Computing and Systems Technology Division 2016

Core Programming Area at the 2016 AIChE Annual Meeting

San Francisco, California, USA
13 - 18 November 2016

ISBN: 978-1-5108-3432-3

Printed from e-media with permission by:

Curran Associates, Inc.
57 Morehouse Lane
Red Hook, NY 12571

CURRAN ASSOCIATES INC. proceedings.com

Some format issues inherent in the e-media version may also appear in this print version.

Printed by Curran Associates, Inc. (2017)

AIChE
120 Wall Street, FL 23
New York, NY 10005-4020

Phone: (800) 242-4363
Fax: (203) 775-5177

www.aiche.org

Additional copies of this publication are available from:

Curran Associates, Inc.
57 Morehouse Lane
Red Hook, NY 12571 USA
Phone: 845-758-0400
Fax: 845-758-2633
Email: curran@proceedings.com
Web: www.proceedings.com

TABLE OF CONTENTS

466153 Smart City, Smart Energy, Smart Decision Making

Abstract:

Research Interests:

Integrated system thinking and analyses are more and more important for sustainable development by addressing challenges in natural resources, economic sectors, and human behaviours in a systematic manner. Process systems engineering has abundant modelling, simulation, and optimization tools to solve these complex problems through a comprehensive 'system of systems' approach. My research focuses on the demand side management of the constantly growing and developing populations, and sustainable provisioning of energy, water and material resources to satisfy needs, as well as the subsequent processing and management of wastes and pollutions. The nexus among energy-water-food is a representative example to adopt system engineering tools for overall economic, environmental and social benefits improvements.

I am developing a decision-making platform that combines comprehensive database, agent based simulation and resource technology network optimization, to deal with the challenges of energy, water, transportation and many interlinked systems in urban development. The open-source, data-driven, integrated systematic tool can gather social, environmental and economic data in a bottom-up approach to inform urban planning, investment and policymaking for city regions globally. Long-term socio-economic scenarios with temporal and spatial feedback loops are addressed to form smart city development strategies.

From macro-level down to detailed systems, sustainable energy and water treatment technologies are studied as fundamental to system analyses. I am researching smart energy management strategies for both generation and demand side with applications to residential and industrial users, as well as emerging technologies. For example, hybrid renewable energy systems (HRES) that consist of photovoltaic thermal, wind, fuel cells, and diesel generator as a backup resource as well as storage components turn to be a promising solution to overcome the drawbacks of single-resource systems. Compared with traditional operation of energy systems, the on-line process I am developing that combines sensors, optimizers and controllers to realize real-time prediction, optimization and demand-responsive scheme will improve system efficiency and take better advantage of all available resources.

Postdoctoral Project: "Water and Energy Systems in Sustainable City Development - ***Resilience.IO*** Resource Economic Human Ecosystem Modeling Platform"

Team under supervision of Professor Nilay Shah, Centre of Process Systems Engineering, Imperial College London.

PhD Dissertation: "***Modeling, Operation and Optimization of Energy Systems Integrating Demand Management***"

Co-advised by Professor Ahmet Palazoglu and Professor Nael H. El-Farra, Department of Chemical Engineering and Materials Science

Teaching Interests:

The philosophy of Knowing, Acting, and Being in education constitute my teaching career to encourage students' engagement, practice, and confidence, during instructing a course or even as simple as teaching a concept, method or software. I look upon a deeper connection between theory and practice when leading a class and creating impressive moments by challenging students and challenging myself in the meantime. Whether for teaching the classic mathematic theories and methods that remain the same, or engineering skills that advance with time, I would put the topics in context, evolve my teaching style and optimize class materials timely to always make the learning and teaching experience satisfactory and pleasant from the feedback of my students. It is the best reward to a teacher to witness how the "WOW" moments positively impact the students' career and life.

I had more than five-year teaching experience giving lectures with excellent ratings, for courses such as computational methods, process design and analysis, separations and unit operations. I was also the instructor for a series of software and tools training such as Aspen Plus and Matlab application. I supervised more than 10 undergraduate and master students' research projects. As a junior faculty member, I bear the mission to foster cooperation and ambition to bridge interdisciplinary researchers and build up a cooperative environment. I would organize series of webinars by inviting prestigious experts in the specific fields to discuss the trending topics by sharing knowledge and insights. With my broad exposure to the research environment of the U.S., UK, Asia, and Africa, I can take the advantage to unite researchers for global collaboration. It is also vital to use research models and results to get involved in government policy making internationally and engage the public to build collaborative communities that expand beyond the academic campus. I had some good practice and will transform the research output through more channels.

472947 Solution Approaches for Large Scale Multistage Stochastic Programs with Endogenous and Exogenous Uncertainty

Abstract:

Research Interests:

Multistage stochastic programs (MSSPs) with endogenous and exogenous uncertainty arise from problems where there is uncertainty in one or more of the problem parameters. In chemical engineering these problems can include the pharmaceutical clinical trial planning problem (Colvin and Maravelias, 2008), the oil field infrastructure planning problem (Gupta & Grossmann, 2011), and the process synthesis problem (Goel & Grossmann, 2006). Multistage stochastic programming formulations are limited computationally by problem size. The growth of the problem is related to the number of uncertain parameters and the number of realizations for each uncertain parameter. Linear increases in the number of uncertain parameters cause the number of scenarios in a MSSP to grow exponentially. Similarly, non-linear growth in problem size is also seen when the number of realizations for each uncertain parameter is increased. Increasing the number of scenarios in a MSSP increases the size of the decision variables and the number of constraints. As a result, most MSSPs considered in literature are limited to a few uncertain parameters with a few realizations for each parameter.

Often real-world size problems have many uncertain parameters with multiple realizations. Therefore, the MSSPs of these problems quickly become computationally intractable. One approach to addressing the computation complexity of MSSPs is to develop improvements for existing solution procedures. Sample Average Approximation (SAA) is a commonly applied approach to determine upper and lower bounds for a MSSPs. One challenge faced in generating the SAA form of the problem is the generation of the constraints. During my PhD I have been able to use my of graph theory to develop a systematic approach to generate

constraints for the SAA problem.

One drawback to the SAA approach is that the approach can only provide bounds on the true solution. In my graduate studies, I have worked on the development of a solution procedure to address the computation challenges presented by the MSSP. The solution procedure uses a branch and bound approach to gradually solve the MSSP formulation. The first step in the development of this procedure was to develop a way to quick generate a feasible solution. To do this I developed two heuristic approaches. The first heuristic approach decomposed the original problem into a series of two-stage stochastic problems (Christian & Cremaschi, 2015).The benefit of this approach is that the number of constraints in each of the problems solved was reduced, however the number of scenarios in the problem caused the approach to quickly become computationally intractable. The second approach decomposed the problem into a series of knapsack problems (Christian & Cremaschi, 2015). Unlike the two-stage decomposition the knapsack did not need the full set of scenarios and was able to generate solutions for very large MSSPs. To complement the feasible solution it was important to develop an approach for solving a relaxed version of the problem. Currently my research focuses on the development of a progressive hedging based approach to solve a relaxed version of the MSSP. The benefit of using progressive hedging is that approach is highly parallelizable.

Continued growth in the computational abilities of computers has opened the door for the use of large datasets in the development of chemical engineering models and optimization. Going forward, an increased understanding of how datasets can be incorporated into existing optimization frameworks will become increasingly important. In my future work I plan to be use my knowledge of statistics, programming, optimization, and graph theory to facilitate the advancement and development of new approaches to incorporating large data sets into chemical engineering applications.

Christian, B., & Cremaschi, S. (2015). Heuristic solution approaches to the pharmaceutical R&D pipeline management problem. *Computers and Chemical Engineering, 74,* 34–47.

Colvin, M., & Maravelias, C. T. (2008). A stochastic programming approach for clinical trial planning in new drug development. *Computers & Chemical Engineering, 32,* 2626–2642. http://doi.org/10.1016/j.compchemeng.2007.11.010

Goel, V., & Grossmann, I. E. (2006). A Class of Stochastic Programs with Decision Dependent Uncertainty. *Mathematical Programming, 108*(2), 355–394.

Gupta, V., & Grossmann, I. E. (2011). Solution strategies for multistage stochastic programming with endogenous uncertainties. *Computers and Chemical Engineering, 35*(11), 2235–2247. http://doi.org/10.1016/j.compchemeng.2010.11.013

Teaching Interests:

During my graduate studies I have had the opportunity to be a teaching assistant. As a teaching assisant I worked under a faculty advisor to develop assignments and lesson plans to engage students. My goal as a teacher will be to develop a program that promotes curiousity and a desire for life long learning. To do this I plan on incorporating real life examples and hands on learning to develop student passion for chemical engineering.

475545 Multi-Scale Process Systems Engineering

Abstract:

Multi-Scale Process Systems Engineering

Bruno A. Calfa

bruno.calfa@wisc.edu

In this poster session, I will discuss my Ph.D. work in the area of Process Systems Engineering (PSE) at Carnegie Mellon University, my Postdoc work at UW-Madison, the research topics and projects I am proposing to pursue, and my teaching interests and experience. For more information, please visit my website: http://bacalfa.com/.

Research Interests:

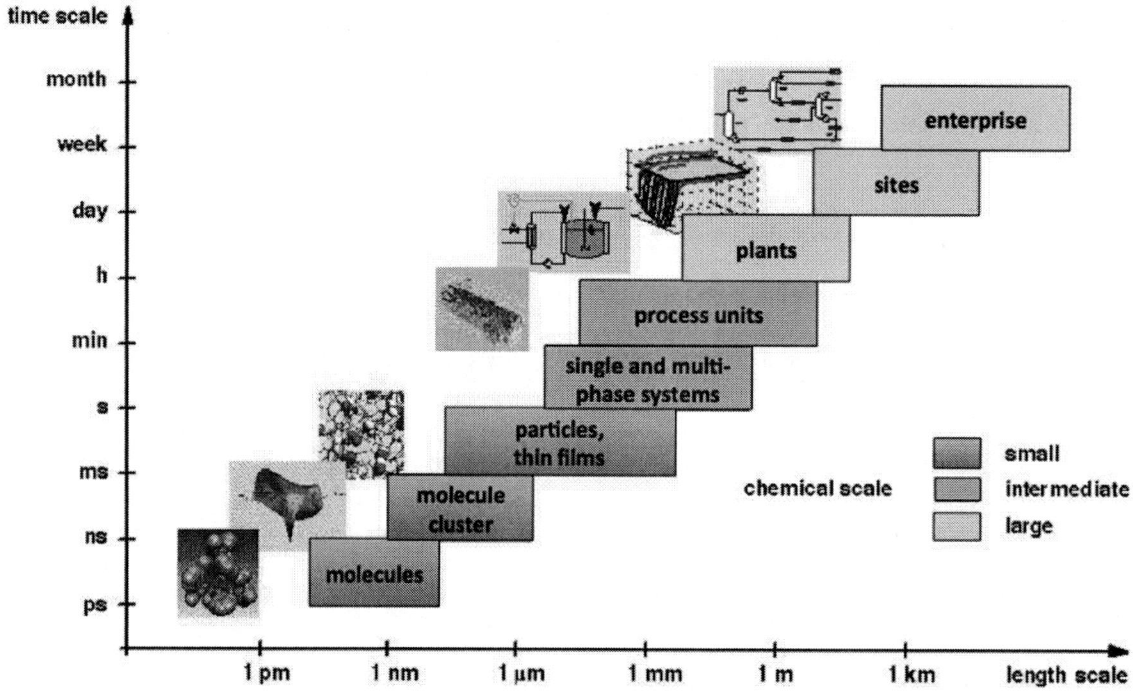

Figure 1. Multiple scales in Process Systems Engineering (PSE) research.

My Ph.D. research focused on the large scale top blocks in Figure 1. I **integrated** and **efficiently solved** planning and scheduling models for a network of batch plants, and developed **data-driven**approaches for modeling uncertainty in Enterprise-wide Optimization (EWO) problems [4-11]. I am interested in developing PSE methods (modeling, simulation, optimization, and control) to solve multi-scale problems of practical importance.

- **Large Scale:** novel *data-driven* models for uncertainty in sales and operations planning; multilevel optimization with contracts and pricing.
- **Intermediate Scale:** reduced-order modeling and optimization; conceptual/operational analysis of *sustainable* technologies, e.g., solar fuels [1].
- **Small Scale:** *property prediction* [3] and computer-aided *materials design* (e.g., crystals) via optimal inverse problems; microkinetics and optimal catalyst design.

I am the lead solver of a winning solution-a systematic crystal design framework-to the Materials Science and Engineering Data Challenge put forward by the Air Force Research Lab. I received the **2015 Ken Meyer Award for Excellence in Graduate Research**, which is given by the Department of Chemical Engineering at CMU in which the faculty base their selection of the student on research quality, productivity, recognition, and impact.

Teaching Interests:

I had extensive teaching experience as a TA at CMU. I prepared several teaching materials, and gave guest lectures and tutorials. I received the **2012 Mark Dennis Karl Outstanding Graduate Teaching Award**, which is given by the Department of Chemical Engineering at CMU to a student judged by the faculty to have done an outstanding job as a teaching assistant.

At UW-Madison, I have developed case studies that are incorporated into a *Chemical Business Simulator* (http://uwchembussim.che.wisc.edu/), which is an education tool that integrates engineering and business best practices [2]. I have also volunteered as a tutor at the Undergraduate Learning Center for freshman and sophomore Chemical Engineering courses.

References

[1] B. A. Calfa and C. T. Maravelias. Conceptual Analysis of Process Alternatives for Solar Thermochemical Methanol Production: The Role of Chemical Storage. In Preparation.

[2] B. A. Calfa and W. F. Banholzer. Web-Based Simulation Games for the Integration of Engineering and Business Best Practices. In Preparation.

[3] B. A. Calfa and J. R. Kitchin. Property Prediction of Crystalline Solids from Composition and Crystal Structure. In: AIChE Journal. (2016). DOI: http://dx.doi.org/10.1002/aic.15251.

[4] I. E. Grossmann, R. M. Apap, B. A. Calfa, P. Garcia-Herreros, and Q. Zhang. Recent Advances in Mathematical Programming Techniques for the Optimization of Process Systems under Uncertainty. In: Computers & Chemical Engineering. (2016). DOI: http://dx.doi.org/10.1016/j.compchemeng.2016.03.002.

[5] B. A. Calfa, A. Agarwal, S. J. Bury, J. M. Wassick, and I. E. Grossmann. Data-Driven Simulation and Optimization Approaches to Incorporate Production Variability in Sales and Operations Planning. In: Industrial & Engineering Chemistry Research. 54.29 (2015), pp. 7261-7272.

[6] B. A. Calfa and I. E. Grossmann. Optimal Procurement Contract Selection with Price Optimization under Uncertainty for Process Networks. In: Computers & Chemical Engineering. 82.1 (2015), pp. 330-343.

[7] B. A. Calfa, I. E. Grossmann, A. Agarwal, S. J. Bury, and J. M. Wassick. Data-Driven Individual and Joint

Chance-Constrained Optimization via Kernel Smoothing. In: Computers & Chemical Engineering. 78.1 (2015), pp. 51-69.

[8] B. A. Calfa. A Memory-Efficient Implementation of Multi-Period Two- and Multi-Stage Stochastic Programming Models. Carnegie Mellon University. Technical Report, 2014. URL: http://repository.cmu.edu/cheme/246/.

[9] B. A. Calfa, A. Agarwal, I. E. Grossmann, and J. M. Wassick. Data-Driven Multi-Stage Scenario Tree Generation via Statistical Property and Distribution Matching. In: Computers & Chemical Engineering. 68.1 (2014), pp. 7-23.

[10] I. E. Grossmann, B. A. Calfa, and P. Garcia-Herreros. Evolution of Concepts and Models for Quantifying Resiliency and Flexibility of Chemical Processes. In: Computers & Chemical Engineering. 70 (2014), pp. 22-34.

[11] B. A. Calfa, A. Agarwal, I. E. Grossmann, and J. M. Wassick. Hybrid Bilevel-Lagrangean Decomposition Scheme for the Integration of Planning and Scheduling of a Network of Batch Plants. In: Industrial & Engineering Chemistry Research. 52.5 (2013), pp. 2152-2167.

[12] B. A. Calfa and M. L. Torem. Bioreagents: their use in the removal of heavy metals from liquid streams by biosorption/bioflotation. (in Portuguese). In: Revista Escola de Minas. 60.3 (2007), pp. 537-542.

[13] B. A. Calfa and M. L. Torem. On the Fundamentals of Cr(III) Removal from Liquid Streams by a Bacterial Strain. In: Minerals Engineering. 21.1 (2007), pp. 48-54.

Formatting Issue - Best Available Paper/Graphic

475589 Petroleum Coke Morphology Mapping: A Mechanistic Approach Using Machine Learning

Abstract:

Petroleum Coke Morphology Mapping: A Mechanistic Approach Using Machine Learning

Amorim, P.H., University of Tulsa

2nd Year PhD Student

Research Interests:

Machine learning is an established and important technique for introducing intelligent systems to chemical engineering and augment the knowledge of a subject-matter. In the field of big data analytics, a full comprehension of data is easily achievable using this powerful tool, which can analyze a gamma of parameters simultaneously, and tell how they correlate with the outputs that one may be interested in. This feature, which is not found in other modeling techniques, gives an advantage to a machine learning user, since he will get the correct interpretation of variables that have complex association with each other. While machine learning can surely be applied to innumerous research fields (such as process safety, process optimization, or any field that contains big data processing), the application within this particular research, which is being advised by Dr. Keith Wisecarver, is to classify and map the morphology of petroleum coke, according to different chemical and physical parameters. Moreover, the mechanistic approach will give an explanation on why each type of coke morphology is formed. It does so by analyzing suspension stability and rheological properties of the feedstock (which can be considered a suspension) and observing and extracting information out of the liquid crystal carbonaceous mesophase generated by thermolysis and polymerization reactions. Further modeling techniques involving large eddy simulations will be used, as an attempt to simulate the different coke morphologies in flow conditions.

Research Experience:

My research background is a blend of diverse experimental methods and computational work. Initially, I focused in microemulsion systems applied to environmental remediation, and process design in the oil & gas industry, where I worked in collaboration with Petrobras and the Brazilian Association of Oil, Gas, and Renewable Fuels (ANP). Currently, I am focused on the structure and rheology of complex fluids, big data analytics using machine learning, with applications in process design and process monitoring. My current research group, the University of Tulsa Delayed Coking Project (TUDCP), gives me the opportunity to conduct research funded by several national and international oil companies.

Teaching Experience:

As an undergraduate student, I was the teaching assistant of Unit Operation Involving Solids and Fluids at the Federal University of Rio Grande do Norte for one and a half years, during which I had the opportunity to be the lecturer and conduct process design projects. Moreover, I created and was the teacher of an English class for low income students for 6 months, also at the same university.

Teaching Interests:

My teaching interests are in Fluid Mechanics, Process Design (HYSYS, Comsol, Ansys), Computational Fluid Dynamics Applied to Chemical Engineering, Heat Transfer, Numerical Methods for Chemical Engineering, and Process Analysis and Optimization Using Machine Learning.

Future Direction:

As a faculty member, I intend to continue applying machine learning in different research projects which have complex association of variables and also conduct research associated with industry. More specifically, these would be conducted in two main areas: Process analysis & optimization, and Process safety. The former will include two tasks, including analyzing variables extracted from energy applications involving solar thermal energy and biomass and applying algorithms, such as Neural Networks, Support Vector Machines, and Naive Bayes to get a better understanding of the association of operating and chemical parameters; and to optimize the design and performance of the process at the same time. The latter will involve troubleshooting an already existing process using the same techniques previously listed, as a way to improve the process safety and prevent hazardous situations in industrial plants, a topic that can attract a variety of fields. Machine learning has the potential to be one of the main tools for data processing, in a world where computers and chemical processes are getting increasingly interconnected.

Additionally, I will conduct studies of turbulent flow using large eddy simulations simulation models, especially with applications involving simulations of the human body systems (respiratory treat simulations to study allergens deposition, drug delivery simulations), particle agglomeration and deposition in suspensions, and even studies of air and water pollution. These studies will establish research on the fields of medicine, fluid suspensions, and environment. Besides these topics, research involving oil & gas industry will still be in my list of interests. Therefore, because of the different research topics, I see myself establishing an interdisciplinary research network with other faculty members, in a way to increase the scientific collaboration among a variety of fields.

Publications:

Amorim ValenÃ§a, P.H., Waturuocha, A., Wisecarver, K. â€œ2D Axisymmetric Temperature Profile Modelling of a Delayed Coking Drum During Pre-Run Warm Up.â€ Proceedings from Comsol Conference Boston, 2015.

475776 Simulation and Optimization of Chemical Processes for CO2 Sequestration and New Clean Energy: Cyclic Adsorption Process, Membrane, and Direct Methanol Fuel Cell

Abstract:

Research Interests:

Because the goal of my research is to solve environmental and energy issues, the focus is the simulation and optimization of CO_2 sequestration processes and a fuel cell system to prepare against global warming. As the methods to reduce the CO_2 emission, gas separation processes such as cyclic adsorption processes (CAPs) and membrane processes are considered in this study.

At first, this study presents the optimization approach (Ko et al., 2005) and the developed simulation model (Ko, 2016) of CAPs to sequester CO_2. The strength of the developed CAP model is to be able to accurately calculate an interstitial gas velocity within adsorption beds resulting in successful scale up designs from a laboratory scale ($0.5Nm^3/h$) to a large commercial scale ($5,000Nm^3/h$).

Then, a dynamic membrane simulation model for the removal of CO_2 is introduced of which mass and energy balances are derived. Optimizations are performed to conveniently perceive the scale up design conditions up to commercial scale ($500Nm^3/h$) by solving mixed integer nonlinear programming (MINLP) problems.

Finally, this work also introduces a non-isothermal dynamic optimization model of DMFC which can generate clean energy (Ko et al., 2008). The optimal operating strategy of DMFC systems was attained through the dynamic simulation and optimization.

In summary, my research includes 1) the development of the scale up design technologies based on the mathematical approaches and 2) dynamic simulation and optimization works for various chemical processes such as the gas treating and energy systems

My research will focus on the optimization technology for the process design and operation. The target processes for the optimization studies will mainly be 1) the gas separation/purification processes to sequester and capture CO_2 and 2) energy systems such as hydrogen plants, fuel cell system, or lithium ion batteries.

Teaching Interests:

What students learn should be practical as well as academic. In other words, I would like to say "to be practical" is same as "to be academic" and the courses of all the chemical engineering area must be helpful for the students to apply to industries. Thus, I would like to teach the core chemical engineering courses including chemical engineering thermodynamics, fluid dynamics, heat transfer, mass transfer, reaction engineering, and process control for the undergraduate level. At the graduate level, I would like to select specialized fields of chemical engineering processes such as process dynamic simulation, gas separation technology, and energy systems. These specialized courses are based on my industrial and academic experiences. The results of my research and teaching must be able to be applied to real industries.

475836 Simulation of the Oil-Treatment Process in the Oil Sands Plant

Abstract:

Research Interests: Process Simulation

Teaching Interests: Process Simulation

The purpose of this paper is to define criteria to be used in the engineering design of an oil sands plant equipped with the steam assisted gravity drainage process. In this effort, the oil treatment process of the oil sands plant on a pilot scale was focused for detailed investigation. The thermodynamic properties of the process fluid, which is mainly composed of bitumen and water, were estimated with the CPA model. The commercial aspen HYSYS was implemented in the analysis throughout this work along with the provided input data and some necessary assumptions. From the simulation results, the heat and mass balances of the process for a 300 BPD oil sands plant were established in order to define standard data for its modular design. In particular, the designing basis for equipment size, heat transfer areas, capital cost and operation cost was extensively discussed.

475841 Effects of the Mixed Refrigerant Composition on the Peformance of the Rankine Cycle Driven By LNG Cold Energy

Abstract:

Research Interests: Process Simulation

Teaching Interests: Process Simulation

The power generation system empowered by the LNG cold energy, which evolves in a large amount during the vaporization process of the liquefied natural gas, was designed in favor of the Rankine cycle with a mixed refrigerant as the working fluid. In this study it is intended to identify the limits of the working fluid composition in respect of equipment safety in the Rankine cycle-type power generation system driven by the cold energy. The thermodynamic properties of the working fluid, which is a hydrocarbon mixture, were calculated with the Peng-Robinson equation. In the steady state simulation of the power generation system by using the commercial simulator Aspen HYSYS, the feed conditions of LNG from Test Bed Train No.1 along with some necessary assumptions were incorporated. The results indicated that deterioration of the mechanical performance of the equipment as well as its safety would be brought about if contents of C_2H_6 and C_3H_8 in the mixture become, respectively, too high or too low.

476056 Process System Engineering for Advanced Modular Continuous Pharmaceutical Manufacturing Platform

Abstract:

Research Interests: My main research area is process system engineering with focus on continuous pharmaceutical manufacturing. I have contributed significantly in the field of process monitoring and control, process modelling and simulation, optimization, PAT, QbD, control hardware/software integration, new methodology and software tools development.

Process system engineering (PSE) is a very advanced field and has been very successful in solving complex problems of chemical based product manufacturing. However, the particulate systems are still poorly understood and there is different level of complexities involved, therefore new techniques and tools need to be developed for pharmaceutical manufacturing through extensive fundamental and applied research.

Currently, major pharmaceutical industries are shifting toward continuous manufacturing technology in which PSE can play a very significant role. Furthermore, there is a need of advanced modular continuous manufacturing platform technology for fast and robust development of solid oral products. Master formulations, process models, standardized sensing and control architecture, superior automation and novel methods and tools need to be developed to enable platform technology. Author's main research area is pharmaceutical process system engineering. This area of research and education are highly encouraged and supported by pharmaceutical companies, academic institutions, regulatory authority (FDA) and national science foundation (NSF). For example, national science foundation of USA has established a 100 million dollar research center for structured organic particulate systems (C-SOPS) at Rutgers University New Jersey. Furthermore, more than 40 companies are supporting this center economically and providing scientific inputs so that the research outcomes can be directly applied to the companies.

In this work, a scientific foundation for the optimal design, optimization and control of the modular continuous process involving structured organic particulate systems has been developed. The mathematical models for continuous manufacturing of active pharmaceutical ingredients and final product tablet, has been developed for virtual experimentation and applied for plant design, optimization and control. Advanced modelling techniques such as PBM, DEM and CFD have been applied for pharmaceutical process modeling. The developed continuous tablet manufacturing plant and process model has been adapted by several pharmaceutical companies. A flexible monitoring and control system incorporating different PAT techniques and tools and control strategies for continuous pharmaceutical manufacturing process has been designed and implemented to Rutgers plant. Different control strategies such as PID, PID coupled with dead time compensator, and MPC has been compared and evaluated practically for pharmaceutical process control in terms of set point tracking and disturbances rejection ability, robustness, and stability/instability. A new hybrid MPC-PID control strategy has been developed and implemented into the pharmaceutical tablet manufacturing plant and its performance has been found better than the other control strategies. For the first time, the closed-loop operation of the tablet manufacturing plant using NIR sensor has been demonstrated, which is a significant advancement in pharmaceutical manufacturing as endorsed by the Food and Drug Administration (FDA) and other regulatory authorities as part of the Quality by Design (QbD) paradigm. For the first time NIR has been applied for real time monitoring of powder bulk density. Feedforward control system has been integrated with feedback control system to take proactive mitigation actions on raw materials and process variability. A systematic generic framework including the methods and tools through which the cost and profit can be estimated and the Dynamic Real Time Optimization (DRTO) can be integrated with hybrid control system has been also developed. The integrated DRTO provides the optimal operational set points for the plant control system in real time optimizing an objective function. DRTO integrated with hybrid control strategy ensures the maximum possible profit irrespective of the market demand fluctuations.

Presently, I am working as an assistant research professor at Department of Chemical and Biochemical Engineering of Rutgers University and currently managing and leading ten projects funded by NSF, industries and regulator with total grants exceeding 18 million USD. I was involved in proposal writing of these projects and I am PI/Co-PI of some of these projects. Prior to this, I worked in a large scale European pharmaceutical project (F^3 Factory project), which involves 25 partners from industry and Academia; where my responsibility was to manage and lead the DTU's parts of the F^3 Factory project and also to coordinate with the other partners associated with it. My doctoral research work, done at Department of Chemical and Biochemical Engineering, Technical University of Denmark, has been awarded for ***prestigious "EFCE Excellence Award 2010", given in Recognition of an Outstanding PhD Thesis, from European Federation of Chemical Engineering***. I have exemplary academic records, previous teaching experiences, and an established trail of valued research accreditations'. I have published several peer reviewed scientific international Journal articles, popular reputed magazine articles, book chapters and presented in more than 80 international conferences. I am actively serving as a conference session chair, guest editor of Journal, manuscripts reviewer and bachelor, master and PhD projects supervisor.

The objective of this presentation is two-fold. First to give the overview of my research and teaching experience, and second to demonstrate the simulation, design, optimization, monitoring and control of advanced modular continuous pharmaceutical manufacturing plant.

Teaching Interests:

My teaching experience that I acquired serving as a teaching assistant, lecturer, and assistant research professor makes me confidant to develop and teach any courses of chemical and biochemical engineering. The industrial invited crash courses that I have given provide me opportunity to extend my teaching skills to cover more applied courses and to interact with highly experienced scientist and operators from manufacturing companies. I demonstrated my skills to instruct hands-on exercise through several workshops that I conducted. Several training courses that I conducted for industrial and academic participants make me instrumented with additional capability of building highly skilled workforce. In my department, I am a lead organizer of postdoctoral training program with the aim of building highly skilled workforce for industry and academia. I developed two online course modules where I learn how to develop and give online courses. My unique research area- applied process system engineering- provides me dual ability to teach theoretical as well as practical courses. I am open to teach any courses that your university will assign me. I am especially interested to teach courses related to process control, process modeling and simulation, process monitoring, chemical reaction engineering, process and tool integration and pharmaceuticals. If needed then I can develop new courses based on my research outcomes.

Grants:

Extensive experience in funding proposals writing. Wrote several grants proposals for national science foundation (NSF), FDA, and pharmaceutical industries. 11 projects with total grant more than 18 million USD got funded in which I have contributed significantly.

Publications:

Scientific articles (43); Book chapters (4); International conference presentations (80); Invited industrial/academic lectures (8); Workshops (5); Thesis (3); Reports (>100).

Activities in AIChE 2015 conference:

Session co-chair

1. 10B08 Process Control Applications I: 32339
2. 10B12 Process Control Applications II: 34890
3. 10C03 Dynamic Simulation and Optimization: 32320
4. 10D01 Advances in Computational Methods and Numerical Analysis: 32306
5. 10E02 Big Data Analytics in Chemical Engineering: 32315

Presenter (Oral presentations)

1. **Singh, R.,** Muzzio, F. J, Ramachandran, R. Ierapetritou, M. **(2016)**. Advanced flexible control system implementation into direct compaction continuous pharmaceutical manufacturing pilot-plant. Oral presentation at AIChE annual meeting, San Francisco, CA, USA, 13 – 18 November. ***Session: Product Quality Attribute Monitoring and Control in Drug Product Manufacturing. Schedule: Thursday, November 17, 3: 59 PM.***
2. **Singh, R.,** Cao, H., Mushnoori, S., Higgins, B., Kolipara, C., Fermier, A., Hausner, D., Jha, S., Ierapetritou, M., Ramachandran, R. **(2016)**. Data management and integration for continuous pharmaceutical manufacturing. Oral presentation at AIChE annual meeting, San Francisco, CA, USA, 13 – 18 November. ***Session: Industrial Applications of Data Analysis, Information Management, and Intelligent Systems. Schedule: Wednesday, November 16, 9.46 AM.***
3. **Singh, R.,** Pereira, G. C., Soni, N., Román-Ospino, A. D., Ierapetritou, M., Ramachandran, R. **(2016)**. Feedforward control of continuous pharmaceutical manufacturing process. Oral presentation at AIChE annual meeting, San Francisco, CA, USA, 13 – 18 November. ***Session: Quality By Design in Drug Product Formulation, Design, & Process Development II. Schedule: Friday, November 18, 8.30 AM.***

Co-author

1. Wang, Z., Escotet-Espinoza, M. S., **Singh, R.,** Muzzio, F. J., Ierapetritou, M. G. **(2016)**. Surrogate-Based Optimization Methodology for Pharmaceutical Tablet Manufacturing Processes. Oral presentation at AIChE annual meeting, San Francisco, CA, USA, 13 – 18 November. ***Session: Design and Operations Under Uncertainty I. Schedule: Thursday, November 17, 8:30 AM.***
2. Wang, Z., Escotet-Espinoza, M. S., **Singh, R.,** Muzzio, F. J., Ierapetritou, M. G. (2016). Feasibility Analysis of Flowsheet Models in Continuous Pharmaceutical Manufacturing Processes Considering the Effects of Noise. Poster presentation at AIChE annual meeting, San Francisco, CA, USA, 13 – 18 November. ***Poster Session: Pharmaceutical. Schedule: Monday, November 14, 3:15 - 5:45 PM.***

Selected publications:

1. International scientific articles:

Year 2016

1. Wu, S., Panikar, S. S., **Singh; R.**, Zhang, J., Donepudi, A., Glasser, B., Ramachandran, R. **(2016)**. Systematic framework to monitor mulling processes using Near Infrared spectroscopy. Advanced Powder Technology. doi:10.1016/j.apt.2016.03.022.
2. **Singh, R.,** Muzzio, F., Ierapetritou, M., Ramachandran, R. **(2016)**. Systematic framework for design and implementation of plant-wide multilayer, sensing and control architecture into continuous pharmaceutical manufacturing plant. Computer Aided Chemical Engineering. Accepted.
3. Román-Ospino, A. D., **Singh, R.**, Ierapetritou, M., Ramachandran, R. Méndez, R., Ortega, C., Muzzio, F. J., Romañach, R. J., **(2016)**. Near Infrared Spectroscopic Calibration Models For Real Time Monitoring Of Powder Density. International Journal of Pharmaceutics. Accepted.

Year 2015

1. **Singh, R.,** Román-Ospino, A. D., Romañach, R. J., Ierapetritou, M., Ramachandran, R. **(2015)**. Real time monitoring of powder blend bulk density for coupled feed-forward/feed-back control of a continuous direct compaction tablet manufacturing process. International Journal of Pharmaceutics, 495, 612-625.
2. **Singh, R.,** Muzzio, F., Ierapetritou, M., Ramachandran, R. **(2015)**. A combined feed-forward/feed-back control system for a QbD based continuous tablet manufacturing process. PROCESSES Journal, 3, 339-356.
3. **Singh, R.,** Sen, M., Ierapetritou, M., Ramachandran, R. **(2015)**. Integrated moving horizon based dynamic real time optimization and hybrid MPC-PID control of a direct compaction continuous tablet manufacturing process. Journal of Pharmaceutical Innovation, 10 (3), 233-253.
4. Simon, L. L., .. **Singh, R.,** et al. **(2015)**. Assessment of Recent Process Analytical Technology (PAT) Trends: A Multi-author Review. Organic Process Research & Development, 19, 3-62.
5. **Singh, R.,**Zhang, J., Ierapetritou, M., Ramachandran, R. (2015). Designing a novel continuous manufacturing plant with superior monitoring and control. European Pharmaceutical Review, 20(6), 37-41.
6. Sen, M., **Singh, R.,** Ramachandran, R. **(2015)**. Model Manufacturing. The Medicine Maker, February, 05, 42-45.
7. Escotet-Espinoza, M. S., **Singh, R.,** Sen, M., O'Connor, T., Lee, S., Chatterjee, S., Ramachandran, R., Ierapetritou, M., Muzzio, F. **(2015)**. Flowsheet Models Modernize Pharmaceutical Manufacturing Design and Risk Assessment, Pharmaceutical Technology 39 (4), 34-42.
8. **Singh, R.,** Muzzio, F., Ierapetritou, M., Ramachandran, R. **(2015)**. Plant-wide control of a continuous tablet manufacturing for Quality-by-Design based pharmaceutical manufacturing. Computer Aided Chemical Engineering, 37, 2183 - 2188.
9. Karry, K. M., **Singh, R.,**Muzzio, F. J. (2015). Fit-for-Purpose Miniature NIR Spectroscopy for Solid Dosage Continuous Manufacturing. American Pharmaceutical Review, 18(4), 64 - 67.
10. **Singh, R.,**Ierapetritou, M., Ramachandran, R. (2015). The scopes of PAT in real-time advanced control of tablet quality. European Pharmaceutical Review 20(2), 76-80.

Year 2014

1. **Singh, R.,** Sahay, A., Karry, K. M., Muzzio, F., Ierapetritou, M., Ramachandran, R. (**2014**). Implementation of a hybrid MPC-PID control strategy using PAT tools into a direct compaction continuous pharmaceutical tablet manufacturing pilot-plant. International Journal of Pharmaceutics, 473, 38–54.
2. **Singh, R.,** Sahay, A., Fernando Muzzio, Ierapetritou, M., Ramachandran, R. (**2014**). Systematic framework for onsite design and implementation of the control system in continuous tablet manufacturing process. Computers & Chemical Engineering Journal, 66, 186-200.
3. **Singh, R.,** Barrasso, D., Chaudhury, A., Maitraye Sen, Ierapetritou, M., Ramachandran, R. (**2014**). Closed-Loop Feedback Control of a Continuous Pharmaceutical Tablet Manufacturing Process via Wet Granulation. Journal of Pharmaceutical Innovation, 9, 16-37.
4. Sen, M., Barrasso, D., **Singh, R.**, Ramachandran, R. (**2014**). A Multi-Scale Hybrid CFD-DEM-PBM

Description of a Fluid-Bed Granulation Process. Processes Journal, 2(1), 89-111.

5. Sen, M., **Singh, R.,** Ramachandran, R. (**2014**). Simulation based design of an efficient control system for the continuous purification and processing of active pharmaceutical ingredients. Journal of Pharmaceutical Innovation, 9, 65-81.
6. Sen, M., **Singh, R.,** Ramachandran, R. **(2014)**. A hybrid MPC-PID control system design for the continuous purification and processing of active pharmaceutical ingredients. PROCESSES Journal, 2, 392-418; doi:10.3390/pr2020392.
7. Sen, M., Chaudhury, A., **Singh, R.,** Ramachandran, R. **(2014)**. Two-dimensional population balance model development and validation of pharmaceutical crystallization processes. American Journal of Modern Chemical Engineering, 1, 13-29.

Year 2013

1. **Singh, R.,** Ierapetritou, M., Ramachandran, R. (**2013**). System-wide hybrid model predictive control of a continuous pharmaceutical tablet manufacturing process via direct compaction. European Journal of Pharmaceutics and Biopharmaceutics, 85(3), Part B, 1164-1182.
2. **Singh, R.,** Godfrey, A., Gregertsen, B., Muller, F., Gernaey, K. V., Gani, R., Woodley, J. M. (**2013**). Systematic substrate adoption methodology (SAM) for future flexible, generic pharmaceutical production processes. Computers & Chemical Engineering Journal, 58, 344 - 368.
3. Sen, M., Dubey, A., **Singh, R.,** Ramachandran, R. (**2013**). Mathematical Development and Comparison of a Hybrid PBM-DEM description of a Continuous Powder Mixing Process. Journal of Powder Technology, http://dx.doi.org/10.1155/2013/843784.
4. Sen, M., Chaudhury, A., **Singh, R.,** John, J., Ramachandran, R. (**2013**). Multi-scale flowsheet simulation of an integrated continuous purification–downstream pharmaceutical manufacturing process. International Journal of Pharmaceutics, 445 (1-2), 29-38.
5. Sen, M., Rogers, A., **Singh, R.,** Chaudhury, A., John, J., Ierapetritou, M., Ramachandran, R. (**2013**). Flowsheet optimization of an integrated continuous purification-processing pharmaceutical manufacturing operation. Chemical Engineering Science, 102, 56 – 66.
6. **Singh, R.,** Ierapetritou, M., Ramachandran, R (**2013**). Hybrid advanced control of a flexible multipurpose continuous pharmaceutical tablet manufacturing process via direct compaction. Computer Aided Chemical Engineering, 32, 757-762.
7. Muzzio, F., **Singh, R.,** Chaudhury, A., Rogers, A., Ramachandran, R. Ierapetritou, M. (**2013**). Model-predictive design, control and optimization of pharmaceutical process. Pharmaceutical Technology magazine, 37(6), 40-41, 77.
8. **Singh, R.,** Sahay, A., Oka, S., Liu, X., Ramachandran, R., Ierapetritou, M., Muzzio, F. (**2013**). Online monitoring, advanced control and operation of robust continuous pharmaceutical tablet manufacturing process. BioPharma magazine Asia, 2(5), 18-25.

Year 2012

1. **Singh, R.,** Ierapetritou, M., Ramachandran, R. (**2012**). An engineering study on the enhanced control and operation of continuous manufacturing of pharmaceutical tablets via roller compaction. International Journal of Pharmaceutics, 438 (1-2), 307-326.
2. Sen, M., **Singh, R.,** Vanarase, A., John, J., Ramachandran, R. (**2012**). Multi-dimensional population balance modeling and experimental validation of continuous powder mixing processes. Chemical Engineering Science, 18, 349-360.
3. **Singh, R.,** Raquel Rozada-Sanchez, R., Dean, W., Perkins, J., Muller, F., Godfrey, A., Gernaey, K. V., Gani, R., Woodley, J. M. (**2012**). A generic process template for continuous pharmaceutical production. Computer Aided Chemical Engineering, 31, 715-719.
4. **Singh, R.,** Boukouvala, F., Jayjock, E., Ramachandran, R. Ierapetritou, M., Muzzio, F. (**2012**). Flexible Multipurpose Continuous Processing: integration of process flow modeling for continuous processing of pharmaceutical solid dosage forms. PharmPro Magazine, Pharmaceutical Processing, 27(6), 22-25.
5. **Singh, R.,** Boukouvala, F., Jayjock, E., Ramachandran, R. Ierapetritou, M., Muzzio, F. (**2012**). Flexible Multipurpose Continuous Processing of Pharmaceutical Tablet Manufacturing Process. GMP news, European Compliance Academic (ECE), http://www.gmp-compliance.org/ecanl_503_0_news_3268_7248_n.html.

Years 2011-2008

1. **Singh, R.,** Gernaey, K. V., Gani, R. **(2010)**. ICAS-PAT: A Software for Design, Analysis & Validation of PAT Systems. Computers & Chemical Engineering Journal, 34(7), 1108-1136.
2. **Singh, R.**, Gernaey, K. V., Gani, R. **(2010)**. An ontological knowledge based system for selection of process monitoring and analysis tools. Computers & Chemical Engineering Journal, 34(7), 1137-1154.
3. **Singh, R.,** Gernaey, K. V., Gani, R. **(2009)**. Model-based computer-aided framework for design of process monitoring and analysis systems. Computers & Chemical Engineering Journal, 33(1), 22-42.
4. Samad, N. A. F. A., **Singh, R.,** Sin, G., Gernaey, K. V., Gani, R. **(2011)**. A generic multi-dimensional model-based system for batch cooling crystallization processes. Computers & Chemical Engineering Journal, 35(5), 828-843.
5. Samad, N. A. F. A., **Singh, R.,** Sin, G., Gernaey, K. V., Gani, R. **(2011)**. Systematic Procedure for Generating Operational Policies to Achieve Target Crystal Size Distribution (CSD) in Batch Cooling Crystallization. IEEE, 1-6. Print ISBN: 978-1-4577-0003-3. DOI: 10.1109/ICMSAO.2011.5775588.
6. Samad, N. A. F. A., **Singh, R.,** Sin, G., Gernaey, K. V., Gani, R. **(2011)**. Integration of Generic Multi-dimensional Model and Operational Policies for Batch Cooling Crystallization. Computer Aided Chemical Engineering, 29, 86-90.
7. Samad, N. A. F. A., **Singh, R.,** Sin, G., Gernaey, K. V., Gani, R. **(2010)**. Control of process operation and monitoring of product qualities through generic model-based in batch cooling crystallization. Computer Aided Chemical Engineering, 28, 613-618.
8. **Singh, R.**, Rozada-Sanchez, R., Wrate, T., Muller, F., Gernaey, K. V., Gani, R., Woodley, J. M. **(2010)**. A retrofit strategy to achieve "Fast, Flexible, Future (F^3)" pharmaceutical production processes. Computer Aided Chemical Engineering, 29, 291-295.
9. **Singh, R.,** Gernaey, K. V., Gani, R. **(2009)**. A software tool for design of process monitoring and analysis systems. Computer Aided Chemical Engineering, 26, 321-326.
10. **Singh, R.,** Gernaey, K. V., Gani, R. **(2008)**. Off-line design of PAT systems for on-line applications. Computer Aided Chemical Engineering, 25, 423-428.

2. Book Chapters

1. **Singh, R.,** Figueroa, CV, Sahay, A., Karry, KM, Fernando Muzzio, F., Ierapetritou, M., Ramachandran, R. **(2014)**. Chapter 7: **Advanced Control of continuous pharmaceutical tablet manufacturing processes**. Book title: Process Simulation and Data Modeling in Solid Oral Drug Development and Manufacture. Publisher: Humana Press, ISBN: 978-1-4939-2995-5, 191 – 223.
2. Ierapetritou, M., Escotet-Espinoza, M. S., **Singh, R. (2016)**. **Process Simulation and control for continuous pharmaceutical manufacturing of solid drug products**. A chapter of book "Continuous manufacturing of pharmaceuticals" edited by Peter Kleinebudde, Johannes Khinast and Jukka Rantanen. Publisher: Wiley-VCH. Accepted.
3. Oka, S, Escotet-Espinoza, M. S., **Singh, R.,** Scicolone, J., Hausner, D., Ierapetritou, M., Muzzio, F. **(2016)**. **Design of an integrated continuous manufacturing system**. A chapter of book "Continuous manufacturing of pharmaceuticals" edited by Peter Kleinebudde, Johannes Khinast and Jukka Rantanen. Publisher: Wiley-VCH. Accepted.
4. **A Book chapter in "Product and Process Modelling: A case study approach":**

- **Singh, R. (2011). Fermentation process modeling**. Chapter 12.2 of book "Product and Process Modelling: A case study approach" edited by I Cameron & R. Gani. Publisher: Elsevier, pp 380-396.
- **Singh, R. (2011). Milling process model.** Chapter 12.4 of book "Product and Process Modelling: A case study approach" edited by I Cameron & R. Gani. Publisher: Elsevier, pp 407-413.
- **Singh, R. (2011). Granulation process model.** Chapter 12.5 of book "Product and Process Modelling: A case study approach" edited by I Cameron & R. Gani. Publisher: Elsevier, pp 413-422.
- **Singh, R. (2011). Pharmaceutical tablet pressing process model.** Chapter 12.6 of book "Product and Process Modelling: A case study approach" edited by I Cameron & R. Gani. Publisher: Elsevier, pp 422-430.
- **Singh, R. (2011). Milk pasteurization process modeling**. Chapter 12.3 of book "Product and Process Modelling: A case study approach" edited by I Cameron & R. Gani. Publisher: Elsevier, pp 396-406.

476073 Optimization-Based Quantification of Performance Limits for Process Networks

Abstract:
Research Interests: Process systems engineering, global optimization, energy systems, process synthesis and design, modeling and simulation, computational fluid dynamics.

Teaching Interests: Transport phenomena, mass transfer, heat transfer, thermodynamics, chemical process computer-aided design and analysis, numerical and mathematical methods, process dynamics and control.

The goal of any synthesis task is to design a feasible network that can deliver desired products from a set of inlet raw material using one or more chemical processes. Computational tools are usually employed to perform this task, aiming to find an optimal solution to a given design problem. Nevertheless, existing tools cannot ensure that the identified solution is a global rather than local optimum, keeping open the possibility of further performance improvements.

In order to find a global optimal solution for the process network synthesis problem, a methodology that employs the Infinite DimEnsionAl State-Space (IDEAS) conceptual framework is presented in this work. The IDEAS framework is based on the techniques of optimization by vector spaces, convex analysis and chemical engineering fundamentals, which guarantee global optimality.

This work presents several examples of globally optimal synthesis of process networks involving reactive/non-reactive distillation of azeotropic mixtures.

476079 Metabolic Modeling for Improved Bioprocess Efficiency

Abstract:

Research Interests:

Microbial fermentation plays a critical role in efforts to enable sustainable manufacturing of commodity chemicals, and the economic feasibility of the integrated biorefinery concept hinges sharply on the productivity, yield, and titers that can be achieved by metabolic hosts. Flux balance modeling has emerged as a fundamental tool in pushing biological transformations towards profitability, as it enables model-based design of gene knockout and overexpression efforts. My interests center on the unbiased search of intervention strategies for models of central carbon metabolism, together with the consideration of the dynamic systems which govern metabolic flows within the network.

Postdoctoral Projects: Metabolic modeling of industrially relevant organisms for improved yield and productivity

Under supervision of Yannick Bomble and Gregg Beckham, Biosciences Center, National Renewable Energy Laboratory, Golden CO

PhD Dissertation: ÒMathematical Approaches to Understanding Mammalian Circadian RhythmsÓ

Under supervision of Francis J. Doyle III, Department of Chemical Engineering, University of California, Santa Barbara.

Future Direction:

As a faculty I would like to work towards standardizing and accelerating model-based design of metabolic interventions, particularly in non-model organisms. Industrial bioprocesses predominantly leverage specialized organisms to achieve the desired operating conditions, and while these microbial platforms offer unique advantages over traditional hosts, their relative obscurity slows the design/build/test cycle for determining optimum metabolic manipulations. To speed this process I aim to develop standardized workflows and new computational techniques that exploit the overlap in metabolic functionality between classes of microbes. By extracting general trends in gene regulation and enzyme kinetics and combining this knowledge with organism-specific results, searches for the optimum overexpression or gene knockout could explicitly consider the underlying uncertainty in the gene regulatory network.

In addition to metabolic engineering, my interests extend to bio- and cheminformatics, where machine-learning algorithms can be used to extend patterns and trends to new, unmeasured points. I hope to apply these techniques throughout the pipeline from feedstocks to commodity chemicals, as well as in applications in synthetic biology and pharmaceuticals.

Teaching Interests:

My primary teaching interests lie in computational and analytical methods in dynamic systems, statistics, and process control. As a teaching assistant I have given several guest lectures in introductory statistics, and have instructed undergraduate and graduate students on how to perform computational analyses in numerous programming languages.

Proposal Experience: Mitsubishi Chemical Fellowship recipient

NSF GRFP Honorable Mention

Selected Publications: 312 Citations, h-index 4.

Abel, J.H., Meeker, K., Granados-Fuentes, D., **St. John, P.C.**, Wang, T.J., Bales, B.B., Doyle F.J. III, Herzog, E.D., and L.R. Petzold. Functional network inference of the suprachiasmatic nucleus (2016) *PNAS*, 113 (16) pp. 4512-4517

St. John, P.C. and F.J. Doyle III. Quantifying stochastic noise in cultured circadian reporter cells (2015), *PLoS Computational Biology* 11(11): e1004451.

St. John, P.C., Taylor, S.R., Abel, J.H., and F.J. Doyle III. Amplitude metrics for cellular circadian bioluminescence reporters (2014) *Biophysical Journal*, 107 (11) pp. 2712-2722

St. John, P.C., Hirota, T., Kay, S.A. and F.J. Doyle III. Spatiotemporal separation of PER and CRY posttranslational regulation in the mammalian circadian clock (2014) *PNAS*, 111 (5) pp. 2040-2045.

St. John, P.C. and F.J. Doyle III. Estimating confidence intervals in predicted responses for oscilla- tory biological models (2013) *BMC Systems Biology* 7:71.

Hirota, T., Lee, J.W., **St. John, P.C.**, Sawa, M., Iwaisako, K., Noguchi, T., Pongsawakul, P.Y., Son- ntag, T., Welsh, D.K., Brenner, D.A., Doyle, F.J. III, Schultz, P.G., Kay, S.A., Identification of small molecule activators of cryptochrome (2012) *Science*, 337 (6098) pp. 1094-1097.

476087 Dynamics of Discrete Systems: At the Interface of Engineering and Medicine

Abstract:

Dynamics of Discrete Systems: At the Interface of Engineering and Medicine

Anwesha Chaudhury, PhD,

Center for Systems Biology, Massachusetts General Hospital; Pathology, Massachusetts General Hospital; Systems Biology, Harvard Medical School.

Email: achaudhury@mgh.harvard.edu, anwesha_chaudhury@hms.harvard.edu

Education

Doctor of Philosophy

December 2014

Department of Chemical and Biochemical Engineering

Rutgers-The State University of New Jersey, NJ, USA

Masters of Technology

August 2010

Department of Chemical Engineering

Indian Institute of Technology, Kharagpur, India

Bachelors of Technology

June 2008

Department of Chemical Engineering

West Bengal University of Technology, India

Research Interests:

Research experience:

I am a Chemical Engineer by training and have worked on a wide variety of research projects in various realms of basic sciences and engineering. The common denominator across all my research experiences is mathematical modeling and systems engineering. The multifaceted areas of my expertise include working on non-linear system dynamics pertaining to reaction engineering, developing mechanistic mathematical models for improving processes operation in pharmaceutical drug manufacturing, and clinical data analyses related to human pathophysiology.

During the course of my research expedition I have been interested in diverse fields in which I feel I may be able to make a significant contribution using the core skill sets of systems modeling. While working as a theoretician during my PhD, I worked closely with my experimental colleagues for developing realistic mathematical models and to better understand the underlying assumptions required for a specific model. Furthermore, I have had opportunities to become adept in various hands-on laboratory techniques including the use of a fluidized bed granulator. Towards the end of my PhD, I got drawn into the enigma of systems biology, which led me to switch fields and broaden the applications of my mathematical modeling skills.

Currently as a postdoctoral research fellow I work in an extremely multi-disciplinary environment with clinicians, biologists, chemists, engineers, and statisticians. I am leveraging my engineering training to develop models that can improve disease prognosis and alleviate the shortcomings of a delayed therapeutic intervention. Moreover, I am collaborating with experts in the field of microfluidic device development with the aim of developing a point-of-care system to detect biomarkers for sickle cell disease using single-cell measurements.

Throughout my PhD and postdoctoral journey I have received several awards and grants that exhibit my ability to secure external funding and inculcate others' interests in my research.

Research Vision:

I plant to integrate my training in various fields of engineering and medicine, with a special focus on systems engineering, by developing tools that describe hematological systems and particulate processes. In the short-term, I aim to develop data-driven population dynamics models for blood cells to predict the occurrence of diseases. Furthermore, by using this approach I intend to develop hypotheses for the underlying mechanisms of diseases. These hypotheses in turn will be investigated by performing laboratory bench experiments. I plan to integrate these models for various lineages of white blood cells in order to investigate the crosstalk within the immune system. Additionally, I plan to develop models (macroscopic, microscopic and mesoscopic) for quantitative systems pharmacology and then consolidate the dynamic behavior of the various components to study the overall pharmacodynamic behavior of the human body.

From a process systems perspective, I aspire to continue developing mathematical models for particulate systems, which can then be utilized for process control and optimization. I have significant experience with developing numerical techniques to solve complex partial differential equations, which I would like to continue incorporating into my research. One of my primary goals is to develop realistic models, which can actually be used for practical applications. With this intent in mind, I plan to continue collaborating closely with experimentalists in order to use systems engineering with a practical application.

Teaching Interests:

In addition to research, I have a keen interest in mentoring students and teaching core as well as advanced engineering topics. During the tenure of my PhD I was a teaching assistant (TA) for several courses including: transport phenomenon-I, Junior design and economics, and process control. I have derived immense pleasure and a sense of accomplishment from teaching and helping undergraduate students learn the chemical engineering curriculum. I have mentored several undergraduate and masters students, who have successfully performed research and have co-authored publications in leading peer-reviewed journals. I have persistently encouraged undergraduate students to pursue research in addition to their coursework. I enjoy teaching the core chemical engineering subjects and am extremely excited to design and teach new courses that combine the knowledge of chemical engineering, mathematics and statistics.

Research appointments:

Research fellow(February 2015-Present), Center for Systems Biology, Massachusetts General Hospital, MA, USA;

Pathology, Massachusetts General Hospital, MA, USA;

Systems Biology, Harvard Medical School, MA, USA.

Advisor: John Higgins, MD.

Project: Mathematical modeling of pathophysiological processes for disease prognosis.

Graduate Student(September 2010-December 2014), Department of Chemical and Biochemical Engineering, Rutgers University, NJ, USA.

Advisor: Rohit Ramachandran, PhD.

PhD dissertation: Mathematical modeling, simulation and optimization of pharmaceutical manufacturing processes.

Exchange research student(June 2014-August 2014), Department of Mathematics and Statistics, University of Konstanz, Germany.

Advisor: Stefan Volkwein, PhD.

Project: Reduced order modeling and optimal control of granulation processes.

Exchange research student(July 2012-October 2012), Research Center Pharmaceutical Engineering, Graz, Austria

Advisor: Johannes Khinast, PhD

Project: Investigations for particulate processes relevant to pharmaceutical drug manufacturing.

Masters student(January 2008-May 2010), Chemical Engineering, Indian Institute of Technology, Kharagpur, India

Advisor: Saikat Chakraborty, PhD

MS thesis: Pattern formation in autocatalytic reactions

Awards and Honors:

February 2016, Finalist for the Life Science Research Foundation (LSRF) postdoctoral fellowship (final result pending).

May 2015, Awarded the prestigious Outstanding Graduate Student Award by the School of Engineering, Rutgers University in recognition of the achievements as a graduate student.

June 2014, Awarded the Baden-Wuerttemberg Scholarship to pursue research at

University of Konstanz, Germany.

July 2012, Awarded the Marshall Plan Fellowship to pursue research at TU Graz, Austria.

May 2012, Awarded the best poster prize at the New Jersey Pharmaceutical Association for Science and Technology graduate student poster competition.

September 2010, School of Engineering Fellowship, Rutgers University, USA.

Selected publications: (total 22, 9 first author publications, 2 book chapters)

A. Chaudhury, A. Tamrakar, M. Sch ngut, D. Smrčka, F. Štěpánek, R. Ramachandran, Multidimensional population balance model development and validation of a reactive granulation process, Industrial Engineering & Chemistry Research, *2015*, 54(3)

A. Chaudhury, A. Armenante, R. Ramachandran, Compartment based population balance modeling of a high shear wet granulation process using data analytics, Chemical Engineering Research and Design, *2015*, 95

A. Chaudhury, H. Wu, M. Khan, R. Ramachandran. A mechanistic population balance model for granulation processes: Effect of process and formulation parameters. Chemical Engineering Science, *2014*,107

A. Chaudhury, I. Oseledets, R. Ramachandran. A computationally efficient technique for the solution of multi-dimensional PBMs of granulation via tensor decomposition. Computers and Chemical Engineering *2014*, 61

M. Sen, **A. Chaudhury**, R. Singh, J. John, R. Ramachandran, Multi-scale flowsheet simulation of an integrated continuous purification-downstream pharmaceutical manufacturing process, International Journal of Pharmaceutics, *2013*, 445(1-2)

A. Chaudhury, A. Kapadia, A.V. Prakash, D. Barrasso, R. Ramachandran. An Extended Cell-average Technique for Multi-Dimensional Population Balance Models describing Aggregation and Breakage. Advanced Powder Technology, *2013*, 24 (6)

A. Chaudhury, A. Niziolek, R. Ramachandran. Multidimensional Mechanistic Modeling of Fluid Bed Granulation Process: An Integrated Approach. Advanced Powder Technology, *2012*, 24

A. Chaudhury, S. Chakraborty, Dynamics of Mixing-Limited Pattern Formation in Non-Isothermal Homogeneous Autocatalytic Reactors: a Low-Dimensional Computational Analysis, Industrial Engineering and Chemistry Research, *2011*, 50 (8)

Book chapters:

A. Chaudhury, D. Barrasso, Daniel A. Pohlman, James D. Litster, R. Ramachandran, Mechanistic modeling of high shear and twin screw mixer granulation processes, Modeling of pharmaceutical unit operations in solid dosage forms, Elsevier, *2016*, In Press

A. Chaudhury, M. Sen, D. Barrasso, R. Ramachandran, Population balance models for pharmaceutical processes, Handbook in Process Simulation and Data Modeling in Solid Oral Drug Development and Manufacture, Humana Press, *2016*, 43-83

Teaching and research mentoring experience:

September 2011-June 2012 and September 2014-December 2014: *Teaching assistant*

Department of Chemical and Biochemical Engineering, Rutgers University

Courses taught:

Transport phenomenon I, Junior design and economics, Process control

Personnel mentored during PhD:

6 Undergraduate students, 2 Masters students.

476137 Application of Modeling and Optimization Methods in Biomedicine and Biorefineries

Abstract:

Application of modeling and optimization methods in biomedicine and biorefineries

Kirti M. Yenkie, University of Wisconsin-Madison Current Position: Second year Postdoctoral Researcher **Research Interests:** My research focuses on development of methodologies for predicting policies or forecasting decisions which can enhance the performance of existing processes as well as act as guidelines for developing novel processes in chemical and biological systems. I plan to incorporate methods such as mathematical modeling, simulation and numerical techniques for the process analysis and then apply principles from optimization and control for prediction of better operating policies and decisions. Policy prediction is particularly advantageous for processes in pharmaceutical industries where the product specifications are stringent and it is essential to maintain the product quality and purity. Also, medical treatments can be benefitted by prediction of drug dosing policies for chronic problems such as diabetes or treatments with adverse side effects such as cancer chemotherapy. These methods are also applicable to selection of best alternatives in a process; for example, selection of the best option among the available alternatives for wastewater treatment, based on initial conditions and final water quality requirements. To ensure the robustness of the results predicted, I intend to apply methods for uncertainty characterization and incorporation into process models through stochastic modeling and control methods. **Postdoctoral Project: Modeling, simulation, analysis and optimization of biorefineries Advisor:** Dr. Christos T. Maravelias, Chemical and Biological Engineering, University of Wisconsin-Madison. - Assessment of separation technologies for efficient recovery of bio-based chemicals from microbial cultures - Collaboration with experimentalists from Dr. Brian Pfleger's group for information about microbial strains, substrates, culture conditions, titers, productivity and yield. - Developing guidelines for selection of best possible separation scheme based on physical properties of selected chemicals by application of mathematical modeling, mixed-integer non-linear programming optimization and high performance computing methods. PhD Dissertation: Stochastic processes from batch crystallization to in-vitro fertilization (IVF) **Advisor:** Dr. Urmila M. Diwekar, Bioengineering, University of Illinois at Chicago (UIC). - Studied effects of uncertainties in batch crystallization, developed stochastic models using Ito processes and applied optimal control methods for predicting robust operating policies for enhanced product quality. - Analyzed clinical data for superovulation (first stage in IVF) cycles. Developed mathematical model for multiple ovulation using differential equations and applied optimization and control for predicting optimal hormone dosing policies. - Developed a new approach for parameter estimation in stochastic differential equations and applied it to develop stochastic models for multiple ovulation and follicle growth. Other Research Projects:

(i) Developed models for cellular systems in terms of stochastic differential equations as an alternative to discrete event simulation methods such as the Gillespie algorithm.

(ii) Application of optimal control methods to research on sustainability by using representative models for earth's ecological-economic interactions in terms of differential-algebraic equations and regulatory policies. Future Research Plans: ***Optimizing cancer chemotherapy*** - I plan to work on modeling the cancer propagation by using deterministic and stochastic models. Using these models, treatment therapies in terms of drug dosing regimens as well the time-gaps between dosing periods will be designed. I would also like to explore the possibility of treatment by combination of multiple drugs or multiple therapies such as chemotherapy, radiotherapy and gene therapy. ***Optimizing wastewater treatment methods*** - Processes such as activated sludge, advanced oxidation, ion-exchange and anaerobic digestion are used to treat wastewater for compliance with the environmental norms. I plan to use optimal control and MINLP optimization methods for prediction of treatment conditions as well as selection of best alternatives. **Teaching Interests:** My B. Tech. and M. Tech. degrees in Chemical Engineering and Ph.D. in Bioengineering have provided me with the opportunity to learn core chemical engineering subjects, like thermodynamics, mass transfer, reaction engineering, transport phenomena, etc. as well as some interdisciplinary topics like metabolic engineering, biochemical engineering and computational biology. Also, my interdisciplinary research facilitated me to develop skills in applied optimization, biomedicine, crystallization, separation processes, sustainability and advanced mathematics. I have been a teaching assistant for several courses such as chemical engineering thermodynamics, plant utility, reaction engineering laboratory, senior process design, and cell and tissue engineering during my graduate studies. I have graded assignments, conducted problem solving sessions and supervised experiments. I have also conducted tutorials for programming tools like MATLAB and GAMS. I have been a teaching associate for the introduction to optimization course at UIC and have taken lectures on genetic algorithms and simulated annealing. I have mentored undergraduate and graduate students for their research projects. I believe that I am capable of teaching most of the chemical engineering subjects, particularly the ones requiring computer programming and numerical methods such as transport phenomena, process design and optimization, bioprocess and biomedical engineering. I am also interested in developing and teaching innovative courses like modeling and control in biomedicine and stochastic processes in chemical engineering. Selected Publications

1) **Yenkie, K. M**.; Diwekar, U. 2013. Stochastic optimal control of seeded batch crystallizer applying Ito process. *Ind. Eng. Chem. Res.*, 52:108-122.

2) **Yenkie, K. M**.; Diwekar, U.; Bhalerao, V. 2013. Modeling the superovulation stage in in-vitro fertilization. *IEEE Trans. Biomed. Eng.*, 60(11): 3003-3008.

3) **Yenkie, K. M.**; Diwekar, U. 2014. Optimal control for predicting customized drug dosage for superovulation stage of in-vitro fertilization. *Journal of Theoretical Biology*, 355: 219-228.

4) **Yenkie, K. M**.; Diwekar, U.; Bhalerao, V. 2014. Modeling and prediction of outcome for the superovulation stage in in-vitro fertilization. *JFIV Reprod.Med.Genet.* 2(2):1000122(1-8).

5) **Yenkie, K. M**.; Diwekar, U. 2014. Comparison of optimal control methods for customized drug dosage prediction in superovulation stage of in-vitro fertilization. *Computers and Chemical Engineering*, 71: 708-714.

6) **Yenkie, K. M.**; Diwekar, U. 2014. Uncertainty in clinical data and stochastic model for in-vitro fertilization.

Journal of Theoretical Biology, 367: 76-85.

7) **Yenkie, K. M.**; Diwekar, U.; Linninger, A. 2016. Simulation-free estimation of reaction propensities in cellular reactions and gene signaling networks. *Computers and Chemical Engineering*, 87: 154-163.

476169 Advanced Adaptive Control Approaches for Complex Batch or Semi-Batch Operations

Abstract:

Research Plan:

Batch/semi-batch processes are an important set of unit operations in industries such as biofuel synthesis, synthesis of active pharmaceutical ingredients (APIs), specialty chemicals, etc. Such applications, in general, have complex dynamics and are operated in conjunction with other unit operations such as distillation, separation, etc. The products of these applications need to meet stringent quality requirements. Given the complex dynamics of such systems, it becomes essential to develop advanced control schemes that can utilize its multivariable dynamics in order achieve the desired control objectives and optimal performance of the process. Developing an all-encompassing control algorithm in order to achieve these goals involves addressing the following main challenges-

- **Adaptive identification of the pdf of the disturbances for optimal estimation of states and parameters**. While the states and uncertain parameters can be estimated using Bayesian estimation algorithms, these algorithms rely on the fact that the system model and pdfs of the unmeasured disturbances and measurement noise are perfectly known. However, this is not the case since there are uncertainties in the model parameters and the pdfs of the unmeasured disturbances and measurement noise are not known. Given the continuously time-varying nature of the batch/semi-batch processes I would like to work on developing online MLE-based approaches for estimation of the pdfs of the disturbances and noise. This would help maintain the optimal performance of the Bayesian state and parameter estimators, which will aid in improved state-feedback control and monitoring.

• **Stochastic iterative learning control and parameter learning for batch/semi-batch operations**. The repeated nature of batch/semi-batch operations means that knowledge of the control moves and parameters used in the previous batches can be used to improve the controller performance and parameter estimates for the current batch. Given the stochastic nature of batch/semi-batch operations and uncertainties in the system model arising from the time-varying nature of the processes, I would like to work on developing iterative learning control and iterative learning parameter estimation approaches that account for the probabilistic nature of these stochastic uncertainties in computing the control moves and parameter estimates respectively. This would also enable the formulation of a probabilistic control objective, thereby allowing for greater control over the final product quality.

Research Experience:

My research experience has been primarily in the areas of nonlinear Bayesian state estimation, system identification and model predictive control. During my PhD, I developed a maximum likelihood estimates (MLE)-based approach for identification of the covariance matrices associated with the Gaussian process and measurement noise from input-output process data for state and parameter estimation using extended Kalman filter (EKF)[1]. I have also developed an MLE-based approach for identification of grey-box models for state and paramter estimation, when the system under consideration is affected by time-correlated noise in the inputs or disturbances[2]. We have validated these approaches on an experimental setup as well. As a post-doctoral fellow at University of Alberta, I have worked on data-based approaches for modeling and classification for welding processes. We used approaches such as principal component analysis (PCA) and partial least squares (PLS) for classification of gas-metal arch welding jobs[3]. This work has led to an industry-academia collaboration and is being taken forward by the concerned PIs. As a post-doctoral fellow at UC Berkeley, my work has focussed on control-relevant model identification for stochastic model predictive control with a focus on cold atmospheric plasma (CAP) system for biomedical applications[4]. As a result I have research experience in multiple fields of Process Systems Engineering, both in computational and experimental studies.

References:

1. **V. A. Bavdekar**, A. P. Deshpande and S. C. Patwardhan, "Identification of process and measurement noise covariance for state and parameter estimation using extended Kalman filter", Journal of Process Control 21 (4) (2011) 585-601.
2. **V. A. Bavdekar** and S. C. Patwardhan, "Development of grey box state estimators for systems subjected to time correlated unmeasured disturbances", Journal of Process Control 22 (9) (2012) 1543-1558.
3. R. Ranjan, A. Talati, M. Ho, H. Bharmal, **V. A. Bavdekar**, V. Prasad, P. F. Mendez, "Multivariate data analysis of gas-metal arc welding process", Proceedings of the 9th International Symposium on Advanced Control of Chemical Processes (AdCHEM 2015).
4. **V. A. Bavdekar** and A. Mesbah, "Model predictive control with integrated input design for nonlinear systems with probabilistic uncertainties", accepted for the 11th International Symposium on Dynamics and Control of Process Systems, 2016.

Select Publications:

1. **V. A. Bavdekar** and A. Mesbah; *Stochastic nonlinear model predictive control with joint chance constraints*, accepted for the 10th IFAC Symposium on Nonlinear Control Systems, 2016.
2. **V. A. Bavdekar** and A. Mesbah; *Model predictive control with integrated input design for nonlinear systems with probabilistic uncertainties*, accepted for the 11th International Symposium on Dynamics and Control of Process Systems, 2016.
3. **V. A. Bavdekar** and A. Mesbah; *A polynomial chaos-based nonlinear Bayesian approach for estimating state and parameter probability density functions*, accepted for the 2016 American Control Conference.
4. **V. A. Bavdekar**, N. Nandola and S. C. Patwardhan; *Estimation of noise covariance matrices for state estimation of autonomous hybrid systems*, under review, Computers and Chemical Engineering.
5. **V. A. Bavdekar**, J. Prakash, S. C. Patwardhan and S. L. Shah; *A Moving window formulation for state estimation of systems with irregular and delayed measurements*, Industrial and Engineering Chemistry Research

53(35) (2014) 13750-13763.
6. **V. A. Bavdekar** and S. L. Shah; *Computing point estimates from a non-Gaussian posterior distribution using a probabilistic k-means clustering approach*, Journal of Process Control 24 (2) (2014) 487-497.
7. **V. A. Bavdekar** and S. C. Patwardhan; *Development of grey box state estimators for systems subjected to time correlated unmeasured disturbances*, Journal of Process Control 22 (9) (2012) 1543-1558.
8. **V. A. Bavdekar**, A. P. Deshpande and S. C. Patwardhan; *Identification of process and measurement noise covariance for state and parameter estimation using extended Kalman filter*, Journal of Process Control 21 (4) (2011) 585-601.

Research Interests:

My research interests are in the areas of nonlinear Bayesian state and parameter estimation, system identification, control-relevant model identification, stochastic MPC, data analysis uses methods such as PCA, PLS, etc. I'm interested in applications involving batch/semi-batch operations such as ethanol synthesis, algal biofuels, batch polymerizations, active pharmaceutical ingredient (API) synthesis, etc.

As a faculty member, I will use my expertise in the above areas of Process Systems Engineering to form a group that conducts leading research at the interface of systems' theory, applied mathematics and process engineering towards the development of advanced control algorithms for batch/semi-batch operations.

Teaching Interests:

At the undegraduate level, I would be interested in teaching the following courses (or their equivalents) and the laboratory/computational hours associated with them: Process Dynamics and Control, Computer-aided Process Design, Numerical Methods in Chemical Engineering and Process Optimization. I would also like to explore the option of introducing a course on Data Analysis and Statistics for Chemical Engineers, which will introduce students to basic experiment design and analyis techniques, with focus on the multivariate nature of chemical engineering data.

At the graduate level, I would be interested in teaching the following courses (or its equivalents): Digital Control and Advanced Numerical Methods for Chemical Engineers. I am interested in offering a course on Bayesian state estimation, given my research experience in this area and a course on System Identification, which can be a nice complement to the course on Digital Control.

476180 New Modeling and Decision-Making Paradigms in Systems Engineering

Abstract:

Research Interests: energy systems, sustainability, mathematical modeling, optimization

Systems engineering is an interdisciplinary field concentrated on the design and management of complex systems. The tools and perspectives of systems engineers are increasingly necessary to establish new links between research fields and drive inter-disciplinary innovation. This poster discusses my research vision in the context of three new paradigms in systems engineering.

Multi-Stakeholder Decision-Making

The design and control of engineering systems involves complex decision-making processes with multiple conflicting objectives of social, economic and environmental natures. Multi-objective optimization techniques have been applied to many chemical engineering domains, including separations, crystallization, chromatography, control systems, and chemical reactors, and have also been used to co-optimize economic and environmental metrics, quantified through life-cycle analysis. Most literature, however, assumes the perspective of a single decision-maker and considers only two or three competing objectives.

We recently proposed a framework for multi-stakeholder decision-making that addresses these two fundamental issues. The conflicting opinions from multiple stakeholders are transformed into individual stakeholder dissatisfactions that are minimized. We generalize previous approaches using the conditional-value-at-risk norm, which provides statistical interpretations to the compromise solutions and guarantees Pareto efficiency. Most notably, the method is computationally scalable. This is critical, as many sustainability studies consider hundreds of conflicting metrics. As a demonstration, we determine locations for waste-to-biogas conversion facilities while considering safety, transportation, water quality and cost priorities. Future work includes developing alternate solution metrics to explicitly address *fairness* and explore the theoretical connections with stochastic programming. I envision novel design methods for *sustainable chemical processes* based on the proposed framework, where the ambiguity of prioritizing different suitability metrics and life-cycle analysis standards is addressed through a multi-stakeholder perspective.

Stochastic Programming and High Performance Computing

In many engineering systems, uncertainty is addressed through safety margins centered on deterministic analysis. Stochastic analysis helps identify overly conservative designs and operating policies, thus further improving the realistic performance. In stochastic programming, the goal is to optimize a statistic (e.g., expected value) of the objective function while considering the probabilistic nature of input data (e.g., weather, future prices) propagated through constraints. Parallelized algorithms tailored for high performance computing platforms are of increased interest, as complex uncertainty structures can result in stochastic optimization problems with *millions* of variables and constraints.

Many energy systems benefit from stochastic analysis, as their performance depends both on design decisions and the operating policy. Consider, for example, sizing an energy storage system to be coupled with a solar or wind farm. If only average annual weather is considered, the system will likely be under-designed for extreme events. We are exploring the use of new stochastic programming algorithms to design concentrated solar power systems with thermal storage and other energy systems, while considering large historical datasets for weather and energy prices. This analysis is essential to properly value the flexibility provided by storage.

Energy Markets

We are studying the incentives embedded in multi-scale price signals from wholesale electricity markets for a diverse set of energy and manufacturing systems. We have determined that 70% of the economic opportunities are only available at fast timescales (seconds to minutes). Most literature, however, focuses on slower timescales, such as hourly prices. These findings inform the design of energy and manufacturing systems, and emphasize the importance of dynamic flexibility. This presents enormous new opportunities for energy intensive industrial systems or facilities with onsite utility plants. For example, providing regulation energy services to the market with a combined heat and power system may introduce low amplitude, high frequency (seconds) variations in steam availability for the coupled chemical process. Large separation systems take hours to days to reach steady-state, however, and may be resilient to these additional disturbances. Such a system has inherent dynamic flexibility that is not well understood, valued or utilized. Future research goals focus on characterizing the dynamic flexibility of industrial systems and developing multi-scale optimization architectures to exploit such flexibility. Long-term goals focus on inter-disciplinary research to redesign electricity markets to more effectively incentivize industrial participation and leverage dynamic flexibility from existing infrastructure to cost effectively support wide-scale adoption of wind and solar intermittent generation.

Additional details and publication record are available at www.alexdowling.net.

Teaching Interests:

I am qualified to teach all undergraduate chemical engineering core courses, and especially interested in teaching control, design, thermodynamics, separations/unit operations, mathematics, engineering problem solving, and statistics. I am also excited to teach graduate courses in optimization, numerical methods and process design, and look forward to developing new courses related to energy systems and sustainability.

Presentations at AIChE Annual Meeting:

A Framework for Multi-Stakeholder Decision-Making and Conflict Resolution
Session: Advances in Optimization I

Economics and Dynamic Flexibility of Concentrated Solar Power Technologies
Session: Sustainable Electricity: Generation and Storage

Exploiting Dynamic Flexibility to Enable Participation in Multi-Scale Electricity Markets
Session: Energy Systems Design and Operations II

476215 Management of Energy Supply Chains Under Uncertainty

Abstract:

Research Interests:

Energy demand, including electricity demand, is directly linked to factors such as economic development, population growth, and urbanization. However, meeting energy demands entails a critical challenge to use the Earth's energy resources in a sustainable fashion, enhancing quality of life, and facilitating human development. Specifically, a number of environmental impacts have been identified associated with the development of energy systems. In particular, emissions of Greenhouse Gases (GHG), the depletion and degradation of water sources, as well as the potential for underground water contamination, are major concerns associated with these systems. Thus, the design and planning of energy systems is a challenging problem where economic and environmental aspects need to be considered at both the process and supply chain levels.

Even though both levels have received significant attention from the engineering research community, there has been comparatively less research on the development of decision-making support tools for the investigation of integrated energy supply chains. For instance, only a few studies have been devoted to the integration of water management and shale gas supply chains design as well as to the integrated planning of generation and transmission expansion in power systems, which are systems that have been playing and will continue to play important roles in supplying our energy needs. I am especially interested in attacking this gap, through the development of optimization tools for supporting the decision-making processes in these two aforementioned energy systems at both national and regional scales. Specifically, I have targeted the development of decision-support optimization frameworks for the strategic design and planning of shale gas supply chains integrated with water management as well as generation and transmission expansion in power systems. In [1] we present a methodology for the simulation and evaluation of different well-pads designs for the exploitation of shale gas resources. Then, in [2] we present the mathematical formulation of a optimization framework for the integration of well-pad design and water management with shale gas supply chain design. The mathematical formulation consists of a novel large-scale Mixed Integer Linear Programming (MILP) model and a Mixed Integer Nonlinear Programming (MINLP) model. Furthermore, the analysis of the results reveals strong interactions between water management and the design of the shale gas supply chain. For instance, it was found that both the Total Dissolve Solid (TDS) concentration in wastewater and the fresh water availability are key factors not only for the selection of the water management strategy but also for the drilling scheme that should be implemented. Additionally, we demonstrated that a more accurate formulation, which results in a MINLP model, could lead to improved designs which increase by about 71% the economic benefit of the exploitation of shale gas plays in comparison with the MILP formulation. While the MILP is much more efficient from a computational viewpoint it is an approximation of the real problem and thus fails to capture some important synergies.

In addition to the shale gas project, I have been working on the development of an optimization framework for the integrated planning of power generation and transmission in interconnected power systems. In [3] we presented the mathematical formulation, MILP models, of the optimization framework as well as some revealing applications including "business as usual" and CO_2 mitigation policy scenarios. The novelty of this framework stems from the integration of power generation and transmission planning along with spinning and non-spinning reserve constraints as well as CO_2 emission constraints and mitigation options. In the case of CO_2 emission mitigation options, we consider the penetration of renewable energies, the integration of Carbon Capture and Sequestration (CCS) technologies, and the implementation of Demand Side Management (DSM) strategies. The results from the "business as usual" scenario revealed that fuel prices have a direct impact not only on the allocation of electricity generation resources but also on the selection of the fossil power plant technologies. Additionally, the analysis of the results reveals an interdependency between the transmission constraints and the decisions regarding the investment schedule and installation of new power plants. Moreover, the results from the CO_2 mitigation policy scenario demonstrated that the reconfiguration of existing power generation technologies is the most cost-effective alternative for the abatement of CO_2emissions. Furthermore, the integration of renewable energy sources was demonstrated to be preferred over the retrofitting of coal power plants with CCS technologies.

My ongoing research is concentrated on the characterization and modeling of the uncertainties inherent to the shale gas supply chain and the power generation and transmission planning problems, as well as on the development of stochastic optimization models. Specifically, global sensitivity analysis is being carried out in order to identify the most impactful among the many parameters in each model. This analysis will allow us to focused on parameters with the most impact on the optimal solution, thus increasing the computational effectiveness when modeling the uncertainty space. Moreover, two-stage stochastic models will be developed in order to solve the optimization problems under uncertainty.

Teaching Interests:

My academic and industrial experiences have offered me the opportunity to not only strengthen my knowledge of mathematical modeling, simulation, and optimization of processes and systems but also to develop deep understanding on the applicability of such tools to deal with real world design and planning problems in different domains, i.e. refinery operations, shale gas supply chains, and power systems. Therefore, my teaching interest pertains to decision making theory and applications, i.e. courses on computer aided process design and planning or optimization theory with applications to industrial processes and systems. However, I am well prepared to teach courses in the chemical engineering core, e.g., material and energy balances, separation processes, and process economics, design, and evaluation, as well as courses on mathematical and numerical methods.

References:

[1] Calderón AJ, Guerra OJ, Papageorgiou LG, Siirola JJ, Reklaitis G V. Preliminary Evaluation of Shale Gas

Reservoirs: Appraisal of Different Well-Pad Designs via Performance Metrics. *Ind Eng Chem Res*2015;54:10334–49. doi:10.1021/acs.iecr.5b01590.

[2] Guerra OJ, Calderón AJ, Papageorgiou LG, Siirola JJ, Reklaitis G V. An Optimization Framework for the Integration of Water Management and Shale Gas Supply Chain Design. *Comput Chem Eng* 2016; 92:230–255. doi:10.1016/j.compchemeng.2016.03.025

[3] Guerra OJ, Tejada DA, Reklaitis G V. An optimization framework for the integrated planning of generation and transmission expansion in interconnected power systems. *Appl Energy* 2016;170:1–21. doi:10.1016/j.apenergy.2016.02.014.

476225 Energy Systems Analysis to Enable a Sustainable Economy

Abstract:

Research Interests:

Current energy and chemical sectors predominantly depend on fossil resources. Limited fossil resources and significant increase in atmospheric greenhouse gases urge to develop and implement renewable energy conversion processes for a more sustainable economy.

The overarching goal of my research is to meet essential human needs including transportation fuel, chemicals, electricity and water, primarily from renewable energy sources, and ensure a smooth transition to a sustainable economy. My research focuses on multiscale systems analysis of energy conversion technologies for synthesis of energy efficient, synergistic processes incorporating process intensification for optimal utilization of resources.

During my PhD, I have focused on determining the optimal synergistic process design to produce liquid fuel from biomass and natural gas, developing a methodology to synthesize efficient solar thermal hydrogen production processes, identifying and optimizing efficient solar thermal hydrogen production process designs based on the developed methodology, synthesis of integrated thermal desalination processes, creating efficient and intensified processes to convert solar thermal energy into electricity and to continuously supply to the grid by integrated energy storage solutions. Moreover, I have worked on policy analysis of large-scale energy storage options for uninterrupted renewable power supply, particularly on closed loop CO_2-driven energy storage solutions.

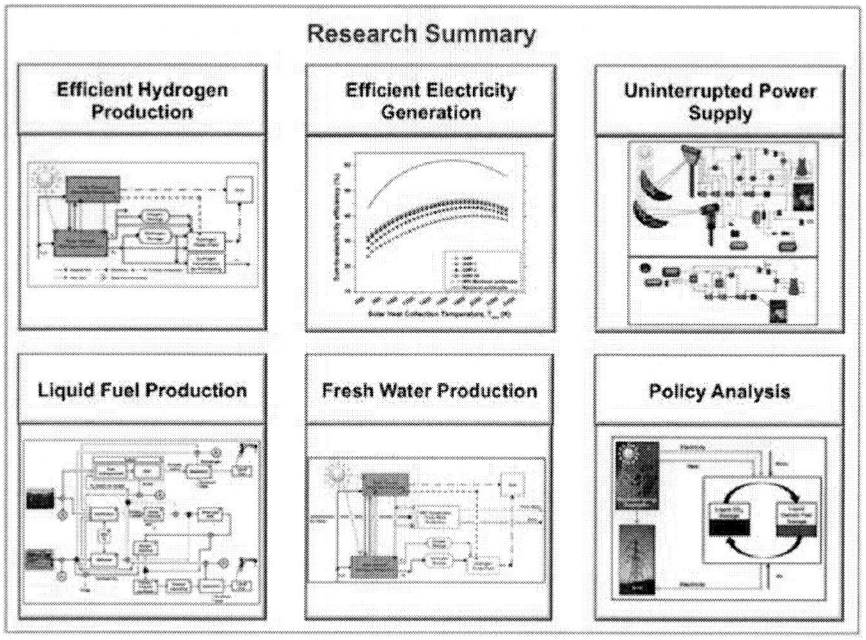

Teaching Interests:

I had extensive teaching experience as a teaching assistant at Purdue University. I gave several lectures and prepared several teaching materials including a new curriculum and a complete set of lab instructions for the newest version of Aspen Plus design class. I am passionate about teaching and I am particularly interested in enhancing students' understanding on energy conversion technologies, process design and advanced mathematical modeling. I am also interested in teaching thermodynamics, separations, heat and mass transfer.

I received several teaching awards including 2016 Graduate School's Excellence in Teaching Award, Purdue University's highest award in recognition of graduate student teachers, Excellence for Undergraduate Teaching Assistant Award as voted by the graduating seniors for Teaching Excellence in Chemical Engineering and 2013 CETA (Committee Education of Teaching Assistants) Award.

References

[1] Gençer E, Mallapragada DS, Marechal F, Tawarmalani M, Agrawal R. Round-the-clock power supply and a sustainable economy via synergistic integration of solar thermal power and hydrogen processes, Proceedings of the National Academy of Sciences (PNAS), 112(52), 15821-15826, 2015.

[2] Gençer E, Al-musleh E, Mallapragada D, Agrawal R. Uninterrupted Renewable Power through Chemical Storage Cycles. Current Opinion in Chemical Engineering, 5, 29-36, 2014.

[3] Gençer E, Agrawal R. A commentary on the US policies for efficient large scale renewable energy storage systems: Focus on carbon storage cycles, Energy Policy, 88, 477-484, 2016.

[4] Gençer E, Tawarmalani M, Agrawal R, Integrated solar thermal hydrogen and power coproduction process for continuous power supply and production of chemicals, Computer Aided Chemical Engineering, 37, 2291-2296, 2015.

[5] Gençer E, Mallapragada DS, Tawarmalani M, Agrawal R. Synergistic biomass and natural gas conversion to liquid fuel with reduced CO2 emissions. Computer Aided Chemical Engineering, 34:525-530, 2014.

476271 Deterministic and Robust Model-Based Strategies for the Online Multi-Level Optimization of Batch Operations

Abstract:

Research Interests:

In the last few decades, there has been a steady growth in applying optimization techniques to the solution of problems of industrial importance, e.g. supply-chain management, operational planning/scheduling and optimal process control/dynamic optimization. This trend has been motivated by both increasing global competition and tighter environmental regulations, which are driving industry towards reducing costs as well as minimizing environmental impacts.

Although extensive process systems engineering research has been reported so far, most of it has addressed continuous processes and only limited attention has been devoted to batch operations, which are widely used to produce high added-value products (specialty chemicals, pharmaceuticals, cosmetics, etc.). Given the opportunities that this application domain provides, I have focused my research on methodologies/applications in operational planning/scheduling and optimal process control/dynamic optimization of batch processes. However, I have also addressed supply-chain problems[1] and continue having interests in such research area.

Most recently, I have focused on developing methods for the optimal online management of both single batch units and entire batch plants. I have formulated and implemented two dynamic optimization/optimal control algorithms applicable to single batch operations (BSMBO&C [2,3], RBSMBO&C [4,5]) and a further one designed for (multi-unit) batch plants (MUBSMBO&C). These strategies are very flexible and can tackle the online optimization/optimal control of both single batch cycles and entire production campaigns. Moreover, RBSMBO&C also implements a novel dynamic scenario selection concept to handle model uncertainty.

My ongoing research involves studying uncertainty propagation/Bayesian inference with the goal of further improving RBSMBO&C. Moreover, I am also planning to extend MUBSMBO&C and make it capable of handling model uncertainty. Longer-term research perspectives include addressing the problem of supply-chain management of batch manufacturing systems and developing integrated frameworks for the simultaneous online optimization of multiple decisional levels.

Teaching Interests:

My research work and graduate studies have provided me with deep knowledge of control theory, numerical methods, dynamic/steady-state modelling and coding/parallel programming (especially in C++). Therefore, I would be interested in teaching any course (broadly) related to these subjects, e.g. process control/design, numerical methods, dynamic/steady-state modelling and process optimization.

References:

1. Rossi F, Manenti F, Reklaitis G. A general modular framework for the integrated optimal management of an industrial gases supply-chain and its production systems. *Computers and Chemical Engineering*. 2015;82:84-104.
2. Rossi F, Manenti F, Buzzi-Ferraris G. A novel all-in-one real-time optimization and optimal control method for batch systems: Algorithm description, implementation issues, and comparison with the existing methodologies. *Industrial and Engineering Chemistry Research*. 2014;53:15639-15655.
3. Rossi F, Copelli S, Colombo A, Pirola C, Manenti F. Online model-based optimization and control for the combined optimal operation and runaway prediction and prevention in (fed-)batch systems. *Chemical Engineering Science*. 2015;138:760-771.
4. Rossi F, Reklaitis G, Manenti F, Buzzi-Ferraris G. Multi-scenario robust online optimization and control of fed-batch systems via dynamic model-based scenario selection. *AIChE Journal*. 2016; DOI: 10.1002/aic.15346.
5. Rossi F, Manenti F, Pirola C, Mujtaba I. A robust sustainable optimization & control strategy (RSOCS) for (fed-)batch processes towards the low-cost reduction of utilities consumption. *Journal of Cleaner Production*. 2016;111:181-192.

476776 Modeling and Control of Hybrid and Nonsmooth Process Systems

Abstract:

Research Interests: Control and systems theory; Optimization; Biological and chemical systems.

Broadly, my interests lie in the modeling, simulation, and control of hybrid and nonsmooth systems. Hybrid systems display a mixture of continuous and discrete dynamics, where the continuous portion is often modeled using differential equations. Nonsmooth systems refers here to dynamic systems whose participating functions exhibit non-differentiability, which can be problematic in conventional simulation and optimization methods. Possible sources of such behavior include thermodynamic phase changes, flow regime transitions, and valve activation, among others. Hybrid and nonsmooth modeling frameworks have become widely applied tools in a variety of real-world problems. Examples are found in pharmaceutical manufacturing, biological systems with changes in population behavior, secure communications via supervisory switching control, and multistream heat exchanger design. My pursuits and contributions are found in this area of research, with an emphasis on sensitivity, stability, and optimization techniques for use in controller and parametric design.

Research Experiences: I am currently a postdoctoral fellow working on dynamic optimization of campaign continuous manufacturing processes, under the supervision of Professor Paul I. Barton (*successful proposal: NSERC Postdoctoral Fellowship*). This collaborative work is a part of the Novartis-MIT Center for Continuous Manufacturing initiative to revolutionize drug manufacturing. I have developed new theory to characterize sensitivities of nonsmooth dynamic systems (e.g., [1, 2]), mirroring the classical results on the topic. These findings lay the theoretical groundwork upon which efficient numerical methods can be designed for large-scale problems, where standard methods can fail. This work thus extends nonsmooth equation-solving and optimization methods to problems with nonsmooth differential-algebraic equations embedded.

I obtained my Ph.D. in Applied Mathematics at the University of Waterloo, under the supervision of Professor Xinzhi Liu (*successful proposal: NSERC Alexander Graham Bell Canada Scholarship*). My Ph.D. studies related to the analysis of complex physical phenomena exhibiting time-delays, which are an inherent feature in a wide range of biological and physical problems. By uncovering underlying mechanisms, I made contributions here in the stabilization, synchronization, and robust control of nonlinear hybrid systems with time-delays (e.g., [3, 4]). I designed switching control strategies to achieve desired performance outcomes where classical methods are either unavailable or inadequate. I connected the theoretical findings to biological systems through the introduction and analysis of hybrid modeling approaches to disease transmission (e.g., [5, 6, 7]).

Future Directions: My goal as faculty would be to combine my training and research experiences in applied mathematics and engineering to uniquely approach high-impact problems in chemical and biological process control. Using my expertise in theoretical and computational mathematics, and by collaborating with experimentalists, I plan to design, develop, implement, and optimize state-of-the-art process control methods. Short-term projects include efforts in characterizing the behavior of high-index differential-algebraic equations and new approaches to nonlinear model predictive control. Long-term projects suitable for Ph.D. students are comprised of problems in stochastic modeling of biological systems, fundamental extensions in the setting of optimal control theory, and the development of global optimization theory and methods for hybrid process systems. Industrial applications in mind include ones in continuous manufacturing processes, power and energy systems, and biological population models.

Teaching Interests: Optimization; Computational methods; Advanced dynamic systems; Process control.

I was the full-time lecturer for two undergraduate engineering courses (approximate class size of 100 each) at the University of Waterloo. I have also acted as a teaching assistant in mathematics and engineering courses (11 undergraduate, 3 graduate level) on topics ranging from computational mathematics to applied functional analysis. My teaching philosophies are instructed by these opportunities, along with the completion of the Kaufman Teaching Certificate Program at MIT. For example, my course design is dictated as follows:

- My lecture and assessment design for first and second-year courses (e.g., differential equations, numerical methods) will aim to solidify the students' foundations going forward in their field of study.
- Upper-year and graduate-level courses (e.g., control theory, optimization) will be designed so that students understand the material in a way that an expert in the field would expect.
- In specialized graduate-level courses (e.g., modeling and simulation of hybrid systems, nonsmooth analysis, topics in mathematical biology), research-oriented goals will be included whenever possible.

Selected Publications (11 accepted peer-reviewed journal articles as lead author):

[1] **P.G. Stechlinski** and P.I. Barton, **In Press**. "*Generalized Derivatives of Nonsmooth Differential-Algebraic Equations.*" Journal of Optimization Theory and Applications, DOI: 10.1007/s10957-016-0988-9.

[2] **P.G. Stechlinski** and P.I. Barton, **Accepted Invited Session Paper**. "*Generalized Derivatives of Optimal Control Problems with Nonsmooth Differential-Algebraic Equations Embedded.*" 55th IEEE Conference on Decision and Control 2016.

[3] X.Z. Liu and **P.G. Stechlinski**, **2016**. "*Hybrid stabilization and synchronization of nonlinear systems with unbounded delays.*" Applied Mathematical Modelling, 280, 140-161.

[4] X.Z. Liu and **P.G. Stechlinski**, **2014**. "*Hybrid control of impulsive systems with distributed delays.*" Nonlinear Analysis: Hybrid Systems, 11, 57-70.

[5] X.Z. Liu and **P.G. Stechlinski**, **2015**. "Application of control strategies to a seasonal model of Chikungunya disease." Applied Mathematical Modelling, 39, 3194-3220.

[6] X.Z. Liu and **P.G. Stechlinski**, **2014**. "SIS models with switching and pulse control." Journal of Applied Mathematics and Computation, 232, 727-742.

[7] X.Z. Liu and **P.G. Stechlinski**, **2013**. "Transmission dynamics of a switched multi-city model with transport-related infections." Nonlinear Analysis: Real World Applications, 14, 264-279.

476958 Energy Security and Environmental Protection: Bridging the Gap Between Theory and Application

Abstract:

Research Interests:

The convergence of recent advances in Computer-aided Engineering and theoretical developments for design and operation of industrial processes has provided unprecedented opportunities to explore new solutions for energy security and environmental protection. This poster presents my recent contributions and future plans in three dimensions; (1) The theory of process design and control, (2) Novel applications for energy security and environmental protection, and (3) Solution algorithms. The justification is that delivering energy and environmental impacts necessitates in-depth theoretical understanding, and solving industrial-scale problems requires tailor-made solution algorithms.

On the theoretical dimension, I demonstrate that integration of the process design and control decisions opens up new venues to energy efficiency and environmental protection. The features of interest include economic performance, dynamic controllability, and flexibility in the presence of disturbances. On the application dimension, I have focused on (1) production of fuel and chemical through biomass pyrolysis, (2) CO_2 capture from power generation systems, and (3) Integration of solar and wind energies into smart electricity grids. The features of interest include whole-system analysis, multi-scale modelling, process integration, and optimization under uncertainty. Finally, applying the advanced theories to address complex applications requires tailored-made solution algorithms. Over the course of my academic studies, I have developed and applied a variety of computer-aided engineering tools spanning over a wide spectrum of off-the-shelf model libraries to customizable simulation tools to low-level optimization programming. Here, the features of interest include multi-scale model integration, optimization with explicit constraints, and exploiting problem logic for improving solution algorithms.

Teaching Interests:

My teaching philosophy is highly inspired by my research vision. It will be based on a balanced combination of theory and application and will employ enabling software tools. I believe the theory of process design and control is an indispensable element of the Chemical Engineering curriculum and will enable future industrial practitioners and academic researchers to close the gap between theory and application. The students should be equipped with a variety of simulation, optimization and programming software tools, that will enable quantification and optimization of economic and environmental performance of industrial processes. I believe students should be inspired by novel applications for energy security and environmental protection and be motivated to deliver impacts to societal wellbeing and economic prosperity.

Publications -The papers are published in highly-ranked journals. The corresponding author is shown by *. The co-authors with equal contribution are shown by ¥. The current bibliometric statistics of my publications are h-index 9, i10-index 7, and Citations >240, according to Google Scholar.

[1] **Sharifzadeh M***, (2013). Integration of process design and control: a review, ***Chemical Engineering Research and Design***, 91 (12), 2515–2549 (Link).

[2] **Sharifzadeh M***, Thornhill NF, (2013). Integrated design and control using a dynamic inversely controlled process model. ***Computers & Chemical Engineering,*** 48, 121–134. (Link).

[3] **Sharifzadeh M***, (2013). Implementation of an inversely controlled process model for integrated design and control of an ETBE reactive distillation, ***Chemical Engineering Science***. 92, 21–39, (Link).

[4] **Sharifzadeh M***, Thornhill NF, (2012).Optimal selection of control structures using a steady-state inversely controlled process model. ***Computers & Chemical Engineering***, 38 (5), 126–138, (Link).

[5] **Sharifzadeh M***, Rashtchian D, Pishvaie MR, Thornhill NF, (2011). Energy induced separation network synthesis of an olefin compression section: a case study. ***Industrial & Engineering Chemistry Research,*** 50 (3), 1610–1623, (Link).

[6] Wang L, **Sharifzadeh M**, Templer R, Murphy RJ*, (2012). Technology performance and economic feasibility of bioethanol production from various waste papers. ***Energy & Environmental Science***, 5, 5717-5730, (Link).

[7] Wang L, **Sharifzadeh M**, Templer R, Murphy RJ*, (2013). Bioethanol production from various waste papers: Economic feasibility and sensitivity analysis. ***Applied Energy,*** 111, 1172–1182, (Link).

[8] Chen L¥, **Sharifzadeh M¥**, MacDowell N, Welton T, Shah N, Hallett JP*, (2014). Inexpensive ionic liquids: [HSO4]–based solvent production at bulk scale. ***Green Chemistry***, 16, 3098-3106. (Link).

[9] **Sharifzadeh M*,** Richards C, Liu K, Hellgardt K, Chadwick D, Shah N. (2015). An integrated process for biomass pyrolysis oil upgrading: the synergistic approach. ***Biomass & Bioenergy***. 76, 108–117, (Link).

[10] **Sharifzadeh M*,** Wang L, Shah N, (2015). Integrated bio-refineries: CO_2 utilization for maximum biomass conversion. ***Renewable and Sustainable Energy Reviews***, 47, 151–161, (Link).

[11] **Sharifzadeh M*,** Wang L, Shah N, (2015). Decarbonisation of olefin processes using biomass pyrolysis oil. ***Applied Energy***, 149, 404–414, (Link).

[12] **Sharifzadeh M***, Cortada Garcia M, Nilay Shah, (2015). Supply chain network design and operation: Systematic decision-making for centralized, distributed, and mobile biofuel production using mixed integer

linear programming (MILP) under uncertainty. Biomass and Bioenergy, 2015, 81, 401–414, (Link).

[13] **Sharifzadeh M***, Shah N, (2015). Carbon capture from natural gas combined cycle (NGCC) power plants: solvent performance comparison at industrial scale. ***AIChE Journal***, 62 (1), 166–179, (Link), Invited paper.

[14] **Sharifzadeh M***, Shah N, (2015). Comparative studies of CO_2 capture solvents for gas-fired power plants: Integrated modelling and pilot plant assessments. ***International Journal of Greenhouse Gas Control***, 43, 124–132, (Link).

[15] **Sharifzadeh M***, Bumb P, Shah N, (2015). Carbon capture from pulverised coal power plants (PCPP): solvent performance comparison at industrial scale. ***Applied Energy***, 163, 423–435, (Link).

[17] Eini S, Shahhosseini HR, Javidi M, **Sharifzadeh M***, Rashtchian D. Inherently safe and economically optimal design using multi-objective optimization: the case of a refrigeration cycle. Accepted for publication at ***Process Safety and Environmental Protection***.

[16] **Sharifzadeh M***, Meghdari M, Rashtchian D. Integrated safe design and operation of Thermal Power System-Solid Oxide Fuel Cell (SOFC) Triple Combined-cycle System. Under review.

[18] **Sharifzadeh M***, Lubiano H, Shah N. Optimal design and operation of integrated renewable energy systems. Under review.

[19] **Sharifzadeh M***, Siirola JD, Grossmann IE. The hybrid reformulation algorithm for generalized disjunctive programs: Implementation in Pyomo. Under review.

[20] **Sharifzadeh M***, Konda M, Wang L, Guo M, Cortada Garcia M, Sadeqzadeh M, Hallet J, Shah N. Biomass pyrolysis and upgrading: the state of art and research frontiers. Under review.

481297 Automatic Exploration of Potential Energy Surfaces: Towards Reaction Mechanisms, Rate Constants, and Product Branching Ratios From First Principles

Abstract:

Research Interests:
My research focuses on computer simulations based on first principles of gas-phase reactions in combustion, atmospheric, and astrochemical systems. With modern computing power and the latest methodologies, ab initio calculations are yielding better agreement with experimental findings yet having the ability to cover a large range of parameters and conditions due to the simulations being done in silico.

In general, the scope of my research can be divided into two categories. The first category comprises of developing methodologies and strategies for automatic transition state (TS) finding and exploring of Potential Energy Surfaces (PES). The second category is applying strategies, either existing or developed by my research, to chemical systems of interests. Later, the methodologies and applications will be extended to condensed phase systems such as catalysis and organometallic systems.

Successful Proposals:
Japan Society for the Promotion of Science (JSPS) Postdoctoral Fellow, Kyoto University, Japan.

Postdoctoral Projects:
"*Sensitivity Analysis on Extinction Strain Rate Calculations of Natural Gas (NG) and Refinement to Detailed Kinetic Models*" and "*Automatic Reaction Discovery of Chemical Reactions*" under supervision of William H. Green, Department of Chemical Engineering, Massachusetts Institute of Technology

"*Automatic Exploration of the Potential Energy Surface from the Pyrolysis and Oxidation of Jet-Fuels*" under supervision of Keiji Morokuma, Fukui Institute for Fundamental Chemistry, Kyoto University

"*Global Investigation of the Pyrolysis of C_1-C_3 Hydrocarbons: Towards the Construction of Detailed Kinetic Models*" under Keiji Morokuma, Cherry L. Center for Scientific Computation, Emory University

Ph.D. Dissertation:
"*Ab Initio Quantum Chemical Studies on Neutral-Radical Reactions of Ethynyl (C2H) and Cyano (CN) with Unsaturated Hydrocarbons.*"

Research Experience:
Succinctly, my research experience has been towards the elucidation of reaction mechanisms, rate parameters, and product branching ratios based on first principle calculations of Potential Energy Surfaces (PES). In graduate school, the reaction mechanisms were constructed used *a priori*knowledge based on the chemical intuition of how molecules react. Then as a postdoctoral associate and postdoctoral fellow, I became a practitioner of automatic reaction search methods that explore the PES without any human intervention or bias of reactivity. This allowed for a systematic and comprehensive search of reaction mechanisms. In each of my postdoctoral appointments, the strategy of automatic reaction search varied, with the Global Reaction Route Mapping (GRRM) strategy employed under the supervision of Keiji Morokuma at Emory and Kyoto University and various chain-of-state methods under the supervision of William H. Green at Massachusetts Institute of Technology. In the latter position, I also did method development towards novel automatic search methods as well as analyzing detailed kinetic mechanisms reaction steps and kinetic parameters.

Teaching Interests:
In the summer and fall semester of 2014, I was an Adjunct Professor at Georgia Perimeter College, a part of the Georgia State University system. There, I taught undergraduate chemistry laboratory courses for Department of Natural Sciences. In graduate school, I was head TA for physical chemistry laboratories for junior and senior level undergraduate students majoring in chemistry. Part of the duties included devising my own curriculum based on given experiments, creating my own grading scale, and physical chemistry laboratory maintenance of equipment and ordering of supply. I would have an open-door office hour, allowing students to come anytime they wish with advanced notice, aside from having a set office hours.

Future Direction:
As a faculty, I would like to divide work in both method development and applications of methodologies. I would focus on method development of automatic search methods of PES, combining my past knowledge and skills in this area. I would also focus on the application of methodologies by elucidating reaction mechanisms and product branching ratios of important chemical reactions in combustion, atmospheric, and astrochemical systems. I would have a research intensive program that includes undergraduate students working on summer research projects towards the application of methods to chemical systems either existing in literature or developed by my group.

Selected Publications:
Ryazanstev, M.*; **Jamal, A.***; Maeda, S.; Morokuma, K. Global Investigation of Potential Energy Surfaces for the Pyrolysis of C_1-C_3 Hydrocarbons: Toward the Development of Detailed Kinetic Models from First Principles. *Physical Chemistry Chemical Physics.* **2015**, *17,* 27789-27805. [*Contributed equally]

Yang, T.; Muzangwa, L.; Kaiser, R. I.; Jamal, A.; Morokuma, K. A Combined Crossed Molecular Beam and Theoretical Investigation of the Reaction of the Meta-Tolyl Radical with Vinylacetylene – Toward the Formation of Methylnaphthalenes. *Physical Chemistry Chemical Physics.* **2015**, *17,* 21564-21575.

Parker, D.; Dangi, B.; Kaiser, R.; Jamal, A.; Ryazantsev, M.; Morokuma, K. Korte, A.; Sander, W. An Experimental and Theoretical Study on the Formation of 2-Methylnaphthalene (C11H10/C11H3D7) in the Reactions of the Para-Tolyl (C7H7) and Para-Tolyl-d7 (C7D7) with Vinylacetylene (C4H4). *Journal of Physical Chemistry A.* **2014**, *118*, 2709-2718.

Dangi, B.; Parker, D.; Kaiser, R.; **Jamal, A.**; Mebel, A. A Combined Experimental and Theoretical Study on the Gas-Phase Synthesis of Toluene under Single Collision Conditions. *Angewandte Chemie International Edition*. **2013**, *52*, 7186-7189.

Jamal, A.; Mebel, A. An Ab Initio/RRKM Study of the Reaction Mechanism and Product Branching Ratio of Ethynyl Radical with 1,2-Butadiene. *Chemical Physics Letters*. **2011**, *518*, 29-37.

Jamal, A.; Mebel, A. Reactions of C2H with 1- and 2-Butynes: An Ab Initio/RRKM Study of the Reaction Mechanism and Product Branching Ratio. *Journal of Physical Chemistry A*. **2011**, *115*, 2196-2207.

Jones, B.; Zhang, F.; Kaiser, R.; **Jamal, A.;** Mebel, A.; Cordiner, M.; Charnley, S. Formation of Benzene in the Interstellar Medium. *Proceedings of the National Academy of Sciences*. **2011**, *108*, 452-457.

Jamal, A.; Mebel, A. An Ab Initio/RRKM Study of the Reaction Mechanism and Product Branching Ratios of the Reactions of Ethynyl Radical with Allene and Methylacetylene. *Physical Chemistry Chemical Physics*. **2010**, *12*, 2606.

459989 Economic Model Predictive Control for Integrating Scheduling and Dispatch of Microgrid Power Systems

Abstract:

Microgrids are small, autonomous power systems that utilize local generation, storage, and controllable loads to serve local energy demands. Microgrids combine the ideas of demand response, renewable power generation, and cogeneration to offer a future energy supply paradigm with a focus on sustainability. However, an important issue in the operation of these microgrid systems is their potential negative impact on the operation and stability of the macrogrid. In short, the integration of distributed generation may exacerbate the stochasticity of the residual and lead to increased reserve requirement and operational cost for the macrogrid [1]. Existing microgrid control literature has addressed this issue in a variety of ways, e.g. penalizing real-time energy purchases [2], designing auction-based markets [3], incentivizing demand response [4], and operating microgrids as islanded systems [5,6]. In this work, we propose an approach based on economic model predictive control (E-MPC) which can be used to integrate scheduling and control decisions so that energy exchange can be carried out in a grid-friendly manner. In particular, energy exchange commitments will be scheduled 24-hours in advance and real-time deviations will be kept small.

A case study is considered for a prototype microgrid consisting of a bi-directional connection to the macrogrid, photovoltaics, microturbines, a battery bank, flexible air conditioning, and an auxiliary electric boiler. This microgrid regulates indoor air temperature, supplies electricity, and generates hot water for a medium office building. A scheduling problem is considered for meeting these energy demands and coordinating energy exchange with the macrogrid. However, the inherent time scales involved in microgrid scheduling (i.e. hours to days) are not significantly separated from the time scales involved in control (i.e. seconds to minutes). Thus, a mixed integer linear E-MPC problem which incorporates low order process models is formulated for scheduling this microgrid on an hourly basis. Chance constraints are used to reduce the probability of commitment violations due to uncertainties in weather processes and local demands. Within each hour, frequent recourse optimization is used to update microgrid dispatch as uncertain conditions are realized.

The performance of this control approach is analyzed with respect to the operational cost, curtailment of renewable power, frequency and magnitude of commitment violations, and satisfaction of thermal demands. Computation time and differences between realized dispatch decisions and scheduling predictions are also considered. In addition, this presentation will highlight some of the important differences between integration of scheduling and control in traditional chemical systems and in these small power systems.

[1] Bayod-Rújula, A. A. (2009). Future development of the electricity systems with distributed generation. *Energy*, 34(3), 377–383.

[2] Atzeni, I., Ordonez, L. G., Scutari, G., Palomar, D. P., & Fonollosa, J. R. (2014). Noncooperative Day-Ahead Bidding Strategies for Demand-Side Expected Cost Minimization with Real-Time Adjustments: A GNEP Approach. *IEEE Transactions on Signal Processing*, 62(9), 2397–2412.

[3] Liu, G., Xu, Y., & Tomsovic, K. (2016). Bidding Strategy for Microgrid in Day-Ahead Market Based on Hybrid Stochastic/Robust Optimization. *IEEE Transactions on Smart Grid*, 7(1), 227–237.

[4] Nguyen, D. T., & Le, L. B. (2015). Risk-Constrained Profit Maximization for Microgrid Aggregators With Demand Response. *IEEE Transactions on Smart Grid*, 6(1), 135–146.

[5] Trifkovic, M., Marvin, W. A., Daoutidis, P., & Sheikhzadeh, M. (2014). Dynamic real-time optimization and control of a hybrid energy system. *AIChE Journal*, 60(7), 2546–2556.

[6] Hosseinzadeh, M., & Salmasi, F. R. (2015). Robust Optimal Power Management System for a Hybrid AC/DC Micro-Grid. *IEEE Transactions on Sustainable Energy*, 6(3), 675–687.

469629 Moving Horizon Closed-Loop Scheduling of Processes Operating Under Dynamic Constraints

Abstract:

Moving Horizon Closed-Loop Scheduling of Processes Operating under Dynamic Constraints

Richard C. Pattison[1], Cara R. Touretzky[1], Iiro Harjunkoski[2], Michael Baldea[1]

[1]McKetta Department of Chemical Engineering, The University of Texas at Austin, Austin, TX

[2]ABB Corporate Research, Ladenburg, Germany

The deregulation of electricity markets has brought new challenges and opportunities in the operation of electricity-intensive chemical processes. Day-ahead/real-time electricity tariffs feature fast changes in energy prices, and it is desirable to impose frequent changes in production targets (e.g., hourly changes in production rate) to capitalize on these fluctuations. In general, production levels (and electricity consumption) are increased at times when prices are low, and the excess product is stored or buffered. As a result, when electricity prices are at a peak, the production levels (and electricity consumption) can be decreased. From the perspective of the electric grid, this operating paradigm effectively amounts to storing energy in the form of chemical products as a means of providing demand response service. However, this constitutes a significant departure from the traditional operation of chemical processes, where setpoint changes are assumed to be infrequent (compared to the process time constant) and the process operates chiefly at steady state. Under the latter conditions, scheduling calculations do not require knowledge of the process dynamics. However, in the more dynamic operating paradigm associated with fast fluctuations in market conditions and prices, transient operations become the norm and the process may never reach steady state in between production target changes. Thus, it is crucial to ensure that a schedule is dynamically feasible, which motivates incorporating a representation of the process dynamics in the scheduling model.

The resulting scheduling problem is in the form of a mixed-integer dynamic optimization (MIDO), and its solution poses several practical challenges. When a first-principles, detailed dynamic process model is used, the problem is often difficult (if not impossible) to solve in a practical amount of time. This challenge is amplified by the fact that, in practical applications, the schedule must likely be recalculated periodically to account for updated price- and demand forecasts, and disturbances. In our previous work [2] we addressed this challenge by constructing low-order dynamic "scale bridging" models (SBMs) [3]. The SBMs predict the closed-loop evolution of variables relevant to the scheduling calculation in response to production target changes, and alleviate the computational challenges associated with embedding the dynamic model in the scheduling calculation.

In this talk, we present a new approach to accounting for changes in market conditions and disturbances to the process [4]. To this end, we close the scheduling loop with feedback based on measurements of the identified scheduling-relevant process variables, in conjunction with an observer structure that is used to update the SBM states. The feedback mechanism is implemented using, i) periodic updates of the schedule over a moving horizon to incorporate the updated price and demand forecasts, and, ii) event-driven updates that account for measurable process and market disturbances.

The theoretical developments are then demonstrated on an industrial-scale cryogenic air separation unit (ASU) model. A system identification routine, based on using transient historical production data, was used to construct SBMs. The ASU is outfitted with a liquid nitrogen storage tank, which enables the process to adjust the production rate in response to variable electricity prices while satisfying product demand by re-gasifying the liquid nitrogen inventory. The optimal schedule is solved using a sequential dynamic optimization framework, and results in a 4.8% savings in electricity cost over the course of 4 days in comparison to a scenario where the production rate is constant. We also evaluate disturbance scenarios comprising, i) a planned, ii) an unplanned, and, iii) a random drop in demand due to a failure at the customer site. In all cases, the moving horizon framework successfully manages the production and inventory levels without any dynamic of product constraint violations.

[1] M. Baldea and I. Harjunkoski. Integrated production scheduling and process control: A systematic review. Comput. Chem. Eng., 71:377–390, 2014

[2] R. C. Pattison, C. R. Touretzky, T. Johansson, I. Harjunkoski, and M. Baldea. Optimal process operations in fast-changing electricity markets: Framework for scheduling with low-order dynamic models and an air separation application. Ind. Eng. Chem. Res., 55(16):4562–4584, 2016.

[3] J. Du, J. Park, I. Harjunkoski, and M. Baldea. Time scale bridging approaches for integration of production scheduling and process control. Comput. Chem. Eng., 79:59–69, 2015.

[4] R. C. Pattison, C. R. Touretzky, I. Harjunkoski, M. Baldea. Moving Horizon Production Scheduling using Dynamic Process Models. Submitted to AIChE J, 2016.

456009 The Control of Self-Interested Agents: Learning from Nature's Wisdom of Crowds

Abstract:

Multi-agent collective dynamics is a prevalent phenomenon. Examples include self-assembly, coupled oscillators, cell migration, quorum sensing, birds flocking, ant colony, etc., which all share something in common: Individuals interact with the crowds that they belong to and benefit from such interactions. "The wisdom of crowds," a century-old discovery in statistics, offers an intuitive explanation: The average opinion formed by a large population can be remarkably accurate. However, supporting theories assume that individual opinions are independent and unbiased. Interactions would thus undermine the wisdom. Those theories, therefore, fail to explain why interacting agents in nature still choose to rely on such wisdom.

In this talk, we formulate this phenomenon as a standard control problem and identify several properties overlooked by existing theories. This work explains the wisdom of *interacting* crowds and sheds light in achieving the elusive goal of controlling self-interested agents.

Our model consists of two processes: First, the agents are self-interested and utility-maximizing. They follow an internal optimization mechanism to achieve the goals of survival or proliferation. In the cell migration system, for example, each cell responds to the concentration gradient signals and adjusts its motion for self-preservation. Sensory noise and cognitive limitation might disturb such process. Second, the agents partially follow the signals from their neighbors, the quorum, or simply the mean-field. We formulate those signals as the control input, which is "softer" than what one might observe in process control. We name such control model "soft regulation."

Through mathematical proofs, simulations, and Amazon Mechanical Turk experiments with human subjects, we show that soft regulation can improve both the individual and the social welfares from the open loop. This result thus agrees with what we observe from nature without restricting us to the condition that requires agents to be independent. Furthermore, we identify how uncertainty, individual heterogeneity, social influence, and network structure affect the efficacy of this mechanism. Such characterizations will help us design and improve control mechanisms for sociotechnical systems.

After all, just like flocking birds, human individuals, organizations, and even nations are also crowds of self-interested agents with common goals of survival and proliferation. Those goals may be improving health and fitness, regulating emerging technologies, or fighting climate change. Learning from nature's wisdom of crowds is the first step of achieving the control of self-interested agents.

469338 Subspace Based Quality Control of Variable Duration Batch Processes

Abstract:

Specialty chemical production makes up an important sector of the chemical processing industry. Examples include armaceutical and high quality polymer products. In these processes, economic gains are realized by achieving tightly specified product quality. This and additional factors, such as the ability to produce a wide range of product grades and the high cost of raw materials, favors using batch reactors for these products. While batch reactors provide benefits, they also introduce a variety of difficulties for closed loop control. In this work we address some of those issues and present a novel, data-driven, model predictive control scheme where the objective is explicitly formed to reach product quality. Furthermore, we demonstrate the ability of our method to incorporate time as a decision variable and appropriately implement corrective discrete events such as mid-batch additions.

In a typical batch process, first the reactor is charged with some ingredients then a transforming process is carried out. As the transformation proceeds, the reactor contents transform exposing a wide range of operating states. Since dynamic behavior of chemical systems is dependent on state, these processes are characterized by transient dynamic behavior. Clearly, the traditional control objective of stabilizing operation around a steady-state condition is not meaningful. Instead, the control objective for this class of processes is to reach a desired product quality by the end of the batch. This objective is particularly important to reject feed stock variations that introduce variance which, left uncorrected, would propagate to the product quality. However, direct feedback control on quality is complicated by a number of factors. Foremost of these, is that the desired quality measurements are often not available online. Furthermore, development of reliable first principal models for these processes is often impracticable because of the complex, nonlinear behavior and the wide range of operating conditions. Traditionally then, indirect approaches such as trajectory tracking have been adopted.

Trajectory tracking control for batch processes involves using closed-loop control to track pre-determined trajectories of measurable process variables. These pre-determined trajectories are often taken from historically successful trajectories. Industrially, tracking is usually achieved using PI control (possibly with some gain scheduling to address the nonlinear process dynamics). Additionally, literature contributions have highlighted the improved trajectory tracking using model predictive control [2, 3, 8, 9]. However, even if perfect tracking is achieved, the desired terminal quality may not be met. This is because the predefined trajectories fail to account for disturbances in initial conditions and throughout the batch.

An alternative to trajectory tracking that can be used for direct quality control is inferential control. In these approaches, a data-driven model is built that relates measurable process variables (temperatures, stirrer-torque, etc.) to product quality. To identify this relationship, a database of historical measurement trajectories and the resulting batch qualities is used. The ability of these methods to correctly drive the process to the desired quality naturally depends on two factors: the predictive ability of the identified model, and the validity of the model when applied to new closed loop input policies.

One highly successful inferential modeling and control method has been latent variable control [6, 7, 1, 4]. In these approaches, latent variable PCA or PLS models are identified from historical batch data. Specifically, the historical batch data is unfolded into a regressor matrix where each column (variable) contains a time indexed measurement from a historical batch. Model parameters are then identified to relate this regressor matrix to the available terminal quality measurements. The resulting model can be used in closed loop control by constructing a vector from new time-indexed measurements as the batch progresses, then applying the identified model to make quality predictions for candidate input trajectories. This method has shown success both in simulation studies and in industrial application. However, one significant restriction on these methods is the dependence on time-indexed model parameters. To address this issue, techniques for aligning batches (for instance using monotonically increasing variables) have been investigated. Nonetheless, the ability of these methods to use time as a decision variable to achieve the desired quality is inherently limited.

In recent work, we presented an alternative, state-space motivated, inferential quality control approach. The key idea is to break the problem of quality modeling and prediction into two parts. In the first part, we identify the dynamic (state-space) behavior of the process. This step is motivated by the understanding that the dynamic evolution of the process is fundamentally state-dependent and time independent. Therefore, by capturing a time independent state-space model of the system, we obtain a model that can be applied at any sampling instant without the need for batch alignment. To identify these dynamic models, we adapt the well studied method of subspace identification for use with historical and identification batch data. The second part of the quality modeling problem is making end-of-batch quality predictions. Fundamentally, we understand that the state of a system captures the entire instantaneous condition of the system. This motivates the claim that the terminal state identified in part one can be used to predict the terminal quality. Since the objective is to produce product in a relatively narrow quality range, a linear model is sufficient to relate terminal state and quality. Taken together, these two models, the dynamic state-space model and the quality model, provide a causal way to relate a candidate input trajectory and the resulting quality and can therefore be implemented in a shrinking horizon MPC.

In the current work, we extend these results to explicitly demonstrate the advantage afforded by the time independence of our proposed approach. Significantly, the time-independence of the model permits batch duration to be readily added as a decision variable in the MPC formulation. This is especially advantageous since the duration of a batch is one of the key factors influencing the resulting product quality. Therefore, enabling batch duration to be a manipulated variable substantially improves the ability of the controller to reach the desired quality set-point. [5]

To demonstrate the efficacy of the proposed modeling and control scheme, a simulation study of a batch PMMA polymerization reactor is presented. First a database of historical and identification batches is constructed. To demonstrate the ability of this methodology to account for discrete events, a discrete addition of reactants half way through the reaction is included in the nominal batch recipe. This database is then used to identify a dynamic state-space model and a quality model. Validation results, for batch data excluded from the training data, demonstrate the predictive ability of the combined dynamic and quality models. Finally, the models are implemented in a shrinking horizon MPC. The resulting quality control ability is demonstrated through quality set-point changes and feed stock disturbance rejection.

References

[1] Siam Aumi, Brandon Corbett, Tracy Clarke-Pringle, and Prashant Mhaskar. Data-driven model predictive quality control of batch processes. AIChE Journal, 59(8):2852–2861, Aug 2013.

[2] Siam Aumi, Brandon Corbett, Prashant Mhaskar, and Tracy Clarke-Pringle. Data-based modeling and control of nylon-6, 6 batch polymerization. IEEE Transactions on Control Systems Technology, 21(1):94–106, Jan 2013.

[3] Siam Aumi and Prashant Mhaskar. An adaptive data-based modeling approach for predictive control of batch systems. Chemical Engineering Science, 91:11–21, Mar 22 2013.

[4] Brandon Corbett, Brian Macdonald, and Prashant Mhaskar. Model predictive quality control of polymethyl methacrylate. IEEE Transactions on Control Systems Technology, 23(2):687–692, Mar 2015.

[5] Brandon Corbett and Prashant Mhaskar. Subspace identification for data-driven modeling and quality control of batch processes. AIChE Journal, 62(5):1581–1601, May 2016.

[6] J Flores-Cerrillo and JF MacGregor. Control of batch product quality by trajectory manipulation using latent variable models. Journal Of Process Control, 14(5):539–553, Aug 2004.

[7] J Flores-Cerrillo and JF MacGregor. Iterative learning control for final batch product quality using partial least squares models. Industrial & Engineering Chemistry Research, 44(24):9146–9155, Nov 23 2005.

[8] J Flores-Cerrillo and JF MacGregor. Latent variable mpc for trajectory tracking in batch processes. Journal of Process Control, 15(6):651–663, Sep 2005.

[9] Masoud Golshan, John F. MacGregor, Mark-John Bruwer, and Prashant Mhaskar. Latent variable model predictive control (lv-mpc) for trajectory tracking in batch processes. Journal of Process Control, 20(4):538–550, Apr 2010.

450737 Optimization Models for Shale Gas Development Planning: A Real-World Marcellus Shale Case Study

Abstract:

Optimization Models for Shale Gas Development Planning:

A Real-World Marcellus Shale Case Study

Markus G. Drouven[1] and Ignacio E. Grossmann[2]

Department of Chemical Engineering

Carnegie Mellon University

Pittsburgh, PA 15213

[1]mdrouven@cmu.edu, [2]grossmann@cmu.edu

Abstract

The production of shale gas from unconventional resource plays is transforming the energy landscape in the United States. Advances in production technologies, notably the dual application of horizontal drilling and hydraulic fracturing, allow the extraction of vast deposits of trapped natural gas that, until recently, were uneconomic to produce. The Energy Information Administration predicts that shale gas will account for 50% of total U.S. natural gas production by 2040 [1]. Natural gas demand is also expected to increase in the electric power and nearly all other industrial sectors. The future development of shale gas resources requires an extensive expansion of the existing gas production, transmission, and processing infrastructure. Virtually all stages of the shale gas supply chain need to be expanded and upgraded to match the ever growing natural gas supply and demand [2]. Since the necessary capital investments for drilling rigs, pipelines, boosting stations and midstream processing facilities are substantial, the long-term planning of upstream production and natural gas transmission is a key challenge.

In this presentation we address a real-world case study for which the general problem can be stated as follows. Within a potential shale gas development area an upstream operator has identified a set of candidate well pads from which shale gas may or may not be extracted. To extract the gas the operator can develop, i.e., drill and fracture a limited number of wells at every candidate pad. Ultimately, the operator wishes to sell extracted gas at a set of downstream delivery nodes which are typically located along interstate transmission pipelines. For this purpose a gathering system superstructure has been identified. This superstructure specifies all feasible, alternative options for laying out gathering pipelines to connect candidate well pads with the given set of delivery nodes. In addition, the superstructure indicates candidate locations for compressor stations as well as the location of existing processing plants.

The long-term shale gas development problem involves planning, design and strategic decisions. In terms of planning decisions the operator needs to decide: a) where, when and how many wells to drill at every candidate well pad, b) whether selected wells should be shut-in and, if so, for how long, and c) how to allocate drilling rigs over time. The design decisions involve: a) where to lay out gathering pipelines, b) what size pipelines to install, c) where to construct compressor stations, and d) how much compression power to provide. The upstream operator's objective is to determine the optimal development strategy by making the right planning and design decisions such that the net present value is maximized.

In this work [3] we present a real-world Marcellus Shale case study that was performed in close collaboration with the EQT Corporation. EQT is one of the largest exploration and production operators in the Appalachian Basin. Their business activities include locating productive natural gas deposits, drilling wells to extract gas and transporting gas through pipelines to transmission and distribution system. The objective of this case study was twofold: (a) validate and refine the proposed modeling framework and (b) attempt to quantify the economic potential of mixed-integer optimization models for long-term shale gas development planning.

As part of a "lookback analysis" the proposed model was applied to an existing gathering system in the Appalachian Basin – owned and operated by the EQT Corporation – using real, historic data. By comparing the development strategy proposed by the optimization with the actual, historic development strategy, we are able to demonstrate and quantify the economic potential of optimization tools for shale gas development. Our findings suggest, in the past, development strategies were primarily driven by trying to drill as many wells as possible at a given location and turning them in line as quickly as possible. However, considering the characteristically steep decline curves of shale gas wells we find that these development strategies led to gathering equipment – including pipelines and compressors – being over-sized and therefore heavily under-utilized over long periods of time.

Our optimization, on the other hand, reveals that so-called return-to-pad operations appear much more suitable and economically promising for shale gas development projects. The idea behind return-to-pad operations is to drill and complete only a small number of wells at a time, but then to "return to the pad" eventually to repeat the process. This strategy allows upstream operators to size gathering pipelines and compressor smaller and to keep them "full", i.e., utilized, over extended periods of time.

Our comprehensive economic analysis reveals that return-to-pad operations and an increased equipment utilization could have improved the profitability of this particular development project by several million U.S. dollars. To the best of our knowledge this is the first case study that provides such deep and realistic insight into past and future shale gas development strategies in collaboration with a major natural gas producer.

References

[1] U.S. Energy Information Administration (EIA). Annual Energy Outlook with Projections to 2040. April 2013.

[2] Goellner, J. F. Expanding the Shale Gas Infrastructure. *AIChE CEP* August 2012, 49-52.

[3] Drouven, M. G.; Grossmann, I. E. Multi-Period Planning, Design and Strategic Models for Long-Term Quality-Sensitive Shale Gas Development. *AIChE J.* 2016 (accepted for publication).

465458 A Theoretical and Computational Study of Continuous-Time Process Scheduling Models in the Context of Adjustable Robust Optimization

Abstract:

The optimal scheduling of multi-purpose batch plant operations has attracted significant attention in the literature [1-2] due to both its economic impact as well as its combinatorial complexity and the unique methodological challenges that are associated with it. As a consequence, a variety of alternative optimization models have been proposed in the literature. In this work we conduct a comprehensive comparison across continuous-time models for multi-purpose batch plant scheduling under decision-dependent uncertainty, evaluating the features and computational performance of these models in the context of our recently introduced Adjustable Robust Optimization framework (ARO) [3].

Uncertainty about process parameters at the time of decision-making is a fundamental reality when scheduling the operation of multi-purpose batch plants. This uncertainty might pertain to parameters such as processing durations, production yields, material prices and demands, as well as availability of raw materials and utilities. It is well-understood that optimizing the schedule deterministically, using nominal values for all these parameters, often leads to a schedule that is suboptimal or even infeasible in view of the actually realized scenario. One of the methods proposed to mitigate the effects of uncertainty is Robust Optimization (RO) [4-7], which guarantees the feasibility of the schedule for any realization of the uncertain parameters in a postulated uncertainty set. The traditional approach, however, treats all of the decisions as "here-and-now," yielding conservative solutions in general. In addition, the "here-and-now" assumption renders RO inadequate to handle some frequent cases such as the case of uncertainty in process yields or the handling of zero-wait materials.

We have previously showed [3] that deferring a subset of the decisions for later ("wait-and-see") can lead to more profitable, yet equally robust, schedules. Furthermore, it was shown that the possibility for time-dependent parameter realizations must be taken into account for a realistic treatment of the problem, leading to the need to consider more involved, decision-dependent uncertainty sets. The application of the ARO framework using affine decision rules in this scheduling context revealed solutions that improved up to 50% compared to the corresponding RO solution, while accounting for the same level of uncertainty in the processing times, and exhibiting the potential for even better solutions when utilizing more complex decision rules. The computational tractability of the approach however, is expected to vary with the selection of deterministic model that is used as the basis for the computations as well as the subset and combinations of the parameters that are considered uncertain.

In this study we discuss the strengths and limitations of various continuous-time deterministic models [8-15] in terms of their amenability to serve as the basis for both a traditional RO as well the ARO framework, and we discuss available strategies to mitigate some of those limitations. A comprehensive computational study using standard literature benchmarks was performed for various levels of uncertainty in the fixed and variable (related to batch size) processing times. Furthermore, the flexibility of the ARO framework allows us for the first time in the open literature to also present robust solutions of problems that involve uncertainty in the production yields, which is a common issue in the chemical industry.

[1] C. Floudas, and X. Lin, 2005. Mixed integer linear programming in process scheduling: Modeling, algorithms, and applications. Annals of Operations Research 139 (1), 131–162.

[2] I. Harjunkoski, C. Maravelias, P. Bongers, P. Castro, S. Engell, I. Grossmann, J. Hooker, C. Mendez, G. Sand, J. Wassick, 2014. Scope for industrial applications of production scheduling models and solution methods. Computers & Chemical Engineering 62, 161–193.

[3] N. Lappas, and C. Gounaris, 2016. Adjustable robust optimization for process scheduling under uncertainty. AICHE Journal 62, 1646–1667.

[4] X. Lin, S. Janak, C. Floudas, 2004. A new robust optimization approach for scheduling under uncertainty: I. bounded uncertainty. Computers & Chemical Engineering 28 (6), 1069–1085.

[5] S. Janak, X. Lin, C. Floudas, 2007. A new robust optimization approach for scheduling under uncertainty: II. uncertainty with known probability distribution. Computers & Chemical Engineering 31 (3), 171–195.

[6] Z. Li, M. Ierapetritou, 2008. Robust optimization for process scheduling under uncertainty. Industrial & Engineering Chemistry Research 47 (12), 4148–4157.

[7] Z. Li, R. Ding, C. Floudas, 2011. A comparative theoretical and computational study on robust counter optimization: I. robust linear optimization and robust mixed integer linear optimization. Industrial & Engineering Chemistry Research 50 (18), 10567–10603.

[8] C. Maravelias, and I. Grossmann, 2003. New general continuous-time state - Task network formulation for short-term scheduling of multipurpose batch plants. Industrial & Engineering Chemistry Research 42 (13), 3056–3074.

[9] P. Castro, A. Barbosa-Povoa, H. Matos, A. Novais, 2004. Simple Continuous-Time Formulation for Short-Term Scheduling of Batch and Continuous Processes. Industrial & Engineering Chemistry Research 43 (1), 105–118.

[10] S. Janak, X. Lin, C. Floudas, 2004. Enhanced Continuous-Time Unit-Specific Event-Based Formulation for Short-Term Scheduling of Multipurpose Batch Processes: Resource Constraints and Mixed Storage Policies. Industrial & Engineering Chemistry Research, 43(10), 2516–2533.

[11] A. Sundaramoorthy, and I. Karimi, 2005. A simpler better slot-based continuous-time formulation for short-term scheduling in multipurpose batch plants. Chemical Engineering Science 60 (10), 2679–2702.

[12] D. Giménez, G. Henning, C. Maravelias, 2009. A novel network-based continuous-time representation for process scheduling: Part I. Main concepts and mathematical formulation. Computers & Chemical Engineering, 33(9), 1511–1528.

[13] N. Susarla, J. Li, I. Karimi, 2009. A novel approach to scheduling multipurpose batch plants using unit-slots. AICHE Journal, 56(7), 1859–1879.

[14] R. Vooradi, and M. Shaik, 2012. Improved three-index unit-specific event-based model for short-term scheduling of batch plants. Computers & Chemical Engineering, 43(19), 148–172.

[15] R. Seid, T. Majozi, 2012. A robust mathematical formulation for multipurpose batch plants. Chemical Engineering Science, 68(1), 36–53.

467328 Decision Making Under Uncertainty in Integrated Planning of Generation and Transmission Capacities in Interconnected Power Systems

Abstract:

Power systems play an important role in the energy sector, supplying a significant share of the total energy demand for the industrial and residential sectors. However, the power sector also accounts for about 90% and 42% of energy related water withdrawal and CO_2emissions, respectively [1, 2]. Therefore, the optimal design and planning of power systems have become one of the most relevant issues in the energy sector. Besides the innate complexity of the system, the handling of the uncertainties in aspects such as electricity demand, fossil fuel cost, and electricity generation rates from renewable sources, have been recognized as one of the biggest challenges of decision making processes in the power sector.

This work presents a methodology for addressing the problem of planning under uncertainty of the integrated generation and transmission expansion of power systems. First, based on the framework developed by Guerra et al. [3], a global sensitivity analysis is carried out in order to rank the sources of uncertainty in the power system. A Monte Carlo sampling method is coupled with Sobol's sensitivity indices [4, 5] to quantify the effect of uncertain parameters on the net present value of the total cost, which is the performance criteria used in the aforementioned framework. Then, two-stage stochastic formulations are derived based on the deterministic Mixed Integer Linear Programming (MILP) optimization model developed by Guerra et al. [3]. First-stage decisions are associated with investment in new power generation and transmission assets, while second-stage decisions are associated with power allocation, non-spinning and spinning reserve allocation, and power transmission flows. Two metrics, i.e. *expected value of perfect information* (EVP) and *value of stochastic solution*(VSS) [6], are used to assess the impact of modeling uncertainty and using the stochastic programming model in power system planning problems. Additionally, scenario reduction strategies are explored to overcome computational challenges. The features of the

proposed methodology are illustrated through a real world case study.

References

[1] IEA, "*Water for Energy: Is energy becoming a thirstier resource?*", *World Energy Outlook*, pp. 1–33, 2012.

[2] IEA, "*World Energy Outlook Special Report 2015: Energy and Climate Change*," 2015.

[3] O. J. Guerra, D. A. Tejada, and G. V. Reklaitis, "*An optimization framework for the integrated planning of generation and transmission expansion in interconnected power systems*," *Applied Energy*, vol. 170, pp. 1–21, May 2016.

[4] I. M. Sobol, "*Global sensitivity indices for nonlinear mathematical models and their Monte Carlo estimates*," *Mathematics and Computers in Simulation*, vol. 55, pp. 271–280, 2001.

[5] A. Saltelli, S. Tarantola, and K. P.-S. Chan, "*A Quantitative Model-Independent Method for Global Sensitivity Analysis of Model Output*," *Technometrics*, vol. 41, no. May, pp. 39–56, 1999.

[6] J. Birge and F. Louveaux, *Introduction to stochastic programming*. 2011.

465020 An Efficient Method for Deriving Normalization Constants for Eigenfunctions of Sturm-Liouville Problems and Its Application to the Graetz Problem for Diffusive and Convection Heat/Mass Transfer

Abstract:

An Efficient Method for Deriving Normalization Constants for Eigenfunctions of Sturm-Liouville Problems and Its Application to the Graetz Problem for Diffusive and Convection Heat/Mass Transfer

Joel A. Paulson[a], T. Alan Hatton[a], and Richard D. Braatz[a,[1]]

[a]Department of Chemical Engineering, Massachusetts Institute of Technology, Cambridge, MA 02139

Sturm-Liouville (SL) problems arise in the solution methods to an important class of partial differential equations (PDEs) relevant in many industrial and scientific applications [1,2,3]. The importance of SL theory is rooted in the fact that the solutions of certain PDEs can be expanded as an infinite series of normalized eigenfunctions of the SL operator multiplied by spectral coefficients [4]. The spectral coefficients can be computed by matching the infinite series at the initial/boundary conditions. Although methods for determining the eigenfunctions (and corresponding eigenvalues) of the SL operator are well-known, few procedures are available for normalizing this set of eigenfunctions (i.e., ensuring the inner product of an eigenfunction with itself is unity). Most normalization procedures require a large amount of time and effort and may not yield an analytic expression in a simple and readily computable form.

This talk presents a new method for analytically determining the normalization constants for the eigenfunctions of regular, singular, and periodic SL problems. The resulting formula only requires derivatives with respect to the eigenvalue parameter, and is much simpler to use than the direct method (whereby the squared eigenfunction must be integrated) or the Greenâ€™s function method (that requires the generation of the full spectral presentation of the SL operator) [5]. This new method offers a gateway for analytically determining normalization constants for some of the more complicated eigenfunctions arising in SL problems that have previously been elusive.

The power of this newly proposed method is demonstrated on the well-known Graetz problem that describes diffusive and convective heat or mass transfer through a pipe in the presence of fully developed laminar flow [4,5]. Combining the proposed method with some other mathematical results, we were able to derive, for the first time (to the best of the authorâ€™s knowledge), an explicit expression for the orthonormal set of eigenfunctions for the Graetz problem that is valid and easily computable over the entire eigenspectrum.

References

[1] I. Stakgold. Boundary Value Problems of Mathematical Physics, volume 11. Macmillian, 1972.

[2] S. S. Bayin. Sturm-Liouville Theory. In Mathematical Methods in Science and Engineering (Chapter 8). John Wiley & Sons, Inc., Hoboken, NJ, 2006.

[3] M. A. Al-Gwaiz. Sturm-Liouville Theory and its Applications. Berlin, Springer, 2008.

[4] W. M. Deen. Analysis of Transport Pheomena (Topics in Chemical Engineering), volume 3. Oxford University Press, New York, NY, 1998.

[5] E. Papoutsakis, D. Ramkrishna, and H. C. Lim. The extended Graetz problem with Dirichlet wall boundary conditions. Applied Scientific Research, 36 (1980): 13-34.

[1]Corresponding author: braatz@mit.edu.

An Efficient Method for Deriving Normalization Constants for Eigenfunctions of Sturm-Liouville Problems and Its Application to the Graetz Problem for Diffusive and Convection Heat/Mass Transfer Joel A. Paulson[a], T. Alan Hatton[a], and Richard D. Braatz[a,[1]]

[a]Department of Chemical Engineering, Massachusetts Institute of Technology, Cambridge, MA 02139

[1]Corresponding author: braatz@mit.edu.

456032 Systematic Framework for Carbon Dioxide Capture and Utilization Processes to Reduce the Global Carbon Dioxide Emissions

Abstract:

In the year 2013, 9.5 billion metric tons of carbon dioxide gas was emitted into the air, and each year this amount is increasing [1]. Carbon dioxide emissions are of particular concern as they represent 80% of greenhouse gas emissions and therefore are a large contributor to global warming. Among the two approaches that are currently being investigated, carbon capture and storage (CCS) and carbon capture and utilization (CCU) [1] to address this issue, the later approach is more promising as it reuses captured carbon dioxide, as a fuel, reactant, solvent, and others, to produce valuable products. There is not only a need for technologies for capture and utilization, via conversion, but also there are numerous questions that need to be resolved. For example, which higher value chemicals can be produced, what are their current demands and costs of production, and, how much of the captured carbon dioxide would be utilized? Also, how much carbon dioxide would be indirectly generated due to the capture and utilization efforts? Can the regulated carbon dioxide emission reduction targets be achieved only through the CCU and/or CCS efforts? Therefore, there is a need for a systematic computer-aided framework through which the issue of global carbon dioxide emissions can be investigated in terms of different available capture-utilization technologies, solution methods, and benefit scenarios, with the objective to determine more sustainable solutions within an appropriate application boundary. The framework would need to provide, amongst other options: useful data from in-house databases on carbon dioxide emission sources; mathematical models from a library of process-property models; numerical solvers from library of implemented solvers; and, work-flows and data-flows for different benefit scenarios to be investigated.

It is useful to start by developing a prototype framework and then augmenting its application range by increasing the contents of its databases, libraries and work-flows and data-flows. The objective is to present such a prototype framework with its implemented database containing collected information-data on various carbon dioxide emission sources and available capture-utilization technologies; the model and solution libraries [2]; and the generic 3-stage approach for determining more sustainable solutions [3] through superstructure (processing networks) based optimization – adopted for global carbon dioxide emission problems. More specifically, the prototype framework and the three-stage approach adopted for systematic and sustainable design of carbon capture and utilization processes incorporates (i) process synthesis stage, the determination of a processing path from a network of alternatives; (ii) process design, the design and analysis of a process or generated processing path in terms of process "hot-spots" or deficiencies to set design targets for improvement; and (iii) innovative and more sustainable design, determination of solutions matching the design targets. In this way, the starting point is an analysis of the current carbon dioxide emission status and the end point is an analysis of the more sustainable solutions in terms of one or more carbon dioxide benefit scenarios.

The developed framework contains a database constructed based on a developed knowledge representation structure, which has collected data on carbon dioxide emission sources, capture technologies, carbon dioxide-based reaction paths, and known utilization technologies leading to various products. The database currently has 5 carbon dioxide emission sources, 3 capture technologies, 10 conversion routes leading to 8 products, and over 50 reaction paths for many more products. The framework also integrates a software tool (the Super-O interface [2]) containing a library of process models and links to numerical solvers that enables the generation of processing networks (superstructure) and the determination of the optimal processing route. Finally, the framework has access to, a collection of tools for analysis, such as economic analysis, sustainability and life cycle assessment, and links to detailed process simulation (process simulators).

Following the work-flow and data-flow implemented in the framework, data for the carbon capture and utilization alternatives are retrieved from the database and linked to form a network using the interface tool, Super-O. For each alternative, models for economic and environmental impacts are available; the promising utilization products considered are methanol, dimethyl carbonate, and succinic acid. The optimal utilization process (or processes) is determined via a superstructure-based method (using solvers in GAMS). Then, the carbon dioxide emission values for the utilization routes are compared with the current industrial production routes for the listed products in order to identify those that have zero or negative emission values (objective function). Subsequently, the identified utilization processes are designed rigorously and further improved with unique technology and integration options. Using this approach, the utilization processes for methanol, dimethyl carbonate and succinic acid give the following interesting result: by converting at least 70% (methanol), 40 % (dimethyl carbonate) and 20% (succinic acid) of the current industrial processes to the corresponding utilization processes, the emissions from these processes can be neutralized, representing over 100 million metric tons of emissions that are neutralized. Note that the issue of product demand increase has not been considered. Also, on a global scale, this reduction of carbon dioxide emissions represents a very small but very valuable reduction. A more complete sustainability analysis is needed to enlarge the boundary of the analysis; for example, the boundary can change in terms of reduction of non-renewable resources and the use of renewable energy sources, as well as more efficient and sustainable designs of current production routes. While a combination of methods is needed to globally address carbon dioxide emissions and other aspects of sustainability, this systematic approach to designing carbon capture and utilization processes shows the capacity for such processes to reduce emissions and improve sustainability while producing valuable products. The current prototype framework with its implemented methods and tools is a small but important step. Collaboration and integration of data, methods and tools is necessary to provide a more sustainable solution to the global carbon dioxide emission problem.

References

[1] Yuan, Z., Eden, M.R., Gani, R., 2016. Toward the Development and Deployment of Large-Scale Carbon Dioxide Capture and Conversion Processes. Ind. Eng. Chem. Res. 55, 3383–3419. doi:10.1021/acs.iecr.5b03277

[2] Bertran, M.-O., Frauzem, R., Zhang, L., Gani, R., 2016. A generic methodology for superstructure optimization of different processing networks, in: Kravanja, Z. (Ed.), Proceedings of the 26th European Symposium on Computer Aided Process Engineering – ESCAPE 26. Elsevier B.V., Portorož, Slovenia.

[3] Babi, D.K., Holtbruegge, J., Lutze, P., Gorak, A., Woodley, J.M., Gani, R., 2015. Sustainable process synthesis–intensification. Comput. Chem. Eng. 81, 218–244. doi:10.1016/j.compchemeng.2015.04.030

462685 Design of Energy Saving CO2 Separation Process Using Circulating Fluidized Bed

Abstract:

By fossil fuel combustion, large amounts of greenhouse gases (GHGs) are generated and emitted into the atmosphere. The reduction of carbon dioxide (CO2) emission has become a major target in efforts to suppress global warming. Carbon capture and storage (CCS) has attracted attention in the past decades to reduce greenhouse gas (GHG) emissions. CO2 separation from gas mixtures requires huge amounts of energy. In fact, CO2 separation from gas mixtures is one of the most energy intensive process in CCS. Therefore it is necessary to reduce the energy consumption of CO2 separation for propagation of CCS.

The most widely used method among CO2 separation is a chemical absorption using amine based solution such as monoethanolamine (MEA). In the conventional CO2 separation by absorption using amine based solution, the thermal energy is supplied from the bottom of regenerator which utilizes reaction heat for regeneration of the solutions and heat of vaporization for stripping. To reduce the supply of the thermal energy, authors have developed CO2 gas separation processes based on self-heat recuperation. Applying the self-heat recuperation to thermal processes, not only the latent heat but also the sensible heat of the process stream can be circulated into the processes without any heat addition, leading to minimizing exergy destruction of heat transfer and reduction of the process energy consumption. However, the absorption using solution physically and chemically requires additional energy for separation of liquid/gas, vaporization for stripping.

In this paper, an innovative energy saving CO2 separation process using chemical absorption was proposed for energy saving point of view. In this process, metal oxide particles were selected as absorbents not for requiring separation energy of liquid/gas. These particles were fluidized by feed gas mixture for increasing segregation effect in the bed and heat transfer rate from chemical reaction to a heat pump system. Heat of absorption was recuperated and supplied to the heat of thermal decomposition through the heat pump by following exergy recuperation. In addition, some of the heat was circulating with absorbents. The energy saving performance of the proposed process was evaluated by simulations.

464620 Systems Design and Economic Analysis of Direct Air Capture of CO2 through Temperature Vacuum Swing Adsorption on Metal Organic Frameworks

Abstract:

The rising levels of CO_2 concentration in the atmosphere poses a serious threat to global climate stability. To mitigate this threat, efforts needs to be made for reduction of CO_2 levels. One of the strategy is to capture CO_2 from stationary, point, sources. These point sources accounts for approximately only one-third of the global emissions. Another strategy is to capture CO_2 directly from ambient air which, if successfully implemented, could result in capture of CO_2 from disperse emission sources. In contrast to CO_2 capture from stationary sources, Direct Air Capture (DAC) plants are location independent and can be scaled more easily. Thus, the DAC plant can be set up near a sequestration or utilization site, eliminating the need for CO_2transportation infrastructure.

Our work proposes comparison of two metal organic frameworks, namely, MIL-101(Cr)-PEI and mmen-Mg_2(dobpdc) [1] for DAC through temperature vacuum swing adsorption (TVSA). Isotherm parameter estimation for CO_2 adsorption was carried out for both the adsorbents using published and original data and a kinetic study was performed through empirical correlations [2] to estimate the mass transfer rates for the process . The adsorbents are grown as films inside 900 cpsi (cells per square inch) monolith. To achieve a temperature swing, steam is used as a stripping agent during the desorption step. In this thermal desorption operation, it is believed that the amines groups present in the metal organic frameworks are prone to oxidative degeneration at higher temperatures[3]. To overcome this challenge, we propose a five step process: adsorption, evacuation, pressurization, desorption and cooling. In the proposed process, evacuation and cooling steps are implemented through vacuum swing to remove the oxygen in the channels when the temperature is high. In order to model this system, partial differential algebraic equations have been implemented in gPROMS, which is a commercial dynamic process modeling and optimization software. The linear driving force (LDF) model is used to approximate the rate of CO_2adsorption [2].

This talk will discuss the energy requirements and overall economics of this DAC process. Different components which contribute to consumption of energy are identified and the TVSA model is simulated to estimate the net energy requirement for MIL-101-PEI-50 and mmen-Mg_2(dobpdc). Operating costs during the five steps (adsorption, evacuation, pressurization, desorption and cooling) and capital costs for metal organic frameworks and monoliths have been estimated for both the adsorbents. It has been shown that lifetime of the adsorbent is critical in estimating the net cost and a sensitivity analysis of the adsorbentâ€™s lifetime has been performed to analyze the DAC economics for both the adsorbents.Â We have identified sensitive parameters such as gas flow rate, cycle time, film thickness etc. which effects the overall energy requirements and net economics of the DAC process.

1. McDonald, T.M., et al., *Capture of carbon dioxide from air and flue gas in the alkylamine-appended metal-organic framework mmen-Mg2(dobpdc).* J Am Chem Soc, 2012. **134**(16): p. 7056-65.

2. Y. Y. Li, S.P.P., B. D. Crittenden, *Zeolite monoliths for air separation Part 2: Oxygen Enrichment, Pressure Drop and Pressurization.* Trans IChemE, 1998. **76**: p. 931-941.

3. Heydari-Gorji, A. and A. Sayari, *Thermal, Oxidative, and CO2-Induced Degradation of Supported Polyethylenimine Adsorbents.* Industrial & Engineering Chemistry Research, 2012. **51**(19): p. 6887-6894.

466081 New Performance Indicators for Adsorbents Used in CO_2 Capture Swing Adsorption Processes

Abstract:

As adsorption process has increasingly received attention in CO_2 capture area, a number of studies have suggested a variety of adsorbents such a metal-organic frameworks (MOFs), zeolitic imidazolate frameworks (ZIFs) and many modified versions of conventional adsorbents. However, because of their limited scope of the studies, they have evaluated their adsorbents by using conventional performance indicators: working capacity and selectivity. The performance indicators rarely consider the situation of the adsorbents' practical application in the adsorption process. In order to evaluate the adsorbents with the consideration of their eventual application, e.g., pressure swing adsorption (PSA) process, dynamic simulations with a rigorous model of the adsorption process are needed, but the simulations cannot be carried out conveniently by the developers of new adsorbents. Furthermore, the high computation cost caused by the nonlinearity and cyclic operation of the PSA system hinders the evaluation of adsorbentsâ€™ suitability for their process-level application. For those reasons, this work derives simple formulas for the process-level performance of an idealized PSA system. The efficiency and the CO_2 purity of produced gas in the idealized PSA system are defined as performance indicators for the evaluation of a target adsorbent. For further simplicity, the performance indicators are derived in the explicit form of an analytic solution. The derived explicit equations are suggested as formulas of the new performance indicators for the evaluation. With the formulas, the performance of a target adsorbent in the PSA process can be obtained with simple calculation only from lab-scale measurement data such as adsorption isotherm data. Because the suggested formulas are based on the idealized PSA process where the theoretically maximum performance of each adsorbent is achieved, the suggested way is fairer than comparing the simulations of different adsorbents with some prior design and operating conditions which may be preferred by particular adsorbents. In order to show the usefulness of the formulas, a case study involving several real adsorbents is presented. The results from the formulas about the new indicators are then compared both to the simulation results of a rigorous dynamic model and to the conventional performance indicators.

Acknowledgement: The authors would like to acknowledge the financial support from the grant (G01160051) funded by Saudi Aramco-KAIST CO2 management center.

466627 Optimal Retrofit of a Post Combustion CO_2 Capture Process Using Reduced Superstructure and Rate-Based Models

Abstract:

We propose a superstructure-based methodology for optimal retrofit of a CO_2 capture pilot plant using rigorous rate-based model for the reactive distillation. The optimal retrofit for existing plants is particular challenging because rigorous process models used for describing the currently operating process often result very complex optimization problems.[1] Herein, we preselect process alternatives implemented in the superstructure based on thermodynamic analysis. Thermodynamic analysis identifies unit operations that have high energy saving potential and process alternatives improving these unit processes are considered in the superstructure. Consequently, the problem size of the superstructure can be substantially reduced and optimization can be performed using commercially available process simulators and optimization solvers. The optimum process configuration and operating conditions are obtained stochastically using a genetic algorithm. Near global optimum solution is obtained by solving the problem multiple times with different crossover fraction. Results indicate that the optimum retrofit process includes three stages solvent recirculation, lean vapor recompression, and mechanical vapor recompression. Thermal energy and total energy consumption in the optimal retrofit process are reduced about 59% and 27%, respectively.

471576 Dynamic Modeling with Uncertainty Quantification of Solid Sorbent Based CO_2 Capture Processes

Abstract:

Post-combustion CO_2 capture is a potential approach to mitigate buildup of CO_2 concentration in the environment. Significant research is currently being carried out all over the world to reduce the penalty for CO_2 capture. Due to the large amount of CO_2 that needs to be captured, significant scale-up is needed, starting from the lab scale investigation of a potential capture technology to its commercial implementation for large-scale power plants. Since scale-up of novel chemical technologies generally extends over 20-30 years, it is imperative that the time required be significantly reduced for the accelerated commercialization of the potential technologies. Rigorous process models of these novel capture processes can be very useful for this purpose. However, process models suffer from uncertainty. The parameters for submodels are typically calibrated deterministically for each submodel. Further, the models are imperfect and the experimental data are not only imperfect, but are often inadequate. The resulting parametric and model-form uncertainties can lead to high discrepancy in the model outputs. Therefore, process models should be developed with quantified uncertainty, which can also be instrumental in identifying where additional experimental data should be collected and/or where different or modified model form should be considered to reduce the discrepancy.

With these motivations, a 1-D dynamic, non-isothermal and non-adiabatic fixed bed adsorption reactor model is developed for novel solid sorbent-based CO_2 capture processes. Rigorous models for mass transfer, heat transfer and chemical kinetics, if applicable, of both physisorption- as well as chemisorption-dominated processes are developed. A Bayesian calibration approach is used for quantifying parametric uncertainty. Model-form discrepancy is described by a set of orthogonal functions. Posterior distribution of the parameters as well as the model-form discrepancies are propagated through the process model to obtain the bounded confidence on the modelâ€™s predictions of variables of interest such as bed pressure drop, breakthrough curve, and bed temperature profile. Our study shows that the output discrepancy strongly depends on the operating conditions and helps to identify approaches that can be used to reduce the scale-up uncertainty of the novel CO_2 capture processes.

468716 CO_2 Capture and Conversion to Chemicals Via Syngas: Rigorous Modeling, Intensification, and Superstructure-Based Process Synthesis

Abstract:

Recent global warming and climate change are attributed to increased CO_2 emission from fossil fuels. Significant efforts have been made in the past to develop CO_2 capture and sequestration (CCS) technologies. However, large-scale CCS is still not deployed in reality for reasons including high cost, technological barriers and uncertainty in geological storage. A promising alternative to CCS is to convert CO_2 to value-added chemicals via syngas (CO and H_2), which is an intermediate for many hydrocarbon-based fuels and chemicals. As much as 2 Gt/yr of fuels and 200 Mt/yr of chemicals can be produced from CO_2 utilization [1].

In this work, we explore alternative technologies and routes for the thermochemical conversion of CO_2 to syngas and then syngas to various oxygenates. We postulate a flowsheet superstructure which includes layers with various alternatives for separation, conversion and upgrading. We allow both pure and dilute raw materials, such as CO_2 from flue gas, methane from stranded sources, oxygen from air and water, and hydrogen from renewables and other sources. A novel feature of our superstructure is that it also includes various intensified alternatives. One such example is the inclusion of a tri-reforming membrane reactor which, instead of using pure CO_2, uses the mixed flue gas as feed, produces syngas, and separates nitrogen from the product at the same time. Therefore, it does not need a separate CO_2 capture section at the upstream. For each alternative, we develop a rigorous model detailing the transport and kinetics to accurately predict the process performance. We use a pseudo-homogenous model for plug-flow reactors (PFR) [2] to describe various alternatives for CO_2 reforming (e.g., steam methane reforming, dry-reforming, tri-reforming, partial oxidation, and variants of combined reforming). Using the rigorous simulation platform, we also develop efficient

surrogate models [3] for conversion as it changes with reactor type, design and operating conditions. The replacement of the ODE models with their algebraic surrogates allows us to pose the overall synthesis problem as a mixed-integer nonlinear optimization (MINLP) problem, which we solve to optimality using ANTIGONE [4]. We consider different objectives such as maximizing the overall utilization of CO_2, minimizing the total annualized cost, and maximizing profit. We also take into account the auxiliary carbon dioxide emissions associated with various processing tasks in the network. In this presentation, we will discuss our overall synthesis framework and the optimization results.

References:

[1] Quadrelli, E. A.; Centi, G.; Duplan, J.-L.; Perathoner, S. Carbon Dioxide Recycling: Emerging Large-Scale Technologies with Industrial Potential. *ChemSusChem*. 2011, 4:1194â€"1215.

[2] Aboosadi, Z. A.; Jahanmiri, A. H.; Rahimpour, M. R. Optimization of tri-reformer reactor to produce synthesis gas for methanol production using differential evolution (DE) method. *Applied Energy*, 88(8):2691â€"2701, 2011.

[3] Bajaj, I.; Hasan, M. M. F. Effective Sampling, Modeling and Optimization of Constrained Black-Box Problems. Accepted for publication in proceedings of ESCAPE26, Slovenia, 2016.

[4] Misener, R.; Floudas, C. A. ANTIGONE: algorithms for continuous/integer global optimization of nonlinear equations. *Journal of Global Optimization*. 59 (2-3), 503-526, 2014.

467184 Efficient Estimation of Maximum Theoretical Productivity from Batch Cultures Via Dynamic Optimization of Flux Balance Models

Abstract:

Production of chemicals from engineered organisms in a batch culture typically involves a trade-off between productivity, yield, and titer. However, strategies for strain design typically involve designing mutations to achieve as high of a yield as possible while maintaining growth viability. Such approaches tend to follow the principle of designing static networks with minimum metabolic functionality to achieve desired yields. While these methods are computationally tractable, optimum productivity is likely achieved by a dynamic strategy, in which intracellular fluxes change their distribution over time. Previous work has investigated using multi-stage fermentations to increase either productivity or yield. Existing methods, however, have tended to assume an initial control strategy (i.e., a single reaction target) in maximizing productivity - but it is unclear how close this productivity comes to a global optimum. The calculation of maximum theoretical yield is well established in the metabolic engineering literature, and its use helps guide strain and pathway selection for static strain design efforts. In this talk, we present a novel method for the calculation of a maximum theoretical productivity of a batch culture system using a dynamic optimization framework. This method follows the traditional assumptions of dynamic flux balance analysis: that internal metabolite fluxes are governed by a pseudo-steady state and external metabolite fluxes are represented by dynamic system including Michealis-Menten or hill-type regulation. The productivity optimization is achieved via collocation on finite elements, and accounts explicitly for an arbitrary number of fermentation stages and flux variable changes. Solution of the dynamic program is accomplished via a large-scale nonlinear programming solver. We have applied our method to succinate production in two common microbial hosts: *E. coli* and *Actinobacillus succinogenes*. The method can be further extended to calculate the complete productivity versus yield Pareto surface. Our results demonstrate that nearly optimal yields and productivities can indeed be achieved with only two discrete flux stages.

470183 Computing Sensitivities for Nonsmooth Differential-Algebraic Equations

Abstract:

Differential-algebraic equations (DAEs) have become a widely applied modeling tool. In particular, nonsmooth DAEs provide a natural framework for dynamic models of chemical processes such as campaign continuous pharmaceutical manufacturing [1]. Sources of nonsmoothness include thermodynamic phase changes (e.g., flash evaporation), flow transitions (e.g., nonreturn valves), and safety devices (e.g., relief valves). In analyzing such models, sensitivity analysis provides useful information in nonsmooth equation-solving (e.g., semismooth Newton methods) and optimization problems (e.g., bundle methods for local optimization). However, these algorithms require computation of an element of some class of generalized derivative (e.g., Clarke's generalized Jacobian [2]), for which theoretical and computational approaches are currently lacking.

In this talk, we detail some new results on obtaining computationally relevant generalized derivatives of DAEs for which existing methods fail. More specifically, a tractable method is provided to furnish lexicographic derivatives of solutions of nonsmooth parametric DAEs using recent advancements in nonsmooth analysis [3-5]. These generalized derivatives are no less useful than elements of the generalized Jacobian in many important applications, including the nonsmooth problem-solving methods detailed above. As in the classical case, the desired parametric sensitivities are given in terms of the solution of an auxiliary nonsmooth DAE system.

References

[1] B. Benyahia, R. Lakerveld and P. I. Barton, "A plant-wide dynamic model of a continuous

pharmaceutical process," Industrial and Engineering Chemistry Research, vol. 51, no. 47, pp. 15393-15412, 2012.

[2] F. H. Clarke, Optimization and Nonsmooth Analysis, Philadelphia: SIAM, 1990.

[3] Y. Nesterov, "Lexicographic differentiation of nonsmooth functions," Mathematical Programming, vol. 104, no. 2, pp. 669-700, 2005.

[4] K. A. Khan and P. I. Barton, "Generalized derivatives for solutions of parametric ordinary differential equations with non-differentiable right-hand sides," Journal of Optimization Theory and Applications, vol. 163, no. 2, pp. 355-386, 2014.

[5] K. A. Khan and P. I. Barton, "A vector forward mode of automatic differentiation for generalized derivative evaluation," Optimization Methods and Software, vol. 30, no. 6, pp. 1185-1212, 2015.

464348 Evolutionary Optimization Environment for Power Plant Control with Dynsim® Interface

Abstract:

Evolutionary Optimization Environment for Power Plant Control with Dynsim® Interface

Ghassan Al-Sinbol, Mario G. Perhinschi, Debangsu Bhattacharyya

Due to the increasing complexity and multi-dimensionality of modern power plants, selecting and optimizing the architecture of control systems within a plant-wide control strategy is a challenging task. The control problems to be solved are highly nonlinear, multidimensional, constrained, and prone to multiple extrema. Therefore, optimal analytical or numerical solutions are in most cases impossible or impractical. For such problems, evolutionary algorithms can provide effective solutions. Recent efforts have identified the artificial immune system (AIS) paradigm as a promising source for mechanisms that can enhance more common evolutionary algorithms and mitigate some of their drawbacks. The development and use of complex power plant simulation environments, such as Dynsim®, is currently receiving significant attention for a variety of applications. However, their use for control system design and optimization has yet to reach full potential. To synergistically merge the capabilities of evolutionary algorithms and the availability of advanced process simulation software, an interactive computational environment for the optimization of modern power plant control using evolutionary techniques is presented in this paper.

The proposed optimization algorithms are implemented in Matlab®, while the power plant may be modeled in Matlab®/Simulink® or Dynsim®. The overall proposed architecture for control optimization including the interface between these computational environments is aimed at facilitating reaching the full potential of advanced methodologies for power plant control system design and optimization.

The general optimization scenario and specific algorithms and parameters are setup by the user through Matlab® graphical user interface menus. The object linking and embedding for process control data access protocol is used to facilitate variable exchange between Matlab® and Dynsim®. The interface integrates Matlab® based models and controllers into Dynsim® models. Dynsim® feature "Scenarios" is used to facilitate further the communication.

The optimization scenario includes two options: a baseline genetic algorithm and an immunity enhanced evolutionary algorithm. The modular architecture of the computational environment allows for the addition of new algorithms into a central library and testing and evaluation of newly developed approaches. The baseline genetic algorithm implemented within the optimization environment features a standard architecture with randomly generated initial population, integer-based chromosome representation, roulette-wheel selection technique and elitist strategy, gene mutation, and single-point crossover. Immunity-inspired enhancements include seeding and vaccination of initial population, hyper-mutated cloning, selection, and population diversification based on affinity to self/non-self. These algorithms are expected to enhance the computational effectiveness, improve convergence, be more efficient in handling multiple local extrema, and achieve adequate balance between exploration and exploitation.

Two major categories of optimization problems can be addressed: optimization of constant control system parameters (controller gains) and optimization of variable setpoints. For the first type of problem, the solution is defined as the set of best performing controller gains, while for the second, the solution consists of a vector of optimal setpoints over pre-defined time intervals.

The functionality of the proposed optimization computational environment is illustrated through optimizing the control laws for an acid gas removal unit as part of an integrated gasification combined cycle plant. The unit selectively removes H_2S and CO_2 using SELEXOL solvent. The unit consists of numerous equipment items including trayed and packed distillation columns, pressure vessels, heat exchangers, pumps, and compressors. These components are strongly coupled through mass and heat exchange resulting in a challenging control problem.

The general architecture of the evolutionary optimization environment is illustrated in Figure 1.

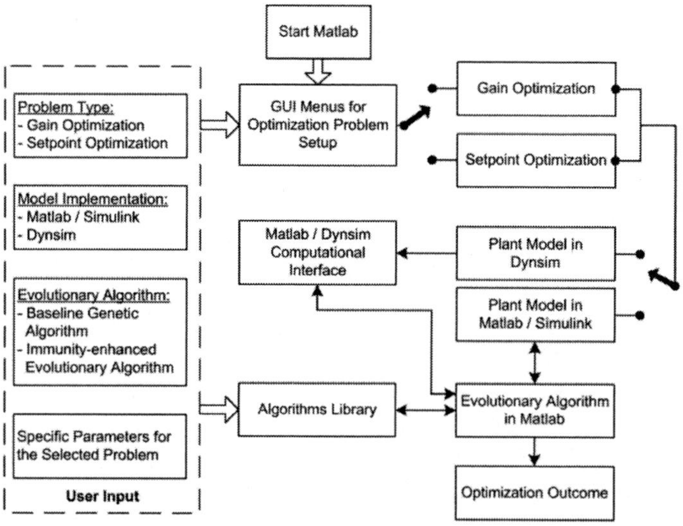

Figure 1. General Architecture of the Evolutionary Optimization Environment

Shale Gas Supply Chain Network Design and Operation Incorporating Rigorous Well Simulations

Jorge Chebeir, Louisiana State University, Baton Rouge, LA, United States
Hope Asala, Louisiana State University, Baton Rouge, LA, United States
Dr. Aryan Geraili, Louisiana State University, Baton Rouge, LA, United States
Dr. Arash Dahi Taleghani, Louisiana State University, Baton Rouge, LA, United States
Dr. Jose Romagnoli, Louisiana State University, Baton Rouge, LA, United States

Introduction

Over the past decade, technological innovations within the oil and gas industry have provided the necessary tools for the optimization of fossil fuels production in the United States. The combination of horizontal drilling and multi-stage hydraulic fracturing has allowed access to enormous volumes of natural gas originally trapped in low permeability shale formations. This has made shale gas development the major factor responsible for the increase in natural gas production observed in recent years. However, the high production levels of natural gas and a near stagnation in demand have led the domestic prices to fall since late 2005 (Figure 1a).[1] These fall in prices has accompanied the sharp decline in production rates that have plagued shale wells after the first three years of production, on average.[2] Figure 1b reports a 3-year average decline in productivity rates ranging between 77% to 89% for the Woodford, Eagle Ford, Barnett, Fayetteville, and Haynesville shale plays. These factors together with the high capital investment involved in a shale gas project, have forced operators to continue drilling and fracturing wells to keep up with the current level of production and profit margins from investment. In this context, re-stimulation of mature wells appears as an affordable option to increase the productivity of existing wells at reduced costs and book additional reserves. Since the local prices for the sale of natural gas still continues to persist, energy companies can take advantage of the price differential between local and external markets. Gas producers may appease the negative conditions in local markets by exporting part of their production as liquefied natural gas (LNG).

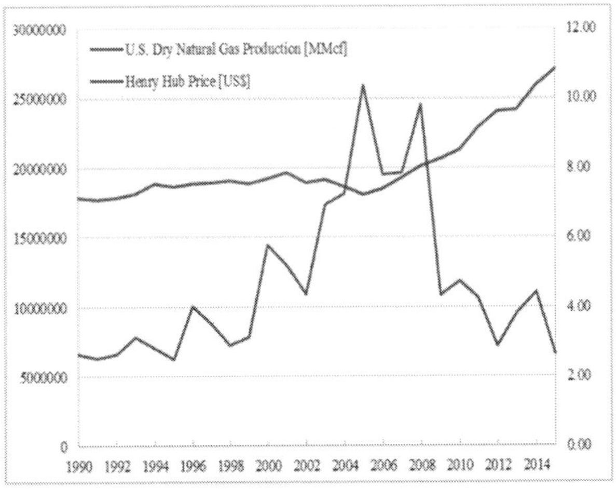

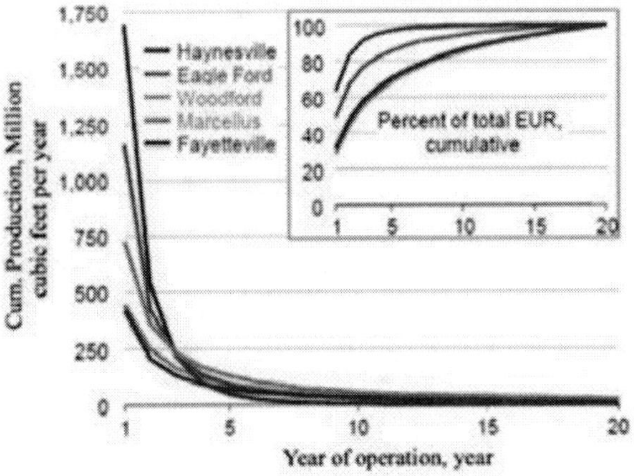

Figure 1. (a) Historical trend of U.S. Natural gas production and the Henry Hub price.

(b) Average production profile for major U.S. Shale plays.[2]

In this work, we propose an optimization framework that integrates the simulation of oil and gas production from a shale reservoir model with the shale gas supply chain structure obtained from a strategic planning model. This framework incorporates the geological characteristics of the shale reservoir as well as the construction of different well designs. This permits a high degree of precision in the determination of oil and gas production rates as well as produced water generated with or without re-fracturing. Results from the reservoir simulations are then incorporated into a strategic planning model that intends to represent a shale gas supply chain network design and operation. The strategic planning model encompasses the different elements of the supply chain including; the drilling, fracturing, possible re-fracturing strategy, mineral rights acquisition policy, quantity of different products transported, processing plant installation and operation, compressors and pipelines procurement and installation, freshwater and wastewater management, and exportation of LNG.

Problem Description

The development of shale gas resources involves strategic and operational decisions including the optimal strategy to drill, stimulate, and re-stimulate wells in their respective pads. Others include: the location and design of the pipeline network required to transport products from subsurface facilities to the final internal and external markets, the location and design of processing plant facilities and compression stations, the management of freshwater for stimulation operations, and the possible reutilization of wastewater generated in the different well pads. A simplified representation of the shale gas superstructure used in this work is presented in Figure 2.

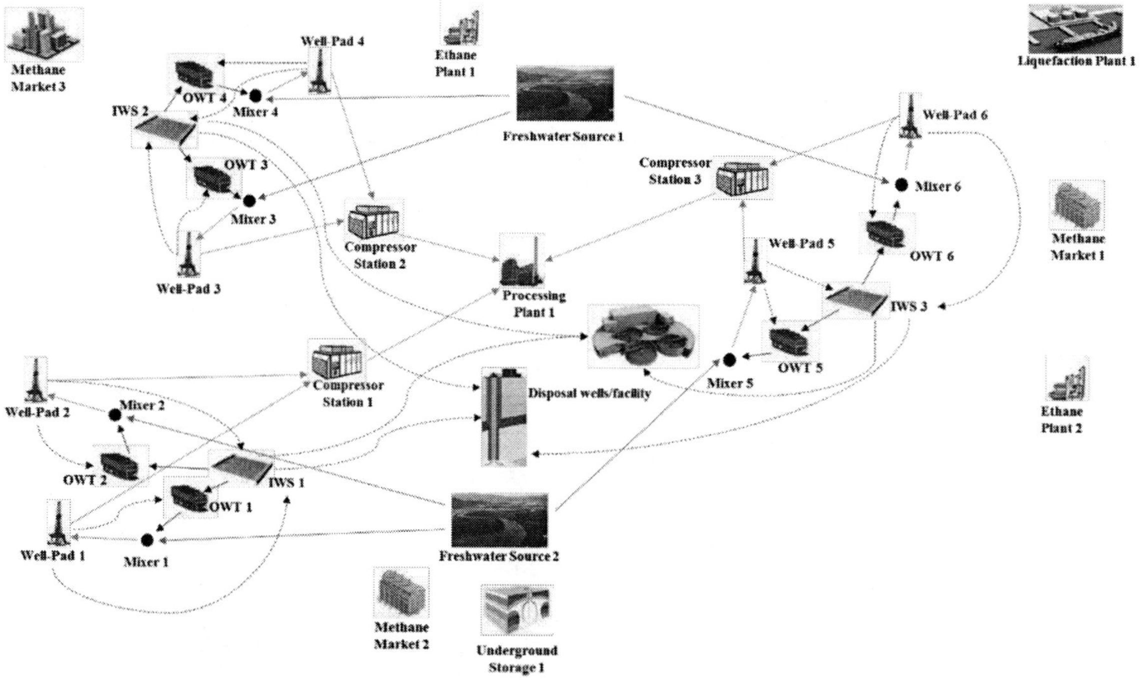

Figure 2. Simplified shale gas superstructure for the proposed framework.

The simplified superstructure includes several freshwater sources that are necessary for obtaining the amount/quality of water requisite for stimulation operations. This freshwater can be mixed in each well pad with produced water that has undergone onsite wastewater treatment (OWT). The produced water generated in each well-pad is transported to the wastewater impoundments for storage or directly treated onsite. The stored wastewater is then transported to disposal wells and/or the centralized wastewater treatment facilities. It can also be returned to the well pad for treatment and reutilization.

Shale gas produced in different well-pads is transported to the compression stations to increase the pressure and make possible its transportation to the processing plants. Once the raw gas reaches the processing plant, it is separated into different products including methane, ethane, propane, butane, and natural gasoline. The methane is then compressed before transportation to different internal markets, underground reservoirs, and/or liquefaction plants for exportation. The ethane is transported to a local petrochemical plant while the heavier hydrocarbons are sold at the processing plant gate. From the underground reservoir, the methane can also be transported to internal markets and/or the liquefaction plant for exportation.

Framework Description

In this section each component of the proposed optimization framework is described in some detail. A schematic of our proposed workflow is presented in Figure 3. As can be observed, the input data required for reservoir simulation includes the reservoir properties of the Marcellus shale rock. The possible well designs are constructed in this model. After simulation is run, output of each design feeds the strategic planning model to determine the optimal network design, drilling, stimulation, and re-stimulation strategies to be implemented. The schedule for the allocation and sale of shale gas components can consequently be determined.

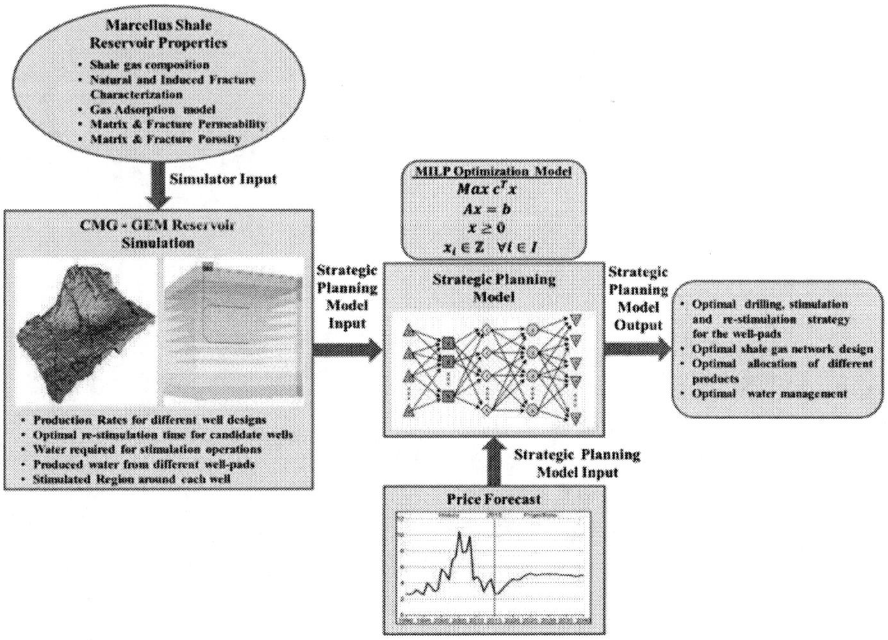

Figure 3. Proposed optimization framework for shale gas supply chain network

Shale Gas Reservoir Model

The Shale Gas Reservoir model used in this work is a logarithmically spaced, locally refined dual permeability (LS-LR-DK) model with 5 x 5 x 1 frac grid refinements. The superstructure containing the well pads is modelled as a 100 x 100 x 10 grid block system in CMG's GEM compositional simulator.

The location of the six pads, labelled A – F, is based on a typical Marcellus shale system of pads owned by a single oil and gas operator. Figures 4 and 5 show the relative location of wells penetrating layer 7 before and after re-fracturing. The relative location of the wells in each pad is such that it maximizes the stimulated reservoir volume (SRV) of the shale reservoir rock so that cumulative production and Estimated Ultimate Recovery (EUR) is maximized. Figure 6 gives an orthogonal representation of the SRV obtained as a result of implementing fracturing on each well. The task of each wells lateral placement, configuration and stimulation design is usually a multi-disciplinary decision that should incorporate rock mechanics, petro-physics, geophysics, reservoir and drilling engineering. According to the GAMS-MILP formulation, each pad can potentially be designed with 17, 20, 25 and 30 stages. The wells were also assumed to be drilled, fractured and flowed back in one-quarter (3-month) time step and this drilling and completion strategy was adopted in this model. Optimal re-fracture time for all wells was designated as when each well has approximately declined to 10-15% of initial well production rate.

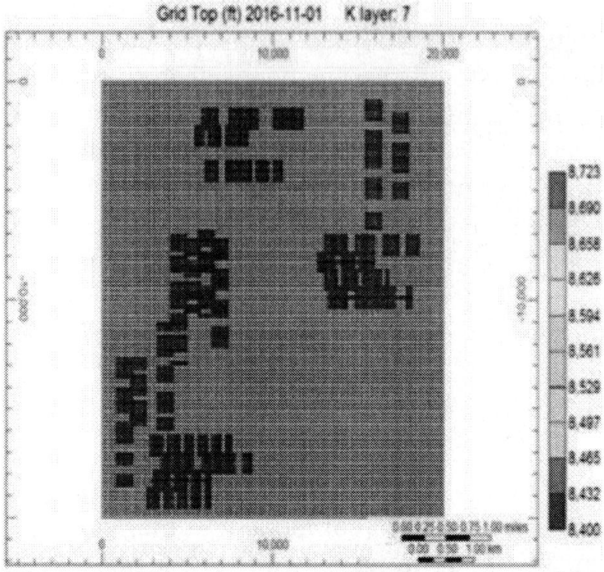

Figure 4. Plan view of wells penetrating grid layer 7 after drilling and hydraulic fracturing

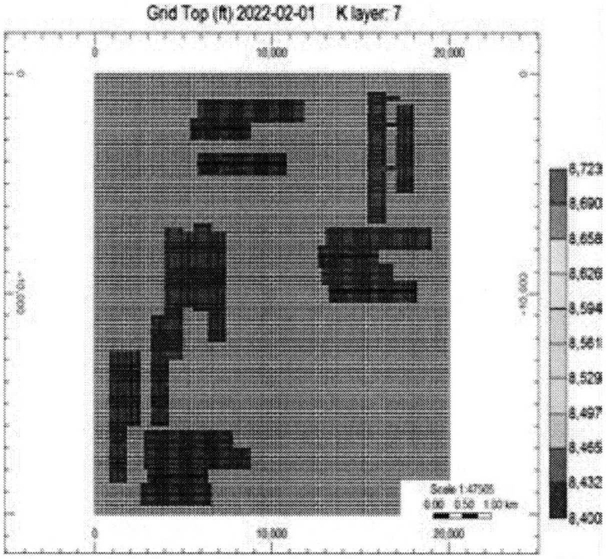

Figure 5. Plan view of wells penetrating grid layer 7 after drilling, fracturing and refracturing

The CMG-GEM Reservoir model used in this work is based on realistic heterogeneous petro-physical and geological properties typical of Marcellus shale reservoir rock. It incorporates an interaction between the natural fracture system in the reservoir and the modelled induced hydraulic fractures. After implementing fracturing and re-fracturing operations, hydrocarbon and water production rate profiles for each well-pad design are obtained and utilized as input for the MILP strategic planning model. Output data for each well pad design is developed for implementations of fracturing or fracturing and re-fracturing. This data provides a best estimate input for GAMS software since historical production data is unavailable at the start of the planning horizon of the project. Realistically, no two wells will experience the same exact decline rate. The well productivity in each pad is a function of: number of fracture stages executed, and the petro-physical properties of each layer in the reservoir. The initial rock content and composition of shale gas is based on an adsorption model. Although more accurate shale gas composition profiles could be obtained from this works formulation, it is assumed that the shale gas composition is constant over time. Figure 7 shows the heterogeneous distribution of porosity in the Marcellus shale reservoir developed in this work.

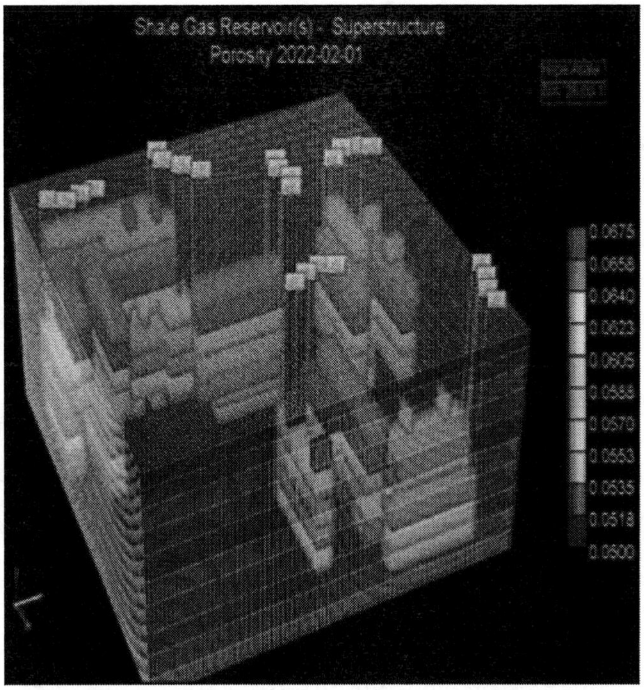

Figure 6. Orthogonal view of the spatial location of wells in a Marcellus superstructure

Figure 7. 3-D view of the porosity distribution of layers in the Marcellus reservoir superstructure

Strategic planning model and Mathematical formulation

The strategic planning model is formulated using mixed-integer linear programming (MILP) with a 10 years planning horizon discretized in 40 quarters. We assume each well-pad can hold multiple wells of certain design, i.e. different number of stimulation stages and horizontal lengths. It is assumed that operators are only able to work in only one well pad in each time period. After performing the drilling and fracturing operations, the wells start to produce in the next time step. Re-fracturing is considered by including an extra index in the production rate variable that pre-establishes if a well with a specific design is only drilled and stimulated or if it is also re-stimulated once it reached a production level below its economic threshold (10-15% of initial production rate). The optimal time step to execute re-stimulation operations for any specific design was therefore a function of decline rate, which was dependent on layer reservoir properties.

In the GAMS model, the non-linear expressions corresponding to the *six-tenth* rule for the capital costs of facilities and pipeline network were linearized by implementing a piecewise method.[4] In the case of the compressor stations, the expression slightly deviates from the *six-tenth* rule. The exponent factor was changed from 0.6 to 0.77.

The objective function of the strategic planning model is an NPV optimization relation given by:

$$NPV = I_s - C_{roy} - C_{fw} - C_{ww} - C_{sgp} - C_{land} - C_{cp} - C_{pp} - C_{pl} - C_{St} - C_{lng} \qquad (1)$$

where I_s is the revenue accrued from the sale of natural gas and natural gas liquids in internal and external markets. The cost terms include capital and operating costs of the following: compression, C_{cp}, processing, C_{pp}, storage facilities, C_{St}, freshwater management, C_{fw}, wastewater management, C_{ww}, LNG production, C_{lng}, pipeline networks, C_{pl}, royalties, C_{roy}, mineral rights, C_{land}, and well pad development, C_{sgp}. The maximization of NPV is subject to:

- The number and design of drilled, stimulated, and/or re-stimulated wells. This constraint is defined by the following equations:

$$\sum_{i}\sum_{n}\sum_{lg}\sum_{q} YW_{i,n,lg,q,t} = 1 \quad \forall t \in T \tag{2}$$

$$N_{i,lg,q,t} = \sum_{n} n \cdot YW_{i,n,lg,q,t} \quad \forall i \in I, lg \in LG, q \in Q, t \in T \tag{3}$$

where $YW_{i,n,lg,q,t}$ is a binary variable that equals 1, if a specific number of wells, n, and number of stages, lg, is selected when operation q is performed at pad i, in each time period t. $N_{i,lg,q,t}$ represents the number of wells drilled in pad i with lg number of stages, when q operation is performed in each time period, t.

- The mineral rights required for the development of wells in the different pads.
- The amount of Freshwater required for fracturing and re-fracturing operations.
- The waste water management options available for the amount of water produced at each well pad (water required for treatment, reuse, and/or disposal).
- The mass balance constraints that determine shale gas flows, natural gas flows, amounts of natural gas liquids sold, and storage of products.
- The capacity of the facilities adopted in the supply chain network.
- The demand for different products in the internal and external markets.

When considering the price uncertainty of products such as natural gas and NGL, the strategic planning model could be adapted to a two-stage stochastic formulation. Here the decision variables would be modelled as first stage here-and-now and second stage wait-and-see variables. Instead of considering product price forecasts of products (unique realization) like in the case of our current model, a scenario planning approach could be utilized to address a possible finite number of realizations of price in the future. The random fluctuation of prices could be described by a process such as Geometric Brownian Motion.

Results

Figure 8 shows the rate decline profiles for fractured wells in pad E while Figure 9 shows the rate decline profile for wells in pad E that have been re-fractured. According to both figures, the production rate of wells increases with the number of stages.

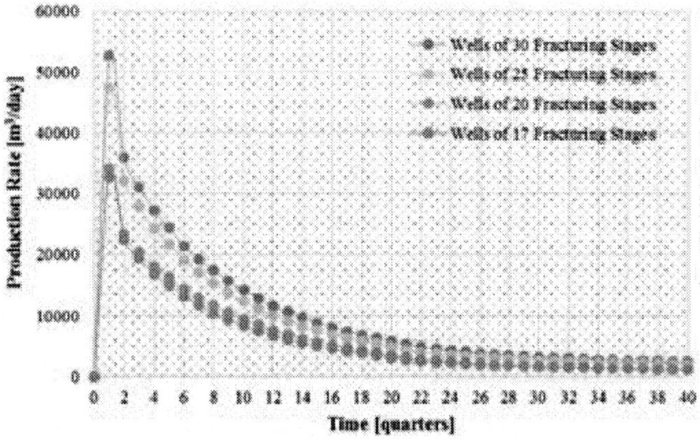

Figure 8. Production decline rate profile for fractured wells in pad E

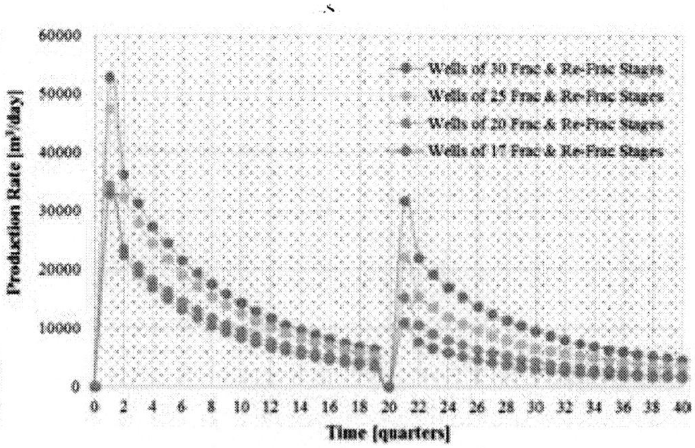

Figure 9. Production decline rate profile for re-fractured wells in pad E

Wells with 17 stages gave the least cumulative production while wells with 30 stages had the highest. The cumulative production realized from re-fractured wells were higher compared to corresponding fractured wells. After re-fracturing, the decline rate in wells with more frac stages was higher than wells with less frac stages. Figure 10 highlights the final proposed drilling strategy for pads A-F.

According to the optimal design, 17.1% of the wells adopted 25 frac stages while 82.9% of the wells adopted 30 frac stages. After implementation of optimization, the total cost of the shale gas project including capital investment, operating costs, and royalties, is estimated at US$ 554.46 MM. Considering the planning horizon of 10 years, the net present value of the shale gas project is US$ 206.74 MM. According to Figure 11, the major costs associated with the optimal supply chain network are related to drilling/completion/stimulation and the processing and transportation of natural gas and other shale gas derivatives. Table 1 highlights the revenue accrued from the project and the contribution of each product.

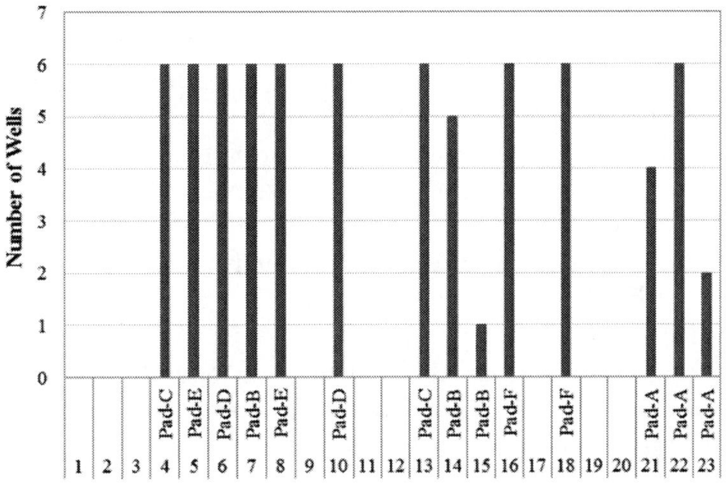

Figure 10. Drilling, fracturing, and re-fracturing strategy for wells in Pad A – F

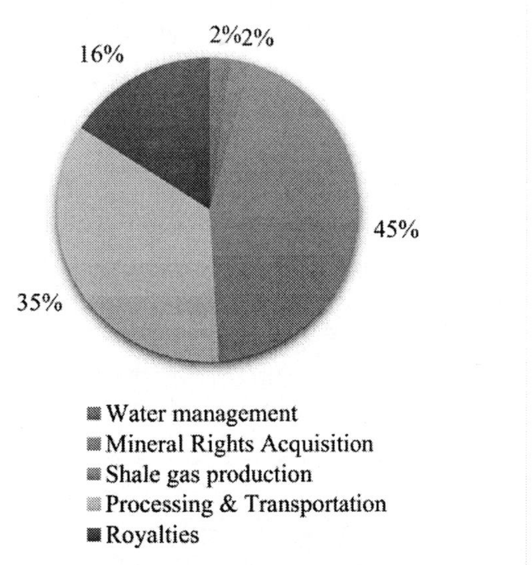

Figure 11. Pie chart showing the relative contribution of different costs associated with the optimal shale gas supply chain network

Table 1. Revenues generated by different products

Products	Revenues [MMUS$]	% Total Revenue
Natural gas	233.54	30.68
Condensates	203.86	26.78
Ethane	42.34	5.56
Propane	49.14	6.46
Butane	35.61	4.68
Natural gasoline	62.24	8.18
LNG	134.46	17.66

Conclusions

Implementing re-fracturing was a necessity for at least 75% of the wells in Pads A to F. The re-fracturing of shale wells decreases the payback time and increases the NPV of shale gas projects. The final well designs adopted considered the maximum number of fracture stages possible. Realistically, the maximum number of frac stages may not be executed because of the length limitations of each well and placement considerations. The total waste water obtained from shale gas operations includes flow back and produced water. The rate decline profiles of waste water obtained from reservoir simulation improves the estimate of total waste water needed for optimizing unconventional shale gas supply chain network operation. The sale of LNG and gas condensates in internal markets as well as their exportation to external markets play a major role in the economic success shale gas projects.

The multi-disciplinary approach utilized in this work introduces a useful economic tool for oil and gas companies to optimize profitability from their shale gas operations. Future work considers the incorporation of a TDS formulation and the application of compositional modelling to the integration of realistic re-stimulation strategies and supply chain optimization.

References

1. Wang Q, Chen X, Jha A, Rogers H. Natural gas from shale formation - The evolution, evidences and challenges of shale gas revolution in United States. *Renew Sust Energ Rev.* 2014;30: 1-28.
2. Asala, H., Ahmadi, M., Dahi Taleghani, A. (2016): Why re-fracturing works and under what conditions. Paper SPE 181516 presented at the SPE Annual Technical Conference and Exhibition held in Dubai, UAE, 26-28 September.
3. Potapenko D.I., Tinkham S.K., Lecerf B., Fredd C.N., Samuelson M.L., Gillard MR, Le Calvez J.H., Daniels J.L. (2009): Barnett shale refracture stimulations using a novel diversion technique. Paper SPE 119636 presented at the SPE Hydraulic Fracturing Technology Conference held in The Woodlands, Texas, USA, 19-21 January.
4. Hugo A., Pistikopoulos E.N. Environmentally conscious long-range planning and design of supply chain networks. *J Clean Prod.* 2005:13: 1471-1491.
5. Calderon, A., Guerra, O., Papageorgiou, L., Siirola, J., Reklaitis, G. Preliminary Evaluation of Shale Gas Reservoirs: Appraisal of Different Well-Pad Designs via Performance Metrics. *J Ind. Eng. Chem. Res.,* 2015: 54: 10334-10349.

Acknowledgments

We thank the Computer Modeling Group Ltd. for providing us with the research and academic licenses for their Winprop, GEM and Builder Simulators.

466157 Dynamic Optimisation of Beer Fermentation: Sensitivity Analysis of Attainable Process Performance Vs. Product Flavour Constraint Levels

Abstract:

To achieve maximum plant throughput and the highest profitability attainable, there is a strong incentive to optimise the fermentation stage in beer production, given the time-consuming and energy-intensive nature of the process (Rodman and Gerogiorgis, 2016). Process complexity prevents the use of comprehensive deterministic models in dynamic optimisation studies: with over 600 chemical species reported (Vanderhaegen et al., 2006), such computational implementations would require prohibitively laborious parameterisations and equally expensive execution. Nevertheless, it is desirable to use a reduced-order kinetic model (considering only the most important species and their interactions within the chemical system) toward dynamic simulation and optimisation of beer fermentation, in order to systematically investigate process performance under varying operating conditions.

The yeast culture suspended within cylindro-conical fermentation vessels is an integral part of the process: the cell enzymatic effect on the sugar content produced in wort pre-processing induces its conversion to ethanol and ensures the desired alcohol concentration level in the beverage. Cell activity is highly sensitive to system temperature (Landaud et al., 2001), influencing the fermentation effective rate and ultimately the process performance. The selection and implementation of an appropriate dynamic temperature profile is thus the most important factor influencing the required processing time, so it is a focus of the present dynamic optimisation study. The system temperature also influences the progression of a multitude of side reactions, many of which lead to the production of undesirable by-product species which are known to have detrimental effects on product flavour beyond certain concentration levels; novel temperature manipulation profiles must thus explicitly incorporate such considerations.

A published and validated kinetic model of beer fermentation considers the dynamic behaviour of a yeast culture (active, latent and dead cells) and its effect on sugar and ethanol concentrations, in addition to two undesirable by

product species: ethyl acetate and diacetyl compounds (de Andrés-Toro, 1998). This model has formed the basis of several prior optimisation studies which considered various systematic protocols to obtain dynamic temperature manipulation profiles capable of achieving higher production efficiency and improving industrial practice (Carrillo-Ureta et al., 2001; Xiao et al., 2003). The novelty of our published optimisation study, however, is the comprehensive visualisation of the four-dimensional attainable envelope of beer fermentation, and the multi-objective optimisation of ethanol production and batch duration under explicit product flavour constraints (Rodman & Gerogiorgis, 2016).

Expert beer tasters and connoisseurs confirm that for a given beer product, the harmful concentrations of both ethyl acetate and diacetyl compounds are actually indistinguishable below certain levels, rendering further reduction towards concentration minimisation redundant or at least less important. Marketing and commercial operations of beer brewing corporations pay particular importance to such thresholds, not only because they may well be subjective and/or local market-dependent, but also because regional, national and/or cultural taste patterns affect desirable flavour perceptions enormously. This experiential reality has motivated the present study, which presents an alternative approach to the formulation of the dynamic multi-objective optimisation problem, in an effort to understand how the arbitrary selection of flavour constraints may impact optimal manufacturing policies.

The sensitivity analysis of attainable process performance (how can we maximise ethanol production while also minimising fermentation duration?) with respect to product flavour constraint levels (how can we ensure that the harmful by-product concentration constraints will be respected?) addresses the importance of technoeconomic optimisation without unnecessary effort and expenditure. Moreover, the novel objective function employed achieves an explicit visualisation of the optimal temperature manipulation profile as a function of the two-dimensional space of by-product constraint levels. A deterministic approach with explicit temperature control vector parameterisation is used to solve the formal optimisation problem; a comparative evaluation of suitable nonlinear programming (NLP) algorithms (Schlegel et al., 2005) will be illustrated. Solving this dynamic optimisation problem over a wide range of realistic threshold values on both by-products allows us to map the attainable concentration of product ethanol and production batch times, thus enabling beer production plant operators to rapidly visualise if and how process performance can be significantly improved as a function of tolerable by-product levels.

LITERATURE REFERENCES

1. Carrillo-Ureta, G., Roberts, P., Becerra, V., 2001. Genetic algorithms for optimal control of beer fermentation. *Proceedings of the IEEE International Symposium on Intelligent Control*, 391-396.

2. de AndrÃ©s-Toro, B., et al. 1998. A kinetic model for beer production under industrial operational conditions. *Math. Comput. Simulat.* **48**(1): 65-74.

3. Vanderhaegen, B., et al., 2006. The chemistry of beer aging â€“ a critical review. *Food Chem.Â* **95**(3): 357-381.

4. Xiao, J., Zhou, Z., Zhang, G., 2003. Ant colony system algorithm for the optimization of beer fermentation control. *J. Zhejiang Univ.*5(12): 1597-1603.

5. Schlegel, M., Stockmann, K., Binder, T. and Marquardt, W., 2005. Dynamic optimization using adaptive control vector parameterization. *Comput. Chem. Eng.*Â **29**(8): 1731-1751.

6. Rodman, A.D., Gerogiorgis, D.I., 2016. Multi-objective process optimisation of beer fermentation via dynamic simulation, *Food Bioprod. Process.Â* (in press, available online).

7. Landaud, S., Latrille, E., Corrieu, G., 2001. Top pressure and temperature control the fusel alcohol/ester ratio through yeast growth in beer fermentation. *J. Inst. Brewing*, **107**(2): 107-117.

458421 Dynamic Optimization of Constrained SEMI-Batch Processes Using Pontryagin'S Minimum Principle – an Effective Quasi-Newton Based Approach

Abstract:

DYNAMIC OPTIMIZATION OF CONSTRAINED SEMI-BATCH PROCESSES USING PONTRYAGIN'S MINIMUM PRINCIPLE – AN EFFECTIVE QUASI-NEWTON BASED APPROACH

Erdal Aydin[a], Kai Sundmacher[a,b,*]

[a]*Max Planck Institute for Dynamics of Complex Technical Systems, Sandtorstraße 1, 39106 Magdeburg, Germany*

[b]*Otto-von-Guericke University Magdeburg, Universitätplatz 2, 39106 Magdeburg, Germany*

[*]Corresponding author: sundmacher@mpi-magdeburg.mpg.de; Tel: +49 391 6110 351, Fax: +49 391 6110 353

Semi-batch processing is employed in many chemical sectors including food, pharmaceuticals, polymers and specialty chemical industries. Semi-batch processes possess a flexible operation scheme for the production of high-value and low volume products. Dynamic optimization of semi-batch processes has been attracting significant attention because of the competing environment in the industry, the requirement of higher product qualities and strict environmental regulations. Direct methods are usually stated as promising techniques for the solution of the dynamic optimization problems. However, depending on the type of the problem, these methods could be inefficient in terms of the time required for the solution. In addition, the usage of the direct techniques may result in unstabilities or infeasible solutions due to required approximations[1]. Indirect methods, such as Pontryagin's Minimum Principle, are the alternative approaches which transcript the problem into a different structure. Although Pontryagin's Minimum Principle was extensively studied for unconstrained problems and considered as an effective method, its application to complex and constrained dynamic optimization problems is limited [2],[3]. In this study, we present an alternative indirect solution technique for the dynamic optimization of constrained batch and fed-batch processes. The algorithm depends on the Pontryagin's Minimum Principle and the necessary conditions of optimality, and uses first order gradients with a Hessian approximation algorithm to update the corresponding control variables at each iteration step. Nonlinear constraints are augmented into Hamiltonian function using indirect adjoining and corresponding constraints are saturated at each iteration step in order to increase the efficiency of the method [3],[4]. The fact that the system equations can be integrated at each iteration step prevents the necessity of any approximation in the model equations. The performance of the proposed strategy was tested through different three case studies, which includes fed-batch reactors and a batch distillation column with complex path and terminal constraints. The performance of the algorithm was also compared with the simultaneous and sequential algorithms. Results show that the proposed solution algorithm is effective in dealing with the complex constraints and it is competing in solving the corresponding problems in small and acceptable computational time.

1. Srinivasan, B., S. Palanki, and D. Bonvin, *Dynamic optimization of batch processes: I. Characterization of the nominal solution*. Computers & Chemical Engineering, 2003. **27**(1): p. 1-26.

2. Chachuat, B., *Nonlinear and dynamic optimization: from theory to practice*. 2007.

3. Hartl, R.F., S.P. Sethi, and R.G. Vickson, *A survey of the maximum principles for optimal control problems with state constraints.* SIAM review, 1995. **37**(2): p. 181-218.

4. Harvey Jr, P.S., H.P. Gavin, and J.T. Scruggs, *Optimal performance of constrained control systems.* Smart Materials and Structures, 2012. **21**(8): p. 085001.

Keywords: Constrained dynamic optimization, optimal control, Pontryagin's Minimum Principle, Gradient-based optimization, semi-batch process

454372 A Parallel Interior Point Solver with Cyclic Reduction for Solving Large-Scale Dynamic Optimization Problems

Abstract:

Dynamic optimization problems directly incorporate detailed dynamic models as constraints within an optimization framework. Model predictive control, state estimation, and parameter estimation are all common applications of dynamic optimization. One dynamic optimization strategy, called direct transcription, is to discretize the continuous dynamics and replace the differential equations with algebraic approximations using some numerical method such as a finite-difference or Runge-Kutta (collocation) scheme. However, for problems with thousands of state variables and discretization points, discretization can lead to nonlinear optimization problems that exceed memory and speed limits of most serial computers. In particular, when applying interior point optimization methods, the computational bottleneck and dominant computational cost lies in solving the linear systems resulting from the Newton steps that solve the discretized optimality conditions. To overcome these limits, we exploit the parallelizable structure of the linear system to significantly accelerate the overall interior point algorithm. Our previous work has shown cyclic reduction (CR) to be a promising parallel linear solver for these problems. In this talk we extend this approach with an implementation of parallel CR within the open-source solver PIPS-NLP (Parallel Interior Point Solver for NLP). We demonstrate the effectiveness and scalability of the resulting optimization framework on several large-scale dynamic optimization applications including challenging distillation and polymerization systems.

453134 Computation of Sensitivities of Dynamic Systems with Lexicographic Linear Programs Embedded

Abstract:

Bioprocesses involving microbial communities have widespread applications in the pharmaceutical, food and biofuels industries. These complex bioprocesses can be modeled accurately using dynamic flux balance analysis (DFBA) [1], [2], [3], which combines genome-scale metabolic network reconstructions with dynamic process models. DFBA models result in dynamic systems with linear programs (LPs) embedded [4]. These LPs are embedded because their right-hand side depends on the dynamic states and the dynamic states depend on the solution vector of the LP. Newly available simulators [4], [5] have made the reliable and efficient implementation of DFBA possible. These simulators transform the LP embedded into a lexicographic LP (LLP) to address complications associated with nonunique and infeasible LP solutions, and they exploit basis information to reformulate these ordinary differential equation (ODE) systems with LPs embedded as differential-algebraic equation (DAE) systems. In order to use DFBA to design optimal bioprocesses involving microbial communities, sensitivity information is required. This paper presents the theory behind the computation of sensitivities for dynamic systems with LLPs embedded and their numerical implementation.

The objectives of a LLP in standard form as a function of its right-hand side are piecewise linear functions [6], and therefore, nonsmooth. This source of nonsmoothness can be propagated to the parametric dependence of the final states of the dynamic system. Therefore, there exist some parameter values for which the Jacobian of the dynamic system may not exist. Computing elements of Clarke's generalized Jacobian for complex nonsmooth functions is challenging [7], but can be done efficiently for piecewise differentiable functions, such as LLPs parameterized by their right-hand side, with lexicographic-directional (LD) derivatives [8]. LD-derivatives of nonsmooth dynamic systems can be computed to obtain elements of the plenary hull of the generalized Jacobian [9] if the right-hand side of the ODE is abs-factorable, which means it can be factored as analytic and absolute value functions. This is the case for ODE systems with LLPs embedded; however, obtaining explicitly the abs-factorable representation of a large LLP is intractable. Therefore, an algorithm that relies on LLP basis information instead is presented. This algorithm enables the efficient computation of LD-derivatives of DFBA systems, which is critical for their systematic optimization.

This paper first introduces the ODE system whose solution gives the LD-derivatives of ODE systems with LLPs embedded. Next, some complications regarding the computation of these LD-derivatives are illustrated. Next, an algorithm that addresses these challenges is presented. Finally, the theory developed is implemented to optimize a DFBA case study.

References

[1] A. Varma and B. Ø. Palsson, "Stoichiometric flux balance models quantitatively predict growth and metabolic by-product secretion in wild-type Escherichia coli W3110," *Applied and Environmental Microbiology*, vol. 60, no. 10, pp. 3724-3731, 1994.

[2] J. D. Orth, I. Thiele and B. Ø. Palsson, "What is flux balance analysis?," *Nature Biotechnology*, vol. 28, pp. 245-248, 2010.

[3] R. Mahadevan, J. Edwards and F. I. Doyle, "Dynamic flux balance analysis of diauxic growth in Escherichia coli.," *Biophysical Journal*, vol. 83, no. 3, pp. 1331-40, 2002.

[4] K. Höffner, S. M. Harwood and P. I. Barton, "A reliable simulator for dynamic flux balance analysis," *Biotechnology and Bioengineering*, vol. 110, no. 3, pp. 792-802, 2013.

[5] J. A. Gomez, K. Höffner and P. I. Barton, "DFBAlab: A fast and reliable MATLAB code for Dynamic Flux Balance Analysis," *BMC Bioinformatics*, vol. 15, p. 409, 2014.

[6] D. Bertsimas and J. N. Tsitsiklis, Introduction to Linear Optimization, Nashua, NH: Athena Scientific, 1997.

[7] F. H. Clarke, Optimization and Nonsmooth Analysis, Philadelphia: Society for Industrial and Applied Mathematics, 1990.

[8] K. A. Khan and P. Barton, "A vector forward mode of automatic differentiation for generalized derivative evaluation," *Optimization Methods & Software*, vol. 30, no. 6, pp. 1185-1212, 2015.

[9] K. A. Khan and P. I. Barton, "Generalized Derivatives for Solutions of Parametric Ordinary Differential Equations with Non-differentiable Right-Hand Sides," *Journal of Optimization Theory and Applications*, vol. 163, no. 2, pp. 355-386, 2014.

460407 Distributed Lyapunov-Based Model Predictive Control with Safety-Based Constraints

Abstract:

Control system designs must ensure process safety. In [1], a control design was proposed that includes explicit constraints on process safety to shrink the region of process operation when required to a smaller region (safety region) around the operating steady-state when a safety logic unit indicates that certain regions of state-space away from the steady-state may lead to process safety concerns due to process disturbances or actuator faults. The safety-based controller design was developed with a centralized model predictive control (MPC) structure; thus, computation time limitations within a sampling period may reduce the effectiveness of such a controller design for promoting process safety. An alternative MPC architecture that is intended to improve the computation time of the MPC algorithm is a distributed model predictive control (DMPC) architecture [2]-[3]. This MPC architecture has been investigated for computation time benefits since it can reduce the number of decision variables in each of the distributed optimization problems and may be able to terminate the optimization problems before the optimal solution is found while maintaining feasibility and closed-loop stability of the controller [4]-[5].

In this work, we propose the integration of a distributed model predictive control architecture with Lyapunov-based model predictive control (LMPC) and Lyapunov-based economic model predictive control (LEMPC) formulated with safety-based constraints. We consider both iterative and sequential distributed control architectures, and the partitioning of inputs between the various optimization problems in the distributed structure based on their impact on process safety. In addition, we investigate the conditions required for guaranteed feasibility and closed-loop stability of the algorithm, and discuss the termination of the algorithm and computation time benefits. Through a chemical process example, we demonstrate the proposed controller design and the effects of the control architecture on process safety considerations, and compare the time required to reach the safety level set and the computation time of the algorithm with that which can be achieved under a centralized safety-based model predictive control scheme.

[1] Albalawi F, Alanqar A, Durand H, Christofides PD. A feedback control framework for safe and economically-optimal operation of nonlinear processes. AIChE Journal. 2016; in press.

[2] Christofides PD, Liu J, Muñoz de la Peña D. Networked and Distributed Predictive Control: Methods and Nonlinear Process Network Applications. London, England: Springer, 2011.

[3] Scattolini R. Architectures for distributed and hierarchical model predictive control- A review. Journal of Process Control. 2009;19:723-731.

[4] Anderson TL, Ellis M, Christofides PD. Distributed economic model predictive control of a catalytic reactor: Evaluation of sequential and iterative architectures. In: Proceedings of the IFAC International Symposium on Advanced Control of Chemical Processes. Whistler, Canada, 2015;26-31.

[5] Venkat AN, Hiskens IA, Rawlings JB, Wright SJ. Distributed MPC strategies with application to power system automatic generation control. IEEE Transactions on Control Systems Technology. 2008;16:1192-1206.

464669 Supervisory Event-Based Control of Networked Process Systems with Limited State Measurements

Abstract:

The design of distributed and supervisory control systems for chemical process networks is a problem of significant and growing interest in both the academic and industrial communities in process control research (e.g., [1]-[5]). Beyond handling the dynamic coupling and interactions between the constituent subsystems in controller design, the management of information flow and communication between the individual subsystems has recently emerged as another important consideration, which is receiving increased attention in light of the growing calls for moving towards smart plant operations [6]-[7]. In this direction, a quasi-decentralized model-based networked control structure that enforces closed-loop stability with minimal periodic communication was developed in [8]. Within each local control system, predictive models of the dynamics of the rest of the plant are used, together with local state measurements, to generate the local control action at times when communication between the plant subsystems is suspended, and the states of these models are updated when communication is permitted at discrete times. Using a time-triggered communication strategy with a fixed communication rate, an explicit characterization of the minimum allowable communication rate necessary for closed-loop stability was obtained. In subsequent work [9], an adaptive event-based communication strategy was used to manage the information flow in the quasi-decentralized control structure. A key idea was for each unit to monitor its local state and prompt the other units to transmit their data to update their models when the local unit is on the verge of instability. The local instability thresholds were obtained with the aid of appropriate local control Lyapunov functions. While this approach provides greater flexibility in responding to unexpected disturbances in plant operations, it could potentially increase the network load and cause delays and data losses when multiple units attempt to access the network at the same time.

An alternative event-based approach is the broadcast-based communication strategy considered in [10], where the local model estimation error is monitored continuously and the local unit transmits its data to the rest of the plant when this error breaches a certain threshold. The communication threshold is determined using a composite control Lyapunov function for the overall system. This approach avoids the need for simultaneous access to the network; however, it focuses only on stability considerations, without taking the performance of the individual subsystems into account. For example, in the event of a transient local disturbance causing the local Lyapunov function to grow for some time – but without causing growth in the composite Lyapunov function and therefore triggering no communication – the deterioration in local performance will go undetected and uncompensated for. The reduction in network load in this approach comes at the expense of sacrificing local process performance.

This limitation was addressed in [11] where a higher-level supervisor was introduced into the control structure to balance the overall plant stability requirement with the local performance needs of the component subsystems, while simultaneously optimizing the extent of information transfer between the distributed control systems. The supervisor kept track of the alarm signals reported by the local control systems in response to the local events, and decided accordingly the time and direction of information flow between the individual subsystems. The supervisory logic took into account the evolution of both the local and global control Lyapunov functions so as to meet both the stability and performance requirements. The supervisory control structure was developed for systems with full-state measurements. In practice, however, and due to problems such as installation difficulty, equipment cost, and slow sampling, only incomplete state measurements are typically available. The lack of full-state measurements poses a difficulty in the implementation of traditional event-triggered control and communication strategies which typically rely on monitoring and evaluating state-dependent thresholds. Moreover, handling the lack of full-state measurements is a more challenging problem in the context of controlling networked process systems due to the interactions between the component subsystems.

In this work, we develop a framework for supervisory event-based control of process networks with uncertain nonlinear dynamics, discrete communication between the local control systems and incomplete state measurements. Initially, a set of quasi-decentralized model-based control systems that can exchange information at discrete times over the network are designed. Each local controller relies on model-estimated states of the local and neighboring units to generate the local control action. To overcome the problem of incomplete state measurements, a suitable high-gain observer is designed and included within each local control system to generate an estimate of the local state based on the available real-time output measurements. The observer-estimated states are used to monitor and assess the stability of the local subsystems, and are also broadcast over the network to update the model estimates used by the local controllers at times when communication is triggered. The choice of a high-gain observer design is motivated by the need to decouple the local observer state estimation error from the influence of the interactions between the plant subsystems. A closed-loop stability analysis taking into account model uncertainties, model and observer estimation errors, is conducted to derive appropriate stability thresholds on both the local model estimation errors and control Lyapunov functions in terms of the local observer states. Each control system continuously monitors the discrepancy between the model-estimated and observer-estimated states, as well as the estimated evolution of the local Lyapunov function using the observer state. Alarm signals are sent to the upper-level supervisor when either threshold is breached. The supervisor collects the alarm signals, and then coordinates based on logic operations the dispatch of data transmissions between the plant subsystems. Finally, the proposed methodology is illustrated using a representative chemical process example, and compared with conventional event-triggered control approaches.

References:

[1] J. Lunze, *Feedback Control of Large Scale Systems*. U.K.: Prentice Hall, 1992.

[2] K. R. Jillson and B. E. Ydsite, "Process networks with decentralized inventory and flow control," *J. Proc. Contr.*, vol. 17, pp. 399–413, 2007.

[3] S. Jogwar, M. Baldea, and P. Daoutidis, "Dynamics and control of process networks with large energy recycle," *Ind. Eng. Chem. Res.*, vol. 48, pp. 6087–6097, 2009.

[4] J. B. Rawlings and B. T. Stewart, "Coordinating multiple optimization-based controllers: New

opportunities and challenges," J. Proc. Contr., vol. 18, pp. 839–845, 2008.

[5] P. D. Christofides, J. Liu, and D. M. de la Pena, *Networked and Distributed Predictive Control: Methods and Nonlinear Process Network Applications*. London: Springer-Verlag, 2011.

[6] P. D. Christofides, J. Davis, N. H. El-Farra, D. Clark, K. Harris, and J. Gipson, "Smart plant operations: Vision, progress and challenges," *AIChE Journal*, vol. 53, no. 11, pp. 2734–2741, 2007.

[7] N. H. El-Farra and P. D. Christofides, "Special issue on control and optimization of smart plant operations," *Chem. Eng. Sci.*, vol. 136, pp. 1–176, 2015.

[8] Y. Sun and N. H. El-Farra, "Quasi-decentralized model-based networked control of process systems," *Comp. & Chem. Eng.*, vol. 32, no. 9, pp. 2016–2029, 2008.

[9] Y. Sun and N. H. El-Farra, "Quasi-decentralized networked process control using an adaptive communication policy," *Proceedings of American Control Conference*, pp. 2841–2846, 2010.

[10] E. Garcia, Y. Cao, H. Yu, P. Antsaklis, and D. Casbeer, "Decentralised event-triggered cooperative control with limited communication," *Inter. J. Contr.*, vol. 86, no. 9, pp. 1479–1488, 2013.

[11] D. Xue and N. H. El-Farra, "Supervisory logic for control of networked process systems with event-based communication," *Proceedings of American Control Conference*, to appear, 2016.

466885 Distributed Extremum Seeking Control over Unknown Network

Abstract:

When it comes to solving large-scale real-time optimization problems involving multi-agents, control approaches employed could either be centralized or distributed (decentralized).
In a centralized approach, the individual agents that make up a system are controlled by a single decision maker. This decision maker monitors, receives and processes the information from all agents, it also sends processed results back to the agents. The main advantage of this control approach is that the decision maker has full knowledge of the state of the system at any given time and can make useful decision(s) that help meet global objective(s). As the number of agents increases, it becomes difficult for the decision maker to perform its task due to increase in computational complexity. This phenomenon could result in the transmission of inaccurate information, loss of information and increase in computation time. In addition, centralized approaches are complex and most importantly lack system robustness as failure of the decision maker could mean failure of the entire system.

Most of the challenges highlighted above can be effectively avoided if one adopts a distributed approach. In a distributed environment, multiple decision makers can be utilized. Each subsystem or agent is controlled by a local decision maker. With the help of the decision maker, each agent's task can be limited to the solution of a simpler local problem. Each local problem can be solved with or without the cooperation of neighbouring agents. In a cooperative environment, agents work with other agents through local communication to find the best solution that meets global objective(s) of the system. Some of the advantages of distributed control include but are not limited to effectiveness, flexibility, scalability and adaptiveness. The greatest advantage of distributed control is system robustness. This implies that the failure of a decision maker does not necessarily mean overall system failure as the system can absorb the effect of a failure and quickly recover through the help of other decision makers.

In this paper, the solution of large-scale real-time optimization problems in the absence of precise knowledge of network connectivity in a distributed environment is considered. The knowledge of the communication network is assumed to be unknown and the agents are faced with the task of ensuring that the system's unknown overall cost (i.e., the sum of the local cost of all the agents) is minimized. Each agent has access to the measurement of two unknown cost functions referred to as the local cost and the local disagreement cost respectively. The minimization of the system's unknown overall cost to its unknown optimum depends on the minimization of the local disagreement cost of all the agents. For the local disagreement cost of the agents to be minimized, dynamic consensus estimation is required. This ensures that the agents reach agreement on their inputs (the decision variables). To help tackle this challenging problem, a distributed proportional-integral extremum seeking control technique is proposed, one that solves both problems simultaneously. A two-time scale approach is employed where consensus is achieved at a faster time-scale than the minimization of the overall cost. Included are two simulation examples, they show the effectiveness and robustness of this technique.

459963 Power Management in Microgrids with Controllable Loads and Energy Exchange Commitments

Abstract:

Microgrids are local energy systems which contain a variety of dispatchable units (e.g. microturbines, batteries), non-dispatchable units (e.g. rooftop photovoltaics), and stochastic loads (e.g. power, space cooling, and hot water demands). The primary operational goal in these systems is to ensure that local energy demands are met in a robust and cost-effective manner. Due to the inherently stochastic nature of this problem, power exchange with the macrogrid is typically used to correct any local power imbalance. However, microgrid operators may want to actively regulate this energy exchange due to demand charges, unappealing real-time prices, or energy exchange commitments [1]. This leads to a challenging control problem since one cannot rely on the grid connection to mitigate stochasticity in renewables and local demands. Previous authors have used approaches like dynamic programming [2], robust optimization [3], bidding on the day-ahead market [4,5], and coordinating intra-microgrid energy transfers [6] to address this challenge.

In this work, we focus on the dispatch of a microgrid given a unit commitment and energy exchange schedule (i.e. energy exchange should lie between some upper and lower bound in each hour). In particular, a case study is considered for a prototype microgrid consisting of a bi-directional connection to the macrogrid, photovoltaics, microturbines, a battery bank, flexible air conditioning, and an auxiliary electric boiler. We propose and formulate a nonlinear economic model predictive control (E-MPC) problem for this system to regulate indoor air temperature, supply electricity and hot water demands, and satisfy the scheduled energy exchange bounds. This E-MPC problem can be resolved frequently so that disturbances due to weather and load forecasting errors are rejected.

Then, a dynamic microgrid and building thermal model is built in the Simulink software environment to test this control approach. We analyze the performance in terms of operational cost, utilization of available PV power, satisfaction of energy exchange bounds, and thermal comfort for occupants. Particular attention is paid to the ability to capture the important nonlinear dynamics in the E-MPC while maintaining a practical solution time. The benefit of incorporating flexible loads (i.e. the AC unit and auxiliary boiler) and leveraging the natural thermal inertia of building elements is also addressed.

[1] Zachar, M. & Daoutidis, P. Microgrid/Macrogrid Energy Exchange: A Novel Market Structure and Stochastic Scheduling. *IEEE Transactions on Smart Grid*, submitted.

[2] Riffonneau, Y., Bacha, S., Barruel, F., & Ploix, S. (2011). Optimal Power Flow Management for Grid Connected PV Systems with Batteries. *IEEE Transactions on Sustainable Energy*, 2(3), 309–320.

[3] Zhang, Y., Gatsis, N., & Giannakis, G. B. (2013). Robust Energy Management for Microgrids With High-Penetration Renewables. *IEEE Transactions on Sustainable Energy*, 4(4), 944–953.

[4] Nguyen, D. T., & Le, L. B. (2015). Risk-Constrained Profit Maximization for Microgrid Aggregators with Demand Response. *IEEE Transactions on Smart Grid*, 6(1), 135–146.

[5] Liu, G., Xu, Y., & Tomsovic, K. (2016). Bidding Strategy for Microgrid in Day-Ahead Market Based on Hybrid Stochastic/Robust Optimization. *IEEE Transactions on Smart Grid*, 7(1), 227–237.

[6] Kuznetsova, E., Li, Y.-F., Ruiz, C., & Zio, E. (2014). An Integrated Framework of Agent-Based Modelling and Robust Optimization for Microgrid Energy Management. *Applied Energy*, 129, 70–88.

460871 Forming Distributed Estimator Networks from Decentralized Estimators

Abstract:

Complex, tightly integrated chemical process networks become more and more common in the modern process industry due to their economic efficiency. Such a process network usually consists of several unit operations (subsystems), which are connected together through material, energy and information flows. Because of the significant interaction between different subsystems, it poses great challenges in the design of automatic control systems. Process networks are traditionally controlled in the decentralized framework. During the past decade, much attention has been given to development of distributed process control systems (in particular, distributed model predictive control). In a distributed control framework, local controllers communicate to exchange information to collaborate their actions. The distributed framework has been demonstrated to be a promising framework for handling the scale and the interaction of these integrated chemical processes. Compared to distributed process control, its dual problem â€" distributed state estimation â€" has received relatively less attention within process control. In this work, we focus on distributed state estimation.

In the literature, the existing algorithms on distributed state estimation are primarily developed in three frameworks: the deterministic observer framework, the Kalman filter (KF) framework and the moving horizon estimation (MHE) framework. Most of these algorithms are developed for linear systems and few for nonlinear systems. A prominent feature of all these algorithms is that they all require the local observers/estimators to be of the same type (e.g., all the local estimators should be MHEs) and each of the developed algorithm can be only applied to a specific type of local observers/estimators (e.g., distributed KF algorithms cannot be used to coordinate distributed MHEs). It is desirable to develop a more general method that is capable of connecting observers/estimators of different types together while ensures overall convergence of the estimates. The motivation of this kind of method can be seen when a process includes subsystems of different types (e.g., linear and nonlinear subsystems) or subsystems posing different process/numerical constraints (e.g., estimation constraints need to be considered in some of the subsystem).

Another feature of the existing distributed state estimation algorithms is that in the design of these algorithms, no consideration is given to the potentially existing (decentralized) implementation of control/estimation algorithms in a process. If a decentralized state estimation algorithm has already been implemented in a process, the above distributed algorithms essentially require a completely redesign of the existing implementation which means high capital investment. This also motivates us to develop a general method that is capable of connecting (decentralized) observers/estimators together to improve the overall estimation performance.

Motivated by above considerations, in this work, we propose a method to connect decentralized observers/estimators (which can be of different types) to form distributed state estimation schemes. One key idea of proposed method is to develop a local compensator for each subsystem, which is used to compensate for the interactions between a subsystem and its neighbor subsystems. Each compensator is connected to the corresponding subsystem observer/estimator designed without considering interaction to form a complete local estimator. Within the proposed framework, the local estimation algorithms for different subsystems can be independently selected and each local estimator can be more easily maintained without deteriorate the structure of the entire estimation system. The proposed method is demonstrated to be effective via the application to a simulated chemical process. It is also demonstrated that the proposed method has the potential to be used in distributed output feedback control which is an issue that has not been addressed in the research of distributed model predictive control.

465373 Coarse Modeling of Circadian Rhythms in Heterogeneous Neural Networks

Abstract:

The suprachiasmatic nucleus (SCN) of the hypothalamus contains about 15,000 neurons in mammals, and is the source of a light-independent 24-hour rhythm which serves as a central clock governing daily rhythms throughout the rest of the body. This rhythm is the result of a set of gene repression/promotion feedback loops occurring in each neuron. Though, in isolation, cells exhibit a range of free-running periods distributed around 24 hours, their coupling in a complex network allows for synchronization, and entrainment to an external zeitgeber.

Our work on coarse-graining the dynamics of synchronized populations uses the emergent smooth dependence of cellsâ€™ states in this synchronized group on the particular parameters of each cell--such as natural period. Such parameters are heterogeneous across the population, but unchanging with time. Under the assumption that this dependence is smooth (and that the cells synchronize at least partially in frequency), this coarse-graining constitutes a significant dimensionality reduction of our description of the system state.

We use this reduced-dimension representation to perform optimizations of drug dosing schedules to treat non-24-hour sleep/wake disorders or to aid with entrainment to an industrial non-24-hour shift schedule.

469960 Dynamic Real-Time Optimization of Distributed MPC Systems

Abstract:

Real-time optimization (RTO) is a supervisory control algorithm that provides set-points to an underlying regulatory control system based on optimizing the plant economics. Industrial RTO systems typically utilize a steady-state model of the process, and are executed every few hours in order to track the economic optimum, as parameters affecting the plant conditions and process economics change (Marlin and Hrymak, 1997; Darby et al., 2011). The limited execution frequency of steady-state RTO systems poses a drawback for applications to processes that exhibit frequent transitions or slowly varying dynamics, which has prompted the development of dynamic RTO strategies. Two key control architectures for addressing plant economics within a dynamic setting are (i) a hierarchical structure, similar to way that current industrial RTO systems are implemented, but with a dynamic model used instead of a steady-state model, and (ii) a single-level control structure in which the economic objective is incorporated within the MPC controller formulation (Engell, 2007; Rawlings and Amrit, 2009; Ellis et al., 2014). The present work follows the former approach due to its compatibility with existing industrial control systems, and builds on recent advances that have considered DRTO in conjunction with a single MPC system.

Previously proposed DRTO strategies that follow a two-layer architecture typically perform economic optimization in an open-loop fashion without taking into account the presence of the plant control system. In this open-loop DRTO approach, the set-points prescribed to the underlying control system are based on the optimal open-loop trajectories under an expectation that the closed-loop response dynamics at the plant level will follow the economically optimal trajectories obtained at the DRTO level. In Tosukhowong et al. (2004), the DRTO framework is designed based on a linear(ized) process model, while Würth et al. (2011) utilize a nonlinear dynamic model. In an earlier study, we proposed the use of a dynamic model at the DRTO level that predicts the effects of the underlying MPC system on the plant dynamics (Jamaludin and Swartz, 2014). Neglecting the effect of the control system in the DRTO optimization tacitly assumes perfect control, which is not achievable in practice and leads to suboptimal economic performance.

However, this new formulation results in a more complex problem, since each MPC control calculation over the future prediction horizon is itself defined by an optimization problem. This is handled by replacing the MPC optimization subproblems by their equivalent first-order optimality conditions, and including the resulting algebraic equations as equality constraints in the DRTO economic optimization problem. In Jamaludin and Swartz (2015), the closed-loop prediction problem was approximated as a bi-level rather than a multi-level problem, resulting in significantly reduced computation times with modest loss in economic performance.

An overall optimal process operation can be achieved via utilization of a centralized model predictive control (MPC) system in which all control inputs are computed simultaneously from a single optimization problem. However, large scale systems, such as refineries, manufacturing plants and power generation networks, typically have a collection of MPC controllers, with each controlling a subset of process units (Qin and Badgwell, 2003). Key considerations in the selection of distributed MPC systems are process scale and complexity, computational tractability, and geographic footprint. Distributed MPC architectures are designed to meet performance specifications that are relatively equivalent to the centralized MPC system, but retaining the modularity, reliability and maintainability of each controller (Pannochia, 2014). A variety of approaches to coordination of distributed MPC systems has been proposed in the literature. Camponogara et al. (2002) and Stewart et al. (2010) present cooperative control strategies in which the MPC modules compute control actions taking into account the actions of the other MPC controllers. Cheng et al. (2008) apply Dantzig-Wolfe decomposition to the calculation of steady-state targets for the MPC controllers based on a linear objective function. In Chen et al. (2012), steady-state optimization is performed, with distributed Lyapunov-based MPC controllers executed sequentially to drive the system to the steady-state optimum.

In this study, the closed-loop DRTO formulation is extended to economic coordination of a distributed MPC system. The DRTO formulation determines optimal set-point trajectories for all MPC controllers simultaneously, based on plant economics. The DRTO plant model represents the overall process dynamics whereas the MPC models represent the local unit dynamics. The plant model used within the DRTO module is consistent with the dynamic models used in the MPC controllers, but with the interactions between the process subsystems captured through appropriate linking relationships, such as material flows. The response of the plant under the action of the MPC subsystems is computed by taking into account the actions of all the controllers, as well as the process interactions between the subsystems. Interaction between local controllers is addressed via output feedback from the DRTO plant model to the MPC model predictions. The MPC optimization subproblems to be used in the DRTO closed-loop response prediction are again be replaced by their first-order optimality conditions to yield a single-level economic optimization problem. The performance of the proposed approach is assessed via case study simulations.

References

Camponogara, E., Jia D., Krogh, B.H. and Talukdar, S. (2002). Distributed model predictive control, IEEE Control Systems Magazine, Feb 2002, 44-52.

Chen, X., Heidarinejad, M., Liu, J., and Christofides, P.D. (2012). Distributed economic MPC: Application to a nonlinear chemical process network. J. Proc. Contr., 22, 689-699.

Cheng, R., Forbes, J.F. and Yip, W.S. (2008). Dantzig-Wolfe decomposition and plant-wide MPC coordination. Comp. Chem. Eng., 32, 1507-1522.

Darby, M.L., Nikolaou, H., Jones, J. and Nicholson, D. (2011). RTO: An overview and assessment of current practice. J. Proc. Contr., 21(6), 874-884.

Ellis, M., Durand, H. and Christofides, P.D. (2014). A tutorial review of economic model predictive control methods. J. Proc. Contr., 24(8), 1156-1178

Engell, S. (2007). Feedback control for optimal process operation. J. Proc. Contr., 17(3), 203-219.

Jamaludin, M.Z. and Swartz, C.L.E. (2014). Effects of closed-Loop dynamics in dynamic real-time optimization. Paper 721f, AIChE Annual Meeting, Atlanta.

Jamaludin, M.Z. and Swartz, C.L.E. (2015). A bilevel programming formulation for dynamic real-time optimization. In Int. Symp. on Advanced Control of Chemical Processes (ADCHEM), Whistler, BC, Canada, 907-912.

Marlin, T.E. and Hrymak, A.N. (1997). Real-time operations optimization of continuous processes. In Kantor, J.C., Garcia, C.E., Carnahan, B. (eds.), Proc. 5th Int. Conf. on Chemical Process Control, AIChE and CACHE, 156-164.

Qin, S.J. and Badgwell, T.A. (2003). A survey of industrial model predictive control technology. Control Engineering Practice, 11, 733-764.

Rawlings, J.B. and Amrit, R. (2009). Optimizing process economic performance using model predictive control. In Magni, L., Raimondo, D., Allgower, F., eds., Nonlinear Model Predictive Control, Lecture Notes in Control and Information Sciences, vol. 384, Springer, 119-138.

Pannocchia, G. (2014). Distributed model predictive control. Encyclopedia of Systems and Control, 1-9.

Stewart, B.T., Venkat, A.N., Rawlings, J.B. and Wright, S.J. (2010). Cooperative distributed model predictive control. Systems & Control Letters, 59, 460-469.

Tosukhowong, T., Lee, J.M., Lee, J.H., and Lu, J. (2004). An introduction to a dynamic plant-wide optimization strategy for an integrated plant. Comp. Chem. Eng., 29(1), 199-208.

Würth, L., Hannemann, R. and Marquardt, W. (2011). A two-layer architecture for economically optimal process control and operation. J. Proc. Contr., 21(3), 311-321.

459912 Overview of CAST Activities and Programming

Abstract:

An overview of activities of the Computing and Systems Technology (CAST) division, including conference technical programming, will be presented. The CAST Executive Committee consists of Ray Adomaitis (Past Chair), Karl Schnelle (Chair), Stratos Pistikopoulos (1st Vice-Chair), Carl Laird (2nd Vice-Chair), Martha A. Grover (Secretary-Treasurer), and the following Directors: Andreas Linninger (2014-2016), Leo Chiang (2014-2016), Michael Nikolaou (2015-2017), Mario Eden (2015-2017), Ruth Misener (2016-2018) and Dimitrios Papavassiliou (2016-2018). The CAST Division Programming Chair is Nick Sahinidis (2013-2016).

459968 Towards a Computational Platform for General Flowsheet Synthesis

Abstract:

Towards a computational platform for general flowsheet synthesis

Qi Chen[1], Ignacio Grossmann[1]

[1]Department of Chemical Engineering, Carnegie Mellon University, 5000 Forbes Avenue, Pittsburgh, PA 15213, USA

ABSTRACT

Synthesis of chemical process flowsheets is a fundamental challenge of the chemical engineering discipline [1]. Recently, low energy prices and new feedstocks introduced by the US shale boom have spurred renewed interest in process design. At the same time, sustainability concerns provide impetus for computer-aided design in the emerging areas of bio-feedstocks, carbon capture systems, and process intensification. Superstructure optimization facilitates a unified basis for tackling synthesis problems across this wide range of interest areas, and from green-field design to retrofit operations [2]. However, despite substantial progress in superstructure optimization, particularly in subsystems like heat exchanger network synthesis and separation sequences, there has been limited adoption of these methods in industry [3]. One driver for this has been lack of a general purpose software platform for performing superstructure synthesis.

In this paper, we present progress towards development of such a platform. A general purpose solution poses both modeling and implementation challenges with a combination of nonlinear, nonconvex equations (frequently arising from mixing/splitting and concave cost functions) and discrete design decisions. We adopt a Generalized Disjunctive Programming (GDP) modeling approach, with optional processing units represented by a disjunction between presence and absence of the unit [4]. We then decompose the problem using logic-based outer-approximation (LOA) [4,5] in order to reduce dimensionality through the reduced-space nonlinear programs (rNLPs) resulting from a fixed configuration of the process network. Furthermore, these rNLPs avoid zero-flow singularities in absent process units by excluding them from the subproblem model. Outer-approximation with equality relaxation (OA/ER) cuts [6] derived from the subproblems communicate information to a mixed-integer linear (MILP) master problem, which selects a new network configuration. The algorithm then repeats until a termination condition is reached. In this work, we propose a modeling distinction between interconnection (e.g. mixers and splitters) and processing nodes in the MILP master problem. We use rigorous underestimators such as polyhedral envelopes for linearizing interconnection nodes rather than OA/ER cuts as before. Since flow rates through interconnection nodes tend to vary more in different process configurations than other types of nodes, rigorous underestimators reduce the likelihood that the optimal solution will be excluded prematurely. This modification yields improved performance both in superstructures predominantly defined by interconnections as well as those driven by complex processing units. We also replace concave cost functions by linear underestimators to predict lower bounds for the case of linear input-output models for the units. Furthermore, in order to improve the robustness of the optimizations, we investigate several strategies for initialization for the rNLPs, the impact of modeling with flows, compositions, and split fractions, and the addition of valid inequalities.

We implement this framework in the high level programming language Python using the Pyomo [7] modeling environment that allows direct formulation of the GDP for superstructure optimization in terms of equations, disjunctions and logic propositions. This combination allows for a true object-oriented modeling paradigm. With the Pyomo Block feature, entire groups of model constraints can be dynamically activated or deactivated for LOA with one line of code. We also demonstrate the ability to perform programmatic application of OA cuts, piecewise McCormick envelopes, multi-parametric disaggregation, and linear underestimators with Python function calls.

The proposed platform is applied to a set of literature case studies, including water treatment network design [8], flowsheet synthesis for the hydrodealkylation of toluene [9], and production of methanol [4]. Computational results demonstrate the ability to readily model superstructure optimization models, and to quickly find near-optimal solutions to a range of synthesis problems.

References

[1] L.T. Biegler, I.E. Grossmann, A.W. Westerberg, Systematic methods of chemical process design, Prentice Hall PTR, 1997.

[2] I.E. Grossmann, G. Guillen-Gosalbez, Scope for the application of mathematical programming techniques in the synthesis and planning of sustainable processes, Comput. Chem. Eng. 34 (2010) 1365–1376. doi:10.1016/j.compchemeng.2009.11.012.

[3] G.J. Harmsen, Industrial best practices of conceptual process design, Chem. Eng. Process. Process Intensif. 43 (2004) 677–681. doi:10.1016/j.cep.2003.02.003.

[4] M. Türkay, I.E. Grossmann, Logic-based MINLP algorithms for the optimal synthesis of process networks, Comput. Chem. Eng. 20 (1996) 959–978. doi:10.1016/0098-1354(95)00219-7.

[5] M.A. Duran, I.E. Grossmann, An outer-approximation algorithm for a class of mixed-integer nonlinear programs, Math. Program. 36 (1986) 307. doi:10.1007/BF02592064.

[6] J. Viswanathan, I.E. Grossmann, A combined penalty function and outer-approximation method for MINLP optimization, Comput. Chem. Eng. 14 (1990) 769–782. doi:10.1016/0098-1354(90)87085-4.

[7] W.E. Hart, C. Laird, J.-P. Watson, D.L. Woodruff, Pyomo – Optimization Modeling in Python, Springer US, Boston, MA, 2012. doi:10.1007/978-1-4614-3226-5.

[8] R. Karuppiah, I.E. Grossmann, Global optimization for the synthesis of integrated water systems in chemical processes, Comput. Chem. Eng. 30 (2006) 650–673. doi:10.1016/j.compchemeng.2005.11.005.

[9] G.R. Kocis, I.E. Grossmann, Computational Experience With Dicopt Solving Minlp Problems in Process Systems-Engineering, Comput. Chem. Eng. 13 (1989) 307–315. doi:10.1016/0098-1354(89)85008-2.

455232 Distributed Modifier-Adaptation Schemes for the Real-Time Optimization of Interconnected Systems in the Presence of Structural Plant-Model Mismatch

Abstract:

The desire to operate chemical processes in a safe and economically optimal way has motivated the development of so-called real-time optimization (RTO) methods [1]. For continuous processes, these methods aim to compute safe and optimal steady-state setpoints for the lower-level process controllers. A key challenge for this task is plant-model mismatch. For example, in the case of a model that is assumed to be structurally identical with the plant but has unknown parameters, the so-called two-step approach [2-4] has been proposed. It repeats two steps: In the first step, plant measurements are used to identify the parameters of the model. In the second step, the economically optimal setpoints for the updated process model are determined by solving an optimization problem. Unfortunately, a structurally correct process model is rarely available in practice. In that case, the optimal setpoints for the model determined by the two-step approach may not be optimal for the plant. To overcome this problem, the so-called modifier-adaptation (MA) methods have been developed [5]. In MA, no structurally correct model is required. Instead, plant measurements are used to formulate and solve a modified optimization problem at each iteration, such that, upon convergence, the first-order optimality conditions of the plant are guaranteed to be satisfied [5].

These and other available RTO methods usually treat the plant as a single entity, and compute the optimal setpoints in a centralized manner. However, this approach may be suboptimal or even infeasible for an increasing number of applications involving so-called interconnected systems. Interconnected systems are here defined as systems composed of subsystems that exchange material, energy or information, such as compressor networks, teams of autonomous vehicles or large industrial parks, in which different business units of a chemical company share certain resources. In these cases, distributed RTO methods can be employed, which utilize the available interconnection variables and exploit the inherent interconnection structure of the particular system.

Only a few distributed RTO methods have been reported in the literature, including the methods proposed by Brdys and Tatjewski [6]. Just as in the two-step approaches, structurally correct models are assumed. In addition to identifying the model parameters, the methods also try to estimate the values of the interconnection variables. Consequently, these methods may not yield the plant optimum in the presence of structural plant-model mismatch.

In this contribution, we propose a set of distributed RTO methods based on the modifier-adaptation framework for interconnected systems in the presence of structural plant-model mismatch. Thanks to the modifier-adaptation framework, all proposed distributed RTO methods are able to reach the plant optimum upon convergence despite possible plant-model mismatch. The proposed schemes employ different types of models, use different measurements, and differ in their algorithmic structure and required controller hierarchy, as well as in their communication topology, as detailed below.

The first method utilizes a model of the local objective function, a model for the dependence of local outputs on the local setpoints and outputs of other subsystems, and a model for the interconnection structure of the system. The algorithm resembles a double-loop structure: In the simulation-based inner loop, the local MA-problems are solved in parallel until the interconnection constraints are satisfied. As soon as the inner loop has converged, the computed setpoints are applied to the plant in the outer loop. When the plant has reached a new steady state, measurements are taken to improve the performance at the next iteration.

The second and third methods do not require a model of the interconnection structure of the system. Consequently, a single-loop algorithmic structure is sufficient. At every iteration, each subsystem computes local setpoints, which are immediately applied to the plant. At the corresponding steady state, plant measurements are taken to improve the performance at the next iteration. At this point, the second and third method proceed differently: The second method uses local measurements of the interconnection variables to each subsystem, whereas the third method additionally measures the local outputs. Consequently, the second method still uses a model describing the dependency of the local outputs on the local setpoints and local interconnection variables, whereas no such relationship is needed for the third method.

Because of their different characteristics, each of these algorithms has its specific advantages regarding applications. For example, the first method requires each subsystem to have a complete model of the full system and its interconnection topology. If these models are good, then fast convergence of this scheme with few setpoint changes can be expected. In applications, where providing a full model of the system to every subsystem is feasible and does not raise any privacy concerns, the first scheme may be the method of choice. The second and third schemes, in contrast, do not require an interconnection model. Therefore, they may be preferred for applications where different subsystems do not want to disclose their models to other subsystems. This could be the case when the different subsystems are owned by competing companies. Moreover, the lack of an interconnection model may be advantageous in certain applications with changing interconnection topologies, such as power system networks. Another advantage of the second and third method is their reduced local modeling effort if accurate measurements are available. On the other hand, these methods may need significantly more iterations to converge than the first method if accurate models are available.

In our contribution, we finally apply the proposed distributed modifier-adaptation schemes to numerical examples. The main features of each method are illustrated, revealing their potential for real-time optimization of interconnected systems with structural plant-model mismatch.

References

[1] C. R. Cutler and R. T. Perry. Real time optimization with multivariable control is required to maximize profits. Comput. Chem. Eng. 7(5):663-667, 1983.

[2] T. E. Marlin and A. N. Hrymak. Real-time operations optimization of continuous processes. In Proceedings of the

5th International Conference on Chemical Process Controls-V: Assessment and New Directions for Research; J. C. Kantor, C. E. Garcia, and B. Carnahan, Eds.; AIChE Symposium Series, No. 316; American Institute of Chemical Engineers (AIChE): New York, 1997, 156-164.

[3] S.-S. Jang, B. Joseph, and H. Mukai. On-line optimization of constrained multivariable chemical processes. AIChE J., 33(1): 26â€“35, 1987.

[4] C. Y. Chen and B. Joseph. On-line optimization using a two-phase approach: An application study. Ind. Eng. Chem. Res., 26:1924-1930, 1987.

[5] A. Marchetti, B. Chachuat, and D. Bonvin. Modifier-Adaptation Methodology for Real-Time Optimization. Ind. Eng. Chem. Res., 48: 6022-6033, 2009.

[6] M. A. Brdys and P. Tatjewski. Iterative Algorithms for Multilayer Optimizing Control. Imperial College Press, Imperial College Press: London, U.K., 2005.

460577 GOSSIP: Decomposition Software for the Global Optimization of Nonconvex Two-Stage Stochastic Mixed-Integer Nonlinear Programs

Abstract:

Stochastic programming provides a natural way of incorporating uncertainty in model parameters, and has been receiving increasing attention in the process systems engineering literature [1–6]. Despite rapid advances in decomposition techniques for solving nonconvex two-stage stochastic mixed-integer nonlinear programs (MINLPs) [1, 7, 8], there is, to the best of our knowledge, no publicly available software framework which implements these techniques. Motivated by the above, we introduce GOSSIP, a decomposition framework for the global optimization of two-stage stochastic MINLPs.

GOSSIP includes subroutines for reformulating user input, detecting special structure, automatic construction of the subproblems required by the decomposition techniques, automatic construction of relaxations, and bounds tightening [9–12]. The decomposition framework includes implementations of nonconvex generalized Benders decomposition (NGBD) [7, 8], Lagrangian relaxation [1, 13], and a modified Lagrangian relaxation algorithm. The option of solving the extensive form of the two-stage stochastic MINLP using a global optimization solver is also included. Solver links to several state-of-the-art optimization software are part of GOSSIP and are used to solve the various subproblems used by the decomposition techniques.

A library of test instances of two-stage stochastic MINLPs from the literature is composed, and the capabilities of GOSSIP are demonstrated over this diverse set of problems.

[1] R. Karuppiah and I. E. Grossmann, "Global optimization of multiscenario mixed integer nonlinear programming models arising in the synthesis of integrated water networks under uncertainty," Computers & Chemical Engineering, vol. 32, no. 1, pp. 145–160, 2008.

[2] S. Rebennack, J. Kallrath, and P. M. Pardalos, "Optimal storage design for a multi-product plant: A non-convex MINLP formulation," Computers & Chemical Engineering, vol. 35, no. 2, pp. 255–271, 2011.

[3] Y. Chen, T. A. Adams, and P. I. Barton, "Optimal design and operation of flexible energy polygeneration systems," Industrial & Engineering Chemistry Research, vol. 50, no. 8, pp. 4553–4566, 2011.

[4] X. Li, E. Armagan, A. Tomasgard, and P. I. Barton, "Stochastic pooling problem for natural gas production network design and operation under uncertainty," AIChE Journal, vol. 57, no. 8, pp. 2120–2135, 2011.

[5] X. Li, Y. Chen, and P. I. Barton, "Nonconvex generalized Benders decomposition with piecewise convex relaxations for global optimization of integrated process design and operation problems," Industrial & Engineering Chemistry Research, vol. 51, no. 21, pp. 7287–7299, 2012.

[6] A. Sundaramoorthy, J. M. Evans, and P. I. Barton, "Capacity planning under clinical trials uncertainty in continuous pharmaceutical manufacturing, 1: mathematical framework," Industrial & Engineering Chemistry Research, vol. 51, no. 42, pp. 13692–13702, 2012.

[7] X. Li, A. Tomasgard, and P. I. Barton, "Nonconvex generalized Benders decomposition for stochastic separable mixed-integer nonlinear programs," Journal of Optimization Theory and Applications, vol. 151, no. 3, pp. 425–454, 2011.

[8] X. Li, A. Sundaramoorthy, and P. I. Barton, "Nonconvex generalized Benders decomposition," in Optimization in Science and Engineering, pp. 307–331, Springer, 2014.

[9] M. Tawarmalani and N. V. Sahinidis, "A polyhedral branch-and-cut approach to global optimization," Mathematical Programming, vol. 103, no. 2, pp. 225–249, 2005.

[10] R. Misener and C. A. Floudas, "GloMIQO: Global mixed-integer quadratic optimizer," Journal of Global Optimization, vol. 57, no. 1, pp. 3–50, 2013.

[11] R. Misener and C. A. Floudas, "A framework for globally optimizing mixed-integer signomial programs," Journal of Optimization Theory and Applications, vol. 161, no. 3, pp. 905–932, 2014.

[12] R. Misener and C. A. Floudas, "ANTIGONE: algorithms for continuous/integer global optimization of nonlinear equations," Journal of Global Optimization, vol. 59, no. 2-3, pp. 503–526, 2014.

[13] C. C. CarøE and R. Schultz, "Dual decomposition in stochastic integer programming," Operations Research Letters, vol. 24, no. 1, pp. 37–45, 1999.

464749 Dimensionality Reduction of Dynamic Networks

Abstract:

A canonical problem in the field of dimensionality reduction is to efficiently capture, in some mathematical sense, the important information contained within a collection of high-dimensional vectors. This leads to a simplified description of the vectors that lends itself to faster computation and easier interpretability. Given the ongoing research into complex networks, we present techniques to extend these dimensionality reduction methods, in particular Diffusion Maps, to operate on a collection of graphs. This allows the researcher to obtain a reduced description of the network using only simulation data, and without analyzing the analytical equations that constitute the model. We illustrate Diffusion Maps' effectiveness in the context of both an epidemiological network model and a dynamic multigraph system. Computational challenges and promising future directions will also be discussed.

468267 Robust Dynamic Principal Component Analysis Method for Modelling Process Data

Abstract:

Advancing sensor and data gathering technology has resulted in a substantial increase in the amount and frequency of data collected from processes, creating a valuable opportunity for data-driven modelling techniques in process monitoring. Reliable data-driven models can lead to online analysis of incoming data so that sensor and process faults can be detected and managed. There are numerous methods for building data-driven models, including the commonly used principal component analysis (PCA) and its extension, dynamic PCA (DPCA) which includes time lagged variables to represent dynamic processes. However, there are several drawbacks to PCA and DPCA, including the inability to handle any missing or corrupted data. Robust PCA methods can address this problem [1].

Two new algorithms were developed in this work by incorporating a recently developed robust PCA method into DPCA to create a new *robust DPCA* method. The new method allows for the recovery of a fault-free matrix even when there are gross sparse errors present, and is therefore more robust to outliers than other methods. The robust PCA method was incorporated into DPCA in two different ways ('Method 1' and 'Method 2') which were tested and compared to each other and to traditional DPCA with real clean and faulty process data.

This work contains two case studies: Case Study 1 used data from a field test which generated two sets of data: a clean data set representing normal operation and a corrupted data set, with known sensor faults. Case Study 2 used operation data from a single piece of process equipment during normal operation to ensure that tags were correlated and the data was largely free of faults. The two case studies allowed for two different approaches for analysis. Since the first data set had a known clean data set available, detection and false alarm rates could be calculated using the squared prediction error (SPE) or Q statistic, as well as the similarity between the normal data set and the one cleaned by the new algorithm. The second data set did not have a basis for comparison, but was normal operation data and was therefore expected to have a low detection rate. Faults were then added to several variables to determine the effectiveness of the algorithm. Data reconstructions based on the models with 'normal' data and 'faulty' data were then compared.

Both the methods presented a significant improvement for fault detection over DPCA models trained on faulty data, which failed to detect most of the faults. Method 1 was also able to detect some faults that Method 2 was not. Furthermore, though Method 2 resulted in high detection rates in all the tests, it also consistently resulted in high false alarm rates, making the results unreliable. Method 1 was effective with proper tuning and sufficient training data; without one or the other, the detection rate and false alarm rates were adversely affected.

Reference

[1] Candes E.J., Li X., Ma Y., & Wright J. (2009) "Robust Principal Component Analysis?" *Journal of the Association of Computing Machinery*, 58 (3), 1-37.

470454 Nonsmooth Models and Methods in Chemical Engineering

Abstract:

Nonsmooth process models occupy a useful niche between discrete models with integer variables and traditional smooth models. Nonsmoothness in a model can be used to encode certain discrete features such as transitions in thermodynamic phase or flow regime, process units that switch between discrete operating modes, and pinch analyses for heat integration.

Compared to models with integer variables, nonsmooth model enforce continuity of design variables, which provides structural information that can be exploited by numerical methods. Moreover, allowing nonsmoothness in a model may provide enough versatility to represent discrete behavior that cannot be expressed in a smooth model, and to permit the use of a wider class of methods such as aggressive penalty approaches for constrained optimization. Numerical methods for nonsmooth optimization and equation-solving have convergence properties between those of methods for smooth problems and methods for problems with integer variables.

This presentation covers recent advances in nonsmooth methods and model formulation, based on pioneering work on generalized derivatives by Clarke and Nesterov. These generalized derivatives are designed to serve a similar function to gradients of smooth systems. Recent advances include an accurate, automatable, computationally tractable method for evaluating generalized derivatives for nonsmooth functions that are compositions of simple, known functions and operations. This method has been extended to nonsmooth dynamic systems, and to implicit quantities such as the unknown time at which a relief valve opens. In turn, these methods yield new, effective approaches for designing and simulating multistream heat exchangers with pinch analyses embedded, for example, and for simulating dynamic bioreactors with pseudo-steady metabolic models embedded. Implementations, future directions, and open challenges will also be discussed.

469917 Perspectives on Predictive Control with Dual Control Feature for Stochastic Systems

Abstract:

Model predictive control (MPC) has demonstrated exceptional success for high-performance control of complex systems in a wide range of applications [1,2]. However, systematic model uncertainty handling under closed-loop conditions remains a key theoretical and practical challenge in MPC. Even though standard MPC implementations on the basis of a nominal system model typically exhibit some degree of robustness to sufficiently small model uncertainty and system disturbances, marginal robust performance may not be deemed satisfactory in many practical situations. This consideration has led to development of robust MPC [3], as well as stochastic MPC when the probabilistic nature of uncertainties can be characterized [4], with the goal of guaranteeing robust stability and performance of the closed-loop system by explicitly accounting for system uncertainties. However, robust MPC and stochastic MPC are generally blind to the system changes and, therefore, their performance is limited by the quality of the a-priori specified nominal model and uncertainty descriptions. The time-varying nature of system dynamics (i.e., changes in plant and/or disturbance dynamics) tends to increase model uncertainty over time, which manifests itself in terms of inadequate model structure and/or parameter estimates. Increased model uncertainty generally results in degradation of the closed-loop control performance, as verified via performance monitoring and diagnosis (e.g., [5,6]). Thus, some form of online model adaptation must often be performed for retaining the control performance at an admissible level.

The problem of model uncertainty handling in model-based control lies at the intersection of the fields of robust control, adaptive control, and system identification for control. This talk draws on the seminal work of Feldbaum, which is widely regarded as the pioneering attempt that addressed the problem of model (parameter) uncertainty in model-based control design [7]. Feldbaum recognized the *dual effect* that input signals controlling an unknown system should have - the *investigating effect* to probe the system dynamics for actively learning about the system and the *directing effect* to effectively control the system dynamics. Control signals are said to have dual effect when, in addition to affecting the system states, they affect the uncertainty associated with the system states. The key notion of dual control hinges on compromising between the investigating and directing effects of the input signals, so that more system information can be gathered at the current time to enable achieving better control performance in future. In this talk, we will discuss the connection of the stochastic optimal control problem with dual feature to the stochastic dynamin programming problem, followed by an overview of the approximate solutions to the dual control problem in the context of predictive control of stochastic systems. We will then offer our perspectives for future research in this area in light of our recent results [8,9,10].

[1] M. Morari and J. H. Lee. Model predictive control: Past, present and future. Computers & Chemical Engineering, 23:667-682, 1999.

[2] J. H. Lee. Model predictive control: Review of the three decades of development. International Journal of Control, Automation and Systems, 9:415-424, 2011.

[3] A. Bemporad and M. Morari. Robust model predictive control: A survey. In Robustness in identification and control, pages 207-226. Springer, London, 1999.

[4] A. Mesbah. Stochastic model predictive control: An overview and perspectives for future research. IEEE Control Systems Magazine, Accepted, 2016.

[5] M. Zagrobelny, L. Ji, and J. B. Rawlings. Quis custodiet ipsos custodes? Annual Reviews in Control, 37:260-270, 2013.

[6] A. Mesbah, X. Bombois, M. Forgione, H. Hjalmarsson, and P.M.J. Van den Hof. Least costly closed-loop performance diagnosis and plant re-identification. International Journal of Control, pages 1-13, 2015.

[7] A. A. Feldbaum. Dual-control theory. I. Automatic Remote Control, 21:874-880, 1960.

[8] C.A. Larsson, M. Annergren, H. Hjalmarsson, C.R. Rojas, X. Bombois, A. Mesbah, and P.E. Moden. Model predictive control with integrated experiment design for output error systems. In Proceedings of the European Control Conference, pages 3790-3795, Zurich, 2013.

[9] V. Bavdekar and A. Mesbah. Stochastic model predictive control with integrated experiment design for nonlinear systems. In Proceedings of the 11th IFAC Symposium on Dynamics and Control of Process Systems (DYCOPS), Accepted, Trondheim, Norway, 2016.

[10] V. Bavdekar, V. Ehlinger, D. Gidon, and A. Mesbah. Stochastic predictive control with adaptive model maintenance. In Proceedings of the 55th IEEE Conference on Decision and Control, Submitted, Las Vegas, 2016.

472288 Mathematical Modeling at the Food-Energy-Water Nexus

Abstract:

Maintaining a sustainable food supply for the foreseeable future without excessive environmental degradation will require novel engineering approaches for remediating the human disruption of the nitrogen cycle. Over the past several decades, food production has been able to keep pace with human population growth only thanks to the development of new high-yielding crop varieties grown with the help of fertilizers. Thus, vast amounts of energy are consumed every year to produce the approximately 200 Mt of ammonia needed to make the nitrogen (N) fertilizers used by farmers around the globe.

Unfortunately, more than half of the N fertilizer applied to a field is not available for plant growth due to losses caused by surface runoff, leaching into surface and ground water, or volatilization. As a result, the increased use of N as fertilizer has been linked to a variety of water pollution problems ranging from the expansion of the hypoxic zone in the Gulf of Mexico caused by eutrophication to the contamination of wells and groundwater with nitrate N. More importantly, denitrification reactions occurring in saturated soils convert nitrate to nitrous oxide, a major greenhouse gas and significant contributor to the depletion of atmospheric ozone

Several studies have reported that the addition of biochars to weathered tropical or temperate region soils decreased leaching losses of nutrients. In other cases, however, biochar amendments did not have the expected beneficial effects. Such contradictory results generated a lot of interest in the production of *designer biochars*engineered and optimized to provide specific environmental performance. However, matching the properties of a biochar to a specific application is a daunting challenge. There are dozens of feedstocks from which biochars can be produced in multiple types of reactors under varying temperature and oxygen conditions, quickly leading to thousands of potential biochars with widely varying properties. Then, the performance of each of these biochars must be tested for a wide range of crops, soil types and climates.

Mathematical models derived from first principles can greatly speed up the process of selecting the most promising candidates for field-testing. By allowing us to isolate and analyze the interactions of key components of our soil-biochar-water system with the microbial populations carrying out the nitrification/denitrification reactions, mathematical modeling provides a systematic way for tuning the properties of a designer biochar to maximize its agricultural benefits. The amount of biochar applied to a field, for example, could be matched to the hydrodynamic characteristics (permeability, pore water velocity, soil type) of a specific field or to the amount of expected rainfall in order to achieve the desired nutrient retention effects. Or, we could select a feedstock and production method that would yield a biochar with optimal adsorption capacity and sorbate affinity.

The amendment of soils with biochar introduces a new stationary phase that necessitates an extension to the classical dual porosity models that considered convection and pore diffusion but only one adsorbent (stationary) phase. Biochars are highly porous materials with interconnected networks of pores that span multiple length scales: from sub-nanometer micropores to macropores with sizes of the order of 10 Î¼m or larger. Moreover, the size and adsorption capacity of biochar particles may be significantly different than the corresponding properties of soil particles.

We will first discuss results from a dynamic convection-dispersion-adsorption model that considers nutrient (fertilizer) transport through beds with two porous adsorbent phases: soil and biochar. After a single fertilizer application, rain or irrigation water flows through the bed and nutrients diffuse in the water-filled pores of both stationary phases, adsorbing reversibly on their pore surfaces. The resulting system of PDE/DAEs is solved numerically using the method of lines. Simulation results will be presented to demonstrate that addition of biochar can effectively slow nutrient transport through the soil if the biochar/soil ratio and crucial biochar properties (like its adsorption capacity) are carefully matched to the soil properties (e.g. soil permeability) and the amount of rainfall or irrigation.

To further investigate the importance of biochar properties, we will focus on the interplay of external mass transfer, intraparticle diffusion and adsorption. We will analyze the dynamic behavior predicted by models considering surface diffusion of nutrients in addition to pore diffusion, pore structures with bimodal size distribution, and adsorption kinetics described by Freundlich or Langmuir isotherms. One of the more surprising findings here will be the sensitivity of the adsorption/desorption dynamics on the Biot number of the biochar particle. Biochar behaves like a â €œcontrolled releaseâ€ medium for adsorbed nutrients primarily because of the small Biot numbers that characterize the flow of rainwater through amended soils.

Finally, we will turn our attention to the problem of nitrogen transport and simultaneous reaction in biochar amended soils. Our equivalent continuum model will assume that ammonium-N is added to the amended soil as fertilizer, dissolves in water and is transported through the bed, while it is reversibly adsorbed on the two solid phases (soil and biochar). Ammonium is oxidized first to nitrite and then to nitrate by soil microbes (nitrification). The nitrate is uptaken by plant roots, immobilized (fixed) by soil microbes, transported along the bed through the convection/adsorption process described by the earlier models or converted to N_2, N_2O (a greenhouse gas) and NO_x (an atmospheric pollutant) by a different family of soil microbes (denitrification reactions). Preliminary simulation results will be presented and analyzed to show how key biochar properties and environmental variables influence the transport and uptake of nutrients, the loss of N fertilizer through leaching and the fluxes of nitrogen oxides.

473979 Towards Systems-Scale Dynamic Metabolic Modeling

Abstract:

Here we present recent efforts to create genome-scale, dynamic metabolic modeling methods that utilize a form of systems-scale data that is of rising importance: metabolomics. Metabolomics data have all too often been omitted from metabolic engineering and other related analyses of metabolism, in part because most systems-scale models of metabolism assume a pseudo-steady state and thus limit the time variability of metabolite levels. Here we explore new ways to include this data that help to drive the development of computational approaches that may ultimately allow for genome-scale dynamic metabolic models. Creating such models would be critical advance for numerous fields including metabolic engineering, allowing in silico modeling even when experimental evolution work is not feasible.

464877 Modeling Convergence of Circadian Clocks and Metabolism

Abstract:

The circadian rhythms orchestrate the mental and physical changes that follow a 24-hour cycle, primarily responding to light/dark schedule. In mammals, the central pace maker resides in the superchiasmatic nuclei (SCN) and takes in light as the entraining signal while synchronizing the clock genes in the periphery. Aside from light/dark cycle, food intake is a strong *zeitgeber*, a cue given by the environment on host to reset the internal body clock. Studies revealed meal timing may have clinical significance, as several groups using mouse models showed that time-restricted-feeding (TRF) during the resting period can suppress tumor growth [1] and reverse liver-specific metabolism abnormalities [2]. This impressive phenomenon is observed because low-level targets and mechanisms do not function in isolation, but rather as a part of broader network of defense mechanisms. Therefore, it is imperative to understand the mechanism behind how TRF affects the hostâ€™s disease progression. In particular, recent studies revealed that nutrient availability has close ties to circadian rhythms, exhibiting bi-directional influence and complex signaling cascades [3-5]. Among the many ways that light/dark cycle and feeding schedule impact each other, nutrient sensors such as AMPK, SIRT1, and PARP1 are gaining attention as they exhibit circadian behavior while also playing key roles in gluconeogenesis[6, 7] and metabolic activities.

Recent studies suggest that SIRT1 is a key candidate for bridging the circadian clocks and metabolism. SIRT1 (human sirtuin 1) is a class III histone deacetylase (HDAC), a homolog of Sir2 (silence information regulator 2) in yeast [4]. Its activity takes place in the nucleus, modulating lipid metabolism and enhancing mitochondrial activity [8], in addition to modulating a variety of biological activities such as oncogenesis and aging [9]. SIRT1 has a clear role in metabolism as its enzymatic activity requires binding of NAD^+ into its catalytic site along with the substrate, effectively acting as a sensor for the energy state of the cell. Modeling the activity of SIRT1 would require an accurate portrayal of the NAD^+ level in the cell, since it directly activates SIRT1. In addition to synthesizing NAD^+ from amino acids, cells can also recover NAD^+ from the NAD^+ salvage pathway [10], where nicotinamide (NAM) is released from NAD^+, NAM is converted to nicotinamide mononucleotide (NMN), and NMN is converted back to NAD^+.

SIRT1 activity has complex relationships with the NAD^+ salvage pathway and the peripheral clocks. First, SIRT1 activity requires binding of NAD^+, thus NAD^+ has an activating effect on SIRT1. Second, SIRT1 activity is inhibited by NAM, the precursor of the rate-limiting step. Third, expression of nicotinamide mononucleotide adenylyltransferase (NAMPT), the rate-limiting enzyme of the NAD^+ salvage cycle, is under the control of SIRT1, which is a co-transcription factor for the Nampt gene along with peripheral clock genes. In summary, SIRT1 functions as a nutrient sensor, being under the influence of the energy state of the cells, represented by NAD^+. It is also under the effect of the circadian rhythmicity presented by NAD^+, NAM, and NAMPT.

From the above observations, we propose a mathematical model that describes the interactions among SIRT1, NAD^+ salvage cycle, and peripheral clock genes, to study the effects of light and feeding schedules on circadian dynamics and metabolism. Our model builds upon earlier works [11] of a semi-mechanical model for light entrainment on the peripheral clock genes through the HPA axis. The oscillations in the HPA axis are generated due to the negative feedback between glucocorticoid and corticotropin-releasing hormone (CRH)/adrenocorticotropic hormone (ACTH) using a modified Goodwin oscillator, entrained by light. The oscillating glucocorticoid is secreted into the periphery, scheduling the peripheral clock genes (Per/Cry, Bmal1, and Clock). The new model captures important interactions between this existing light-entrained oscillator and new feeding entrainment. The Clock/Bmal1 complex binds to SIRT1 and activates transcription of NAMPT, which facilitates the NAD^+ salvage cycle. NAD^+, in turn, activates SIRT1, as well as entraining glucocorticoids through a transit compartment that represents neural input through the ventromedial arcuate nucleus [12]. SIRT1 also facilitates the degradation of PER/CRY proteins in the nucleus, exerting more influence on the peripheral clock genes.

Current model successfully predicts some observations from experimental time-restricted feeding studies. Some basic predictions include that restricting feeding during the active phase results in stronger oscillations of peripheral clock genes [13]. In contrast, restricting feeding to the rest phase results in declined oscillation amplitude for peripheral clock genes. The model also exhibits the â€œphase jumpâ€ behavior often observed in two-*zeitgeber* systems [14], ensuring that both light and feeding entrains the periphery in a significant manner. More interestingly, the model agrees with the recently proposed hypothesis that feeding is a strong *zeitgeber* for the peripheral clock genes, and light-controlled SCN interferes the entrainment to feeding schedule [15]. This proposal counters the traditional belief that the light/dark cycle most strongly entrains the peripheral clock genes. Our model can demonstrate that removing the light schedule can reduce the time required for the system to equilibrate to a changed feeding schedule, supporting the new hypothesis. Furthermore, the new model provides the foundation for mathematically modeling hepatic gluconeogenesis in future works, in an attempt to potentially elucidate the mechanism of light and feeding *zeitgebers* resetting the circadian and metabolic clocks.

1. Wu, M.W., et al., *Effects of meal timing on tumor progression in mice.* Life Sciences, 2004. **75**: p. 1181-1193.

2. Maywood, E.S., et al., *Disruption of peripheral circadian timekeeping in a mouse model of Huntington's disease and its restoration by temporally scheduled feeding.* The Journal of neuroscience : the official journal of the Society for Neuroscience, 2010. **30**: p. 10199-204.

3. Asher, G. and U. Schibler, *Crosstalk between components of circadian and metabolic cycles in mammals.* Cell Metab, 2011. **13**(2): p. 125-37.

4. Bellet, M.M., et al., *The time of metabolism: NAD+, SIRT1, and the circadian clock.* Cold Spring Harb Symp Quant Biol, 2011. **76**: p. 31-8.

5. Eckel-Mahan, K. and P. Sassone-Corsi, *Metabolism control by the circadian clock and vice versa.* Nat Struct Mol Biol, 2009. **16**(5): p. 462-7.

6. Luna, A., et al., *Predicted Role of NAD Utilization in the Control of Circadian Rhythms during DNA Damage Response.* PLoS Comput Biol, 2015. **11**(5): p. e1004144.

7. Shi, M. and X. Zheng, *Interactions between the circadian clock and metabolism: there are good times and bad times.* Acta Biochim Biophys Sin (Shanghai), 2013. **45**(1): p. 61-9.

8. Thakran, S., et al., *Role of sirtuin 1 in the regulation of hepatic gene expression by thyroid hormone.* J Biol Chem, 2013. **288**(2): p. 807-18.

9. Bishop, N.A. and L. Guarente, *Genetic links between diet and lifespan: shared mechanisms from yeast to humans.* Nat Rev Genet, 2007. **8**(11): p. 835-44.

10. Sahar, S. and P. Sassone-Corsi, *Regulation of metabolism: the circadian clock dictates the time.* Trends Endocrinol Metab, 2012. **23**(1): p. 1-8.

11. Mavroudis, P.D., et al., *Mathematical modeling of light-mediated HPA axis activity and downstream implications on the entrainment of peripheral clock genes.* Physiol Genomics, 2014. **46**(20): p. 766-78.

12. Yi, C.X., et al., *Ventromedial arcuate nucleus communicates peripheral metabolic information to the suprachiasmatic nucleus.* Endocrinology, 2006. **147**(1): p. 283-94.

13. Hatori, M., et al., *Time-restricted feeding without reducing caloric intake prevents metabolic diseases in mice fed a high-fat diet.* Cell Metabolism, 2012. **15**: p. 848-860.

14. Oda, G.A. and W.O. Friesen, *Modeling two-oscillator circadian systems entrained by two environmental cycles.* PLoS One, 2011. **6**(8): p. e23895.

15. Angeles-Castellanos, M., et al., *The Suprachiasmatic Nucleus Participates in Food Entrainment: A Lesion Study.* Neuroscience, 2010. **165**(4): p. 1115-1126.

464747 Pore-Pore and Pore-Edge Interactions in Graphene Sheets and Nanoribbons

Abstract:

Defect engineering of graphene is a promising approach toward fabrication of carbon-based two-dimensional materials with unique properties and function. The assembly of nanopores, formed by irradiation or lithographically, is of particular interest for developing patterning strategies in graphene sheets and graphene nanoribbons (GNRs). Toward this end, a fundamental understanding is required of the interactions between such pores in graphene sheets and between pores in graphene nanoribbons and the nanoribbon edges and how these interactions mediate pore migration and coalescence processes.

In this presentation, we report a systematic analysis of pore-edge interactions in GNRs and pore-pore interactions in graphene sheets based on atomic-scale simulations according to reliable interatomic potentials. Based on molecular-statics (MS) computations, we have found that when two pore edges or a pore edge and a GNR edge are in close proximity (a few bond lengths), their interactions are strongly attractive and lead to pore coalescence or drive the pore to the GNR edge, thus changing the GNR edge morphology. At longer distances between pore edges or between pore and GNR edges, elastic interaction potentials based on elastic inclusion theory can describe the MS simulation results very well, and can be used for coarse-grained modeling of pore assembly in graphene and GNRs. We have characterized systematically the effects of pore size and GNR edge type on these interaction potentials. We have found that the GNR edge type has a very weak effect on the pore-edge interaction potential, while the pore size and, thus, curvature has a stronger effect on the interaction. With this understanding of pore-pore and pore-GNR edge interactions, we designed and conducted molecular-dynamics (MD) simulations of nanopore dynamics at high temperature in the vicinity of both GNR edges and other pores in graphene sheets. We found that the pore-edge and pore-pore interactions provide thermodynamic driving forces for nanopore migration toward the GNR edge or toward a larger pore through a sequence of carbon ring reconstructions and facilitate pore coalescence and GNR edge morphological evolution. Finally, we have conducted nudged elastic band calculations to construct the optimal pore migration paths and obtain a fundamental understanding of the mechanisms for driven pore migration and coalescence.

474750 Mathematics for Data-Driven Modeling - the Science of Crystal Balls

Abstract:

In mathematical modeling one typically progresses from observations of the world (and some serious thinking!) to equations for a model, and then to the analysis of the model to make predictions. Good mathematical models give good predictions (and inaccurate ones do not) - but the computational tools for analyzing them are the same: algorithms that are typically based on closed form equations. While the skeleton of the process remains the same, today we witness the development of mathematical techniques that operate directly on observations -data-, and "circumvent" the serious thinking that goes into selecting variables and parameters and writing equations. The process then may appear to the user a little like making predictions by "looking into a crystal ball". Yet the "serious thinking" is still there and uses the same -and some new- mathematics: it goes into building algorithms that "jump directly" from data to the analysis of the model (which is never available in closed form) so as to make predictions. I will present a couple of efforts that illustrate this
new path from data to predictions. It really is the same old path, but it is travelled by new means.

461842 Decoding Common Features of Protein-Nanoparticle Interactions

Abstract:

Nanoparticles have been recognized as key to revolutionizing the healthcare industry. However, excitement about the applications envisioned for nanoparticles is tempered by concerns about their potential toxicity. Both healthcare applications and potential toxicity require knowledge of nanoparticleâ€™s interactions with proteins. Our goal is to discover the mechanisms that govern protein-nanoparticle interactions and to develop a tool that can predict interaction strengths based on knowledge solely of the proteinâ€™s sequence and the nanoparticleâ€™s chemicophysical properties. We calculate the interaction energies between 2315 proteins and three bare gold nanoparticles (diameter = 1.0, 2.0 and 4.0 nm) and identify the binding sites using computer simulations. This allows us to rank the protein-nanoparticle interaction strengths, explore the chemicophyiscal properties of the binding sites, and develop models that can predict nanoparticle-protein interaction energies. Our results show that glutamic acid and lysine have the highest chance to appear on nanoparticle binding sites. We develop several models that can qualitatively predict the interaction energy.

471936 Design of Optimal Experimental Probes for Protein Dynamics Using Machine Learning and Variational Approach to Modeling Conformational Kinetics

Abstract:

In this talk, we plan to answer a broad question about design of experiments and validation of protein conformational ensembles, what is the minimum experimental information required for describing the conformational dynamics of a protein? Markov state models of protein dynamics describe the transitions between the conformational states of proteins using a transition probability matrix built from the raw simulation data. The eigenvalues and eigenvectors/eigenfunctions of the transition matrix provide the timescales and the population shift between states due to a particular dynamic process respectively. Considering only the m-slowest processes, an ideal set of experimental observations would unequivocally describe these m-slowest processes using only m-experimental observables. Our machine learning protocol involves using the variational approach for building Markov state models that provides approximation of the first m eigenfunctions of the molecular dynamics propagator using all possible experimental probe positions on the protein and picks the optimal set of probe locations that best describe the m-slowest processes. In this work, we combine two central ideas from machine learning and chemical physics - hyperparameter selection via cross-validation and variational approaches for linear operator eigenproblems; to create a new method for predicting observables that optimally discriminate between alternative models for molecular kinetics constructed from MD simulations. This protocol could also be used with a broad range of experimental techniques or a combination of experimental techniques. We present results from application of this protocol to three systems of biological importance, conformational change in membrane peptide transporter PepT$_{SO}$ and Î²2-Adrenergic receptor, and folding of WW domain.

469029 Guiding Experiments Towards New Functional Materials with Informatics

Abstract:

In this talk, we will discuss our informatics approach that integrates machine learning, symmetry analysis and density functional theory (DFT) to guide experiments in search of novel $BaTiO_3$-based ferroelectric materials with high Curie temperature (T_C). $BaTiO_3$-based materials have attracted significant attention as a potential alternative for Pb-free piezoelectrics. In the development of $BaTiO_3$-based piezoelectrics, one of the common strategies involves mixing two end member compositions that have Cubic (*Pm-3m*) to Tetragonal (*P4mm*) and Cubic to Rhombohedral (*R3m*) transformations. Our design objective is to identify suitable chemical substitutions at the Ba- and Ti-sites such that the T_C of the end members are high. We perform DFT calculations on undoped $ATiO_3$ and $BaBO_3$ perovskites in *P4mm* and *R3m* symmetries, respectively; A and B are divalent and tetravalent cations, respectively. From symmetry-mode analysis, we decompose the two ferroelectric phases (*P4mm* and *R3m*) in terms of irreducible representations (irreps). These irreps, along with the unit cell volume from DFT, serve as *feature sets* for understanding $ATiO_3$ and $BaBO_3$ crystal chemistries and their role in impacting the ferroelectric distortions. We then link these features with known experimental T_C data using machine learning methods and predict a new $CdTiO_3$-$BaTiO_3$ solid solution for experimental synthesis. We performed experiments and found that our predictions do not agree with the measurements. The reason for the discrepancy is attributed to the anomalous behavior of the $CdTiO_3$-$BaTiO_3$ solid solution, the phase diagram of which is found to be different from any of the $BaTiO_3$-based solid solutions studied so far.

472588 Pushing the Frontiers of Atomistic Modeling Towards Predictive Design of Materials

Abstract:

The ability to perform accurate calculations efficiently is crucial for computational materials design. In this talk, we will discuss our stream-lined approach to force field development using first principles density functional theory training data and machine learning algorithms. Our objective has been to develop new, first-principles based, more accurate and more robust inter-atomic potentials for accurate simulations of dynamical processes at reactive interfaces and low dimensional systems such as clusters and molecules. The procedure involves several steps including (a) generation and manipulation of extensive fitting data sets through electronic structure calculations, (b) defining functional forms, (c) formulating novel highly optimized fitting procedures, (d) dual-Hamiltonian optimization to leverage classical FFs with more accurate approaches, and (d) subsequently coding and implementing these algorithms on high performance computers (HPCs). We will also discuss the validation of this approach on several diverse material systems ranging from precious metal nanocatalysts to newly discovered two dimensional materials such as stanene and silicene.

469177 Design of Ternary Transparent Conducting Oxides

Abstract:

Transparent conducting oxides (TCOs) are well-developed and commercialized class of wide-bandgap semiconductors that are crucial for the function of many electronic devices. Recent experimental work has demonstrated bandgap engineering in ternary $(Al_xGa_yIn_{1-x-y})_2O_3$ over several ~ 3 eV by adjusting the ratio of In/Ga[1] and Ga/Al.[2]

The focus of this work is an examination of the phase diagram for ternary $(Al_xGa_yIn_{1-x-y})_2O_3$ materials using DFT-based cluster expansion models combined with fast stochastic optimization techniques (*e.g.*, nested sampling). This combined computational approach allows for both an efficient search of the stable and metastable configurations for $(Al_xGa_yIn_{1-x-y})_2O_3$ at various lattice types and the consideration of entropy on the relative stability of ternary TCOs. Statistical learning, in particular compressed sensing, is used to efficiently identify a structure-property relationship between the targeted properties (*e.g.*, mobilities and optical transparency) and the fundamental chemical and physical parameters that control these properties.

[1] Zhang et al., *Solid State Commun*, 186, 28 (2014).

[2] Ito et al., *Jpn. J. Appl. Phys.*, 51, 100207 (2012); Zhang et al., *Appl. Phys. Lett.*, 105, 162107 (2014).

472938 Development of Empirical Charge Transfer Interatomic Potential for Tantalum Oxide Nanostructures from First Principle Calculations

Abstract:

There has been considerable interest over the years in metal oxides and their nanostructures owing to their exotic properties making them ideal candidates for several applications, notably in electronics, thermal management, catalysis and biochemistry. In particular, mesoporous (2-50 nm pore sizes) tantalum oxide nanostructures can be used for photocatalytic decomposition of water, thermal insulation and molecular adsorption, while Ta_2O_5nanoparticles have applications in bio-imaging due to their inertness and non-toxicity. It is essential to obtain a clear fundamental understanding of the structure and chemistry of metal oxide nanostructures to optimize the material performance in these applications.

Atomistic molecular dynamics (MD) simulations provide an ideal tool to investigate the underlying mechanisms at play in the nanoscale processes around these nanostructures and to decode the structure-property relationship. The key to a reliable MD simulation is the underlying empirical force field that describes the interatomic interactions. While robust interatomic potentials have been developed for several metal oxides, namely, alumina, titanium oxide, iridium oxide etc., there is no available empirical charge transfer force field to simulate the behavior of tantalum oxide nanostructures and metal-oxide interfaces. This is primarily due to a lack of comprehensive structural and energetics information, both experimental and ab initio theory, on the different polymorphs of Ta_2O_5.

In this talk, we present a detailed tabulation of ab initio density functional theory (DFT) calculations of structure, cohesive energies and elastic properties for the three main experimentally observed polymorphs of Ta_2O_5, namely, hexagonal, orthorhombic and monoclinic. The DFT-generated dataset is used to parameterize an EAM+QEQ (Streitz-Mintmire) charge transfer interatomic potential for tantalum oxide via a genetic algorithm based framework.

Streitz-Mintmire force field is a variable charge potential, which can account for dynamic charge transfer, and has been demonstrated to be a reasonable potential for single-metal oxides. In particular, oxidation of metallic nanoparticles can be investigated using this charge transfer formalism. The developed force field can also be used to investigate Ta/TaO interfaces, the properties of which are significantly affected by the nanostructure of the interface.

In this work, the developed interatomic potential is used to investigate the oxidation of tantalum thin films and nanoparticles. The thermal and mechanical stability of tantalum oxide nanoparticles and mesoporous nanostructures, and the atomic scale processes (oxidation and sintering) that occur in the vicinity of these nanoclusters are also investigated.

455131 Machine Learning for Advancing Discovery of Novel Thermoelectric Materials. the Thermoel

Abstract:

Based on current estimates 59% of all primary energy produced in the United States is lost as waste heat, recovery of even 1% of this heat by a novel thermoelectric materials would amount to nine times the energy currently produced by photovoltaics. Thermoelectric materials are materials that exhibit a strong coupling between electric and thermal responses, producing an electric potential in the presence of a temperature gradient and vice versa.
The physics governing the material systems is described as a network of models and functional relationships between different scales and modalities (i.e. processing/structure/properties). However, the disjointed nature of materials property databases and relative immaturity of the thermoelectrics multi-scale functional relationships make integrated optimization across numerous length and time scales particularly difficult in this case. A data-driven design of advanced thermoelectrics is possible with tools capable of integrating materials databases into a consistent framework and finding the multi-scale underlying processing/structure/property functional relationships responsible for thermoelectrics performance.
The ThermoEl could be an effective solution to discovery of novel thermoelectrics. It is a predictive toolkit developed based on a supervised machine learning algorithms and known experimental measurements of various materials. Aside from other tools on the market, ThermoEl is based on regression models that predict numerical value of such parameters as Seebeck coefficient for non-stoichiometric materials and Bulk modulus of crystalline compounds. Based on our experience, we would like to discuss tricks and challenges in using data mining approaches with materials databases.

468062 Identifying Descriptors for Dielectric Breakdown Strength Using Genetic Programming

Abstract:

The development of phenomenological theories by identifying relevant descriptors for material properties enables rapid screening large numbers of materials and facilitates the design of new materials. One of the leading challenges for computational researchers is determining the best ways to analyze large material data sets to identify the best descriptors for a given property. In this presentation, we demonstrate the use of genetic programming to identify relevant descriptors of dielectric breakdown based on 82 representative materials. We identified band gap E_g and phonon cut-off frequency w_{max} as the two most correlated features, and new classes of phenomenological models featuring exponential functions of E_g*w_{max} were uncovered. The genetic programming model was found to outperform other models for descriptor identification, and we discuss some of the advantages of the genetic programming approach.

462881 Machine Learning with Structural Fingerprints of Local Particle Environments

Abstract:

As the computational power available to scientists increases, researchers themselves — rather than hardware or algorithms — become the bottleneck in scientific discovery. The computational study of colloidal self-assembly is one area that keenly feels this effect: even after computers generate massive amounts of raw data, performing an exhaustive search to determine what, if any, ordered structures occur in a large parameter space of many simulations can be a long, manual process. Here we demonstrate how "off-the-shelf" machine learning algorithms can be applied to results of self-assembly studies both to find interesting regions in a phase diagram and identify characteristic local environments in simulations in an automated, high-throughput manner for both simple and complex crystal structures. These methods can form the foundation of an intelligent exploration of parameter space — a key component in the process of creating new materials by design.

454766 Using Semi-Supervised Machine Learning to Map the Phase Diagrams of Open Materials Data Sets

Abstract:

The Materials Genome Initiative (MGI) is a government program intended to expedite materials discovery, optimization, and deployment by combining theory, experiments, and data science in an integrated workflow. This effort requires that large quantities of high quality experimental data be made publicly available via minable materials databases for the purpose of validating new data mining techniques and validating theoretical results. High-throughput experimentation has been identified as an important technology for creating large-scale experimental databases that delineate the impact of synthesis, processing and composition on crystal phase stability and figure of merit. To date, however, relatively few of these rich data sets are freely available and those that are available are primarily restricted to unlabeled as-obtained structural data.

Such data is critically important to identifying the materials genome, as it creates the linkage between composition, structure and property. Algorithmic approaches to automated phase diagram mapping have been a hot issue in the high-throughput field for a number of years and most studies have focused on an open FeGaPd data set that has been available for about 10 years. Here, we demonstrate a semi-supervised machine learning technique, SS-AutoPhase, which uses a two-step approach to automatically identify phases within structural data sets. In the first step, clustering analysis is used to automatically select a representative sub-set of samples to be manually analyzed by a human expert. In the second step, these labeled samples are used by an AdaBoost classifier to identify the presence of the different phases in the FeGaPd diffraction data. SS-AutoPhase was used to identify the metallographic phases in 278 diffraction patterns from a FeGaPd sputtered composition spread sample. The accuracy of SS-AutoPhase was greater than 82.6% for all phases when 15% of the diffraction patterns were used for training. Furthermore, the predicted phase diagram of SS-AutoPhase was determined and compared to phase labels from a human expert and other algorithmic approaches. This comparison showed that not only did SS-AutoPhase have very high agreement with the expert phase labels, but

that it was able to determines and correctly identify a previously unreported phase. Finally, I will report on a first-of-its-kind identification of a novel ferromagnetic shape memory alloy via the data mining of an open materials database.

475423 Sustainability Concepts in the Food-Energy-Water Nexus: Chemical Engineering Perspective

Abstract:

The idea of sustainability and/or sustainable development emerged from the World Commission on Environment and Development (WCED), generally known as the Brundtland Commission. It defined sustainable development in "Our Common Future" as that which meets the needs of the present without compromising the ability of future generations to meet their own needs. Environmental sustainability and/or sustainable development "occurs when we maintain or improve the material and social conditions for human health and the environment over time without exceeding the ecological capabilities that support them". As such, sustainability has three underlying bases: (1) economic viability - the cost and business aspects of a project; (2) social concern - human health and social welfare; and (3) natural or ecological issues - depletion of natural capital and environment. Sustainable development has become a major driving initiative in engineering businesses throughout the world today. Our focus is on all aspects of sustainable development including environmental, industrial, business, economical, ethical, and social sustainability as they impact the availability of food, energy, and water. This is important because it ***has been generally acknowledged that the increasing use of non-renewable resources to support an increasing population has created an unsustainable or nearly unsustainable situation where the ability to provide adequate food, energy, and water is in question. Should this rate of use be maintained, two dire consequences will follow: first, the future generations will be unable to maintain a good standard of living, and second, developing countries will have less of an opportunity to bring their living standards to a level comparable to that of the affluent countries. This is important because the availability of food, energy, and water is an indispensable necessity for maintaining a good standard of living. The practice of chemical engineering, perhaps more than any other technical discipline, involves the use of both non-renewable and renewable materials, and energy and exergy (available energy) resources for the production of chemicals, value-added goods, services of commerce, and waste as***

pollution. This paper primarily deals with the ideas of environmental sustainability and/or sustainable development, methodologies to redesign sustainable chemical and allied processes and products development using the LCA approach. These principles are illustrated with a case study for the production of ethanol, an energy source, from molasses, a human food, in a combined agricultural and chemical process that requires substantial amounts of water.

461670 Optimizing Spatio-Temporal Sensor Placement for Nutrient Monitoring: Algorithmic Framework

Abstract:

INFEWS is a new initiative in NSF which deals with Innovations at the Nexus of the Food, Energy and Waters Systems. This project addresses the INFEWS: N/P/H2O solicitation. The food production system generates waste streams that are characterized by high concentrations of organic matter, nitrogen- and phosphorus-containing species in water. Therefore, monitoring nitrogen and phosphorous species is important for water quality requirements for agricultural as well as energy production. Currently these species are monitored via stationary monitoring stations. However nitrogen and phosphorous species move via agricultural run-off to other water systems, and requires portable sensors which can change the positions in real time. This type of dynamic sensing requires novel algorithms which decide sensor locations in real time in the face of inherent uncertainties in the fate and transport of the species. To develop such an algorithmic framework to solve the problem of sensor placement in real time is the objective of this proposal.

This paper presents the first part of the work which is the algorithmic framework for optimal spatial-temporal locations of the sensors in real time. The algorithmic framework is based on a novel algorithm called Better Optimization of Nonlinear Uncertain Systems (BONUS). We present a case study of an agent based model from social sciences for testing these framework. This model is based on behavior of cows in a field where there are grass patches. The cows move in a herd towards the greener pasture and eat grass till it finishes. The grass growth and cow movements are probabilistic. We are using four sensors for object tracking (e.g. cows) and these sensor move spatially and temporally in order to track maximum number of cows. Results show that the optimal movement of sensors allows to track maximum number of cows.

In the future, the algorithmic framework will be coupled with nutrient and fate transport models. A case study of a small watershed will be selected from literature for nutrient monitoring sensor location problem. Optimal spatial-temporal locations will be the result of this new framework for this case study.

Bandwidth Study on Energy Use and Potential Energy Saving Opportunities in Manufacturing Food and Beverages

Presenter:
Caroline Dollinger
Engineer, Energy Analysis
Energetics Incorporated

American Institute of Chemical Engineers
Annual Meeting
San Francisco, CA
November 14, 2016

Collaboration

Caroline Dollinger (Speaker)
Energetics Incorporated
cdollinger@energetics.com

Sabine Brueske
Energetics Incorporated
sbrueske@energetics.com

Energetics Incorporated is an energy consulting company, providing subject matter expertise to private and public sector clients

Joseph Cresko
U.S. Department of Energy
Advanced Manufacturing Office
joe.cresko@ee.doe.gov

Alberta Carpenter
National Renewable Energy Laboratory
Alberta.carpenter@nrel.gov

2

Overview

- An *Energy Bandwidth Study* is currently being performed to analyze energy use and energy saving opportunities in energy intensive manufacturing sectors
- This presentation focuses on the work being done for a *Food and Beverage Manufacturing Energy Bandwidth Study*
- Energy data will broken down by sub-sector, product, and unit operation to identify processes and energy-saving technologies with the greatest opportunities for energy savings

Results presented in this talk are draft data; the work is still ongoing and a final report is expected to be published next year.

Energy bandwidth studies can be a useful tool for assessing energy savings opportunities in manufacturing

Energy bandwidth studies frame the range (or *bandwidth*) of potential energy savings in manufacturing, and technology opportunities to realize those savings.

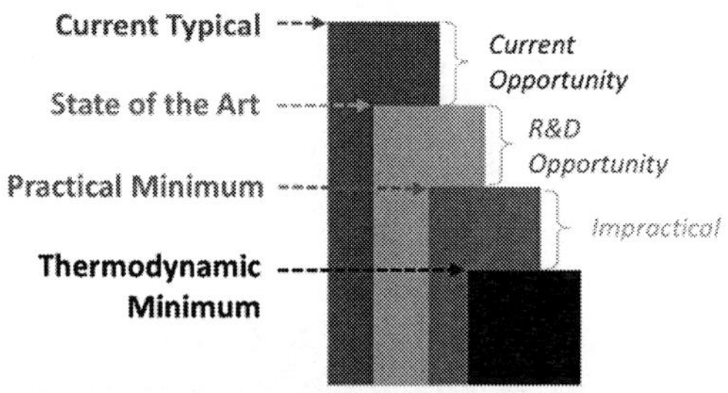

Measures of energy intensity studied:

Current Typical (CT)	State of the Art (SOA)	Practical Minimum (PM)	Thermodynamic Minimum (TM)
Determined from a literature review and stakeholder outreach, based on current typical manufacturing processes in the U.S.	Determined from a literature review and stakeholder outreach, based on the most energy-efficient technologies and practices available worldwide	Modeled based on plausible energy savings from identified R&D technologies under development worldwide	Calculated analytically using a Gibbs free energy approach assuming ideal conditions

Six new energy bandwidth studies on lightweight materials were recently completed in draft form

- **These studies have previously provide a deep-dive analysis into manufacturing energy use and potential energy savings for:**
  - **Four energy-intensive manufacturing sectors**
  - **Manufacturing of six lightweight materials**
- **This talk will focus primarily on a new bandwidth on food and beverage manufacturing.**
- **Published bandwidths (including this one, once completed) are available on the DOE Advanced Manufacturing Office website**

Summary thumbnail for bandwidth study on petroleum refining

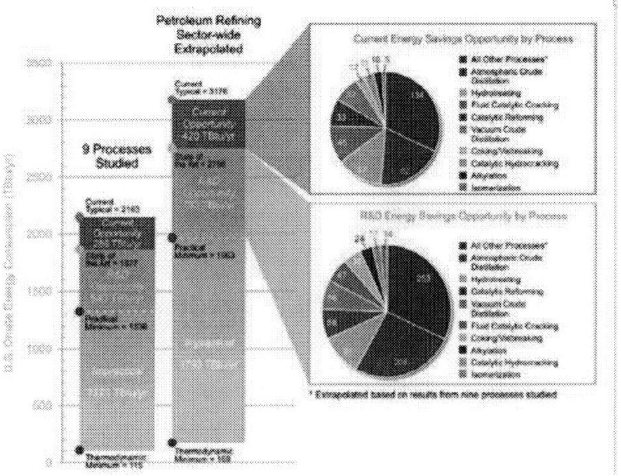

5

ENERGETICS
A Subsidiary of VSE Corporation

First Four Industry Bandwidths – US Results

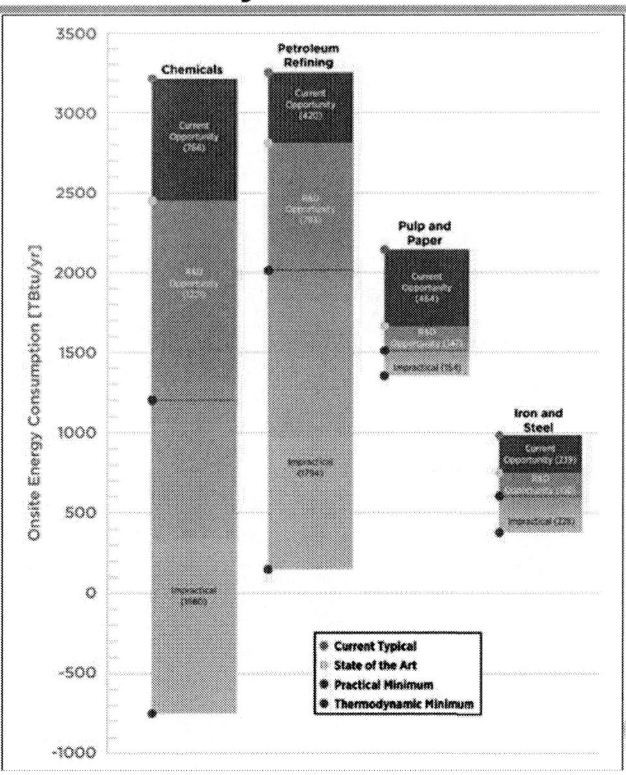

ENERGETICS
A Subsidiary of VSE Corporation

Challenges in Researching Energy Use and Savings in the Sector

- Highly diversified sector that produces thousands of different products
- Processing facilities range from small to large units, normally producing more than one product
  - As a result energy intensities will vary from facility to facility and from product to product
- Less active in the R&D, with a focus on energy savings technologies, compared to other sectors
- Competitive marketplace, where energy efficiency may help to offer a competitive advantage
- Largely mechanized processes, difficult to ascertain and source energy data at the sub-process level

7

ENERGETICS
A Subsidiary of VSE Corporation

Food & Beverage Manufacturing Bandwidth
Study Coverage

Food and Beverage Sub-sectors Studied	Food and Beverage Products Studied
Grain and Oilseed Milling	Wet corn milling- corn oil and by-products Soybean oil and meal Other vegetable oils and meals
Sugar Manufacturing	Beet sugar Cane sugar
Fruit and Vegetable Preserving and Specialty Food Manufacturing	Frozen fruits, vegetables, juice Canned fruits, vegetables Pickled vegetables Dried and dehydrated fruits and vegetables
Dairy Product Manufacturing	Milk Butter Cheese Dry, condensed, evaporated milk Ice cream/frozen desserts
Animal Slaughtering and Processing	Beef, pork, chicken, lamb Animal fat, lard, edible tallow
Beverage Manufacturing	Soft drinks Beer, wine, distilled liquors

8

ENERGETICS
A Subsidiary of VSE Corporation

Food & Beverage Manufacturing Bandwidth
Study Coverage

Food and Beverage Sub-sectors Studied	Energy Intensive Sub-processes
Grain and Oilseed Milling	Crushing Extraction Evaporation Dewatering and drying
Sugar Manufacturing	Dewatering and drying Evaporation Vacuum pans (crystallization)
Fruit and Vegetable Preserving and Specialty Food Manufacturing	Heat sterilization/retorting Cooking/blanching Freezing Concentration Evaporation Drying Canning
Dairy Product Manufacturing	Pasteurization Freezing Concentration
Animal Slaughtering and Processing	Rendering
Beverage Manufacturing	Beer processes (brewhouse, packaging) Wine processes (refrigeration)

Food and Beverage Manufacturing Bandwidth
Scope Description

What's Out- Agriculture Sector NAICS 11

- Establishments primarily engaged in growing crops, raising animals, from a farm, ranch, or their natural habitats
- 111-Crop production
- 112-Animal production

What's In- Manufacturing Sector NAICS 31

- Industries in the Food Manufacturing subsector transform livestock and agricultural products into products for intermediate or final consumption (basically, any food that is processed)
- 311 – Food manufacturing

What's Out- NAICS 42, 44, 72

- 424 – Food Wholesalers
- 445 – Food and Beverage Stores
- 722 – Food Services and Drinking Places

ENERGETICS
A Subsidiary of VSE Corporation

10

Food and Beverage Sector
U.S. Energy Consumption Overview

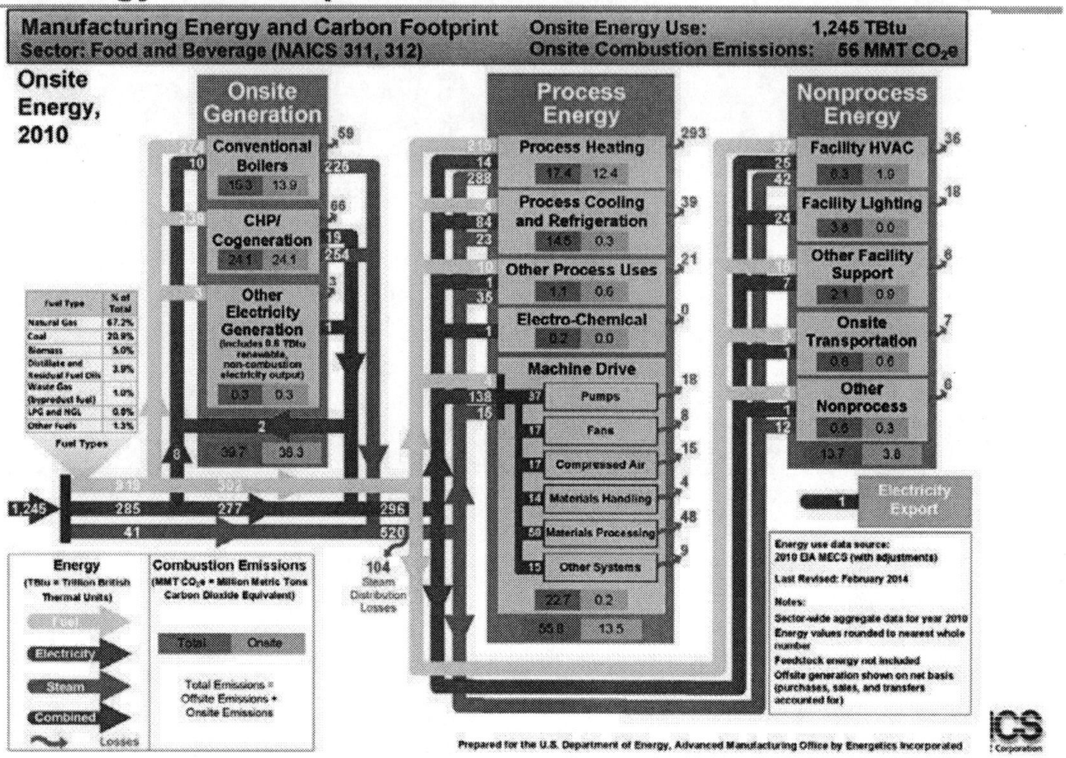

Food and Beverage Bandwidth
U.S. Baseline Energy Consumption, 2010

MECS Food & Beverages Energy Consumption Data by Sub-sector, 2010 1,245 TBtu

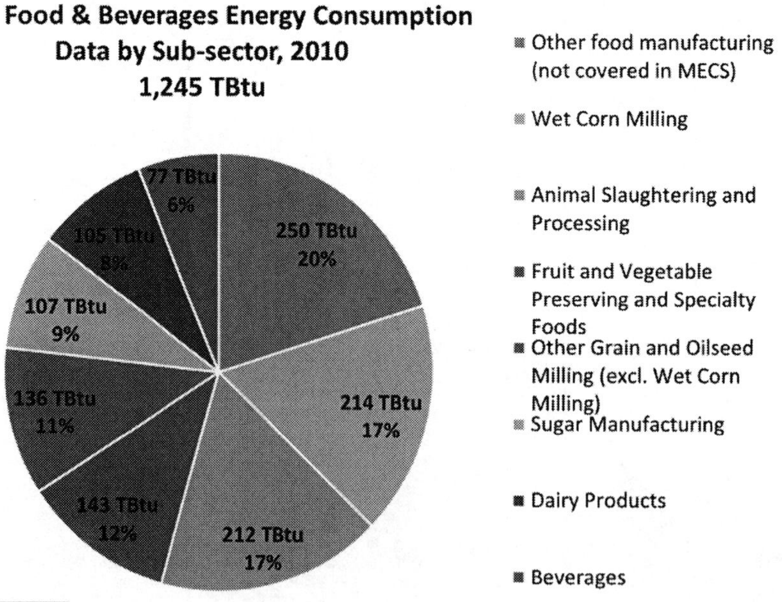

Source: EIA 2014, Manufacturing Energy Consumption Survey (MECS)

12

ENERGETICS
A Subsidiary of VSE Corporation

Food and Beverage Bandwidth
U.S. Production for Sub-sectors Studied, 2010

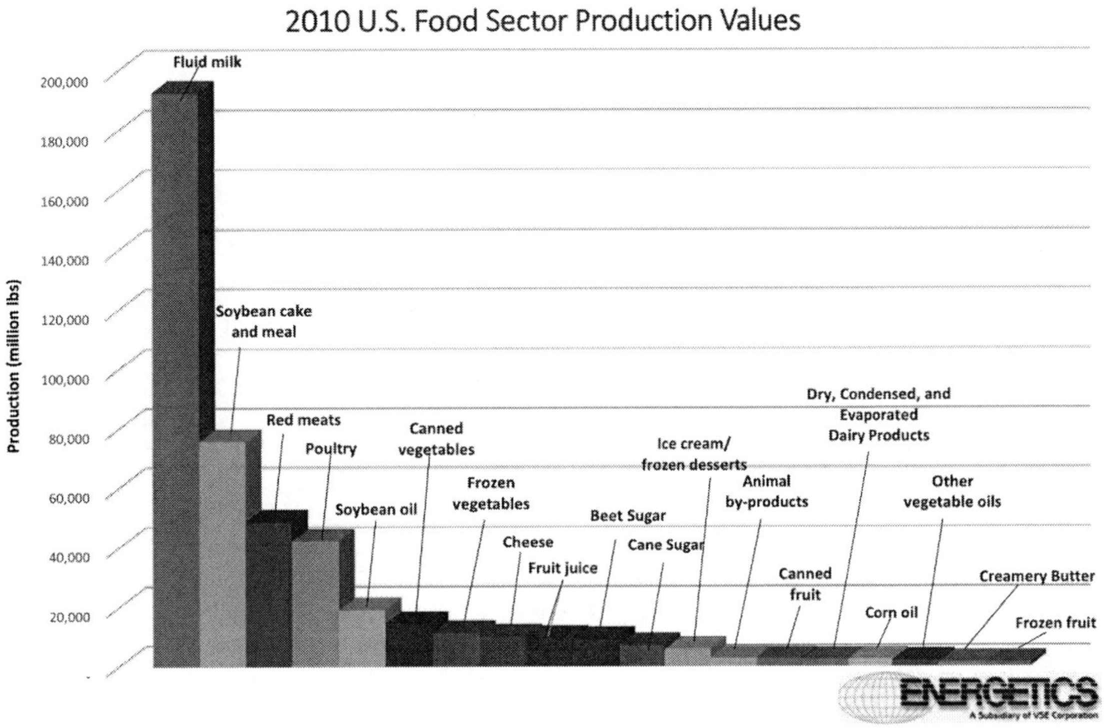

Dairy Products Manufacturing
Energy intensive sub-processes being analyzed

- Pasteurization
- Cooling/refrigeration
- Concentration
- Freezing
- Motors/pumps
- Spray Drying
- Canning

Manufacturing Process	CT (TBtu/year)	% of sub-sector
▶ **Pasteurization**	20	18.6%
▶ **Cooling**	16	15.6%
▶ **Motors, pumps**	9	8.4%

14

Grain and Oilseed Milling

Energy intensive sub-processes being analyzed

Wet Corn Milling

- Starch dewatering and drying
- Gluten thickening and drying
- Gluten feed drying
- Steeping
- Steepwater evaporation
- Fiber and gluten recovery
- Germ dewatering and drying
- Grinding

Other vegetable oils/meals

- Crushing and extraction
- Neutralization
- Oil refining
- Deodorization

15

Fruit and Vegetable Preserving

Energy intensive sub-processes being analyzed

Canned Fruits & Vegetables

- Washing
- Retorting
- Blanching/cooking
- Canning

Frozen Fruits & Vegetables

- Washing
- Peeling
- Blanching/cooking
- Freezing

Other fruit/ vegetable products

- Pasteurization (fruit juice)
- Dehydrating
- Retorting (fruit juice)

16

Animal Slaughtering and Processing
Energy intensive sub-processes being analyzed

Red Meat Products

- Dressing and cutting
- Refrigeration
- Processing (curing, smoking, cooking)
- Packaging
- Splitting
- Scalding
- Singeing
- Blood processing
- Edible rendering

Poultry Products

- Scalding, slaughtering, defeathering
- Evisceration and cooling
- Refrigeration
- Liquid effluent treatment
- Packaging

17

Current Typical (CT) energy intensities and consumption for red meat products, U.S. (2010)

Manufacturing Process	CT (Btu/lb)	TBtu/year (% of sub-sector)
Chilling	468	23 (10.8%)
Processing (curing, smoking, cooking)	320	16 (7.4%)
Edible rendering	4,573	14 (6.5%)
Dressing and cutting	181	9 (4.2%)
Packaging	93	5 (2.1%)
Scalding*	149	3 (1.6%)
Splitting*	117	3 (1.2%)
Singeing*	82	2 (0.9%)

* Pork products

18

ENERGETICS
A Subsidiary of VSE Corporation

Animal Slaughtering and Processing
Example of challenges in data and analysis

- **Lack of recent analysis done on energy usage in U.S. manufacturing facilities, must rely on internationally available data**
- **Most estimates of energy intensity are presented in terms of total energy/product or input (e.g., kWh/tonne), not broken down by sub-process**
- **Studies done on individual plants may not be applicable to be generalized sector-wide**
- **Energy data presented in different forms (consumption per head, carcass weight, dressed weight, product weight)**
  - **Involves accurate conversions to calculate total energy consumption using production values from USDA, other assumed factors**
- **If data is provided for sub-processes, only provided for a few, not allowing for calculation of balance**

19

ENERGETICS
A Subsidiary of VSE Corporation

Current Opportunity and *R&D Opportunity*

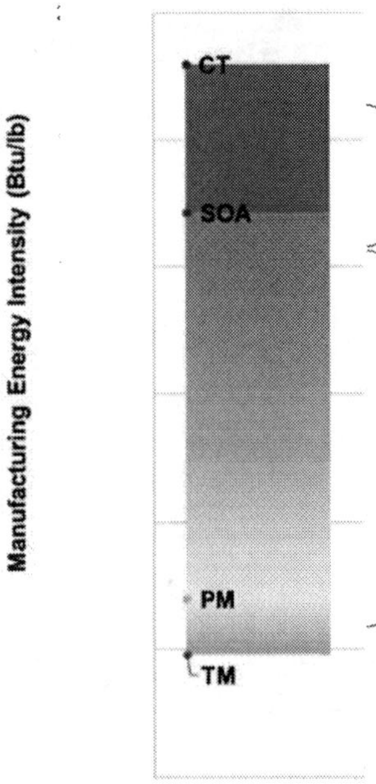

Current Opportunity

Energy savings if the best technologies and practices available were used to upgrade production

R&D Opportunity

Additional energy savings if applied R&D technologies under development worldwide were successfully deployed

SOA Technology Examples:
- **Steam system improvements**
- **Adjustable-speed drives**
- **Air compression improvements (reduce leaks, efficient vacuum pumps)**
- **Alternative technologies**
- **...**

R&D Technology Examples:
- **Membrane filtration**
- **Microwave heating**
- **Advanced Ohmic Heating**
- **High pressure processing**
- **Cost efficient thermal management in motors using nanomaterials**
- **...**

20

ENERGETICS
A Subsidiary of VSE Corporation

What about the Water part of the Nexus?

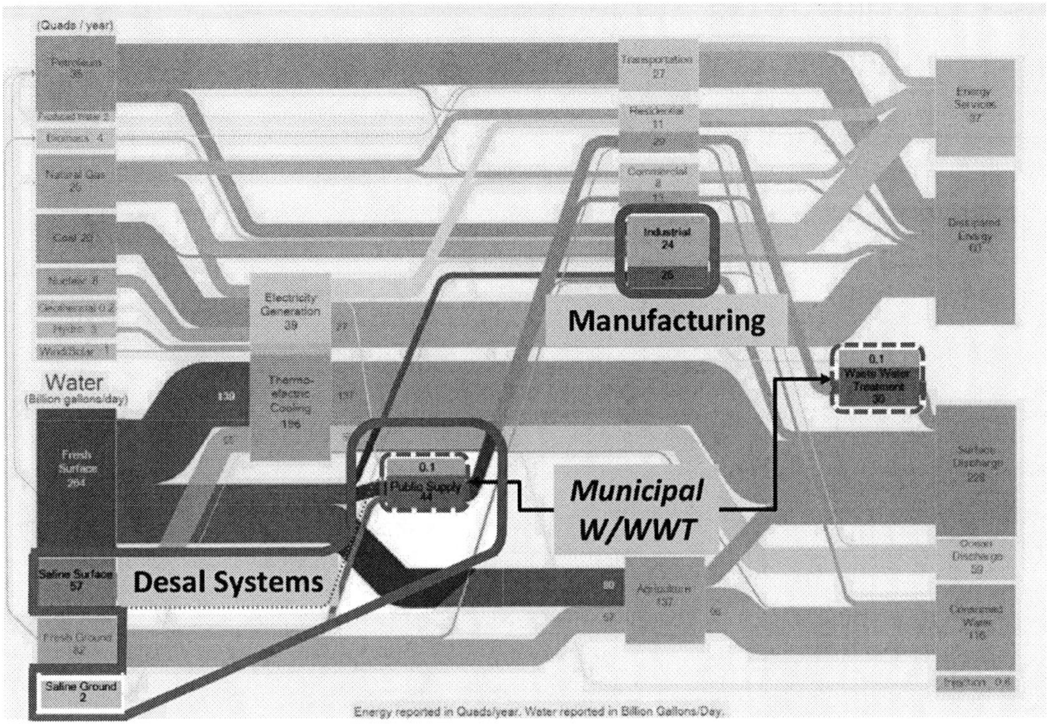

Conclusions

- An *Energy Bandwidth Study* is being performed to analyze manufacturing energy use and energy saving opportunities for the food and beverage manufacturing industry
- Manufacturing energy data are broken down by sub-sector, product, and sub-process to identify processes and energy-saving technologies with the greatest opportunities for energy savings
- This type of study has not been done in the past, so data must been gathered carefully; also there is a lack of data in certain areas
- While more research is to be done, research to date implies that there may be more current rather than future energy savings opportunity, due to the fact that there are few areas with active R&D in food and beverage, while more traditional areas for energy efficiency savings (steam systems, etc.)

Results presented in this talk are draft data; studies are currently being peer reviewed. If you are interested in participating, please contact me!

Thank you!

Energetics Incorporated
Caroline Dollinger
(CDollinger@energetics.com)

A Subsidiary of VSE Corporation

23

468656 Insight-Based Design of Local Integrated Systems for Food, Energy and Water

Abstract:

The importance of Food-Energy-Water nexus (also referred to as Water-Energy-Food nexus) has been widely recognised by national governmental bodies and international organisations such as the United Nations, for its significance to sustainable development and national security. Though these three sectors are inextricably linked, as actions in one sector have impacts in one or both of the others, these areas have too often been considered individually. Thus, there is a need to look at them from a holistic perspective and develop tools that take into consideration their interdependencies. This opens up opportunities for process system engineering research. As such, the aim of this study is to develop a systematic insight-based approach for the design of an integrated local production system particularly to meet local demands for food, energy and water. One of the benefits of the insight-based approach is that it allows decision variables to remain tractable throughout the design process. Unlike mathematical optimisation approaches, intermediate results from insight-based approaches can be analysed, interpreted and used to inform the implications of a given design for the operation of the whole system, thus offering a more practical tool for decision-makers and local planners to use. Insight-based approaches which include techniques such as pinch analysis for heat integration and mass integration have proved particularly useful for the integration of energy and material streams in industrial production processes within eco-industrial parks and industrial symbiosis systems. They have been successfully implemented for enhancing resource efficiency and sustainability of processes involving heat, water and hydrogen, to name a few applications. Although the proposed study also pursues higher resource efficiencies through integration of resources, the design of the local integrated production system has a distinctive emphasis on local resources and demands and on the holistic consideration of all types of agricultural, industrial and municipal processes to take place at the locality of concern. This work will combine existing methods and adapt them to enable the identification of the efficient use of various locally available resources. A net gain indicator will be defined to quantify the avoided cumulative effort that would have been spent for importing resources to satisfy the demand.

Using the net gain indicator, parallel processing and utilisation options within and between different subsystems can be compared and the optimal actions identified. The proposed insight-based approach for the design of local integrated system of food, energy and water will be illustrated through a case study based on an eco-town in the UK.

468765 Recovering Runoff Particulate-Bound Phosphorus Via Fungal Bioextraction

Abstract:

Phosphorus (P) loss is a key issue affecting food-energy-water nexus sustainability. P in fertilizers is critical for ensuring food security for the continuously growing world population. Currently, P is obtained via energy-intensive mining and processing of phosphate rock, which is becoming more scarce and concentrated in a few geographical areas. Excessive fertilizer application due to slow diffusion and heavy fixation of P in soils have led to massive P losses amounting to almost half of all mined P due to agricultural soil erosion and runoff. P-rich runoff streams discharging into surface water bodies can cause eutrophication, toxic algal blooms, and overall quality deterioration of freshwater resources. In this work, we propose capturing and processing runoff sediments using a low-energy and inherently benign fungal bioprocessing method to unlock a previously untappped source of recoverable phosphorus for nutrient recycling and closing the loop of P use. This can potentially reduce the demand for mined sources while at the same time help protect watersheds from the devastating effects of eutrophication.

This paper presents the preliminary findings of our work involving the use of phosphorus-solubilizing fungi (PSF) for extracting and recovering particulate-bound phosphorus in erosion and runoff streams. PSF species can generate organic acids and phosphatase enzymes that can induce mobilization of matrix-associated or surface-adsorbed phosphate ions on inorganic minerals in runoff particulates, and degrade phosphate ester bonds in sedimentary organic phosphorus compounds to facilitate the release of phosphate ions, respectively. Samples of runoff sediments discharging into a eutrophic lake in Kalamazoo, MI were collected and characterized for total P content and P speciation. P extraction experiments on runoff sediment samples were performed using model organic acids and spent culture broth *Aspergillus niger* ATCC 15475. Phosphate solubilization and changes in P speciation of the runoff sediments during treatment were measured using a sequential chemical extraction method and colorimetric analysis. Changes in runoff sediment mineralogy was analyzed using X-ray Diffraction (XRD) spectroscopy. The results indicate that oxalic and citric acids can extract and solubilize a significant fraction of phosphates bound to the iron, aluminum, and calcium mineral components of the runoff sediments. A kinetic model describing phosphate mobilization via ligand-exchange and induced desorption mechanisms was developed and fitted into the experimental data. We also observed a significant reduction in the organic P fraction of sediments treated with spent fungal culture broth, indicating the possibility of phosphatase enzyme production and activity. Thus, changes in organic P speciation in the sediments were monitored using solution ^{31}P nuclear magnetic resonance (NMR) spectroscopy. The findings demonstrate the feasibility of the proposed technology for P recovery from nonpoint nutrient pollution sources such as runoff sediments. It will help develop an effective, efficient, economical, and environmentally benign system that will benefit the community, agriculture and food production systems, and the environment in terms of resource security and environmental sustainability.

469658 Preliminary Study on Detroit's Urban Food-Energy-Water (FEW) Nexus

Abstract:

With 66% of the worldâ€™s population expected to live in urban areas by 2050, cities around the world will face numerous challenges in sustainable provision of food, energy, and water. These challenges are particularly complex due to the interconnectedness of the food system, energy system, and water system. For example, water is required for food and energy production; energy is needed for food production as well as water and wastewater treatment and distribution; and industrial agriculture and food production lead to oversupply of nitrogen and phosphorus in the urban water system through fertilizer use. Therefore, policy and technology solutions addressing urban food-energy-water (FEW) challenges need to be evaluated through the lens of FEW nexus to identify co-benefits and avoid unintended consequences. This calls for innovative approaches to model the urban FEW nexus as an integrated whole, instead of examining them individually.

Existing studies primarily investigated urban FEW systems individually (e.g., nutrient flows, energy flows, and water flows) or focuses on two of the three FEW systems (e.g., water-energy nexus and food-water nexus). Existing studies that consider all FEW systems primarily aim to understand the impact of one system on the other two, such as how urban water-sector innovations change food and energy flows. There lacks an integrated approach that characterizes the comprehensive interdependence of urban FEW systems.

In this study we developed a systems modeling framework for the urban FEW nexus. We also applied this framework to the Detroit Metropolitan Area (DMA) as a preliminary study. The urban FEW systems modeling framework we proposed is based on material and energy flow analysis (MEFA) which quantifies flows and stocks of a group of substances within a system during a given period of time. This framework includes three steps:

Step 1: Individual FEW systems are modeled as resource flows across major sectors (e.g., water treatment, wastewater treatment, food production) of the urban system. Given the variety of food crops and products, we used nitrogen and phosphorus flows as the currency to represent the food system. Quantities of resource flows (water, energy, nitrogen, and phosphorus) are estimated based on a variety of data such as population, water withdrawal, and wastewater generation that were collected for the DMA case study.

Step 2: We identified and measured interdependence of individual FEW systems. This was done through literature review for common FEW interactions and quantitative estimation of resource flows between individual FEW systems (e.g., the amount of water used for food production and energy production).

Step 3: Each individual FEW system is characterized as a network of resource flows. Based on the interconnections between individual resource flow networks obtained from Step 2, we integrated the individual FEW resource flow networks as a network of four networks (water, energy, nitrogen, and phosphorus). This provides a mathematic representation of the urban FEW nexus, which will be the base of further analysis and modeling.

462067 Closed-Loop Scheduling: Major Considerations, Paradoxes and Remedies

Abstract:

Process industries employ a complex chain of operations (processes) which compete for limited resources. Hence a resource allocation problem arises wherein decisions with regards to the start time of operations, how much to process, the equipment to utilize and several associated decisions need to be taken on a repetitive basis. In the last two decades, there has been an increasing thrust towards using advanced computational tools to achieve more profitable schedules. Hence, working with scheduling models to optimally plan operations has become an important problem in process systems engineering [1,2].

Although the process systems engineering (PSE) community has worked on building accurate models and better solution methods, the aspect of repetitive scheduling has received limited attention. Rescheduling has been emphasized in some works [3-9], but in most cases scheduling is still thought to be a static open loop problem, wherein, if rescheduling is carried out, the emphasis is only on restoring feasibility or optimality to the current static schedule. Quantifying the quality of actual implemented schedule obtained by rescheduling has not been addressed.

In a dynamic environment multiple disturbances, such as task delays, yield losses, unit breakdowns or rush order arrival can render a previously computed schedule, suboptimal or even infeasible. In addition, Subramanian et. al. [10] emphasized, that rescheduling also needs to be performed due to advancement of the scheduling horizon, over which the schedule is computed, even when there are no disturbances.

In this work, we first and foremost show that open-loop and closed-loop scheduling are two different problems, even in the deterministic case, when no uncertainty is present. We also show how equally good open-loop schedules can translate to very different closed-loop schedules, so much so, that it could mean a difference between, no production versus production at full capacity. In addition, we stumble upon a paradox, wherein, solving a well-defined open-loop problem to optimality in every iteration, leads to a worse closed-loop schedule, than if this same open-loop problem was to be solved to a suboptimal solution.

Thereafter, we investigate various design attributes of the open-loop problem that affect the quality of the resulting implemented closed-loop schedule. The design attributes we study, are scheduling horizon length, rescheduling frequency and optimality gap of each open-loop optimization. We choose illustrative production networks, and extensively study combinations of the aforementioned attributes, over a reasonably exhaustive set of short term demand patterns, production load, and scheduling objectives. From these test cases, we identify trends, and problem characteristics which can facilitate in carefully choosing the three attributes.

First, we find that it is important to reschedule periodically, even when there are no "trigger" events, something that is in contrast with the current approaches to rescheduling. Second, we show that finding suboptimal open-loop schedules does not necessarily "accumulate" in the closed-loop. On the contrary, suboptimal solutions are "corrected" through repetitive revisions due to feedback. Third, we observe that there exist lower and upper thresholds for appropriate rescheduling frequency and moving horizon length, operating outside of which leads to substantial deterioration in closed-loop performance. Fourth, we show that the various design attributes work in conjunction with each other, and hence it is important to study them simultaneously rather than individually. The aforementioned observations lead to some interesting questions:

1) Should we reschedule on encountering trigger events (e.g., new order arrival, unit breakdown) or at a predetermined frequency or both?

2) Is it better to reschedule fast, even if we obtain suboptimal solutions, or less frequently but obtain optimal open-loop schedules?

3) Is it better to consider models over short scheduling horizons solved to optimality; or long horizons, which require larger models possibly leading to suboptimal solutions?

4) How can we choose the rescheduling frequency and moving horizon length? Can we trade one (e.g., long horizon) for the other (e.g., higher rescheduling frequency)? If yes, what are the trade-offs?

We address the following questions through a series of cases studies and then draw some general conclusions.

Lastly, we explore objective function modifications and addition of constraints to the open-loop problem as methods to improve closed-loop performance. Specifically, we show (1) how the former, among others, can help us "choose" among many equivalent open-loop solutions; and (2) how the latter can be used to enforce solution characteristics which can possibly lead to suboptimal open-loop schedules but ultimately result in higher quality closed-loop (implemented) schedules.

In this work, we use state-task network representation [11] and the discrete time-grid state-space model proposed by Subramanian et al. [10]. However, the framework and analysis presented, is also applicable to all discrete-time models and can be extended to continuous-time models represented in state-task network as well as resource-task network [12] representations.

References

[1] Méndez, C.A., Cerdá, J., Grossmann, I.E., Harjunkoski, I. and Fahl, M., 2006. State-of-the-art review of optimization methods for short-term scheduling of batch processes. *Computers & Chemical Engineering*, *30*(6), pp.913-946.

[2] Harjunkoski, I., Maravelias, C.T., Bongers, P., Castro, P.M., Engell, S., Grossmann, I.E., Hooker, J., Méndez, C., Sand, G. and Wassick, J., 2014. Scope for industrial applications of production scheduling models and solution methods. *Computers & Chemical Engineering*, *62*, pp.161-193.

[3] Kanakamedala, K.B., Reklaitis, G.V. and Venkatasubramanian, V., 1994. Reactive schedule modification in multipurpose batch chemical plants. *Industrial & engineering chemistry research*, *33*(1), pp.77-90.

[4] Rodrigues, M.T.M., Gimeno, L., Passos, C.A.S. and Campos, M.D., 1996. Reactive scheduling approach for multipurpose chemical batch plants. *Computers & chemical engineering*, *20*, pp.S1215-S1220.

[5] Vin, J.P. and Ierapetritou, M.G., 2000. A new approach for efficient rescheduling of multiproduct batch plants. *Industrial & engineering chemistry research*, *39*(11), pp.4228-4238.

[6] Cui, J. and Engell, S., 2010. Medium-term planning of a multiproduct batch plant under evolving multi-period multi-uncertainty by means of a moving horizon strategy. *Computers & Chemical Engineering*, *34*(5), pp.598-619.

[7] Novas, J.M. and Henning, G.P., 2010. Reactive scheduling framework based on domain knowledge and constraint programming. *Computers & Chemical Engineering*, *34*(12), pp.2129-2148.

[8] Nie, Y., Biegler, L.T. and Wassick, J.M., 2012. Integrated scheduling and dynamic optimization of batch processes using state equipment networks. *AIChE Journal*, *58*(11), pp.3416-3432.

[9] Kopanos, G.M. and Pistikopoulos, E.N., 2014. Reactive scheduling by a multiparametric programming rolling horizon framework: a case of a network of combined heat and power units. *Industrial & Engineering Chemistry Research*, *53*(11), pp.4366-4386.

[10] Subramanian, K., Maravelias, C.T. and Rawlings, J.B., 2012. A state-space model for chemical production scheduling. *Computers & Chemical Engineering*, *47*, pp.97-110.

[11] Shah, N., Pantelides, C.C. and Sargent, R.W.H., 1993. A general algorithm for short-term scheduling of batch operations—II. Computational issues. *Computers & Chemical Engineering*, *17*(2), pp.229-244.

[12] Pantelides, C.C., 1994, July. Unified frameworks for optimal process planning and scheduling. In *Proceedings on the second conference on foundations of computer aided operations* (pp. 253-274). New York: Cache Publications.

460885 Integration of Scheduling and Control Under Process Uncertainties

Abstract:

Scheduling and control problems are closely related and, when addressed simultaneously, can lead to an improved objective function and avoid suboptimal solutions (Flores-Tlacuahuac and Grossmann 2006). Meanwhile, the integration of this problems often results in MIDO problems, and the discretized MINLP requires excessive computation effort to get solved (Flores-Tlacuahuac and Biegler 2005). Furthermore, dynamical models of the systems used in the integration problems are usually considered to adequately describe the system behavior, and it is known that this assumption is not valid for physical system (Campo and Morari 1987), which are subject to uncertainties in process operations and models.

An effective framework for the scheduling and control problem involves two control loops, where the inner loop is a nominal model predictive control and the outer loop is an integrated schedule and control problem, which generates both the production scheduling and the state reference for control problem, and it is solved every time the system suffers a disturbance that cannot be handle by the inner control (Zhuge and Ierapetritou 2015). However, this is a reactive approach, which can become computationally expensive if the uncertainties occur frequently.

In this work, a preventive strategy for the integration of scheduling and control problems under uncertainty in process operations is proposed. First, the design of robustly stabilizing model predictive control (RMPC) of nonlinear systems is presented. Process uncertainties are treated as variables and the original process dynamics is approximated using a piece-wise affine model (PWA). The uncertain parameters are confined by upper and lower bounds, and the model is formulated aiming to optimize the worst case performance. For uncertainty descriptions that are linearly related to the state variables, the minmax optimization can be recasted as a quadratic program. Second, the PWA approximation is integrated with the scheduling level and an integrated problem is formed. Solving the integrated problem results in the scheduling and state references to be used by the robust MPC. Several case studies are presented to illustrate the

proposed approach and provide comparisons with the deterministic case.

Campo, P. J. and M. Morari (1987). "Robust Model Predictive Control." Proc. American Contr. Conf.. **2**: 6.Flores-Tlacuahuac, A. and L. Biegler (2005). "A robust and efficient mixed integer non-linear dynamic optimization approach for simultaneous design and control." Computer Aided Chemical Engineering **20**: 5.Flores-Tlacuahuac, A. and I. E. Grossmann (2006). "Simultaneous cyclic scheduling and control of a multiproduct CSTR." Industrial Engineerign Chemistry Research **45**(20): 15.Zhuge, J. and M. G. Ierapetritou (2015). "An Integrated Framework for Scheduling and Control Using Fast Model Predictive Control." Aiche Journal **61**(10): 16.

464526 A Framework for Integration of Design, Scheduling and Control for Chemical Processes Under Disturbances

Abstract:

Chemical process design is key for the optimal operation and control of chemical processes under the effect of external perturbations and process uncertainty. In addition to this, process scheduling also plays a critical role in facilities from various sectors, e.g. semiconductor manufacturing, electricity generation, chemical process industry, etc. The most common approach to address optimal design is known as the sequential approach, where the design, control, and scheduling are all considered separately. This approach is often inadequate, as it assumes steady state optimization for the specification of design and scheduling decisions, which are often suboptimal, as the dynamic interactions of design, control, and scheduling are not taken into account. An alternative emerging approach is to perform simultaneous optimization of design, control, and scheduling. This approach has the potential to produce improved solutions; however, it is a challenging task because it greatly increases the complexity of the problem to be addressed. Some of the challenging tasks associated with this approach are finding suitable approximations for process disturbances, accounting for parameter uncertainty, and reducing the computational time required to solve the resulting optimization formulations.

The aim of this research is to develop a mathematical framework for integrated optimization of design, control, and scheduling of chemical engineering systems under process disturbances. At this stage, the proposed framework optimizes design, control, and scheduling of a non-linear chemical system by using an iterative decomposition algorithm. At each iteration, the design, control, and scheduling are optimized under critical realizations of the process disturbances. Following that, the specified design, control, and scheduling scheme is fixed, and a dynamic feasibility problem is solved to determine the worst-case realization in the process disturbances (identified from a pre-specified set of possible disturbance realizations). The identified critical disturbance realization is then added to the set of worst-case disturbance realizations encountered from previous iterations in the algorithm. The loop then continues to the next iteration, where a new design, control and scheduling scheme is sought in the presence of the updated set of critical realizations in the disturbances. The decomposition algorithm continues up until no further critical realizations in the disturbances that produce process constraint violations are detected. This framework has been applied to a case study, which involves a single CSTR continuously switching between production of multiple different product grades. The design, control, and scheduling provided by the framework have been proven to be optimal when compared to the sequential design approach.

In future work, this framework will be expanded to include multiple process units and parameter uncertainty. This framework will be of great value to the chemical process industry, as the integrated optimization of design, control, and scheduling facilitates improvements in overall efficiency and process economics.

463434 Design, Scheduling and Control: A Simultaneous Approach By Multi-Parametric Programming

Abstract:

The long term operation of a plant requires decision making in several different levels; decisions that significantly affect the initial capital investment, optimal operation and economic, environmental and energy impact over its life span. The inherent uncertainty of a plant's long term operation is among the critical aspects associated with its overall performance. A variety of approaches have been considered including in particular, the consideration of (i) operability and design under uncertainty [1] as well as (ii) the system controllability with regulatory control structure selection and its tuning [2] and (iii) the design and model predictive control with stability and robustness aspects taken into account in a unified framework [3]. During the last few years there has also been great interest in the field of optimal operation, with several studies in literature exploring the interactions between shorter term operational decisions (regulatory and model-based control) with their longer term counterparts (scheduling and planning techniques) thus attempting to bridge the gap between the two [4-7]. The aim of this approach targets the long-term optimal operation of a process while taking into account its control aspects and the inherent handling of uncertainty [8].

In this work we consider decisions in three different time scales spanning from the lifetime operation timescale to the 1-second operation timescale of the plant.

I. The long term decisions include the optimal size of the plant equipment. The design of the equipment is a here-and-now decision that affects the process operability and controllability as well as its economic assessment throughout its lifespan. More, specifically, the design of the plant influences heavily the operational strategies that result into optimal operation. The operating bounds and performance metrics are determined by the design.

II. In processes where multiple products are considered, the design affects the overall production structure as well. In such processes, the production strategies focus on the alternation of production focus based on uncertainty associated with the raw material acquisition cost, product demand and resources availability. We enhance the multi-parametric Rolling Horizon Optimization (mp-RHO) policies presented in [9] by considering the design of the plant as a unknown but bounded parameter. Furthermore, we take into consideration the interactions of the control policies with the schedule within the mp-RHO formulation.

III. Model based control strategies, and more specifically their multi-parametric programming counterparts, are developed to ensure efficient set-point tracking. The control formulation takes into account the design of the plant in a similar way to the scheduling formulation. Based on the production scheduling different control schemes and structures are considered.

IV. The development of the receding horizon policies follow the principles presented in [10]. Starting from a high fidelity model developed in gPROMS®, different multi-parametric optimization problems are designed and validated. In terms of the optimal scheduling policies we derive actions as a function of (i) the system operational state, (ii) the uncertain demand and pricing, (iii) the process design and (iv) the control interactions. Similarly, the optimal control policies are a function of (i) the system operational state, (ii) the process design and (iii) the operating set-points. The design of the system is determined by the formulation and solution of a (mixed integer) dynamic optimization problem based on the process at hand, the optimal operation of which is ensured by the multi-parametric optimization based policies.

V. The approach is demonstrated on a domestic cogeneration unit that provides usable heat and power upon demand [11]. More specifically, the size of the internal combustion engine of the CHP is optimized under optimal operation which is ensured by design dependent scheduling and control policies. This multi-product process requires the derivation of different decentralized control schemes to account for the different operating policies as well as design dependent scheduling policies that determine the operational focus under uncertainty. We show that the simultaneous optimization of the design and operation of the process at hand achieves maximum economic benefits to the user while ensuring optimal operation.

References:

1. van Schijndel, J.M.G. and E.N. Pistikopoulos, Towards the integration of process design, process control and process operability - Current stares and furore trends. Foundations of Computer-Aided Process Design, 2000. 96: p. 99–112.

2. Bansal, V., et al., Simultaneous design and control optimisation under uncertainty. Computers & Chemical Engineering, 2000. 24(2 7): p. 261–266.

3. Sakizlis, V., J.D. Perkins, and E.N. Pistikopoulos, Parametric Controllers in Simultaneous Process and Control Design Optimization. Industrial & Engineering Chemistry Research, 2003. 42(20): p. 4545–4563.

4. Baldea, M. and I. Harjunkoski, Integrated production scheduling and process control: A systematic review. Computers & Chemical Engineering, 2014. 71: p. 377–390.

5. Zhuge, J. and M.G. Ierapetritou, Integration of scheduling and control for batch processes using multi-parametric model predictive control. AIChE Journal, 2014. 60(9): p. 3169–3183.

6. Subramanian, K., C.T. Maravelias, and J.B. Rawlings, A state-space model for chemical production scheduling. Computers & Chemical Engineering, 2012. 47(0): p. 97–110.

7. Engell, S. and I. Harjunkoski, Optimal operation: Scheduling, advanced control and their integration. Computers and Chemical Engineering, 2012. 47: p. 121-133.

8. Patil, B.P., E. Maia, and L.A. Ricardez-Sandoval, Integration of scheduling, design, and control of multiproduct chemical processes under uncertainty. AIChE Journal, 2015. 61(8): p. 2456-2470.

9. Kopanos, G.M. and E.N. Pistikopoulos, Reactive scheduling by a multiparametric programming rolling

horizon framework: A case of a network of combined heat and power units. Industrial and Engineering Chemistry Research, 2014. 53(11): p. 4366-4386.

10. Pistikopoulos, E.N., et al., PAROC-An integrated framework and software platform for the optimisation and advanced model-based control of process systems. Chemical Engineering Science, 2015.

11. Diangelakis, N.A., S. Avraamidou, and E.N. Pistikopoulos, Decentralized Multiparametric Model Predictive Control for Domestic Combined Heat and Power Systems. Industrial and Engineering Chemistry Research, 2016. 55(12): p. 3313-3326.

458986 Coordinating Production Scheduling and Process Operation Via Economic Model Predictive Control

Abstract:

Supply chain optimization has been an active area of research for improving the overall profit from a chemical process [1]-[2]. Although supply chain optimization in general addresses many aspects of the production pipeline from raw material manufacturers to customers, an important aspect of the supply chain is planning and scheduling of plant production to determine what will be produced and when it will be produced. The integration of production management and control has been investigated to improve process profits [3]-[4]. Generally, only a subset of the process states are required to track a schedule; the additional states may be varied in time for economic benefit if the production schedule is simultaneously met.

In this work, we develop a Lyapunov-based economic model predictive control (LEMPC) formulation that can maintain closed-loop stability of the process states while causing states that are required to meet a schedule to track the scheduled values. By enforcing the schedule through a soft constraint, feasibility of the LEMPC is maintained at all times even in the presence of disturbances when the Lyapunov level sets used in the stability constraints of the LEMPC intersect at the time that the schedule is changed. The states that are not required to meet a schedule are permitted to vary in time to maximize the process economics. The proposed LEMPC is applied to a chemical process example in which the heating rate is to be minimized while the product concentration tracks a desired schedule. This chemical process example demonstrated that the proposed method was able to cause the product concentration to meet the desired schedule while maintaining closed-loop stability and minimizing the process energy cost.

[1] Subramanian K, Rawlings JB, Maravelias CT, Flores-Cerrillo J, Megan L. Integration of control theory and scheduling methods for supply chain management. Computers & Chemical Engineering. 2013;51:4-20.

[2] Perea-López E, Ydstie BE, Grossmann IE. A model predictive control strategy for supply chain optimization. Computers & Chemical Engineering. 2003;27:1201-1218.

[3] Baldea M, Harjunkoski I. Integrated production scheduling and process control: A systematic review. Computers & Chemical Engineering. 2014;71:377-390.

[4] Gutiérrez-Limón MA, Flores-Tlacuahuac A, Grossmann IE. A reactive optimization strategy for the simultaneous planning, scheduling and control of short-period continuous reactors. Computers & Chemical Engineering. 2016;84:507-515.

466691 Travelling Salesman Problem (TSP) Based Integration of Planning, Scheduling and Control for Continuous Processes

Abstract:

Advanced decision making in the process industries requires efficient use of information available at the different hierarchical decision levels [1]. Traditionally, planning, scheduling and control problems are solved in a decoupled way, neglecting the strong interdependence among them. Integrated Planning, Scheduling and Control (iPSC) aims to address this issue [2,3]. In the last few years, a considerable amount of research work has focused on the integration of scheduling and control in order to enhance feasibility, robustness and profitability of the operations in the chemical industries while planning was until recently dealt hierarchically. Formulating the iPSC problem results in a complex optimisation problem involving continuous (algebraic and differential) and integer variables. Since planning, scheduling and control problems consider different time scales (varying from seconds for control to weeks or months for scheduling and planning), the final optimisation problem can be computationally intensive [4,5]. Until now, most of the research done in the context of integrating the control with operations has used as basis for the planning and scheduling, time-slot based formulations and little to none attention has been given to immediate precedence formulations such as the one presented in Liu et al. (2008) [6]where the authors proposed a Travelling Salesman Problem (TSP) formulation.

In the present work, we propose a new approach for the iPSC problem of continuous processes aiming to reduce model and computational complexity. The original nonlinear dynamics, which arise in the control of the process, are approximated with piece wise affine (PWA) dynamics within specific regions of validity. For the planning and scheduling we use a TSP based formulation [6] where the planning periods are modelled in discrete time while the scheduling within each week is in continuous time. Another key feature of the proposed iPSC framework is that backlog and multiple orders from different customers are allowed. The resulting iPSC problem is a Mixed Integer

Programming (MIP) problem and different solution strategies are employed and analysed. Model Predictive Control (MPC) is employed to account for the stability and recursive feasibility of the underlying control problem. Finally, the proposed iPSC approach is tested on a number of case studies and compared with existing approaches.

References

1. Grossmann, I. E. (2012). Advances in mathematical programming models for enterprise-wide optimization. Computers & Chemical Engineering, 47, 2-18.
2. Chu, Y., & You, F. (2015). Model-based integration of control and operations: Overview, challenges, advances, and opportunities. Computers & Chemical Engineering, 83, 2-20.
3. Baldea, M., & Harjunkoski, I. (2014). Integrated production scheduling and process control: A systematic review. Computers & Chemical Engineering, 71, 377-390.
4. Chu, Y., & You, F. (2014). Integrated planning, scheduling, and dynamic optimization for batch processes: MINLP model formulation and efficient solution methods via surrogate modeling. Industrial & Engineering Chemistry Research, 53(34), 13391-13411.
5. Gutiérrez-Limón, M. A., Flores-Tlacuahuac, A., & Grossmann, I. E. (2014). MINLP formulation for simultaneous planning, scheduling, and control of short-period single-unit processing systems. Industrial & Engineering Chemistry Research, 53(38), 14679-14694.
6. Liu, S., Pinto, J. M., & Papageorgiou, L. G. (2008). A TSP-based MILP model for medium-term planning of single-stage continuous multiproduct plants. Industrial & Engineering Chemistry Research, 47(20), 7733-7743.

468543 Optimal Scheduling of Advanced Energy Plants with CO_2 Capture

Abstract:

Pre- and post-combustion carbon dioxide (CO_2) capture from fossil fuel-fired power plants are potential approaches to mitigating build-up of CO_2 concentration in the environment. However, state-of-the-art capture processes have high operating costs, mainly attributable to the energy penalties for regeneration and CO_2 compression. In addition, the high penetration of renewable generators into the grid is expected to force advanced fossil energy plants with CO_2 capture to follow the fluctuating load. Therefore, optimal scheduling of power plant dispatch and the CO_2 capture rate are critical for maximizing plant profitability without sacrificing optimal control performance. While control of energy plants with carbon capture has been reported in the literature, there is little work that has investigated optimal scheduling of advanced power plants from the perspective of plant economics and controllability in the face of expected penalty/carbon tax scenarios.

It is anticipated that the CO_2 capture targets will need to be satisfied over a period of time, denoted here as â€˜base timeâ€™, which is expected to vary from region/country to region/country. The base time for CO_2 capture may be several months to a year or more. Therefore, any discrepancy in the CO_2 captured and the target during the previous time instances can be accounted for in the future time instances. This is unlike typical control objectives where the deviation from the set point in the past might be completely neglected or be considered with lower weights in the future. Thus inclusion of CO_2 capture in the scheduling problem requires the solution of an optimization problem with a large horizon. One challenge for this class of problems is the large size of the online optimization problem. Efforts to reduce the horizon can lead to undesirable closed-loop characteristics such as bang-bang actuation or inventory creep. In addition, the input sequence for a reduced horizon formulation may be suboptimal when compared to an infinite horizon control formulation. Our work has evaluated the optimal scheduler profile for CO_2 capture rate over time by

considering three penalty/tax scenarios including the possibility of a cap-and-trade policy. The presentation will include a proof for the
stability of the developed scheduler algorithm and highlight its application to an integrated gasification combined cycle (IGCC) power plant with CO_2 capture.

465615 Optimal Design and Control of a Catalytic Distillation Column. Case Study: Ethyl *Tert*-Butyl Ether (ETBE) Synthesis Column

Abstract:

Optimal design and control of a catalytic distillation column. Case study: Ethyl *tert*-butyl ether (ETBE) synthesis column.

David E. Bernal and Jorge M. Gómez

Most of the chemical processes include two of the most important operations, chemical reaction and thermodynamic separation. These two operations are used to be carried out in different equipment. The reactions take place in different equipment, called reactors (continuously stirred tank reactors -CSTR-, tubular reactors or batch reactors, among others), and are operated under different conditions. On the other hand, the separation is usually made by different unitary operations (distillation, extractions, crystallization, absorption, among others) being the distillation by far the most common one [1]. Distillation is the most popular liquid mixture separation technique in the chemical and in the oil industry. The energy requirement of this operation can represent up to 40% of the whole plant's energy requirement [2], [3].

Recycle streams are used between the reaction and separation sections to increase the yield and conversion, minimize the undesired products synthesis, improve the energetic efficiency and guarantee the controllability of these processes. Instead of carrying the reaction and separation in independent units, these operations can be carried out in a single equipment [1]. The implementation of these combined processes represents one of the most promising methodologies of process intensification. Economic and environmental considerations have led the industry to develop this kind of processes [4], which offer considerable benefits compared to the traditional multi-unit scheme.

Integrating reaction and product purification in a single multifunctional unit leads to considerable improvements compared to the traditional sequential approach such as: overcoming of the chemical equilibrium limitations, increment in product selectivity and the use of the reaction heat for the separation [5]. A limitation of this integration is that the "operational window" is reduced considerably as the conditions of the reaction and the separation must be satisfied simultaneously [1]. The general concept this process intensification is referred as Reactive Distillation (RD) and when a heterogeneous catalyst is involved it is referred as Catalytic Distillation (CD).

The operation of the distillation with a chemical reaction is important in the process, becoming into a critical unit and in certain cases a limiting unit, therefore its stable behavior must be guaranteed. This unit is very sensitive to perturbations, which means that a change in the operational conditions may affect the process dynamics compromising significantly the steady state operation, affecting the product purity, the energy consumption or the produced quantity [6].

Traditionally, the design of a process has been made in a sequential manner. First, a stationary analysis determines the process design and then a dynamic analysis determines its control law. Ziegler and Nichols [7] identified a direct relationship between the design of a process and its dynamic response under uncertainty, which compels to consider the process controllability in the first stages of the design [8], [9]. From this fact, the optimal control arises as a field where mathematical optimization tools are used to find the optimal profiles of operational variables for a certain process guaranteeing the product quality and the process profitability and security facing a periodic change or a perturbation to the system. Coupling the optimal design and optimal control results in a process that maximizes its profitability, assuring the specifications of its product even when the system is subjected to dynamic disturbances.

Taking into account these facts, the optimal design and control of a CD column becomes important. This kind of problems have a considerable mathematical complexity because of the interactions that exist between the coupled processes of separation and reaction. The modeling of these phenomena is highly nonlinear, which generates complications while solving it and the possibility of multiple solutions [10]. The mathematical complexity of this type of problems is mainly that the set of equations that describe the dynamic behavior of the system is a differential algebraic system of equations (DAE), which requires certain mathematical reformulations for solving it. The DAE problems are reformulated into large-scale NonLinear Programming (NLP) problems using orthogonal collocation.

The optimal design and control of a CD column designed for the production of Ethyl *tert*-butyl ether (ETBE) has been proposed as the case study of this work. The ETBE is a chemical compound used as oxygenate for fuels and it is classified as semi-renewable, as it can be synthetized from the etherification of bioethanol and isobutene in the presence of an acid catalyst [11]. The optimal control problem had included tracking objectives and an economic objective, for instance a weighted sum of the objectives was used as objective function. A methodology for determining the weighting parameters of each objective was proposed, based on an offline multi-objective utopia tracking optimization [12].

A comparison is made between the sequential and the simultaneous approach in the optimal design and control for this equipment. The results show that both approaches result in different design parameters that affect the economic profit of the process, its controllability and the satisfaction of the operational constraints of the system.

Several formulations of the dynamic model were proposed based on the description level and the assumptions that support them and resulted in different index DAE problems. This models were compared in terms of resulting NLP problem size (e.g. number of equations and variables) and results by solving and OCP problem. Finally, a first approach to an advanced control strategy called Economic-Oriented Nonlinear Model Predictive Control (EO-NMPC) was implemented and compared to a PI (Proportional and Integral) controller, showing advantages of the EO-NMPC in terms of economic profit.

References

[1] W. L. Luyben and C.-C. Yu, *Reactive Distillation Design and Control*. 2005.

[2] D. C. White, “Optimize Energy Use in Distilation,” *Am. Inst. Chem. Prog.*, no. March, p. 7, 2014.

[3] P. Fact, “Distillation Column Modelling Tools,” *Chem. Proj. Fact Sheet*, 2001.

[4] T. Keller, B. Dreisewerd, and A. Górak, “Reactive distillation for multiple-reaction systems: Optimisation study using an evolutionary algorithm,” *Chem. Process Eng. - Inz. Chem. i Proces.*, vol. 34, no. 1, pp. 17–38, 2013.

[5] K. Sundmacher and A. Kienle, *Reactive distillation: status and future directions*, vol. 3. 2006.

[6] R. Ross, J. D. Perkins, E. N. Pistikopoulos, G. L. M. Koot, and J. M. G. Van Schijndel, “Optimal design and control of a high-purity industrial distillation system,” in *Computers and Chemical Engineering*, 2001, vol. 25, no. 1, pp. 141–150.

[7] J. G. Ziegler and N. B. Nichols, “Optimum settings for automatic controllers,” *InTech*, vol. 42, no. 6, pp. 94–100, 1995.

[8] V. Bansal, J. D. Perkins, E. N. Pistikopoulos, R. Ross, and J. M. G. Van Schijndel, “Simultaneous design and control optimisation under uncertainty,” *Comput. Chem. Eng.*, vol. 24, no. 2–7, pp. 261–266, 2000.

[9] R. L.-N. de la Fuente and A. Flores-Tlacuahuac, “Integrated Design and Control Using a Simultaneous Mixed-Integer Dynamic Optimization Approach,” *Ind. Eng. Chem. Res.*, vol. 48, no. 4, pp. 1933–1943, 2009.

[10] M. G. Sneesby, M. O. Tadé, and T. N. Smith, “Multiplicity and Pseudo-Multiplicity in MTBE and ETBE Reactive Distillation,” *Chem. Eng. Res. Des.*, vol. 76, no. 4, pp. 525–531, 1998.

[11] H. E. P. and its Offices, “Bio-ETBE: The Right Road to High Quality 21st Century Motor Fuels,” Washington, 2012.

[12] V. M. Zavala and A. Flores-Tlacuahuac, “Stability of multiobjective predictive control: A utopia-tracking approach,” *Automatica*, vol. 48, no. 10, pp. 2627–2632, 2012.

464945 A Novel and Systematic Method for Process Intensification

Abstract:

Process intensification combines multiple operations in a single unit, and leads to substantially smaller, cleaner, safer, and more energy-efficient technologies [1]. It has been shown that intensification represents a limit case of tight integration through significant material recycling [2]. Although there are established methods for process synthesis and integration, few methods exist for the systematic identification and selection of intensification opportunities. Current superstructure-based process synthesis relies on pre-specified equipment configurations and is unable to automatically construct novel equipment configurations for intensification. Phenomena based process intensification [3-5] allows a bottom-up approach for the identification, screening and incorporation of intensification opportunities, but this approach often requires sequential or decomposition based solution strategies to solve large models, which can lead to suboptimal solutions.

In this work, we propose a unified process synthesis and intensification approach that leverages on an original representation of the flowsheet superstructure. Unlike a classical flowsheet superstructure which encompasses all feasible sequences of unit operations (e.g., reactors, separators, movers, etc.), our proposed superstructure is an ensemble of building blocks, where each block represents a unit use of a material and nexus between the blocks are achieved by intra-block mass and energy transfer. An intensified unit is realized by selecting and assembling multiple neighboring blocks with different functionalities. While reaction, mixing, heating and cooling are achieved within the block, separation related phenomena (e.g., phase separation, phase-contact, etc.,) are achieved via the boundaries between two neighboring blocks. This enables to include many intensification alternatives, including divided wall column distillation, reactive distillation, reactive absorption and membrane reactor, within the same superstructure without any *a priori* postulation.

The overall process intensification model is formulated as a single mixed-integer nonlinear optimization (MINLP)

problem and is solved to optimality. Through a range of applications, we demonstrate that the proposed intensification method is capable of generating innovative and intensified configurations and obtaining optimal process conditions for given feed and product specifications.

References:

[1] Reay, D.; Ramshaw, C.; Harvey, A. Process intensification: engineering for efficiency, sustainability and flexibility. Waltham, MA: Butterworth-Heinemann, 2013.

[2] Baldea, M. From process integration to process intensification. *Computers & Chemical Engineering*2015, 81:104â€"114.

[3] Arizmendi-Sanchez, J. A., and Sharratt, P. Phenomena-based modularisation of chemical process models to approach intensive options. *Chemical Engineering Journal* 2008, 135(1):83-94.

[4] Lutze P, Babi DK, Woodley JM, Gani R. Phenomena based methodology for process synthesis incorporating process intensification. *Industrial & Engineering Chemistry Research* 2013, 52(22):7127â€"44.

[5] Babi, D. K., Holtbruegge, J., Lutze, P., Gorak, A., Woodley, J. M., and Gani, R. Sustainable process synthesisâ€"intensification. *Computers & Chemical Engineering* 2015, doi:10.1016/j. compchemeng.2015.04.030.

469723 High-Resolution Process Design: Flowsheet Optimization with Embedded High Fidelity Unit Models

Abstract:

High-resolution process design: Flowsheet optimization with embedded high fidelity unit models

R.C. Pattison and M. Baldea

McKetta Department of Chemical Engineering, The University of Texas at Austin, Austin, TX 78712

Optimal process design is key to maximizing the profitability of chemical plants. Due to the large upfront capital investments required to construct new processes or retrofit existing plants, it is imperative that the design is optimized using modern gradient-based optimization tools coupled with detailed models of the processing units present in the flowsheet.

Equation-based simulation and optimization have witnessed substantial progress, and most current simulators offer equation-oriented simulation options as well. In this presentation, we illustrate a novel equation-oriented simulation tool which allows for high-fidelity models of select unit operations (including reactors, multi-stream heat exchangers) to be easily embedded into the process flowsheet model. These models capture phenomena occuring over multiple length scales, including smaller dimensions (e.g., diffusion and reaction in a catalyst pellet) which are typically not considered in flowsheet simulation or optimal design calculations.

Our framework is based on a pseudo-transient representation of the unit operation models developed in our previous work [1-3]. The pseudo-transient modeling framework relies on a systematic reformulation of the material and energy balances into a differential and algebraic equation model having the same steady state solution to that of the original algebraic equation model. The (typically) partial differential equations associated with detailed unit models lend themselves naturally to a pseudo-transient reformulation which can then be incorporated into the process flowsheet model. This, in turn, enables the detailed steady state unit designs to be optimized simultaneously with the design of the entire process flowsheet. To this end, we use a time relaxation-based optimization algorithm previously developed in our group [1].

The framework will be illustrated with industrial case studies, including the design optimization of an intensified dimethyl ether production process featuring a dividing wall distillation column and the optimal design of a natural gas liquefaction process.

References

[1] Pattison, R.P.; Baldea, M. Equation-oriented flowsheet simulation and optimization using pseudo-transient models. *AIChE J.*, **2014**, 60, 4104-4123.

[2] Pattison, R.P.; Baldea, M. Multistream heat exchangers: Equation-oriented modeling and flowsheet optimization. *AIChE J.*, **2015**, 61, 1856-1866.

[3] Pattison, R.P.; Gupta, A.M.; Baldea, M. Equation-oriented optimization of process flowsheets with dividing-wall columns. *AIChE J.*, **2016**, 62, 704-716.

456412 Process Intensification of Multicomponent Distillation Configurations Using Minimum Additional Number of Heat and Mass Integration Sections

Abstract:

We present a powerful and simple-to-use fact about heat and mass integration to consolidate distillation columns in a multicomponent distillation configuration. First, thermal couplings are introduced to all intermediate transfer streams in the distillation column. And then, distillation columns associated with a lighter final product reboiler and a heavier final product condenser are consolidated to produce a heat and mass integrated configuration. We call this novel set of configurations the HMP configurations.

Coupled with the already developed techniques[1] that enumerate all basic and thermally coupled distillation configurations, a systematic way of synthesizing all HMP configurations for *n*-component separations is introduced here. Also, we compare the energy savings of HMP configurations with the well-known fully thermally coupled (FTC) configuration using the state-of-the-art Global Minimization Algorithm (GMA)[2] developed based on Underwood equations. GMA provides a global optimization based rank-list of basic and thermally coupled configurations according to minimum total vapor duty requirement under minimum reflux conditions for separating any ideal or near-ideal multicomponent mixture. We demonstrate through case studies that HMP configurations can have very similar and sometimes even the same minimum total vapor duty requirement as the FTC configuration, while using far less number of column sections, intermediate transfer streams, and thermal couplings than the FTC configuration.

These findings not only intensify the search space of distillation configurations, but also provide industrial practitioners with insights in designing energy efficient distillation configurations that are easy to operate and control.

[1] Shah VH, Agrawal R. A matrix method for multicomponent distillation sequences. AIChE J. 2010;56(7):1759â€"1775.

[2] Nallasivam U, Shah VH, Shenvi AA, Huff J, Tawarmalani M, Agrawal R. Global optimization of multicomponent distillation configurations: 2. Enumeration based global optimization algorithm. AIChE J. 2016;62(6):2071â€"2086.

469729 Novel Optimization-Based Approach for Process Design and Intensification of High-Dimensional Modular Systems

Abstract:

In the recent years, due to the advent of the shale gas revolution, the US has incremented the annual production of processed natural gas by approximately 21% from 16.3 to 19.8 trillion cubic feet[1]. To accelerate the potential benefits brought by the abundant amounts of shale gas in the country, approaches are needed for the optimization and techno-economic viability analysis of novel natural gas utilization processes that produce heat, power, fuels and chemicals. However, the design of such processes is a challenging task as these systems are typically operated in a highly constrained and integrated environment that is represented by complex large-scale and nonlinear models. In this presentation, a novel optimization-based operability approach for process design and intensification of high-dimensional and modular energy systems is proposed. These modular units if enabled have the potential to transform the US economy by utilizing local stranded gas without the need of building expensive pipelines.

Process operability was originally developed as an approach for the design and control of complex chemical processes[2]. This approach has been applied to processes that are described by high-dimensional linear models[3,4]. For nonlinear systems, an operability-based approach for design and intensification of low-dimensional membrane reactor systems has also been proposed[5,6]. However, operability methods for nonlinear systems are currently limited by the problem size that they can address. The proposed approach bridges this gap in the literature by addressing the challenges of process nonlinearity and model size. The proposed method also broadens the scope of the traditional path of the operability approaches for design and control, mainly oriented to obtain the achievable output set (AOS) from the available input set (AIS), and compare the computed AOS to a desired output set (DOS). In particular, an optimization algorithm based on nonlinear programming tools is formulated for the high-dimensional calculations of the desired input set (DIS) that is feasible considering process constraints, performance levels and intensification

targets.

To illustrate the effectiveness of the developed method, two modular systems are addressed: i) a membrane reactor (MR) for the direct methane aromatization (DMA) conversion to benzene and hydrogen; and ii) a natural gas combined cycle (NGCC) system for heat and power generation. Results on the application of this novel method as a tool for process intensification show reduction of the DMA-MR footprint (â‰ˆ77% reactor volume and 80% membrane area reduction) for an equivalent level of performance, when compared to the base case. These results indicate that the novel approach can be a powerful tool for enabling the realization of the concept of the modular economy.

References

1. EIA (2016). U.S. Energy Information Administration. Natural Gas Report. Available at: http://www.eia.gov/dnav/ng/ng_sum_lsum_dcu_nus_m.htm
2. Vinson D. R., Georgakis C. A new measure of process output controllability. J. Proc. Cont. 2000, 10(2-3), 185-194.
3. Lima F. V., Georgakis C. Design of output constraints for model-based non-square controllers using interval operability. J. Proc. Cont. 2008, 18(6):610-620.
4. Lima F. V., Georgakis C. Input-output operability of control systems: the steady-state case. J. Proc. Cont. 2010, 20(6):769-776.
5. Carrasco J. C. and Lima F. V. Nonlinear operability of a membrane reactor for direct methane aromatization. In Proceedings of the 2015 IFAC ADCHEM Symposium, Whistler, Canada, June 2015.
6. Carrasco J. C. and Lima F. V. Novel operability-based approach for process design and intensification: application to a membrane reactor for DMA. Submitted for publication, 2016.

470719 Towards Optimal and Sustainable Operation of Separation Processes: The Computational Approach

Abstract:

Separation processes (such as Simulating Moving Bed, Pressure Swing Adsorption Systems and chromatographic processes) are widely applied in the chemical/biochemical industry for the separation and/or purification of various components [1-3]. Commonly, chromatographic separations are assessed by their product recovery rate, as well as the purity of the end product, with the latter being of vital importance. Consequently it is essential to guarantee that the process runs under optimal operating conditions, achieving high recovery rates of products that meet the desired specifications. However, downstream processing is coupled with high capital and operational costs that are significantly increased as a function of the feed stream produced by the upstream [4]. Current industrial needs suggest the development of sustainable processes, characterized by lower investment and/or operation costs. To answer to global demand and competition arising from novel solutions such as single-use technologies [5, 6], separation processes and particularly preparative chromatography, requires significant advancements.

Along this line, in this work we focus on the development of advanced control strategies for a semi-continuous, chromatographic separation process [7] aiming to increase both product purity and yield. In this framework we examine a highly nonlinear, periodic system and we design novel control strategies to ensure that optimal operation is maintained. The controllers are designed and tested following our recently presented PAROC framework that comprises four steps: (i) development of a high-fidelity process model, (ii) approximation of the complex, process model, (iii) design of a multi-parametric controller and (iv) â€˜closed-loopâ€™, in-silico validation of the controller against the process model. The controllers are tested in-silico against the high-fidelity process model and demonstrate an overall satisfactory performance, tracking the predefined setpoints and providing input profiles of reduced utility consumption. In addition, the proposed control scheme accounts for variable feeding composition, thus enabling the downstream processing to operate independently from the mixture composition resulting from the upstream.

Acknowledgements

The authors would like to thank Mr R. Oberdieck and Miss A. Quiroga-Campano for their contribution in the assessment of the controller validation. The authors would also like to thank Mr. Fabian Steinebach & Prof. M. Morbidelli from ETH Zurich, as well as Dr. Thomas Mueller-Spaeth from ChomaCon AG for their valuable input on the understanding of the MCSGP process. Financial support from the European Commission (OPTICO/G.A. No.280813) & Texas A&M University are also gratefully acknowledged.

References

1. Subramani, H.J., K. Hidajat, and A.K. Ray, *Optimization of Simulated Moving Bed and Varicol Processes for Glucoseâ€"Fructose Separation.* Chemical Engineering Research and Design, 2003. **81**(5): p. 549-567.

2. Khajuria, H. and E.N. Pistikopoulos, *Dynamic modeling and explicit/multi-parametric MPC control of pressure swing adsorption systems.* Journal of Process Control, 2011. **21**(1): p. 151-163.

3. Müller-Späth, T., et al., *Chromatographic separation of three monoclonal antibody variants using multicolumn countercurrent solvent gradient purification (MCSGP).* Biotechnology and Bioengineering, 2008. **100**(6): p. 1166-1177.

4. Gronemeyer, P., R. Ditz, and J. Strube, *Trends in Upstream and Downstream Process Development for Antibody Manufacturing.* Bioengineering, 2014. **1**(4): p. 188-212.

5. Shukla, A.A. and U. Gottschalk, *Single-use disposable technologies for biopharmaceutical manufacturing.* Trends in Biotechnology, 2013. **31**(3): p. 147-154.

6. Xenopoulos, A., *A new, integrated, continuous purification process template for monoclonal antibodies: Process modeling and cost of goods studies.* J Biotechnol, 2015. **213**(0): p. 42-53.

7. Krättli, M., F. Steinebach, and M. Morbidelli, *Online control of the twin-column countercurrent solvent gradient process for biochromatography.* Journal of Chromatography A, 2013. **1293**: p. 51-59.

454543 Multi-Scale Computer Aided Synthesis–Design–Intensification Method for Sustainable Hybrid Solutions

Abstract:

Process synthesis-design refers to identification of the processing route to reach a desired product from a specified set of raw materials, design of the operations involved in the processing route along with calculations of utility requirements, waste generations and emissions to the environment. Process intensification refers to strategies by which new processes can be designed or existing processes can be redesigned to be more efficient, compact and sustainable.

Therefore, in order to achieve sustainable process design it would be logical to perform process synthesis-design and process intensification together.

In this work, a four stage hierarchical decomposition solution strategy is proposed to generate sustainable hybrid solutions. The framework operates at different scales, the unit operations scale, the task scale and the phenomena scale. In stage1, process synthesis is performed (at the unit operations scale) using a newly developed computer-aided flowsheet design method (CAFD) (Tula et al, 2015), which applies principles of computer-aided molecular design to the synthesis and design of process flowsheets. That is, use process-groups representing different unit operations (reactor, distillation, flash, crystallization, etc.), bonds representing streams and/or recycles, rules for chemical feasibility also representing process flowsheet feasibility and sum of group contributions representing the performance of the flowsheet that can later be used to quickly screen the alternatives and to generate the optimal processing route. In stage 2, the base case design is first established based on the generated processing route. This base case design is further analyzed to identify process hotspots (process bottlenecks) using a comprehensive indicator based analysis incorporating process economics, sustainability measures (Carvalho et al., 2013) and life cycle assessment factors (Kalakul et al., 2014). These indicators corresponding to the hotspots are translated to design targets that are targeted in the next stage to achieve more sustainable and intensified designs. In stage 3,

integrated task phenomena based synthesis-intensification method is applied (Babi et al., 2015). The main idea here is to generate all the phenomena involved in the process by breaking the tasks involved into corresponding phenomena. Later these phenomena scale units are combined using combination rules in order to generate new and/or existing unit operations that constitute the (more sustainable) flowsheet alternatives which satisfy, the process design targets. In this way, truly predictive and innovative hybrid solutions are generated much in the same way as atoms are combined to form molecules with desired properties (that is, analogous to computer-aided molecular design) which are otherwise could not be found from the higher scales. In the final stage, detailed model based simulations and/or experiments are performed to verify and compare the new solutions.

In this presentation, a multi-scale computer-aided framework will be presented together with the corresponding databases, computer-aided methods and tools needed to implement and apply the synthesis-design-intensification method. The application of this framework will be highlighted through case studies generating hybrid solutions for energy efficient processing schemes involving downstream separations and intensified reaction-separations.

References:

[1] Babi, D. K., Holtbruegge, J., Lutze, P., Gorak, A., Woodley, J. M., & Gani, R. 2015. Computers & Chemical Engineering.

[2] Carvalho, A., Matos, H. A., & Gani, R. 2013. Computers & Chemical Engineering, 50, 8–27

[3] Kalakul, S., Malakul, P., Siemanond, K., & Gani, R. 2014. Journal of Cleaner Production

[4] Tula, A. K., Eden M. R., Gani, R. 2015. Computers & Chemical Engineering.

458409 Microwave Assisted Crystallization: Intensification of Direct Nucleation Control

Abstract:

Direct nucleation control (DNC) is a feedback control strategy used to control the size distribution of crystals (one of the many possible control objectives) during a crystallization process. DNC has been shown for maintaining a narrow size distribution by removal of unwanted fine particles. Application of DNC for obtaining desired size distribution is based on alternating dissolution and growth cycles, achieved via temperature cycling and triggered as a result of deviations in the particle count set point (measured by in-situ measurement PAT tool).

However, a major draw-back of the DNC implemented in conventionally heated/cooled reactors is that the temperature cycles needed to dissolve the nuclei considerably increases the process/batch time. However recently it has been demonstrated that by application of microwave fields for heating, the same control performance can be achieved with a 50% reduction in process/batch time. The study showing the intensification of the DNC in terms of process time by integrating microwave fields into a crystallizer was performed in a 1 litre jacket vessel in which the slurry was circulated via an external loop through a microwave cavity. However a better application would be when microwave fields are truly introduced inside the crystallizer thus eliminating external loops for slurry circulation which can be problematic.

In this study we present DNC experiments using a microwave integrated crystallizer. The crystallizer utilizes internal transmission line technology made available by Sairem SAS which enables direct introduction of microwave fields into the slurry. In this study we demonstrate cooling crystallization with DNC control and address key challenges in process integration and implementation, namely: Evaluation and optimization of the microwave heating to ensure homogeneous temperature distribution in the system, Optimization of the reactor configuration for safe introduction of PAT tools essential for DNC feedback (e.g FBRM), by avoiding the exposure to high intensity MW field and evaluation of the performance of microwave assisted DNC in a cooling crystallization process.

467819 A Systems Approach for the Development of Osn Membrane Cascades

Abstract:

The processes involved in separating the components of chemical mixtures into pure or purer forms, such as distillation, account for 10-15% of the worldâ€™s energy consumption. Methods to purify chemicals that are more energy efficient could, if applied to the U.S. petroleum, chemical and paper manufacturing sectors alone, save 100 million tonnes of CO2 emissions and US$4 billion in energy costs annually [1]. Membrane-based separation methods, or other non-thermal ones, can be an order of magnitude more energy efficient than heat-driven separations that use distillation, but these methods are underdeveloped [2].

Given the recent advances in organic solvent nanofiltration (OSN) membranes [3], the focus of this presentation is on OSN membrane cascades for separating organic/hydrocarbon mixtures. As a case study, we investigate the design of a lab-scale membrane cascade for purifying a binary mixture of heptane and hexadecane, with a 75-25 weight fraction. The main objective is to maximize heptane purity in the permeate stream by using a maximum of 10 identical membrane units, each having an effective area of 14 cm.

In a first step, we use a superstructure optimization approach based on mixed-integer nonlinear programming (MINLP) to determine the most promising cascade configurations, in terms of the number of stages in the cascade, the number of membrane units in each stage, and the interconnections thereof, along with certain operating parameters [4,5]. Constraints are imposed for limiting the transmembrane pressure drop and the stage-cut in each unit. The permeate and retentate flux and composition for each membrane unit are predicted by a classical solution-diffusion model, where the permeability coefficients for heptane and hexadecane are estimated based on dedicated experiments for the membranes at hand, and the activity coefficients of heptane and hexadecane are approximated using the UNIFAC method for different mixture compositions. An optimal 3-stage cascade obtained with this approach is shown on the figure below, with GAMS 24.5.6 and the global optimizer BARON 15.9 used to solve the MINLP problem.

In a second step, we implement the optimised OSN membrane cascades experimentally in order to verify the predictions. For the 3-stage cascade shown on the figure, the mismatch between the model predictions and the measurements remains small, within a few percents, thereby validating the model-based optimization approach.

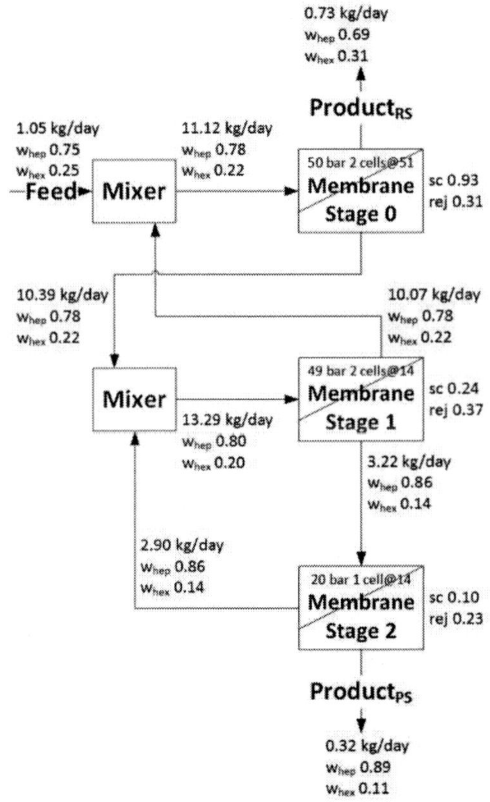

References:

1. U.S. Department of Energy, Advanced Manufacturing Office, "Bandwidth Study on Energy Use and Potential Energy Saving Opportunities in U.S. Petroleum Refining", June 2015.

2. D.S. Sholl, R.P. Lively, "Seven chemical separations to change the world," Nature 532:435-43, 28 April 2016. U.S. Petroleum Refining (US Dept. Energy, 2015).

3. P. Marchetti, M.F.J. Solomon, G. Szekely, A.G. Livingston, "Molecular separation with organic solvent nanofiltration: A critical review," Chemical Reviews 114:10735-10806, 2014.

4. R.H. Qi, M.A. Henson, "Membrane system design for multicomponent gas mixtures via mixed-integer nonlinear programming," Computers & Chemical Engineering 24:2719-2737, 2000.

5. V.S.K. Adi, M. Cook, L.G. Peeva, A.G. Livingston, B. Chachuat, "Optimization of OSN membrane cascades for separating organic mixtures," Proc. 26th European Symposium on Computer Aided Process Engineering (ESCAPE), 12-15 June 2016, PortoroÅ¾, Slovenia.

452857 Solution Strategies for the Dynamic Warehouse Location Under Discrete Transportation Costs

Abstract:

A Mixed-Integer Linear Programming model is proposed to determine the optimal number, location and capacity of the warehouses required to support a 5-year forecast for a business with seasonal demand. Discrete transportation costs, dynamic warehouse contracting, and handling of safety stock are the three main features that make the problem different from the previous reported work in the facility location problem. The most effective modeling alternative for each of these complicating issues was identified. To solve large-scale problems, tightening constraints, approximating reformulations and multistage heuristics are proposed. Solution algorithms that exploit the problem structure were developed. A 2-stage heuristic based on products grouping is able to obtain the optimal solution in 95% of the instances tested. A 3-stage heuristic, using a single mode model as intermediate step, was found to be better suited for very large-scale problems yielding only a 1% deviation from the optimal solution. To take advantage of the feasible solution found using the developed heuristics, decomposition strategies are explored to yield proven optimal solutions.

References:

Melo, M. Teresa, Stefan Nickel, and Francisco Saldanha-Da-Gama. "Facility location and supply chain management– A review." European journal of operational research 196.2 (2009): 401-412.

Bravo, Juan José, and Carlos Julio Vidal. "Freight transportation function in supply chain optimization models: A critical review of recent trends." Expert Systems with Applications 40.17 (2013): 6742-6757.

Daskin, Mark S., Collette R. Coullard, and Zuo-Jun Max Shen. "An inventory-location model: Formulation, solution algorithm and computational results." Annals of operations research 110.1-4 (2002): 83-106.

454275 Multiscale Production Routing in Multicommodity Supply Chains with Complex Production Facilities

Abstract:

In order to increase the efficiency of industrial supply chains, the simultaneous optimization of production and distribution operations is becoming increasingly important. This problem is especially relevant in the industrial gas industry where multiple products (e.g. liquid and gaseous oxygen and nitrogen) are produced in multiple plants, and distributed to a large set of customers. Among the integrated supply chain planning problems, the so-called production routing problem (PRP) (Adulyasak, Cordeau, & Jans, 2015) is the most comprehensive one as it considers production, inventory, distribution, and routing decisions simultaneously. The PRP in its classical form can be formulated as a mixed-integer linear program (MILP) and integrates two well-known problems, namely the lot-sizing problem (LSP) and the inventory routing problem (IRP). The PRP is notoriously difficult to solve; so far, good solutions to instances of relevant sizes have only been obtained by applying tailored metaheuristics.

In this work, we introduce an extension of the PRP, the multiscale production routing problem (MPRP), which considers multicommodity supply chains with complex production facilities. The complexity in the production processes requires the consideration of operational transitions and time-sensitive prices, which are not modeled in a simple LSP formulation. As a result, the MPRP is an even more difficult problem than the classical PRP.

For modeling the MPRP, we propose an MILP formulation involving two different time grids. While a detailed mode-based production scheduling model captures all critical operational constraints on the fine time grid, vehicle routing is considered in each time period of the coarse time grid. In order to solve large instances of the MPRP, we propose an iterative MILP-based heuristic approach that solves the MILP model with a restricted set of candidate routes at each iteration and dynamically updates the set of candidate routes for the next iteration. The results of an extensive computational study show that the proposed algorithm finds high-quality solutions in reasonable computation times

(within an hour), and in large instances, it significantly outperforms a standard two-phase heuristic approach and a solution strategy involving a one-time heuristic pre-generation of candidate routes (Marchetti, et al., 2014). Similar results are achieved in an industrial case study, which considers a real-world industrial gas supply chain with 2 plants, approximately 240 customers, 20 vehicles, and a planning horizon of 4 weeks, resulting in 168 time periods on the fine grid and 56 time periods on the coarse grid.

References

Adulyasak, Y., Cordeau, J. F., & Jans, R. (2015). The production routing problem: A review of formulations and solution algorithms. Computers & Operations Research, 55, 141-152.

Marchetti, P. A., Gupta, V., Grossmann, I. E., Cook, L., Valton, P.-M., Singh, T., Li, T., & André, J. (2014). Simultaneous production and distribution of industrial gas supply-chains. Computers & Chemical Engineering, 69, 39-58.

461232 Supply Chain Planning and Scheduling Approach for Multiproduct Multistage Continuous Plants under Uncertainty

Abstract:

This paper presents a novel MILP-based (Mixed Integer Linear Programming) approach for the planning and scheduling of multiproduct multistage continuous plants in a supply chain network with product demands and price elasticity under uncertainty. For this, we consider a supply chain network where multiple echelon nodes, such as, plants, distribution centers (DCs) and markets, are interconnected to satisfy those customer's demands (Liu et al., 2012). In here, multiproduct continuous plants are in charge to produce several products in an efficient way to supply the DCs, while DCs are in charge to store and deliver these products to the final markets. In general continuous plants, several products must be processed in simultaneous by following a series of production stages. In this particular problem we are considering multiproduct multistage continuous plants, with a single production unit per stage where sequence-dependent changeovers occur in each processing unit between different products. Another important issue of the real supply chain networks is the uncertain behaviour of the system. In this work, two uncertain parameters, the demand and the price elasticity, are taken into account. In this context, planning decisions must be done across the supply chain to deal with the production, storage and delivery issues of the final products.

A hybrid discrete/continuous precedence-based model was proposed for this problem by using the main ideas of the Traveling Salesman Problem (TSP) and global precedence representation (Mendez et al., 2006). For this, the original TSP of Liu et al. (2008, 2009) was reformulated to easily tackle changeover issues without generating additional binary variables. In order to deal with the uncertainty, we adapt the Model Predictive Control (MPC) approach proposed in Liu et al. (2012) for this particular problem. Despite of the efficiency of this solution technique, the final solution reported by this method could be far from the global optimal when many of the decision variables are successively fixed iteration by iteration. Due to this, a Local-Search (LS) approach was developed to improve the solution of the MPC by rescheduling successive

products from the current schedule (Castro et al., 2011). The effectiveness of this alternative solution technique has been demonstrated by solving large scale instances of the problem.

References:

1. Castro, P. M., Harjunkoski, I., & Grossmann, I. E. (2011). Greedy algorithm for scheduling batch plants with sequence dependent changeovers. *AIChE journal, 57*(2), 373-387.

2. Liu, S., Pinto, J. M., & Papageorgiou, L. G. (2008). A TSP-based MILP model for medium-term planning of single-stage continuous multiproduct plants. *Industrial & Engineering Chemistry Research, 47*(20), 7733-7743.

3. Liu, S., Pinto, J. M., & Papageorgiou, L. G. (2009, June). Medium-term planning of multistage multiproduct continuous plants using mixed integer optimisation. In *19th European Symposium on Computer Aided Process Engineering: ESCAPE-19: June 14-17, 2009, Cracow, Poland* (Vol. 26, p. 393). Elsevier.

4. Liu, S., Shah, N., & Papageorgiou, L. G. (2012). Multiechelon supply chain planning with sequence dependent changeovers and price elasticity of demand under uncertainty. *AIChE Journal, 58*(11), 3390-3403.

5. Méndez, C. A., Cerdá, J., Grossmann, I. E., Harjunkoski, I., & Fahl, M. (2006). State-of-the-art review of optimization methods for short-term scheduling of batch processes. *Computers & Chemical Engineering, 30*(6), 913-946.

462437 Safety Stocks Revisited: Terminal Constraints for Closed-Loop Scheduling

Abstract:

Safety stocks have been used in supply chain (SC) management to prevent stockouts, and it has been shown that they also play an important role in scheduling, especially closed-loop scheduling. In the optimal solution for a scheduling model without any safety stock as terminal constraints, inventory levels tend to be very low at the end of horizon, so that the production, transition and inventory holding cost can be minimized [1]. However, if such a solution is implemented, the problem may be infeasible for the following horizon, because the initial inventory is almost depleted, and the demand at the start of the following horizon cannot be satisfied. In the SC literature, a safety stock is generally required for the inventory level of each product at the end of the horizon, based on the statistics of the lead time of SC arcs and demand rate of SC nodes [2]. In some scheduling works, the terminal inventory levels are required to be equal to the initial value or to one of the values in a cyclic solution, and these would lead to recursive feasibility [3]. However, all of the aforementioned approaches neglect the relationship of the inventory levels among products, or among materials in different stages (echelons). For instance, in a single-stage, two-product network, if the terminal level of one product is higher, we can possibly decrease the other. Thus, in this work, we present methods to generate terminal constraints based on the specific network structure. The proposed methods account for the interdependencies among products or materials in different stages. These terminal constraints are linear, which can be easily incorporated in any mixed integer programming (MIP) model.

We generalize our methods from simple network structures to the more complex ones. First, we focus on the single-unit, multi-product problem. By solving an auxiliary linear programming model, we obtain the frequencies of tasks in a campaign mode. Using these frequency numbers, we formulate the linear terminal constraints, and we present two different formulations. We also prove that these constraints ensure feasibility for the next optimization problem.

Second, we generalize the first network to the single-stage, multi-unit, multi-product network, in which both homogeneous and heterogeneous units are considered. Third, we study the multi-stage single-product network. Due to the integrality requirement of batches, the original terminal constraints have round-down operators leading to nonlinearities. We show different options to linearize the constraints, either by introducing auxiliary integer variables or by writing stronger terminal constraints. Fourth, based on the logic of the aforementioned simpler networks, we discuss how to formulate terminal constraints for the multi-stage, multi-product network. For the simpler networks, we rigorously prove that the terminal constraints are valid, i.e., the constrained inventory levels ensure the feasibility for future horizons. For the complex networks, we present results from a case-study to show the effectiveness of the proposed terminal constraints. By recursively solving the scheduling model starting from different initial inventory levels and checking the feasibility, we characterize the true feasible region of the terminal inventory levels. Comparing the true feasible region to the feasible region from terminal constraints, we show that most, if not all, of the inventory levels satisfying the terminal constraints lead to recursive feasibility.

References:

[1] Lima, R. M.; Grossmann, I. E.; Jiao, Y. Long-term scheduling of a single-unit multi-product continuous process to manufacture high performance glass. *Comput. Chem. Eng.* **2011**, *35*(3), 554â€"574.

[2] Eppen, G.D.; Martin, R.K. Determining safety stock in the presence of stochastic lead time and demand. *Manag. Sci.* **1988**, *34*(11), 1380-1390.

[3] Subramanian, K.; Maravelias, C.T.; Rawlings, J.B. A state-space model for chemical production scheduling. *Comput. Chem. Eng.* **2012**, *47*, 97-110.

465506 Adaptive Robust Optimization for Tactical-Level Distribution Problems Under Customer Order Uncertainty

Abstract:

Tactical-level distribution operations over a short-term horizon are central to many transportation problems. In such operations, customer orders are often received dynamically over the planning horizon. Each order consists of a demand quantity and a set of time periods (e.g., days) during which service can take place. The supplier must then assign a visit day to each customer order and design routes for each day so as to minimize the overall transportation cost. The plan is usually implemented in a rolling horizon fashion: routes of the first day are executed while new requests are recorded; unfulfilled orders at the end of the first day and new requests accumulated during the day constitute the new portfolio of orders to be considered for scheduling the following day. In contrast to typical vehicle routing problems (VRPs), where the planning horizon consists of one period during which all customer requests must be serviced, in the tactical level scheduling and routing decisions must be made simultaneously over a multi-period horizon. This problem has been referred to as the Dynamic Multi-Period VRP [1] or as the Tactical Planning VRP [2] and is highly relevant to systems in which logistics and other services are provided by appointment. For example, applications arise in food distribution [1], automobile transportation [3] and in maintenance scheduling [4].

In most existing approaches to address this problem, no knowledge is assumed about potential customers that may call in to request service in future time periods; in other words, the uncertainty in future customer requests is completely ignored. While this setup may be unavoidable in some contexts, companies often have large amounts of historical data, which can be used to obtain suitable order forecasts. It is thus possible to take advantage of this information and generate better, risk-averse solutions. In fact, not accounting for this uncertainty may lead to situations which can either be (i) infeasible because too many vehicles were required on some day, or (ii) too expensive in terms of routing costs. In this regard, [4] consider uncertainty in future requests; they utilize a probabilistic description where, at any time period, they assume the probability of demand in subsequent time periods of every possible customer that does not have a pending request. They consider an adaptive policy by solving a VRP subproblem to identify the set of pending customers that should be serviced on the first day of the planning horizon, along with the actual vehicle routes.

In this work, we propose a novel Robust Optimization approach in which the uncertainty in future customer requests is captured via a purely discrete uncertainty set. This uncertainty set consists of a finite (but possibly very large) number of discrete scenarios and admits a polyhedral representation. Each scenario represents a possible combination of customer requests that may potentially realize during the planning horizon. This model of uncertainty is especially attractive in cases where historical record is not rich enough to regress out detailed probabilistic information. It is also very flexible, as it allows us to capture various practically meaningful correlations that link customer requests in temporal, geographical or other terms.

Our solution approach views the problem as a two-stage robust optimization problem, in which the routing and scheduling decisions for the known customers (i.e., customers who have a pending service request) are first-stage decisions while the assignment of future customer requests to the routing plan constitute second-stage decisions. We design routes for each day of the planning horizon such that customers who have already placed an order are visited exactly once throughout the planning horizon. Moreover, any realization of future customer requests from the uncertainty set can be assigned to the designed routing plan in a way that all vehicle capacities are respected. We guarantee that the routing decisions to which we commit to here-and-now will not render the overall operation infeasible when an admissible realization of future customers materializes.

We remark that a static robust policy, where all assignment decisions are made here-and-now, would be infeasible or overly conservative in this setting for all but the most favorable instances. To that end, we consider an adaptive strategy where the assignment decisions are allowed to adjust on the value of the observed realization, and we present algorithms to solve the resulting robust models to guaranteed optimality. We investigate the tradeoffs in using an adaptive strategy in terms of cost savings and computational time. Our study allows the decision-maker to quantify the cost increase in implementing a robust plan and to decide the level of robustness that is appropriate to the application. For the instances we considered in our computational experiments, robust routing plans can be obtained with a modest cost increase of less than 5%.

[1] M. Wen, J.-F. Cordeau, G. Laporte, and J. Larsen (2010). The dynamic multi-period vehicle routing problem. *Computers & Operations Research*, 37:1615—1623.

[2] R. Baldacci, E. Bartolini, A. Mingozzi, and A. Valetta (2011). An exact algorithm for the period routing problem. *Operations Research*, 59(1):228—241.

[3] J.-F. Cordeau, M. Dell'Amico, S. Falavigna, and M. Iori (2015). A rolling-horizon algorithm for auto-carrier transportation. *Transportation Research Part B: Methodological*, 76:68—80.

[4] H. Tang, E. Miller-Hooks, and R. Tomastik (2007). Scheduling technicians for planned maintenance of geographically distributed equipment. *Transportation Research Part E: Logistics and Transportation Review*, 43(5):591—609.

[5] M. Albareda-Sambola, E. Fernandez, and G. Laporte (2014). The dynamic multiperiod vehicle routing problem with probabilistic information. *Computers & Operations Research*, 48:31—39.

470506 Optimal Integrated Water Management and Shale Gas Supply Chain Planning Under Uncertainty

Abstract:

The production of unconventional fossil fuels, i.e. shale gas, has gained increasing importance in the energy sector. For instance, under the New Policy Scenario (the International Energy Agency (IEA) base line scenario), the IEA estimated that shale gas production will increase by almost two-fold (from 331 billion cubic meters (bcm) to 941 bcm) from 2013 to 2040 [1]. However, the development of shale gas resources imposes important challenges and potential risks to water resources. Specifically, the depletion, potential contamination, and degradation of both underground and surface water sources have been raised as impediments to the wider development of shale gas resources [2, 3]. Additionally, the design and planning of both water management and shale gas supply chains involve uncertainties that are inherent to the system: these include shale play productivity and wastewater composition, as well as uncertainties associated with market conditions, i.e. gas demand and prices.

This work deals with the development and implementation of a two-stage stochastic optimization approach for the design and planning of the integrated water management and shale gas supply chain. First, a global sensitivity analysis is carried out using the framework developed by the authors [4, 5] to assess and rank uncertain parameters in the integrated supply chain. Then, a Monte Carlo sampling technique is combined with Sobol's sensitivity indices [6, 7] to compute the effect of uncertainties on the net present value (NPV) performance metric used in the aforementioned framework. Based on the outcomes of the sensitivity analysis, a two-stage stochastic model involving a Mixed Integer Linear Program (MILP) is developed. The first-stage decisions in this model consist of the investment in drilling and fracturing operations as well as in transportation and processing facilities for both water and shale gas. The second-stage decisions are associated with operational issues related to water management and gas delivery. The potential benefits of modeling uncertainty and implementing stochastic models are quantified through two metrics: *expected value of perfect information* (EVP) and *value of stochastic solution* (VSS) [8]. Additionally, scenario

reduction approaches are evaluated to mitigate computational challenges.

References

[1] IEA. "*World Energy Outlook*" 2015. doi:10.1787/weo-2014-en.

[2] Vidic RD, Brantley SL, Vandenbossche JM, Yoxtheimer D, Abad JD. "*Impact of shale gas development on regional water quality*". Science 2013;340:1235009. doi:10.1126/science.1235009.

[3] Olmstead SM, Muehlenbachs LA, Shih J, Chu Z, Krupnick AJ. "*Shale gas development impacts on surface water quality in Pennsylvania*". Proceedings of the National Academy of Sciences of the United States of America 2013;110:4962–7. doi:10.1073/pnas.1213871110.

[4] Calderón AJ, Guerra OJ, Papageorgiou LG, Siirola JJ, Reklaitis G V. "*Preliminary Evaluation of Shale Gas Reservoirs: Appraisal of Different Well-Pad Designs via Performance Metrics*". Industrial & Engineering Chemistry Research 2015;54:10334–49. doi:10.1021/acs.iecr.5b01590.

[5] Guerra OJ, Calderón AJ, Papageorgiou LG, Siirola JJ, Reklaitis G V. "*An Optimization Framework for the Integration of Water Management and Shale Gas Supply Chain Design*". Computers & Chemical Engineering 2016;(*in press*, doi: 10.1016/j.compchemeng.2016.03.025).

[6] I. M. Sobol, "*Global sensitivity indices for nonlinear mathematical models and their Monte Carlo estimates,*" Mathematics and Computers in Simulation, vol. 55, pp. 271–280, 2001.

[7] A. Saltelli, S. Tarantola, and K. P.-S. Chan, "*A Quantitative Model-Independent Method for Global Sensitivity Analysis of Model Output,*" *Technometrics*, vol. 41, no. May, pp. 39–56, 1999.

[8] J. Birge and F. Louveaux, *Introduction to stochastic programming.* 2011.

472035 Surrogate-Based Derivative Free Optimization of a Multi-Enterprise Supply Chain

Abstract:

Surrogate-based Derivative Free Optimization of a Multi-Enterprise Supply Chain

Nihar Sahay, Lisia Dias and Marianthi Ierapetritou

Department of Chemical and Biochemical Engineering, Rutgers University, Piscataway, NJ

In the present scenario, with the growing interactions among the different enterprises constituting supply chain networks, it is rare for a single enterprise to operate its supply chain independently. The advancements in web technologies have also helped get rid of the limitations with multi-enterprise global supply chain networks. Multi-enterprise supply chain operations can be significantly different from that in a monolithic supply chain belonging to a single enterprise. It is important to take into consideration the effects of interactions among the different enterprises when modeling such networks.

Supply chain optimization strategies that consider centralized networks might not necessarily be suitable for the individual entities in the network. Such factors gain more importance when the network consists of more than one enterprise and the optimal solution for the overall network is not beneficial for all the participating enterprises. Under such circumstances, the model needs to take into consideration the interaction among the entities in terms of transfer prices, coalition formation, negotiations. Among the initial studies discussing multi-enterprise networks, D'Amours et al.[1] proposed a network approach with multiple firms. The firms are selected and scheduled with the objective of fulfilling demand at minimum cost and collaborative approach is shown to be more profitable. During the late 1990's, many new supply chain paradigms evolved which were studied by many authors.[2-4] Among the different ways enterprises may interact with each other for transactions, auctions have been gained some emphasis recently.[5-8]Although buyers usually rely on contracts with suppliers for the critical products, auctions have become an efficient and effective means to procure non-critical products. They help achieve lower acquisition costs as well as enable new suppliers to enter the market. Another advantage such marketplaces offer is that they allow the buyers and sellers to determine price elasticity and dynamically adjust their prices based on supply and demand. In the recent past, there has been a growing body of literature in the field of operations management.

In this work we consider a multi-enterprise supply chain network. The production sites and warehouses are considered to belong to one particular enterprise while the retailers are different enterprises. The transaction between warehouses and retailers happens through auctions. A multi-attribute double auction is considered where the attributes of the auction are price and quantity. The auction mechanism consists of two phases where the first stage determines the buyer and the seller while the second stage uses the Nash Bargaining solution to determine the final trade. An agent-based simulation model is used to depict the representation of the supply chain dynamics and study the interactions among the enterprises. Additionally optimal warehouse capacities are obtained for the minimization of total cost. A kriging based derivative free optimization methodology is proposed. Different adaptive sampling methods including Expected Improvement and Goal Seeking are compared. Local search is performed using the trust region method starting from the different initial points obtained by clustering.

References:

1. D'Amours S, Montreuil B, Lefrançois P, Soumis F. Networked manufacturing:: The impact of information sharing. *International Journal of Production Economics.* 1999;58(1):63-79.

2. Bartezzaghi E. The evolution of production models: is a new paradigm emerging? *International Journal of Operations & Production Management.* 1999;19(2):229-250.

3. Lehtinen U. Subcontractors in a partnership environment:: A study on changing manufacturing strategy. *International Journal of Production Economics.* 1999;60–61:165-170.

4. Lindsey JH, Samuelson W, Zeckhauser R. Selling Procedures with Private Information and Common Values. *Management Science.* 1996;42(2):220-231.

5. Chen F. Auctioning Supply Contracts. *Management Science.* 2007;53(10):1562-1576.

6. Beil DR, Wein LM. An Inverse-Optimization-Based Auction Mechanism to Support a Multiattribute RFQ Process. *Manage. Sci.* 2003;49(11):1529-1545.

7. Gallien J, Wein LM. A Smart Market for Industrial Procurement with Capacity Constraints. *Management Science.* 2005;51(1):76-91.

8. Moyaux T, McBurney P, Wooldridge M. A supply chain as a network of auctions. *Decision Support Systems.* 2010;50(1):176-190.

472522 A Leader-Follower Game-Based Life Cycle Optimization Framework and Shale Gas Supply Chain Application

Abstract:

In the past decade, accompanying the rapid development of shale gas industry, concerns are raised regarding the environmental impacts of shale gas production, among which the greenhouse gas (GHG) emissions is of special interest to both academia and industry. There have been a huge body of literature focusing on evaluating the life cycle GHG emissions of shale gas, most of which are based on life cycle assessment methodology [1, 2]. Meanwhile, some publications focus on optimization of shale gas supply chains; issues such as supply chain planning [3], process design [4], water management [5], uncertainties [6], and GHG emissions [7] were addressed. However, all of these literature consider a centralized model, where a single decision-maker is assumed with absolute power in the supply chain design and management. Nevertheless, in practice the management of a shale gas supply chain is normally decentralized and run by different companies [8]. As a consequence, the optimal strategy obtained from a centralized model can be practically infeasible [9]. Therefore, there is an urgent need to properly address the non-cooperative relationship between multiple stakeholders in the life cycle optimization of a shale gas supply chain [10].

To address this challenge, we propose a novel leader-follower game-based LCO modeling framework, which integrates both the leader-follower Stackelberg game and LCO approach [9, 11, 12]. This holistic modeling framework enables us to simultaneously address the trade-offs between conflicting objectives as well as the interactions between different players. As the initiator in the shale gas supply chain, the upstream shale gas producer is identified as the leader. Due to the key role of leader in a game, the producer not only enjoys the priority to make decisions first, but senses the responsibility to mitigate the life cycle GHG emission embedded in the final product in addition to pursuing profit. The midstream player shale gas processor is identified as the follower, who will take actions rationally according to the leader's decisions to pursue its own profit. The resulting problem is a multiobjective mixed-integer bilevel linear programming (MIBLP) problem that cannot be solved directly by any off-the-shelf solver. To address this computational challenge, we preent a novel projection-based reformulation and decomposition algorithm for efficient solution of this problem.

To illustrate the application of proposed modeling framework and solution algorithm, we present a case study based on Marcellus shale play, where a "cradle-to-gate" system boundary is chosen covering the whole shale gas production system from upstream shale sites to midstream processing plants. Multiple strategic decisions are taken into account, including the well drilling schedule, water management, installation of gathering pipelines, allocation, capacity and design of processing plants, and selection of processing contract. In order to demonstrate the advantage of the game-based model, we further present a detailed comparison with classical centralized models. Based on a case study of Marcellus shale play, the leader's net present value (NPV) ranges from $43.0 M to $154.7 M, and the corresponding total greenhouse gas (GHG) emissions range from 1.03 Mt CO_2-eq to 2.37 Mt CO_2-eq. Moreover, the centralized model gives over-optimistic solutions. The highest total NPV obtained from centralized model is 10.2% higher than that in the game-based model, and the lowest GHG emission is 3.4% lower.

References

[1] C. L. Weber and C. Clavin, "Life Cycle Carbon Footprint of Shale Gas: Review of Evidence and Implications," *Environmental Science & Technology,* vol. 46, pp. 5688-5695, 2012.

[2] G. A. Heath, P. O'Donoughue, D. J. Arent, and M. Bazilian, "Harmonization of initial estimates of shale gas life cycle greenhouse gas emissions for electric power generation," *Proceedings of the National Academy of Sciences,* vol. 111, pp. E3167-E3176, 2014.

[3] D. C. Cafaro and I. E. Grossmann, "Strategic planning, design, and development of the shale gas supply chain network," *AIChE Journal,* vol. 60, p. 21, 2014.

[4] C. He and F. You, "Toward more cost-effective and greener chemicals production from shale gas by integrating with bioethanol dehydration: Novel process design and simulation-based optimization," *AIChE Journal,* vol. 61, pp. 1209-1232, 2015.

[5] L. Yang, I. E. Grossmann, M. S. Mauter, and R. M. Dilmore, "Investment optimization model for freshwater acquisition and wastewater handling in shale gas production," *AIChE Journal,* vol. 61, pp. 1770-1782, 2015.

[6] J. Gao and F. You, "Deciphering and handling uncertainty in shale gas supply chain design and optimization: Novel modeling framework and computationally efficient solution algorithm," *AIChE Journal,* vol. 61, pp. 3739-3755, 2015.

[7] J. Gao and F. You, "Shale Gas Supply Chain Design and Operations toward Better Economic and Life Cycle Environmental Performance: MINLP Model and Global Optimization Algorithm," *ACS Sustainable Chemistry & Engineering,* vol. 3, pp. 1282-1291, 2015.

[8] G. Cachon and S. Netessine, "Game Theory in Supply Chain Analysis," in *Handbook of Quantitative Supply Chain Analysis*. vol. 74, D. Simchi-Levi, S. D. Wu, and Z.-J. Shen, Eds., ed: Springer US, 2004, pp. 13-65.

[9] H. Von Stackelberg, *Market structure and equilibrium*: Springer Science & Business Media, 2010.

[10] D. J. Garcia and F. You, "Supply chain design and optimization: Challenges and opportunities," *Computers & Chemical Engineering,* vol. 81, pp. 153-170, 2015.

[11] D. Yue and F. You, "Game-theoretic modeling and optimization of multi-echelon supply chain design and operation under Stackelberg game and market equilibrium," *Computers & Chemical Engineering,* vol. 71, pp. 347-361, 2014.

[12] D. J. Yue, M. A. Kim, and F. Q. You, "Design of Sustainable Product Systems and Supply Chains with Life Cycle Optimization Based on Functional Unit: General Modeling Framework, Mixed-Integer Nonlinear Programming Algorithms and Case Study on Hydrocarbon Biofuels," *Acs Sustainable Chemistry & Engineering,* vol. 1, pp. 1003-1014, 2013.

2016 AIChE Annual Meeting
189 CAST Rapid Fire Session : II

Simulation of Propane Dehydrogenation (Catofin) Process

Byeonggil Lyu and Il Moon*
Yonsei University, Seoul, Korea
longs25@yonsei.ac.kr

Monday, November 14, 2016
Carmel II (Hotel Nikko San Francisco)

16AIChE
Annual Meeting, San Francisco, CA

Introduction

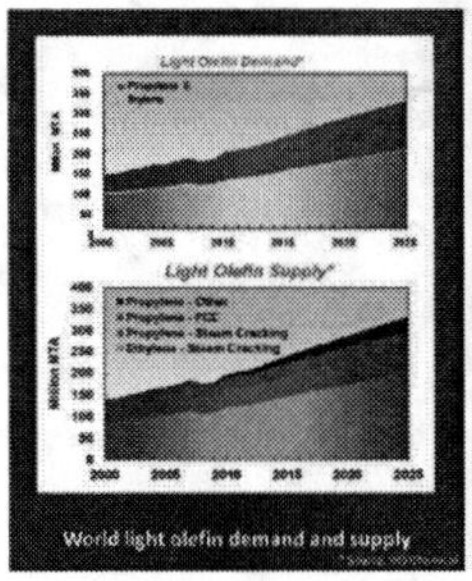
World light olefin demand and supply

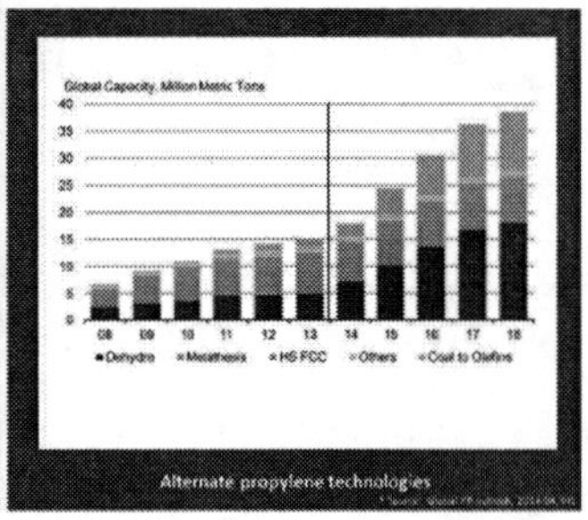

Alternate propylene technologies

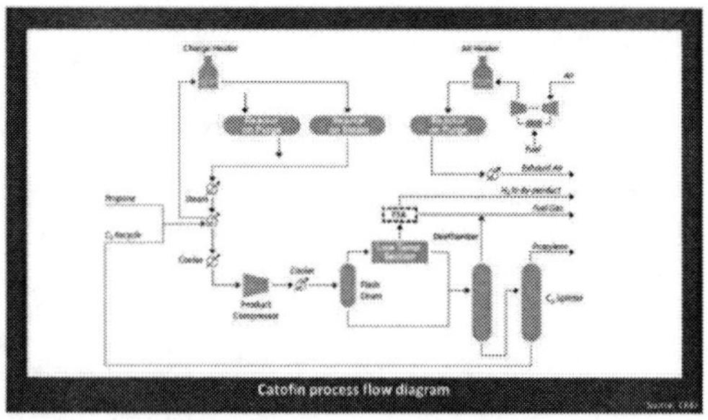
Catofin process flow diagram

Growing propylene supply-demand gap

➡ **On-purpose propylene production**

➡ **Propane dehydrogenation is the most**

- ✓ **Continuous process with multiple fixed bed reactor**
- ✓ **Operating higher temperature(500-700℃), lower pressure(≤1atm)**
- ✓ **Coke formation deactivates catalysts, reduces selectivity**
- ✓ **Cyclic process includes [1]reaction period, [2]discharge of the reactor, and [3]regeneration of the catalyst**

Objective

- ✓ **Cyclic steady state simulation of Catofin process, 4 multi reactor system**
- ✓ **Find optimal reaction condition for maximizing profit**

Cyclic steady state simulation

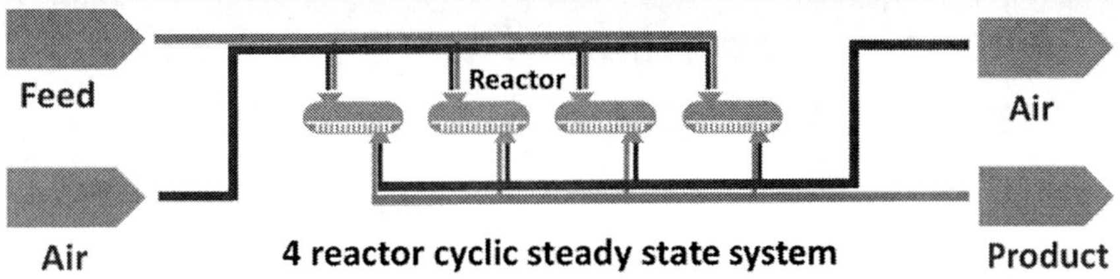

4 reactor cyclic steady state system

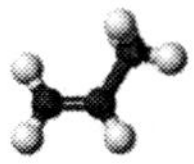

Reaction Kinetics

❑ ***Dehydrogenation Reaction :*** $C_3H_8 \Leftrightarrow C_3H_6 + H_2$

✓ Langmuir-Hinshelwood-Hougen-Watson reaction mechanism
✓ Assumes negligible propane adsorption

$$(-r_{C_3H_8}) = \frac{k_1 a[P_{C_3H_8} - (\frac{P_{C_3H_6}P_{H_2}}{K_{eq}})]}{1 + K_{C_3H_6}P_{C_3H_6}}$$

$$k_1 = k_{01}\exp\left(\frac{-E_{a_1}}{R}\left(\frac{1}{T} - \frac{1}{T_m}\right)\right)$$

a : catalyst activity

$$K_{C_3H_6} = k_{02}\exp\left(\frac{-\Delta H}{R}\left(\frac{1}{T} - \frac{1}{T_m}\right)\right)$$

$k_{01} = 0.18576\ kmol/(kg \cdot h)$
$E_{a_1} = 3.55E4\ J/mol$
$k_{02} = 3.45E-3\ kmol/(kg \cdot h)$
$\Delta H = -5.95E5\ J/mol$
$T_m = 823.15\ K$

❑ ***Coke formation reaction :*** $C_3H_8 \rightarrow CH_{0.5} + 3.25H_2$

✓ Monolayer-multilayer-coke growth model

$C = C_m + C_M$
(Total coke = monolayer coke + multilayer coke)

$$\frac{dC}{dt} = \frac{dC_m}{dt} + \frac{dC_M}{dt}$$

$$\frac{dC_m}{dt} = k_2(C_{max} - C_m)$$

$$\frac{dC_M}{dt} = k_3 C_m$$

$$k_2 = k_{02}\exp\left(\frac{-E_{a_2}}{R}\left(\frac{1}{T} - \frac{1}{T_m}\right)\right)$$

$$k_3 = k_{03}\exp\left(\frac{-E_{a_3}}{R}\left(\frac{1}{T} - \frac{1}{T_m}\right)\right)$$

$k_{02} = 0.00242s^{-1}$
$k_{03} = 0.000357s^{-1}$
$E_{a_2} = 221kJ/mol$
$E_{a_3} = 325.8kJ/mol$

❑ ***Catalyst activity***

$$a = \exp(-\alpha_1 C_m)\exp(-\alpha_2 C_M)$$

$$k_4 = k_{04}\exp\left(\frac{-E_{a_4}}{R}\left(\frac{1}{T} - \frac{1}{T_m}\right)\right)$$

$E_{a_4} = 308kJ/mol$
$\alpha_1 = 813$
$\alpha_2 = 289$
$C_{max} = 0.000682\ mg\ coke\ /\ mg\ catalyst$

❑ ***Propane cracking reaction :*** $C_3H_8 \rightarrow CH_4 + C_2H_4$

$$(-r_{C_3H_8}) = k_4 P_{C_3H_8}$$

(Gascón, Téllez et al. 2008)

❖ Operation schedule and simulation result

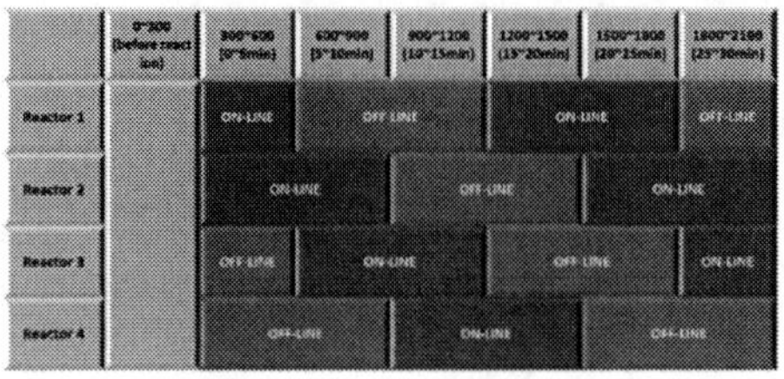

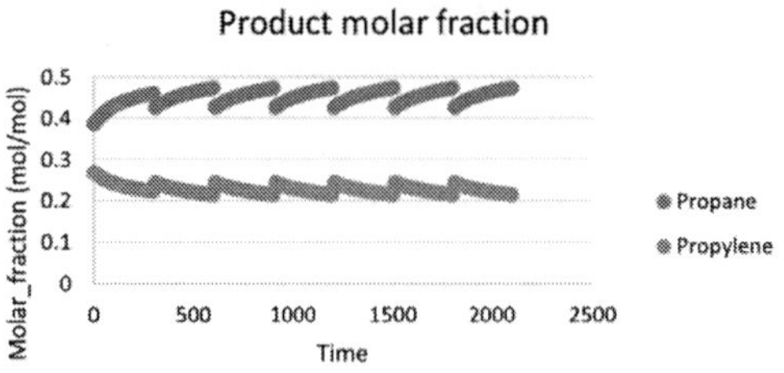

Optimization&conclusion

❖ Sensitivity analysis

Propane Conversion	0.5atm	0.75atm	1atm
575°C	16%	17%	18%
600°C	36%	38%	40%
625°C	56%	61%	64%

Propene Selectivity	0.5atm	0.75atm	1atm
575°C	90%	85%	82%
600°C	89%	86%	82%
625°C	83%	81%	78%

9 case studies with changing reaction temperature and pressure

✓ **Propane conversion increases with higher temperature and pressure**
✓ **Propylene selectivity decreases with higher temperature and pressure**

Need to find optimal operating condition

❖ Optimization

❑ **Objective function**
✓ *profit = propylene_price * propylene_produced – propane_price * propane_consumed*

❑ **Control variables**
✓ *Reaction temperature : [848.15 : 898.15] (K)*
✓ *Reaction pressure : [0.25 : 1] (atm)*

❑ **Constraint**
✓ *Conversion : higher than 30%*
✓ *Selectivity : higher than 85%*

	Initial value	Final value
Pressure	0.75 atm	0.40 atm
Temperature	873.15 K	886.35 K
Profit	492.53 $/kg	642.48 $/kg

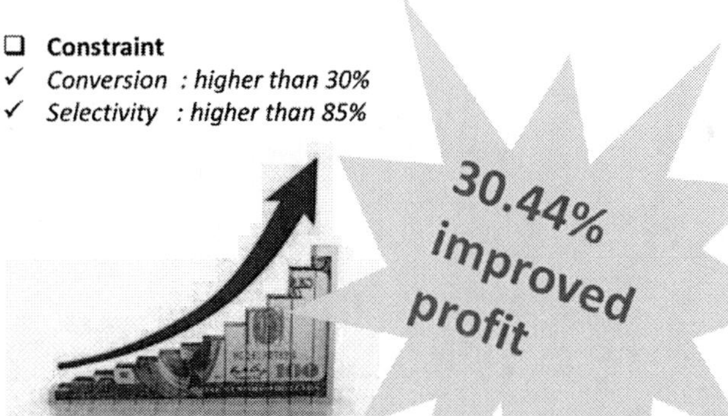

463134 Product Dynamic Transitions Using a Derivative-Free OptimizationÂ Trust-Region Approach

Abstract:

Traditionally the optimization of processing systems has relied on the availability of an explicit
model together with the corresponding gradient information. However, there are some practical scenarios such as (a) non-differentiable systems, (b) physical experimental systems, (c) simulation environments and (d) reduced order systems where such a model and its gradient are not available. Under these scenarios the deployment of derivative-free optimization strategies provide an alternative manner to cope with the optimization of such systems. In particular, in this
work we deploy a derivative-free optimization trust region approach to deal with the product dynamic optimization problem of processing systems. To this aim, we use a closed-loop model predictive control strategy where the system to be optimized is embedded
in a black-box dynamic simulation environment. The results demonstrate that black-box dynamic models can be dynamically optimized assuming that the number of decision variables is not large. The first-principles dynamic model of
a binary distillation column embedded in the ASPEN dynamic simulation environment was deployed as our black-box dynamic model, to demonstrate the advantages of solving product dynamic transition problems when an explicit model of the dynamic model and/or its gradient information are not available.

470591 Advanced Modeling of Tissue:Blood Partition Coefficients for Industrial Chemicals

Abstract:

The application of Physiologically Based BioKinetic (PBBK) models in the risk assessment arena is limited due to the lack of the generic character of these models. A critical limiting factor of describing ADME processes for a large chemical space is the proper parameterization for "data poor" compounds. In order to expand the applicability of PBBK models, so that the chemical space is covered as much as possible, input parameters of these models are predicted using advanced Quantitative Structure-Activity Relationships (QSARs). In silico approaches, including QSARs, are widely used for the estimation of the above physicochemical and biochemical properties, biological effects as well as understanding the physicochemical features governing a biological response. QSARs are described as regression or classification models, which form a relationship between the biological effects and chemistry of each chemical and comprise the activity data to be modeled, the data with which to model and a method to formulate the model.

Several approaches incorporating QSARs have been proposed for the prediction of partition coefficients for PBBK modeling, including (a) the Peyret, Poulin and Krishnan algorithm, which is based on the fractional content of cells, interstitial fluid in tissue, plasma in blood, erythrocyte in blood, tissue lipids and the lipophilicity of the compound of interest; (b) the molecular fractions algorithm proposed by Béliveau et al. that takes into account the frequency of occurrence of the several molecular fragments of the compounds and (c) Abraham's solvation equation for estimating biological properties, which takes into account the excess molar refraction, the compound dipolarity/polarizability, the solute effective or summation hydrogen-bond acidity, the solute effective or summation hydrogen-bond basicity and the McGowan characteristic volume that can trivially be calculated for any solute simply from a knowledge of its molecular structure.

The methodological approach presented in this study proposes the modeling of tissue/blood partition coefficients for five main human tissues (muscle, kidney, adipose, liver, brain) using PaDEL Descriptor and QSARINS. PaDEL Descriptor is an open source software for the calculation of 1D, 2D, 3D molecular descriptors and fingerprints of chemical compounds. QSARINS is used for the development of QSAR models, based on Multiple Linear Regression (MLR) by Ordinary Least Squares (OLS) as modeling method and Genetic Algorithm (GA) for descriptors' selection. In QSARINS, models are analysed using tools such as Principal Component Analysis (PCA), fitting, internal and external validation criteria and applicability domain procedure. Users can browse through different options, ending up with a robust and reliable model according to the OECD principles.

The first step of QSAR modeling was the preparation of input data, which included the experimental values of the tissue/blood partition coefficients and the molecular descriptors of the corresponding chemical compounds. The dataset was consisted of 33 environmental chemical compounds, which were randomly splitted to a training and a prediction set. The splitting was based on random selection through property sampling, performed by ordering the chemicals according to their descending experimental values. The prereduction process was followed for the derived PaDEL descriptors in order to avoid the semi-constant and intercorrelated ones. A set of 435 descriptors for each chemical compound was used for the development and analysis of QSAR models.

The next step was the dataset analysis and the development of QSAR models. The distribution of chemical compounds in the chemical space was explored using PCA. The score plot indicated that the molecules were clustered by structures and the loading plot showcased the most influential descriptors for the chemicals' categorization. As mentioned before, the statistical method of MLR, combined with OLS, was used for the development of the models. Variable selection was done by means of a genetic algorithm, which aimed to find the best combination of variables for the derived models. A large number of models was developed and ordered according to their fitting performance. In order to evaluate the models' validity, internal (Leave One Out (LOO) and Leave Many Out (LMO) technique, Y-scrambling) and external validation methods were applied. The selection of the best model for each tissue/ blood partition coefficient was based on the Multi-Criteria Decision Making (MCDM) value, which summarized the fitting, cross validation and external validation criteria.

The fitting performance (R^2) of the selected models for predicting muscle, kidney, adipose, liver and brain/ blood partition coefficient was 0.92, 0.92, 0.97, 0.94 and 0.96, respectively. The LOO technique ( ) indicated that models' performance in predictions was equal to 0.88, 0.90, 0.96, 0.92 and 0.94, while the LMO technique ( ) resulted in 0.86, 0.89, 0.95, 0.92 and 0.93, respectively. The external validation value ( ) was found to be 0.56, 0.62, 0.98, 0.81 and 0.81 for muscle, kidney, adipose, liver and brain/ blood partition coefficients, respectively. The absence of chance correlation was confirmed by the low values, obtained from the Y-scrambling method. The Root Mean Square of Error (RMSE) for the training set was calculated and ranged from 0.08 to 0.16, while for the prediction set ranged from 0.18 to 0.27. The Applicability Domain (AD) analysis showed that there were not outliers, verifying the reliability of each of the developed QSAR models.

The proposed models for the estimation of tissue/blood partition coefficients were checked for their fitting, validity and applicability. It was found that they are stable, reliable and capable to predict physicochemical parameters of "data poor" chemical compounds that fall within the applicability domain. The developed predictive models could serve as a tool to fill in data gaps of environmental chemicals with unknown values of tissue/blood partitioning. In this way, the animal testing and experiments could be reduced and the wide use of PBBK models could be reinforced. In conclusion, the "safe by design" concept for environmental chemicals is supported, by allowing the successful prediction of toxicokinetic behavior based on molecular parameters, promoting green chemistry and cost saving of product development.

Optimization of Dynamic Systems Including Ordinary and Fractional Differential Equations

Urmila M. Diwekar (with Julio C. Barrera-Martinez, Edgar O. Calderon-Gonzalez and Vicente Rico-Ramirez)

urmila@vri-custom.org

Vishwamitra Research Institute
Instituto Tecnologico de Celaya, Mexico

Fractional Calculus as a Modeling Tool

- ***Fractional Calculus (FC) is a generalization of ordinary differentiation and integration to an arbitrary NON INTEGER order.***

$$\frac{d^\alpha f}{dx^\alpha} = ? \qquad if\ \alpha = \frac{1}{2}, \frac{1}{3}, 0.89, \ldots any\ non\text{-}integer$$

- ***FC has been applied as a useful modeling tool and is generally used to represent the behavior of state variables presenting memory effects.***
- ***The several definitions of fractional integration and differentiation involve convolution integrals (non local property of operators) .***

$${}_aI_t^\alpha f(t) = \frac{1}{\Gamma(\alpha)} \int_a^t (t-\tau)^{\alpha-1} f(\tau) d\tau \qquad {}_aD_t^\alpha f(t) = \frac{d}{dt} {}_aI_t^{1-\alpha} f(t) = \frac{1}{\Gamma(1-\alpha)} \left(\frac{d}{dt}\right) \int_a^t (t-\tau)^{-\alpha} f(\tau) d\tau$$

- ***Literature reports successful applications of FC to dynamic systems in chemical engineering: Process Control (Fractional controllers), Transport Phenomena (Mass and momentum diffusion processes), Fractional Kinetics (Non-conventional chemical reactions)***

Fractional and Ordinary Optimal Control Problems

- *Dynamic systems including differential equations within an optimization framework result in Optimal Control Problems (OCP)*
- *Literature reports the use of Calculus of Variations and the Maximum Principle to derive the Optimality Conditions (Euler-Lagrange Equations) for both Fractional and Ordinary OCP*
- *Numerical techniques for the integration of Fractional Differential Equations have also been developed*

Our Basic Ideas:

- *Large scale dynamic systems will SIMULTANEOUSLY involve both ordinary and fractional differential equations (Fractional-Ordinary OCP)*

$$y=\begin{bmatrix} y_1 \\ y_2 \end{bmatrix} \qquad \frac{dy_1}{dt}=f(y_1,y_2,t) \qquad y_1(t_0)=y_{1,0} \qquad {}_0^C D_t^\alpha y_2 = g(y_1,y_2,t) \qquad y_2(t_0)=y_{2,0}$$

- *Numerical techniques for fractional problems can be applied to ordinary cases (by simply using integer orders), but the calculations would involve unnecessary significant computational efforts due to the non-local definitions of fractional operators*

⇓

- *An efficient integrated theoretical/numerical strategy is needed to approach fractional-ordinary OCP*

16AIChE

Fractional-Ordinary Optimal Control Problems

This work: An integrated approach has been developed and implemented

- *Theoretical optimality conditions result in a two-point boundary value problem including both fractional and ordinary differential equations*

Objective $J(u)=\varphi(y_1(t_f),y_2(t_f))+\int_{t_0}^{t_f} L(y_1(t),y_2(t),u(t),t)dt \qquad y=\begin{bmatrix} y_1 \\ y_2 \end{bmatrix}$

Fractional-Ordinary State Equations

$$\frac{dy_1}{dt}=f(y_1,y_2,t) \qquad y_1(t_0)=y_{1,0}$$

$${}_0^C D_t^\alpha y_2 = g(y_1,y_2,t) \qquad y_2(t_0)=y_{2,0}$$

Fractional-Ordinary Adjoint Equations

$$-\frac{d\lambda_1}{d}=\frac{\partial L}{\partial y_1 t}+\lambda_1^T\frac{\partial f}{\partial y_1}+\lambda_2^T\frac{\partial g}{\partial y_1} \qquad \lambda_1^T(t_f)=\left.\frac{\partial\varphi}{\partial y_1}\right|_{t_f}$$

$${}_tD_{t_f}^\alpha \lambda_2=\frac{\partial L}{\partial y_2}+\lambda_1^T\frac{\partial f}{\partial y_2}+\lambda_2^T\frac{\partial g}{\partial y_2} \qquad {}_tD_{t_f}^{\alpha-1}\lambda_2^T(t_f)=\left.\frac{\partial\varphi}{\partial y_2}\right|_{t_f}$$

Optimal Control

$$\frac{\partial L}{\partial u}+\lambda_1^T\frac{\partial f}{\partial u}+\lambda_2^T\frac{\partial g}{\partial u}=0$$

Fractional-Ordinary Hamiltonian

$$H(y_1,y_2,\lambda_1,\lambda_2,t)=L+\lambda_1^T f+\lambda_2^T g$$

- *An integrated numerical strategy combining the gradient method and the Predictor-Corrector technique for ordinary and fractional integrations has been implemented*

The approach has been tested with a simple dynamic system representing the tequila production process

16AIChE

THIN FILM DEPOSITION USING RAREFIED GAS JET

S. Pradhan

Department of Chemical Engineering
Indian Institute of Science
Bangalore-560 012
India

ABSTRACT

The rarefied gas jet of aluminum is studied at Mach number $Ma = (U_j / \sqrt{kb\ T_j / m})$ in the range $0.01 < Ma < 2$, and Knudsen number $Kn = (1 / (\sqrt{2} \pi d^2\ n_d\ H)$ in the range $0.01 < Kn < 15$, using two-dimensional (2D) Direct Simulation Monte Carlo (DSMC) simulations [1, 2, 3, 4, 5, 6, 7, 8, 9, 10, 11, 12, 13], to understand the flow phenomena and deposition mechanisms in a physical vapor deposition (PVD) process. Here, H is the characteristic dimension, U_j and T_j are the jet velocity and temperature, n_d is the number density of the jet, m and d are the molecular mass and molecular diameter and kb is the Boltzmann constant.

In 2D DSMC simulations we find that the jet velocities both before and after the substrate reduces as the Knudsen number is decreased (increasing the number density). As Knudsen number decreases, the gas jet became much more concentrated on the centerline, and the gas depleted region behind the substrate (flat plate) decreases due to the decrease in the streamwise velocity and hence increased residence time. The decreased Knudsen number (smaller mean free path) confined the gas jet to the centerline by reducing the rate of lateral diffusion. Increasing the jet Mach number also leads to a narrowing of the gas jet. This results from the higher jet velocity, and hence higher momentum of the gas jet along the streamwise direction, which reduces the transverse scattering effects of collisions. We examine the jet streamlines in the vicinity of the substrate by determining the average trajectories along the simulation cells.

The variation of the local fluxes along the streamwise direction away from the jet are studied. The qualitative nature of the local flux at high Mach number $(Ma = 2)$ is similar to those in the incompressible limit $(Ma = 0.01)$. These include the initial fast decay, then slow variation, and finally rapid decay near the substrate. However, there are important differences. The amplitudes of the local flux increase as the Mach number increases. The flux decreases by an order of magnitude before reaching to the substrate. There is significant velocity and temperature slip ((Pradhan & Kumaran, *J. Fluid Mech*, 2011); (Kumaran & Pradhan, *J. Fluid Mech*, 2014)) at the solid surfaces of the substrate. In a compressible rarefied jet flow, we determine the local flux deposited on the substrate surface with the thermal accommodation coefficient equal to 1. The gas atom concentration in the jet slowly decreases due to the infrequent scattering, and deposition occurs due to the reduced diffusion distance. The gas flux profile at the front and rear surfaces of the substrate is determined, and show significant dependence upon the Knudsen number.

An important finding is that the capture width (cross-section of the gas jet deposited on the substrate) is symmetric around the centerline of the substrate, and decreases with increased Mach number due to an increase in the momentum of the gas molecules. DSMC simulation results reveals that at low Knudsen number $((Kn = 0.01)$; shorter mean free paths), the atoms experience more collisions, which direct them toward the substrate. However, the atoms also move with lower momentum at low Mach number, which allows scattering collisions to rapidly direct the atoms to the substrate. At high Knudsen number $((Kn = 15)$; longer mean free paths), the atoms travel greater distances without depositing onto the substrate. Changing the Mach number influences the flux profile by controlling the momentum of the incident gas atoms. At high Mach number $(Ma = 2)$, the gas atoms have large momentums, which increases the distance atoms traveled before depositing onto the substrate. A very small Mach number $(Ma = 0.01)$ resulted in atoms depositing quickly near the centerline of the substrate.

DSMC SIMULATION

A molecular simulation technique, the direct simulation Monte-Carlo (DSMC) algorithm (Bird (1994), is used here to understand the flow phenomena and deposition mechanisms in a physical vapor deposition (PVD) process. The computational domain consists of the rectangular region, in the *(x-y)* plane. A specific aspect ratio, $L_x:L_y = 1:0.6$ is chosen to facilitate the DSMC simulations. A gas jet of aluminum at $((x/x_{max}) = 0.0, (y/y_{max}) = 0.29)$, and width along y-direction $(h = 0.26 - 0.32)$, is considered. The fluid is a hard sphere gas with the molecular diameter $d = 4.151 \times 10^{-10}$ *m*, and molecular mass $m = 4.4828 \times 10^{-26}$ *kg/molecule* and specific heat ratio $\gamma = 1.6,6$ corresponding to aluminum atoms. The hard sphere (HS) molecular model is used in the simulations. It is assumed that the gas molecules undergo diffuse reflection from all the solid surfaces with the thermal accommodation coefficient equal to 1. The gas jet temperature is maintained at 523 K. The dimensionless parameters affecting the fluid flow are the Knudsen number, *Kn*, Mach number, *Ma*, the Prandtl number, *Pr*, and the ratio of specific heats, γ.

The Direct Simulation Monte Carlo method (DSMC) is a numerical method for modeling rarefied gas flows. Here, the Boltzmann equation for the probability distribution function for gas molecules is solved using a probabilistic (Monte Carlo) simulation technique. The Boltzmann equation is more fundamental than the Navier-Stokes equations, since the it describes the evolution of the velocity distribution function of the gas molecules. The Navier-Stokes equations are obtained by taking the first, second and third moments of the Boltzmann equation with respect to the molecular velocities. While taking moments, the assumption of `local equilibrium' (velocity distribution function close to the Maxwell-Boltzmann distribution) is made, thereby restricting the Navier-Stokes equations to near-equilibrium flows. Since there is no assumption of local equilibrium in the Boltzmann equation, it is valid even for far-from equilibrium flows, provided the molecular chaos approximation is satisfied.

The DSMC method uses 'simulated molecules' which represent a large number of real gas molecules. A basic assumption is that the inter-molecular collisions and molecular motion between collisions can be decoupled over time scales comparable to the time between collisions. The translation of molecules between collisions is deterministically modeled, while the inter-molecular collisions and molecular-surfaces are incorporated using probabilistic models. The most common model is the hard-sphere model, but other models such as the variable hard sphere and variable soft sphere models are also used in order to accurately reproduce the density and temperature dependence of the viscosity. The hydrodynamic fields (density, velocity, temperature) are obtained by averaging over the molecular distribution functions in the DSMC simulations.

At the boundaries, it is necessary to prescribe interaction laws between the molecules and the solid surfaces. Accurate specification of the interaction laws requires a detailed knowledge of the molecule-surface interaction. However, there are standard interaction laws based on the surface velocity and temperature which are sufficient for predicting the hydrodynamic fields accurately. The most commonly used is the `diffuse reflection' conditions, where a molecule incident on the surface gets reflected with a Gaussian velocity distribution with the same mean velocity and temperature as that of the surface. Another commonly used condition is the specular reflection condition, that the velocity normal to the surface is reversed, while that tangential to the surface is unchanged. It is important to note that these boundary conditions are different from the no-slip conditions in the continuum models. In high Knudsen number flows, the diffuse and specular reflection conditions do predict a slip velocity at the surface.

There are four sub-cells in each cell in 2D in most of the calculations, though the number of sub-cells per cell has been varied up to a maximum of 25 in some cases to check the dependence of the mean free path on the sub-cell size. In DSMC simulations, the time step is chosen to be small compared to mean collision time in order to decouple the molecular motion and collisions. The mean collision time is calculated as, $t_c = \lambda / C_{av}$, where λ is the mean free path and C_{av} is the average molecular velocity. The minimum value of mean free path for the range of parameters used in the simulation is 1.306×10^{-2} m. The average molecular velocity, $\sqrt{kb\ T / m}$, is 401.30 m/sec. This gives the mean collision time of 3.255×10^{-5} *sec*. The time step used in the simulations, 1.0×10^{-6} *sec*, which is about an order of magnitude smaller than the smallest mean collision time.

RESULTS

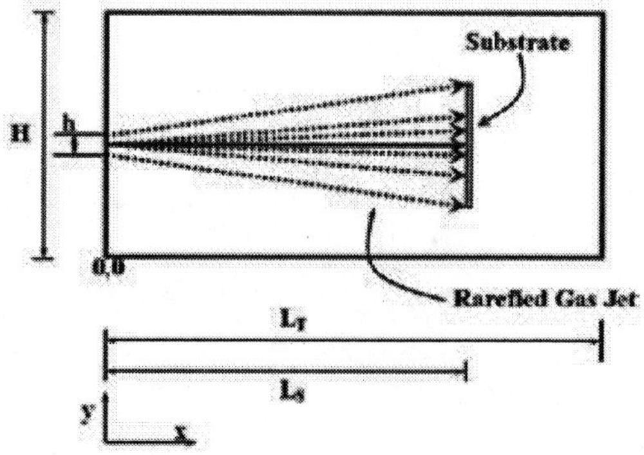

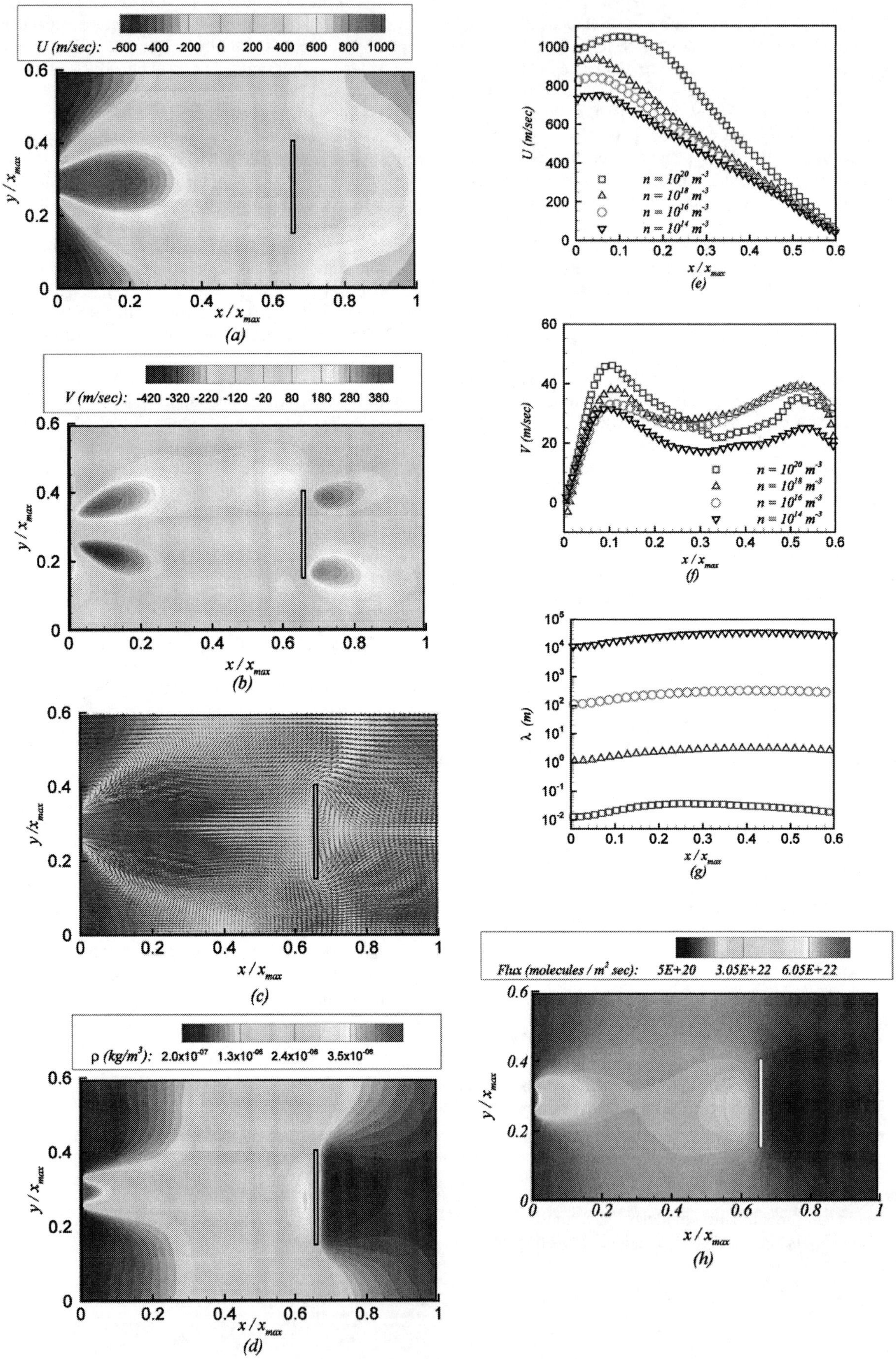
U (m/sec): -600 -400 -200 0 200 400 600 800 1000
y / x_max
x / x_max
(a)
V (m/sec): -420 -320 -220 -120 -20 80 180 280 380
y / x_max
x / x_max
(b)
y / x_max
x / x_max
(c)
ρ (kg/m³): 2.0x10⁻⁰⁷ 1.3x10⁻⁰⁶ 2.4x10⁻⁰⁶ 3.5x10⁻⁰⁶
y / x_max
x / x_max
(d)
U (m/sec)
n = 10²⁰ m⁻³
n = 10¹⁸ m⁻³
n = 10¹⁶ m⁻³
n = 10¹⁴ m⁻³
x / x_max
(e)
V (m/sec)
n = 10²⁰ m⁻³
n = 10¹⁸ m⁻³
n = 10¹⁶ m⁻³
n = 10¹⁴ m⁻³
x / x_max
(f)
λ (m)
x / x_max
(g)
Flux (molecules / m² sec): 5E+20 3.05E+22 6.05E+22
y / x_max
x / x_max
(h)

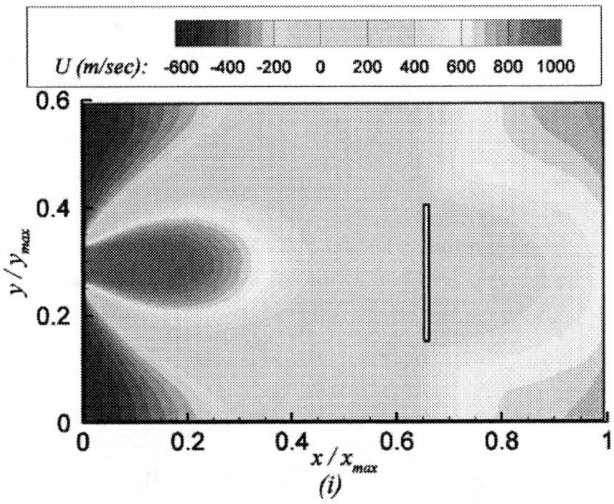

(i)

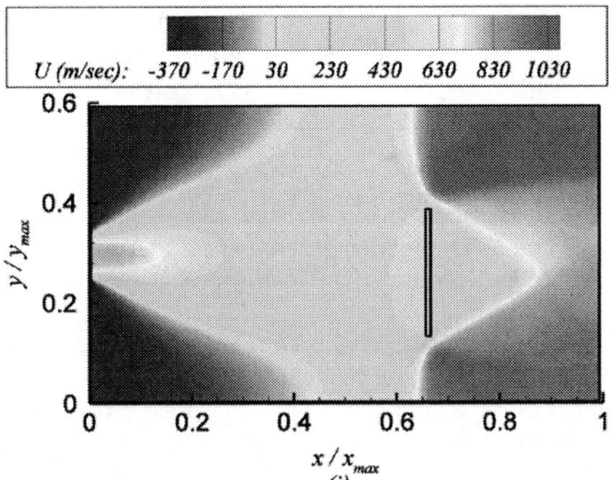

(j)

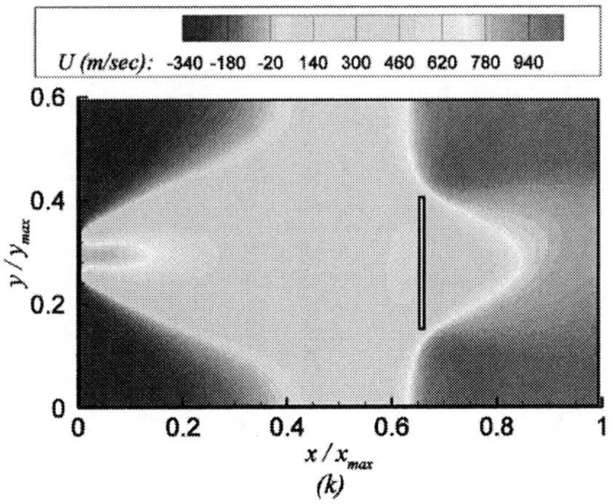

(k)

CONCLUSIONS

The objective of the present work was to carry out a comprehensive study of the rarefied gas jet of aluminum at Mach number *Ma = (U_j / \sqrt{\kb T_j / m}) in the range 0.01 < Ma < 2,* and Knudsen number *Kn = (1 / (\sqrt{2} \pi d^2 n)) in the range 0.01 < Kn < 15,* using 2D direct simulation Monte Carlo (DSMC) simulations to understand the fundamental phenomena involved and to identify the role of the process conditions on the coating's uniformity, ahdesion, as well as the hydrodynamic fields like streamwise and cross-streamwise velocity, number density, mean free path, and flux distribution of the gas jet inside the coating chamber for the development of the thin film of aluminum. The DSMC simulations utilize nearly collision-less vapor transport to the substrate. In this study, we have studied the rarefied gas jet based deposition processes of the thin film of aluminum to the stainless steel substrate.

One major influence on the flow patterns in PVD is, obviously enough, the chamber geometry. Amongst the biggest aims of reactor design is to get this right to minimize the costs of operation and process control. It must be quite obvious by now that the reactor design exercise calls for a better understanding and quantification of the extent to which each process variable can exert an influence. This in turn, requires a rigorous study of the chamber flow phenomena and deposition mechanisms. Minor variations in any of the process variables can cause significant changes in the flow patterns in the chamber, affecting the deposition rate and uniformity. Extensive experimental study is not viable, as generating high vacuums requires technology that is not cheap. Real-time flow measurements in these high vacuum, high temperature regimes also requires non-trivial tooling. A computational tool for the analysis and design of such equipment is, hence, of great value.

Such a capability will provide a significantly cheaper and shorter design iteration cycle.

The one unifying feature for most PVD techniques is their requirement of an evacuated (low-pressure) chamber. This is because a low background pressure ensures that the sputtered atoms have high mean free paths, and thus form a molecular beam from the target to the substrate. This achieves higher deposition rates. High chamber pressures will result in more atomic collisions, a significantly higher role for diffusion processes, and hence, a drastic reduction in the incident vapor phase flux at the substrate. Apart from this, the heating effects on the target, due to evaporation or sputtering, generally cause steep temperature gradients. Additionally, in several real deposition environments, the ionic plasmas that participate in the vapor phase creation contribute heavily to thermal non-equilibrium effects. Such factors result in the failure of continuum flow simulations within the deposition chamber. The flow exhibit distinct non-continuum characteristics like thermal non-equilibrium, non-trivial vibrational energy components, low or

counter-intuitive diffusion behaviors. The numerical study of PVD flow phenomena, hence, requires molecular simulations like direct simulation Monte Carlo (DSMC) simulations. The bulk of the numerical approaches used in vapor deposition simulations are particle based methodologies. This is because the applicability of continuum methods is severely limited.
Coating uniformity is determined by the binary scattering conditions in chamber regions near the substrate surface. The uniformity is controlled through both the rate of scattering collisions near the substrate and the momentum imparted from the collisions. The higher jet momentum causes the atoms to travel further in the streamwise direction between collisions, and results in the gas atoms experiencing fewer collisions as they travel past the substrate. Our results show the importance of two fundamental quantities that determine coating uniformity: (a) the gas atom concentration and (b) the mean free path. Gas atom concentration is of great importance for deposition onto macroscopic substrates. With larger mean free paths, scattering collisions are less frequent, allowing the gas atoms to travel further in a straight line. A large mean free path will allow deep penetration into the substrate, but may prevent the necessary collisions for eventual deposition onto the surface. Thus, ideal deposition conditions are found through balancing these two parameters, which occurs at moderate Mach number and moderate Knudsen number.

The simulations point to a rich variety of phenomena in rarefied gas jet assisted PVD technique. And DSMC simulations provide an invaluable tool for analyzing these phenomena.

REFERENCES

[1] Pradhan, S. & Kumaran, V. 2011 The generalized Onsager model for the secondary flow in a high-speed rotating cylinder. J. Fluid Mech. 686, 109.

[2] Kumaran, V. & Pradhan, S. 2014 The generalized Onsager model for a binary gas mixture. J. Fluid Mech. 753, 307.

[3] Balakrishnan, J. & Boyd, I. D. & Braun, D. G. 2000 Monte Carlo simulation of vapor transport in physical vapor deposition of titanium. *Journal of Vacuum Science Technology* 3, 907.

[4] Bird, G.A. 1994 Molecular gas dynamics and the direct simulation of gas flows. *Clarendon Press.Oxford.*

[5] Fan, J. & Boyd, I. D. & Shelton, C. 2000 Monte Carlo modeling of electron beam physical vapor deposition of yttrium. *J. Vacuum Science Technology.* 18 (6), 2937.

[6] Powell, C. F. & Oxley, J. H. 1966 *Thin Film Device Applications.* Vapor deposition, New York, USA.

[7] Chapman, S. & Cowling, T.G. 1952 The Mathematical Theory of Non-Uniform Gases, 2nd edition. Cambridge, Cambridge University Press.

[8] Chopra, K. L. 1983 Thin Film Device Applications. Plenum Press. New York.

[9] Hass, D. D., Marciano, Y. & Wadley, H. N. G. 2004 Physical vapor deposition on cylindrical substrates. Surface and Coatings Technology, 185, 283.

[10] Kersch, A., Morokoff, W., Werner, C., Restaino, D. Vollmer, B. 1992 Modeling of a Sputter Reactor Using the Direct Simulation Monte Carlo Method. International Electron Devices Meeting (Tech. Digest)., 181.

[11] Quinto, U. T. 1987 High temperature microhardness of hard coatings produced by physical and chemical vapor deposition. Thin Solid Films, 153, 19.

[12] Youchison, D. L., Gallis, M. A., Nygren, R. E., McDonald, J. M. \& Lutz, T. J. 2004 Effects of ion beam assisted deposition, beam sharing and pivoting in EB-PVD processing of graded thermal barrier coatings. Surface and Coatings Technology, 177, 158.

[13] Quinto, U. T. 1987 Comparison study of physical vapor-deposited and chemical vapor-deposited titanium nitride thin films using X-ray photoelectron spectroscopy. Applied Surface Science, 158, 246.

456810 Non-Convex MINLP for Utility Optimization

Abstract:

Non-convex MINLP for Utility Optimization

Rajkumar Vedam, Detong Zhang, Purt Tanartkit, Gareth Hiller, Schneider-Electric, 10900 Equity Drive, Houston, TX 77041, USA

Sankararao Boddupalli, Mallikarjun Lavate, Manju Murali, Cognizant Technology Solutions, Hyderabad, India

*(Author for correspondence: rajkumar.vedam@schneider-electric.com)*Extended Abstract

Utility optimization in the process industry offers the potential for significant energy cost savings. Among the technologies used for utility optimization are mixed integer linear programming (MILP) [1], nonlinear programming (NLP) [2], and case-study enumeration. In this paper, we present utility optimization using mixed-integer nonlinear programming (MINLP) [3], and highlight practical difficulties in the solution of the same. We highlight the typical non-convexities in operation, and present strategies to address them. We present novel modeling methods that encode plant operational logic as constraints and strategies for performance optimization.

Utility optimization involves calculating the best economic solution to meet the energy requirements of a process, while operating the underlying process optimally with desired operational specifications. The utility optimization problem is relevant when there are multiple sources that can supply the energy to do work in a process. Some sources can be endemic to the process – such as re-captured steam, and others, extraneous to the process – such as utility electricity. An additional requirement driving utility optimization is the dynamic nature of fuel costs, which can vary based on peak-pricing as well as other demand-related free-market forces. Thus an optimal solution for a given period might not be optimal at a different period where the energy costs have changed, necessitating frequent re-optimization.

Conventionally, the energy optimization problem has been solved as a nonlinear programming (NLP) problem, wherein using nonlinear models of switchable units such as steam and gas turbines, motors, generators, boilers, fuels, process streams and other process equipment, optimal values of independent process variables are computed, exploiting the degrees of freedom in the process, such that the energy costs are minimized while the throughput is maximized. Case-studies help to select between candidate switched configurations.

Typically, units in a process are operated between lower-bounds and upper-bounds, operation of which is infeasible beyond those limits. There is also a disjoint feasible point at the zero-state of the process unit (which can either be zero for power-off or non-zero for processes that are always-on, such as turbine units that always have a base-loading). This is a source of non-convexity in practical systems. When optimized using an NLP solver, the more operationally-expensive units are driven to their lower-bounds, whereas optimality might require the unit to be turned off (or operate at the disjoint zero-state). Designed to search in convex regions, conventional NLP solvers cannot turn off devices, because that requires crossing a non-convex boundary. NLPs therefore finding sub-optimal solutions. An additional drawback is that conventional NLP solvers seek the strongly optimal local minimum, thereby restricting their outlook to a neighborhood of operation. Depending on the size of the locally-convex regions, this sometimes precludes the ability of such solvers to find significantly different solutions that require large excursions from the current minimum.

Utility optimization requires the capability to switch in and out energy-supplying units, depending on the economics, energy and material balance, and underlying nonlinear-process optimality. Such switching could be done for example, using selective enumeration of case-studies where each case study involves solving an NLP problem that has encoded a particular switching of the energy-units. Due to exponential growth, the enumerative approach cannot be scaled up easily with increasing switchable units, even after screening out infeasible switching.

In this paper, we propose utility optimization using a Mixed Integer Nonlinear Programming (MINLP) branch and bound solver [3], using a reduced gradient sequential quadratic programming (SQP) NLP solver. Our novel methods include models for switching, the ability to handle a class of non-convexities in the process, and methods to control the complexity of state-space search and the performance and robustness of the solution.

The utility optimization problem can be expressed as follows:

min $\mathbf{c}\ (\mathbf{x},\mathbf{y},\mathbf{z})$, s.t. $\mathbf{f}(\mathbf{x},\mathbf{y},\mathbf{z})=\mathbf{0}$ and $\mathbf{x}_{min} <= \mathbf{x} <= \mathbf{x}_{max}$ and $\mathbf{y}_{min} <= \mathbf{y} <= \mathbf{y}_{max}$ and $\mathbf{z}$is Boolean (1)

In (1), the nonlinear process is modeled with equality constraints, $\mathbf{f}(\mathbf{x}, \mathbf{y}, \mathbf{z}) = \mathbf{0}$, where the function "**f**" is nonlinear in the vector variables **x**, **y** and **z**. Let $\mathbf{f}$: $\mathbf{R}^N \times \mathbf{R}^S \times \mathbf{B}^B \text{->} \mathbf{R}^N$. The variables "**x**" are free-dependent, and can be varied between some operational bounds, of dimension Nx1. The variables "**y**" are free-independent variables, and are usually referred to as "specifications", and also varied between bounds, and is of dimension S x 1. Normally, $S < N$. The variables "**z**" are free-independent, and are the binary-switches in the problem, and is of dimension B x 1. The cost function **c** is linear in the variables **x**, **y**, **z**.

Starting from an initial relaxed NLP solution, the branch-and-bound solver selects a relaxed binary variable for expansion based on a heuristic measure such as most-fractional, and performs a depth-first-search. When a binary variable is held at zero, the NLP solver is presented with a nonlinear problem that has the corresponding unit in its zero-state, thus circumventing the non-convex region between the bounds and zero-state. In continuum, a guarded relaxation of the lower-bound permits the obtaining of sensitivity of the solution in a direction towards zero. The branch and bound solver prunes branches on infeasibilities, as well as on branches that have worse cost than a "best-so-far" candidate solution – both relying on assumptions of convexity and globalism of the NLP solutions. While the worst-case complexity is exponential in the number of binary variables, we have observed that the branch and bound solver obtains a solution by examining a small percentage of the NLP nodes, indicating good performance of the heuristics.

Plant operational constraints sometimes require some switching to be synchronous, i.e., some switches need to operate coherently or complementarily when a switching occurs. A group is defined as a set of binary switches that switch coherently or complementarily. In such a set, we have just one independent variable, termed the group-leader. All other binary members of the set have constraint equations describing their relationship to the group-leader, and are included in (1). The use of groups results in two advantages:

(a) The number of independent binary switches that the MINLP solver must see reduces to the number of group-leaders. Let there be B binary switches, and G groups, where, $G <= B$. In this case, the worst-case branches of MINLP reduce from 2^{B+1} to 2^{G+1}, resulting in a provably improved worst-case performance.

(b) By encoding plant operational wisdom into grouping logic, we are reducing the state-space search to a smaller search-space, due to considering only operationally-feasible switching. This results in fewer infeasible NLP solutions, thereby improving the robustness and performance of MINLP, and time-to-result.

The use of logical cuts encodes known unit switching in the plant, to include "must-run" units, or to take out units that are down for maintenance. This has the effect of reducing the number of binary solver variables, and reducing the infeasible search-space, thereby improving both the performance and robustness.

The price sensitivity around an MINLP solution is explored by performing "What-If" analysis, where all binary variables are fixed either at 0 or 1, as set by the user, and by performing an NLP run. "What-If" analysis allows performing case-studies and the selection of an operating point from a candidate set of case-studies.

The table below shows results obtained with an illustrative example of a generic utility optimization problem, where the switchable devices are nonlinear models of motors, steam turbines, and generators [4], whose numbers are doubled in each run. The number of NLP nodes visited by the MINLP solver and the number of NLP iterations are scalable linear functions of the number of binary variables in this example (but not always, owing to NP-Hard nature of the problem), using a heuristic of branching on the most-fractional node.

Binary Vars	Eqns	Total Variables	NNz	NLP Nodes	NLP Iterations	CPU Time (sec)
4	145	153	325	6	31	2.66
8	242	257	560	10	50	5.73
16	436	465	1030	18	113	17.2
32	824	881	1970	34	224	49.5
64	1602	1715	3855	66	378	225

Conclusion

A branch and bound MINLP solver is proposed with a nonlinear SQP solver for the utility optimization problem. The non-convexity of the problem is discussed and addressed. Performance and robustness improvements are proposed. An illustrative example of utility optimization is presented. Reference

[1] Operational optimization of the utility system of an oil refinery, Sandra R. Michelettoa, Maria C.A.

Carvalhoa,José M. Pinto, *Computers & Chemical Engineering*, Volume 32, Issues 1–2, January 2008, Pages 170–185

[2] Modeling and Optimization of Utility Systems, PS Varbanov, S Doyle, R Smith, *Chemical Engineering Research & Design*, Vo.82, Issue 5, May 2004, pp.561-578.

[3] An algorithmic framework for convex mixed integer nonlinear programs, P Bonami, LT Biegler, AR Conn, G Cornuéjols, IE Grossmann, CD Laird, J Lee, A Lodi, F Margot, N Sawaya, A Wachter, *Discrete Optimization*, Vol 5, Issue 2, 2008, pp. 186-204.

[4] SimSci ROMeo Process Optimization, Retrieved from http://software.schneider-electric.com/products/simsci/optimize/romeo-process-optimization/, April 2016.

458494 Novel Sampling Technique for High Dimensionalstochastic Optimization/Stochastic Programmingproblems

Abstract:

Uncertainty is essence of real world optimization problem. Computational speed is critical in optimizing large scale stochastic problems. The major bottleneck in solving large scale stochastic optimization/stochastic programming problems is the computational intensity of scenarios or samples. To this end, this paper presents a novel sampling approach. There are two important properties for sampling randomness and uniformity. It is the uniformity that is important for accuracy of the sampling. Latin Hypercube sampling (LHS) provide good convergence rate as compared to random Monte Carlo simulations but LHS shows uniformity in single dimension only. For Multi Dimension it computes stratied sampling for each uncertainty but randomly pairs to form a K dimensional sample set. Due to this random combination; good uniformity for single dimension is lost for K dimension. Furthermore quasi random sequences like Hammersley, sobol, and Halton sequences show better uniformity in k-dimensions but are not one dimensionally uniform. For Quasi Monte Carlo sequence sampling pairings are done as per Hammersley or Sobol sequence points to form a sample set of K dimension. For higher dimension (more than 40) some of these quasi-random sequences like Hammersley and Halton show clustering and correlation pattern eects due to exponential increase in prime base. This pattern can be broken by coupling LHS with HSS. Similarly Sobol sequence sampling is not linked to the prime base so for dimension more than 40 there are no clustering or correlation patterns upto 100 variables. In this work, we combine one dimensional uniformity of LHS with k-dimensional uniformity of sobol to generate a new sampling called LHS-Sobol sampling. This sampling breaks the correlation patterns for sobol and

shows superior convergence for larger dimensions. This sampling plays an important role innancial modeling.

References
[1] U. Diwekar, Introduction to Applied Optimization, 2nd Edition, 2008, Springer.

[2] J. Kalagnanam and U. Diwekar, An ecient sampling technique for oine quality control, Technometrics, 43, 440, 1997.
[3] F. Akesson and J. Lehoczky, Path generation for quasi-Monte carlo simulation of morgage backed securities, Management Science, 46(9), 1171, 2003.

457199 Monte Carlo Simulation-Based Method for Process Performance Assessment in Sterile Manufacturing of Biopharmaceuticals

Abstract:

The fast expansion of biopharmaceuticals such as monoclonal antibodies is calling for more extensive research in the manufacturing processes of sterile drug products, such as injectables. In this work, we present a Monte Carlo simulation-based method, which enables the assessment of process performance under operational uncertainty in sterile drug product manufacturing. The study is supported by an industrial case study.

Our method enables the systematic performance assessment of processes in commercial operation with a probability-based approach. It also permits the multi-facet assessment of the operations, such as design of the process, human factor and predictive maintenance, in order to ensure high industrial applicability. The method, which consists of five main steps, supports use of Monte Carlo-based mathematical models, and enables its application in an industrial environment. The five steps are: define key performance indicators (KPIs), collect and adapt data, create process model, assess operational properties and identify sensitive task. In the first step KPIs are defined accordingly to the industrial needs and interests, e.g., increase of capacity, yield or robustness of the process. Subsequently, data related to the KPIs are collected from data management systems and adapted to the purpose, resulting in a hierarchical data set. In the third step the process model is created, which covers all tasks inside the process, and considers the KPIs as well as uncertainties such as failures, repetition or human factors. Next, the operational properties are assessed by taking into account historical information on production stops, duration of tasks, remedy times and correlations. During this step, the creation of fitted process durations and empirical event probabilities lays the basis for the Monte Carlo simulation. Stochastic simulation enables reflection of operational uncertainty, which is a prominent element on level of manufacturing shop-floor, into the simulation results. Finally, significant tasks, which would require improvement, are identified by the performance of sensitivity analysis. Because of the non-deterministic nature of the models, variance-based sensitivity analyses approaches, such as Sobolâ€™s indices and partial rank correlation coefficients (PRCC), are applied to rank tasks by their importance. Furthermore, for enhancing the industrial applicability, the combination of the KPIs with feasibility indicators, such as implementation time, is considered as an additional feature. This approach allows incorporation of good manufacturing practice (GMP) in the assessment, an omnipresent and critical factor in the pharmaceutical industry.

The method was applied in an industrial case study, and more specifically, on key processes of the sterile manufacturing of biopharmaceuticals termed clean-in-place (CIP) and sterilize-in-place (SIP). These processes have been rarely the research object because of their supportive functions, but are known as the bottleneck of manufacturing sterile drug products in the industry. In the plant which was the object of the case study, the CIP and SIP processes occupy nearly half of the production timeline. In this case study we applied the above mentioned method in view of a future process optimization of the entire CIP and SIP procedure. Our goal was to identify the tasks, which have the highest impact on the production time, the KPI determined in the case study. Generally CIP and SIP are semi-automated processes, consisting of around 200 tasks with different durations and dealing with fluid dynamic laws, which render their complete modelling unrealistic. Also, the operations environment leads to the presence of failures and time distribution due to technical or human issues. There a quasi-Monte Carlo method was applied to simulate the duration of an entire CIP and SIP procedure, which was supported by Latin hypercube sampling (LHS) of task durations generated by fitted non-parametric distributions. The pattern, describing failures and repetitions of process tasks, was randomly generated at each iteration on the base of empirical probabilities. For the validation of process model, Kolmogorov-Smirnov test was applied assuming the identical distributions between actual and simulated production time as the null hypothesis. The identification of the important tasks, which have the highest influence on the CIP/SIP duration, was performed by PRCC analysis. To enhance the applicability, the result is combined with feasibility evaluation regarding various aspects, e.g., GMP, time duration or resource availability, which is provided by the industrial experts. For the first twenty most important tasks, Sobolâ€™s indices were calculated to quantify the taskâ€™s contribution to the total variance of CIP and SIP durations.

Our method permits the evaluation of the current process with an intention for redesign, as was demonstrated in the case study. Moreover, it enables considerations on the quality of the operations as well as the effects of failures. In practice, this last point represents a failure-mode-effect analysis with a quantitative criticality assessment, and finds great application in a GMP environment. Since the method was developed in an industrial environment, it shows high compatibility with the process improvement methodologies currently in use, i.e., Lean Six Sigma (LSS). In fact LSS procedures can be used for the interpretation of the simulation results, e.g., Pareto chart and root cause analysis (RCA). In future studies we intend to develop a model based on machine learning to be coupled with RCA, which will enable considerations on predictive maintenance and predictive planning.

Modeling and Optimization of Industrial Ammonia Synthesis Process

Stanislav Y. Ivanov, Ajay K. Ray
Western University, London, ON, Canada

Introduction

Ammonia is one of major chemicals produced in the industry. It has a variety of applications: for manufacturing of inorganic salts, polymer fibers, explosives, etc. Its biggest role it plays for agricultural fertilizers: as an intermediate in urea production or used directly as liquid. It is hard to diminish the importance of ammonia for agricultural sector all over the world.[1]

The use of rigorous mathematical models in process design and optimization has been extensively used for a large-scale chemical engineering processes boost up the performance of existing units. Model simulations allows for fast and accurate investigation of the decision domain providing an engineer/researcher with valuable knowledge on a process behaviour while avoiding expensive and time consuming lab scale or industrial experiments. Therefore, one can perform a comprehensive study of a process for the numerical optimization to boost up process performance.

Conceptually, there are two approaches in optimization - single and multi-objective (SOO and MOO respectively). The former formulates a sole objective function to be optimized into a single point solution, while latter treats two or more objective objective functions to find (likely) a set of equally good solutions.

However, SOO approach is still extensively used, but MOO has gained its popularity in chemical engineering. Large-scale industrial processes are very complex system, so often it is troublesome to single out the only objective. Thus, it is especially beneficial to use multi-objective approach for optimization of industrial processes.

The main outcome of the problem solution is obtained set (known as Pareto set) of non-dominating points, which resemble best solution with respect to all objectives. Additionally, they reflect the trade-off between objectives, thus illustrating wherther one can significantly improve one objective while not worsening other beyond reasonable.

In order to perform a proper optimization study, a valid and adequately accurate model is needed. Modelling of ammonia synthesis has been drawing attention over the years and a number of attempts has been made. Baddour et al.[2] developed a simple one-dimensional plug-flow model for autothermal ammonia converter. They studied effect of process parameters such (*i.e.* feed temperature and composition) on ammonia production rate and temperature profile along catalyst bed. Shah[3] discussed two-bed adiabatic model with recovery heat exchange for the process control. He found best stable operating point as balance between heat generation and consumption. Gaines[4] used mathematical model for adiabatic bed with empirical correlation for effectiveness factor to perform analysis on model parameters. Singh and Saraf[5] modelled converters with both types - adiabatic and non-adiabatic - catalytic beds. They used pseudo homogeneous model with effectiveness factor based on reactants diffusion. Mansson and Andresen[6] did an optimization study for tubular ammonia converter with one-dimensional pug low model aiming to maximize ammonia content in the effluent and obtain optimal temperature profile. Elnashaie et al.[7] compared the performance of homogeneous and heterogeneous models for auto-thermal converter showing advantages and better accuracy for the latter. Later, Upreti and Deb[8] performed a design optimization study with genetic algorithm by maximizing overall economic return of the synthesis having total converter length as a major decision variable. Babu and Angira[9] performed a similar study but could obtain more stable model solution in wider temperature range.

However, none of the works, to the best of the authors' knowledge, utilizes multi-objective optimization principles and performs optimization search for more than one objective. Hence, it is beneficial for the complex system as industrial converter.

In this work we developed a first-principle mathematical model of an industrial ammonia synthesis converter. The model incorporates heterogeneous reaction kinetics with intra-particle diffusion resistances. As process being highly exothermic, the heat recovery model is also embedded into the overall converter model. The model is able to predict gas temperature and concentration profiles along converter's bed, and the amount of heat which is recovered from a gas stream and recycled to supply for the pre-heat.

Then, the model was validated with industrial data and showed a good accuracy. Then the model was used for in two MOO studies for industrial ammonia synthesis. One can find a broad range of optimal solutions with a trade of between objectives.

Overview of industrial ammonia synthesis

Industrial ammonia synthesis is done through Haber process[1]. It is a stoichiometric reaction of hydrogen and nitrogen (Eq.1). The reaction is slow at ambient conditions as well as limited by equilibrium, thus to achieve significant yields is carried out in a presence of catalyst under elevated temperatures (350 - 500 °C) and pressure (100-250 atm.)

The object of study is a three bed ammonia converter. The simplified layout is provided at Fig. 1. The synthesis gas is split into two flows - for main feed and quench. The main feed enters prior the first bed, while quenching is done in between. The process gas passes though a bed and experiences a temperature rise. This is where quenching and heat recovery is used - to remove the excess heat from the gas stream.

Converter modelling and validation

Industrial ammonia synthesis is done through Haber process[1] (Eq. 1). The Eq. 1 is a gas-solid reaction occurring in presence of iron or ruthenium catalyst. The catalytic ammonia converter is made of few packed beds. The reaction mixture is sequentially fed into beds to produce ammonia in a number of steps. The intermediate cooling is essential as the reaction 1 is exothermic. It prevents catalyst from exposure to high temperatures and shift reaction equilibrium the excess heat

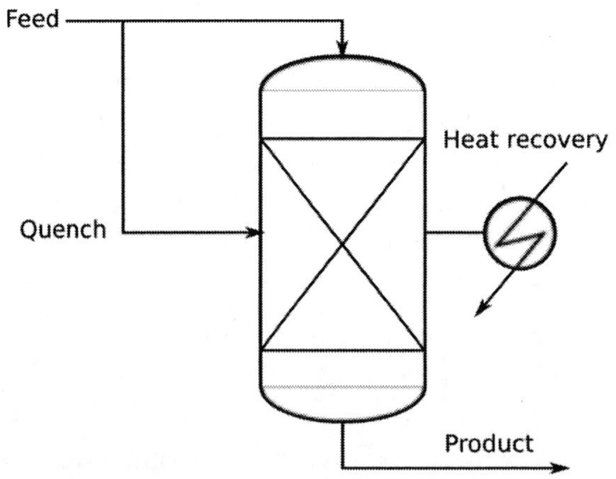

Figure 1. Ammonia converter schematic

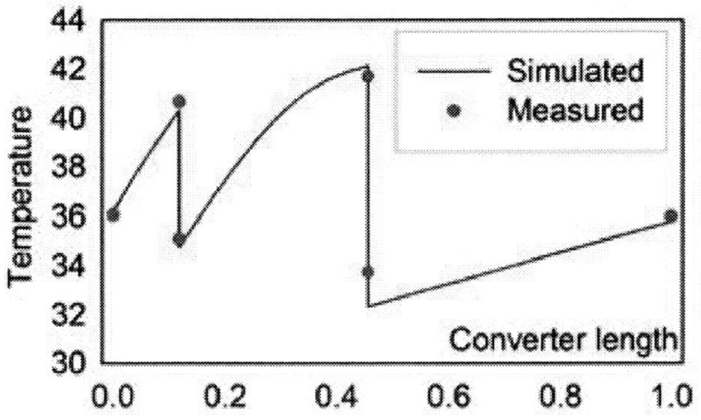

Figure 2. Comparison between simulated and observed temperatures along reactor

is removed between beds.

$$N_2 + 3\,H_2 \longleftrightarrow 2\,NH_3, \quad \Delta H^0_{298} = 46.22 \quad \text{kJ/mol} \tag{1}$$

Considering the process desctiption and converter layout, The model is comprosed of 4 following parts:

1. Synthesis reaction kinetics by Temkin equation[10]
2. Intrapacticle diffusion of species inside porous pellet[11]
3. Fixed bed hydrodynamics as plug flow system
4. Heat recovery model

Due to proprietary reasons authors cannot provide a layout for the heat recovery system in the converter, thus making it impossible to disclose details of the heat recovery model as well. So this part will remain unknown for the reader. Also, all numbers related to process operating parameters or the converter dimensions are given in dimensionless form. To assess validity, one can compare temperatures between simulated profile and on bed boundaries, measured by thermocouples. One can find that the model proved a good overall match. The relative error for the most cases lies within 4 %. Moreover, one can look into comparison of gas compositions in Table 1. Even though components concentrations were not included into fitting, but along with temperatures they have a close match.

Table 1. Comparison between simulated and experimented gas composition

	Gas stream composition				
	N_2	H_2	NH_3	CH_4	Ar
simulated	3.77	9.02	2.13	-	-
experimental	3.70	9.04	2.17	-	-

Multiobjective optimization cases

Case 1 : The two objective discussed above were included into two-objective optimization problem. The optimization problem is constrained by maximum bed temperature as it is limited by catalyst manufacturer in order to prevent damaging the catalyst and loss of activity. Industrially feasible domain for process parameters is a subject to minimum and maximum bounds as decided in accordance with industrial practice. The MOO problem is summarized as:

$$\begin{aligned} maximize \quad & I_1(\mathbf{x}) = X_{\mathrm{NH_3}}^{out} \\ maximize \quad & I_2(\mathbf{x}) = \mathrm{HR} \end{aligned}$$

subject to:

$$\begin{aligned} T_{bed}^{out} \leq T_{max} \quad & for \quad each \quad bed \\ \mathbf{x} = T_{feed}, QFR, P \in & [min; max] \end{aligned}$$

Case 2 : Also, we considered MOO problem involving converter's design parameters. Beside ammonia production and HR, another important process parameters is a total catalyst loading within converter and it distribution among beds. The lower total loading will allow for reduction of catalyst price, while its distribution might potentially improve ammonia production as well. Therefore, another multi-objective optimization problem to investigate possibility of lowering total catalyst volume in reactor while keeping ammonia production high to be solved as follows:

$$\begin{aligned} maximize \quad & I_1(\mathbf{x}) = X_{\mathrm{NH_3}}^{out} \\ minimize \quad & I_2(\mathbf{x}) = \sum_{i=1}^{3} V_i \quad i = 1 - 3 \end{aligned}$$

subject to:

$$\begin{aligned} T_{bed}^{out} \leq T_{max} \quad & for \quad each \quad bed \\ \mathbf{x} = T_{feed}, QFR, V_i \in & [min; max], P = 10 \end{aligned}$$

Results and discussion

Pareto set of optimal solution for Case 1 was obtained for two different converter pressures. As the converter pressure being the parameter not easily changeable in industry (contrary to feed temperature and quench-to-feed ratio), we solved Case 1 for pressures 10 and 15 atm. independently. Former pressure is found to be most common for the converter, while latter is intended to show the potential changer in converter operation and is possible to achieve within the unit. The results are shown on Fig. 3. Overall, higher temperatures and QFR favour higher ammonia production. If one is striving for it, it is necessary to control both parameters simultaneously to stay near optima. However, at higher pressure the impact of quenching is becoming less important thus more attention has to be paid to feed temperature.

One can note that higher pressure is much in favour of higher ammonia production. Any point of pressure 15 set is providing higher molar fraction of ammonia in product. On contrary, lower pressure on average allows for higher heat recovery within converter. Moreover, it also provides a wider range of Pareto optimal solutions for both objectives. If one willing to come from low to high heat recovery (*i.e.* from most right to most left point at pressure 10), it will allow for 8% in heat saving. Considering large scale of production it will result in significant heat duty reduction. The range is narrower at higher pressure due to faster reaction rate and heat generation, thus,

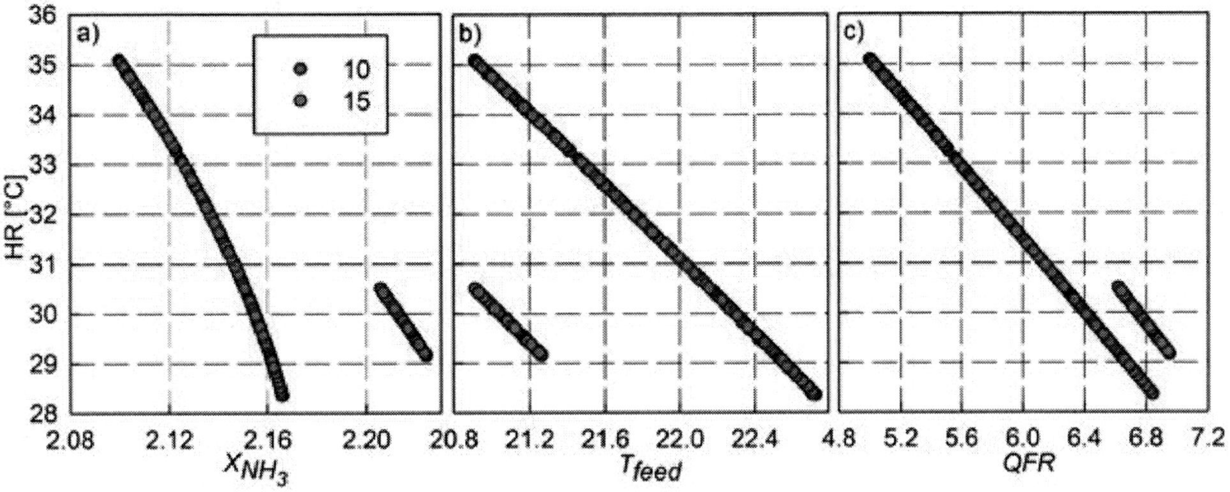

Figure 3. Pareto optimal set of solutions for Case 1
Figure a) shows both objectives. Figures b), c) shares the same Y-axis with figure a) and shows relation of process parameters to Pareto-optimal set of objectives.

consequently, more bounded by upper limit on catalyst bed temperature.

The solution for Case 2 is given at Fig. 4. As for Case 1, Fig. 4 a) provides a range of objectives. One can find that a range of $X_{\mathrm{NH_3}}^{out}$ is much wider than in case 1 (0.22 vs 0.08). Also, for every point on the figure ammonia content in the product stream is higher than for any point in Case 1. Therefore, catalyst distribution is another important parameter to consider when designing a converter or loading fresh catalyst at maintenance. As for process parameters, QFR and V_1 hit their upper and lower bound respectively. The relation of other process parameters to Pereto optimal points is given on Fig. 4 b) and c). Feed temperature and two other bed volumes are experiencing more complex relations in order to stay Pareto optimal. Firstly, higher total catalyst volume will allow for lower inlet temperature due to higher space times. Secondly, the major decrease in feed temperature (by 1.7) and rise in ammonia (77 %) is achieved through the increase in third bed catalyst loading. The further boost in ammonia (33%) is through increase of the loading in second bed, while keeping T_{feed} and V_3 constant.

To take a closer look into importance of catalyst distribution on converter performance, one can consider three cases from Fig. 5 a) highlighted as red dots. A and C correspond to marginal cases with minimum and maximum loading while B depict solution "in the middle". In fact, total loading of catalyst at point B is the same as in the converter studied but with different distribution. The existing catalyst distribution is labelled as "original" at Fig. 5. Firstly, one can note a major contribution of the second and third beds into overall ammonia production. While first bed loading remains constant, the change in other two can drastically improve ammonia production. Secondly, even the lowest catalyst loading provides higher ammonia production by far ("original" vs. A). Needless to say that if one can properly redistribute existing amount of catalyst among beds, it can bring even better ammonia production ("original" vs. B).

For all considered cases operating parameters remain in the same range, while pressure is kept at 10. One can recall from optimization of Case 1 that at higher pressure ammonia production increases. Thus, if ammonia production is the first priority, there's even more room for improvement at higher pressures if engineers are willing to increase compression duties and/or if there's no other restrictions to do so.

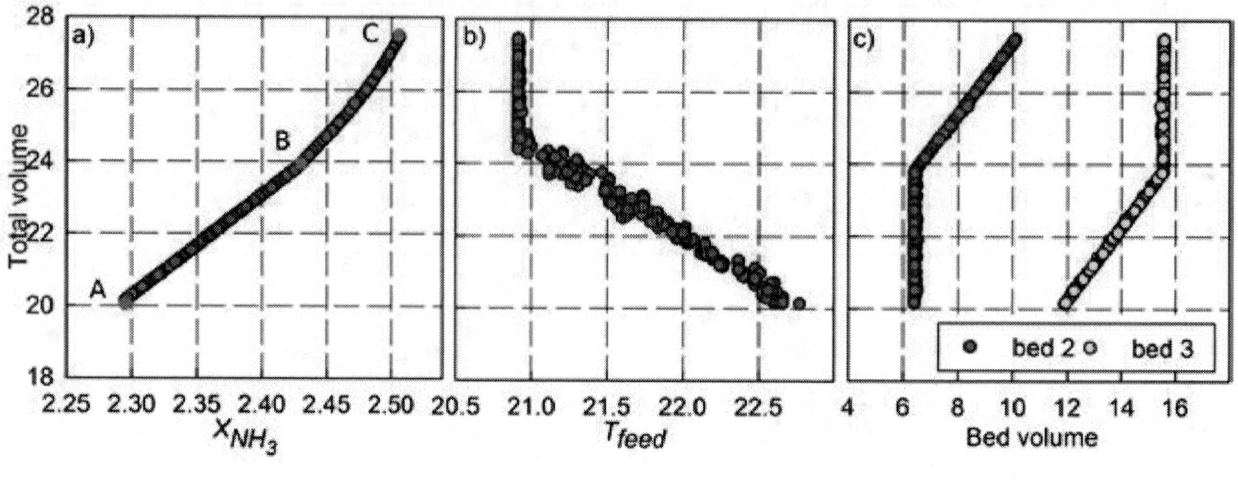

Figure 4. Pareto optimal set of solutions for Case 2
Figure a) shows both objectives. Figures b) shares the same Y-axis with figure a) and shows relation of feed temperature to Pareto-optimal set of objectives. QFR is at upper bound, V_1 is at lower bound.

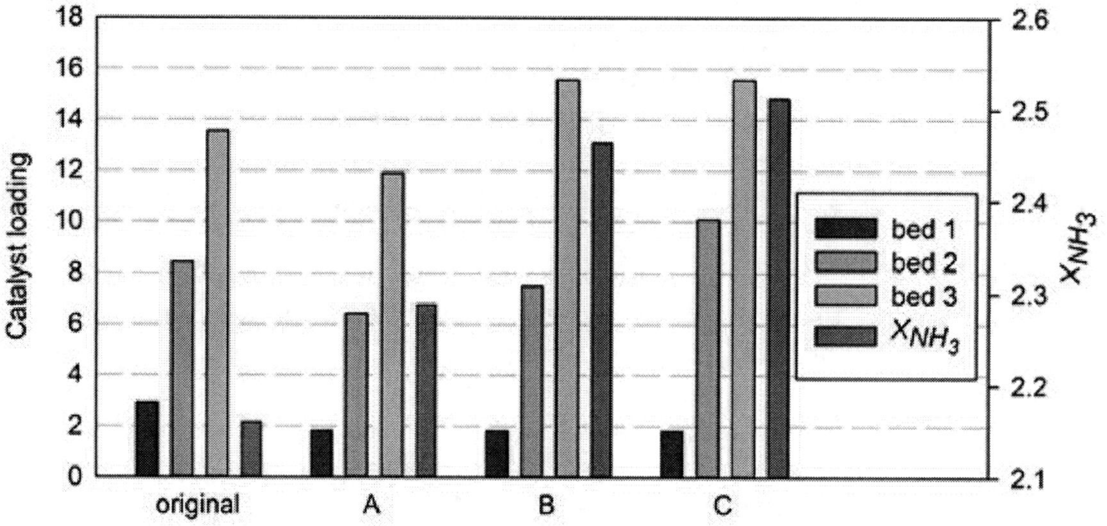

Figure 5. Comparison of catalyst loading distributions
Distribution of catalyst among beds at points A, B and C from Fig. 4 are compared. "Original" denotes current catalyst distribution in the object of study. Case B in total gives the same amount of catalyst as in "original" case.

Summary and conclusion

Multi-objective study of the industrial ammonia converter was performed. The MOO study was done with use of surrogates trained on simulated data from first-principle model. Two MOO cases - operation and design optimization - were solved with NSGA-II. The former case yielded a Pareto set of solutions compromising between ammonia production and heat recovery depending on control parameters. The latter case showed that for any point in Pareto set for design MOO ammonia production is higher that in operation MOO. Therefore one can benefit even more with change of total catalyst loading in the converter and/or proper distribution of catalyst among beds. The total ammonia production at any point of Pareto set for design optimization was higher that for operation case even for lower total catalyst loading in converter.

Nomenclature

bed	refers to catalyst bed
HR	recovered heat
P	pressure
Q_i	volumetric flow rate
QFR	quench-to-feed ratio
T	temperature
V	bed volume

References

[1] Max Appl. *Ammonia : principles and industrial practice.* Wiley-VCH, Weinheim ; New York, 1999. ISBN 3527295933.

[2] R.F.F Baddour, P.L.T. Brian, B.A Logeais, and J.P.P Eymery. Steady-state Simulation of an Ammonia Synthesis Converter. *Chemical Engineering Science*, 20(4):281–292, 1965. ISSN 0009-2509. doi: 10.1016/0009-2509(65)85017-5.

[3] Shah. Control simulation in ammonia production. *Industrial and Engineering Chemistry*, 59(1):72–83, 1967.

[4] Larry D. Gaines. Ammonia synthesis loop variables investigated by steady-state simulation. *Chemical Engineering Science*, 34(1):37–50, 1979. ISSN 00092509. doi: 10.1016/0009-2509(79)85176-3.

[5] C.P.P. Singh and D.N. Saraf. Simulation of Ammonia-Synthesis Reactors. *Industrial & Engineering Chemistry Process Design and Development*, 18(3):364–370, 1979. ISSN 0196-4305. doi: 10.1021/i260071a002.

[6] Bengt Mansson and Bjarne Andresen. Optimal Temperature Profile for an Ammonia Reactor. *Industrial & Engineering Chemistry Process Design and Development*, 25:59–65, 1986. ISSN 0196-4305. doi: 10.1021/i200032a010.

[7] S.S.E.H Elnashaie, A.T. Mahfouz, and S.S. Elshishini. Digital Simulation of an Industrial Ammonia Reactor. *Chemical Engineering and Processing*, 23(3):165–177, 1987.

[8] Simant R. Upreti and Kalyanmoy Deb. Optimal design of an ammonia synthesis reactor using genetic algorithms. *Computers & Chemical Engineering*, 21(1):87–92, 1996. ISSN 00981354. doi: 10.1016/0098-1354(95)00251-0.

[9] B. V. Babu and Rakesh Angira. Optimal design of an auto-thermal ammonia synthesis reactor. *Computers and Chemical Engineering*, 29(5):1041–1045, 2005. ISSN 00981354. doi: 10.1016/j.compchemeng.2004.11.010.

[10] Anders Nielsen, Jorgen Kjager, and Bennie Hansen. Rate Equation and Mechanism of Ammonia Synthesis at Industrial Conditions. *Journal of Catalysis*, 3:68–79, 1964.

[11] S.S.E.H. S Elnashaie, M.E. Abashar, A.S. Alubaid, and A.S. Al-Ubaid. Simulation and optimization of an industrial Ammonia Reactor. *Industrial & Engineering Chemistry Research*, 27(8):2015–2022, 1989. ISSN 08885885. doi: 10.1021/ie00083a010. URL `http://pubs.acs.org/doi/abs/10.1021/ie00092a024`.

458760 A New Proactive Scheduling Methodology for Front-End Crude Oil and Refinery Operations Under Uncertainty of Shipping Delay

Abstract:

Since 1950s, mathematical programming has been implemented in long-term decision making for crude planning to optimize both production and crude purchasing plan. Crude planning and scheduling is very important to a petroleum refinery due to the potential realization of large cost savings. It has been studied and practiced for a long time, especially in the last two decades driven by increasingly intensive global competitions, more volatile feedstock and product markets, as well as stricter environmental regulations. It is not easy task due to the domino effects with the crude flows from upstream to downstream units and the differences between their densities, sulfur contents, product yields etc. And in reality, the crude scheduling activities are highly vulnerable to disruptions brought by various uncertainties. For example, the shipping delay may disrupt the associated activities and cause the originally well set schedule operations suboptimal or even infeasible.

In this paper, a new proactive large-scale continuous scheduling model has been developed for crude unloading, transferring, and processing (which we called CUTP) system to simulate and optimize the front-end and refinery crude-oil operations simultaneously. The scope of the scheduling problem includes crude oil unloading from vessels to storage tanks at onshore berths, transfer of crude from these tanks to charging tanks, charging crude distillation units and further processing of crude inside the refinery, which includes crude distillation, cracking, coking, reforming, hydrotreating, product blending and component recovery. The general objective of the proposed CUTP scheduling methodology is to maximize the total operational profit; meanwhile, it should meet inventory and production demands, and try to maintain refinery plantâ€™s normal operations under uncertainty of shipping delay.

The profit maximization and the production flexibility maximization are generally two contradictory aspects that should be well balanced. After developing a CUTP scheduling model, a *CFI* model need to be added to quantify the shipping delay tolerance. The upper and lower bounds of the combined flexibility index (*CFI*) need to be identified. When the modeling solution is identified as feasible and the range of *CFI* is obtained, then the profit maximization and the customized minimum *CFI* will be considered simultaneously. This makes the optimization more applicable in terms of having taken into account shipping delay uncertainty. For comparison, the optimization results with and without customized *CFI* are presented and discussed. OA (outer-approximation) based decomposition method has been employed to obtain the optimal solution of the developed MINLP scheduling model. The efficacy of the proposed rescheduling model has been demonstrated by different case studies. There are three major contributions in this paper: (i) the simultaneous scheduling of front-end crude transfer and refinery processing has been achieved; (ii) RPST has been considered in the CUTP model for seamlessly connecting both front-end crude transfer and refinery processing models; and (iii) uncertainty of shipping delay has been taken into account to avoid refinery shut down events by introducing an inventory-related time flexibility index. Based on these new features, the developed methodology has greatly increases the potential profitability and production flexibility of refineries.

461823 A Generic MILP Modelling Framework for the Systematic Design of Lignocellulosic Biorefining Supply Chains

Abstract:

Abstract

The development of sustainable biobased economies could help overcome the high societal dependency on fossil resources. Therefore, research has focused on the study of advanced biorefining networks. The complexity of such production systems requires the use of efficient decision-making tools, enabling a full exploitation of biomass (and its macrocomponents, i.e. cellulose, hemicellulose and lignin) for the production of biobased products and platform chemicals (Kokossis and Yang, 2010). Therefore, it is also essential to identify the most promising pretreatment process that fractionates biomass into cellulose, hemicellulose and lignin and usually represents the highest cost part of the entire biorefining system. In addition, the deployment of second-generation technologies is still hindered by high capital costs as well as the existence of uncertainties (e.g. demand and price of biobased products) in the so far immature biobased market. Consequently, one of the most important and challenging aspects in the quest of producing a set of sustainable biobased products, is the design of an integrated and economically viable biorefinery supply network (Akgul et al., 2011; Martín and Grossmann, 2010; Čuček et al., 2014). Optimisation tools could play a powerful role supporting decision in such novel production systems, through the identification of the major cost drivers, the performance of sensitivity analysis as well as the assessment of economic and technical uncertainties (Kim et al., 2013).

The aim of this work is the modelling and optimization of biorefining chain systems using an integrated approach to the modelling of all the entities involved across the technology chain, with the purpose of achieving a long-term, decision-making regarding the systematic design and planning of advanced biorefining networks.

Methodology

The design problem is formulated assuming as given, sets of lignocellulosic biomass types including their temporal and geographical availability, biobased product types and production facilities of different scale, transport logistics, economic and technical parameters as a function of biomass and product type as well as conversion technology and plant scale, biobased products market characteristics, demand over a fixed time horizon and geographical distribution.

The objective is the identification of the optimal network configurations that satisfy the target regional demand of the selected biobased products over the entire planning horizon, while maximising the overall financial profitability of the system. Therefore, the decision variables of the problem refer to the planning and operation of the biorefining network.

The biobased supply chain optimization problem is formulated as a mixed-integer linear programming (MILP) model, based on previous research works (Giarola et al., 2012) (Giarola et al., 2011) (Zamboni et al., 2009). In particular, the model represents a spatially-explicit, multi-feedstock, multi-period and multi-echelon lignocellulose-based supply chain. A maximum profit-based objective function is considered, including the capital investments, the operating costs and the revenues evaluated across the entire biorefining network.

A binary variable is introduced, accounting for whether a conversion facility is installed within a territorial element or not. Logical constraints depending on the value of the binary variable, impose that only one plant can be built in each geographical cell and production can only take place if a plant is already established.

In addition to the system, several other equality (e.g. mass balance) and inequality constraints were defined in the mathematical formulation of the optimization problem to characterize each supply chain node.

Case study

The proposed MILP model was implemented in the GAMS® software tool using the CPLEX solver (Rosenthal, 2014). In particular, the model was used to solve a European case study involving the development of biorefining systems in the South-West of Hungary. The case study considered the arable land availability as well as agronomical factors in four Hungarian counties, i.e. Tolna, Baranya, Somogy and Zala. The examined region is discretized, for computational reasons, into 102 square grids, of 225 km^2each. According to the Corine Land Cover database, truck was considered the most suitable transportation mode for this study. The model is optimized over a 1-year planning horizon, divided into 12 months, assuming seasonal availability of biomass for July, August, September and October. Additionally, the technical (i.e. technological yields) and economic (i.e. capital and operating costs) inputs, used in this work, are derived from the BIOCORE project (O'Donohue, 2014).

Concluding remarks

Results show that the development of optimization models considering all the supply chain entities can shed the light onto the systematic design and planning of novel production infrastructures.

The model embedding the definition of the process technology superstructure for a portfolio of selected biobased products and platform chemicals, enables the evaluation of environmental impacts along with costs as well as the analysis of the key sources of uncertainty and their evolution over time.

Acknowledgement

The EC (Reneseng-607415 FP7-PEOPLE-2013-ITN) is gratefully acknowledged for supporting this work.

Prof. Kohlheb Norbert and the Szent Istvan University (Hungary) are thanked for providing support in the agronomical and economic characterization of the examined region.

References

Akgul, O., Shah, N., & Papageorgiou, L., G. (2011). ESCAPE21 Journal, 1799-1803.

Čuček, L., Martín, M., Grossmann, I., E., Kravanja, Z. (2014). Computers and Chemical Engineering, 66, 57-70.

Giarola, S., Shah, N., & Bezzo, F. (2012). Bioresource Technology, 107, 175-185.

Giarola, S., Zamboni, A., & Bezzo, F. (2011). Computers and Chemical Engineering, 35, 1782-1797.

Kim, J., Sen, S., M., & Maravelias, C., T. (2013). Energy Environ. Sci., 6, 1093-1104.

Kokossis, A., C., & Yang, A. (2010). Computers and Chemical Engineering, 34, 1397-1405.

Martín, M., & Grossmann, I., E. (2010). ESCAPE20 Journal, 943-948.

O'Donohue, M. (2014). BIOCORE – FINAL REPORT, Institut National de la Recherche Agronomique (INRA), Toulouse, France.

Rosenthal (2014). GAMS – A User's Guide. GAMS Development Corporation, Washington, DC, USA.

Zamboni, A., Shah, N., & Bezzo, F. (2009). Energy & Fuels, 23, 5121-5133.

463402 Application of an Artificial Bee Colony Algorithm for Constrained Optimization Problems

Abstract:

The goal of this project is to develop a stochastic optimization algorithm, Artificial Bee Colony (ABC) for the computer-aided molecular design of novel antibiotics. The design of a novel antibiotic can be formulated as a non-convex, non-linear combinatorics problem, which has a very large search space and fairly inexact constraints, such the use of deterministic methods is inefficient. To think of this problem in a simpler manner, the optimization problem resulting from the CAMD approach can be mathematically represented by the Knapsack Problem (KP), in which the total value of items, each with their own value and weight, in a knapsack is optimized without exceeding the weight constraint. This project focuses on the testing and tuning of the ABC algorithm to find near-optimal solutions of the Knapsack Problem. The 0-1 KP is solved first, followed by the Bounded and Unbounded KPs. The algorithm is written in C++, and was first tested on two sets of 10 and 20 items each before moving on to larger search spaces. By varying only which items were included in or excluded from the knapsack, many potential solutions were found and the best solution was recorded for each instance that it was run. The algorithm consistently found solutions within 95% of the upper bound of the LP relaxation. To represent molecular design more accurately, the algorithm has been expanded and applied to the antibiotics design problem, requiring additional constraints so that multiple chemical and biological properties can be considered. The complete 2D structure of the novel antibiotics is obtained by the use of topological indices as structural descriptors.Â These initial results show the ABCâ€™s potential not only for antibiotics design, but for many other large scale problems in chemical product optimization as well.

Infrastructure Maintenance and Inspection Scheduling Under Partial State Observability

Jong Woo Kim, Go Bong Choi, Jong Min Lee*
skjw1224@snu.ac.kr

Nov. 14, 2016

School of Chemical and Biological Engineering,
Seoul National University, South Korea

Infrastructure Scheduling

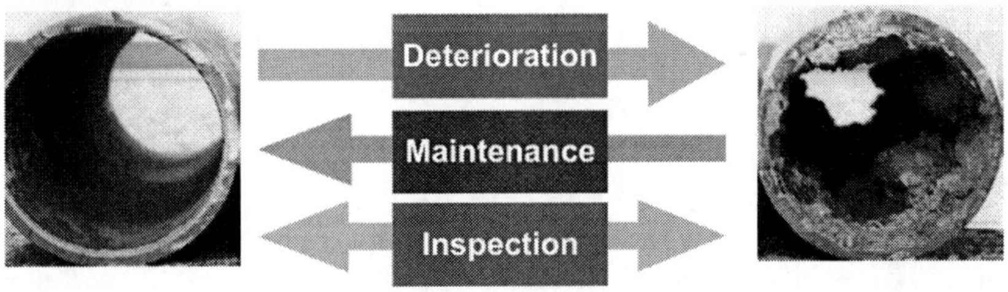

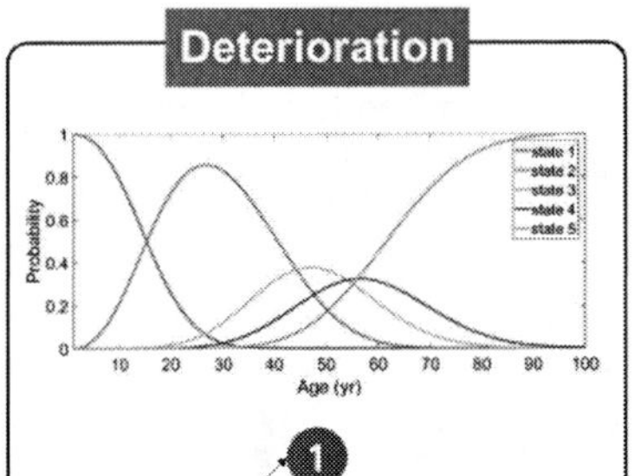

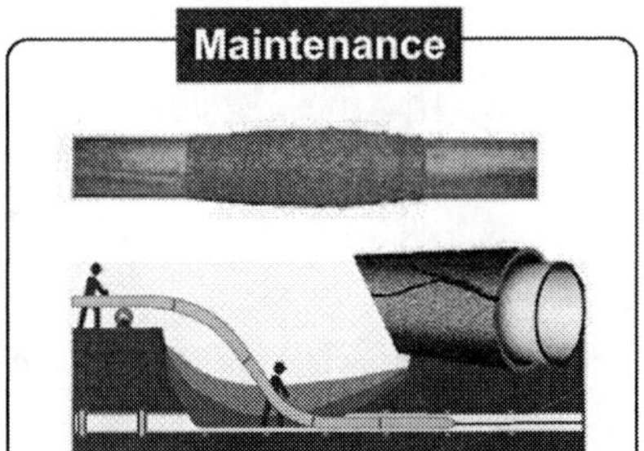

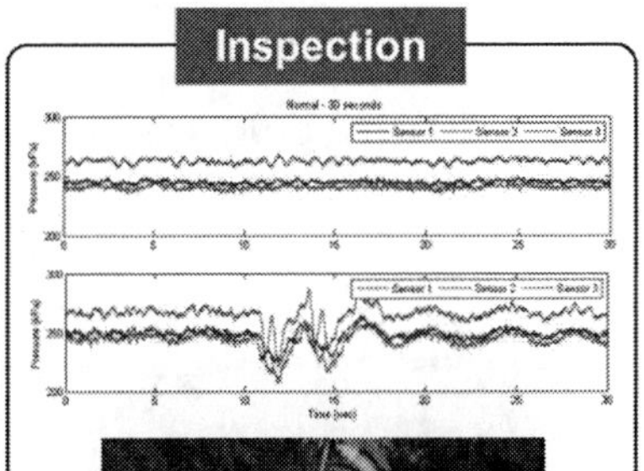

1

POMDP for Joint Action Scheduling

● Why POMDP?

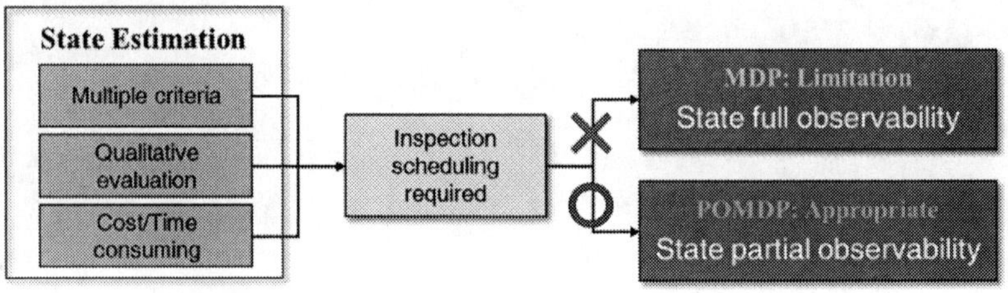

● POMDP framework

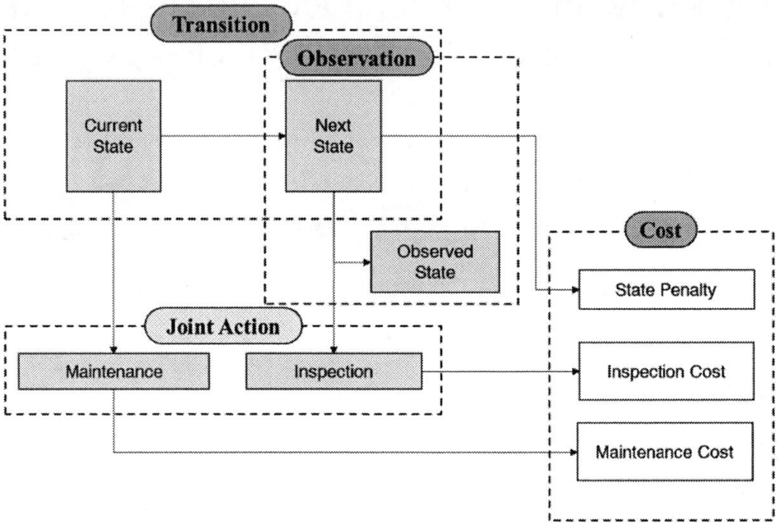

2

Simulation Result

● The optimal action sequence within finite horizon

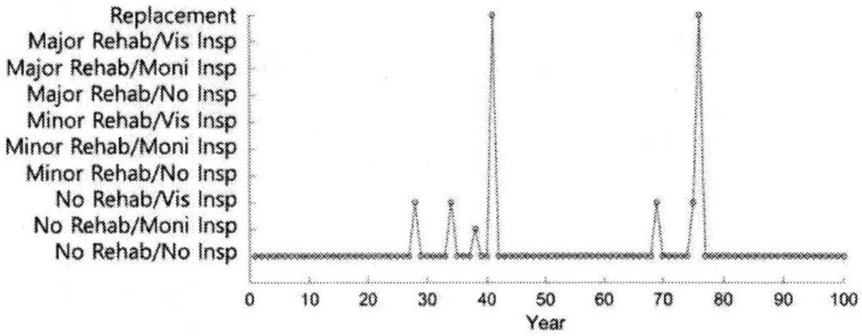

● Comparison with other policies

POMDP policy is ***optimal***

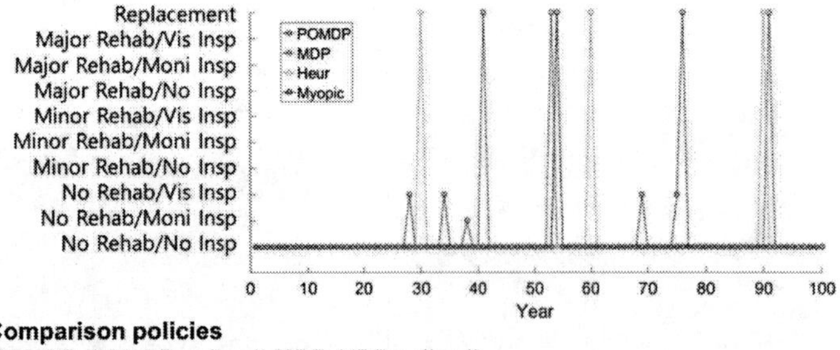

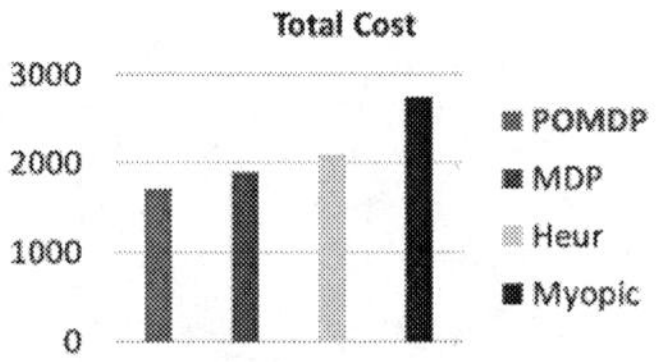

Comparison policies
POMDP: POMDP policy // ***MDP***: MDP policy //
Heuristic: Replace every 30 years // ***Myopic***: Replace after failure occurs

3

Proactive-reactive scheduling under uncertainty using approximate dynamic programming

Go Bong Choi, Jong Woo Kim, Jong Min Lee

Seoul national university

November. 13. 2016

School of Chemical and Biological Engineering

Seoul National University

16 AICHE

neocompty@snu.ac.kr

Introduction

○ Integration of planning and scheduling

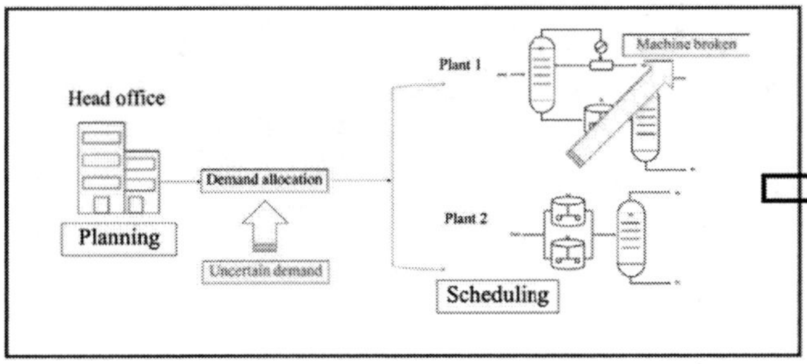

Very large size stochastic optimization

Simultaneously solving : Computational complexity

Separately solving : Non-optimal

○ Surrogate model using value function (Introduce function that connect with planning and scheduling)

⇒ This inherent computational complexity : approximate dynamic programming approach

(a recursive manner with Monte Carlo simulation and newly observed data)

Formulation

○ Framework

Maximum expected amount of production ⇒ represented by value function	Planning : Minimizing total cost under meeting demand Scheduling : Meeting demand at appointed time	Assignment of each amount of production ⇒ Give a parameter for low level optimization

If Uncertain event occurs,
→ If only infeasible solution, planning can assign new schedule for each plant.
→ If feasible, new schedule is made.

○ Application

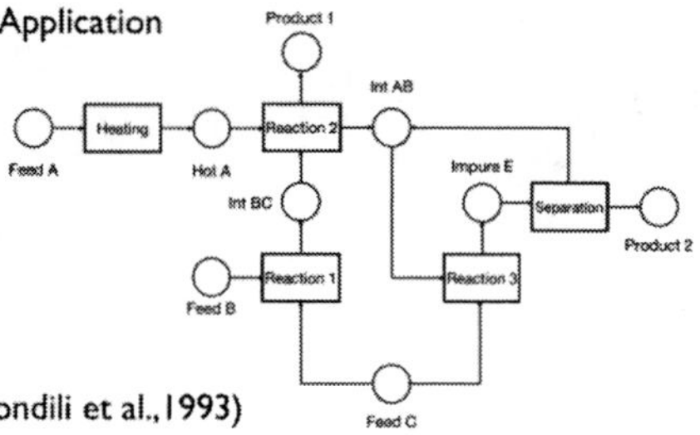

(Kondili et al.,1993)

○ 2 Case

Case 1) Machine breakdown

Case 2) Maximizing Product 2 under meeting demand of Product 1

Result & Discussion

Online simulation using ADP policy : Approximated value function vs real amount of product

○ Case 1 : Machine breakdown

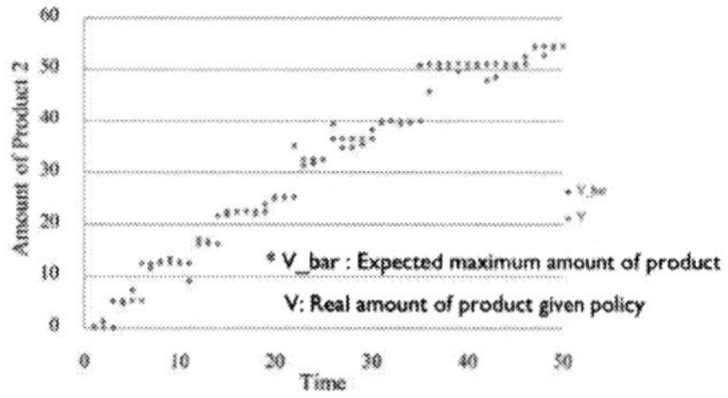

○ Case 2 : Maximizing Product 2 under meeting demand of Product 1

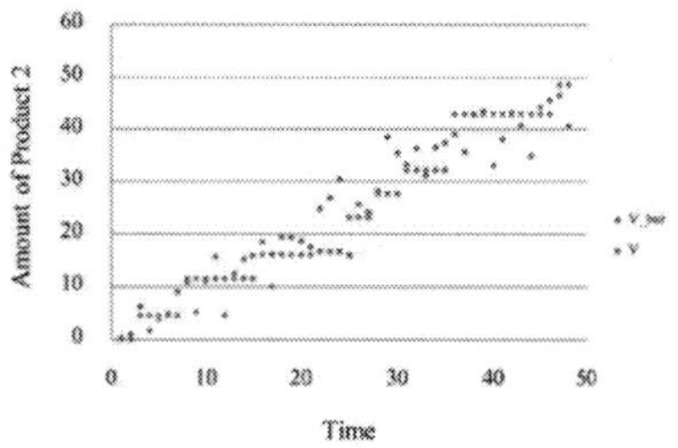

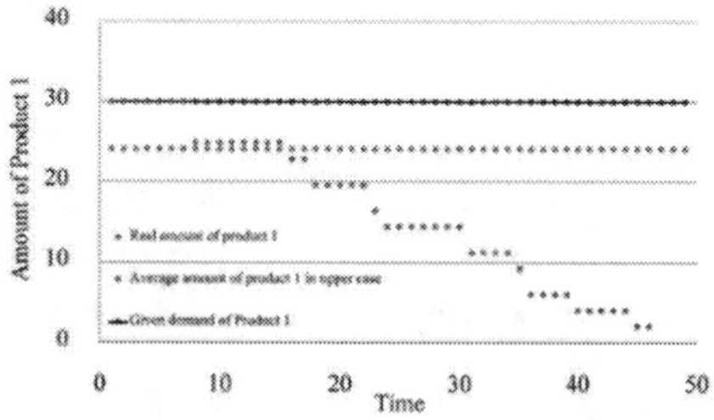

2016.11.14

Performance and dynamic behavior of Shell entrained-flow gasifier by coals with different moisture and rank in a 300 MW IGCC power plant

Youngsan Ju[1]
(Yonsei University, wndudtks@yonsei.ac.kr)

Hyeon-Hui Lee[1], **Min Oh**[2] **and Chang-Ha Lee**[1]*

[1]Department of Chemical and Biomolecular Engineering, Yonsei University, Seoul, Korea
[2]Department of Chemical Engineering, Hanbat National University, Daejeon, Korea

Gasification process with various applications

- **Gasification process** with various fuel sources

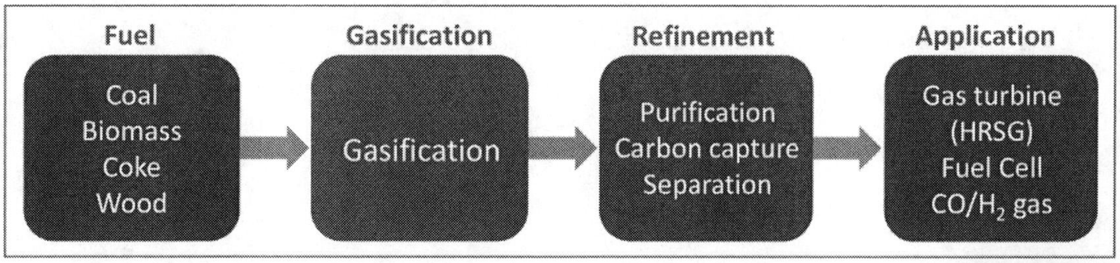

  - Gasification is a process to convert organic or fossil fuel based carbonaceous materials into CO, H_2, CH_4 and CO_2.
  - It is achieved at harsh condition(>700°C, high pressure) without combustion.
  - It offers advantages over **easier emissions control** (NO_x, SO_x, mercury, arsenic, ...) and **efficient CO_2 removal**
- **Coal gasification process** in integrated gasification combined cycle
  - Due to harsh condition of gasifier, dynamic simulation is necessary before process simulation.
- **Dynamic model** for gasification process

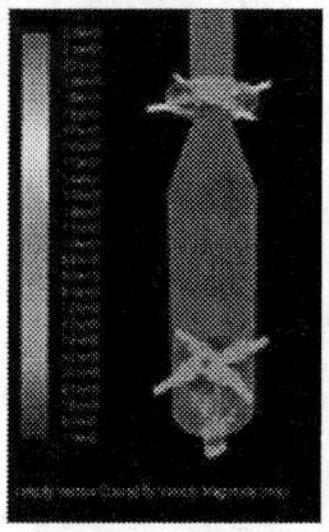

Fig. CFD simulation of Shell gasifier

Separation & Purification LAB.

16AIChE

2

Dynamic model of Shell entrained-flow gasifier

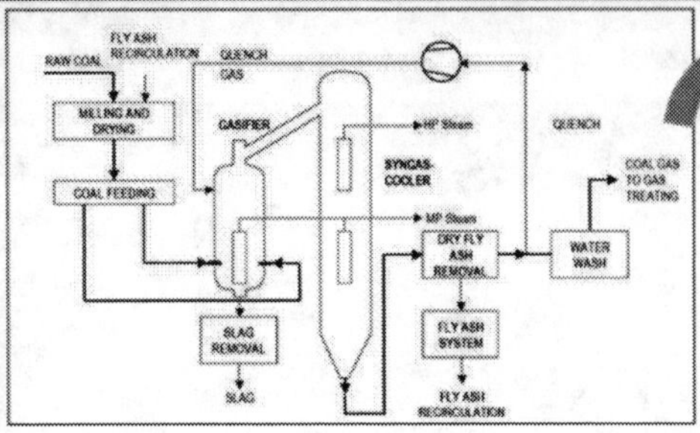

Fig. Simplified SCGP Flow Scheme
Source: National Energy Technology Laboratory – Department of Energy

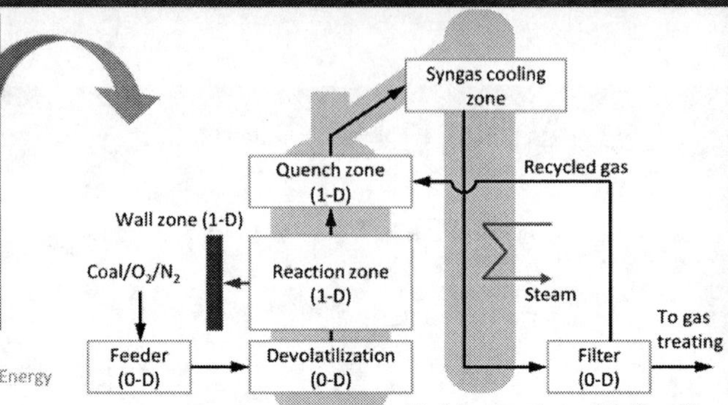

Fig. Modeling topology of Shell gasifier

44 Main governing equations with >25 sub-equations

Example of the governing equations in reaction zone
- Mass/energy/momentum balance for Gas phase species i

$$\frac{\partial(A_{cs}\varepsilon_g C_{gi})}{\partial t}=\frac{\partial}{\partial z}\left(A_{cs}D_{g,eff}\frac{\partial(\varepsilon_g C_{gi})}{\partial z}\right)-\frac{\partial(A_{cs}u_g\varepsilon_g C_{gi})}{\partial z}+A_{cs}(MS_{gi}^{Hetero}+MS_{gi}^{Homo})$$

$$\frac{\partial(A_{cs}\varepsilon_g C_{g,tot}c_{p,g}T_g)}{\partial t}=\frac{\partial}{\partial z}\left(A_{cs}k_{g,eff}\frac{\partial T_g}{\partial z}\right)-\frac{\partial(A_{cs}u_g\varepsilon_g C_{g,tot}c_{p,g}T_g)}{\partial z}+A_{cs}(HS_g^{Homo}+HT_g)$$

$$\frac{\partial(A_{cs}\varepsilon_g\rho_g u_g)}{\partial t}=-\frac{\partial(A_{cs}u_g^2\varepsilon_g\rho_g)}{\partial z}+A_{cs}\left(\frac{\partial P}{\partial z}+\rho_g\varepsilon_g g-F_{g,w}-F_{g,p}\right)$$

Example of the sub-equations in reaction zone
- Heat transfer between gas, char, gasifier wall

$$Q_{conv,pg}=A_p h_{pg}(T_p-T_g)\qquad Q_{conv,gw}=A_w h_{gw}(T_g-T_w)$$

$$Q_{rad,pw}=A_p(\varepsilon_p\sigma T_p^4-\varepsilon_w\sigma T_w^4)$$

- Friction between gas, char, gasifier wall

$$F_{g,p}=\frac{3\varepsilon_p C_D\rho_g\varepsilon_p^{-2.65}(u_g-u_p)^2}{8(d_p/2)}\frac{(u_g-u_p)}{|(u_g-u_p)|}\qquad F_{g,w}=\frac{f\rho_g u_g^2}{16(d_p/2)}$$

Separation & Purification LAB.

16AIChE

Detailed model description and further application

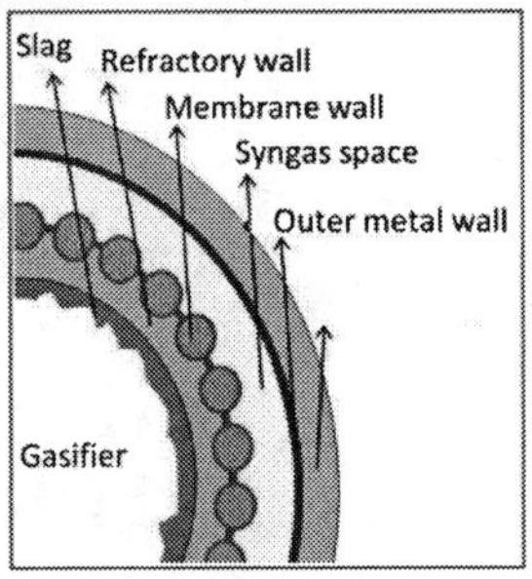

Wall zone with 5 sub-models

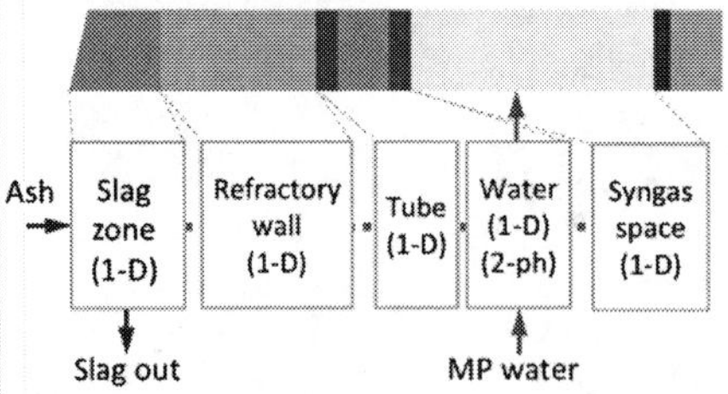

Fig. Schematic diagram of the wall model

Example of the equations in 2-phase water wall
- Energy balance equation with 2-phase heat transfer coefficient

$$A_{cs,j}\cdot\rho_{H_2O}\frac{\partial h_{mass,H_2O}}{\partial t}=-\frac{\partial(\dot m_{H_2O}h_{mass,H_2O})}{\partial z}+2h_{tp}P_{surf,34}(T_3-T_{H_2O})$$

Where, $h_{tp}=\left((h_L F_{tp})^3+(h_{nb,p}F_{nb})^3\right)^{1/3}$

$$F_{tp}=\left[(1-x^{vap})^{1.5}+1.9(x^{vap})^{0.6}\left(\frac{\rho_L}{\rho_G}\right)^{0.35}\right]^{1.1}\Bigg|_{x\le0.6}$$

$$F_{nb}=\left(1.73(p_r)^{0.27}+\left[6.1+\frac{0.68}{1-p_r}\right](p_r)^2\right)\left(\frac{q_{flux}}{q_{flux,0}}\right)^{(0.9-0.3(p_r)^{0.15})}$$

Dynamic model → **Validation** → **Applications**

Main models
Sub-models
- Reaction kinetic model
- Heat transfer model
- Fluid dynamic model
- Slag model
- 2-phase water model
- ...

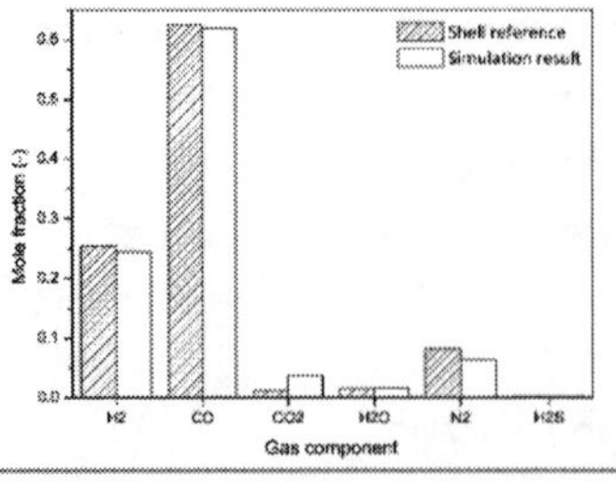

Fig. Validation with syngas composition at the outlet of quenching zone in a real 300MW gasifier

- Improving the gasification process operation more efficient or environment-friendly
- Evaluation of the gasifier performance by varying coal ranks, operating condition and drying level of the coal
- Optimization of operating conditions, depending on the coal types

Poster in #249 session
Abstract No.:189g

Separation & Purification LAB.

16AIChE

466912 Understanding Rare Safety and Reliability Events Using Transition Path Sampling

Abstract:

In the chemical manufacturing industry, processes and their control systems are typically well-designed to mitigate events that have the potential to lead to plant shutdowns or safety events – those that often result in human health losses or environmental impacts. Clearly, there is strong motivation to understand how these events develop and propagate. Effective operator training, safety system design, and safety analysis, all benefit from a full understanding of such events. A major challenge in the study of events that propagate to process shutdown or safety incidents is their sparsity – typically these events occur so rarely that statistical techniques alone are incapable of describing and characterizing them – especially when they have not yet occurred. Simulation of these events could be useful to understand them, however, a daunting computational challenge exists. Typical rare events occur on the order of years or decades apart, yet the events occur within minutes or hours. Thus, the bulk of the computational effort in simulating rare events is allocated to normal operation, making the event computationally infeasible to simulate with meaningful frequency.

A rare-event sampling technique, Transition Path Sampling (TPS), has been developed by the molecular dynamics community. While the time and length scales between molecular dynamics and process dynamics differ greatly, the ratios of the time of the rare events and the waiting time between them are similar. This Monte-Carlo based technique relies on the simulation of perturbed rare-event trajectories – an initial rare-event trajectory is randomly modified such that large numbers of trajectories are generated. Clusters of rare safety-event trajectories can be the basis for alarm and safety-system design, assuring that TPS-generated clusters are preventable. Important modifications to the TPS technique are needed to apply it to process dynamics. The backwards integration, a key attribute of TPS, is not possible for most process simulations – instead a boundary-value optimization technique is used. Furthermore, process models use of the vast amount of process data available for model verification and to estimate the relative likelihood of

one trajectory to another. The application of TPS will be demonstrated using a simple jacketed exothermic CSTR, as well as a more complex air separation unit process. This innovative approach allows for a quantitative rationalization of alarm and safety systems to prevent rare, yet serious, safety events.

467480 Chance-Constrained MINLP Optimization for the Process Synthesis of the Oxidative Coupling of Methane

Abstract:

Uncertainty in process models is of vital importance for most process synthesis tasks. This is especially the case when a close interrelation between process unit outputs, such as required product purities, and parametric uncertainty exists. In recent contributions, such as (Quaglia et al., 2013) and (Steimel et al., 2014), frameworks have been presented, which incorporate uncertainty in mixed-integer (non-)linear programming problems associated with superstructure optimization in chemical engineering. Most approaches published so far follow along the lines of a two stage technique with an outer stage in charge of making the superstructure decisions on the design and the inner stage reacting to a set of different scenarios generated from a description of the uncertainty. Consequently, these frameworks require a great number of simulations and the structural optimization reacts only indirectly to the effect of the uncertainty. As an alternative approach to tackle the aforementioned class of optimization problems, Esche et al. (2016) introduced a chance-constrained MINLP approach. The framework incorporates uncertainty into process synthesis tasks by requiring the adherence to probabilistic process constraints, namely chance constraints.

In this contribution, a comparison of the results of the chance-constrained MINLP framework to deterministic, robust, and other sampling-based methods (e.g. Monte Carlo sampling) for the introduction of uncertainty into the optimal process synthesis task is carried out. The oxidative coupling of methane process is employed as an exemplary process synthesis task. The process concept considers choices on different reactor feeding policies and various downstreaming options. The latter consist of combined pressure and temperature swing adsorption, combinations of gas permeation membranes, and absorption desorption processes using different scrubbing liquids for the removal of carbon dioxide from the product stream.

In addition to the comparison of the results, options for the extension of the chance-constrained optimization

framework for larger scale systems, greater numbers of uncertain variables and chance constraints, as well as the interfacing with additional solvers is discussed.

ReferencesÂ

Quaglia, A., Sarup, B., Sin, G., Gani, R. (2013). A systematic framework for enterprise-wide optimization: Synthesis and design of process networks under uncertainty. Computers & Chemical Engineering, 59:47â€"62.

Steimel, J., Harrmann, M., Schembecker, G., Engell, S. (2014). A framework for the modeling and optimization of process structures under uncertainty. Chemical Engineering Science, 115:225-237

Esche, E., MÃ¼ller, D., Werk, S., Grossmann, I.E., Wozny, G. (2016). Solution of Chance-Constrained Mixed-Integer Nonlinear Programming Problems. In: Proceedings of the 26th European Symposium on Computer Aided Process Engineering, June 12th-15th, 2016

467416 Health-Aware Operation of a Subsea Gas Compression Station Under Uncertain Operating Conditions

Abstract:

Subsea oil and gas production systems have stringent requirements to availability and reliability due to the very high intervention cost, and safety and environmental consequences of failure. The need for specialized lifting vessels and possibly harsh climate conditions can lead to production stops for months, leading to significant economic losses. At the same time, low oil prices require high efficiency and small margins in order to ensure competitive production. The current trend in industry is to move production and processing equipment from topside to the seabed, which will increase the complexity of the system and the available degrees of freedom by a manifold. In order to ensure optimal operation, a systematic approach is needed to include information about the health state of the system in the operation strategy.

Traditionally, prognostics and health monitoring (PHM) systems are used by the operators to schedule maintenance. However, the information is seldom used to adjust the optimal operation strategy. Typically, the operators will choke back production until the estimated remaining useful life (RUL) is comfortably within the acceptable range. While this approach might work when there are only a few degrees of freedom that can be adjusted, it becomes difficult to ensure optimal for more complex processes.

Escobet et al. propose a control structure that combines PHM and process control, resulting in a health-aware control structure (Escobet, Puig, & Nejjari, 2012). The proposed control structure is able to ensure optimal operation without compromising the integrity of the system. Their approach involves increasing or decreasing the set-point of a PI controller by a fixed value until the predicted RUL is within an acceptable range, essentially making it an automated version of the current practice. While the algorithm is faster and more accurate than an operator, the set-point selection algorithm can result in cases where the proposed set-point is suboptimal or infeasible.

Pereira et al. propose a model predictive control (MPC) structure that utilizes prognosis and health monitoring of actuators to ensure optimal operation (Pereira, Galvão, & Yoneyama, 2010). By using an MPC, multivariate systems can also be treated. Reliability requirements are included by adding RUL constraints in the optimization routine. A similar approach is found in (Salazar, Weber, Nejjari, Theilliol, & Sarrate, 2016)

We extend the results of Escobet, Pereira, Salazar and coworkers by considering a stochastic degradation model and a nonlinear process model, with a nonlinear MPC formulation. We utilize the concept of health-aware control to derive a control structure for a subsea production system. A statistical model is used to predict the evolution of the remaining health of the system. Model uncertainty is treated by linearizing around worst case realizations, resulting in a robust control structure (Diehl, Bock, & Kostina, 2006). We also consider the case where the maintenance horizon is a decision variable, rather than optimizing over an a priori horizon. As an example, we consider a subsea gas compression station.

References

Diehl, M., Bock, H. G., & Kostina, E. (2006). An approximation technique for robust nonlinear optimization. *Mathematical Programming*, 213-230.

Escobet, T., Puig, V., & Nejjari, F. (2012). Health Aware Control and model-based Prognosis. *Mediterranean Conference on Control & Automation (MED)* (pags. 691-696). Barcelona: IEEE.

Pereira, E. B., Galvão, R. K., & Yoneyama, T. (2010). Model Predictive Control using Prognosis and Health Monitoring of actuators. *International Symposium on Industrial Electronics* (pags. 237-243). Bari: IEEE.

Salazar, J. C., Weber, P., Nejjari, F., Theilliol, D., & Sarrate, R. (2016). MPC framework for system reliability optimization. En Z. (. Kowalczuk, *Advanced and Intelligent Computations in Diagnosis and Control* (pags. 161-177). Springer.

470754 Optimal Grade Transitions for a Propylene Polymerization Loop Reactor

Abstract:

Polyolefin plants operate under continuous production, but they need to supply many types of polymer grades (polymer with different specification, used in different applications). The operating conditions are maintained constant during the production of a certain grade and need to change periodically to produce another grade. Grade transitions allow the production of different polymer grades in a single reactor, but as they require changes in the operating conditions, there is a large production of off-specification polymer. Therefore, a desired transition drives the polymer properties to the new grade in a short period of time, producing a small amount of off-specification polymer. Efficient strategies to operate the transition can be obtained by solving a problem of dynamic optimization. In this work, a quadratic integral objective function was minimized by using dynamic sequential techniques for solving optimization. The results were optimized for the copolymerization of propylene with ethylene. The results show that to increase theÂ *MIÂ* was necessary to reduce the concentration of hydrogen in the feed and increase the comonomer flow rate; and to decrease the density was necessary to increase the comonomer flow rate. It was observed that the instantaneous properties have faster and aggressive dynamics compared to the cumulative properties. The trajectories of the transition properties towards grade 1 - grade 2 showed different behavior compared to grade 2 â€“ grade 1 transition, concluding that the profile of a transition depends strongly on the direction of change in the properties.

472504 Dynamic Global Sensitivity Analysis on Wastewater Stabilization Pond Networks

Abstract:

In this work, a wastewater treatment network consisting of a system of ponds is modeled rigorously, taking into account dynamic mass balances for the main groups of bacteria, together with different types of organic load, algae biomass and nutrients. A dynamic global sensitivity analysis (DGSA) is performed on the kinetic parameter of the model system for the optimal configuration of three stabilization ponds, obtained in a previous work (Ochoa et al., 2016). The system is comprised of two aerobic ponds in series followed by a facultative one. The main objective is the determination of the most influential parameters of the model and parameter ranking, considering the whole range of parameters variation, allowing the identification of the uncertaintyÂ sources and thus determining the needed experimental measures.

DGSA is implemented using a variance based technique proposed by Sobolâ€™ (1993). Other first order and total order sensitivity estimators are also studied (Saltelli et al., 2010). An important characteristic of the method is the possibility to detect interactions between uncertain parameters. The technique is implemented within gPROMS platform, a differential algebraic equation oriented environment where stochastic simulations are performed. Temporal profiles for the first order, total order and interactional sensitivity indices are calculated for 18 differential variables taking into account 20 parameters as uncertain.

Results show a great influence of the temperature coefficient (Î¸) over all differential variables along the entire horizon of time, not only by its first order sensitivity effects but also throughÂ its interactional effect. There is also a notable influence of the mass transfer coefficient between layers (k_m) in the facultative pond. Global sensitivity analysis results provide a valuable insight into model features.

References

Ochoa, M.P., Estrada, V., Hoch, P.M., (2016) MINLP Wastewater Stabilisation Ponds Synthesis using Rigorous Models under Different Scenarios, Computer Aided Chemical Engineering.

Saltelli, A., Annoni, P., Azzini, I., Campolongo, F., Ratto, M., Tarantola, S., (2010) Variance based sensitivity analysis of model output. Design and estimator for the total sensitivity index, Computer Physics Communication 181, 259-270.

Sobol', I. M. (1993) Sensitivity estimates for nonlinear mathematical models. Mathematical Modelling & Computational Experiment, 1, 407-414.

Optimization of a Dual PSA Process for the Separation of Mixture Gas consisting of Methane and Carbon Dioxide

Seungnam Kim[a], Daeho Ko[b], Il Moon[a*]
[a]Department of Chemical and Biomolecular Engineering, Yonsei University; [b]GS E&C

16AIChE Annual Meeting, San Francisco, CA

To support CCS,

- Pressure swing adsorption (PSA) has been widely used in separating or purifying mixed gases.
- Generally available for carbon capture, operates over a wide range of pressure and temperature, and offers a competitive power consumption rate in the separating field.

The Dual-PSA process

- Suggested by Grande and Blom (2012)
- Consists of two units: (1) a rectifying unit and (2) a stripping unit.
- The additional PSA (stripping) unit purifies the waste flow from the conventional PSA (rectifying) unit.

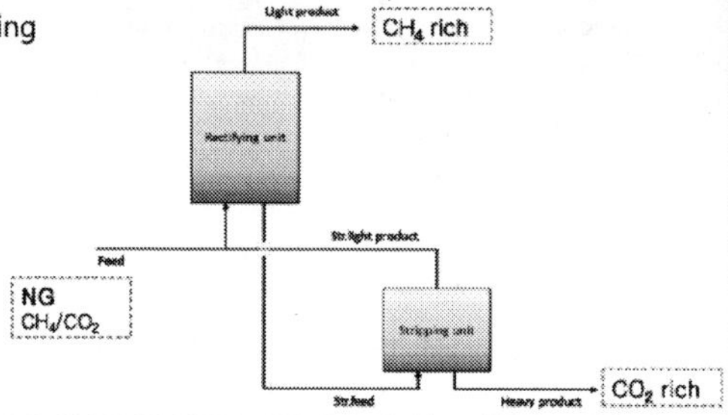

System description

Adsorbent: CMS
Adsorption isotherm: LF

Feed composition: CH4(85%)/CO2(15%)
Feed pressure: 10 atm

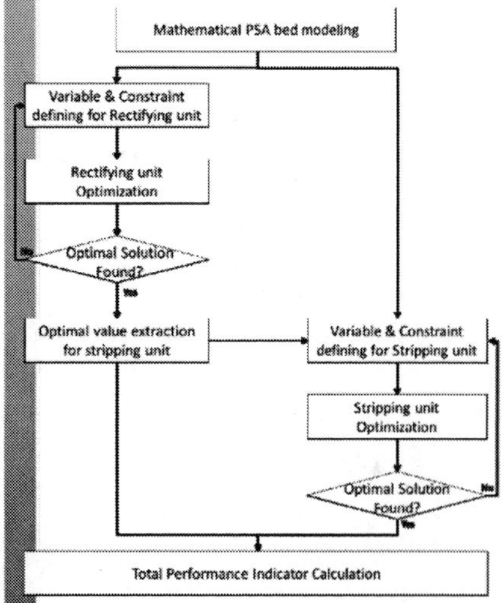

Objective

- Max. CH_4 recovery for rectifying unit
- Max. CO_2 purity for stripping unit

Decision variable

t_{PR} (s) t_{AD} (s)
t_{DP} (s) t_{RG} (s)
u_{ad} (m/s)
P/F ratio
$STR.L_{bed}$

Retrofit design for stripping unit

Constraint

$-0.00001 \le y_{prod,CH_4} - target\_purity \le 0.00001$

$-0.0001 \le t_{PR} - t_{DP} \le 0.0001$

$-0.0001 \le t_{AD} - t_{RG} \le 0.0001$

$-\varepsilon \le C_{ini,i,z} - C_{final,i,z} \le \varepsilon$

$-\varepsilon \le T_{ini,z} - T_{final,z} \le \varepsilon$

$-\varepsilon \le q_{ini,i,z} - q_{final,i,z} \le \varepsilon$

$-0.0001 \le Mole\_feeding - Mole\_waste\_Rec \le 0.0001$

$-0.0001 \le Cycletime_{STR} - Cycletime_{REC} \le 0.0001$

Optimization results

	Rectifying unit	Stripping unit
Objective	CH_4 recovery	CO_2 purity
	78.5%	76.3%
t_{PR} (s)	9	8
t_{AD} (s)	187	188
t_{DP} (s)	9	8
t_{RG} (s)	187	188
t_{cycle}	394	394
P/F ratio	0.109	0.045
u_{ad} (m/s)	1.160	0.339
L_{bed} (m)	1.55	0.659

LP: Light Product HP: Heavy Product		Conventional PSA	Dual PSA	△
Product Recovery	CH4 (LP)	78.5%	95.4%	16.9%
	CO2 (HP)	86.2%	83.3%	-2.97%
Product Purity	CH4 (LP)	97.0%	97.0%	-
	CO2 (HP)	41.4%	76.3%	34.9%

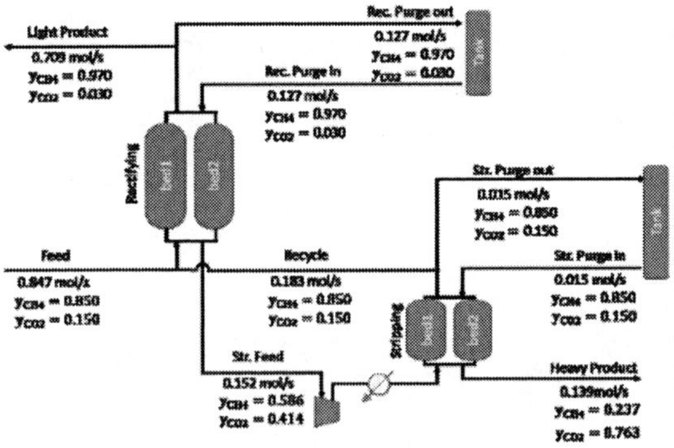

Interval No.		1	2	3	4	5	6
Time (sec)		8	1	187	8	1	187
Rectifying Unit	Bed 1	PR	PR	AD	DP	DP	RG
	Bed 2	DP	DP	RG	PR	PR	AD
Stripping Unit	Bed 1	PR	AD	AD	DP	RG	RG
	Bed 2	DP	RG	RG	PR	AD	AD

Conclusion

- The performances of Dual-PSA are considerable improved by the dynamic optimization with operational variables.
- New Optimization framework for multi-bed and multi-stage PSA processes

469123 Optimal Decision-Making of Decentralized Multi-Participant Multi-Period Supply Chains under Uncertainty

Abstract:

Current Supply Chains (SCs) decision-support models focus on monopolistic situations leaded by one centralizing decision-maker. However, in practice, SCs comprise independent enterprises interacting with their contrasting objectives in a decentralized uncertain environment. Each enterprise decision-maker seeks to optimize its own revenues, by making decisions without complete information about the other enterprises' policies or their uncertain reaction. Many works have been carried out to optimize the tactical decision-making of decentralized SCs based on cooperative (Cao et al., 2013; Hjaila et al, 2016a) and non-cooperative systems (Yue and You, 2014; Hjaila et al., (2015, 2016b). However, to the best of our knowledge, there is still need to explore the strategic decision-making of decentralized SCs. Since the SC strategic level is a long-term investment, any strategic decision affects the participating enterprise bottom lines. The tradeoff between the payoffs of the different enterprises involved may lead to fundamental changes in the strategic decisions of the participating enterprises, which is a challenge to be explored and analyzed. Such an analysis requires the consideration of the uncertain reaction of any of the participants to the dynamic market.

Accordingly, this work proposes a generic bi-level model based on a non-cooperative non-zero sum non-symmetric role negotiation approach which has been extended from Game Theory. The main purpose is to state the best conditions for long-term expected win-win coordination/collaboration contracts. Different net payoff scenarios which compare the principal game player with the standalone case are analyzed. This analysis is carried out considering the capacity uncertainty of one of the partners, and their effect on the optimal strategic and tactical decisions of each enterprise are evaluated until the best compromise between their long term payoffs is reached. A penalty cost is proposed as if one of the contract options (e.g., quantity demanded at the short-term) reaches a maximum, so as to avoid risks or disruptions in the other player's SC. As a result, a set of long-term coordination contracts, including the annual resources (flows among the game players, the unit transfer price, and the penalty) is proposed to cope with the different risk-behavior decision-makers.

The proposed approach and its potential are demonstrated using a case study based on a multi-product multi-period global SC, which is analyzed from the points of view of different participants. The proposed generic collaboration model is flexible enough to allow dealing with all possible links between the enterprises SCs nodes. It supports strategic and tactical decisions such as best locations and capacities for new facilities, capacity expansions, equipment technology, raw material acquisition, production, storage, and distribution programs, etc. The results show how the tradeoff between the payoffs of the enterprises involved affects the strategic and tactical decisions of the whole system, thus stressing the necessity of considering this wider view of the different enterprises' options in order to improve the enterprise-wide coordination.

Acknowledgements: Financial support from the Generalitat de Catalunya (AGAUR FI program and grant 2014-SGR-1092-CEPEiMA) is fully appreciated.

References:

Cao E., Wan C., and Lai M. (2013). Coordination of a supply chain with one manufacturer and multiple competing retailers under simultaneous demand and cost disruptions. Int. J. Prod Econ, 141,425–433.

Hjaila K, Laínez-Aguirre JM, Puigjaner L, and Espuña A. (2015). Decentralized manufacturing supply chains coordination under uncertain competitiveness. Procedia Eng, 132, 942–9.

Hjaila K, Laínez-Aguirre J.M., Zamarripa M, Puigjaner L. and Espuña A. (2016a). Optimal integration of third-parties in a coordinated supply chain management environment, Computers and Chemical Engineering, 86, 48–61.

Hjaila K, Laínez-Aguirre J.M., Puigjaner L. and Espuña, A. (2016b). Scenario-Based Dynamic Negotiation for the Coordination of Multi-Enterprise Supply Chains under Uncertainty. Computers and Chemical Engineering, 91, 445–470.

Yue D, and You F. (2014). Game-Theoretic Modelling and Optimization of Multi-echelon Supply Chain Strategic and Operation under Stackelberg Game and Market Equilibrium. Computers and Chemical Engineering, 71, 347–361.

453130 Optimal Design of Feasible Clinical Tests for the Identification of Physiological Models of Type 1 Diabetes Mellitus

Abstract:

Type 1 diabetes mellitus (T1DM) is a metabolic disorder characterised by inadequate insulin secretion, causing damaging imbalance to the glucoregulatory system. Cutting edge therapies involve the use of a wearable artificial pancreas (WAP), that is a device capable of maintaining physiological glucose concentration through accurate insulin infusions. A WAP consists of three elements: a glucose sensor, an insulin pump and a control algorithm. Customisation of this device to the individual needs may be accomplished provided that a precise identification of the diabetes physiological model [1] used in the control algorithm is performed [2]. The identification procedure is typically carried out by means of clinical tests, such as the oral glucose tolerance test (OGTT) or the intravenous glucose tolerance test (IVGTT).

Model-based design of experiment (MBDoE) techniques have already been applied to the design of clinical tests for the identification of individual parameters in subjects with T1DM [3,4], however, the feasibility requirements are not always satisfied. Considering the possibility of improving standard clinical tests, this work applies MBDoE techniques to devise a new physiological test, whose pivotal feature is its implementation feasibility. In order to satisfy this condition, a novel approach based on heuristic indices is hereby introduced to assess several aspects that could impact the most on the practical applicability of the designed protocol. Therefore, the overall evaluation of the test is performed considering 5 indices, taking into account the following features: information, invasivity, safety, duration and number of samples.

A constrained MBDoE strategy, applied to the design of a modified OGTT lasting 5h [5], where sampling times are optimised while controlling insulin infusion and imposing constraints on safety, generates the optimal test shown in Figure 1a. In the MBDoE formulation, the lower bound on glucose concentration is treated as a hard constraint (hypoglycaemic conditions are harmful even in the short period) while the upper bound is treated as a soft constraint (hyperglycaemic conditions can be tolerated for short periods of time). The designed protocol (MOGTT) is easy to implement both in terms of insulin administration and glucose measurement and also guarantees safety conditions throughout the execution. Evaluation of the designed protocol through in silico studies attests its statistically satisfactory parameter estimation capabilities and its high robustness. Beyond its parameter identification performance, the MOGTT complies with the requirements imposed by feasibility. In fact, the effectiveness of this MBDoE approach is proved by a comparison between the OGTT and the MOGTT protocols in terms of the indices introduced (Figure 1b): safety is deeply improved and information obtainable is maximised.

Finally, the significant enhancement of the information index reflects on the MOGTT exhibiting excellent parameter estimation even upon utterly different parametric sets.

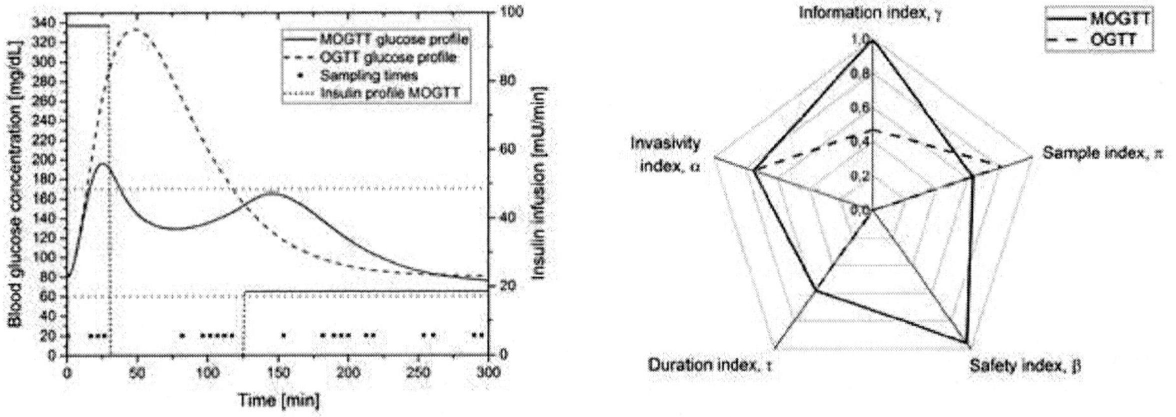

(a) *(b)*

Figure 1: Glucose profiles of the MOGTT and OGTT, along with sampling times and insulin infusion profile, horizontal dots represent glucose bounds imposed in the design (a). Indices radar chart comparing MOGTT and OGTT performance (b). References

[1] Balakrishnan, N. P.; Rangaiah, G. P.; Samavedham, L. *Review and analysis of blood glucose (BG) models for type 1 diabetic patients*. Ind. Eng. Chem. Res. 2011; 50(21):12041-12066.

[2] Cobelli, C.; Renard, E.; Kovatchev, B. *Artificial pancreas: Past, present, future*. Diabetes 2011; 60(11): 2672-2682.

[3] Galvanin, F.; Barolo, M.; Macchietto, S.; Bezzo, F. *Optimal design of clinical tests for the identification of physiological models of type 1 diabetes in the presence of model mismatch*. Med. Biol. Eng. Comput. 2011; 49(3): 263-277.

[4] Galvanin, F.; Barolo, M.; Macchietto, S.; Bezzo, F. *Optimal design of clinical tests for the identification of physiological models of type 1 diabetes mellitus*. Ind. Eng. Chem. Res. 2009; 48, 1989-2002.

[5] Man, C. D.; Campioni, M.; Polonsky, K. S.; Basu, R.; Rizza, R.A.; Toffolo, G. et al. *Two-hour seven-sample oral glucose tolerance test and meal protocol: Minimal model assessment of β-cell responsivity and insulin sensitivity in nondiabetic individuals*. Diabetes 2005; 54(11): 3265-3273.

Advances in DEM Computing Which Improve Predictive Capability for Processes

Howard Stamato[1], Aditya Vanarase[1], Wei Chen[1],
Preetanshu Pandey[1], Rahul Bharadwaj[2], Martin Hack[3], Lucilla Almeida[2]

(1) Bristol-Myers Squibb (2)Rocky DEM, Inc., (3)L.B. Bohle, LLC

Previous Efforts and Limitations

- **DOE:** limited prediction, experiments wasted if out of range, poor fidelity of response surface
- **Mechanistic Equation:** adjustable and difficult to measure parameters requires training set, limited fidelity
- **Computation DEM:** spheres or composites, limited number of particles, simplified geometry, slow computation

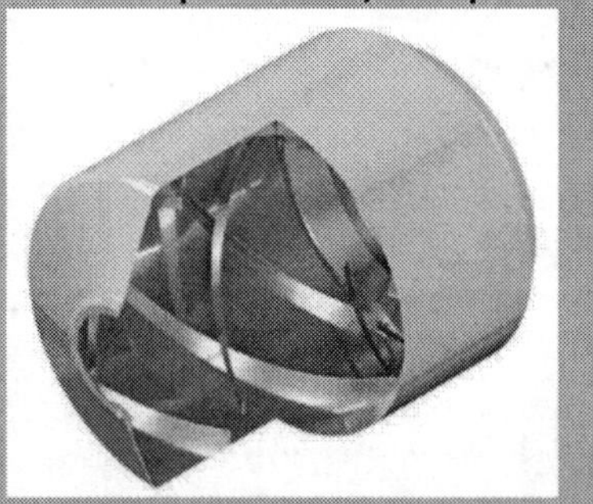

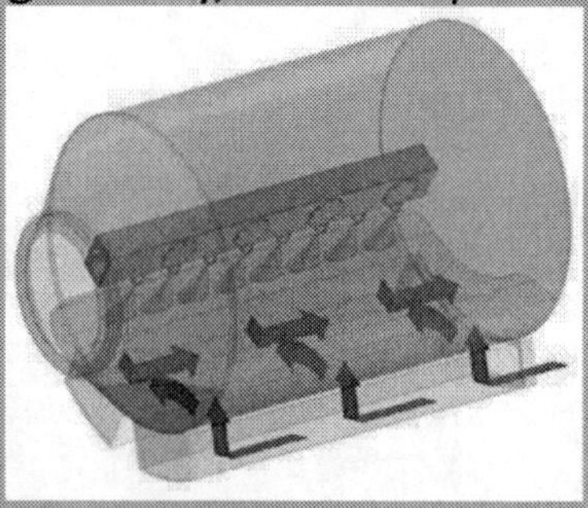

Model Challenges

- Number of factors considered
- Number of scales
- Transfer of data among model components
- Computing

"all models are wrong, but some are useful" - George E. P. Box

Rocky DEM Advantages

Geometry

Glued spheres

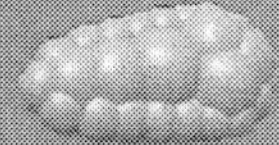

Polyhedral shape

sharp edges

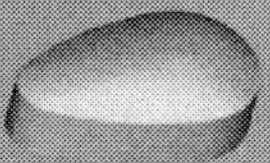

Polyhedra allow

- more accurate friction
- open to use of breakage models.

Solver Speed – GPU & CPU

Property	Value
Tablet size	0.88 X 0.41 X 0.30 [in]
Tablet vertices	222
Pan Load	350 kg
Number of particles	241379
Simulation per day	5 [s]

Post -Processing Tools

- Easy-to-use interface
- Time plots, histograms, residence time…
- Discrete to continuum analysis
- Volume fraction, velocity, stress tensor invariants
- Python integration – additional post-processing scripts
- Stream wise velocity plots

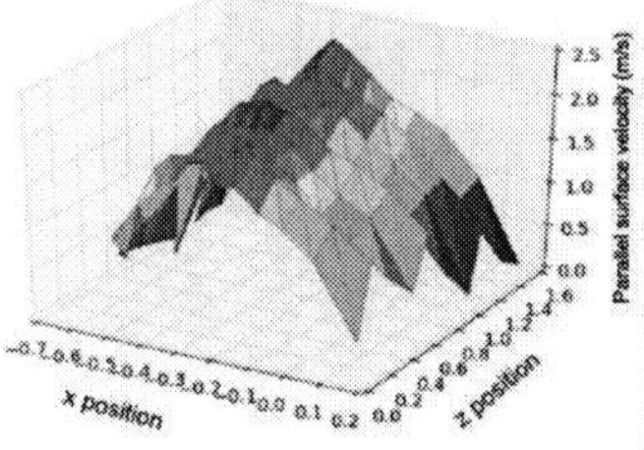

Rocky DEM Output

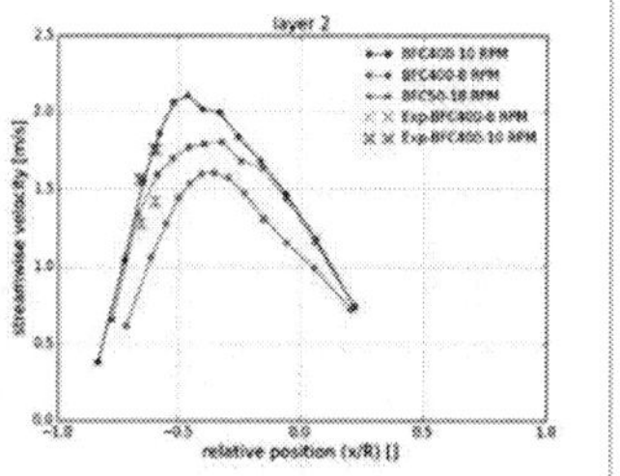

SHEAR STRESS

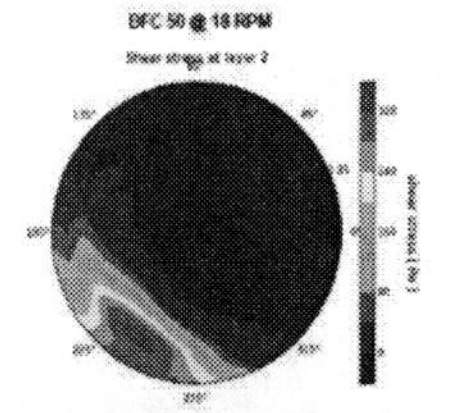

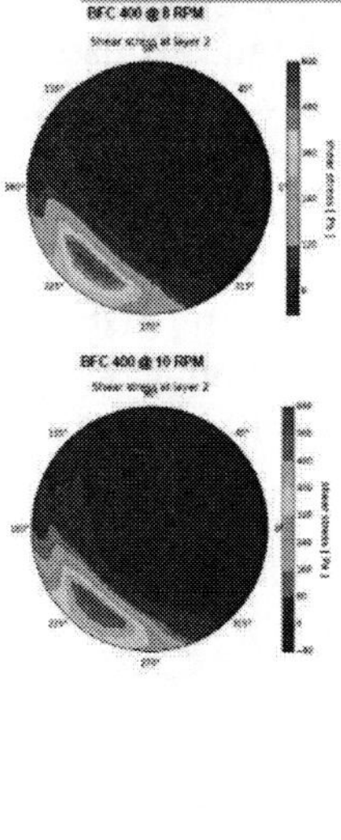

Coater	Max Shear Stress
BFC 50 @ 18 RPM	~ 320 Pa
BFC 400 @ 8 RPM	~ 600 Pa
BFC 400 @ 10 RPM	~ 640 Pa

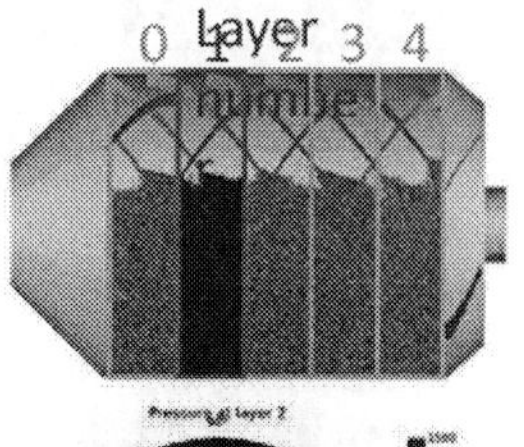

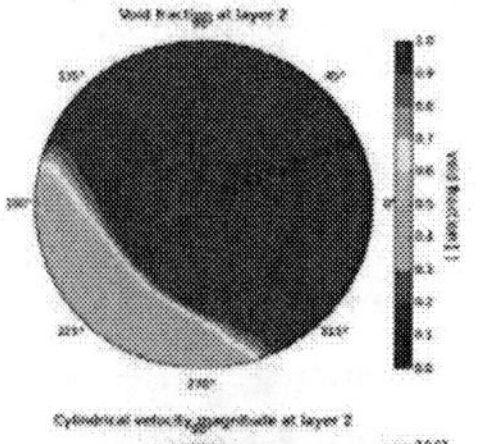

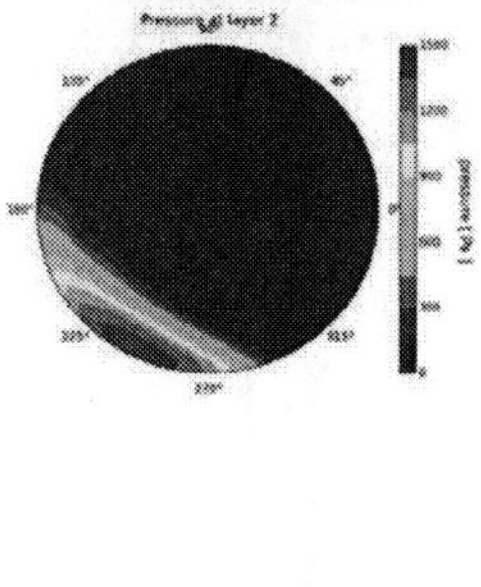

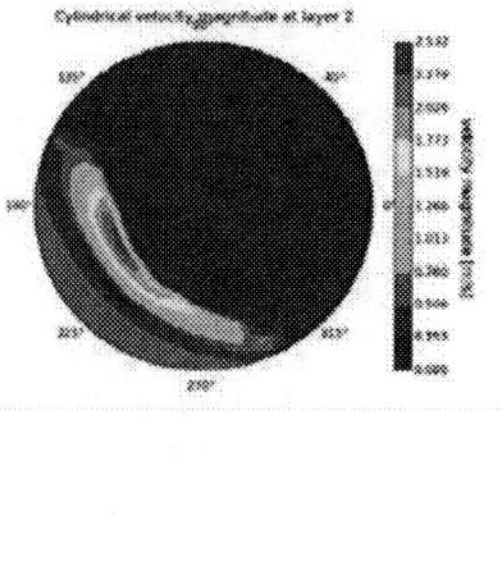

459784 Travelling Traders Exchange Problem: Stochastic Simulation and Sensitivity Analysis

Abstract:

In the traveling salesman problem (TSP) [1], given a list of cities and the distances between each pair of cities, the problem is to find the shortest possible route that visits each city exactly once and returns to the city of origin. The TSP problem is here recast and revisited as a travelling traders' exchange problem (TTEP), in order to analyse a population of N traders when the traders can move in space and interact with each other. The following assumptions are introduced for describing the TTEP: 1. The N traders travel across a country over time (i.e. the problem is dynamic); 2. The traders initially start with a certain amount of money M; 3. During a trading season lasting τ the traders are free to move over a territory (i.e. the problem is space-dependent); 4. Each time they meet they exchange money 5. The total amount of money is conserved.

The TTEP is a fundamental problem arising in a number of relevant applications including epidemiology [2], chemistry [3] and physics [4]. In all these studies, computational methods based on stochastic simulations [5] are frequently employed for the study and characterisation of these systems in which uncertainty exists. This method is widely employed when it is difficult to describe and analyse the system in a deterministic way. In particular, Bansal et al. [2] studied a stochastic simulation of a compartmental model in epidemiology; stochastic modelling was also discussed for thermal conductivity in harmonic lattices [4]. Stochastic simulation approaches with various extensions and modifications for chemical reaction processes are presented by a number of researchers [3] [6] [7]. Within a stochastic simulation model, some variables are randomly changing in time while the entire system evolves dynamically and presents a highly time-dependent outcome. The objective of this paper is to develop a stochastic simulation model for describing the TTEP to get insights in the emergent properties and evolution over time of the distribution of traders' accounts.

The TTEP stochastic simulation model developed here is a model within a bounded region (i.e. city or country), in which a number of travelling traders, each with an initially assigned amount of money, stochastically migrate. Both the migration of the traders and the potential money exchange will influence the amount of money every trader has in time. A sketch of a 1-D TTEP model is given in Figure 1a. The definitions of (*i*) initial allocation of amount of money to each trader and (*ii*) interaction mechanism of any two traders who collide are assumed during the construction of the stochastic simulation model. In the model, these influences and definitions are presented mathematically by a number of parameters including money exchange location L_i, money exchange direction D_i and money transfer coefficient k_i^a and k_i^p.

In the TTEP model, complexity is added sequentially for describing the mechanisms at different levels of detail and the computational results are related to the underlying assumptions 1-5 so as to provide expectations and compare various scenarios. Instead of studying the individual stochastic trajectories of each trader and his money, the time evolution of the probability distribution of the amount of money the traders hold in this stochastic system is of significant interest (Figure 1b). The probability distribution is described by a set of distribution parameters whose values and precision depends on the analysis of probability density function (PDF) and derivation of stochastic differential equations (SDEs). Results show how the impact of several parameters on the system can be mathematically described; a sensitivity analysis on each distribution parameter is carried out to evaluate the impact of the exchange mechanism on the outcomes.

Future work aims at extending and modifying the stochastic model so that an inverse estimate of the model parameters, based on optimisation techniques, can be carried out to relate model output and input. It is also of great interest to relate the parameters describing the exchange on a trader-by-trader basis to the emergent properties of the entire distribution by analytical techniques.

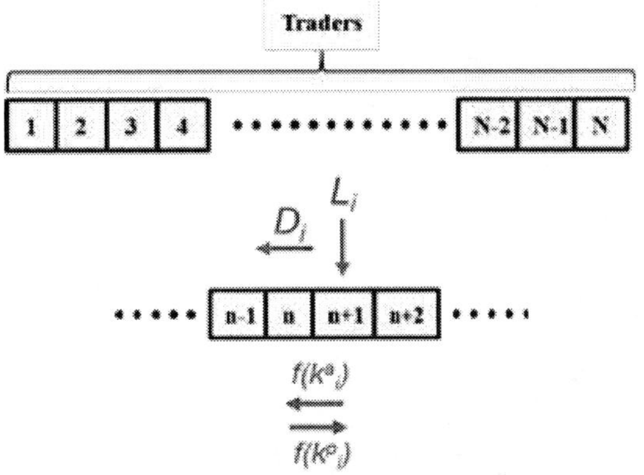

Figure 1a. Sketch of 1-D position-fixed TTEP model and i-th money exchange

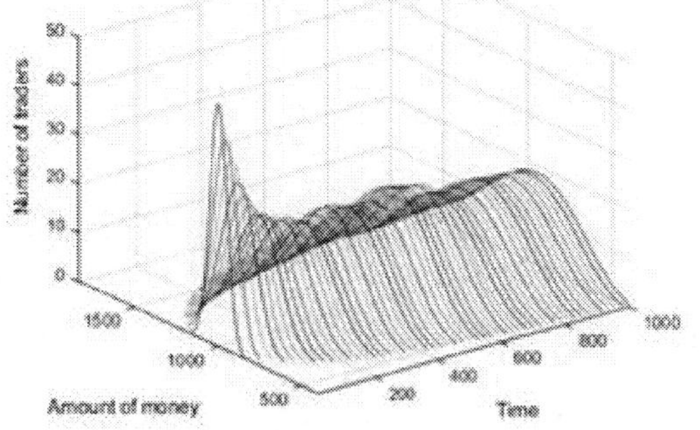

Figure 1b. Time evolution of the fitted money distribution in 1-D position-fixed TTEP model.

References

[1] D. L. Applegate, R. E. Bixby, V. Chvátal and W. J. Cook, The Traveling Salesman Problem: A Computational Study, Princeton University Press, 2011.

[2] S. Bansal, B. T. Grenfell and L. A. Meyers, "When individual behaviour matters: homogeneous and network models in epidemiology," *Journal of the Royal Society Interface,* vol. 4, no. 16, pp. 879-891, 2007.

[3] M. A. Gibson and J. Bruck, "Efficient Exact Stochastic Simulation of Chemical Systems with Many Species and Many Channels," *J. Phys. Chem.,* vol. 104, no. 9, pp. 1876-1889, 2000.

[4] G. Basile, C. Bernardin, M. Jara, T. Komorowski and S. Olla, "Thermal Conductivity in Harmonic Lattices with Random Collisions," in *Thermal Transport in Low Dimensions: From Statistical Physics to Nanoscale Heat Transfer*, Cham, Springer International Publishing, 2016, pp. 215-237.

[5] B. Nelson, Foundations and Methods of Stochastic Simulation, New York: Springer US, 2013.

[6] D. T. Gillespie, "Approximate accelerated stochastic simulation of chemically reacting systems," *The Journal of Chemical Physics,* vol. 115, no. 4, p. 1716, 2001.

[7] E. L. Haseltine and J. B. Rawlings, "Approximate simulation of coupled fast and slow reactions for stochastic chemical kinetics," *The Journal of Chemical Physics,* vol. 117, no. 15, p. 6959, 2002.

463759 Population Interaction Based on Occupation: Agriculturists and Industrialists

Abstract:

Population interaction based on occupation: Agriculturists and Industrialists

Babita K.Verma and S.Pushpavanam*

Department of Chemical Engineering, IIT Madras, Chennai 600036, India

*Corresponding author:spush@iitm.ac.in, Tel: +91-44-22574161

There are three classical kinds of population interaction in ecological modeling: Prey-predator, competitive and mutualism [1]. These three interactions between various species results in a rich behavior in population dynamics. These interactions occur in both chemical systems and ecological systems.

Societies in the world today are moving towards being industrially developed. This has resulted in a shift in the occupation of people from agrarian to industrial. The rapid growth of economies with industrialization has raised questions on the sustainable nature of the growth. Industrial growth can have a positive as well as a negative influence on the agricultural development [2]. For instance at low levels of industrialization farming practices can become more efficient because of mechanized tools being available, while at higher levels of industrialization the land resources available for agriculture could become strained. Keeping this in mind the objective of this work is to study the population interaction in society based on their occupation. Two broad categories of occupation have been chosen i.e. agriculturists and industrialists. It is observed that this system has all the three classical kinds of interaction discussed above. An analysis of the dynamical system can help in formulation of policies by agencies to direct growth in a particular manner.

We study this model both as lumped and distributed system. A lumped model is developed by taking into account the interaction in the two populations through the products manufactured in each sector. The food that is produced by agriculturist is shared by both the population. The Industrial product in return helps to improve agricultural productivity. A two dimensional model is formulated and a linear stability analysis is performed. We also study spatiotemporal dynamics by introducing the diffusion in this model. We further investigate the type of patterns formed due to diffusion and Allee effect [3]. General conclusions are drawn from the models based on the phase portrait analysis and the physical interactions. This study helps understand the interaction between the agriculturists and industrialists and initiate policies which would result in sustainable growth.

References:

[1] J. D. Murray, *Mathematical Biology I. An introduction*, vol. 17. 2002.

[2] R. Tamura, "Human capital and the switch from agriculture to industry," *J. Econ. Dyn. Control*, vol. 27, pp. 207–242, 2002.

[3] W. Wang, Y. Cai, Y. Zhu, and Z. Guo, "Allee-Effect-Induced Instability in a Reaction-Diffusion Predator-Prey Model," vol. 2013, pp. 24–27.

Ashish Kumar

Department of Chemical Engineering
University of Massachusetts, Amherst

ashishkumar@umass.edu

Current-Driven Dynamics of Single-Layer Epitaxial Islands on Crystalline Conducting Substrates

Ashish Kumar, Dwaipayan Dasgupta, and Dimitrios Maroudas
Department of Chemical Engineering, University of Massachusetts, Amherst

Financial Support: U. S. DOE, Office of Basic Energy Sciences (Award No: DE-FG02-07ER46407)

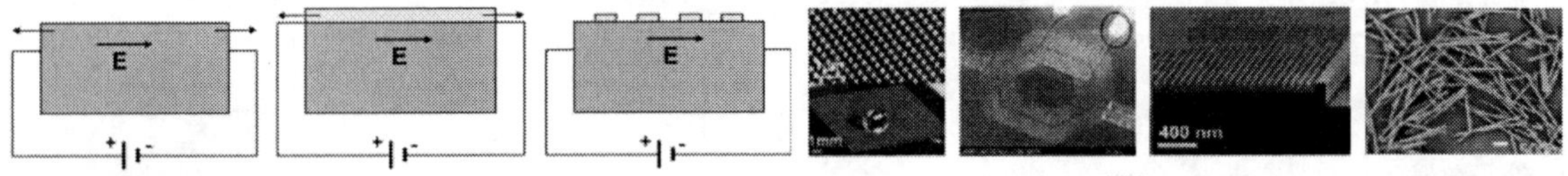

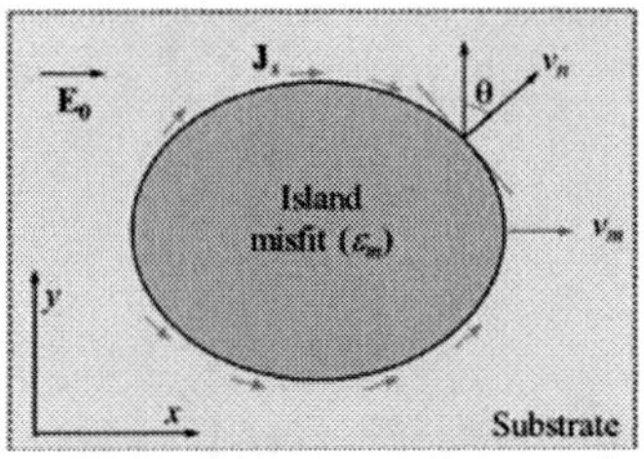

Model: $v_n = -\Omega \nabla_s \cdot \mathbf{J}_s$

$$J_s = \frac{D_s(\theta)}{k_B T}\left[-\frac{\partial \mu}{\partial s} + q_s^* E_t\right]$$

$$D_s(\theta) = D_{s,\max}\left(\frac{1 + A\cos^2\left[m(\theta+\varphi)\right]}{1+A}\right)$$

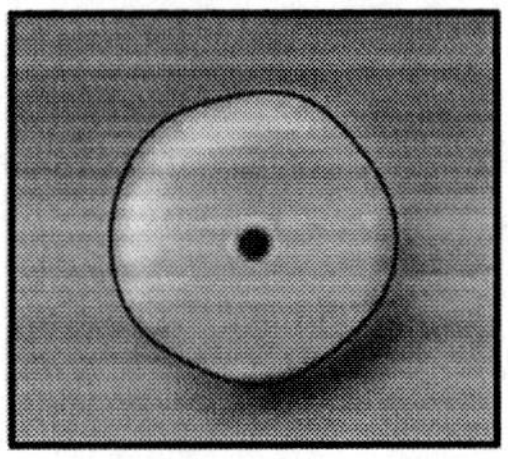

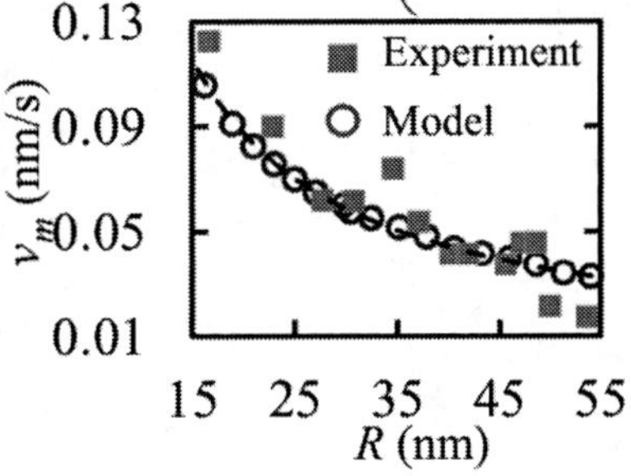

[C. Tao, W. G. Cullen, and E. D. Williams, *Science* **328**, 736 (2010)]

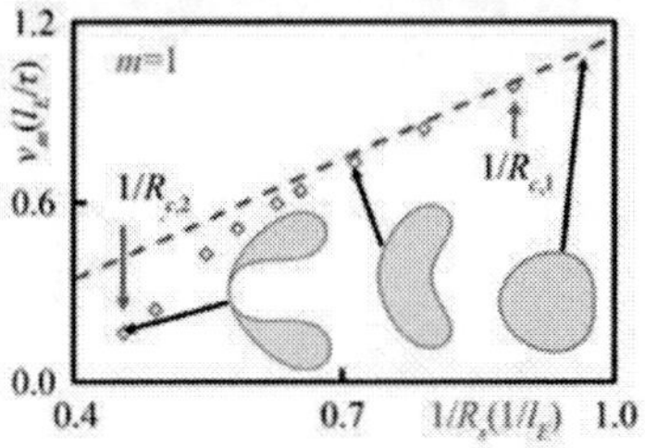

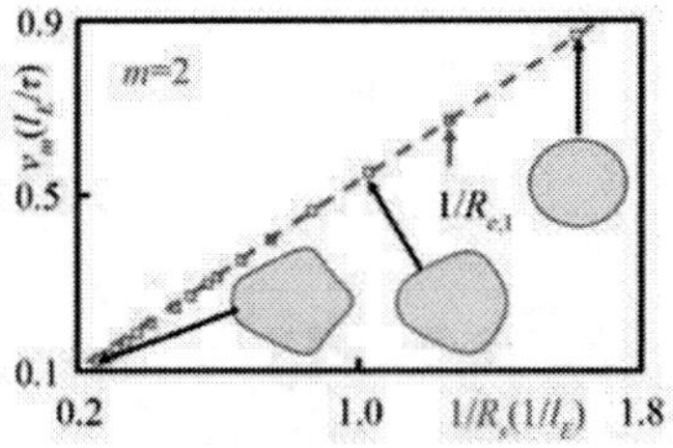

[D. Dasgupta, G. I. Sfyris, and D. Maroudas, *Surf. Sci.* **618**,L1-L5 (2013)]

Complex Island Shape Dynamics: Oscillations and Nanowire Formation

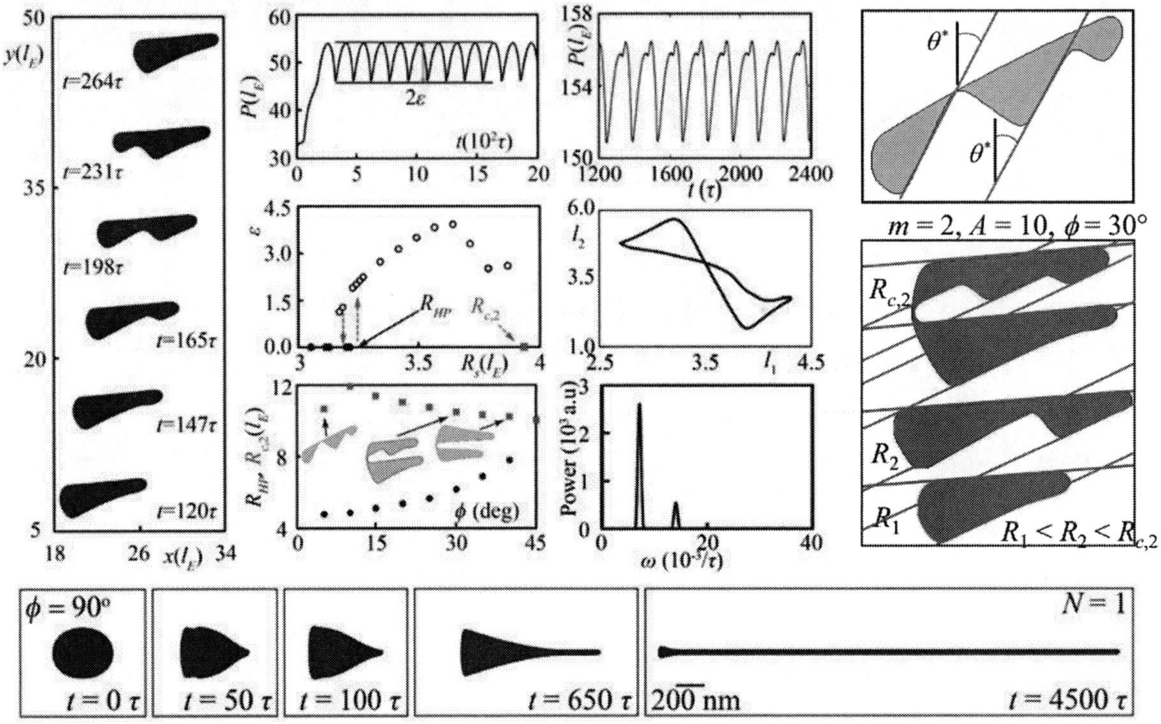

[A. Kumar, D. Dasgupta, C. Dimitrakopoulos, and D. Maroudas, *Appl. Phys. Lett.* **108**, 193109 (2016); A. Kumar, D. Dasgupta, and D. Maroudas, *Appl. Phys. Lett.* **109**, 113106 (2016)]

Complex Nanopattern Formation from Current-Driven Large-Island Dynamics

- Electric field directed along fast edge diffusion direction

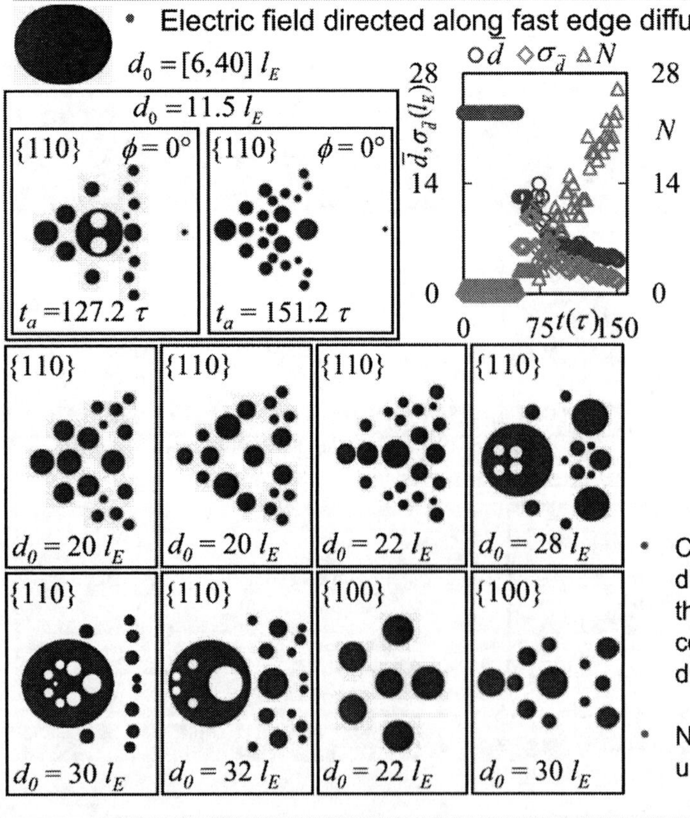

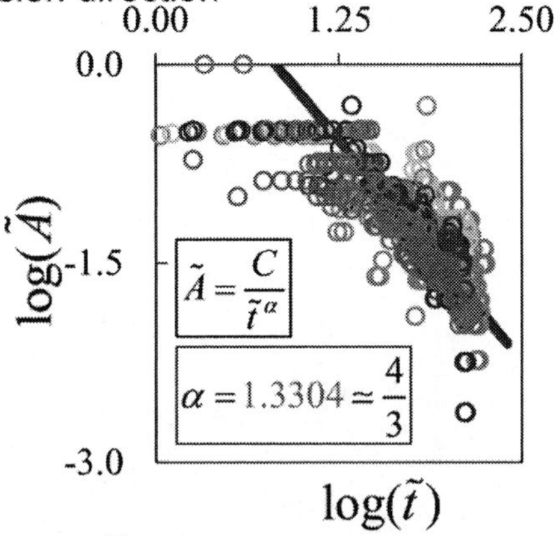

$$\tilde{A} = \frac{\bar{A} - A_{\min}}{A_0 - A_{\min}}, \qquad \tilde{t} = \frac{t - t_{\text{necking}}}{\tau}$$

- Complex patterns with islands distributed symmetrically with respect to the electric field direction and tuned by controlling original island size and duration of electric field action
- Nanopatterning kinetics follows the universal scaling relation $\tilde{A}(\tilde{t}) \sim \tilde{t}^{-4/3}$

454554 A Systematic Procedure for Analysis and Modification of Nonlinear Equation Sets to Enhance Convergence and Reduce Computational Effort

Abstract:

Many practical problems in chemical engineering give rise to the need to solve systems of non-linear algebraic equations (NLE). Typical examples include phase equilibrium and chemical equilibrium computations, steady state process simulation, flow distribution in pipeline networks etc. Nowadays powerful mathematical software packages are available for solving systems of NLEs. Shacham and Cutlip[1] evaluated and compared six state of the art software packages. In this work only the POLYMATH[2] and MATLAB[3] packages are employed. Based on our experience, these packages can serve as good representative of the additional ones.

Some of the library functions for solving NLEs employ "globally convergent" methods (see, for example p. 176 in Press et al[4]), but even such methods are often unable to find the solution from certain sets of initial estimates. In such cases some trial and error with regard to the initial estimates may be required in order to enable the algorithm to converge to the correct solution. A common cause of such difficulties is the presence of discontinuities and/or regions where some of the functions are undefined (Shacham and Brauner[5]). The impact of such discontinuities becomes most severe if they lie in the vicinity of the solution. Consider, for example the "Combustion of Propane in Air" test problem of Hiebert[6] shown in Table 1.

Table 1. Hiebert's[6] "Combustion of Propane in Air" NLE Solver Test Problem

No.	Equations	Variable	x_0	x*	f(x*)
1	R = 40				
2	f1 = x1+x4-3 = 0	x1	2	2.9976	-3.55E-05
3	f2 = 2*x1+x2+x4+x7+x8+x9+2*x10-R = 0	x2	5	3.9664	3.37E-04
4	f3 = 2*x2+2*x5+x6+x7-8 = 0	x3	40	80	-5.34E-05
5	f4 = 2*x3+x5-4*R = 0	x4	1	0.0023645	6.04E-04
6	f5 = x1*x5-0.193*x2*x4 = 0	x5	0	6.04E-04	-1.99E-08
7	f6 = x6*sqrt(x2)-0.002597*sqrt(x2*x4*xs) = 0	x6	0	0.0013659	-1.00E-07
8	f7 = x7*sqrt(x4)-0.003448*sqrt(x1*x4*xs) = 0	x7	0	0.064573	2.57E-08
9	f8 = x8*x4-1.799e-5*x2*xs = 0	x8	0	3.5308	-3.59E-08
10	f9 = x9*x4-0.0002155*x1*sqrt(x3*xs) = 0	x9	0	26.432	1.45E-06
11	f10 = x10*x4^2-3.846e-5*xs*x4^2 = 0	x10	5	0.0044998	-1.56E-13
12	xs = x1+x2+x3+x4+x5+x6+x7+x8+x9+x10			117.00021	

This example comprises of 10 implicit algebraic equations with 10 unknowns (x_i, $i=1...10$), requiring 10 initial guesses. The x_i's represent the number of moles of the various components, thus all x_i must be ≥ 0. Four variables are constrained to non-negative values (x_1, x_2, x_3, x_4 and x_s). Using a constrained NLE solver several solutions to this problem can be obtained, some of them are physically infeasible. The solution shown in Table 1 (**x***) was obtained by CONLES, the constrained NLE solver algorithm of Shacham[7] , which is available in POLYMATH and also for MATLAB (see pp. 198-199 in [8]).

To alleviate the solution of NLE systems which contain discontinuities the following systematic procedure is proposed.

1. Identify expressions with discontinuities (like logarithm, square root, division by unknowns, etc.). Modify such expressions as to eliminate the discontinuities (replacing log by exp., for example), if possible.

2. Select a unique "output" variable from each of the equations, where an output variable only appears on the left hand side (l.h.s) of one (and only one) of the equations. Leave the equations from which no output variable can be selected in an implicit form.

3. Mark the unknowns that could not be selected as output variables as of "known" (value) and arrange the explicit equations in a sequence so that only "known" variables appear in the r.h.s. of the equations and the "output" variable on the l.h.s. Explicit equations that cannot be arranged in this form should be rewritten as implicit equations.

4. Solve the revised system of equations with an NLE solver program.

5. Substitute the solution obtained into the original NLE system to verify that the reformulated and rearranged system is identical (in terms of the solution obtained) to the original system.

.Applying the proposed procedure to the example problem yields the results of Table 1

Table 1. Revised Formulation of the Example Problem

No.	Equations	Variable	x_0	x*	f(x*)
1	R = 40				
2	x1 = 3 - x4	x1	-	2.9976	-
3	x5 = 4 * R - (2 * x3)	x2	-	3.966427	-
4	x2 = x1 * x5 / (.193 * x4)	x3	40	79.9997	-
5	x6 = .002597 * sqrt(x4 * xs)	x4	1	0.0023645	-
6	x7 = .003448 * sqrt(x1 * xs)	x5	-	6.04E-04	-
7	x8 = .00001799 * x2 * xs / x4	x6	-	0.0013659	-
8	x9 = .0002155 * x1 * sqrt(x3 * xs) / x4	x7	-	0.0645727	-
9	x10 = 0 - ((2 * x1 + x2 + x4 + x7 + x8 + x9 - R) / 2)	x8	-	3.530816	-
10	f1 = 2 * x2 + 2 * x5 + x6 + x7 – 8 = 0	x9	-	26.43155	1.15E-10
11	f2 = x10 * x4 ^ 2 - (.00003846 * x4 ^ 2 * xs) = 0	x10	-	0.0044998	-3.02E-16
12	f3= xs - (x1 + x2 + x3 + x4 + x5 + x6 + x7 + x8 + x9 + x10) = 0	xs	100	116.9995	-5.409E-11

Observe that the use of the proposed procedure alleviated the problem solution by reducing the dimension of the problem (3 implicit equations, instead of 10), reduction of the number of initial estimates that need to be provided (3 instead of 10) that enables providing good initial estimates based on physical considerations, and reducing the number of variables that are associated with discontinuity (x_2 is removed from the list).

In the presentation the proposed procedure for analysis and modification of NLE systems will be described in more detail and several examples where the use of this procedure alleviates considerably the solution of NLE systems that contain regions where some of the functions are undefined, will be presented.

References

1. Shacham, M. and M.B. Cutlip, “A Comparison of Six Numerical Software Packages for Educational Use in the Chemical Engineering Curriculum”, *Computers in Education Journal*, IX(3), 9-15 (1999)

.(POLYMATH is a product of Polymath Software (http://www.polymath-software.com .2

.(MATLAB is a trademark of The Math Works, Inc. (http://www.mathworks.com .3

Press, W.H., P.B. Flannery, S.A. Teukolsky and W.T. Vetterling, *Numerical Recipes*, 2nd Ed., Cambridge .4
(University Press, Cambridge (1992

5. Shacham, M. and N. Brauner, ”Numerical Solution of Nonlinear Algebraic Equations with Discontinuities”, *Comp. & Chem. Eng.*, 26, 1449-1457 (2002)

6. Hiebert, K. L. "An evaluation of mathematical software that solves systems of nonlinear equations". *ACM Transactions on Mathematical Software*, 8, 5–20 (1983).

Shacham, M. Numerical solution of constrained nonlinear equations. *International Journal of Numerical Methods* .7
..(*in Engineering*, 23, 1455–1481 (1986

Cutlip, M. B., and M. Shacham, Problem solving in chemical and biochemical engineering with POLYMATH, .8
(Excel and MATLAB. 2nd Ed., Prentice Hall, Upper Saddle River, N. J. (2008

Daniel P Howsmon

Department of Chemical & Biological Engineering
Rensselaer Polytechnic Institute
howsmd@rpi.edu

Rensselaer

Rensselaer

Poster Number: **464869**

Multivariate Data Analysis of Markers of Oxidative Stress and DNA Methylation in Children with Autism Spectrum Disorder

Autism Spectrum Disorder (ASD) encompasses a large group of early-onset neurological diseases characterized by

1. difficulties with social communication/interaction and
2. expression of restricted repetitive behaviors and interests

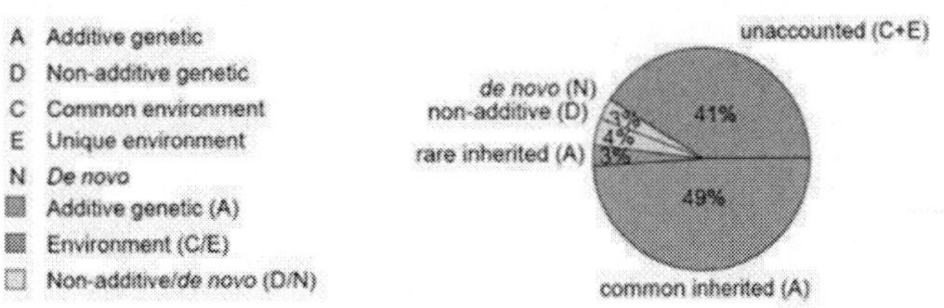

Folate-Dependent One Carbon Metabolism

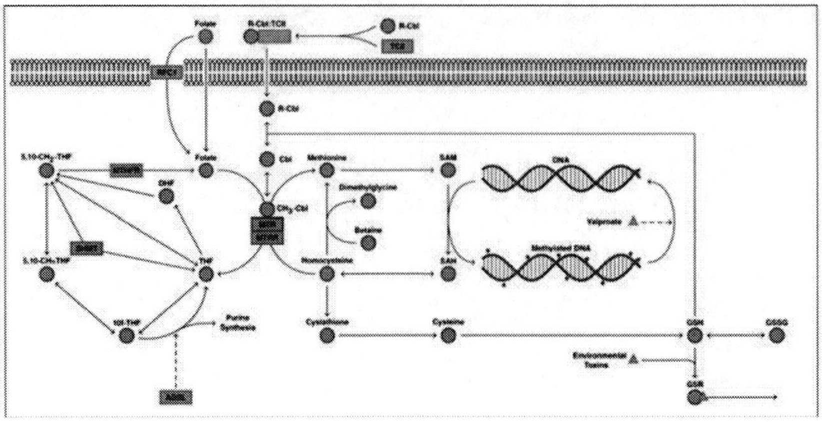

Univariate, linear statistics show differences in these metabolites, but do not exhibit separability or regression acceptable for diagnostic purposes

Rensselaer

Multivariate Data Analysis

Data from the autism IMAGE study conducted at Arkansas Children's Hospital

83	47	**76**
ASD	SIB	**NEU**

Methionine	SAM	SAH
SAM/SAH	% DNA methylation	8-OHG
Adenosine	Homocysteine	Cysteine
Glu.-Cys.	Cys.-Gly.	tGSH
fGSH	GSSG	fGSH/GSSG
tGSH/GSSG	Chlorotyrosine	Nitrotyrosine
Tyrosine	Tryptophane	fCystine
fCysteine	fCystine/fCysteine	% oxidized glutathione

Classification (FDA)

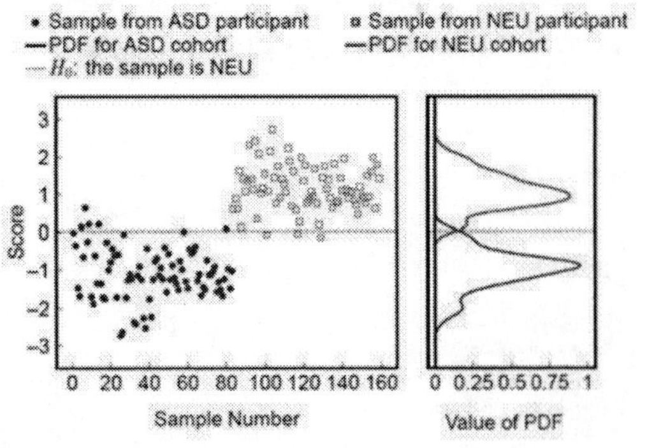

Regression (Kernel PLS)

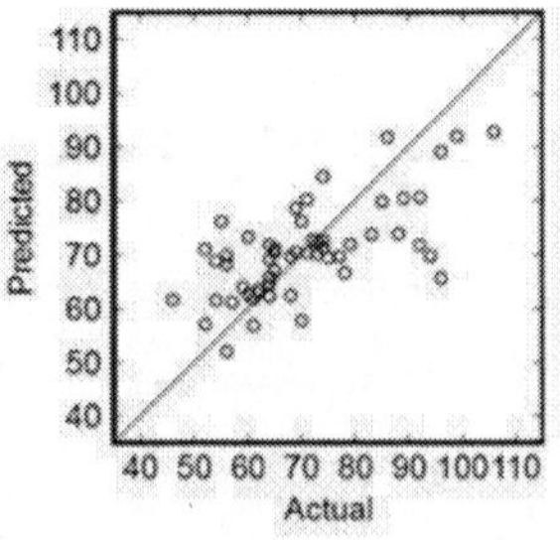

Rensselaer

465306 Nonlinearity Analysis of Periodically Forced Bioreactors

Abstract:

Nonlinearity Analysis of Periodically Forced Bioreactors

Chi Zhai[a], Ahmet Palazoglu[b], Wei Sun[a*], Zengzhi Du[a]

[a] China Beijing Key Lab of Membrane Science and Technology, College of Chemical Engineering, Beijing University of Chemical Technology, 100029 Beijing, CHINA

[b] Department of Chemical Engineering, University of California, Davis,

CA 95616, USA

Abstract

Nonlinearity is the underlying characteristic in process systems that allows for time average performance enhancement by periodic forcing of external parameters[1, 2]. The well-known pi-criterion provides a sufficient condition for optimal periodic operation[3], and is based on the linearization of the system and second-order truncation of the performance index, which is valid only for weakly nonlinear systems with infinitesimal forcing amplitudes. On the other hand, bioreactors often exhibit highly nonlinear dynamics which poses difficulties on the systematical analysis of periodic operation. Hence, it becomes desirable to explore how inherent nonlinearities would contribute to performance improvement if periodic forcing is applied on the process. In this paper, a two-compartment structured model for ethanol production is studied[4]. By judiciously setting the scope of the design/operation parameters and implementing bifurcation analysis on the unforced model, Hopf bifurcation points are detected which separate the parameter space into a hyperbolically stable region and a self-oscillatory region.

For a hyperbolically stable point, analysis of the periodic operation can be accomplished by applying the center manifold theory to the forced nonlinear system. With a multivariable expansion of the center manifold equations, higher-order terms of the nonlinear system are evaluated[5]. This method is a higher-order correction of the pi-criterion, which will reduce to the pi-criterion by truncation up to 2nd order terms. But this method is only suitable for sinusoidal-like forcing inputs. Typically, when pulsed periodic inputs are applied on the system, Fourier transform of the inputs would complicate this approach significantly[6]. Previous studies[7] used Carleman linearization as another method for pulsed periodic optimization problems. This method is carried out by Carleman linearization on the nonlinear model and solving it analytically, and then the state terms are substituted into the Taylor expansion of the performance index. Since Carleman linearization is cumbersome and suffers from lack of parsimony, a more compact functional expansion method using the Laplace-Borel transform is applied[8]. This method is found to be comparatively easy for dealing with both pulsed inputs and sinusoidal-like inputs[9]. These three methods are compared with a case study on the two-compartment bioreactor model, and the superiority of the functional expansion method is addressed.

This bioreactor may exhibit self-oscillatory behavior by adjusting the parameters properly. We show that for a self-oscillatory point, external periodic forcing may bifurcate the system to more complex dynamic behaviors. Bifurcation analysis can be implemented on the forcing waveform that is continuously differentiable, i.e., when the sinusoidal input is exerted on the self-oscillator, complex dynamics such as invariant torus or chaos would emerge. In this regard, the route to chaos by period-doubling cascades is discussed in this case study. While, intermittency[10] may also lead the system to chaos, and one may only concentrate on the onset of chaos in industry as chaotic behavior is viewed as an operational state to be avoided. An algebraic criterion on the onset of chaos is adopted in this case study, which is the Laplace-Borel transformation of the autocorrelation function. Unlike other criterion for the formation of chaos, which is usually simulation-based, this criterion is analysis-based and offers better insight.

Keywords: Carleman linearization; functional expansion; Laplace-Borel transform; central manifold; the onset of chaos.

References:

1. Douglas, J.M., Rippin, D.W.T., 1966. Unsteady state process operation, Chem. Eng. Sci., 21, 305-315.

2. Sterman, L.E., and Ydstie, B.E., 1991. Periodic forcing of the CSTR: an application of the generalized ¦Đ-criterion. AIChE. J. 37, 986-996.

3. Bittanti, S., Fronza, G., and Guardabassi, G., 1973. Periodic control: a frequency domain approach. IEEE Trans. automat. Contr. 18, 33-38.

4. Abashar, M.E.E., Elnashaie, S.S.E.H., 2010. Dynamic and chaotic behavior of periodically forced fermentors for bioethanol production. Chem. Eng. Sci. 65, 4894-4905.

5. Kravaris, C., Dermitzakis, I., Thompson, S., 2012. Higher-order corrections to the Pi criterion using center manifold theory. Eur. J. Contr.1, 5-19.

6. Dermitzakis, I., Kravaris, C., 2009. Higher-order corrections to the pi criterion for the periodic operation of chemical reactors. Control Applications, (CCA) & Intelligent Control, (ISIC), IEEE, 376-381.

7. Hatzimanikatis, V., Lyberatos, G., Pavlou, S., 1993. A method for pulsed periodic optimization of chemical reaction systems. Chem. Eng. Sci. 48, 789-797.

8. Batig¨¹n, A., Harris, K. R., Palazoglu, A., 1997. Studies on the analysis of nonlinear processes via functional expansions-I. Solution of nonlinear ODEs. Chem. Eng. Sci. 52, 3183-3195.

9. Harris, K. R., Palazoglu, A., 1997. Studies on the analysis of nonlinear processes via functional expansions-II. Forced dynamic responses. Chem. Eng. Sci. 52, 3197-3207.

10. Shivamoggi, B. K., 2014. Nonlinear dynamics and chaotic phenomena: an introduction. Springer.

471515 Convenient Prediction of Steady State Multiplicity Patterns - Multiple Autocatalytic Reactions

Abstract:

Chemical and biological reactions with nonlinear kinetics when carried out in well-mixed reactors give rise to the possibility of reactor operation at multiple operating points at steady state. When multiple steady states are physically realizable, identification of various multiplicity regions in the space of reactor operating conditions is important for design, optimization and controlled operation of chemical and biological reactions. Prediction of steady state multiplicity patterns in terms of profiles of concentrations of species influencing reaction kinetics is important for better understanding of the reaction behavior and design, optimization and control of reactors used to conduct these reactions. These patterns are simpler with respect to certain operating variables, such as the feed concentration of a reactant, and complex with respect to other operating variables, such as the reactor space time. With reactor space time as the variable parameter, exotic patterns, such as isolas, mushrooms, and multi-stability, have been reported for isothermal and non-isothermal reactor operation. In prior studies, these have been obtained via extensive iterations involving sweeps through parameter spaces. In the present work, a convenient procedure for analyzing and predicting steady state multiplicity patterns is developed and illustrated considering one or more autocatalytic reactions as specific examples. Autocatalytic reactions and processes are commonly encountered in growth of all living cells, processes involving free radicals, polymerization processes, many inorganic and organic reactions, and crystallization processes. An analysis of steady state multiplicity of two (parallel, series and mutating) autocatalytic reactions occurring in a well-mixed reactor is presented. The generation of an autocatalyst (single for parallel/series autocatalysis and two for mutating autocatalysis) from competing or consecutive resources by cubic autocatalysis is followed by its decay. Parallel and series autocatalysis is observed in growth of living cells on multiple substitutable resources (nutrients) and co-metabolism of primary and secondary nutrients (simultaneous utilization of multiple nutrients via independent metabolic pathways inside living cells). A secondary autocatalyst may be generated from a primary autocatalyst by mutation, with both autocatalysts competing for a common resource. Examples of primary autocatalyst are (i) plasmid-bearing (recombinant) cells subject to instability and (ii) healthy cells in the case of variety of cancers. A single well-mixed reactor may operate at up to five steady states for two parallel/series autocatalytic reactions and seven steady states for mutating autocatalysis. For each example, steady state multiplicity patterns are predicted in a non-iterative fashion by a judicious choice of parameter combinations. The space of ratios of the kinetic parameters for the autocatalytic reactions and ratios of supply of resources is divided into different regions depending on the maximum number of steady states admissible. The space of the remaining kinetic and operating parameters is divided into multiple regions based on the number and identity of physically realizable steady states. This division allows exact determination of appearance and disappearance of particular steady states. Continuation curves for limit point (LP) bifurcation are identified in the multi-dimensional kinetic and operating parameter space. There is a wide variety of dispositions of these for each example of multiple autocatalytic reactions. Concentration profiles of species participating in the reactions are predicted conveniently using the LP continuation diagrams. These are simple when feed composition is varied and the same is the case when reactor space time is varied in the absence of autocatalyst decay. With autocatalyst subject to decay, exotic patterns such as single and multiple isolas and mushrooms of different varieties, are possible with reactor space time as the variable parameter. The LP continuation diagrams enable convenient and precise prediction of emergence and extinction of isolas and mushrooms. A reversible sweep in a parameter combination as reactor space time is varied leads to these exotic patterns. Two or more consecutive reactions are therefore necessary for admission of isolas and mushrooms. The reaction systems exhibit rich varieties of steady state patterns, specific illustrations being provided for some of these. The present approach enables a much easier identification of the multitude of rich steady state multiplicity patterns.

471873 Mesoscale Effects in Heat Conduction through Crystalline Solids

Abstract:

Heat conduction in crystalline solids occurs through the motion of molecular-scale vibrations, or phonons. At large enough time and length scales there are sufficient phonon-phonon interactions for local equilibrium to be established, and heat conduction is described by Fourier's law. However, at length scales comparable to the mean-free path of the phonons, Fourier's law becomes inaccurate, and more fundamental "mesoscale" descriptions of heat transfer are required. We are using the phonon Boltzmann Transport Equation (BTE) to describe heat conduction in the high-energy material ß-HMX. Using a recently derived Green's function for the BTE, we calculate phonon distribution functions and temperature changes for stationary and moving heat sources. The results are interpreted in terms of continuum-scale simulations of crack formation following the shock-induced collapse of air-filled pores in ß-HMX. The latter simulations were performed by using an Arbitrary Lagrangian-Eulerian finite element method.

472254 A Nonlinear Programming Framework for Estimating Spatial Coupling in Disease Transmission

Abstract:

This work addresses the issue of estimating disease transmission parameters using a flexible, scalable modeling framework. Highly infectious pathogens can cause significant outbreaks that are followed by regional extinctions as they run out of susceptible hosts within a population. Rescue effects due to spatial transmission of these pathogens can once again spark epidemics and lead to disease persistence within subpopulations [1]. Spatial coupling parameters are a measure of the level of mixing of individuals between these subpopulations [2]. Computational models have become essential for understanding epidemiological patterns and developing the evidence base for decision-making, but require accurate estimates of the parameters that enter into the models. Furthermore, it is important that the computational modeling framework be flexible and readily scalable to larger problem sizes [3]. This work focuses on quantifying rates of disease transmission for a network of subpopulations by estimating their spatial coupling parameters with a nonlinear programming approach.

A framework is presented for efficient estimation of city-to-city spatial transmission rates by inferring transport information from localized disease case data using a statistical, hazard-based SIR model [4]. First, a stochastic model is constructed to predict spatio-temporal disease dynamics and accurately match existing datasets. A statistical hazard-based approach focusing on disease fade-out periods provided the basis for the estimation [2]. Subsequently, the strength of city-to-metapopulation spatial transmission is estimated using disease fade-out information between outbreaks. Then, the model is extended to estimate the strength of city-to-city spatial transmission [5]. The estimation is demonstrated using various records of datasets for measles outbreaks. The proposed approach is found to be feasible for this large-scale estimation, and accurately reproduces existing literature values. Additionally, the proposed approach for estimating disease transmission parameters is flexible, and allows for further investigation of larger model spaces.

[1] Heesterbeek, H., Anderson, R. M., Andreasen, V., Bansal, S., De Angelis, D., Dye, C., Eames, K. T. D., Edmunds, W. J., Frost, S. D. W., Funk, S., Hollingsworth, T. D., House, T., Isham, V., Klepac, P., Lessler, J., Lloyd-Smith, J. O., Metcalf, C. J. E., Mollison, D., Pel- lis, L., Pulliam, J. R. C., Roberts, M. G., and Viboud, C. (2015). Modeling infectious disease dynamics in the complex landscape of global health. *Science*, 347(6227):aaa4339–aaa4339.

[2] Bjø rnstad, O. N. and Grenfell, B. T. (2008). Hazards, spatial transmission and timing of outbreaks in epidemic metapopulations. *Environmental and Ecological Statistics*, 15:265–277.

[3] Word, D. P., Cummings, D. a. T., Burke, D. S., Iamsirithaworn, S., and Laird, C. D. (2012). A nonlinear programming approach for estimation of transmission parameters in childhood infectious disease using a continuous time model. *Journal of the Royal Society, Interface / the Royal Society*, 9:1983–97.

[4] Grenfell, B. T. (2000). Time series modelling of childhood diseases: a dynamical systems approach. *Appl. Statist.*

[5] Xia, Y., Bjø rnstad, O. N., and Grenfell, B. T. (2004). Measles metapopulation dynamics: a gravity model for epidemiological coupling and dynamics. *The American naturalist*, 164(2):267–281.

465177 Scheduling and Purchasing Optimization for Olefin Cracking Complex

Abstract:

Olefin cracking takes an important role in chemical industry. Efforts are made to develop new technology to increase its production. Due to the large production volume, olefin plants can make hundreds of thousands of profits by optimal scheduling without any investment. The optimal production scheduling can further give direction for raw material purchasing. In this paper, a MILP model was built for scheduling and purchasing optimization of an olefin plant with consideration of downstream constraints, non-simultaneous cleaning constraints, and recycling ethane relaying cracking constraints. The efficacy of the model is demonstrated by case studies.

466670 Kinetic Parameter Estimation Including Uncertainty Under Mass Transfer Limited Conditions

Abstract:

Kinetic models are commonly used to describe chemical processes (1) to allow gaining insight in the underlying reactions, and (2) for process optimisation. In the former application, the model needs to be calibrated using experimental data. The main aim of such a calibration exercise is to obtain kinetic parameter values which are independent from process settings such as the mixing speed, the reactor geometry, etc. This can only be achieved if the reactor is not suffering from mass transfer limitations and, hence, no major spatial heterogeneities exist. This "homogeneous" reactor behaviour is aimed for by mixing. However, it is often not possible to have non-limiting mass transfer, and thus it becomes impossible to use kinetic models to estimate the intrinsic parameter values using only a kinetic model and assuming a CSTR. However, by coupling computational fluid dynamics (CFD) with the kinetic model, it is possible to account for these spatial heterogeneities and calibrate the kinetic model.

Recently, Verbruggen et al. (2016) [1] coupled a Langmuir model with CFD, and stated that they were able to accurately retrieve the intrinsic kinetic parameter values. This approach proved to be very powerful, but requires the evaluation of the CFD model at multiple parameter values to find the minimum of the objective function, and thus is computationally very demanding. When overcoming the computational burden, however, it might seem that the problem of estimating the intrinsic parameter values has been solved. But this is not the case as one can only prove that the obtained parameter values really represent the intrinsic parameter values when the measurements are not noise corrupted and an unlimited amount of such data is available (the so-called theoretical identifiability). In reality, only a limited amount of noise-corrupted measurements is available, and thus the reliability of the estimated parameter value will depend on the measurement uncertainty and the experimental design (e.g. the sampling locations and the initial substrate concentrations). As a consequence, only parameters with a sufficiently important (and uncorrelated) impact on the model predictions will be estimated properly. Therefore, it is important to assess the uncertainty of the parameter estimates in such a calibration context. This is often referred to as practical identifiability analysis [2].

It is clear that even for parameter estimations performed with CFD, assessing the parameter uncertainty remains an important aspect, since uncertain parameter estimates will yield models with low predictive power. The application of uncertainty analysis for CFD-based kinetic model calibration has not yet been performed to our knowledge. A new methodology is developed, and is applied *in silico* to a second-order reaction (Eq. (1)) for a microreactor where the enzyme is immobilised at the wall (Figure 1).

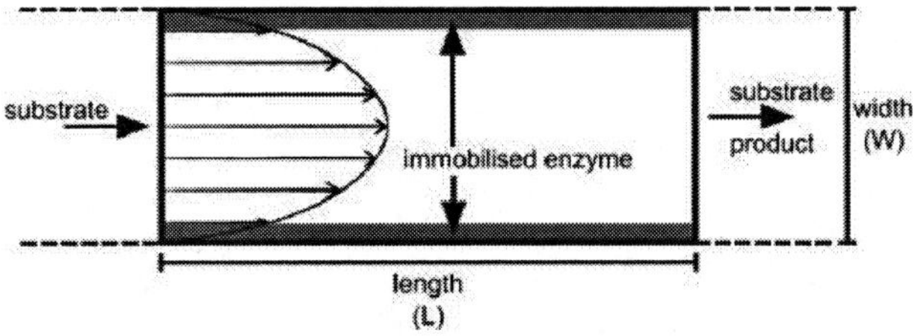

Figure 1: Microreactor (W=200 μm, L=0.10 m) with enzyme immobilised at the walls.

$$E + S \xrightarrow{k_{\text{cat}}} E + P \quad (1)$$

The corresponding kinetic model is given in Eq. (2), and was implemented in OpenFOAM 2.2.2:

$$\frac{\mathrm{d}P}{\mathrm{d}t} = -\frac{\mathrm{d}S}{\mathrm{d}t} = k_{\text{cat}}[E][S] \quad (2)$$

The uncertainty of k_{cat} was estimated as a function of the "optimal" k_{cat} value, using the likelihood confidence region method [3]. Since the parameter uncertainty is affected by the mass transfer, the sampling locations, and measurement noise, a Monte Carlo strategy was used to determine the impact of these degrees of freedom on the parameter uncertainty. The use of the developed methodology showed that in this case k_{cat}, could be properly estimated when using the CFD-based model calibration, even when mass transfer limitations were high. Moreover, it allows estimating the uncertainty of the parameter values.

References

[1] Verbruggen, S.W., Keulemans, M., van Walsem, J., Tytgat, T., Lenaerts, S., Denys, S., (2016) CFD modeling of transient adsorption/desorption behavior in a gas phase photocatalytic fiber reactor, *Chem Eng J*, 292:42-50

[2] Dochain, D., Vanrolleghem, P.A., (2001) Dynamical Modelling and Estimation in Wastewater Treatment Processes

[3] Seber G.A.F., Wild C.J., (1989) Nonlinear Regression

470228 A Kinetic Model of Lignin Biosynthesis in *ArabidopsisÂ thaliana* for Improved Biofuel Production

Abstract:

A Kinetic Model of Lignin Biosynthesis in *Arabidopsis thaliana* for Improved Biofuel Production

Rohit Jaini[1]*, Longyun Guo[2], Peng Wang[2], Natalia Dudareva[2], Clint Chapple[2], John A. Morgan[1,2]

[1]*School of Chemical Engineering, Purdue University, West Lafayette, Indiana 47907,* [2]*Department of Biochemistry, Purdue University, West Lafayette, Indiana 47907*

Plants naturally channel almost 20-30% of the carbon fixed via photosynthesis towards the phenylpropanoid pathway for its conversion to lignin, a hetero-aromatic polymer that imparts mechanical support and hydrophobicity to plant vasculature while impedes cellulosic biofuel production. The past two decades witnessed several genetic engineering efforts targeting the phenylpropanoid pathway to manipulate lignin content and composition to improve forage digestibility and saccharification efficiency. Despite the progress, several questions pertaining to control and regulation of carbon flux in this pathway remain unanswered, revealing a significant gap in the mechanistic understanding of flux control in this pathway. In an attempt to elucidate this underlying regulatory framework, we are developing a kinetic model for the phenylpropanoid network in *Arabidosis thaliana*. In this study, we present a holistic strategy for kinetic parameter estimation and validation using targeted metabolomics data obtained by stable-isotope labeling experiments conducted on wild-type (WT) and transgenic plants with knockdown of the gene encoding caffeoyl-shikimate esterase (CSE) respectively. Arabidopsis *cse* lines, compared to WT, exhibit a phenotype with reduced lignin content that is enriched in *p*-hydroxyphenyl units presenting an interesting system to investigate.

As a first step, we developed a novel and comprehensive analytical method based on liquid chromatography and tandem mass spectrometry (LC-MS/MS) for accurate quantitation of the phenylpropanoid pathway intermediates. Employing two separate extraction strategies allowed successful profiling of all the intermediates of the pathway, including the previously un-detectable and labile hydroxycinnamyl CoA esters. Soluble extracts from basal section of stems were analyzed on a reverse phase LC-MS/MS platform in the negative ion mode harnessing the sensitivity conferred by multiple reaction monitoring (MRM). Chromatography and compound ionization was optimized for achieving low limits of detection (LODs) for the metabolites of interest. WT and transgenic Arabidopsis plants were fed with three different concentrations (100, 300 Î¼M & 1 mM) of [ring-$^{13}C_6$]-phenylalanine as part of our stable isotope labeling experiments. Concentrations and label enrichments of phenylpropanoid metabolites measured at seven different time points after feeding constituted the metabolomics data sets for parameter estimation.

The analytical method allowed the detection and quantification of 15 of the 19 compounds of the pathway, for which standards were available, in WT tissue. This data was used to extend our previously presented kinetic model on the first three reactions of the lignin biosynthesis pathway catalyzed by phenylalanine ammonia-lyase (PAL), cinnamic acid 4-hydroxylase (C4H), 4-(hydroxy) cinnamoyl CoA ligase (4CL) to now include a key section consisting of enzymes hydroxycinnamoyl-CoA: shikimate hydroxycinnamoyl transferase (HCT), *p*-coumaroylshikimate 3′-hydroxylase (C3′H), and CSE. Kinetic parameters were identified by simultaneously fitting fractional label and pool sizes of metabolites using a covariance matrix adaptation evolution strategy (CMA-ES), a stochastic and derivative-free numerical optimization method widely applied to ill-conditioned systems. Our current model captures pathway dynamics over the range of feeding treatments considered for the study. Parameter uncertainties were evaluated using a Markov Chain Monte Carlo (MCMC) sampling scheme. The model was validated on Arabidopsis *cse* lines by comparing dynamic metabolite-profiling data to predicted simulations. We believe our study has laid foundation to extend the kinetic model to the remaining metabolic pathway showing promise as a predictive tool to aid in rational metabolic engineering of lignocellulosic feedstock for efficient biofuel production.

473025 Minimization of Entropy Generation in Multi-Pressure Reactive Distillation Networks

Abstract:

In the synthesis of reactive distillation networks, a promising process intensification technology, different pressure levels are allowed to be used to overcome the azeotropes, which can enhance the separation process. Nevertheless, the use of reactions or pressure swing to execute a separation can lead to different levels of entropy generation for the system. The amount of entropy generated in a process can be used as a measure the energetic efficiency of the system. Looking from the angle of availability analysis, the amount of entropy generated in a process is directly related to the destruction of available work.

A methodology for the globally optimal synthesis of a reactive distillation network with minimum entropy generation, under certain conditions, is presented in this work. The proposed methodology employs the Infinite DimEnsionAl State-Space (IDEAS) conceptual framework, which is shown to be applicable to the problem under consideration. The IDEAS framework considers an infinite number of units with all physically possible mixing and splitting streams existing between them. The resulting mathematical formulation is an infinite-dimensional program (ILP). Using the proposed formulation, the objective function of minimum entropy generation is function of only the outlet streamsâ€™ composition, regardless of the number of units or interconnections the reactive distillation network contains. The method is demonstrated on a case study involving the multi pressure reactive distillation of methanol/isobutene/MTBE azeotropic mixture.

COUPLED AUTOCATALYTIC REACTIONS : EXTINCTION OF A SPECIES

Autocatalysis or self-replication is an important phenomena which arises across various disciplines like chemistry, ecology (*animal reproduction*), economics (*sustained growth of goods and services*), biology (*DNA replication, RNA polymerization*).

In the context of chemical reaction engineering, the dynamics of single autocatalator in an open system were studied in [4] [5]. These autocatalytic reaction systems are non linear and are characterized by a positive feedback. As a result, they exhibit complex behavior like multiple steady states and oscillations. Coupled systems in which two species are produced autocatalytically are also found in the literature: Lotka-Volterra predator-prey equations [7] [3], Complementary metabolism [8], Hypercycles [1] [2].

In this work, we analyze another coupled autocatalytic system (0.1) where each species catalyzes the autocatalysis of the other species. Here, each species acts as a substrate and gets consumed in the autocatalysis. The system we propose here is comprised of two species A and B both undergoing autocatalysis. One undergoes cubic autocatalysis while the other quadratic autocatalysis. Further, the system is coupled as each species catalyzes the autocatalysis of the other. The asymmetry in this interconversion is introduced by the different order of the autocatalysis. The species B is converted to A in the presence of two molecules of A, while A is converted into B in the presence of just one molecule of B. Thus, the conversion of A into B is easier than that of B to A.

$$\begin{aligned} B + 2A &\rightarrow 3A \\ A + B &\rightarrow 2B \end{aligned} \tag{0.1}$$

The model (0.1) can capture interconversion of two isomers. It is distinct from asymmetric autocatalysis [6]. Here, instead of a substrate, an enantiomer in the presence of its optical isomer can get converted into its isomer and vice versa. Also, (0.1) can represent interconversion of two social or religious groups. It is analogous to the threshold ethnokinetics model proposed by Turchin in [9].

In this work, in addition to the two reactions in (0.1), uncatalyzed generation and decay of these species are included. The system doesn't remain a simple threshold model in the presence of these additional factors. It is found to exhibit multiple steady states and oscillations. We capture the non linear behavior of this system by applying Singularity theory and Bifurcation theory.

An important question to ask is in a system where both species exist under what conditions will only one of the species remain. This situation corresponds to one ethnie being wiped out by getting completely converted to another. In a system of enantiomers reacting autocatalytically, it corresponds to conversion of a racemic mixture to pure enantiomer. The analysis helps understand the operating conditions to obtain a desired enantiomer from an undesired enantiomer.

In the absence of uncatalysed generation or decay of one of the species, the system behavior differs considerably. To capture this, all the combinations of generation and decay reactions are considered. A comprehensive analysis is performed by analyzing a total of twelve cases. Although the system doesn't remain a simple threshold model, but it is still found to be very sensitive to the initial value of quadratic autocatalator A in all these cases. Also, it is found that even though only one species is produced by uncatalysed generation,it can get wiped out and the steady state may converge to the other species.

---

[1] Eigen, M., Naturwissenschaften **58**, 465 (1971).

[2] Eigen, M. and Schuster, P., *The hypercycle: a principle of natural self-organization* (Springer Science & Business Media, 2012).

[3] Goel, N. S., Maitra, S. C., and Montroll, E. W., Reviews of Modern Physics **43**, 231 (1971).

[4] Gray, P. and Scott, S., Chemical Engineering Science **38**, 29 (1983).

[5] Gray, P. and Scott, S., Chemical Engineering Science **39**, 1087 (1984).

[6] Islas, J. R., Lavabre, D., Grevy, J.-M., Lamoneda, R. H., Cabrera, H. R., Micheau, J.-C., and Buhse, T., Proceedings of the National Academy of Sciences of the United States of America **102**, 13743 (2005).

[7] Lotka, A. J., The Journal of Physical Chemistry **14**, 271 (1910).

[8] Meyer, J. S., Tsuchiya, H., and Fredrickson, A., Biotechnology and Bioengineering **17**, 1065 (1975).

[9] Turchin, P., *Historical dynamics: why states rise and fall*, Vol. 41 (Princeton University Press Princeton, NJ, 2003).

473996 Computational Approaches in Systems Modelling for Environmental Impacts of Industries: Automating Physical Input-Output Tables (PIOTs) Via Process Modelling

Abstract:

Systems modeling for environmental impacts of industrial systems is currently limited by the available computational approaches, thus leading to slow and inefficient decision making for the development of sustainable industrial ecosystems. One such systems model is based on Physical Input Output Tables (PIOTs), an important part of the proposed system of environmental accounts, with the potential to aid in the understanding of industrial metabolism, economic structure, interdependencies, waste generation and management, and tracing of scarce resources. PIOTs have not been adopted on a wide scale, mostly due to the high costs related to data collection[1]. We propose using a hybrid approach including process modeling and computations to partially automate the process of constructing the PIOT. Once the process model is complete, this will be combined with readily available statistical data (a combination of bottom up and top down approaches) to create a relevant PIOT for the systems under consideration with much less effort than traditional methods. The process modeling segment is most relevant for the sectors that are particularly intensive in the physical component being studied. Our work looks at the industrial systems that process nitrogen intensive products using data from the state of Indiana. We use process modelling (Aspen Plus) to model corn through the production chain combined with statistical census data on total flows in Indiana to generate the PIOT. We intend to use this methodology to create a complete PIOT for Indiana for nitrogen. Challenges include the treatment of solids in traditional process modeling software, like Aspen Plus.

1 Hoekstra, R. (2010, February). Physical Inputâ€"Output Tables: Developments and Future. In *18th International Input-Output Conference* (pp. 20-25).

459931 Process Systems Lifecycle Management Using a Model Based Engineering Approach

Abstract:

The motivation of this work is the constant evolution in industry. Nowadays we are in what is called the fourth industrial revolution (called Industry 4.0 or Connected Enterprise or Industrial Internet depending on the country or consortium). This revolution is being fostered in many countries to get a more competitive industry. Industry 4.0 target is to make more efficient and flexible plants, reduce times and costs of projects and products lifecycle.

This revolution tries to converge the physical and the virtual plant, this is achieved by means of developing a digital model of the plant before constructing it. Industrial Internet tackles the current challenges of the industry: increased functionality in products, more connectivity, high interdisciplinarity, design for X, more product/process complexity, very software intensive, digitalization, needs of simulation and visualization, compliance with more demanding regulations, etc. It is an integrated engineering approach were different software tools are interconnected during the whole lifetime of the project, reducing inconsistencies and developing times.

To fulfil the new requirements, traceability throughout the entire system lifecycle is needed, traceability from the initial requirements to the final physical product

The concept of Model Based Systems Engineering provides methods to guide the cross-disciplinary, virtual product development process and to achieve the required traceability. Model Based Systems Engineering (MBSE) is a multi-disciplinary engineering paradigm propagating the use of models instead of documents to support analysis, specification, design and verification of the system being developed. Using models instead of documents, a discipline-neutral view of the system specification is created. The resulting coherent system model helps to understand and to overview the complexity of the developed system.

System models are created by application of the Systems Modelling Language (SysML). This is a general-purpose graphical modelling language for specifying, analysing, designing, and verifying complex systems that may include hardware, software, information, personnel, procedures, and facilities. In particular, the language provides graphical representations with a semantic foundation for modelling system requirements, behaviour, structure, and parametrics, which is used to integrate with other engineering analysis models.

Under this framework models appear as a core component in every new development. Using a systems engineering methodology the developed model will be the one that guarantees the consistency and derives the different applications needed in every stage of the lifecycle, from simulation, to risk assessment or even documentation maintenance.

The objective of our work is to develop a model of a process plant using SysML. This model will follow a systems engineering approach, starting from the requirements and will cover the whole lifecycle of the project. This model will be the core of the project allowing for the different applications.

In this paper, we present first a Model Based Systems Engineering architecture for process systems and its relationship with the different stages of the lifecycle process. The core of the architecture is the neutral model of the system. This model is generated once the requirements are clearly stated. The generation of the general system model starts with the development of the functional model that refines the requirements to more specific functions that have to be fulfilled by the system. Once the functional model is built the behavioural and structural models are developed. These models reify the functions to physical equipment and to a topology and defines the behaviour of the components that achieve the functions. This general model has many parts that can be reusable and a model library is also constructed to store them. The final step is to build the specific models, these are models related to different views of the process and related with specific disciplines. The generic model is developed using SysML but as for specific models more efficient languages and tools exist it is necessary to do some model transformations from and back to SysML. Different applications are needed for the design and operation phases of the project (steady state simulation for process design, risk assessment tools, fault diagnosis, alarm management, etc.)

After the architecture has been introduced we create a SysML model of a process plant (the production of ethylbenzene) starting with the requirements (package diagram). This model includes the behavioural (block definition and activity diagrams) and structural (internal and block definition diagrams) aspects of the model. Once the model is created an automatic transformation (bidirectional) from it to some applications, where specific and more efficient environments for those purposes exist, is illustrated. The applications addressed in this work are:

-Process design. Transformation to a steady state model used in a process simulation environment, in this case Aspen Plus. This model is used for plant design calculating process operating conditions, equipment design and different production routes.

-Risk assessment. Transformation to a functional model suited to perform risk analysis (HAZOP like analysis). In this case the transformation is made to D-higraphs, a graphical functional language.

-Process optimization. Automatic generation of the optimization objective and constraints. In this case we do it using Matlab. The model used in the optimization will be the one generated for Aspen Plus that can be embedded in Matlab.

In summary, this work shows some of the potential of having models at design time. Models that are neutral and that can be used for different applications. Current industry trends look for models that generate virtual plants that can be "operated" and that allow to discover design mistakes in this early design phase. Thus providing, when constructed, more robust, consistent and efficient plants.

References

Eigner,M., Dickopf,T., Apostolov,H., Schaefer,P., Faißt,K.-G., Keßler,A.,2014. System lifecycle management: Initial approach for a sustainable product development process based on methods of model based systems engineering. In: Product Lifecycle Management for a Global Market. Springer, pp. 287–300.

Friedenthal, S., Moore, A., Steiner, R., 2014. A Practical Guide to SysML, 3rd Edition. Morgan Kaufmann.

Kagermann, H., Wahlster, W., 2013. Recommendations for implementing the strategic initiative INDUSTRIE 4.0, Final report. Tech. rep., Industrie 4.0 Working Group

Evans, P. C., Anninziata, M., 2012. Industrial internet: Pushing the boundaries of minds and machines. Tech. rep., General Electric

Paredis, C., 2012. Why model-based systems engineering? benefits and payoffs. In: 4th PLM Future. Tagung, Mannheim.

Rodriguez, M., de la Mata, J. L., 2012. Automating hazop studies using d-higraphs. CACE 45, 102 – 113.

460837 Comparison of Two Identification Methods for Identifying Sequential Alarms in Plant Operation Data

Abstract:

The advance of distributed control systems (DCS) in the chemical industry has made it possible to inexpensively and easily install numerous alarms in DCS. While most alarms help operators detect an abnormality and identify its cause, some are unnecessary. A poor alarm system might cause sequential alarms, which are a collection of numerous alarms that almost always occur simultaneously with specific time lags within a short time (EEMUA, 2007). Sequential alarms reduce the ability of operators to cope with plant abnormalities because critical alarms are buried under many unnecessary ones. Sequential alarms are usually caused by poor alarm rationalization (Hollifield, *et al.*, 2007). Therefore, it is very important to identify sequential alarms in a plant operation data for improving the alarm system and process operation.

The grouping correlated sequential alarms in accordance with their degree of similarity helps to reduce sequential alarms more effectively than by analyzing individual sequential alarms. Event correlation analysis (Nishiguchi, *et al.*, 2010) was proposed to identify sequential alarms in noisy plant operation data. Event correlation analysis was applied to the operation data of an industrial ethylene plant and was able to correctly identify similarities between correlated sequential alarms (Higuchi, *et al.*, 2010). However, event correlation analysis occasionally failed to detect similarities between two physically related sequential alarms when deletions, substitutions, and/or transpositions occurred in the alarm sequence.

In previous study (Wang, *et al.*, 2015), we proposed a new identification method of sequential alarms by applying the dot matrix analysis to a plant operation data (Kurata *et al.*, 2011). Dot matrix analysis (Gibbs, *et al.*, 1970) is one of the sequence alignment methods for identifying similar regions in DNA or RNA. Similar regions in DNA or RNA may be a consequence of functional, structural, or evolutionary relationships between the sequences. In this research, in order to demonstrate the usefulness of the previous study, we compare it with the event correlation analysis. Two methods were applied to the plant operation data of an azeotropic distillation column. The results revealed that the dot matrix analysis is able to identify similar sequential alarms in plant operation data, when deletions, substitutions, and/or transpositions occurred in the alarm sequence..

References

The Engineering Equipment and Material Users' Association (EEMUA): Alarm Systems A Guide to Design, Management and Procurement Publication No. 191 Edition 2,EEMUA, London (2007)

Gibbs, A. J., McIntyre, G. A.: The diagram method for Comparing Sequences. Its Use with Amino Acid and Nucleotide Sequences, *Eur. J. Biochem.*, 16, 1-11 (1970)

Hollifield, B. R., Habibi, E.: Alarm Management: Seven Effective Methods for Optimum Performance, ISA, Research Triangle Park (2007)

Higuchi, F., Noda, M., Nishitani, H.: Alarm Reduction of Ethylene Plant using Event Correlation Analysis (in Japanese), *Kagaku Kogaku Ronbunshu*, 36(6), 576-581 (2010)

Kurata, K., Noda, M., Kikuchi, Y., and Hirao, M.: Extension of Event Correlation Analysis for Rationalization of Plant Alarm Systems (in Japanese), *Kagaku Kogaku Ronbunshu*, 37, 338-343 (2011)

Nishiguchi, J., and Takai, T: IPL2&3 Performance Improvement Method for Process Safety Using the Event Correlation Analysis, *Computers & Chemical Engineering*, 34, 2007-2013 (2010)

Wang, Z., Noda, M. : Identification of Sequential Alarms in Plant Operation Data by using Dot Matrix Analysis (in Japanese), *Kagaku Kogaku Ronbunshu*, 41, 333-339 (2015)

463834 Heterogeneous Supercomputing with Multi-Scale Modeling - Towards Virtual Reality in Process Engineering

Abstract:

Dynamic simulation in process engineering can be very time-consuming but still of low accuracy due to the strong coupling between different scale (e.g., molecular, elemental and systematic) and deferent processes (e.g., reaction, diffusion and flow). A natural and effective approach to address this challenge is multi-scale modeling which combines or cascades models at different scales in a single simulation so that the effects of smaller-scale behaviors on larger scales can be accounted without describing their full details. However, multi-scale modeling may also complicate the computation by mixing different numerical methods and algorithms, e.g., continuum-based and discrete methods. Heterogeneous computing is, therefore, especially suitable for simulation using multi-scale models, with processors of different architectures for different algorithms accordingly.

In this talk, we will demonstrate this possibility by introducing a series of multi-scale simulations on particle-fluid systems such as fluidized bed, pneumatic conveyor, mixing tanks etc. Highly parallel codes for continuum-based finite volume method, discrete element method and lattice Boltzmann method are implemented on the central processing units (CPUs), many-integrated cores (MICs) and graphics processing units (GPUs), respectively, using the petaflops heterogeneous supercomputer Mole-8.5E [1,3]. The accuracy, speed and efficiency of the simulation can be improved dramatically to quasi-realtime level for lab-scale systems with particle-scale resolution. In contrast, traditional simulation may cost weeks for seconds of physical process. Based on this capability, interactive simulation and visualization with online control of the operating conditions has been realized also, suggesting a more exciting prospect of virtual-reality (VR) style simulation in process engineering. Some preliminary results can be found in our recent publications [2,3] and the most recent developments will be presented in the talk.

[1] Wang, X. & Ge, W. 2013. Chapter 5: The Mole-8.5 Supercomputing System. In: Vetter, J. S. eds. *Contemporary*

High Performance Computing from Petascale toward Exascale. pp. 75-98. Boca Raton: Chapman & Hall / CRC.

[2] Ge, W. et al. 2011. Meso-scale oriented simulation towards virtual process engineering (VPE)-The EMMS Paradigm. Chemical Engineering Science 66, 4426-4458.

[3] Ge, W. et al. 2015. Multiscale discrete supercomputing - a game changer for process simulation? Chemical Engineering & Technology 38, 575-584.

465847 Developing a Strategy for Renewable Energy System in Process Industry with a Focus on Employing Energy Storage Systems

Abstract:

Energy plays an important role in process industry. Energy is an essential utility resource of which impact is bigger than any other. Process industries are also heavily dependent on reliable and sufficient energy supply. It is important to supply enough amount of energy in a cheap way. Managing the energy system in process industry is a challenge from both of supply and demand perspective. In terms of supply, the introduction of renewable energy sources that is subject to variation makes the control of supply difficult. The increasing pressure to employ clean and sustainable energy sources enforces the introduction of renewable energies. At the same time, the varying demand cannot be underestimated caused by the frequent change in operation schedule. The fluctuating energy price makes the problem of energy supply, energy demand and their integration even further complicated. Recently attention has been given to energy storage system that is expected to be an affordable tool in mitigating the difficulties of the energy system. Many kinds of energy storage systems are under development such as capacitor, flying wheel, etc. It is expected that the deployment of ESS have a positive impact on the energy system. ESS allows us to use the redundant power that would have been wasted by storing them in the physical facility. Peak energy loads can thus be flattened. The amount of energy generation can be reduced. In terms of cost, the cheap energy generated at off peak can be used at the peak time. Individual energy storing performances will be improved due to the active R&D. The existing energy supply and demand network should be re-designed by incorporating the presence of ESS in the network. There has been little reported in terms of transforming the expectation into practice by a systematic manner. In order to utilize the full potential of ESS, it should be managed in a systematical framework based on rigorous decision-making tools. This paper proposes a mathematical programming based operation strategy for employing energy storage systems in the renewable energy supply and demand networks. After mathematical modeling of ESS and its energy system is presented, numerical case studies are addressed with some remarks.

466884 A Tree Ontology for Multi-Scale Muti-Disciplinary Process Model

Abstract:

We present an new approach to multi-disciplinary model representation and generation, which is based on an ontology that forms a tree with the branches shaped by the involved disciplines and the leaves being the detailed multi-facet models.

Each leave in the tree represents a network of capacities linked by transfers. What is "living" in these networks depends on the discipline. In physical systems this will be mass, energy, species detailed into phases that may be detailed even further. In control this will be signals and information, in material models it can be particles, molecules, atoms. The structure is completed by linking the leaves together where the discipline-specific networks are linked through connection networks. All networks build on a very small set of meta terms, the set of which is the result of intense research over the last couple of years not at least in connection with a European project that has the objective to model polyurethane foams from quantum to mechanical properties (http://modena.units.it).

The process of building the ontology start with generating the tree structure. One of the keywords provides the branching the rest provides the key construction elements for the equation/variable system that eventually will represent the behaviour of the modelled process in a mathematical form.

The ontology is used in conjunction with a graphical user interface that implements a hierarchical network representation. The hierarchy being used here is different from the hierarchy of the ontology. In the graph model designer, the hierarchy serves the purpose of managing the model complexity, in terms of a large number of nodes and arcs in the respective discipline-dependent network.

The system generates automatically code for target applications that have the necessary solution methods implemented.

The presentation will focus on the philosphy of the approach and the range of applications with the example of materials modelling and controlled plants.

469393 Data-Mining Assisted Coarse-Grained Optimization

Abstract:

The design of complex engineering systems usually leads to high-dimensional optimization problems, where the objective function is expensive to evaluate and may possess a rugged surface due to the presence of multiple local minima. In such cases gradient-based minimization methods won't be able to locate the global minima unless very specific initial conditions are chosen. We assume that some "coarse relaxation" of the objective function is smooth and we can use this underlying structure to perform coarse-grained optimization. We show that short runs of a stochastic optimization method, Simulated Annealing [1] can be described by an effective stochastic differential equation (SDE). Using appropriate subsampling we can estimate the drift and diffusion coefficients of this SDE [2] and extract information about the coarse gradient to speed up the "dynamics" of the optimization method. In addition, we show that if the data lies on a low-dimensional hyperplane in the parameter space, we can uncover it using data mining techniques such as PCA or diffusion maps [3], making the exploration of the objective function surface even more efficient.

References

[1] S. Kirkpatrick, C. D. Gelatt, and M. P. Vecchi, "Optimization by Simulated Annealing," Science, vol. 220, no. 4598, pp. 671–680, 1983.

[2] G. A. Pavliotis and A. M. Stuart, "Parameter Estimation for Multiscale Diffusions," Journal of Statistical Physics, vol. 127, pp. 741–781, mar 2007.

[3] R. R. Coifman, S. Lafon, A. B. Lee, M. Maggioni, B. Nadler, F. Warner, and S. W. Zucker, "Geometric diffusions as a tool for harmonic analysis and structure definition of data: diffusion maps.," Proceedings of the National Academy of Sciences of the United

States of America, vol. 102, pp. 7426–31, may 2005.

469680 Data Analytics Applied to Reduce Hazards during Process Transitions

Abstract:

A data analysis approach based on machine learning is used to identify and characterize process data with standard transition processes. Transitions are a ubiquitous part of any continuous process occurring every time a valve is opened manually, a pump is started, or equipment is taken in or out of service. A data analysis framework is proposed to identify recurring transitions and detect deviations from normal transitions due to faults. We use classification, regression, clustering, dimensionality reduction, and other machine learning techniques to extract information from process data. The information extracted can then be used to identify clusters of data and detect similar patterns and be used as a basis to characterize the transition procedure. It is also possible to detect different trend relationships and characterize the trend responses.

450818 Dissimilarity-Based Fault Diagnosis through Ensemble Filtering of Informative Variables

Abstract:

Despite the fact that fault diagnosis, similar to pattern recognition, has been widely studied in recent years, two key challenges remain: insufficient training samples and overlapping characteristics faced by reference fault classes. Recognition of these challenges motivate this study. First, an ensemble filtering of informative variables, also serving as a dimensionality reduction step, is proposed to address the challenge of insufficient training samples vs. high dimensionality. Second, to characterize the difference among overlapped fault classes, a dissimilarity analysis, that detects changes in a distribution of two datasets, is proposed. A moving window technique with incrementally increasing window sizes is used to gather data from online abnormal samples as well as each reference fault class. The dissimilarity for a pairwise set of data windows is then computed using the informative variables. The fault class recognition depends on the minimum dissimilarity achieved by the reference fault classes at each moving window step. The comparisons demonstrate that the recognition performance of the proposed approach is considerably better than that of discriminate models as well as other pattern matching methods.

Systems Analysis of Ensemble of Decision Trees for Modeling and Process Control

Zelimir F. Kurtanjek
University of Zagreb, Croatia

Abstract

Considered are methodologies for systems analysis of processes based on ensemble (forest) of decision tree models. The ensembles are derived from multivariate variables records of numerical and categorical data from plant monitoring systems and can be considered as an application of production plant big data analysis. The ensembles are optimized by the gradient boosting algorithm. Structure and dimensions of individual trees is minimized by the regularization of the objective function defined by minimization of the model prediction error with penalty function for number of trees, tree complexity and crispness of classification. Dimension of the ensembles are minimized by cross validation on training sets and validated on independent training set. Systems properties are evaluated on systemic random search of process variables by the numerical procedure for global Fourier Amplitude Sensitivity Test (FAST) simulation. Derived are the variable importance ranks, confidence intervals of the variables for the ensemble models, global sensitivity and inference of process variable synergism. The methods are demonstrated on a set of data from a communal waste water treatment plant (WWTP). Monitored are 37 chemical, biological and physical variables continuously during the period of two years. In view of process control evaluated are accuracies for prediction of output control variables depending on the length of prediction horizon and inclusion the decision trees into model predictive control (MPC) structure.

Key words: Process control, decision trees, systems analysis, WWT

Introduction

The basic paradigm of mathematical modeling in chemical engineering based on differential and algebraic equations is being extended to include modeling based on artificial intelligence (AI) and machine learning (ML) techniques. The driving forces for the changes are result of "big data" introduction into industrial environment and significant advances in software tools with sophisticated application programming interfaces (API). Large amount of on-line and historical data from a production plant enables use of complex statistical methodology and development of validated and reliable models for process control and design. Especially important departure from the basic paradigm is use of qualitative information, images, heuristic reasoning and unstructured text. Such models have shown potential for control decision support on large scale, whole plant or corporation, and extended time horizons.

Here is presented, as an example of the opportunities of ML modeling, analysis of a communal biological waste water treatment plant during continuous monitoring during two years[1]. Wastewater treatment process are complex and are difficult to inclusion of qualitative information and learned experience into a differential equation based models, such as activation sludge model (ASM) developed and coordinated by International Water Association (IWA). Applied is an ensemble of decision trees models for auto regression moving averages (ARMA) applicable for model predictive control (MPC). Based on the model evaluated are systems properties such as process variable importance analysis, variable global sensitivity and levels of variable interactions (synergy between process variables).

Model

Applied is ensemble of related, i.e. nonrandom, decision trees DT_i obtained by the tree boosted algorithm[2]. Prediction of output response is un-weighted average of the whole ensemble[3-4]:

$$\hat{y} = DTF(X) = \frac{1}{N}\sum_{i=1}^{i=N_T} DT_i(X) \tag{1}$$

The model evolution is regularized for ensemble size N, tree complexity (depth) and crispness of prediction of output leafs. Ensemble growth is a sequential minimization problem of the sum of square errors of prediction SSE and the penalty function ϕ for the model complexity (2-3):

$$\min F_{object.}(X, DTF, \omega) = SSE(y, \hat{y}) + \phi(DTF, \omega) \tag{2}$$

$$\phi(DTF, \omega) = \gamma \times N_T + \frac{1}{2} \times \lambda \times \sum_{i}^{N_T} \omega_i^2 \tag{3}$$

At each iteration minimization is determined by the maximum gradient based on analytical expansion of the objective function to the corresponding Hessian[5]. Number of needed iterations is determined by five or ten fold cross validation until minimal validation SSE is reached. To use the model for MPL control the decision trees are trained with ARMA time series data:

$$y(t+1) = DTF(y(t), X(t)) \tag{4}$$

Properties of the derived model(s) are analyzed for process variable importance, global sensitivity of each individual process variable and corresponding levels of interaction among the variables. The analysis harnesses very efficient simulation by ensemble of decision trees. The process variable space is "scanned" by uncorrelated perturbation mapped by a single Lissajous curve[6]. The model response is analyzed by Fourier amplitude sensitivity test (FAST). Application of FAST algorithm provides sensitivities to large and simultaneous changes of the complete set of the model parameters.[6] The method is based on transformation of the multidimensional parameter space to a one dimensional space of a single parameter, s, by the nonlinear transformations of the normalized parameters x_i:

$$x_i = \frac{1}{2} + \frac{1}{\pi} \times \arcsin(\sin(\pi \times \omega_i \times s + \varphi_i)) \qquad (5)$$

The s parameter space covers the range $s \in [-1,+1]$. With each parameter transformation are associated two random parameters, frequency ω_i and phase angle φ_i. Variations of the process variables are assumed independent, i.e. joint probability density function is given by the product:

$$\rho(X) = \prod_{i=1}^{N} \rho_i(x_i) \qquad (6)$$

Applied are uniform density function corresponding to ranges (global) between first and third quintiles and normal distribution functions with corresponding mean and variance:

$$\rho_i(x_i) = \begin{cases} U(x_i Q_{i1}, x_i Q_{i3}) \\ N(x, \sigma_i(x_i)) \end{cases} \qquad (7)$$

Individual global sensitivities S_i are based on the relative variance contributions:

$$S(x_i) = \frac{V(E(y|x_i))}{V^T(y)} \qquad (8)$$

Total effect S^T of each variable perturbation are given by the relative expected value of variance evaluated for the whole set of variables except the considered (marked as $\neq$):

$$S^T(x_i) = \frac{E(V(y|_{\neq} x_i))}{V^T(y)} \qquad (9)$$

The difference between the total effect and corresponding individual determines synergy:

$$Syn(x_i) = S^T(x_i) - S(x_i) \qquad (10)$$

Process and data

For decision tree modelling considered is a process of communal continuous biological waste water treatment depicted in Fig. 1. The plant constitutes the pre-treatment processing unit, followed by the primary settler and biological processing reactor. Output from the biological reactor is processed in the second settler in the recycle closed loop. The model is based on data collected during two years of continuous monitoring and evaluation of 37 variables[1]. Most of the process variables can be considered statistically independent, as depicted by the auto correlation matrix in Fig. 2. The modelling data set has 2.6 K numerical variables and for systems analysis applied is simulation data set with 2 M (7 Gbyte file) numerical data. As the output variable considered is biological oxygen demand (BOD) from the biological reactor. Model predictive control is envisioned by the feedback loop with bypass regulation of the inlet row feed. Process dynamics is accounted by auto moving averages for one day delay. The model test set is 60 days of continuous processing. All numerical and statistical evaluations are performed by use of open source R software[4].

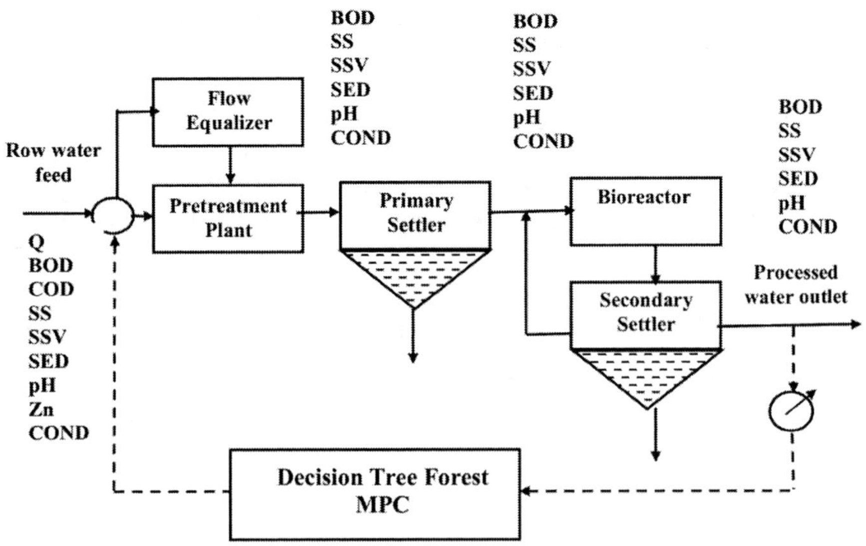

Fig. 1. Schematic diagram of the biological waste water treatment plant with a decision tree based model predictive (MPC) feedback control loop

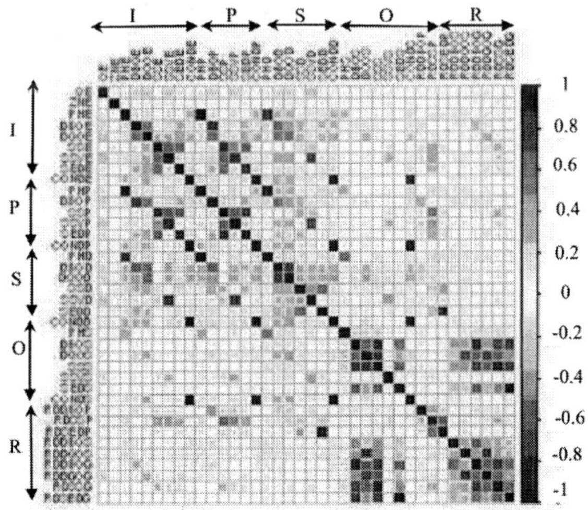

Fig. 2. Correlation matrix of the 38 process data and performance indices measured continuously daily during two years. I=input, P=primary settler, S=secondary settler, O= bioreactor output, R=performance indices.

Results and Discussion

Systemic properties of the decision tree model of a dynamic system, WWT plant, are compared to the linear moving averages with multivariate exogenous inputs ARMAX. The decision tree model is ensemble of 49 individual trees with two depth levels. Number of trees is determined by minimization of 10 fold root mean square error (RMSE). The linear ARMAX is obtained with the test data set by unweighted and autoscaled least square method. The two models are applied for prediction of untrained data set for 60 days operation. BOD responses are depicted in Fig. 3. The improved accuracy by the DT model can be observed by correct prediction of the BOD transient increase outside the process regular range. The same conclusion can be obtained by determination of significant increase of the correlation coefficients between the model predictions and process observation, for DT model R=0.71 while for LS model R=0.56.

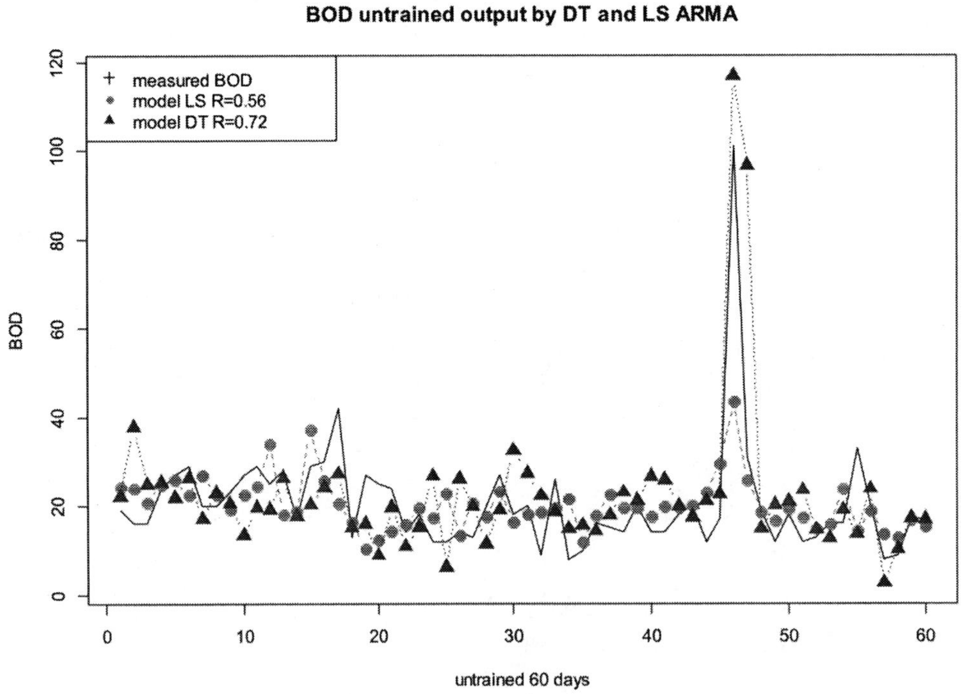

Fig. 3. Comparison of untrained predictions of the outlet BOD by linear least squares (LS) and decision tree forest (DTF) autoregression moving averages (ARMAX) for 60 days period of the plant operation.

Importance of individual parameters for LS ARMAX model are evaluated by the absolute values of corresponding coefficients for autoscaled variables. Variable importance for DT ARMAX model are evaluated by corresponding frequencies of the tree branch splitting. The comparison is presented in Fig. 4A. The linear model predicts sedimentation in the bioreactor as the most important variable effecting outlet BOD with relative importance of 35 %.

However, contrary to the linear model, the DT ARMAX predicts BOD in the second with 60% of the total account of importance.

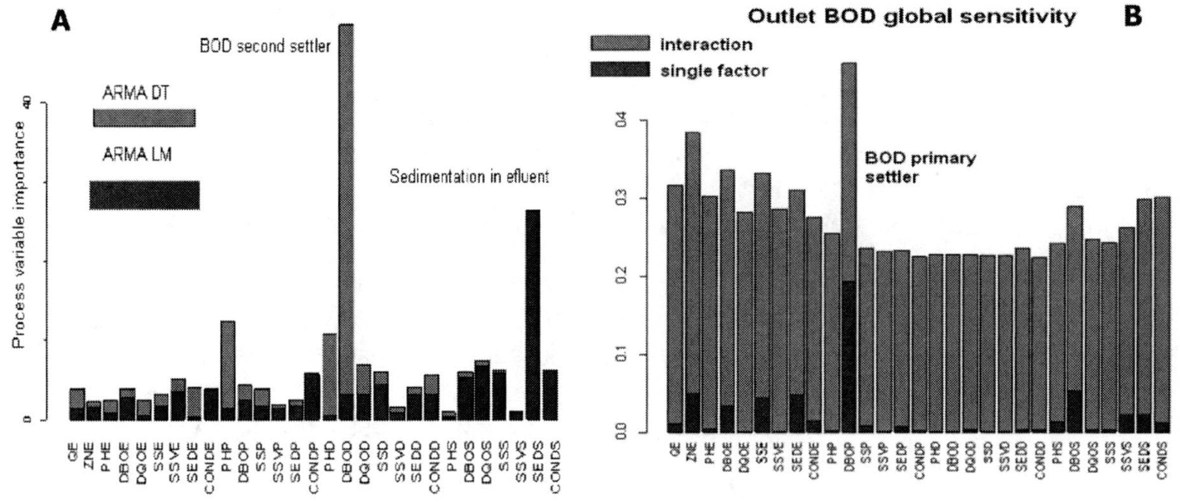

Fig. 4. A) Comparison of the model process variable importance by the least squares and decision tree forest ARMA models. B) Comparison of global sensitivities by the single process variable and their global interaction (process variable synergism).

Obtained variable importance ranks and the variable interaction levels (synergism) are evaluated for the sub-ranges of the observed process variables. Evaluated are the ranks under the assumption of uniform cumulative probability density functions (CPDF) between the first Q_{1i} and third Q_{3i} quintile corresponding to each of the variable x_i, and with the assumption of the corresponding normal probability density distributions $N(\mu_i,\sigma_i)$ in the ranges of $x_i \pm \sigma_i$. The results for the uniform distributions are depicted as stacked barplot in Fig. 4B. Similar results of global analysis are obtained for the assumption of the corresponding normal distributions. From the analysis of BOD outlet reveals that about 90 % process interaction results from of the variable mutual interactions, while only 10 % corresponds to the individual variable effects. The important outcome of the global analysis is the shift of the sedimentation in the second settler as the most important variable, obtained from local (variations around the process working point), to BOD in the first settler obtained by global analysis. The individual global effect of BOD in primary settler accounts for 40 % of all individual effects, and also has the highest level of interactivity (about 30 % of all interactions).

Conclusions

Analyzed is application of decision tree forest DTF for modeling of time series of a complex industrial process and is compared to linear least squares (LS) ARMAX modeling. The two models are compared for prediction error with untrained data obtained during sixty days of continuous process operation.

Due to property of DTF to model nonlinearity accuracy of prediction is greatly improved (DFT R=0.72 versus LS R=0.5). The increased accuracy of DTF, transparency of decision rules, and numerical efficiency makes them the best modeling choice for model predictive based control.

Applied is process variable importance analysis needed for the task of process control structure design. Variable importance for LS ARMAX are evaluated by magnitude of the model coefficients with auto-scaled process variable data, and are compared with variable importance list obtained by splitting frequency during DTF development. There is a significant difference between the lists of the most important variables. The LS model points to the variables measured at the last process unit where the output variable is measured. However, the DTF model predicts measured state at mid process unit as the most important variable.

The importance variable lists are compared with variable global analysis based on variance analysis of the model simulation data with "complete" search of the whole (expected between first and third quintiles) state space. The same process variable is selected by the global analysis of the simulation data and local DTF obtained from the training data. The difference is that the global analysis predicts that the same most influential variable is in the first process stage (unit).

Most importantly for complex processes, evaluated are effects of mutual process variables interactions. Such is the selected process (biological WWT plant) which is complex with high process variable interactions and difficult to model based on the first principles. The DTF simulations reveal that variable interactions are more pronounced than effects associated with individual variables. About 90 % of variance of WWT data corresponds to variable interactions (synergism).

In conclusion, for control of complex processes difficult to model on the first principles, regularized DFT models provide the most accurate predictions on untrained data. DFT transparency and numerical efficiency and simplicity makes the most suitable for process control.

References

1) J. Bejar, Ulises, UCI Machine Learning Repository [http://archive.ics.uci.edu/ml]. Irvine, CA: University of California, School of Information and Computer Science (1996)
2) L. Breiman, J. H. Friedman, R. A. Olshen, C.J. Stone, „Classification and Regression Trees", CRC, Boca Raton, 1998.
3) T. Hastie, R. Tibshirani, J. Friedman, The Elements of Statistical Learning, Data Mining, Inference and Prediction, Springer Verlag, Second Edition, New York, 2009.
4) R Development Core Team, , R: A language and environment for statistical computing. R, Vienna, Austria,www.R-project.org 2015.
5) T. Chen, T. He and M. Benesty (2016). xgboost: Extreme Gradient Boosting. R package version 0.4-3. https://CRAN.R-project.org/package=xgboost
6) G. Pujol *et al.*, https://cran.r-project.org/web/packages/sensitivity/

Želimir Kurtanjek

University of Zagreb,
Faculty of Food Technology and Biotechnology
CROATIA

zelimir.kurtanjek@gmail.com

Systems Analysis of Ensemble of Decision Trees for Modeling and Process Control

Poster number: 455980

Final Paper##: 190e

Session: Interactive Session: Systems and Process Control
scheduled for Monday, November 14, 2016: 3:45 PM - 5:45 PM, Hilton, Grand Ballroom B

* Retired professor of chemical engineering

Goal

The goal of the presentation is to demonstrate principles and advantages of regularized decision tree forests (DTF) for information synthesis and process modelling , control and systems analysis.

Model goal function of the regularized DTF:

$$\min F_{object.}(X, DTF, \omega) = SSE(y, \bar{y}) + \phi(DTF, \omega)$$

Process: WWT, 38 variables, 2 years continuous monitoring data

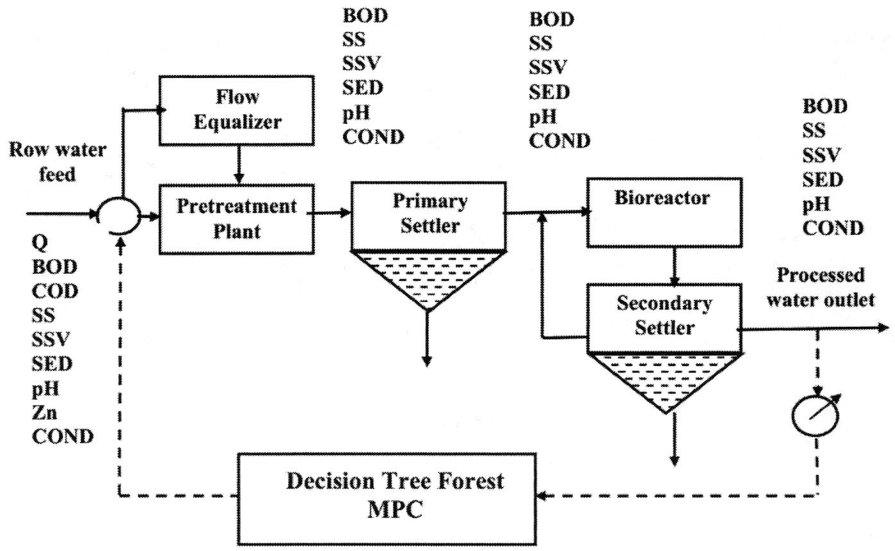

Advantages

Significantly improved, over OLS, predictivity on untrained data (2 months of operation)

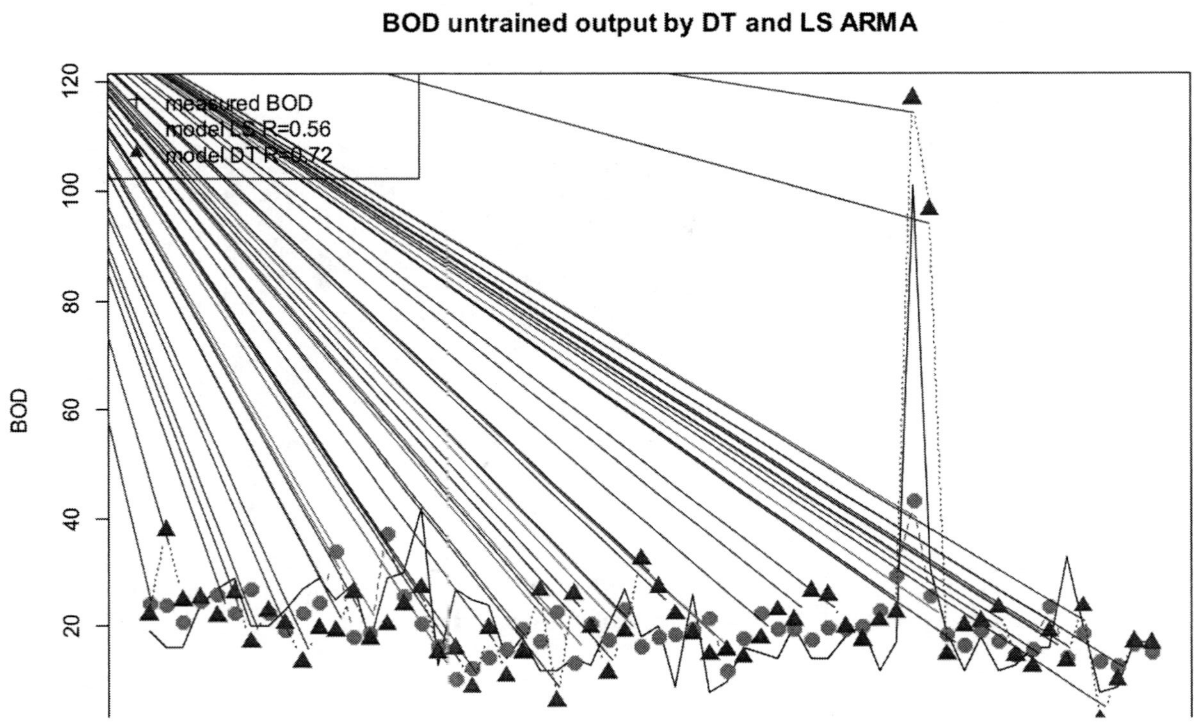

Systems analysis

Global analysis in the ranges between 1-3rd quintile

1) Variable importance

2) Process variables interactions (synergism)

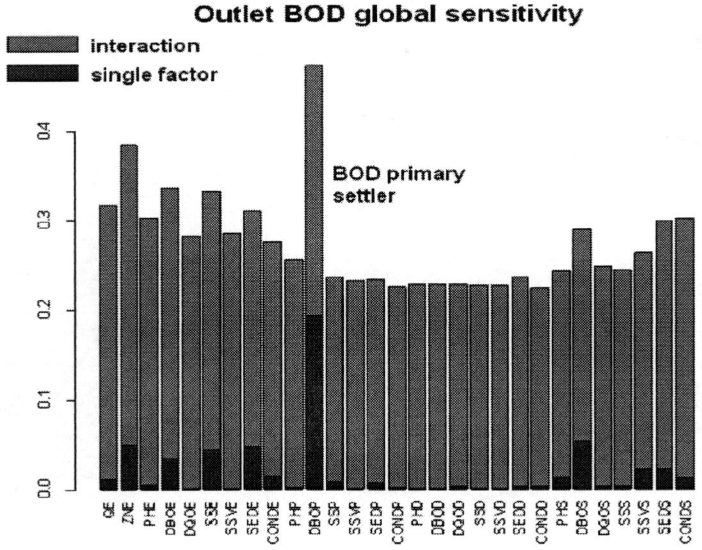

Conclusions

1) DTF enables integration of heterogeneous process data

2) Regularization yields model transparency and inference of system information

3) Improved process control is enabled

462208 Modeling Emergent Phenomena in Complex Sociotechnical Systems

Abstract:

A number of systemic failures, such as the Global Financial Crisis, BP Deepwater Horizon oil spill, and the Northeast Power Blackout, share surprising similarities in the failure mechanisms. Such similarities call for a unified framework that can describe, model, and more importantly, manage systemic risks across domains. One of the challenges in developing such a framework is creating a universal representation, which can describe components, relationships, and emergent behaviors of a complex system. The representation should also take into consideration the human factors. Existing methods are generally suitable only for modeling a subset of complex systems. For example, equations describe system dynamics while digraph-based techniques deal with causal relationships.

In this work, we propose such a unified modeling framework by combining system teleology with axiomatic set theory. The framework abstracts elements of sociotechnical systems as sets and models system teleology as nonlinear relations. Using this framework, we can demonstrate the underlying process of emergence -- an essential element of systemic risk. We illustrate our approach using several case studies.

462353 On the Robust Explicit Model Predictive Control of Hybrid Discrete-Time Linear Systems

Abstract:

In robust model predictive control, the aim is to control a system in the face of model uncertainties [1-3]. For the case of discrete-time linear systems, one of the most challenging issues is encountered when the state-space matrices are uncertain, a problem which is termed multiplicative/parametric uncertainty. In the open literature, this type of problem has been tackled using the concepts of open-loop and closed-loop control [4]. In the open-loop approach, the aim is to identify a set of control actions which will guarantee constraint satisfaction over the entire horizon. However, this approach is quite conservative, as it does not consider the fact that future control actions will have future measurements at their disposal. Thus, the idea of closed-loop robust MPC has been formulated, which aims at finding a control action to apply '*now*' such that there will exist a '*future*', feasible control action. While such robust approaches have been applied extensively to the case of discrete-time continuous systems, only few approaches have been proposed regarding hybrid systems, i.e. systems featuring both continuous and discrete components [5].

In this work, we consider the development of novel closed-loop robust explicit MPC strategies for hybrid discrete-time linear systems. It extends our results for continuous systems [6], which is based on the following components: (i) a robust counterpart formulation for the special case of box-constrained uncertainty, guaranteeing constraint satisfaction, (ii) a multi-parametric linear programming (mp-LP) problem for each stage (starting from the final stage in a dynamic programming setting), and (iii) the recursive solution of the mp-LP problem yielding the initial set of states for which robust constraint satisfaction can be guaranteed. For the case of hybrid systems, we show that (i) a similar type of robust counterpart formulation can be derived under mild assumptions, (ii) this robust counterpart results in a multi-parametric mixed-integer linear programming (mp-MILP) problem which can be formulated for each stage, and (iii) using state-of-the-art methods the mp-MILP problem can be solved recursively. This then enables the formulation of the robust explicit hybrid MPC problem as a multi-parametric mixed-integer quadratic programming problem, for which the authors recently proposed the first exact solution algorithm [7]. Using a series of example problems, the applicability and scalability of this novel approach will be highlighted.

References

[1] Mayne, D. Q.; Rawlings, J. B.; Rao, C. V.; Scokaert, P. O. M. (2000) Constrained model predictive control: Stability and optimality. Automatica, 36(6), 789 – 814.

[2] Kerrigan, E. C.; Maciejowski, J. M. (2004) Feedback min-max model predictive control using a single linear program: robust stability and the explicit solution. International Journal of Robust and Nonlinear Control, 14(4), 395 – 413.

[3] Wan, Z.; Kothare, M. V. (2003) An efficient off-line formulation of robust model predictive control using linear matrix inequalities. Automatica, 39(5), 837 – 846.

[4] Bemporad, A.; Borrelli, F.; Morari, M. (2003) Min-max control of constrained uncertain discrete-time linear systems. IEEE Transactions on Automatic Control, 48(9), 1600 – 1606.

[5] Nascu, I.; Oberdieck, R.; Pistikopoulos, E. N. (2015) Offset-free Explicit Hybrid Model Predictive Control of Intravenous Anaesthesia. In Proceedings of the IEEE International Conference on Systems, Man and Cybernetics (SMC), 2475 – 2480.

[6] Oberdieck, R.; Misener, R.; Pistikopoulos, E. N. (2016) Robust explicit/multi-parametric control. AIChE Journal, in revision.

[7] Oberdieck, R.; Pistikopoulos, E. N. (2015) Explicit hybrid model-predictive control: The exact solution. Automatica, 58, 152 – 159.

Data Based Process Monitoring of Industrial Processes Using Multiscale Nonlinear Multivariate Statistical Methods

Chiranjivi Botre, Dr. M. Nazmul Karim

Texas A&M University
College Station, Texas

Email ID: chirubotre@gmail.com

1

Dwight Look College of ENGINEERING TEXAS A&M UNIVERSITY

2016 AIChE Annual Meeting, San Francisco, CA

Introduction

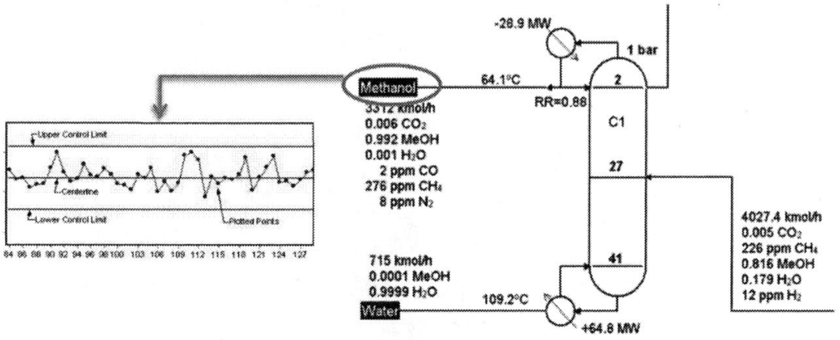

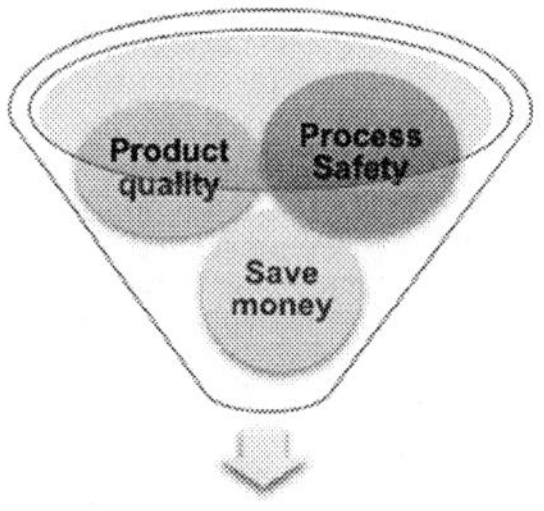

- Data based process monitoring technique have been successfully applied where accurate process models are not available

- Obstacles with industrial process data
  - ✓ Complex nonlinearity in data set
  - ✓ Noise in data set
  - ✓ Auto-correlated variables in data set

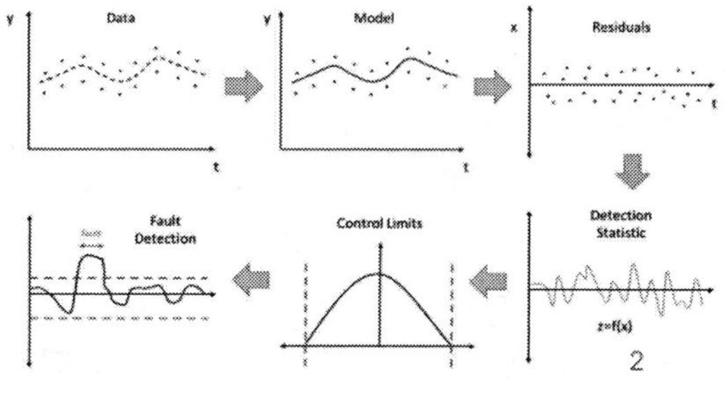

2

Dwight Look College of ENGINEERING TEXAS A&M UNIVERSITY

16AIChE Annual Meeting, San Francisco, CA

Methodology

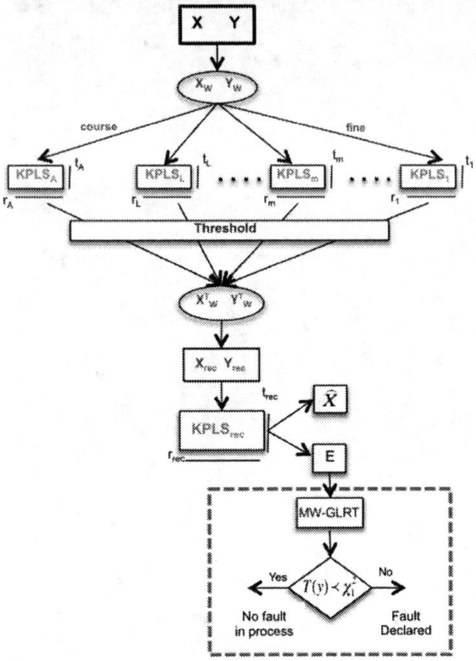

Kernel Partial Least square (KPLS)

Map the input data to a feature space, where the non-linear structure in the input space is more likely to be linear.

$$\Phi: \ \chi \rightarrow F \qquad\qquad k(x_i, x_j) = \Phi(x_i)\Phi(x_j)$$

Multiscale representation

Wavelet decomposed coefficients are

1. De-noised 2. De-correlated 3. Normalized

Moving window GLRT

1. Composite hypothesis problem is solved
2. θ is computed by maximizing likelihood ratio function

$$\begin{cases} H_0 = \{E \sim N(0, \sigma^2 I_N)\} \\ H_1 = \{E \sim N(\theta, \sigma^2 I_N)\} \end{cases}$$

Optimization of the kernel parameter

$\underset{h}{\text{Min}}$ *(Missed detection rate + false alarm rate)*

s.t.

Kernel NNPLS algorithm

GLRT statistic algorithm

$$K(x,y) = \exp\left(-\frac{\|x-y\|^2}{2 \times h^2}\right)$$

3

Dwight Look College of ENGINEERING TEXAS A&M UNIVERSITY

16AIChE Annual Meeting, San Francisco, CA

Result and Take Away Points

Moving window GLRT

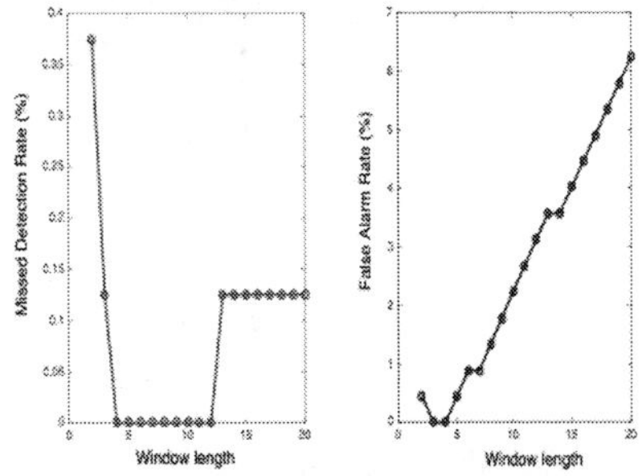

MS-KPLS based MW-GLRT

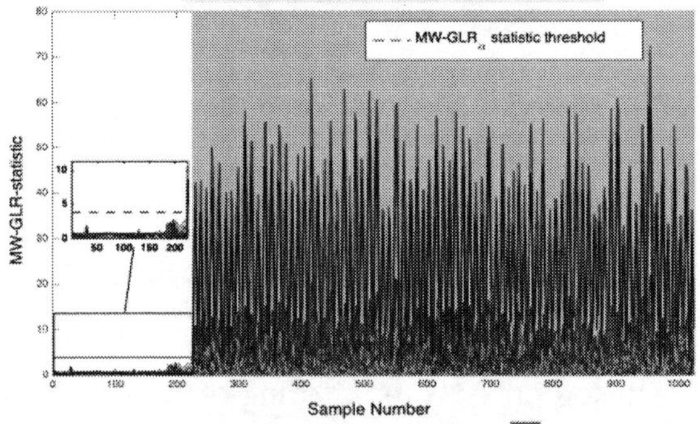

Missed Detection rate ⇩ False Alarm rate ⇩

- Take away from presentation
  - ✓ Applicability of data based multivariate statistical techniques
  - ✓ Improved fault detection performance of KPLS model by use of multiscale representation and optimized kernel parameter
  - ✓ Composite hypothesis testing statistic (GLRT) over conventional statistic methods

4

462783 Beyond Foptd Models in Tuning PI Controllers

Abstract:

The first-order plus time delay (FOPTD) model is widely used to design proportional-integral (PI) controllers.[1-2] The FOPTD model

$$G_F(s) = \frac{k\exp(-\theta_F s)}{\tau_F s + 1}$$

has three parameters (process gain, time constant, and time delay) and many PI controller tuning rules based on these FOPTD model parameters are available.

When high-order process models are available, their FOPTD models can be obtained by applying model reduction techniques. Otherwise, they can be obtained from process responses. Traditional simple methods to obtain FOPTD models are based on the several measurements of process step responses as shown in Fig. 1.[3]

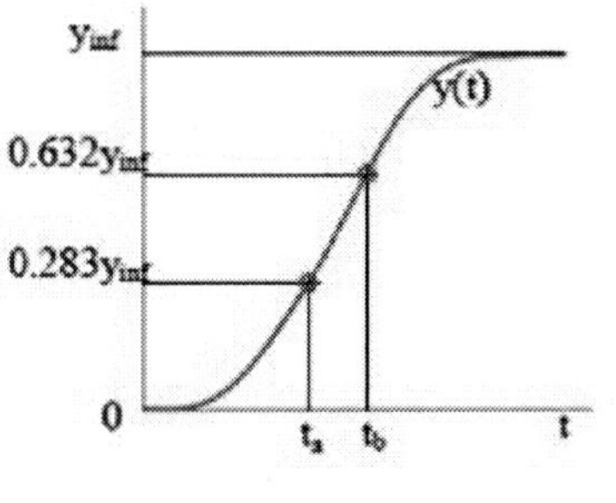

Fig. 1. A graphical method to identify the FOPTD model from open-loop step response.

For some overdamped processes, even though their step responses are very similar to that in Fig. 1, FOPTD models fail to provide acceptable PI controller parameters. For such processes, a half-order plus time delay (HOPTD) model

$$G_H(s) = \frac{k\exp(-\theta_H s)}{\sqrt{\tau_H s + 1}}$$

can be used to approximate their dynamics and tune PI controllers.[4] Field data for the three model parameters of $G_H(s)$ are just the same as those for the FOPTD model, i.e., only two point data with the final steady state are the field data required.

For processes showing overshoots in step responses as shown in Fig. 2, FOPTD models also fail to describe step responses and provide acceptable PI controller parameters. For such processes, instead of finding high order models that fit step responses, step responses are divided into two parts. FOPTD models for two regions are obtained and they are combined as

$$G_{dFOPTD}(s) = \frac{k_1}{\tau_1 s + 1}e^{-\theta_1 s} + \frac{k_2}{\tau_2 s + 1}e^{-\theta_2 s} + k_1 e^{-(t_p - \theta_1)/\tau_1}\left(1 - \frac{1}{\tau_1 s + 1}\right)e^{-t_p s}$$

Here the first two terms are for $G_I(s)$ and $G_{II}(s)$ in Fig. 2 and the last term is to cancel the effects of $G_I(s)$ after t_p. Each model of $G_I(s)$ and $G_{II}(s)$ can describe well each part of the response in Fig. 2. This combined model (dFOPTD model) can be used to tune PI controller.[5]

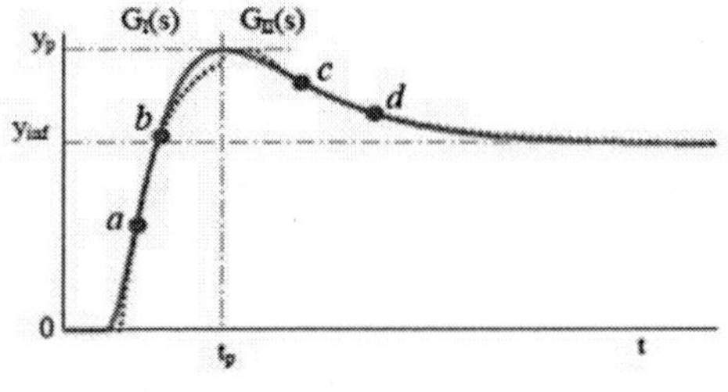

Fig. 2. Typical processes where dFOPTD models are applied.

For the above two types of processes, FOPTD models suffer from approximating step responses due to the structural limitations and fail to provide acceptable PI controller parameters. Here merits and demerits of two models of HOPTD and dFOPTD are studied.

References

[1] Seborg DE, Mellichamp DA, Edgar TF, Doyle FJ. *Process Dynamics and Control*, 3rd ed. New York: Wiley, 2010.

[2] O'Dwyer A. *Handbook of PI and PID Controller Tuning Rules*, 3rd ed. London, UK: Imperial College Press, 2009.

[3] Liu T, Wang QG, Huang HP, A tutorial review on process identification from step or relay feedback test. *J. Process Control*, 2013, *23*, 1597-1623.

[4] Lee J, Lee Y, Yang DR, Edgar TF, Half Order Plus Time Delay (HOPTD) Models to Tune PI Controllers, submitted to *AIChE J.*, 2016.

[5] Lee Y, Yang DR, Lee J, Edgar TF, Double First Order Plus Time Delay Models to Tune PI Controllers, submitted to *I&EC Research*, 2016.

463245 Hybrid Algorithms for the Design of Sensor Networks of Nonlinear Systems

Abstract:

The design of the instrumentation system of a chemical plant is a complex multilevel task, which comprises the definition of the global objectives, the selection of the measured variables, and the specification of the implementation details, such as measurement intervals, sample procedures, interfaces, maintenance activities, etc.

During the on-line operation, the quality and availability of process knowledge essentially depends on the instrumentation selection performed on the second design stage. The structure of a sensor network (SN) is defined by the type, amount, precision, reliability and location of its instruments.

To determine the optimal SN that minimizes the total instrumentation cost and simultaneously satisfies the information requirements, a combinatorial optimization problem is stated that involves binary variables. The difficulties associated to the solution of that problem come from the dimension of the search space, which increases significantly for large scale processes, and the type of mathematical relationship among the process variables.

Nguyen and Bagajewicz (2008) addressed the design of SNs that provide information about the operation of processes described by a set of non-linear algebraic equations. They proposed the equations-based tree search method, which guarantees to find an optimal solution and is computationally efficient for small and middle scale problems. However, its performance is not always satisfactory for dealing with large scale problems (47 binary variables). At the same time Kelly and Zyngier (2008) presented a MILP formulation for designing sensor structures that satisfy observability, software/hardware redundancy and precision. But to deal with a problem that involves 34 binary variables, some heuristics were used. Then Nguyen and Bagajewicz (2011) exploited certain cost properties of the different nodes in the tree search to efficiently prune nonoptimal nodes using a breadth-first/level traversal tree search method to obtain the global optimum, and proved the efficiency of the algorithm using a benchmark that comprises 24 binary variables. Finally the aforementioned authors combined the tree search procedure with a local search heuristics to reduce the computational time (Nguyen and Bagajewicz (2013).

Regarding the application of stochastic optimization problems to solve the design problem, Gerkens and Heyen (2005) presented two ways of parallelizing the classic GA to reduce the solution time, and applied them on large scale problems (212 and 618 binary variables), but a solution reproducibility analysis is missed.

In this work, the metaheuristic proposed by Carnero et al. (2013) is extended to the design of SNs of processes whose operation is described using the set of non-linear algebraic equations associated to the mass, component and energy balances. The approach uses an Estimation of Distribution Algorithm (Hauschild and Pelikan, 2011), i.e. a global search engine, in combination with a local search heuristic. This is Strategic Oscillations within the framework of Tabu Search. Application results of the hybrid technique have demonstrated a rewarding performance for solving the same problem for linear systems of incremental size. It achieves a high reproducibility of the good solutions using a controlled computational time.

The incorporation of the process knowledge contained in the balance equations is extremely useful to guide the search. In this work, a technique is presented to form the initial population of the Estimation of Distribution Algorithm in such a way that at least the observability constraints of the key variables are satisfied. The performance of the proposed design methodology is evaluated and compared with the corresponding ones to other existing strategies for small and medium size benchmarks. Regarding a large scale case study, best known solutions are provided, and their reproducibility is analyzed.

References

Baluja S, Caruana R. Removing the genetics from the standard genetic algorithm. Technical Report CMU-CS-95-141; Carnegie Mellon University, 1995.

Carnero M, HernÃ¡ndez J, SÃ¡nchez M. A new metaheuristic based approach for the design of sensor networks. Comp. Chem. Eng. 2013; 55: 83-96.

Gerkens C, Heyen G. Use of parallel computers in rational design of redundant sensor networks. Comput. Chem. Eng. 2005; 29: 1379-87.

Hauschild M, Pelikan M. An introduction and survey of estimation of distribution algorithms. Swarm and Evolutionary Computation 2011,1: 111â€"28.

Nguyen D, Bagajewicz M. Design of nonlinear sensor networks for process plants. Ind. Eng. Chem. Res. 2008; 47: 5529-42.

Nguyen D, Bagajewicz M. New efficient breadth_first/level traversal tree search method for the design and upgrade of sensor networks. AIChE J. 2011; 57: 1302-9.

Nguyen D, Bagajewicz M. Efficient Approximate Methods for the Design and Upgrade of Sensor Networks. Ind. Eng. Chem. Res. 2013; 52: 83-90.

Kelly JD, Zyngier D. A new and improved MILP formulation to optimize observability, redundancy and precision for sensor network problems. AIChE J. 2008; 54: 1282-92.

464300 Power Plant Abnormal Condition Detection Using the Artificial Immune System Paradigm

Abstract: Power Plant Abnormal Condition Detection Using the Artificial Immune System Paradigm

Ghassan Al-Sinbol, Mario G. Perhinschi, Debangsu Bhattacharyya

Modern power plants are expected to function safely and efficiently under both normal and abnormal operating conditions. Due to the complexity, multi-dimensionality, and strong interaction among various sections of these plants, health monitoring requires comprehensive and integrated methodologies. In this paper, the artificial immune system (AIS) paradigm is used to develop a detection scheme for abnormal operation of advanced power plants.

The functionality of the biological immune system and its capability to discriminate between "self" (normal conditions) and "non-self" (abnormal conditions) inspired the AIS paradigm as a novel artificial intelligence technique with promising capabilities for abnormal condition detection. An abnormal situation or failure affecting a dynamic system is considered similar to an antigen invasion. Using a positive selection-type of algorithm, failures are declared or detected when the current configuration of "features" does not match with any configuration from an exhaustive set known to correspond to normal situations or the self. Building the self requires a substantial amount of data collected at normal operational conditions; however, it does not require modeling of the system.

In this paper, a novel approach, denoted as 'the partition of the universe' approach, for self/non-self generation is presented. In this approach, the n-dimensional feature space is divided into uniform partitions with predefined centers, shape, and size. The raw self points are then tested against the universe partitions. Selfpartitions are then identified and saved as integer strings that can be produced and used with reduced computational effort.

An abnormal condition detection scheme is designed based on a positive-selection-type of algorithm in conjunction with a dendritic cell (DC) mechanism that is expected to allow extension towards abnormal condition identification and evaluation. The artificial DC is a computational module inspired from the interaction between the innate and adaptive immune systems. The proposed artificial DC mechanism is expected to provide the detection outcome based on the current and past discrimination results and overcome any imperfections in the definition and generation of the self. The general flowchart of the detection process is presented in Figure 1 and the flowchart of the proposed abnormal condition detection scheme using the partition of the universe approach is presented in Figure 2.

The proposed approach is demonstrated with promising results using a rigorous model of an acid gas removal (AGR) unit as part of the integrated gasification combined cycle power plant developed in Dynsim® environment. A total of 150 features was selected to build the self/non-self of the AGR unit, including pressure, temperature, flow rate, and composition measurements across the unit. The AGR unit is divided into 22 subsystems and lower dimensional projections of the self are considered for each subsystem within the hierarchical multi-self strategy. For the purpose of demonstrating the operation of the proposed detection scheme, a limited number of abnormal conditions that include deposition of solids, such as flyash, and leakages in the pipes or equipment items has been considered. These abnormal conditions affect five main sub-systems characterized by sixty features.

The proposed detection scheme provides excellent performance with high detection rates and zero false alarms for all cases considered. The detection rates vary between 94.6% and 99.3% with detection times between 2.5 and 10.5 seconds.

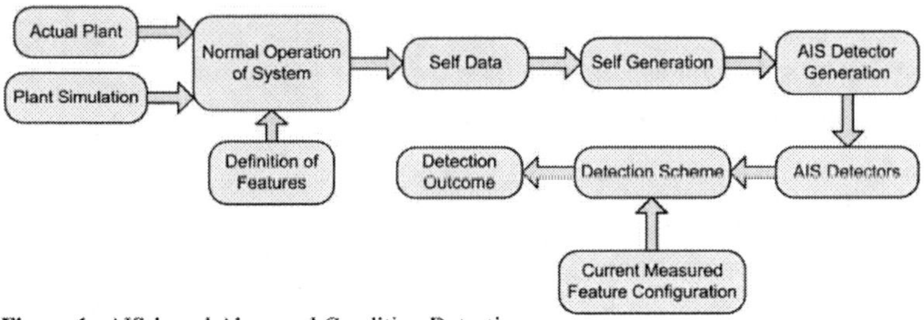

Figure 1. AIS-based Abnormal Condition Detection

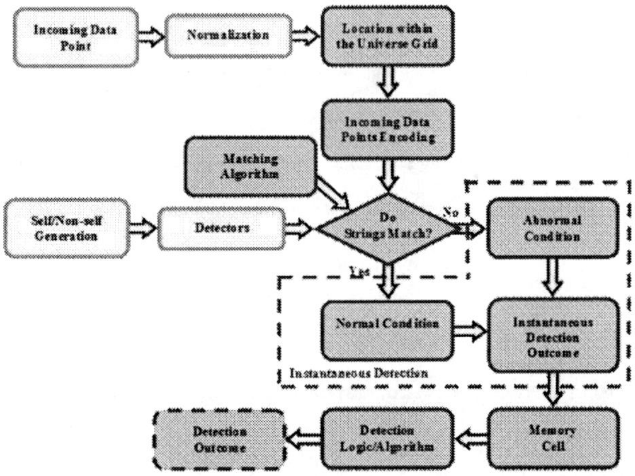

Figure 2. Abnormal Condition Detection Using the Partition of the Universe Approach

465078 Delay-Timer Alarm Design for Uncertain Chemical Systems

Abstract:

Modern process industries are equipped with highly automated fault detection and diagnosis (FDD) systems to maintain process safety and reliability standards. The primary objective of an alarm system is to alert the operator every time a fault orccurs or the process deviates from its normal behaviour. According to the Abnormal Situation Management Consortium (ASM), the US petrochemical plants alone lose more than $10 billion per year due to ab-normal plant situations and lack of ecient alarm management strategies [1]. This is due to the complex tasks performed by the plant operators, which include presentation, processing, and handing of hundreds and thousands of alarms from dierent units in a process plant. Without an ecient alarm management system, alarm monitoring becomes a daunting task. The alarm design is a critical rst step in setting up an ecient alarm management system. The design and performance evaluation of alarm systems have received considerable attention from both academic and industrial communities [2, 3, 4]. Although it has been extensively studied and applied primarily to linear, Gaussian, and stationary systems with uncorrelated signals [5], the real systems often tend to be nonlinear, non-Gaussian and exhibit complex correlations in the dynamics. This is particularly true for closed-loop systems, since they introduce strong system correlations even if the open-loop system has uncorrelated states. We consider a limit-checking based alarm design problem for complex chemical processes represented by discrete-time nonlinear stochastic state space models (SSMs). Given a sto-chastic model representation of a process, we are interested in designing a univariate delay timer alarm conguration for the process states. Once a delay-timer is designed, we evaluate its performance based on the false and missed alarm rates. The ecacy of the proposed alarm design is illustrated on a continuous stirred tank reactor (CSTR) reaction system.

References

[1] E. Cochran, C. Miller, and P. Bullemer, "Abnormal situation management in petrochemical plants: can a pilot's associate crack crude?" in Proceedings of the IEEE 1996 National Aerospace and Electronics Conference, Dayton, Ohio, May 20-23 1996.

[2] J. Ahnlund, T. Bergquist, and L. Spaanenburg, "Rule-based reduction of alarm signals in industrial control," Journal of Intelligent and Fuzzy Systems, vol. 14, no. 9, pp. 73-84, 2003.

[3] R. Brooks, R. Thorpe, and J. Wilson, "A new method for dening and managing process alarms and for correcting process operation when an alarm occurs," Journal of Hazardous Materials, vol. 115, no. 1, pp. 169-174, 2004.

[4] D. Rothenberg, Alarm Management for Process Control: A Best-Practice Guide for Design, Implementation, and Use of Industrial Alarm Systems. Momentum Press, 2009.

[5] N. A. Adnan, Y. Cheng, I. Izadi, and T. Chen, "Study of generalized delay-timers in alarm conguration," Journal of Process Control, vol. 23, no. 3, pp. 382-395, 2013.

467430 Monosilane Production: Analysis and Control of a Reactive Distillation Column to Reduce Cooling Utilities

Abstract:

Alternative fuels and renewable energy have been subjects of interest in recent years. With the depletion of fossil fuels, several investigations have focus their attention to solar energy. Polysilicon is the principal raw material by the solar photovoltaic and electronics industry. The evolution of applications in these areas has demanded an increase in polysilicon production.

One route to obtain ultra-high purity of polysilicon is achieved first by the preparation of a volatile silicon hydride i.e. silane and its subsequently purification, generally using fractional distillation. This is followed by the decomposition of this hydride to hyper-pure elemental silicon by reductive pyrolysis or chemical vapor deposition. The preparation of the volatile Si compound involves external reactants and its decomposition generates by-products, which need to be recycled.

The Union Carbide Corporation (UCC) process used to produce polysilicon is based on the thermal decomposition of monosilane on a heated silicon rod or filament placed inside a deposition chamber. This process requires two reactors and more distillation columns for producing monosilane. Unreacted reactants need to be separated and recycled back for further reaction, therefore utility and capital costs are extremely large.

Alternative processes to produce monosilane have been investigated and patented. Reactive distillation is a promising alternative to conventional reactor-distillation processes, so that capital and operating costs can be reduced [1]. Reactive distillation was first proposed by Bakay [2], since then, several research have been conducted to improve the process and reduce costs. One of the drawbacks of using reactive distillation is the expensive refrigeration services needed in the top of the column for silane production. Moreover, it is important to avoid thermal decomposition of the catalyst in the reactive trays. Block [3] proposed to use a two column scheme to increase temperature at the top of the reactive tower and reduce the temperature in the dome of the auxiliary purification tower by increasing operating pressure generating lower operation cost.

This work focuses on the analysis and control of such scheme and the comparison to the conventional reactive distillation column regarding economic and dynamic performance. Steady state of the configurations is simulated in AspenPlus. Simulations are then exported to AspenDynamics and temperature controllers are configured by means of a sensitivity analysis. The steady state results show that the two configuration scheme reduces operating cost but capital cost is increased. The dynamic responses are analyzed and compared under feed disturbances. The results show that the closed loop response for the two column configuration is slower, however it brings the controlled variables back to their nominal values.

[1] Muller, D., Ronge, G., Fer, J. S., & Leimkuhler, H. J. (2002). Development and economic evaluation of a reactive distillation process for silane production. *Distillation and Adsorption: Integrated Processes, Bayer AG, D-51368 Leverkuse.*

[2] Bakay, C. J. (1976). *U.S. Patent No. 3,968,199.* Washington, DC: U.S. Patent and Trademark Office.

[3] Block, H. D., Leimkühler, H. J., Müller, D., Schäfer, J. P., & Ronge, G. (2005). *U.S. Patent No. 6,905,576.* Washington, DC: U.S. Patent and Trademark Office.

467959 Active Fault Detection and Isolation and False Alarm Elimination By Constrained Optimization of Built-in and Maintenance Test Conditions

Abstract:

In this presentation, we will outline workflows and methods for the improvement of system reliability, in terms of capability and robustness in fault detection and isolation (FDI). We will present methods that improve fault identifiability and reduce or eliminate false alarms that otherwise cannot be correctly diagnosed or detected under standard operational or maintenance testing. This work is driven by the increasing need to eliminate false alarms and incorrectly diagnosed or intermittent faults that lead to the event called Fault Not Found (FNF) [1]. We present data from the literature and simulations that illustrate the problems of current methods of FDI in complex systems operating in uncertain environments. In particular, this work explores the issues of FDI and FNF in systems supporting aircraft operations.

A general framework is provided for the design of tests for fault detection and isolation, which is based on model-based methods for active fault diagnostics. We present the standard work and corresponding methods to treat fault detection and isolation as a set of constrained optimization problems. We consider the issues of uncertainty caused by system operations and modelling error and address them through the formulation of mathematical problems to minimize their impact on fault detection and isolation capabilities. We improve the identifiability of faults by maximizing the sensitivities of the system outputs with respect to faults and uncertain conditions through the adjustment of input trajectories. This method is based on optimal experimental design methods that improve precision and reduce correlation of estimated model parameters [2,3]. After calculating the optimal test design for fault detection and isolation, we analyze the system at its optimal operating point(s) to assess the identifiability of faults and determine if false alarms can occur using methods of model structural identifiability [4,5]. We show that this analysis is effective at determining if there are fault-free conditions that produce similar outputs to faulty conditions, which could lead to false alarms.

We will present two separate case studies that compare the identifiability of faults at nominal and optimal operating points. In each case study, we examine a virtual system that represents part of a typical aircraft environmental control system. The first case study will focus on a built- in test for a cross-flow plate fin heat exchanger that is prone to particulate fouling. The heat exchanger analysis is based on previous work on fouling identifiability with uncertain operating conditions [6]. We show how the estimation and precise quantification of heat exchanger fouling can be affected by system uncertainty and how false alarms can be eliminated by optimizing the built-in test design. The fouling extent of the nominal and optimal built-in tests is estimated from heat exchanger exit temperature measurements and then these estimates are compared to each other in terms of accuracy and confidence in the estimation. The second case study focuses on the maintenance of a cabin air compressor (CAC) system that has bias in its outlet mass flow sensor (sensor fault). We show how uncertain environmental conditions can cause false alarms when detecting faults in the CAC. We also show that the impact of these conditions can be minimized by adjusting the CAC admissible inputs, making the detection and isolation of sensor faults feasible and reliable during operation.

Acknowledgment: This work was sponsored by the UTC Institute for Advanced Systems Engineering (UTC-IASE) of the University of Connecticut and the United Technologies Corporation. Any opinions expressed herein are those of the authors and do not represent those of the sponsor. This document does not contain any export controlled technical data.

References:

[1] Soderholm P. A system view of the No Fault Found (NFF) phenomenon. Reliability Engineering and System Safety 2007;92:1–14.

[2] Fedorov V. Optimal Experimental Design. Wiley Interdisciplinary Reviews: Computational Statistics 2010;2:581–9.

[3] Han L, Zhou Z, Bollas GM. Model-based analysis of chemical-looping combustion experiments. Part II: Optimal design of CH 4 -NiO reduction experiments. AIChE Journal 2016;in press.

[4] Asprey SP, Macchietto S. Statistical tools for optimal dynamic model building. Computers and Chemical Engineering 2000;24:1261–7.

[5] Galvanin F, Boschiero A, Barolo M, Bezzo F. Model-Based Design of Experiments in the Presence of Continuous Measurement Systems. Industrial & Engineering Chemistry Research 2011;50:2167–75.

[6] Palmer KA, Hale WT, Such KD, Shea BR, Bollas GM. Optimal design of tests for heat exchanger fouling identification. Applied Thermal Engineering 2016;95:382–93.

468744 Exploiting Connectivity Structure for Online Selection of Primary Controlled Variables

Abstract:

While there are numerous works on controller design, research in the area of optimal control structure design is still very limited. Optimal selection of the primary controlled variables by considering plant economics as well as controllability is critical for optimal control structure design. However, due to change in the operating conditions, control objectives, as well as operating constraints, optimal primary controlled variables can change over time. Due to large number of candidate variables that lead to significant number of combinatorial possibilities even for small plants, online selection of primary controlled variables is computationally prohibitive. In this work, we propose a novel approach that exploits the structural connectivity of the plant to decompose the large-scale problem into smaller subsets. This approach makes parallel solution of the controlled variable selection problem feasible thus achieving solution within the target execution time.

For determining the changing connectivity among plant sections, we draw from the neuroscience literature where the connectivity among cortical/sub-cortical areas are determined using the fMRI responses. At a cortical level, the neuronal populations can be modelled as states that dynamically evoke brain responses as a function of inputs. Thus, by modeling islands of sub processes in a chemical plant as cortical/sub-cortical areas, the effective connectivity of these islands of sub processes can be extracted using bilinear approximation. These connectivity matrices can then be employed as a basis for controlled variable selection.

In this approach, a bilinear model of the process is developed where the model parameters represent intrinsic coupling among the states, describe the influence of extrinsic inputs on the states and as well as capture the effect of inputs on coupling. These parameters are identified using a Bayesian framework. The algorithms developed are implemented on a model of an integrated gasification combined cycle plant with CO_2 capture. This plant includes significant mass and energy interactions with strong change in the coupling as the inputs are changed making it a prefect test case for the developed algorithm. Using an expectation maximization algorithm with uninformed priors, the algorithm provides information about the connectivity strengths among various plant sections. It is observed that the algorithm correctly captures the â€˜trueâ€™ connectivity as would be expected from the first-principles model, plant configuration and operating conditions. The presentation will include algorithmic details as well as detailed results that provide insights into change in connectivity induced by external disturbances.

Control Configuration Selection Using Agglomerative Hierarchical Clustering

Graph theoretical approach

Lixia Kang, Wentao Tang, Yongzhong Liu, Prodromos Daoutidis.
Journal of Process Control (2016) 46: 43-54.

Wentao Tang
Department of Chemical Engineering & Materials Science
University of Minnesota

UNIVERSITY OF MINNESOTA
Driven to Discover℠

Background

- Distributed control for large-scale, complex networks.
- Hierarchical clustering for control architecture design.
- Optimization formulations → ***Graph-theoretic formulations.***

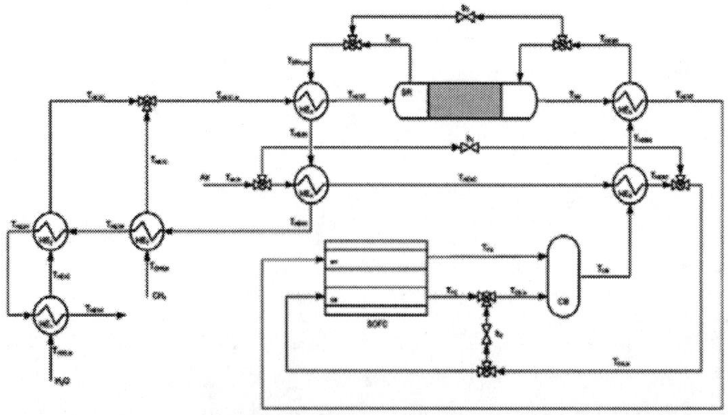

Fig. 1. Solid oxide fuel cell (SOFC) system

UNIVERSITY OF MINNESOTA
Driven to Discover℠

Control architecture design

Steps	Relative Degree Calculation	Input-Output Pairing
Goal	Characterization of response directness or input-output connectivity	Assignment of inputs to outputs to minimize interactions
Interaction measure	Relative degree	Structural decoupling index
Graph representation	Equation graph	Input-output bipartite graph
Algorithm	***Shortest path search***	***Maximum matching***

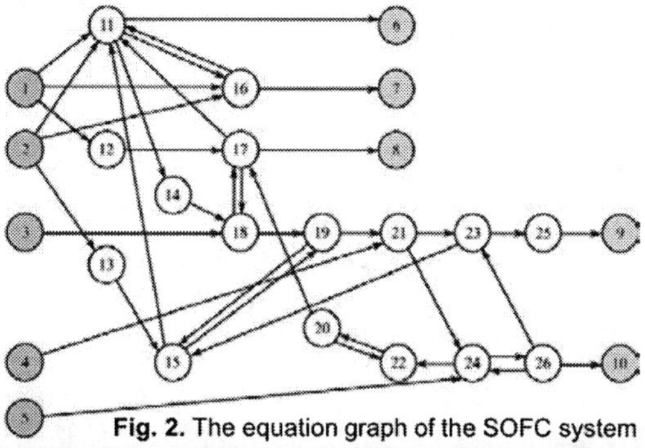

Fig. 2. The equation graph of the SOFC system

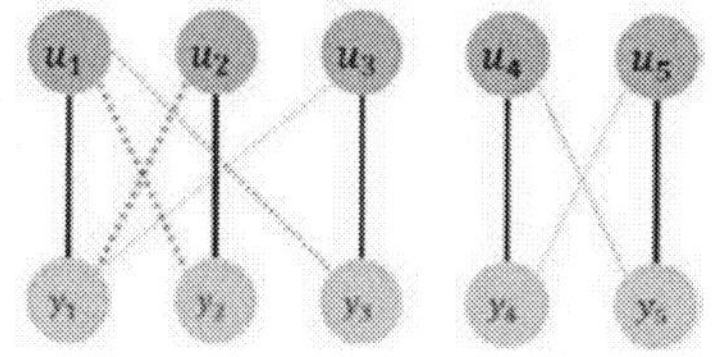

Fig. 3. Input-output pairing by bipartite matching

Control architecture design

Steps	Hierarchical Clustering	Configuration Evaluation
Goal	Clustering of weakly interacting input-output pairs to form control subsystems	Selection of the optimal control architectures based on the difference between inter-cluster and intra-cluster interactions
Interaction measure	Distance between input-output pairs	Pairwise modularity measure
Graph representation	Input-output pair complete graph	Input-output bipartite graph
Algorithm	***Minimum spanning tree***	***Modularity calculation***

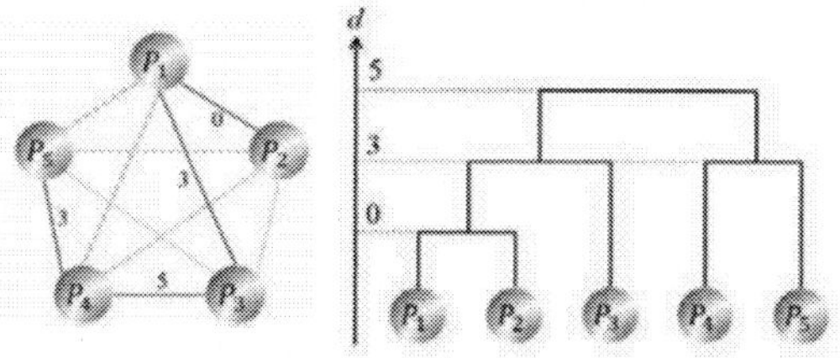

Fig. 4. Minimum spanning tree in i/o pair complete graph

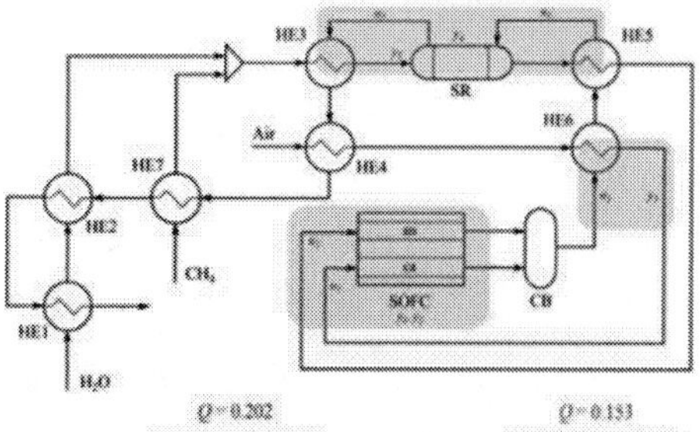

Fig. 6. Optimal control architecture for SOFC system

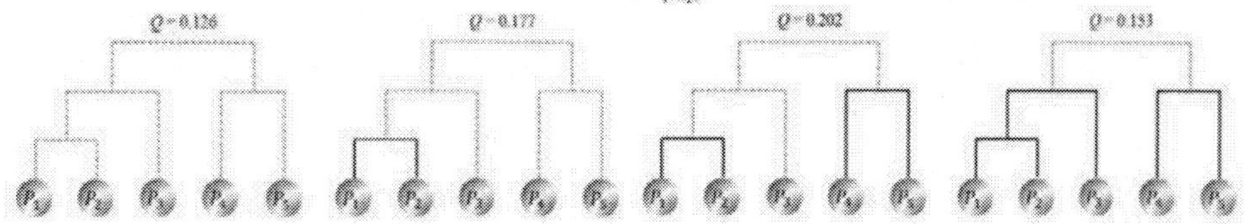

Fig. 5. Optimal architecture selection based on modularity

459275 Steam Methane Reforming Furnace Control: Design and Implementation on a CFD Model of an Industrial Furnace

Abstract:

The steam methane reforming (SMR) process is one of the most effective and economical processes for hydrogen production on a commercial scale, i.e., it alone accounts for more than 40% of the worldwide production of hydrogen [1]. Due to the importance of hydrogen as a raw material for many essential processes in petroleum refineries, e.g., the desulfurization process and hydrocracking process, much research has been performed on the optimization of the operating conditions of the SMR process and on formulating feedback control algorithms to compensate for the effect of process disturbances. It is noteworthy that the SMR process is an overall endothermic process in which methane reacts with superheated steam in a tightly packed nickel-based catalyst network to produce hydrogen, carbon dioxide and carbon monoxide. Therefore, a higher outer reforming tube wall temperature theoretically results in a higher hydrogen conversion. Nevertheless, operating at excessively high temperature decreases the reforming tube expected life span, i.e., an increase in outer reforming tube wall temperature of 20 K relative to its nominal operating temperature can reduce the reforming tube expected lifetime by half [1] [2]. In addition, it might cause the reforming tube wall to lose its integrity and to rupture, which results in production and capital losses. Specifically, the required total capital investment to replace a typical industrial-scale reformer is estimated to be 5 â" 8 million USD [2]. This motivates the development of robust feedback control schemes to maintain the SMR process at the optimized operating conditions in the presence of process disturbances.

From previous work, a CFD model of an industrial-scale reforming tube and a CFD model of a pilot-scale reformer have been successfully developed and validated with the available typical plant data and the publically available experimental data reported in literature [3] [4]. Additionally, our previous work to design feedback control schemes for the industrial-scale reforming tube assumes that the outer reforming tube wall temperature can reach the optimized profile instantaneously and does not account for the dynamics of the temperature of the outer reforming tube wall. This

work aims to develop a more realistic single industrial-scale reforming tube CFD model by accounting for the dynamics of the outer reforming tube wall temperature when it is controlled by the feedback control scheme. Specifically, the pilot-scale reformer CFD model is employed, and the furnace-side feed flow rates to the individual burners are subjected to three different step change inputs. The evolution of the outer reforming tube wall temperature is recorded as the pilot-scale reformer reaches the corresponding steady-state condition, and the dynamics of the outer reforming tube wall temperature are obtained and implemented into the industrial-scale reforming tube CFD model. Finally, classical (i.e., proportional control scheme and a proportional-integral control scheme) as well as optimization-based control schemes (i.e., model predictive control), for which the manipulated input and controlled output are chosen to be the outer reforming tube wall temperature and outlet area-weighted average hydrogen mole fraction, are used to form the closed-loop CFD simulation to track the desired set-point in the presence of process disturbances.

[1] Latham D. Masters Thesis: Mathematical Modeling of an Industrial Steam Methane Reformer. Queenâ€™s University, 2008.

[2] Pantoleontos G, Kikkinides ES, Georgiadis MC. A heterogeneous dynamic model for the simulation and optimisation of the steam methane reforming reactor. International Journal of Hydrogen Energy. 2012;37:16346-16358.

[3] Lao L, Aguirre A, Tran A, Wu Z, Durand H, Christofides PD. CFD modeling and control of a steam methane reforming reactor. Chemical Engineering Science. 2016;148:78-92.

[4] Aguirre A, Tran A, Lao L, Durand H, Crose M, Christofides PD. CFD Modeling of a Pilot-Scale Steam Methane Reforming Furnace. Advances in Energy Systems Engineering, Kopanos G, Liu P and Georgiadis M (Eds.), Springer, in press.

464845 Modeling and Control of Proppant Bank Height to Achieve Uniformity of a Hydraulic Fracturing System

Abstract:

Typically, the term shale gas refers to natural gas trapped in rock of low porosity (2% or less) and ultra-low permeability (0.01 to 0.0001 md or even less) [1]. Therefore, if a well were drilled into such a rock, it would take extremely long time to reach economic amount of production. What has made the recovery of shale gas economically attractive is the extensive use of two crucial technologies: directional drilling and hydraulic fracturing. From a control engineering viewpoint, hydraulic fracturing has been traditionally viewed as an open-loop problem. Well logs and mini-frac test results are interpreted prior to operation in order to obtain petrolphysical and rock-mechanical properties of the formation. The operation is designed based on the properties and then is conducted accordingly. However, the open-loop operation may lead to poor performance, which has motivated this work that considers the closed-loop operation of hydraulic fracturing.

With respect to applying automatic control strategies to oil and gas production, it is important to point out that the literature over the last ten years has been focused on applying MPC to the drilling process, which is known as managed pressure drilling (MPD) that enhances pressure control flexibility, efficiency, and safety of the process [2, 3]. However, applying MPC to improve the uniformity of the proppant bank height while optimizing the fracture geometry, both of which significantly affect the productivity of the treated well, has not been considered because of the following reasons: (1) limited access to real-time measurements, (2) presence of uncertainties in the measurement data, (3) time-dependent spatial domain, and (4) large computational burden when solving high-fidelity models of hydraulic fracturing systems. While some attempts to employ model-based control schemes have been made [4, 5], there are a number of unresolved fundamental as well as practical implementation issues to make MPC a truly real-time control technique.

Motivated by these considerations, we will first address the development of a first-principles model for the hydraulic fracturing process. Second, a novel numerical scheme will be developed to deal with the high computational requirement caused by coupling of multiple partial differential equations (PDEs) defined over a time-dependent (evolving) spatial domain. Third, a reduced-order model will be constructed by using these simulation results. Lastly, nonlinear MPC theory will be utilized for the design of the feedback control system that provides a foundation for the online control of proppant bank height in order to achieve uniformity at the end of the treatment, which is directly related to the overall efficiency of the operation.

References:

[1] M. Nikolaou. Computer-aided process engineering in oil and gas production. *Computers & Chemical Engineering*, 51:96–101, 2013.

[2] O. Breyholtz, G. Nygaard, and M. Nikolaou. Automatic control of managed pressure drilling. *Proceedings of American Control Conference, Baltimore, MD*, 442 – 447, 2010.

[3] R. A. Shishavan, C. Hubbell, H. D. Perez, J. D. Hedengren, D. S. Pixton, and A. P. Pink. Multivariate control for managed pressure drilling systems by use of high-speed telemetry. *SPE Annual Technical Conference and Exhibition (SPE 170962), Amsterdam, Netherlands*, 2014.

[4] Q. Gu and K. A. Hoo. Evaluating the performance of a fracturing treatment design. *Ind. & Eng. Chem. Res.*, 53:10491–10503, 2014.

[5] Q. Gu and K. A. Hoo. Model-based closed-loop control of the hydraulic fracturing process. *Ind. & Eng. Chem. Res.*, 54:1585–1594, 2015.

466495 Linking Production and Utilities a Holistic View on Plants and Energy Supply

Abstract:

In chemical engineering the use of commercial process simulation software and detailed models for each individual unit of a chemical plant is state of the art, especially during plant design and as a basis for tuning of operating points.

For cross-plant optimization tasks in a complex compound system of several plants, where each plant comprises numerous unit operations, this detailed approach is often limited because substantial interdependencies between the plants are not considered, in particular the connections among the production plants and from production plants to the energy supply chain.

An oil refinery is a typical example of a complex compound system in which several production plants are interconnected in order to produce numerous product streams such as fuels and petrochemicals from crude oil. Residue-fired power plants as the main utilities provide the production plants with steam and electrical power.

Due to the tight interconnection between production plants and utilities, a simultaneous treatment is obvious. In doing so, the usually huge number of streams, recycles, apparatuses and chemical compounds makes it difficult to combine detailed individual models of several plants while maintaining model stability at reasonable amounts of simulation time. The latter aspect suggests to reduce the modeling-depth in an appropriate way.

This paper presents a comprehensible methodology to reduce the modeling-depth of a compound system using a top-down approach. This is achieved by decomposing the real system into an aggregation of appropriate sub-systems. Each plant is regarded as a deterministic black box, in which product streams and steam consumption are described as linear functions of the feed. These model equations are obtained by fitting the linear functions to measured mass flow data of the real plants, using the fulfilment of the mass balance for each plant as constraint. The overall static model of the refinery is represented by mixers, splitters and the fitted functions.

This procedure results in a set of functions in which all product streams and steam streams only depend on 1.) the crude feed to the refinery and 2.) a set of several split fractions.

All links with the associated utilities, like steam network, electric supply network, hydrogen network and fuel network can simply be considered by summation of all imports and exports of each plant. The same procedure is carried out for the power plants, where steam production as well as power generation are described by linear or polynomial functions of fuel or steam flows, combined with several decision variables. The resulting set of functions describing the main characteristics of each process step in terms of product yield and energy consumption can be used to formulate one target criterion or a weighted sum of several target criteria that can be influenced by both production and utilities.

This approach is compared to more detailed models using commercial process simulation software. Different model concepts, representing different modeling depths, are investigated in terms of their ability to represent the interconnections between the energy demand of the production units and the utilities. In this context, the modeling depth is varied between rigorous thermodynamics and tray-to-tray simulations, shortcut methods, component splitters and the presented data-driven approach, which represents the most simplified model concept.

Proposals for an optimal choice of the modelling depth are given based on examples from industry, ensuring a well-balanced trade-off between complexity, stability and accuracy of the model.

469688 Estimation of Gasifier Wall Profile Using Measurements from a Wireless Sensor Network

Abstract:

Temperature is the most important variable in efficient operation of the entrained-flow gasifiers. In addition, slag penetration into the refractory wall of these gasifiers leads to stress buildup and spalling of the refractory bricks reducing the plant availability. An estimation of the slag penetration profile and stress profile can be very useful for monitoring, maintenance planning, and determining operating conditions for enhancing the refractory life. However, the state-of-the-art measurement technologies are inadequate for measuring any variable inside the gasifier due to the high-temperature, high-pressure gasifier environment and corrosive properties of the slag. In addition, the existing technologies need open access port in refractory lining, which provides a penetration path for slag and reduces lifetime of the refractory further. With this motivation, novel smart refractory bricks with embedded sensors are being developed at West Virginia University. However, as the slag penetrates into these smart bricks, their electrical and thermal properties change. In addition, a steep temperature gradient exists along the length of these sensors. Therefore, the existing approach where the sensor output is directly correlated to the measurement variable is inadequate for this system. In addition, slag penetration depth cannot be directly correlated to the sensor output due to complex effect that the slag penetration has on the temperature profile as well as the refractory and sensor properties. To address these issues, a model-based estimation approach has been developed. The measurements from the embedded sensors are noisy. Due to temporal change in the material properties, it is likely that the discrepancy of the process and measurement models would also evolve with time. In addition, the measurement data are available from a wireless sensor network that suffers from communication constraints, packet dropouts, and synchronization errors. Due to the limitations in the communication bandwidth, only limited packets can be transmitted to the estimation algorithm at any transmission instant. In addition, random packet dropouts can occur in wireless network caused by the unavoidable transmission error. Furthermore, the estimation algorithm may receive out-of-sequence measurements. Therefore, a modified nonlinear unscented Kalman filter (UKF) algorithm for this differential-algebraic equation (DAE) system has

been developed for estimating the temperature profile, stress profile, and slag penetration profile with due consideration of the issues with the wireless sensor network. Our study shows that the existing approach can increase the refractory life considerably.

468895 Robust Classification of Systematic Measurement Errors

Abstract:

Today's chemical plants have increased the capacity and quality of their data base, as result of the exponential growth of the computers capacity and the quantity of process variables which can be measured on-line. The observations provided by different instruments must be analyzed in order to take advantage of them and obtain relevant information of the process state.

Data Reconciliation (DR) is a well known methodology that considers the measurements and the model of the plant. This tool works in the sense that measurements and conservation equations have the less possible discrepancy. It consists in solving an optimization problem which minimizes a non linear function of the error subject to the satisfaction of a set of process constraints (Romagnoli and Sánchez, 2000).

Classical estimators provide optimal estimates when measurements follow exactly a probability distribution function, while robust estimator aims at giving goods results when the measurements recollected just make this approximately (Maronna et al., 2006). Many researchers have made comparisons between these families of estimators and reported results that verified the estimation quality of the robust ones (Ozyurt and Pike, 2004; Martinez Prata et al., 2010; Chen et al., 2013; Llanos et al., 2015).

The simultaneous presence of bias and outliers was treated by Chen et al. (2013). They applied the correntropy estimator to solve the DR problem and analyzed the results of performance's parameters. The reduction of gross errors effect was also studied by Nicholson et al. (2014). They employed the Hampel's redescending estimator plus an advanced step moving window for the resolution of the dynamic DR problem. Those authors considered that measurements are contaminated by outliers, biases and drifts.

Regarding the robust classification of systematic errors, Martinez Prata et al. (2010) addressed the simultaneous identification of bias and outliers using the Welsch estimator and a Particle Swam Optimization algorithm. Recently Zhang and Chen (2015) presented a robust strategy to distinguish among outliers, biases and drifts, which makes use of the Correntropy estimator.

In Llanos et al. (2015), a comparison of recently appeared robust methodologies was made taking into account performance parameters that measure the capacity of detection, the power of reconciliation and the computational requirement. It was concluded that one of their proposed approaches, called the Simple Method (SiM) is an efficient alternative for solving the type of problems in consideration. The procedure is made up of two steps. First, the robust median of the measurements contained in a moving window is calculated using the Bi-square function. Then an optimization problem is solved applying the Huber estimator and the solution of the previous step as initial point.

In this work, a new strategy for the robust classification of systematic measurements errors is presented. The DR stage is based on the SiM. The identification stage uses the Measurements Test (MT) to detect outliers or suspicious variables. When the MT is applied to the last two observations of the moving window, two cases may arise. If the statistical test of only the penultimate measurement is greater than its critical value, an outlier has been identified. In contrast if the statistical values of the two observations are greater than the critical ones, then the next measurements are saved to make a robust linear regression. The relation between the estimated slope and the variance of the observations allows distinguishing between biases and drifts.

Application results of the methodology for linear and nonlinear benchmarks, extracted from the literature, are used to evaluate its performance. Three cases are analyzed for different lengths of the data horizon. The first one involves the presence of outliers and biases; the second one considers that measurements may contain outliers and drifts, and the last one studies the possible existence of all the systematic observation errors. The performance of the proposed methodology is measured in terms of the Percentage of Right Classification for each type of systematic error and a global index. Results show that there is a window length which allows obtaining the best classification of the simulated errors.

References

1) Maronna, R. A.; Martin, R.D.; Yohai, V. *Robust Statistics: Theory and Methods*; John Wiley and Sons Ltd.: Chichester, 2006.

Chemical Processes. *Comput. Chem. Eng.* **2004,** *28,* 381-402.

3) Romagnoli, J.; Sánchez, M. *Data Processing and Reconciliation for Chemical Process Operations*; Academic Press: San Diego, 2000.

4) Martinez Prata, D.; Schwaab, M.; Lima, E.L.; Pinto, J.C. Simultaneous Robust Data Reconciliation and Gross Error Detection through Particle Swarm Optimization for an Industrial Polypropylene Reactor. *Chem. Eng. Sci.* **2010,** *65,* 4943-4954.

5) Chen J, Peng Y, Munoz J. Correntropy Estimator for Data Reconciliation. *Chem. Eng. Sci.* **2013**, 104, 10019-27.

6) Nicholson, B.; López-Negrete, R.; Biegler, L. T. On-line State Estimation of Nonlinear Dynamic Systems with Gross Errors. *Comput. Chem. Eng.* **2014,** *70,* 149-159.

7) Zhang, Z.; Chen, J. Correntropy Based Data Reconciliation and Gross Error Detection and Identification for Nonlinear Dynamic Processes. *Comput. Chem. Eng.* **2015,** *75,* 120-134.

8) Llanos C. E.; Sánchez, M. C.; Maronna, R. A. Robust estimator for Data Reconciliation. *Ind. Eng. Chem. Res.*, **2015**, *54* (18), pp 5096–5105

469306 A Multivariate Data Driven Approach for Modelling the Cognitive Behaviour of Control Room Operator Using Eye Tracking

Abstract:

In an attempt to improve process safety, todayâ€™s plants utilize automation and control strategies along with sophisticated safety management regimes. Despite these advancements, accidents continue to occur, indicating that process safety is at least as important today as it was in the 1970s and 80s. Statistics along with numerous surveys indicate that human error is one of the major cause for accidents in process industries. Advancements in automation transfers the responsibilities to various individuals and does not eliminate the need for human operators. Therefore, it is important to understand the role of human operators in modern plants and develop methodologies to prevent human errors.

Traditionally, human errors are qualitatively considered during HAZOP analysis and human failures quantified using error probabilities. However, this view-point abnegates the cognitive abilities of operator in handling a situation. Recent studies in process industries have utilized eye tracking measures to qualitatively understand the behaviour of control room operators. Results from these studies not only revealed the existence of a specific eye gaze pattern in certain participants (completed the task using only one manipulated variable) but also patterns in the eye gaze behaviour indicated that not all successful operations were alike; nearly 12% of the successful operators closely resembled operators who had failed to control the task [1]. Also, gaze patterns and pupillometry can be used to identify the various cognitive tasks performed during different phases of the task [2]. These works however provide only qualitative information about the eye gaze behaviour.

Visual entropy is one of the widely used measures in several safety critical domains (aviation, healthcare) for quantification of eye gaze movements. Visual entropy reveals the degree of randomness in the scanning behaviour. Relatively low values of entropy e represents highly focused visual attention while scattered gaze behaviour results in higher value of visual entropy. This quantitative measure has been used understand the mental workload of operators, their expertise level and situation awareness [3,4,5]. Apart from visual entropy, there are several other quantitative measures (known in the literature) that could provide different perspectives on the cognitive behaviour of the operators. For instance, pupil dilations has the potential to provide real-time information about the mental workload of control room operators [6]. Visual entropy provides indication about the scanning behaviour of the operator (on the HMI) while pupil size variations help understand their perceived workload. Also, both these measures are correlated and provide information about the cognitive behaviour of operator. For instance, during high workload the operator will focus on large number of variables (high visual entropy) as the cause of abnormality is not known while little/no pupillary dilations with low visual entropy indicates the high confidence level of the operator in handling the situation.

This work attempts to develop a multivariate approach to model the behaviour of control room operator using visual entropy and pupillary dilations. Experimental studies using a simulated ethanol plant were conducted on 44 participants with control engineering background. Process variables, alarm details, mouse clicks, eye movement and pupil measurements were conducted during these experiments. Eye gaze data and pupil measurements were utilized to compute visual entropy and pupil dilation measure. A multivariate data driven measure is developed from these measures to quantify the cognitive behaviour of control room operators. Results from application to experimental studies reveal the potential of the propose measure in modelling the cognitive behaviour of process control room operators.

References

[1]Sharma C, Bhavsar P, Srinivasan B, Srinivasan R. Eye gaze movement studies of control room operators: A novel approach to improve process safety. Computers & Chemical Engineering. 2016 Feb 2;85:43-57.

[2]Kodappully, Madhu, Babji Srinivasan, and Rajagopalan Srinivasan. "Towards predicting human error: Eye gaze analysis for identification of cognitive steps performed by control room operators." Journal of Loss Prevention in the Process Industries (2015).doi:10.1016/j.jlp.2015.07.001.

[3]Tole JR, Stephens AT, Vivaudou M, Ephrath AR, Young LR. Visual scanning behavior and pilot workload.

[4]Hermens, F., Flin, R., Ahmed, I., 2013. Eye movements in surgery: A literature review. Journal of Eye Movement Research 6 (4), 1â€"11.

[5]Gilland J. Driving, eye-tracking and visual entropy: Exploration of age and task effects. ProQuest; 2008.

[6]Bhavsar P, Srinivasan B, Srinivasan R. Pupillometry based Real-time Monitoring of Operatorâ€™s Cognitive Workload to Prevent Human Error during Abnormal Situations. Industrial & Engineering Chemistry Research. 2015.

470662 Improved Soft Sensors for Mixed Culture System Monitoring

Abstract:

Mixed cultures are biological systems with more than one type of organism sharing or competing for available nutrients. Due to many advantages associated with mixed cultures, their applications in biotechnology has expanded rapidly in recent years with newer applications for producing solvents, acids, plastics, hydrogen, antibiotics, and other valued commodities. At the same time, many challenges remain for effective mixed culture applications. One of them is how to efficiently and accurately monitor the individual cell populations in a mixed culture. The traditional approaches on individual cell mass quantification are suitable for off-line, in frequent characterization of mixed cultures.

To address this challenge, spectra-based soft sensor was proposed [1] to estimate individual biomass concentration in a mixed culture. Although this approach is quick and sufficiently accurate, models are built from data with large number of variable compared to the number of samples, which makes models less accurate and less robust. This work is intended to address this issue.

In this work, models were developed to predict individual biomass concentration for co-culture mixtures of *Scheffersomyces Stipitis* & *Methylomicrobium buryatense*, and *Escherichia coli* & *Saccharomyces cerevisiae* using partial least squares (PLS). For outlier detection, principal component analysis (PCA) was applied. The main goal of this work was to investigate the effect of variable selection and experimental design on the performance of the PLS soft sensor. Variable selection methods studied in this work are variable importance in projection (VIP), synergy-iPLS, genetic algorithm, and recursive-VIP or R-VIP. Performances of soft sensors with different variable selection methods and different model training scenarios are compared and discussed.

References:

[1] Stone K & Shah D., â€œA novel soft sensor approach for estimating individual biomass in mixed culturesâ€. Annual AIChE Meeting, Salt lake city, UT(2015)

Review of Flare Requirements and Challenges in Implementation

Emily Manternach and Erick Mertooetomo

ABSTRACT:

Over time the EPA has increased requirements on flares through regulation, consent decrees, and additional new rules. Meeting and maintaining the flare requirements has multiple challenges in order to comply with data processing, recordkeeping, and reporting requirements that can be completed using different combinations of instruments/analyzers and data processing methods (DAHS, DCS, Microsoft Excel, Microsoft Access). Instruments often have QA/QC requirements suggested by the EPA and technical specifications, installation instructions, and operational instructions per the manufacturer. Processing and transforming the raw data collected from the instruments into data for required recordkeeping, data for determining compliance, and data for required reports also has challenges. Flare compliance requires a large volume of raw data for processing, which includes identifying bad data, whether from instrument downtime or other reasons, and following data substitution procedures. The advantages and disadvantages of the different hardware and software combinations for accomplishing the data transformation and meeting the challenges involved are discussed.

In addition, the compliance standards and limits pose their own challenges. In order to comply with the standards and limits, the process control for collaboration between instruments and analyzers becomes important to help the board operator understand and operate the flare. The process control system has to cope with a wide range of operating conditions which creates difficulty in tuning the controller to be effective. For example from zero flow while waste gas is going to a flare gas recovery system to large flows during emergency flaring or for a range of waste gas composition. Furthermore, there are timing issues to contend with, such as instrument lags, valve response times or gain, and travel time for added steam and supplemental gas to reach the flare tip (or the combustion zone). A degrees of freedom analysis and discussion of the control disturbances is presented for troubleshooting the compliance limits and standards.

472715 Modeling and Prediction of Behavior of Mammalian Cell Culture for Monoclonal Antibody Production - Dual Rate Approach

Abstract:

The demand for high-value biopharmaceuticals is on a rapid rise due to their therapeutic and diagnostic applications. Monoclonal antibodies (MABs) are important reagents used in biomedical research and diagnosis and treatment of diseases and account for nearly 40% of the sale of biologics in the U.S. and are produced in animal cell bioreactors at a variety of scales. Chinese Hamster Ovary (CHO) cells and hybridoma cells have been the popular cell types for production of MABs. The simplest approach for producing MABs in vitro is to grow the mammalian cells in batch and fed-batch cultures, since these are easy to operate and scale up, and recover and purify the MABs from the culture medium. However, under batch conditions, mammalian cells exhibit an inefficient glucose and glutamine metabolism, causing increased production of lactate and ammonia, major waste products of metabolism of glucose and glutamine, respectively. Both lactate and ammonia inhibit growth of mammalian cells and deteriorate MAB production. To reduce formation of these toxic by-products, fed-batch operations with intermittent supplementation of glucose and glutamine have been performed with these cell lines, resulting in a higher volumetric productivity and extended cell viability. Although fed-batch operations have been successfully employed for microbial systems, kinetics of mammalian cell cultures is still under investigation to determine quantitatively as well as qualitatively cost-effective production strategies. Traditional operations of these culture systems at production scales are mainly based on empirical knowledge. There has been limited effort on development of first principle-based mathematical models for these cultures at laboratory scale operations. Due to complexity of the models and uncertainties associated with estimation of model parameters, the first principle models are not suitable for online prediction and control of mammalian cell cultures. Simple data driven models may be more suitable for online prediction and control.

Multi-rate systems, where inputs and/or outputs have different sampling rates, are encountered in many chemical and biological engineering applications. Typically, this occurs in systems where multiple variables are measured. Process variables such as temperature, pressure and pH are measured more frequently, at times nearly continuously, while other variables, such as species concentrations in a reaction mixture, are measured less frequently. An example of such processes is batch and fed-batch mammalian cell cultures, where sampling rates for viable hybridoma cells, nutrients glucose and glutamine, by-products lactate and ammonia, and the target product, a MAB, are significantly different. The simplest way to handle multi-rate systems is to neglect excess data from fast sampling signals and synchronize the signals with the slowest sampling rate. With this method, great amount of data will be discarded, the data that may contain crucial information regarding system dynamics. Therefore, some effort has been devoted to modeling, analysis and identification of multi-rate systems to take advantage of the rich information available in the experimental databases. In all the techniques proposed, multi-rate systems are generally simplified to multiple dual rate systems where the slower sampling rate is a positive integer () multiple of the faster sampling rate. One of the most successful techniques, polynomial transformation technique, can successfully transfer single rate models into dual rate models with ease. The dual rate model, after transformation, can utilize dual rate signals simultaneously. For such model structures, multiple identification techniques have been proposed with good estimation accuracy and convergence rate.

In this work, a second order discrete time system is considered first to demonstrate convergence of the proposed parameter estimation algorithm for a dual rate model. The convergence rates and estimation accuracy for various values are compared to examine efficacy of the dual rate parameter estimation technique. Use of a dual rate model with parameter estimation technique for modeling and prediction of mammalian cell cultures is discussed next. The faster sampling rate corresponds to viable hybridoma cell concentration and the slower sampling rate corresponds to glucose and glutamine concentrations. The sampling rate for monoclonal antibody is much slower than that for glucose and glutamine. Recursive time series models are developed for key culture variables. The model parameters are recursively estimated by using a least square estimation algorithm. Model stability is evaluated and confirmed by converting the time series model into a state space counterpart. The most rapidly measured variable, viable cell concentration in cell culture is expressed by a recursive ARMAX model. The performance of the dual rate model coupled with frequent parameter estimation in representation and prediction of mammalian cell culture producing a MAB is examined in considerable detail.

472980 Port Hamiltonian Approach to Modeling and Control of Coupled Chemical Reactors

Abstract:

Models of chemical processes alter in details from ordinary differential equations (ODEs) for the modeling of a Continuous Stirred Tank Reactor (CSTR) to a large set of Partial Differential Equations (PDEs) for tubular reactors. Usually, models of chemical reactors are developed on the basis of the conservation of extensive quantities, namely balance of energy, mass, momentum and entropy. Due to the coupling between reaction and transport phenomena, these models usually exhibit very complex behavior.

Port Hamiltonian framework has been widely applied in finite-dimensional processes in chemical, electrical, mechanical domains, e.g. CSTR (see [1]). Recently, Port Hamiltonian theory was developed and generalized to infinite-dimensional systems, see [2] [3] [4]. In [4], it is shown that many infinite-dimensional processes in chemical and mechanical domains falls into the category of Port Hamiltonian systems, e.g. most of hyperbolic systems. Moreover,

under the Port Hamiltonian framework, the energy-based control approaches are investigated.

In this work, we propose a Port Hamiltonian representation of a distributed reactor for control purpose. The Hamiltonian representation provides powerful analysis methods (e.g. for stability), and it enables the use of Lyapunov-stability theory and passivity-based control.

In particular, Hamiltonian representation is powerful when addressing the boundary control of distributed parameter systems. We discuss how the state variables are chosen in such a way that geometric properties of the model are emphasized. The chemical reactors given in [5] are studied in this work and the presentation of the considered reactors is given for one dimensional spatial domain by utilizing the internal energy.

[1] Hoang, H., Couenne, F., Jallut, C., and Le Gorrec, Y. (2011). The port Hamiltonian approach to modeling and control of Continuous Stirred Tank Reactors. Journal of Process Control, 21(10), 1449--1458.

[2] Villegas, J. A., Zwart, H., Gorrec, Y. L., and Maschke, B. (2009). Exponential stability of a class of boundary control systems. Automatic Control, IEEE Transactions on, 54(1), 142--147.

[3] Wu, Y. (2015). Passivity preserving balanced reduction for the nite and innite dimensional port Hamiltonian systems (Doctoral dissertation, Universit Claude Bernard-Lyon I).

[4] Jacob, B., and Zwart, H. (2012). Linear port-Hamiltonian systems on innite-dimensional spaces (Vol. 223). Springer Science & Business Media.

[5] Ramkrishna, D., and Amundson, N. R. (1974). Stirred pots, tubular reactors, and self-adjoint operators. Chemical Engineering Science, 29(6), 1353--1361.

456350 Optimal Reactor-Separator-Network Synthesis

Abstract:

Main challenges in process design is finding the optimal pathway to form a specified product in the most economical way and building a process which is able to approximate this theoretical pathway best. For latter one can basically choose between two main operating modes - discontinuous and continuous processes. First is often associated with smaller scale productions like for pharmaceuticals and fine chemicals and latter for larger scale productions of e.g. bulk chemicals. Due to that the design approaches for both kinds of processes are often distinguished as well. However, it would be more promising to figure out the potential of different process pathways or realizations without restricting it in advance to either discontinuous (batch) or continuous process types. The concept of elementary process functions (EPF) allows in particular for that, the analysis of the maximum potential of a process without preliminary restrictions to the optimal process [1]. Core of the EPF methodology is the optimal manipulation of mass and energy fluxes of matter elements during their travel through the process. These fluxes are determined by means of dynamic optimization calculations in an analogous batch process.

Within this scope it is intended to develop an approach to find an optimal design for any arbitrary process by means of the EPF methodology, which corresponds to optimizing it within a batch process optimization with optimal dosing of all attending components. Thereby any characteristics and beneficial control of the reaction progress can be identified, e.g. advantages of mixing, back-mixing, extraction, heating, cooling, etc. The resulting time-dependent set of fluxes is directly related to aforementioned functions and/or the corresponding unit operations, or at least allow for concluding realization options. Analyzing these fluxes provides basic information for optimal discontinuous and continuous realization of the process: A discontinuous process can be realized by optimal time control of a single batch reactor or as optimal control or scheduling of several interconnected batch reactors and separator steps; a continuous process can be developed by translation of the fluxes into an optimal network of continuous reactor and separator steps, whereby also the back-mixing characteristics of the reactor steps and the benefits of recycles can be revealed. For the realization of those optimization results in a batch process several interesting work has already been published, see e.g. [2]. Within this work the focus lies on using those optimization results for developing an optimal continuous reactor-separator-network.

The translation of the fluxes into an optimal reactor-separator-network results in a superstructure of aforementioned continuous reactor types, dosing/mixing spots and extraction spots together with the corresponding cooling/heating strategies for all units. The dosing/mixing spots can thereby be also realized by closed recycle loops. Since this translation only provides a qualitative structure of the process, this superstructure has to be optimized then on its own to obtain a quantitative design of the process. This can be done for different target compositions resulting in an area in the concentration space in which for each point an optimal reactor-separator-network is associated. This is closely connected to the idea of determining the Attainable Region (AR) of a process and then to find the optimal operation point in this region depending on the overall process, the specified process requirements and the objective of the design. But the AR approach is limited to the combination of only reaction and mixing in continuous reactors, i.e. PFR, CSTR and DSR, and mixing units, respectively [3].

Within this work the proposed reactor-separator-network synthesis method is exemplified on two processes. First on an extended van-de-Vusse reaction, which is often taken as example process in literature, as well for the Attainable Region approach. Therefore it allows for a direct comparison of our method to the state-of-the-art Attainable Region concept and other network synthesis approaches. The second example is the hydroformylation of long chain olefins in thermomorphic solvent systems [4]. The hydroformylation is a very important homogeneously catalyzed process in chemical industry. For a profitable, highly selective hydroformylation of long chain olefins a Rhodium/Biphephos catalyst complex is used which is supposed to be recovered by an innovative solvent system, whose separation characteristics can be controlled by temperature.

References:

[1] Freund, Hannsjörg; Sundmacher, Kai (2008): Towards a methodology for the systematic analysis and design of efficient chemical processes. In *Chemical Engineering and Processing: Process Intensification* 47 (12).

[2] Peschel, Andreas; Hentschel, Benjamin; Freund, Hannsjörg; Sundmacher, Kai (2012): Design of optimal multiphase reactors exemplified on the hydroformylation of long chain alkenes. In *Chemical Engineering Journal*188, pp. 126–141.

[3] Glasser, David; Crowe, Cameron; Hildebrandt, Diane (1987): A geometric approach to steady flow reactors. The attainable region and optimization in concentration space. In *Ind. Eng. Chem. Res.* 26 (9).

[4] Schäfer, Elisabeth; Brunsch, Yvonne; Sadowski, Gabriele; Behr, Arno (2012): Hydroformylation of 1-Dodecene in the Thermomorphic Solvent System Dimethylformamide/Decane. Phase Behavior–Reaction Performance–Catalyst Recycling. In *Ind. Eng. Chem. Res.* 51 (31), pp. 10296–10306.

Targeting Multi-Phase Chemical Reactor Networks in Biochemical Processes: A Superstructure Approach with a View to Innovation and Novel Development

Georgios P. Panayiotou[1], Aikaterini D. Mountraki[1,2], Antonis C. Kokossis[1]
[1]School of Chemical Engineering, NTUA, Greece, [2]CIMV, Toulouse, France

Challenges

- Chemical reactor is perhaps the most important unit to optimize within a chemical process plant.
- The growing demand for "green products" has contributed to the rapid development of biorefineries. The fractionation of biomass and its conversion to products is the objective of recently developed chemistries which are still under research in order to be optimized.
- The new objectives are centered around clean technology and are namely: (1) to give as far as possible only the desired product, (2) to reduce by-products and pollution, (3) to minimize the inhibition factors, and (4) to intensify processing in order to improve economics.
- The design of contemporary biochemical reactors is still mainly based on the existing experience and heuristics in conventional reactors.
- Bioreactor design based on conventional rules of reactor design could fail; apart from the fact that the kinetic 'constants' are no longer constants but functions of everything that characterize the biochemical system, mass transfer and the general hydrodynamic behaviour of the system becomes crucial.
- The diversity of biomass feedstock, along with the complicated activity of the microorganisms used, imposes new uncertainty factors for the design of the bioreactors.
- The feed distribution patterns of enzymes can be viral for economics and yield and can assist in reverse engineering the process and help with the design of the biocatalyst.
- This study proposes a **complete superstructure framework** for the full conceptual design and optimisation of different integrated biorefinery processes.

Superstructure Representation

Systems approach: Shadow Reactor Superstructure Model

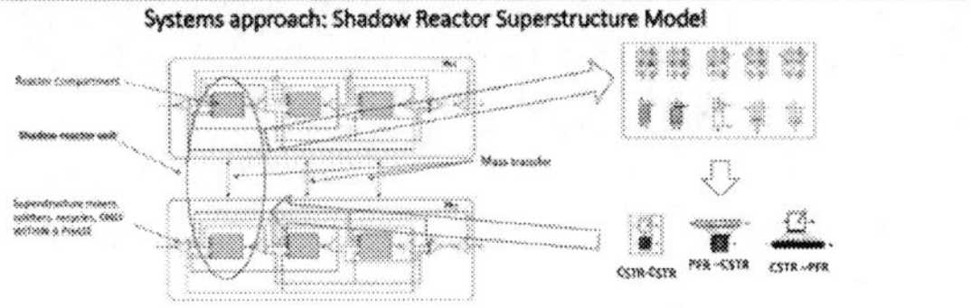

An improved version of the **systematic framework** for the optimization of a superstructure representation of ideal reactor units involving CSTRs and PFRs, firstly introduced by Mehta (1998), for the full conceptual design of multiphase chemical and biochemical reactors.

Degrees of freedom and objective:

- ✓ Type of mass transfer favored - Reactor configuration
  - ✓ currently a list of 12 options each suggesting an experimental rig
  - ✓ possible inclusion of experimental, bespoke correlations
- ✓ Type of mixing in the shadow reactor compartment (CSTR, PFR, macro-mixed PFR)
- ✓ Comprehensive macro-mixing and micro-mixing options among shadow reactor compartments
- ✓ Reactor hold ups, Reactor volumes, flowrates, connectivities, feed distribution patterns, in-situ product recovery (ISPR) patterns
- ✓ **Optimization objectives** (as usual): yields, selectivity, (occasionally) cost

Mass Transfer

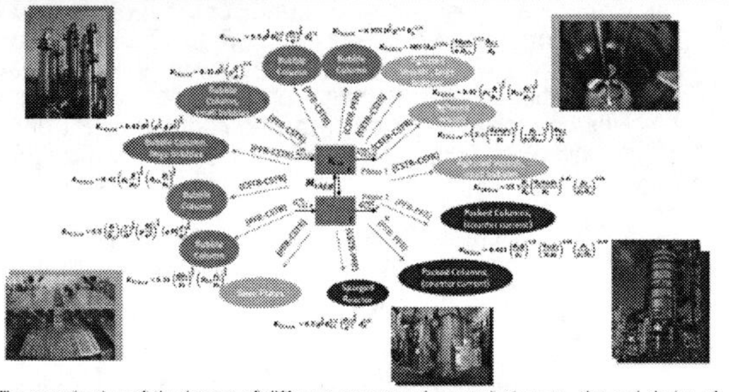

The examination of the impact of different mass transfer correlations on the real design of bio-reactors running different biorefineries examples based on recently developed or under research technologies using the synthesis framework embedded with appropriate mass transfer correlations for the best approximation of mass transfer between phases for each type of reactor.

Optimisation Technology

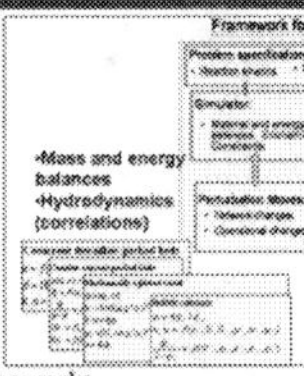

Synthesis framework:

1. **Targeting stage:** creation of initial structure of the superstructure representation using the shadow reactor superstructure
2. **Screening and design stage:** evaluation of the results of the targeting stage and screening of the options that have a positive impact on the system
3. **Analysis stage and validation:** the implementation of **a generic probabilistic metaheuristic global optimization method**

Evolutionary optimisation is employed utilising **Simulated Annealing (SA)** algorithm, on FORTRAN programming language, for the synthesis, design, optimization and superstructure representation of complex reaction network systems (Kokossis & Floudas, 1990).

The approach as a software

➢ ***INPUT:***
- *Reaction system*
- *Reaction kinetics*
- *Design objective*

➢ ***OUTPUT:***
- ✓ *Yield, Selectivity*
- ✓ *Reactor configurations, Reactor volumes*
- ✓ *Macromixing and micro-mixing patterns*
- ✓ *Enzymes/catalyst feed distribution patterns*
- ✓ *In-situ product recovery (ISPR) patterns*
- ✓ *Actual performance limits of the system*

➢ ***CAPE - OPEN Link:***
Link with CAPE-OPEN and Aspen Plus for a very accurate chemical properties calculation and fast feed re-characterization.

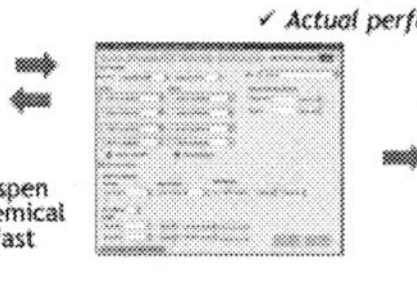

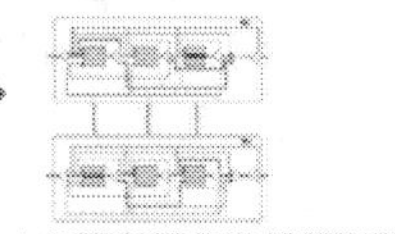

➢ ***INTERPRETATION OF RESULTS:***
(Illustration of superstructure)

✓ *Recommendation for the real design*

✓ *Strong Link to innovative (non-conventional/novel) designs*
(Significant evidence that non-conventional reactors bring key benefits)

Conclusions and Work in Progress

- ✓ The proposed method can be applied for the synthesis, design, and optimization for both, conventional and non-conventional processes since it is fully adjustable to any type of process.
- ✓ Reliability of the framework to adapt with flexibility and high level of accuracy reconfirmed by the results of the case studies.
- ✓ Enables reactor design of the conventional chemical processes with linear simple kinetics
- ✓ Facilitates the synthesis and design of biochemical processes with complex reaction mechanisms.
- ✓ Optimizing various microorganisms & enzymatic reaction systems was proven the Michaelis -Menten kinetics & Monod equation which applies to most biochemical processes can be accurately simulated.
- ✓ The latest version of the framework proposed enables the real design of biochemical reactors with interactive mass transfer coefficients adapted for each case.
- ✓ The great importance of the framework is that it identifies the maximum achievable efficiency of the resulting optimum reactor network, providing useful information about: (1) The number and type of reactors, (2) The desired mass transfer correlation for each type of reactor, (3) The volumes of reactors, (4) The appropriate feeding, mixing, recycling and bypassing strategy and (4)The optimal values of the flowrates, compositions and yields.
- ✓ Work in progress is aiming to make available online this framework as web tool enabling the user to input fast and easy any set of chemical/biochemical reaction equations and kinetics choosing the objectives.

References

A.C. Kokossis and C.A. Floudas, 1990, Optimization of complex reactor networks: isothermal operation. Chem. Eng. Sci., 46, 1361.

V. Linek, T. Moucha, F. J. Reji, M. Kordač, F. Hovorka, M. Opletal, & J. Haidl, 2012, Power and mass transfer correlations for the design of multi-impeller gas-liquid contactors for non-coalescent electrolyte solutions. Chemical Engineering Journal, 209, 263-272.

P. Linke, & A.C. Kokossis, 2003, On the robust application of stochastic optimisation technology for the synthesis of reaction/separation systems. Computers & chemical engineering, 27(5), 733-758.

L.V. Mehta, 1998, Synthesis and Optimization of Multiphase Reactor Networks, University of Manchester, Institute of Science and Technology, Dept. of Process Integration, PhD Thesis.

Acknowledgement

The EC (Reneseng-607415 FP7-PEOPLE-2013-ITN) is gratefully acknowledged for supporting this work.

Contact

National Technical University of Athens, Greece
akokossis@mail.ntua.gr

Automatic Generation of Optimal Structure for Distillation Processes Using Stepwise VLE Description

Hiroshi Takase, Shinji Hasebe*

Kyoto University, Kyoto, Japan

*Email: hasebe@cheme.Kyoto-u.ac.jp

Background

Many designing methods for heat integrated distillation processes have been proposed. However,

the variety of process structures examined in the optimizations is limited!

Ex. Superstructure-based method of Harwardt & Marquardt*

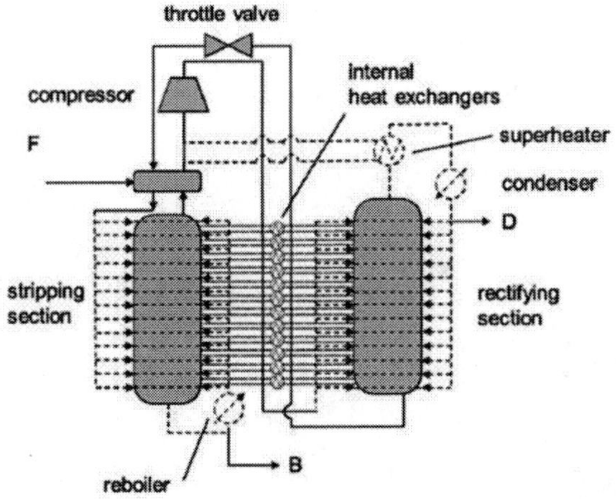

Limitation of the model

- Preassigned combinations of heat exchange
- Pressurization in vapor phase only
- Limited flexibility regarding vapor and liquid flow structure

* A. Harwardt, & W. Marquardt, *AIChE Journal*, 58, pp3740-3750, 2012

Feature of Our Research

Development of a process synthesis method which **does not introduce any structural assumptions**

Decompose to the min. element, "a tray"

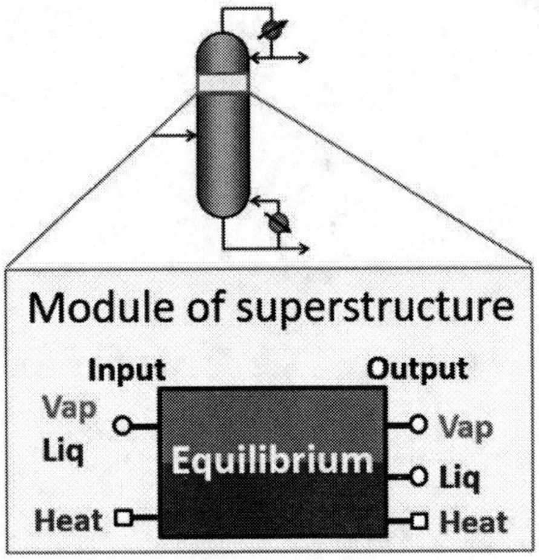

Superstructure containing all possible vapor, liquid, and heat paths

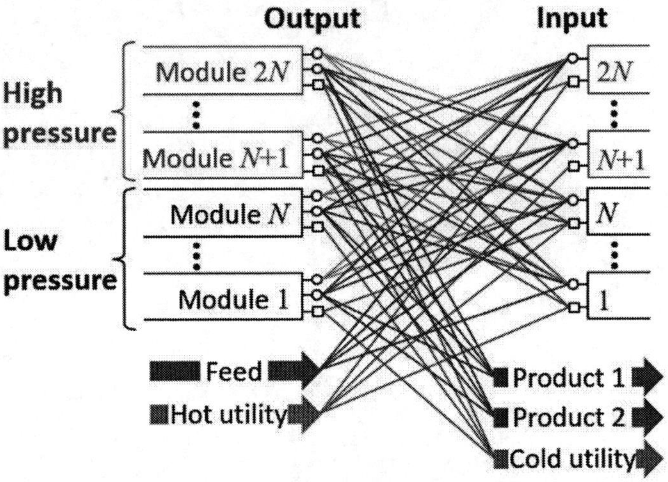

Formulation and Result

Generally, such a detailed superstructure results in **large scale NLP/MINLP** problems, because of

- Highly nonlinear Equilibrium functions
- Bilinear terms in balance equations

→ **Large computational burden**

Approximate the Equilibrium function and assign the VLE conditions in modules in advance

LP / MILP problems

→ **Less computational burden**

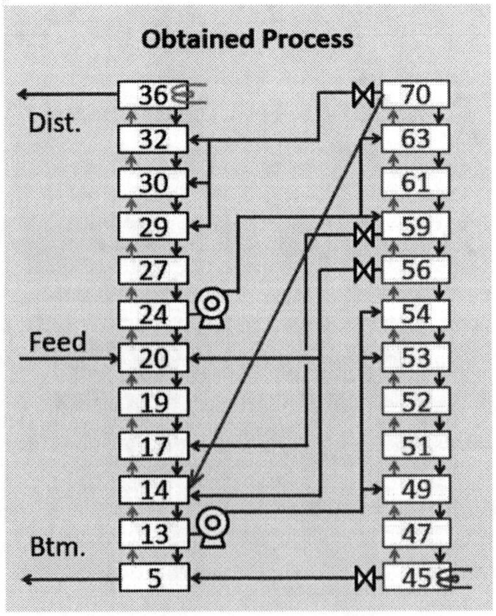

For more details...

Poster #188b

Grand Ballroom B, Hilton, 15:45-17:45, Today

461097 Design and Simulation of Biofuel Production from Pyrolysis of Brown Macroalgae

Abstract:

In recent years, thermochemical conversion processes such as pyrolysis have been greatly studied for biofuel production. Traditional feedstocks which have been considered in this research include various kinds of terrestrial biomaterials such as food crops (1st generation biofuels) and woody biomass (2nd generation biofuels). However, pyrolysis of 3rd generation biomass i.e., microalgae and macroalgae has only been recently studied. As the chemical composition of algae and their seasonal variation differ greatly from terrestrial biomass, especially in large water and ash content, the process of pyrolysis has to be redesigned and evaluated for this particular application. Conceptual process design coupled with computer assisted process simulation is a widely accepted approach for techno-economical evaluation of new production processes for novel feedstocks.

This study aims to simulate and evaluate the pyrolysis conversion pathway of brown macroalgae and subsequent biofuel production on an industrial scale (14,400 tons year^{-1} feed), using Aspen Plus V8.6 simulator. Validity of the simulation is backed up by bench-scale experimental data for indirectly heated, non-catalytic pyrolysis process of brown marine alga *Saccharina japonica* as well as with experimental data for upgrading steps. This study presents the extent of modifications for applying this particular feedstock to thermochemical conversion and biofuel production. Primary differences, compared to similar processes using terrestrial biomass include: (i) pretreatment (mineral content removal and drying), (ii) pyrolysis and optimization of combustion, as well as (iii) the number and severity of stabilization and upgrading steps. Additionally, simulation results demonstrate the overall energy requirement, as well as provide indications of economic viability of the process. Ultimately, the strategic points of integration with bioconversion platform are determined.

461718 Computationally Efficient Evaluation of Energetically Improved Distillation Processes for the Separation of Non-Ideal Mixtures

Abstract:

The majority of fluid separations in the chemical industry are still performed by means of distillation columns, accounting for about half of the total energy consumption in this industry and 10-15% of the US industrial energy use [1]. Besides the pure quantity of these processes, with approximately 40,000 distillation columns in the US [2], another reason for this high energy consumption is to be found in the low thermodynamic efficiency of 10-15% for conventional distillation processes [3]. Consequently, energetically improving these processes provides a huge potential for energy savings [4]. While a variety of different options for heat integration, thermal coupling and heat pump configurations have been proposed in order to improve the energy efficiency of distillation processes, even academic investigations are almost always restricted to specific options, such as dividing wall columns (DWC) [5], or the implementation of heat-pump designs up to a fully heat-integrated distillation column (HIDiC) [6]. The same holds true for the investigation of hybrid separation processes, such as membrane-assisted distillation or the combination of crystallization and distillation [7], for which the achieved improvements are almost always related to conventional distillation processes as benchmark.

One of the reasons for this non-compliance of energetically improved distillation processes is the limited availability of efficient means for their evaluation. For the separation of zeotropic mixtures, powerful synthesis methods to derive all simple and thermally coupled configurations have been developed (see e.g. [8]) and superstructure-based optimization approaches have been proposed to determine the optimal design of these configurations [9]. However, in the way these approaches have been described the application of the methods was limited to the separation of ideal mixtures.

Similar statements can be given for the design of heat-integrated distillation processes, for which elegant superstructure optimization approaches have been presented [10-11], but the applicability was limited to the separation of ideal mixtures as well. Despite this limitation, these methods are also not available in commercial process modeling and simulation software, such that complex configurations like DWCs have to be assembled from the connection of a thermodynamically equivalent combination of conventional distillation column models [12]. Current approaches for screening thermally coupled configurations rely on shortcut methods like the Fenske-Underwood-Gilliland method (FUGM) for the determination of the most suitable options. These are further verified by means of rigorous simulation [13]. However, the FUGM is limited in applicability and accuracy by the underlying assumptions of constant molar overflow and constant relative volatilities. The rigorous simulation of the thermally coupled configurations is complicated by the absence of readily available models, while the complex configurations, which are required in order to depict the thermally coupled configurations by means of conventional column models, have to be initialized manually [14].

In order to efficiently address the problem of evaluating a large set of process configurations, including heat-integration, thermal coupling, heat pumps and multi-effect distillation, a systematic screening procedure is proposed, which operates in two steps. In a first step a full enumeration of all process variants based on reliable shortcut methods is proposed. In order to account for non-ideal thermodynamic models, the rectification body method (RBM) [15] is integrated in an automated evaluation procedure which automatically investigates various ways of heat-integration between single columns by means of variating the operating pressures in the different columns. Thermally coupled configurations and the thermodynamically equivalent DWC configurations are evaluated based on the decomposition method proposed by Carlberg and Westerberg [16], while the distribution of the intermediate boiling component is optimized by means of a simple bisection algorithm. The application of vapor recompression is performed based on isentropic compression computation. For the separation of a mixture into three products more than 30 different configurations are evaluated, including the non-improved direct, indirect and intermediate split. Due to the computationally efficiency of the shortcut-based approach even uncertainty associated to the feed composition, utility costs and the thermodynamic model can be addressed, based on a screening of these configurations for several hundred different scenarios. Based on the full enumeration of all the configurations and potentially all resulting scenarios the most efficient configurations as well as the most relevant scenarios can be identified, and can further be investigated by means of a rigorous optimization approach based on equilibrium-tray models.

In order to evaluate the cost optimal process design from the various configurations, an efficient optimization-based method is proposed. The method relies on a superstructure approach including rigorous thermodynamic models for equilibrium and enthalpy computations [17], making it applicable to ideal as well as non-ideal mixtures. Optimal designs are determined on the basis of the total annualized costs (TAC), considering investment and operating costs. All configurations associated to one sequence of splits are initialized by means of the common column sequence, while subsequently the optimization of each energetically improved configuration is performed by means of an automated adaption of the superstructure. The different process options are optimized for minimum energy demand and TAC. The application of the overall method is demonstrated for the separation of ternary non-ideal mixtures.

REFERENCES

[1] D.S. Scholl and R.P. Lively, Nature, Vol. 532 (2016), 435–437

[2] P.C. Wankat, "Separation Process Engineering", Prentice Hall, 3rdedn. (2012)

[3] G.M. Koeijer and S. Kjelstrup, IJOT, 3 (2010), 105-110

[4] A.A. Kiss, J. Chem. Technol. Biot. 89 (2014), 479–498

[7] A. Harwardt and W. Marquardt, AIChE J., 58 (2012), 3740-3750

[6] I.J. Halvorsen and S. Skogestad, J. Nat. Gas Sci. Eng., 3 (2011), 571-580

[7] M. Skiborowski, A. Harwardt and W. Marquardt, Annu. Rev. Chem. Biomol. Eng., 4 (2013), 45-68

[8] V.H. Shah and R. Agrawal, AIChE J., 56 (2010), 1759-1775

[9] G. Dünnebier and C.C. Pantelides, Ind. Eng. Chem. Res., 38 (1999), 162-176

[10] H.Y. Yeomans and I.E. Grossmann, Comp. Chem. Eng., 23 (1999), 1135-1151

[11] M. Barttfield and P.A. Aguirre, Ind. Eng. Chem. Res., 42 (2003), 3441-3457

[12] A.C. Dimian, C.S. Bildea and A.A. Kiss, Computer-Aided Chem. Eng., 35, Elsevir, (2014)

[13] M. Errico, B.-G. Rong, G. Tola and I. Turunen, Chem. Eng. Proc.: Proc. Int. 48 (2009), 907-920.

[14] I. Dejanović, L. Matijašević and Ž. Olujić, Chem. Eng. Proc.: Proc. Int. 49 (2010), 559-580.

[15] J. Bausa, R. von Watzdorf and W. Marquardt, AIChE J. 44 (1998) 2181–2198.

[16] N.A. Carlberg and A.W. Westerberg, Ind. Eng. Chem. Res., 28 (1989), 1379

[17] M. Skiborowski, A. Harwardt and W. Marquardt, 2015, Comput. Chem. Eng. 72, 34–51.

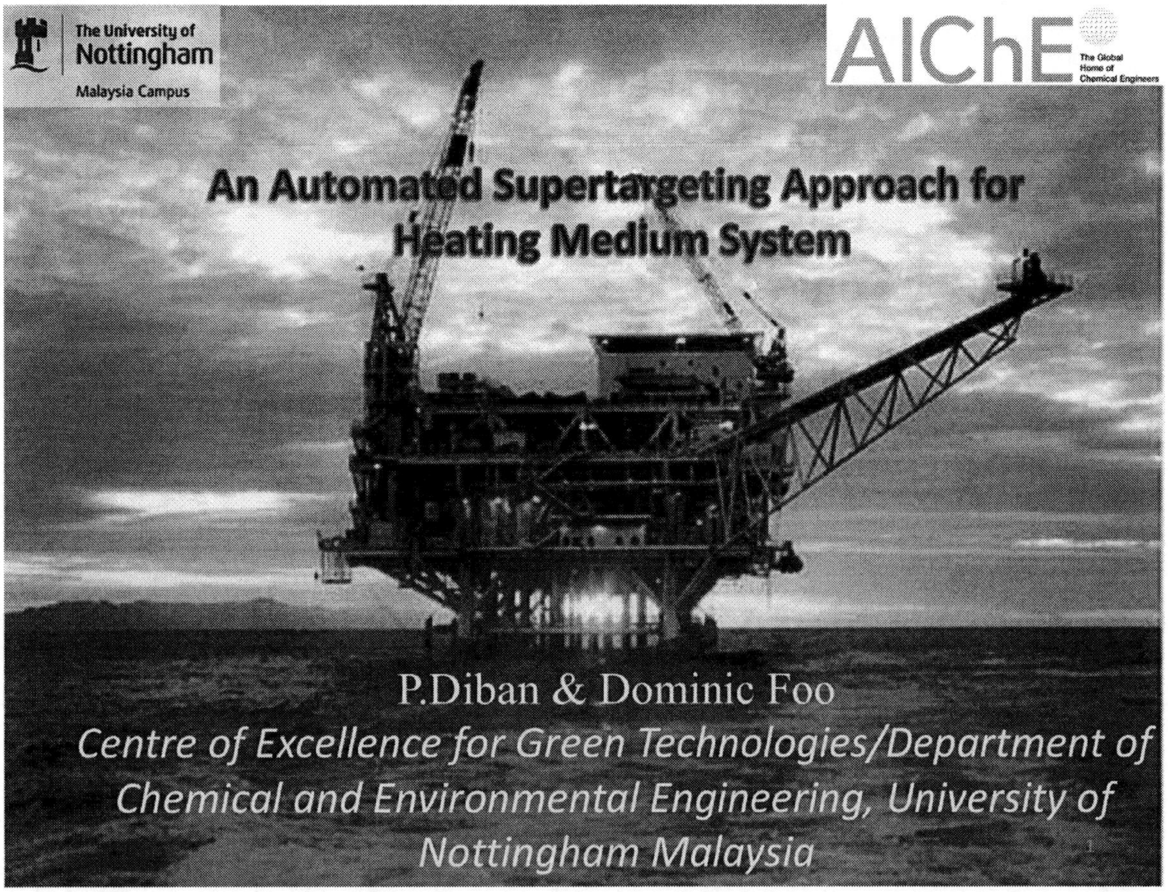

Heating Medium Utility System Description

- Used when heavy crude oil produced from reservoir
- Highly viscous, longer residence time for oil water separation
- Providing heating helps to reduce its viscosity and hence lead smaller size and lower cost separator

- Process safeguard from overheating

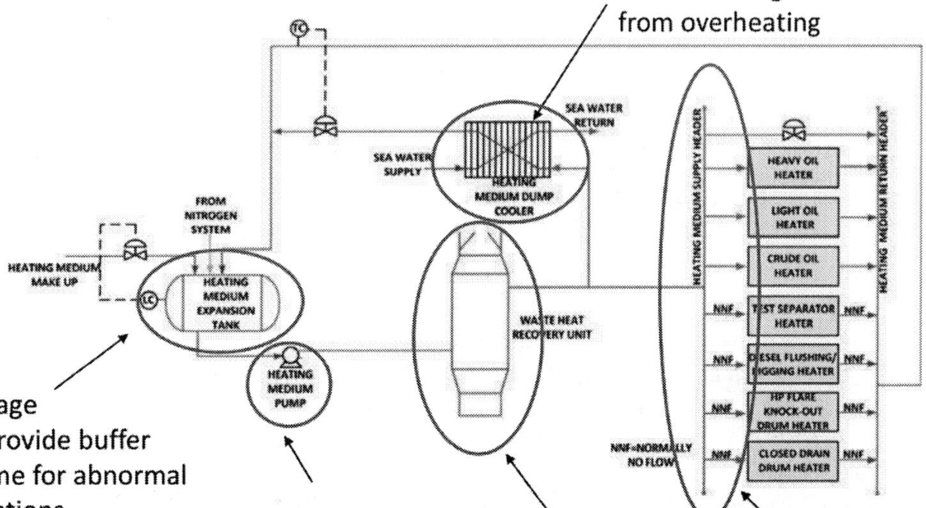

- Storage
- To provide buffer volume for abnormal operations

- To pressurize heating medium

- Uses flue gas from GTG & GTC to heat up the HM

- Common supply header
- Parallel design
- All user receive HM at supply temperature

2

ATM Framework for Heating Medium System

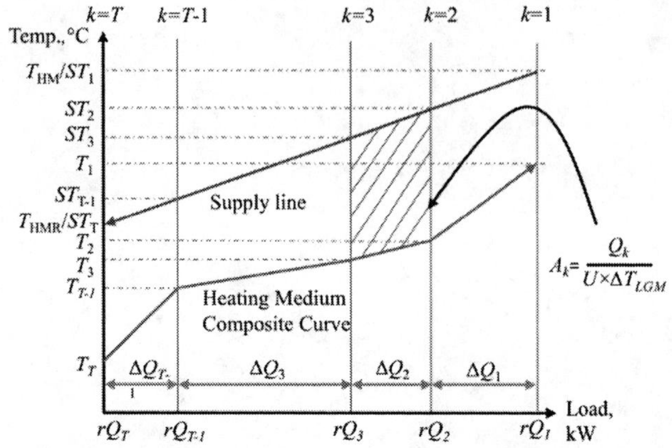

FIG.1. Supertargeting approach in determining the heat transfer area required for HEN in the heating utility system

Heat Exchanger Targeting

$$rQ_t = \begin{cases} dQ_T & t = 1 \\ rQ_{t-1} - \Delta Q_{t-1} & t = 2,3 \ldots . T-1, T \end{cases}$$

$$ST_t = \frac{(rQ_t - dQ_T\ )}{F_{min}} + T_{\text{HM}} \qquad \forall t$$

$$\Delta T_{min} = \text{MIN}\{ST_2 - T_2, ST_3 - T_3, \ldots \ldots . ST_{T-1} - T_{T-1}, ST_T - T_T\}$$

$$T_{\text{LGM}t} = \frac{[(T_t\_ST_t) - (T_{t+1}ST_{t+1})]}{LN[(T_t\_ST_t) - (T_{t+1}ST_{t+1})]} \qquad t = 1,2,3 \ldots T-1$$

$$A_{t,\text{h}} = \frac{F_{min}(ST_t - ST_{t+1})}{h_{\text{HM}}T_{\text{LGM}t}} \qquad t = 1,2,3 \ldots T-1$$

$$A_{t,c} = \sum_m \frac{F_m(T_t - T_{t+1})}{h_m T_{\text{LGM}t}} \qquad t = 1,2,3 \ldots T-1$$

$$A_{HEN} = \sum_t A_{t,\text{h}} + A_{t,c}$$

Flowrate Targetting

$$\Delta Q_t = \sum_m F_m(T_t - T_{t+1}) \qquad \text{for } t = 1,2,3 \ldots T-1$$

$$dQ_t = \begin{cases} 0 & t = 1 \\ dQ_{t-1} + \Delta Q_{t-1} & t = 2,3 \ldots . T-1, T \end{cases}$$

$$F_{min} \geq F_t \qquad \forall t$$

$$F_t = dQ_t/(T_t - T_{\text{HM}} - \Delta T_{\text{min}}) \qquad \forall t$$

$$T_{\text{HMR}} = T_{\text{HM}} - \frac{dQ_T}{F_{min}}$$

WHRU & Pump Sizing

$$T_{WHRU} = \frac{(T_{in,g} - T_{HM}) - (T_{out,g} - T_{HMR})}{LN[(T_{in,g} - T_{HM}) - (T_{out,g} - T_{HMR})]}$$

$$A_{WHRU} = \frac{Q}{U_e T_{WHRU}}$$

$$P_s = \frac{1.1 F_{min} gd}{1 \times 10^3 CP_{HM}\eta}$$

Total Anualised Costing

$$AF = \frac{i(1+i)^n}{(1+i)^n - 1}$$

$$C_{\text{U}} = P_S WG$$

$$C_{HEN} = N\,AF\left[a + b\left({A_{HEN}}/{N}\right)^c\right]$$

$$C_{WHRU} = C_B AF\left(\frac{A_{WHRU}}{D_B}\right)^M$$

$$Total\ Cost = C_{HEX} + C_{WHRU} + C_{\text{P}} + C_{\text{U}}$$

$$C_P = C_B AF\left(\frac{P_S}{D_B}\right)^M$$

3

Industrial Case Study

Description

- An FPSO platform in Angola
- Uses water as heating medium (HM)
- 3 process unit that requires HM
- Min. approach temp. 10°C for HEX design

Result

- ➢Optimum annualised cost is 645657 \$/y
- ➢Optimum ΔT_{min} =37°C
- ➢Saving 3% per year total annualised cost compared to conventional design

Table 1. Comparison of cost between optimum design from supertargeting and conventional design.

Description	Optimum design from supertargeting (this work)	Conventional design
Process HEX (\$/y)	326,696	259,974
WHU (\$/y)	190,266	212,642
Pump (\$/y)	50,804	66,206
Total Capital Cost(\$/y)	567,766	538,822
Total Utility Cost (\$/y)	77,891	126,059
Total Overall Cost (\$/y)	645,657	664,880

4

Thank You

5

463097 Systematic Process Design of Cumene Process

Abstract:

Cumene is produced industrially by the alkylation of benzene with propylene, and is one of the primary uses of benzene. Production of cumene is gaining more interest due to increasing market demands, and is used almost exclusively as a chemical intermediate for production of phenol and acetone [1].

In this work, a sustainable conceptual design of a cumene process has been developed, using a systematic, hierarchical, twelve step approach [2]. The process design has an annual cumene production of $4.1 \cdot 10^5$ metric tons, with a purity of 99.97 wt%. The process includes an alkylation reactor, and a transalkylation reactor. Benzene and propylene (which contains a 5% propane impurity) are fed to the alkylation reactor in a ratio of 3:1, where the main alkylation reaction occurs, as well as two side reactions producing di-isopropylbenzene (DIPB) and 1-hexene. In the transalkylation reactor, DIPB reacts with benzene to form additional cumene, thus increasing the yield [3]. The alkylation reactor outlet consists of a multi-component mixture of propane, propylene, 1-hexene, benzene, cumene and DIPB. First, a flash separation is performed for the separation of propane and 1-hexene. The overhead of the flash also contains benzene, and is sent to a distillation column where benzene is recovered in the bottom product and recycled to the alkylation reactor, and the overhead is further sent to another distillation column to separate propane from 1-hexene. Propane is sold as LPG and 1-hexene is purged. The bottom product of the flash is sent to a distillation column to separate benzene from cumene, and benzene is recycled back to the alkylation reactor. The bottom product is sent to another distillation column to separate cumene from DIPB, and DIPB is sent to the transalkylation reactor, where additional benzene is also fed in a ratio of 1:1. The effluents of the transalkylation reactor consists of benzene, cumene and DIPB, and is recycled back into the separation train, where it is mixed with the bottom product of the flash. This design will be considered the base case design of the process, and has been evaluated for further improvement.

The base case process design was subjected to an analysis including economics, sustainability indicators and life cycle assessment factors, where targets of improvement were identified. The targets were addressed through targeted process re-design, process optimization and heat-mass integration, in order to improve the economics of the process. Through a sustainability analysis and life cycle analysis, targets of improvement related to sustainability of the process were identified. Based on these targets, the sustainability of the process design was improved. The final process design has thus been improved w.r.t. both economic viability and sustainability, including an assessment of the environmental impacts. This poster will present systematic design of the process, including economic and sustainability results of the base case design as well as optimized cases.

References:

[1] R. M. Pandia, "the Phenol-Acetone Value Chain: Prospects and Opportunities," 2013.

[2] D. K. Babi, *Teaching Sustainable Process Design Using 12 Systematic Computer-Aided Tasks*, vol. 37, no. June. Elsevier, 2015.

[3] M. Weber, W. Pompetzki, R. Bonmann, and M. Weber, "Acetone," *Ullman's encyclopedia of industrial chemistry*. INEOS Phenol GmbH, p. 6, 2013.

464340 A Methodology for Process Debottlenecking By Process Intensification: Application to Ethylene Oxide Production

Abstract:

Improvements in process economics and sustainability are key challenges for the chemical industry. Process Intensification (PI), which has broadly been defined as the development of innovative apparatuses and techniques that offer drastic improvements in chemical manufacturing and processing [1], offers potential solutions to this challenge. PI aims at significant enhancement of physical and chemical phenomena, achieved in process-intensifying equipment, such as e.g. microreactors, advanced heat exchangers or by integration of functions (reaction-separation), introduction of new energy sources or process control methods. As a result, PI technologies may replace large, expensive and energy-intensive apparatuses with smaller, more efficient and economically competitive ones [1]. Moreover, PI technologies may overcome process bottlenecks, which cannot be tackled with the classical equipment. Therefore, the interest in PI is increasing steadily, documented by numerous experimental demonstrations of intensified equipment [2]. Nevertheless, up till now, the number of successful PI implementations on an industrial level is limited [3]. One of the reasons for this is the complexity in identifying the most suitable PI option for a given process, necessitating means for an efficient determination of the possible improvement. Due to time and resource constraints detailed modelling or experimental investigations for all existing PI equipment is generally not feasible. Therefore, this paper presents a systematic methodology enabling process analysis, bottleneck identification and selection of the most promising PI options for a detailed investigation.

Numerous PI options for tackling a bottleneck are identified on the basis of a broad database of PI technologies, such as intensified mass or heat-exchange reactors, reactive separations and hybrid processes. The database is an extended version of the one previously presented by Lutze et al. [4]. The number of feasible options is systematically reduced by consideration of technological and process specific constraints. Application of several short-cut methods for fast quantification of possible process improvement enables the pre-selection of the most promising equipment options.

The most suitable option for overcoming a given bottleneck is modelled in detail and investigated by means of rigorous process simulation. The methodology is illustrated for the case study of ethylene oxide production.

Before the methodology for process debottlenecking can be applied, the process that is to be improved needs to be analysed, which either can be done on the basis of process plant measurements or a detailed flowsheet simulation model. Subsequently, the targeted process improvement is defined and formulated as an objective function to be optimized, such as e.g. minimizing operational cost per unit of product. Furtherly, the process bottlenecks need to be identified with respect to the defined objective function. The determination of relevant bottlenecks is achieved based on energy and mass efficiencies calculated for every piece of equipment (what corresponds e.g. to selectivity for a reactor). The values of efficiency factors for the base-case design are subject to sensitivity analysis and manipulated to some value of uncertainty. Based on the results of the sensitivity analysis, the mass or energy efficiency factors having the most significant influence on the improvement of the objective function are identified. Consequently, the equipment and related mass or energy efficiency is selected as the major target for intensification.

Subsequently, the PI database is screened for possible intensified options for the identified bottleneck. The database includes more than 150 intensified technologies previously reported in the literature, which are described by the type of unit operation (e.g. reaction and separation processes, heat exchange) operating parameters (pressure, temperature, available substrate and product phases), and the specific phenomena that present the limitation causing the bottleneck (e.g. mass transfer, heat transfer). Moreover, the database contains characteristic parameters describing the apparatus, as e.g. heat transfer coefficients for advanced heat exchangers, and referenced relevant literature on experimental demonstration and possible applications. First screening of the database is performed in order to identify feasible technologies for tackling the target bottleneck, considering criteria such as feed and product stream phases, as well as constraints on process operating conditions. Additionally, different methods and tools are applied for the determination of the specific physical/ chemical phenomena responsible for the bottleneck (e.g. heat transfer, mass transfer, reaction equilibrium as possible reasons of the bottleneck- reaction selectivity). For instance, based on simplified correlations proposed by Klaewkla et al. [5], the influence of mass transfer limitations on the reaction selectivity is evaluated. If the mass transfer is not limiting selectivity, options for mass transfer improvements are further omitted. One potential option for intensification that may be found by means of such analysis is a reaction with in-situ product removal. In order to identify feasible separation processes that can be coupled with the reaction the thermodynamic insight approach [6] is applied. In case a mass separating agent is required, potential solvents are screened by means of a computer aided molecular design method (CAMD) [7], taking into account restrictions from the reaction and separation.

Based on the results the bottleneck is characterized by specific phenomena responsible for it. On this basis feasible process options are again determined from a database screening. The remaining options are further evaluated by means of short-cut models to evaluate the feasibility and performance improvement of the intensification. Based on these estimates a preselection of the most promising intensified alternatives is performed in an early design phase, at which available data is commonly sparse.

The proposed debottlenecking methodology is applied for the case study of ethylene oxide production. At first, the limitations of the analyzed process are identified based on a detailed simulation of a base case design for the oxygen-based process variant, representing the current state-of-art technology [8]. The major bottlenecks of the process are identified in respect to improvements in operational cost per unit of product. Applying the sensitivity analysis of mass and energy efficiency factors for every piece of equipment, the main process bottleneck is identified to be the selectivity of the reactor towards production of the ethylene oxide. The second process bottleneck is the ethylene loss on the purge stream, being part of the downstream processing. Taking into account the given bottlenecks, the PI database is screened for feasible options. For reactor improvement, the process conditions (temperature> 200° C, pressure> 10 bar, reacting phase- gas) are specified to identify suitable equipment. Furtherly, the reasons for the limited selectivity are studied by evaluation of the impact of heat transfer, mass transfer, reactants concentrations, and reaction equilibrium limitations.

By means of this screening procedure, reaction with in-situ separation is identified as a promising alternative for tackling the selectivity bottleneck. Absorption is identified as potential separation technique based on the application of the thermodynamic insight method [6]. A possible mass separation agent is derived by means of CAMD [7]. The search procedure is repeated for the second bottleneck, which is the ethylene loss on the purge stream. A membrane separation unit is determined as one of promising options for the reactant recovery. The potential of process improvement is quantified by means of simplified models and in the end validated by detailed modelling.

Overall, the presented approach provides a systematic tool for identifying the relevant process bottlenecks and the determination of the most promising intensification options, combining database search, thermodynamic and kinetic studies, as well as a quantitative assessment for the most promising options. Experimental effort and detailed modelling can consequently be focused on such options, offering the highest potential for improvements. The method can therefore be interpreted as support tool for the identification of the PI alternatives which should be investigated in more detail.

References:

[1] Stankiewicz, A.; Moulijn, J.; Process intensification: transforming chemical engineering, Chem. Eng. Prog. 2000, 96 (1), pp. 22-33.

[2] Keil, F; Modeling of process intensification. Weinheim, Chichester: Wiley-VCH; John Wiley 2007, pp. 1-9

[3] Becht, S.; Franke, R.; Geißelmann, A.; Hanh, H.; An industrial view of process intensification, Chem. Eng. Process 2009, 48, pp. 774-780

[4] Lutze, P.; Gani, R.; Woodley, J.; Process intensification: A perspective on process synthesis, Chem. Eng. Processing: Process Intensification 2010, 49, pp.547-558

[5] Klaewkla, .; Arend M.; Hoelderich, W. F.; A Review of Mass Transfer Controlling the Reaction Rate in Heterogeneous Catalytic Systems, in: Mass Transfer - Advanced Aspects, Ed, Hironori Nakajima, InTech,2011, pp. 667-684

[6] Jaksland, C.; Gani, R.; Lien, K.; Separation process design and synthesis based on thermodynamic insights, Chem. Eng. Sci. 1995, 50, pp. 511-530

[7] Constantinou, L.; Bagherpour, K.; Gani, R.; Klein, J.A.; Wu, D.T.; Computer aided product design: problem formulations, methodology and applications, Comput. Chem. Eng. 1996, 20(6-7), pp. 685-702

[8] Rebsdat S, Mayer D. Ethylene Oxide. Ullmannâ€™s Encyclopedia of Industrial Chemistry, Wiley-VCH Verlag GmbH & Co. KGaA, Weinheim, 2012, vol. 13, pp. 547-572

466662 Sustainable Production of Styrene By Conversion of Benzene and Ethylene through Carbon Dioxide

Abstract:

The demand of styrene is increasing more than the capacity, thus new, sustainable, and energy-saving processes are highly requested. When ethylbenzene is dehydrogenated with CO2 to produce styrene, the process is more ecological and economical than when water is used[1].
The aim of this project is to develop sustainable design for styrene production. The economic feasibility, the sustainability issues together with the ethylbenzene:carbon dioxide ratio are investigated. A systematic sustainable process design method consisting of 12 hierarchical tasks are performed to obtain the final sustainable process design.
The starting point is a survey to collect data from which the raw material and the corresponding conversion reaction are selected for the product to be made, which is styrene. Benzene and ethylene are selected as raw material and a carbon dioxide route instead of the well-known water route is selected for the conversion steps. A quick economic potential calculation is made to ensure a profitable starting point. The next set of tasks help to identify the processing route (solve the process synthesis problem) and to obtain the base case design. To reach this point, first mass balance with simple models; then mass and energy balance to establish the operating conditions, also with simple models; then simulation with rigorous models; and finally, sizing and costing calculations are performed. The base case design is found to be economically infeasible. However, the next set of tasks determine the sustainable design through targeted improvement of the base case design through heat-mass integration, design optimization as well as evaluation of environmental impacts. The final design gives a more sustainable process design.
The standard computational tools â€“ process simulator (PROII), economic analysis (ECON[2]), sustainability analysis (SustainPro[2]) and lifecycle assessment (LCSoft[2]) together with well-known methods for heat-mass integration and sustainable process design[3].

The main features of this design are: a high conversion of benzene and ethylene to a higher value product, more energetically favorable reaction conditions through carbon dioxide utilization, and removal of byproducts for a 99.8% pure styrene product.
The work was done as part of a chemical process design course4 and extended after to reach the final more sustainable design.

1 Andressa H. de Morais Batista, et al. Ethylbenzene to chemicals: Catalytic conversion of ethylbenzene into styrene over metal-containing mcm-41. Journal of Molecular Catalysis A-chemical, 315(1):86-98, 2010.
2 Kalakul, S, Malakul, P, Siemanond, K, Gani, R, â€œIntegrated of Life Cycle Assessment Software with Tools for Economic
and Sustainability Analyses and Process Simulation for Sustainable Process Designâ€, Journal of Cleaner Production, 2014, 71, 98-109
3 Babi, DK, Teaching Sustainable Process Design using 12 systematic Computer Aided Tasks, Computer Aided Chemical
Engineering, 2015, 37, 173-178.
4 Process Design: Methods and Principles, MSc-course (C28350), Technical University of Denmark, Lyngby, Denmark

The University of Kansas
Chemical & Petroleum Engineering

Visualization and Modeling of Chemical Processes via MOSAIC

Bridgette Befort[1]
Coauthors: Kyle V. Camarda[1], Kenneth Bishop[1], Erik Esche[2], Sandra Fillinger[2], Jens-Uwe Repke[2], Gregor Tolksdorf[2]
(1) Chemical and Petroleum Engineering, The University of Kansas, Lawrence, KS
(2) Process Dynamics and Operations Group, Technische Universität Berlin, Berlin, Germany
Email: Bridgette.Befort@ku.edu

The University of Kansas
Chemical & Petroleum Engineering

Exploring MOSAIC

Computer modeling and simulation of chemical processes provides the opportunity to visualize, solve, and optimize complex chemical engineering systems

- Free, web-based, open-source modeling, simulation, and optimization environment
- Tools
  - Modeling
  - Simulation
    - Visualization—Flowsheeting
  - Optimization
- LaTeX-Style Entry
- Solves and creates code in many languages
  - Import/Export
- CAPE-Open Compliant

The University of Kansas
Chemical & Petroleum Engineering

MOSAIC Capabilities

- Modeling and Simulation of Chemical Processes, steady state or dynamic
  - Williams Otto Process
  - Plant Operations and Control
- Create Flowsheet Simulations
- Combine Modeling, Simulation, and Optimization Tools with Visualization
  - Importing Visuals and Developing Equations
- Optimize Processes
- Formulating Equation Systems Produces Code to Export
  - GAMS
  - Matlab
  - C++
  - Python
- High Performance Computing
- Design and Build Algorithms

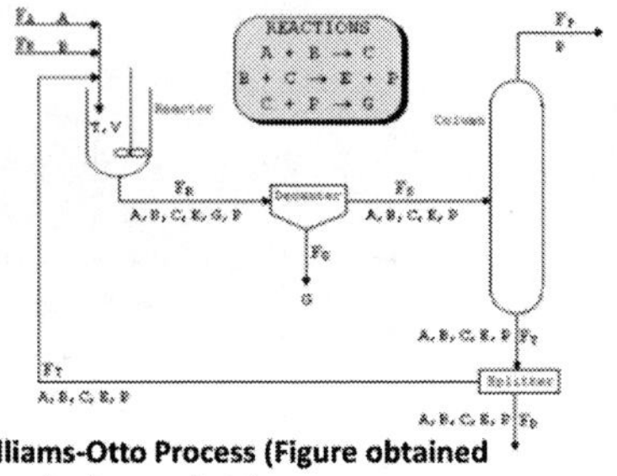

Williams-Otto Process (Figure obtained from che.boun.edu.tr)

The University of Kansas
Chemical & Petroleum Engineering

Future Work

- Visualization, Modeling, and Simulation
  - Models Generate Visuals
  - Visuals Generate Models
- Multi-Objective Optimization
  - Decision making involving many, often conflicting, criteria
  - Applied to Sustainable Chemical Process Plant Design
    - Goal: Achieve both maximum energy efficiency *and* industry profit

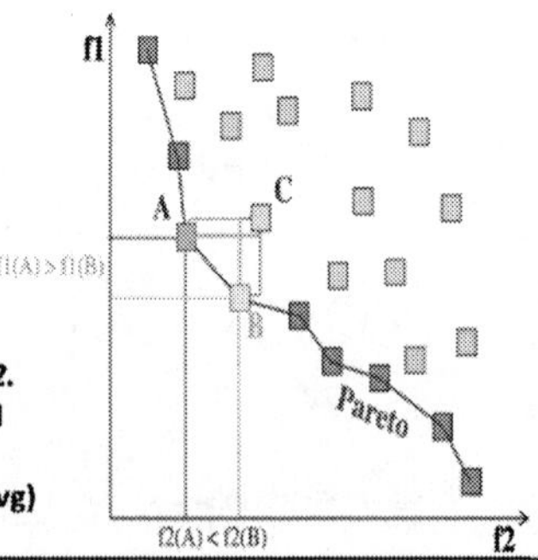

Pareto Optimality Frontier for minimizing *f1* and *f2*. Squares define feasible solutions. (Figure obtained from commons.wikimedia.org/wiki/File:Front_pareto.svg)

Synthesis of Dimethyl Ether from Natural Gas: CO_2 Utilization Process

Alessandro Duso[1], Júlía Rós Hafþórsdóttir[2], Yiyi Cao[3],

Emmanouil Papadakis and Olivia Ana Perederic

Department of Chemical and Biochemical Engineering,

Technical University of Denmark

DTU

[1]*s151118@student.dtu.dk*
[2]*s151246@student.dtu.dk*
[3]*s151096@student.dtu.dk*

Project Framework

- CO_2 has major environmental impact
- Need to identify a process that consumes CO_2 producing a valuable product, DME

Methodology

Synthesis

- Collection of data
- Flowsheet generation
- Mass and energy balance

Design & Analysis

- Equipment sizing
- Economic & sustainability analysis
- Heat integration
- Optimization

PROII, ICAS, ECON

Innovation

Hot-spots identification analysis

PROII, SustainPro, ICAS

Process Flowsheet

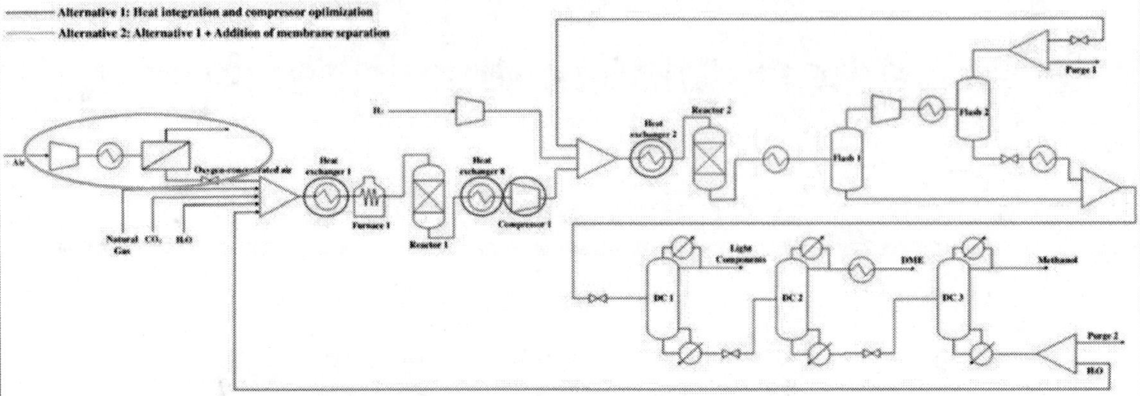

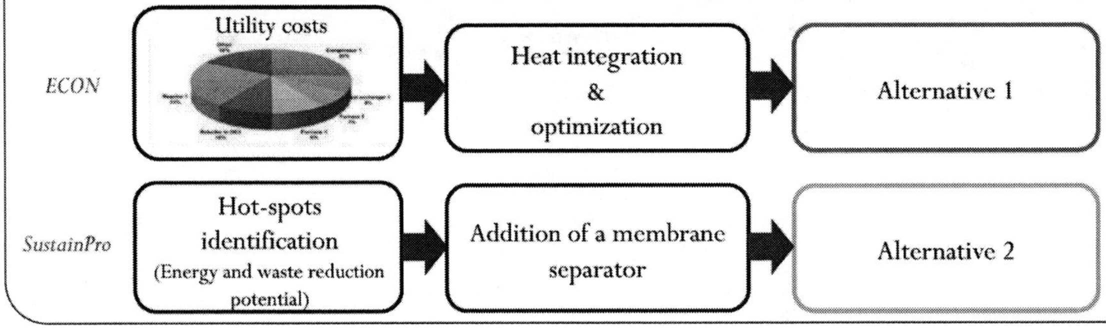

Results

- Alternative 1
  - Production cost reduction of 5.4%
  - Payback period decrease by 18.3%
- Alternative 2
  - Production cost reduction of 6.5%
  - Payback period decrease by 30.7%

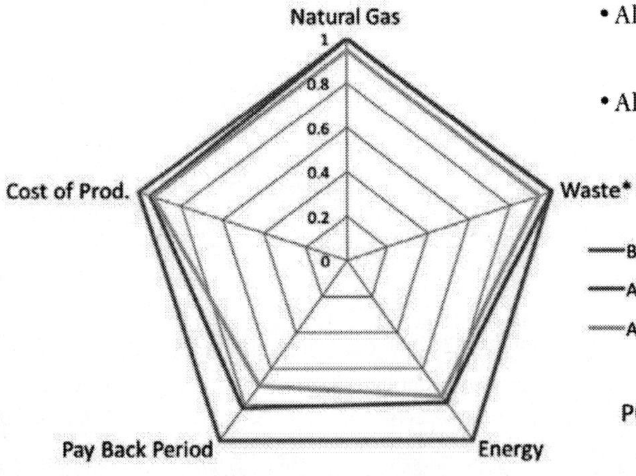

* Evaluated without nitrogen

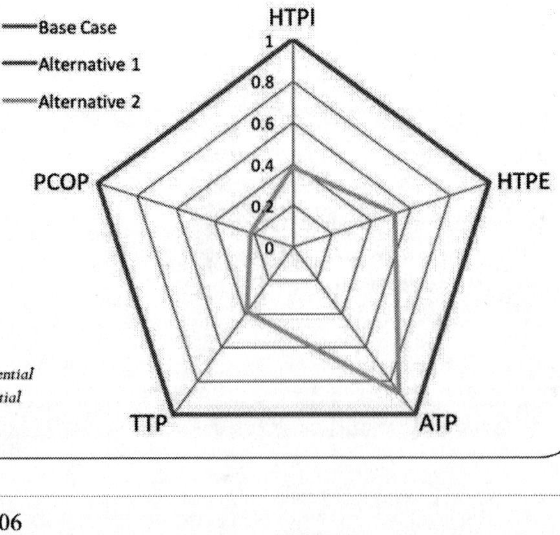

HTPI = Human toxicity potential by ingestion
HTPE = Human toxicity potential by exposure
PCOP = Photochemical oxidation potential
TTP = Terrestrial toxicity potential
ATP = Aquatic toxicity potential

466549 Integrated Biorefineries Using Ionic Liquids: Application to Macroalgae Feedstock

Abstract:

Fossil fuels are estimated to be exhausted in a few decades due to the increase in human population and living standards. A very important challenge is to find an environmentally friendly and sustainable new source of energy. Biofuels, energy from biomass is one of the most promising alternatives. Depending on the type of feedstock, biofuels have been classified as first generation (food crops), second generation or non-edible biomass (some crops, forest residues, animal fat, waste) and third generation biomass (microorganisms as microalgae). The importance of this alternative is reflected in the recent research projects funded by the European Commission like: Eurobioref (European Multilevel Integrated Biorefinery), Biocore (Biocommodity Refinery), Suprabio (Sustainable Products from Economic Processing of Biomass in highly Integrated Biorefineries) or Miracles (Multiproduct Integrated Biorefinery of Algae). One of the main outcomes of these projects is that future biorefineries must be highly integrated, flexible enough to allow the conversion of different feedstock (2nd and 3rd generation biomass) and multiproduct (biofuels, chemicals, polymers, energy, etc.)

One of the issues to be improved in biorefineries is the use of solvents, organic solvents are generally needed in different stages of the production process. These solvents are usually non-environmentally friendly, and its separation and recovery increase the overall cost significantly.

Ionic liquids are salts that are liquid at room temperature. Their properties make them a more environmentally friendly and less costly alternative to conventional species. Ionic liquids are usually used as substitutes of solvents but they can be also used as catalysts or catalyst supports among other uses. One of the main benefits of using ionic liquids, that makes them green solvents, is that they can be recovered almost completely.

In this work we present the process flowsheet of a whole multifeedstock / multiproduct integrated biorefinery. For each feedstock the different available processes are shown, for each process the routes to produce multiple products (chemicals like the antioxidant astaxanthin, polymers like PHB –polyhydroxybutyrate-, specialty chemicals from levulinic acid and esters, biodiesel -mainly jet fuel-, bioethanol, etc.) and for each route some alternatives to achieve the desired product. Biorefinery integration is also depicted indicating what units can be used in several places of the plant. Finally, each operating unit that can use ionic liquids is marked commenting on the specific ionic liquids that are suitable for that unit and its benefits. Some examples of their use in the biorefinery: as solvents for lignocellulosic biomass, for the deconstruction of lignin, as catalysts in the transesterification reaction or in the enzymatic hydrolysis of sugars, as extraction agents of lipids from algae or other biomass, as entrainers in azeotropic distillation, etc.

To illustrate the potential of this approach we have selected a specific biomass, brown seaweed. This is the single lartest macroalgae resource and a very good candidate for energy processing. The carbohydrates present in these seaweed are mannitol, laminarin and alginates. Different products (biogas, bioethanol, value added chemicals like organic acids, etc) can be achieved depending on the carbohydrate used and the technology selected (anaerobic digestion, thermochemical conversion, hydrothermal liquefaction, fermentation, etc.).

Using this feedstock to analyse the benefits of each alternative a superstructure optimization as well as a simulation of the biorefinery is being developed. In this work we present the developed superstructure and the simulation of one of the routes remarking where the ionic liquids are used.

References

Brennan, L., P. Owende P., Biofuels from microalgae—A review of technologies for production, processing, and extractions of biofuels and co-products, Renewable and Sustainable Energy Reviews 14 (2010) 557–577

Bruton, T., H. Lyons, Y. Lerat, M. Stanley and M. B. Rasmussen (2009). A review of the potential of marine algae as a source of biofuel in Ireland. Dublin, Ireland, Sustainable Energy Ireland.

Nautiyal P., Subramanian K.A., Dastidar M.G., Production and characterization of biodiesel from algae, Fuel Processing Technology 120 (2014) 79–88

Peterson, A. A., F. Vogel, R. P. Lachance, M. Froling, M. J. Antal and J. W. Tester (2008).

Thermochemical biofuel production in hydrothermal media: A review of sub- and supercritical

water technologies. Energy & Environmental Science 1(1): 32-65.

Ross, A. B., J. M. Jones, M. L. Kubacki and T. Bridgeman (2008). Classification of macroalgae as fuel and its thermochemical behaviour. Bioresource Technology 99(14): 6494-6504.

Stark, A., Ionic liquids in the biorefinery: a critical assessment of their potential. Energy Environ. Sci., 2011. 4(1): p. 19-32.

Waldron K. (Ed.), Advances in Biorefineries: Biomass and Waste Supply Chain Exploitation, 2014, Woodhead Publishing Series in Energy: Number 53, Elsevier

466928 Real-Time Dynamic Efficiency Optimization of Coal-Fired Steam Power Plants

Abstract:

Physics-based dynamic models of power plants play a substantial role in the design of system configuration, development of control strategies, and optimization of system operation. Due to the seasonal and daily fluctuations of the electricity demand, the plants are subject to frequent load changes and partial shutdowns.[1] Extensive research has been performed on dynamic power plant models, with focus on dynamic simulation,[2,3] real time optimization,[4,5] start-up optimization,[5] and model predictive control.[6,7] In this work, coal-fired power plant configurations are studied, because coal is a more reliable energy source with attractive pricing than other resources.[8]

This work presents a dynamic model of a 600 MW subcritical coal-fired steam power plant and the real-time optimization for power plant efficiency. In this presentation, the steps for the model development and steady-state validation of an integrated steam cycle-boiler model are discussed. The model is validated against steady-state data from a reference subcritical-pressure power plant.[9] Conventional control designs are successfully incorporated in the system model. Transient analyses of the response of the power plant to varying the coal loads show that the model provides a robust test-bed for dynamically changing power demand. The optimization capabilities of the complementary tool chain are demonstrated in case studies of nominal efficiency optimization of the integrated plant with respect to admissible plant inputs. The objective of real-time power plant optimization is to maximize the efficiency of power plants, operating in a transient fashion. This is done by calculating optimum time-varying input trajectories, which satisfy operability and safety constraints during the transition between steady states.

Acknowledgements

This material is based upon work supported by the National Science Foundation under Grant No. 1054718.

References

1. Ticǎ, A., Guéguen, H., Dumur, D., Faille, D. & Davelaar, F. Design of a combined cycle power plant model for optimization. *Appl. Energy* **98,** 256–265 (2012).

2. Lu, S. Dynamic modelling and simulation of power plant systems. *Proc. Inst. Mech. Eng. Part A J. Power Energy* **213,** 7–22 (1999).

3. Chen, C., Han, L. & Bollas, G. M. Dynamic simulation of fixed bed chemical looping combustion reactors integrated in combined cycle power plants. *Energy Technol.* (accepted) (2016).

4. Åkesson, J., Årzén, K.-E., Gäfvert, M., Bergdahl, T. & Tummescheit, H. Modeling and optimization with Optimica and JModelica.org—Languages and tools for solving large-scale dynamic optimization problems. *Comput. Chem. Eng.* **34,** 1737–1749 (2010).

5. Lind, A., Sällberga, E. & Velutb, S. Start-up Optimization of a Combined Cycle Power Plant. *Model. 2012 Conf. Proc.* (2012). at <http://www.control.lth.se/documents/2012/5900_pop.pdf>

6. Franke, R., Babji, B. S., Antoine, M. & Isaksson, A. Model-based online applications in the ABB Dynamic Optimization framework. in *Model. 2008* 279–285 (2008).

7. Johnsson, A. Nonlinear Model Predictive Control for Combined Cycle Power Plants. (2013).

8. Rubin, E. S., Chen, C. & Rao, A. B. Cost and performance of fossil fuel power plants with CO2 capture and storage. *Energy Policy* **35,** 4444–4454 (2007).

9. Singer, J. G. *Combustion Fossil Power: A Reference Book on Fuel Burning and Steam Generation.* (Combustion Engineering, 1991).

467075 Exergy Targeting for Complicated LNG Processes

Abstract:

Exergy is the maximum available work obtained by bringing a system to equilibrium with its environment. Such analysis, which is the combination of first and the second law of thermodynamics, is based not only on the condition of a system but also the environment. These features make exergy efficiency more reliable than energy efficiency, showing the real performance of processes. Thus, exergy analysis has been regarded as a valuable post design tool.

However, exergy can also be used in an early stage of design by calculating the minimum available work required in a process, especially in low-temperature processes. Refrigeration in cryogenic processes is produced by compression power and such power is pure exergy. Traditionally, pinch analysis has been widely used for designing systems related to heat integration. This analysis only contains temperature as a variable. Therefore, pinch analysis is inadequate for designing a refrigeration system where streams experience changes in both temperature and pressure [1].

Exergy targeting for designing a refrigeration process can be performed by combining pinch analysis and exergy analysis, having both temperature and pressure as variables. Previous work in our research group has suggested a new graphical diagram for exergy targeting, based on so-called exergetic temperatures [2]. The advantage of these new temperatures (actually new "lumped" temperature functions) is that there is a linear relation between exergetic temperature and exergy. Simple piecewise linear graphical diagrams similar to composite curves in pinch analysis can easily be drawn and used for exergy targeting.

Previous work studied only simple processes to test the new graphical diagram, such as N_2 gas expander processes assuming the ideal gas model. Thus, the main objective of this paper is to extend this exergy targeting method to multi-component refrigerant LNG processes, while using a real gas model. A discussion is also made on the practicality of the exergy targeting approach for design and optimization of complicated

cryogenic processes.

References

[1] Aspelund, A., Berstad, D., Gundersen, T. "An Extended Pinch Analysis and Design procedure utilizing pressure based exergy for subambient cooling", *Applied Thermal Engineering*, vol. 27, pp. 2633-2649, 2007.

[2] Marmolejo-Correa, D. and Gundersen, T. "New Graphical Representation of Exergy Applied to Low Temperature Process Design", *Industrial & Engineering Chemistry Research*, vol. 52, pp. 7145-7156, 2013.

A Methodology to Generate Modular Equipment for an Equipment Database in Module-Based Plant Design

Martin Eilermann
TU Dortmund University
Laboratory of Plant and Process Design
martin.eilermann@bci.tu-dortmund.de

technische universität dortmund

apt Laboratory for Plant and Process Design

Motivation & Objective

Conventional plant design

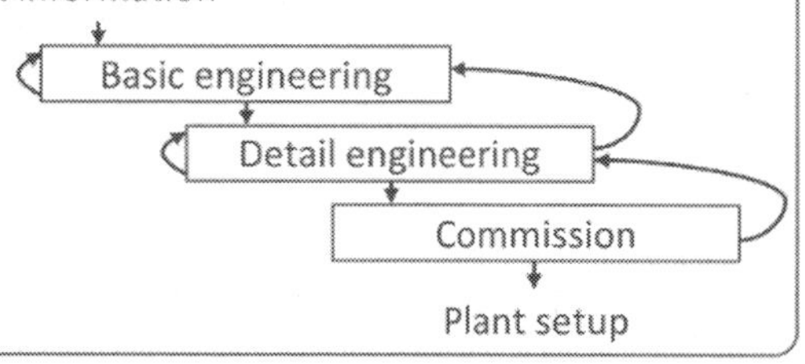

Module-based plant design

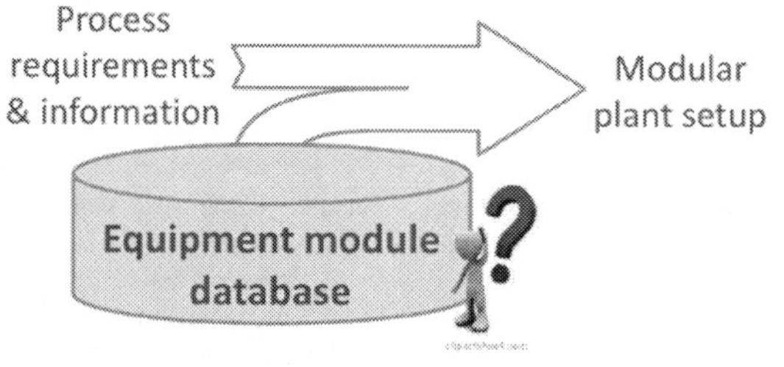

Objective

Generate equipment module database with the least number of equipment modules covering as many industrial applications as possible

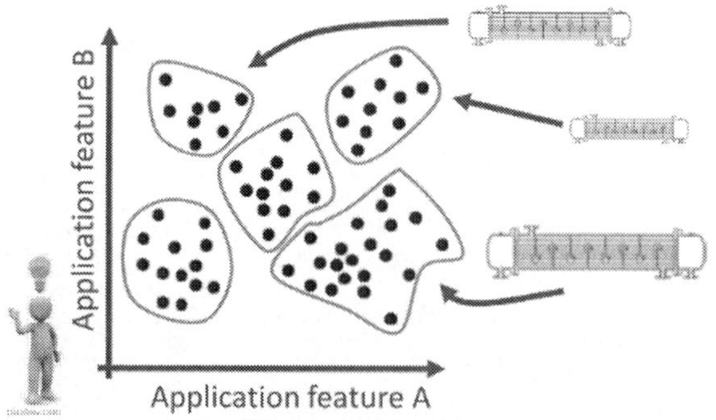

technische universität dortmund

apt Laboratory for Plant and Process Design 2

Approach for Generating Equipment Module Database

Grouping of existing industrial applications

- Hierarchical cluster analysis
  - Number of clusters can be chosen
  - Resulting tree structure advantageous

Selection of representative equipment

- Design for dedicated application not possible
- Sampling of equipment designs by Sobol sequence

Simulation of all applications in all equipment designs

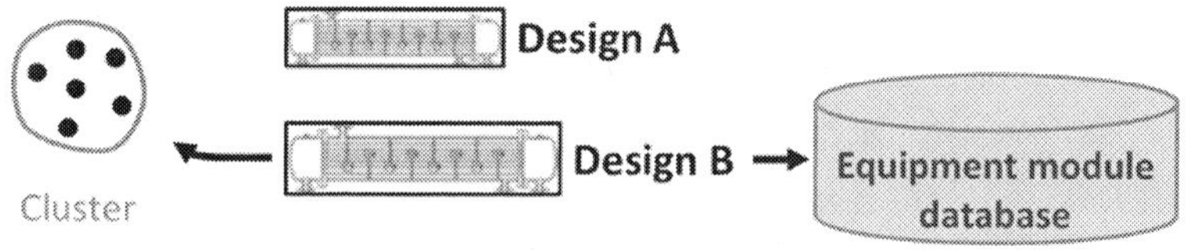

technische universität dortmund

apt Laboratory for Plant and Process Design

3

Results

- Successful application for liquid/liquid heat transfer applications
- Process data sheets of industrial applications provided by EVONIK INDUSTRIES

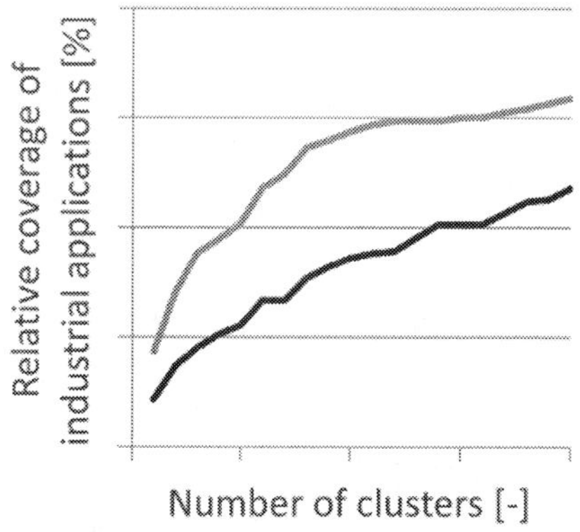

- Vast majority of applications can be covered by highly reduced number of heat exchangers
- Proof of concept in heat exchanger network

For more information or questions meet me at
Poster 188j
Grand Ballroom B (Hilton)
3:45 PM to 5:45 PM

technische universität dortmund

apt Laboratory for Plant and Process Design

4

IPESE
INDUSTRIAL PROCESS AND ENERGY SYSTEMS ENGINEERING

European Commission

EPFL
ÉCOLE POLYTECHNIQUE FÉDÉRALE DE LAUSANNE

Combined water and energy integration in industrial processes with restricted connections

Maziar Kermani [a], Ivan D. Kantor [a], François Maréchal [a]

[a] EPFL, Industrial Processes & Energy Systems Engineering Group (IPESE), Sion, Switzerland

maziar.kermani@epfl.ch

AIChE 2016 Annual Meeting, November 13-18, San Francisco, USA

Motivation

Background

- Strong interconnectivity of water and energy
- Simultaneous optimization of water and energy
- Several methodologies have been proposed [1]:
  - Conceptual (insight-based)
  - Mathematical (MINLP in nature)
- Drawbacks:
  - Complexity for application in real industrial cases
  - Assumption on unrestricted heat/mass exchange

Goal

- Addressing these challenges by combining two methodologies:
  - MILP approach for water and energy optimization [2]
  - MILP approach for energy integration with restrictions [3]
- Proposing a methodology for industrial applications

[1] E. Ahmetović, N. Ibrić, Z. Kravanja, and I. E. Grossmann, *Computers & Chemical Engineering*, vol. 82, pp. 144–171, 2015.
[2] M. Kermani, Z. Périn-Levasseur, M. Benali, L. Savulescu, and F. Maréchal, in *Comp. Aided Chem. Eng.*, vol. 33, Elsevier, 2014, pp. 1561–1566.
[3] H. Becker and F. Maréchal, *Computers & Chemical Engineering*, vol. 37, pp. 104–118, 2012.

Methodology

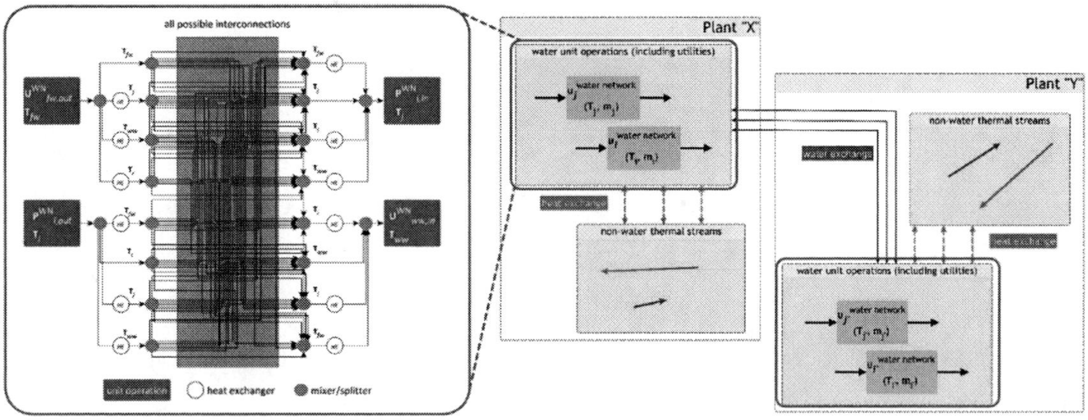

MILP superstructure:

min {total annualized cost}

s.t. *heat cascade constraints*
mass balances
unit sizing
heat/mass restrictions

Remarks

Case study

- Two industrial plants: pulp and paper, petrochemical.
  - No data on contamination. Requires heuristics to avoid infeasible mass exchanges.
- Individual optimization of each plant:
  - 50% reduction in utility consumption.
- Integrated approach
  - Complete elimination of hot utility requirement by recovering waste heat in an integrated water network.

Remarks

- Extending the superstructure for simultaneous optimization of water and energy (SOWE) to also incorporate water as an intermediate heat transfer medium.
- Temperature of storage tanks can be further optimized to lower the penalty of the restrictions.

"poster presentation: #188f Grand Ballroom B, Hilton San Francisco Union Square"

Thank you for your attention.

188f Grand Ballroom B (Hilton San Francisco Union Square)

Acknowledgement: This project has received funding from the European Union's Horizon 2020 research and innovation programme under grant agreement No 679386. This work was supported by the Swiss State Secretariat for Education, Research and Innovation (SERI) under contract number 15.0217, as well as CTI within the SCCER EIP (KTI.1155000150).

Comparison of Capacity Expansion Strategies for Chemical Production Plants and Consideration of Alternative Modular Reactors with a Larger Operating Window

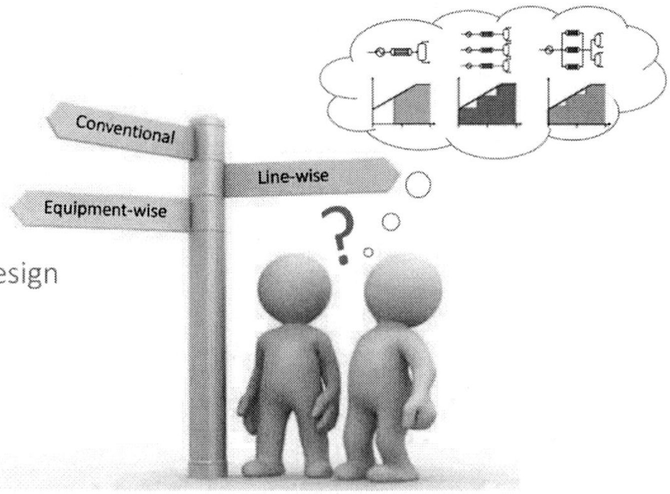

Heiko Radatz
TU Dortmund University
Laboratory of Plant and Process Design
heiko.radatz@bci.tu-dortmund.de

K. Kühne
Prof. Dr.-Ing. G. Schembecker
Dr.-Ing. C. Bramsiepe

technische universität dortmund
Department of Biochemical and Chemical Engineering

apt Laboratory of Plant and Process Design

Today´s world requires new approaches in plant design

Conventional large scale plants profit from economy of scale[1]

- Capital-intensive investment
- Optimized for a single operating point[2]
- Narrow operating window

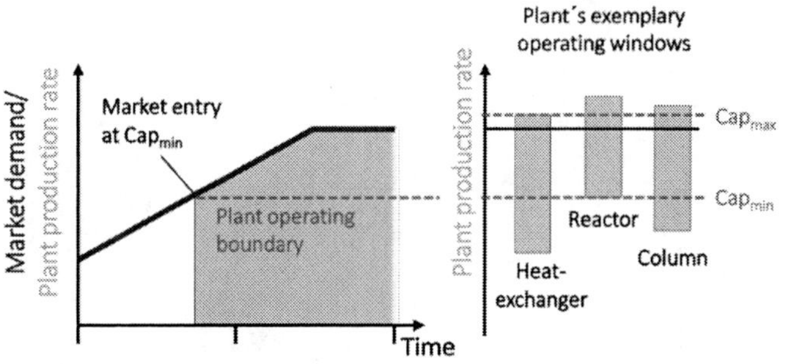

Requirements for production

- Adaptability
- Reduced investment risk
- Early market entrance

New approaches in plant design

⇨ Step-wise capacity expansion

⇨ Use of modular equipment

[1] B.P. Priddat, "*Die Modernität der Industrie*", 2012
[2] H. Mothes, "*No-Regret-Lösungen - Modulare Produktionskonzepte für komplexe, unsichere Zeiten*", 2015

Types of step-wise capacity expansion

Conventional design

- Designed and optimized for specific production capacity[2]
- Late market entry due to large Cap_{min}

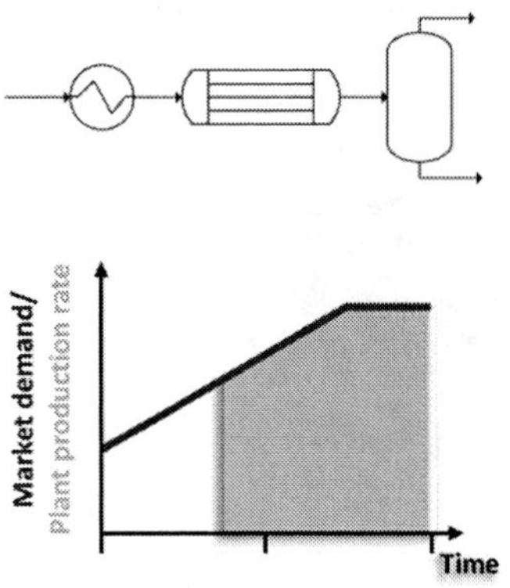

Line-wise capacity expansion

- Multiple smaller designs
- No full operating window utilization of every unit

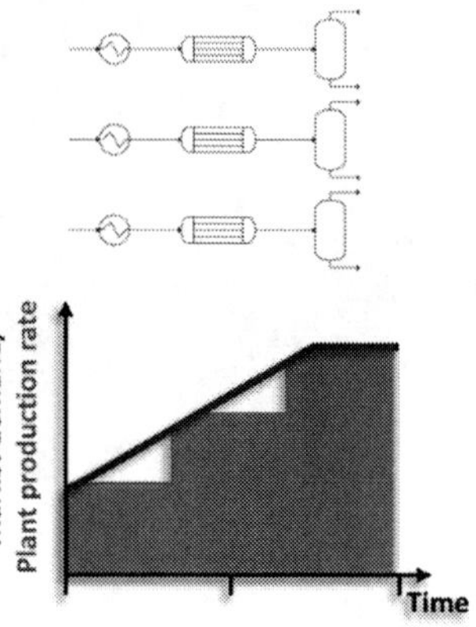

Equipment-wise capacity expansion

- Expand capacity of single units
- Utilizes the operating window of every process unit

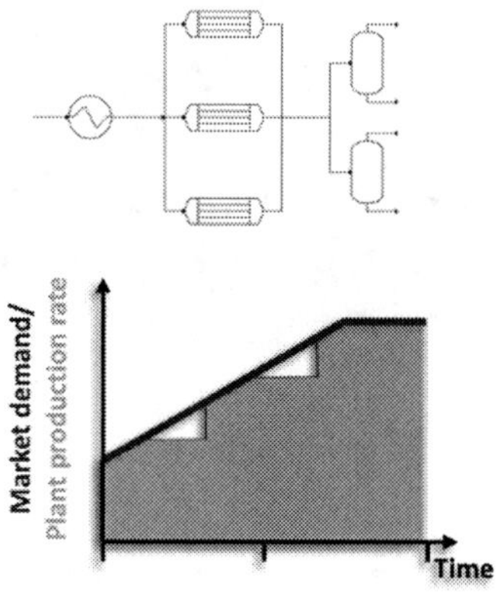

[2] H. Mothes, *"No-Regret-Lösungen - Modulare Produktionskonzepte für komplexe, unsichere Zeiten"*, 2015

technische universität dortmund

3

What you can expect?

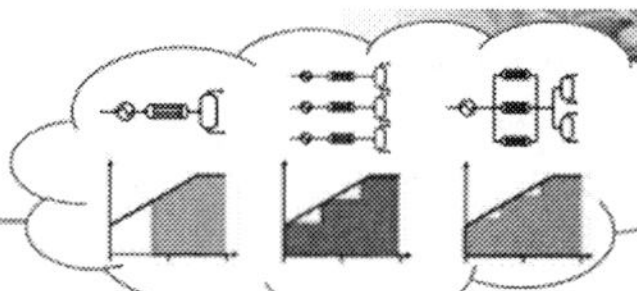

Equipment Module Database

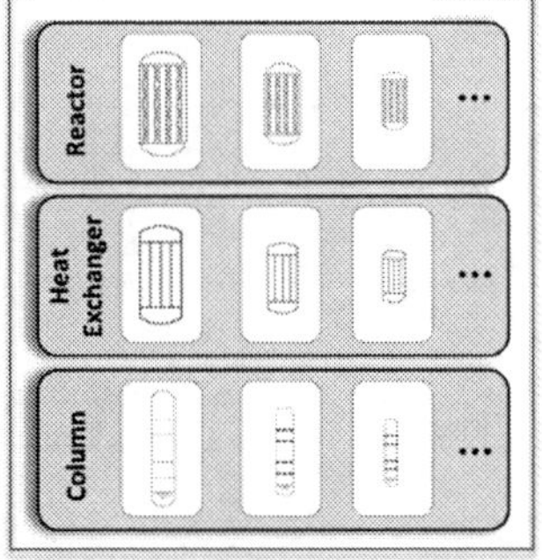

Methodology

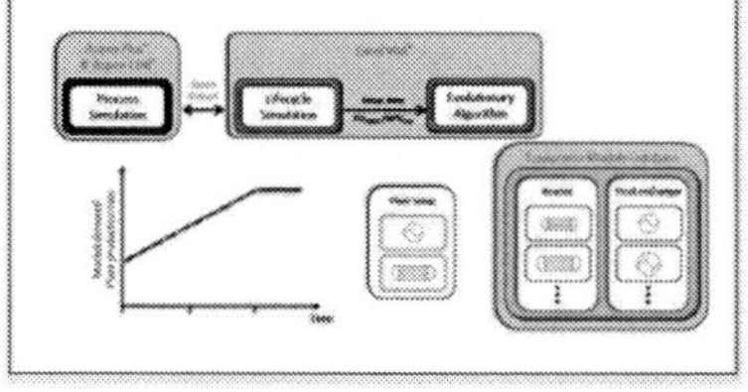

Example Process

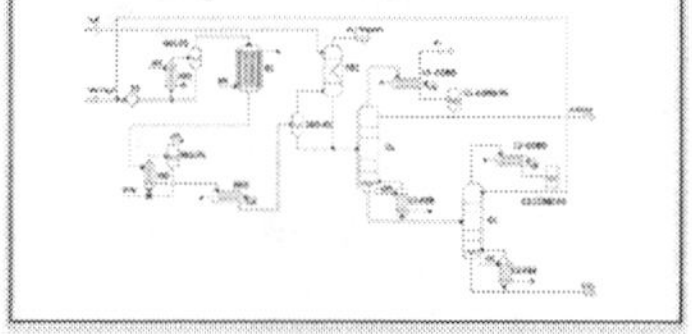

Objectives

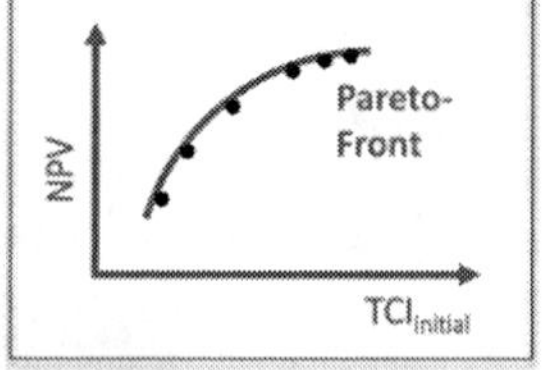

Reactor Modules with larger operating windows

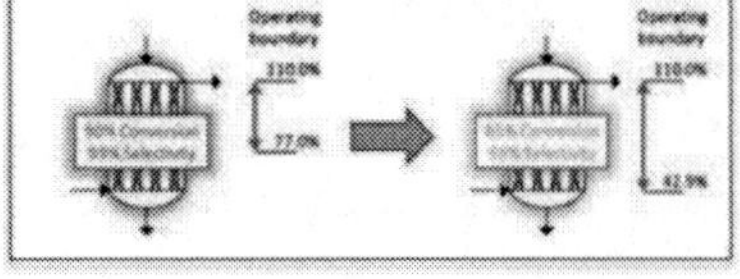

Results

Meet me at **Poster 188h**
Grand Ballroom B (Hilton)
3:45 PM to 5:45 PM

technische universität dortmund

4

468894 Safety Evaluation of Shale Gas Monetization Processes

Abstract:

Exploitation of shale gas has increased in the past few years, raising the interest in developing and analyzing chemical production technologies that use shale/natural gas as feedstock. Different processes have been analyzed and compared from technical, economic and environmental aspects, among them those to produce ethylene and methanol. Another aspect of primary importance to rank technologies is that related to safety. This has encouraged the search for practical methods to quantify the inherent safety levels of different processes at early design stages. Indexes such as the Fire Damage and Explosion Index and the Process Stream Index are used to rank the inherent safety levels of a process using data obtained at a preliminary design stage. In this work the analysis of different processes for the production of ethylene and methanol from shale gas is presented. The results obtained from the selected indexes are used to compare the processes from a safety point of view. These efforts represent a first approach towards the development of a global index that could account for safety as applied to a complete process flowsheet from the conceptual step of a design project.

469597 Optimal Design of LNG Regasification Terminals

Abstract:

Natural gas (NG) is the cleanest and the fastest growing fossil fuel. It has become a key global player in the face of rising energy and climate change concerns. Liquefied Natural Gas (LNG) is the most preferred option to transport natural gas over long distances. Since LNG is a cryogenic liquid (around â€“160) at atmospheric pressure, heat ingress into LNG during its storage and transport is unavoidable in spite of much insulation [1]. This vaporizes a portion of the LNG to form Boil-Off Gas (BOG). LNG regasification terminals which import and store huge amounts of LNG are concentrated sources of BOG. Unless reliquefied and/or reused, BOG represents an economic loss, and changes LNG composition and quality over time. The power needed to reliquefy BOG is a prominent expenditure for a regasification terminal, and minimizing this undesirable power consumption is worthwhile. As LNG is becoming a global commodity, more and more export/import terminals are being built around the world. Therefore, optimizing the process configuration of a regasification terminal is very important.

While some studies in the literature have addressed the retrofit design of a typical regasification terminal, most have focused on energetic or thermodynamic targets, and no study exists on grass roots design. Liu et al. [2] presented multi-stage recondensation schemes to decrease the energy needed for reliquefying BOG. However, they ignored the powerful cooling effect of high pressure LNG. Park et al. [3] considered some degree of cooling via high pressure LNG to minimize power consumption and operating cost. However, their work focused on retrofitting an existing terminal, and did not consider all process synthesis options. The amount of BOG in and energy consumption of a regasification terminal depends heavily on the composition, flow, temperature and pressure of LNG at the terminal. In addition, it depends strongly on the ambient conditions and other processes (e.g. pumping, recirculation, regasification) at the terminal. Therefore, isolated optimization studies such as those existing in the literature comparing selected process configurations may not yield the best results. Hence, none of the existing works is sufficiently comprehensive to provide the best process configurations for energy/cost targets. They also fail to provide reliable lower bounds from energy/cost perspective. There is a need for a fresh and comprehensive process synthesis approach.

In this talk, we will present a comprehensive superstructure for the BOG reliquefaction process at an LNG regasification terminal. The proposed superstructure includes all the previously investigated process options, but also a few that have not been considered before. It applies to all flow conditions and compositions under which it is practical to recondense BOG without the use of any external refrigerant. We highlight the fact that LNG storage tanks are very dynamic in nature, and assuming perfect equilibrium as done in the existing literature [2] is not a realistic assumption. Our superstructure yields the best process configuration for any given flows, properties, and conditions of LNG and BOG. In addition to the superstructure, we have developed an ingenious simulation-based solution strategy incorporating fundamental thermodynamic and process synthesis insights to solve the superstructure optimally without any undue assumptions. Model results not only provide the best designs with minimum cost/energy under realistic considerations, but also gives reliable lower bounds under idealistic assumptions. The latter provide very useful benchmarks for assessing existing regasification terminals. We illustrate our novel process synthesis approach on a South Korean terminal studied by Park et al. [3] for a range of BOG scenarios. We present a superior regasification scheme that offers about 12% reduction in power consumption and significant reduction in costs. We perform a series of sensitivity analyses on BOG conditions and various realistic operating scenarios to illustrate their profound impacts on terminal design and operation. Our work lays the foundation for future studies on the design and optimization of LNG terminals.

References:

1. Hasan MMF, Zheng AM, Karimi IA. Minimizing Boil-Off Losses in Liquefied Natural Gas Transportation. Ind Eng Chem Res. 2009; 48(21):9571-80.
2. Chaowei Liu JZ, Qiang Xu, and John L. Gossage. Thermodynamic-Analysis-Based Design and Operation for Boil-Off Gas Flare Minimization at LNG Receiving Terminals. Industrial & Engineering Chemistry Research. 2010; 49:7412â€“20.
3. Park C, Song K, Lee S, Lim Y, Han C. Retrofit design of a boil-off gas handling process in liquefied natural gas receiving terminals. Energy. 2012; 44:69-78.

471890 A Systematic Workflow for the Design of Robust Batch Processes

Abstract:

Batch processing is an important operation mode in the chemical industry, commonly employed in the production of added-value products such as pharmaceuticals, flavor and fragrance chemicals, and pesticides [1]. The design of such processes is facilitated by using dynamic models, which aim to capture the performance of the unit operations used in the production of such products. In general, all the models involve parameters that i) are obtained experimentally and ii) distinguish processes one another, even if the same type of unit operation model is used to describe them. As an example, kinetic parameters to describe the rate of a reaction, or phase equilibrium coefficients for use in separation models, are obtained by conducting appropriate experiments and fitting the data to postulated models, taking into account measurement errors. The reliability of those numerical values will greatly impact the accuracy of the process simulations and therefore the capability of the final process design. For this reason, considering both the workflow followed in the determination of process parameters and their use in simulation and optimization routines is of great importance to develop robust designs for batch processes.

This contribution illustrates a systematic workflow leading to the optimal design of a batch process. The approach combines experimental procedures, parameter estimation, process modeling and simulation, global sensitivity analysis and process optimization . The synthesis of nopol, which involve reaction, filtration and distillation unit operations, is used as a case study to illustrate the application of the proposed workflow. Nopol is an oxygenated compound produced by the liquid-phase reaction between β-pinene and formaldehyde [2]. It can be produced via heterogeneous catalysis using a SnMCM-41 material [3]. The SnMCM-41 catalyzed process is an environmentally-friendly alternative to homogeneous synthesis procedures for nopol production [4], since in the heterogeneous process, catalyst can be reused and the utilization of toxic chemicals is greatly diminished.

As a first step, an experimental study of nopol chemical reaction rate was carried out in order to determine kinetic and adsorption parameters [5], [6]. Furthermore, a process flowsheet comprising the aforementioned stages and including a recycle stream was defined, and mathematical models consisting of both ordinary differential equations and algebraic equations [7]–[9], were developed for each unit operation. These mathematical models were implemented in a Matlab® process simulation. Next, a Monte-Carlo-based global sensitivity analysis [10], [11] was employed to rank the effect of the process variables and parameters on the model outputs. The sensitivity analysis identifies the model parameters that may require more attention at the experimental level, and prioritizes the process variables and thus reduce the complexity of the process optimization, by focusing attention of the variables that have the highest impact on the performance metrics, i.e. the net present value (NPV) and the break-even cost.

References

[1] A. Cybulski, J. A. Moulijn, M. M. Sharma, and R. A. Sheldon, Fine Chemicals Manufacture. Elsevier B.V., 2001.

[2] E. Kirk and D. F. Othmer, Encyclopedia of Chemical Technology, 4th Ed. New York: Wiley - Interscience, 1998.

[3] A. L. Villa, E. Alarcón, and C. Montes de Correa, "Synthesis of nopol over MCM-41 catalysts." Chem. Commun. (Camb)., no. 22, pp. 2654–5, 2002.

[4] J. P. Bain, "Nopol . I. The Reaction of p-Pinene with Formaldehyde," J. Am. Chem. Soc., vol. 68, pp. 638–641, 1946.

[5] A. L. Villa, L. F. Correa, and E. Alarcón, "Kinetics of the nopol synthesis by the Prins reaction over tin impregnated MCM-41 catalyst," Chem. Eng. J., vol. 215–216, pp. 500–507, 2013.

[6] D. Casas-Orozco, E. Alarcón, and A.L. Villa, "Kinetic study of the nopol synthesis by the Prins reaction over tin impregnated MCM-41 catalyst with ethyl acetate as solvent," Fuel, vol. 149, pp. 130–137, 2015.

[7] S. Fogler, Elements of Chemical Reaction Engineering, Fourth Ed. New York: Prentice Hall, 2006.

[8] S. Tarleton and R. Wakeman, Solid-Liquid Separation: Equipment Selection and Design, 1st Ed. Oxford: Elsevier Inc., 2007.

[9] I. M. Mujtaba and S. Macchietto, "Holdup issues in batch distillation-binary mixtures," Chem. Eng. Sci., vol. 53, no. 14, pp. 2519–2530, 1998.

[10] I. M. Sobol, "Global sensitivity indices for nonlinear mathematical models and their Monte Carlo estimates," Math. Comput. Simul., vol. 55, no. 1–3, pp. 271–280, 2001.

[11] A. Saltelli, M. Ratto, S. Tarantola, and F. Campolongo, "Sensitivity analysis for chemical models," Chem. Rev., vol. 105, no. 7, pp. 2811–2827, 2005.

METABOLIC NETWORK DESIGN FOR ETHANOL PRODUCTION BY
Synechocystis sp. PCC 6803

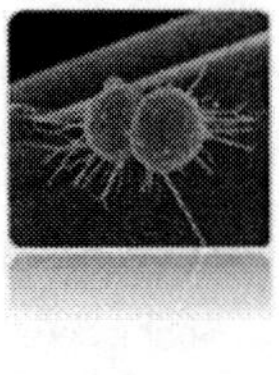

R. Lasry Testa, C. Delpino,
V. Estrada, M. Soledad Diaz

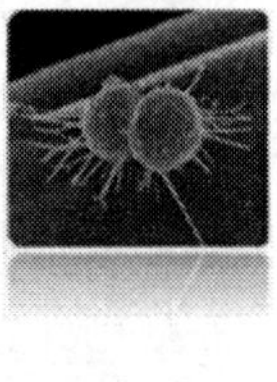

Planta Piloto de Ingeniería Química (PLAPIQUI), Universidad Nacional del Sur-CONICET
Bahía Blanca, Argentina

PLAPIQUI CONICET U N S

16AIChE Annual Meeting, San Francisco, CA

GENETIC ENGINEERED *SYNECHOCYSTIS SP.* PCC 6803

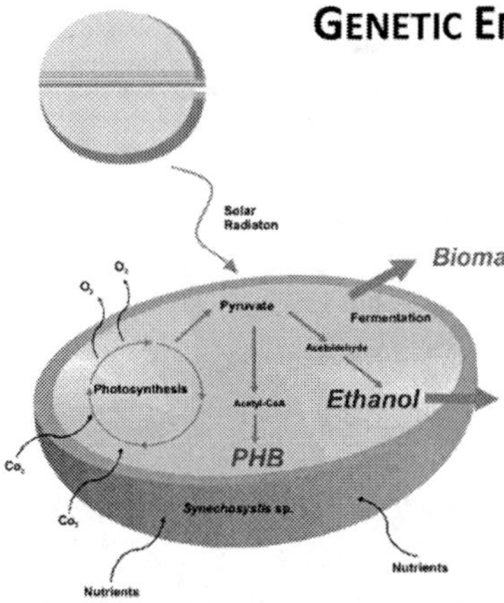

- Photoautotrophic growth
- Minimize carbon source cost
- Reduced greenhouse gases emissions
- Do not compromise food and products derived from terrestrial crops
- First photosynthetic organism completely sequenced
- Naturally competent to take up exogenous DNA
- Oxygenic photosynthesis
- Photosynthetic efficiency greater than higher plants
- High density cell cultures

Challenge

- Increase cell and process yield

Pyruvate decarboxilase (*pdc*)

$$PYR \xrightarrow[PDC]{} ACAL + CO_2$$

Zymomonas mobilis

Alcohol dehydrogenase B (*adh*)

$$ACAL + NADH \xleftrightarrow[ADHII]{} ETH + NAD$$

Under control of gen ***petE*** promoter, Cu-activated

Stoichiometric matrix S: 526x613
- 526 metabolites
- 613 reactions
  - 161 reversible
  - 452 irreversible

Extended model based on (Knoop et al., 2013)

Mathematical Formulation

Bilevel optimization problem

$$
\begin{aligned}
&\text{maximize } v_{ethanol}\\
&s.t.\\
&\quad \text{maximize } v_{biomass}\\
&\quad s.t.\\
&\qquad \sum S_{ij}v_j = 0 \qquad \forall i \epsilon met\\
&\qquad LB_j y_j \leq v_j \leq UB_j y_j \qquad \forall j \in rxn\\
&\quad \sum(1-y_j) = \kappa\\
&y_j \in \{0,1\},\ v_j \in \Re \qquad \forall j \in rxn
\end{aligned}
$$

Single level optimization problem

$$
\begin{aligned}
&\text{maximize } v_{ethanol}\\
&s.t.\\
&\quad \sum S_{ij}v_j = 0 \qquad \forall i \epsilon met\\
&\quad LB_j y_j \leq v_j \leq UB_j y_j \qquad \forall j \in rxn\\
&\quad \sum \lambda_i S_{i,biomass} + \mu_{biomass}^{UB} - \mu_{biomass}^{LB} = 1\\
&\quad \sum \lambda_i S_{i,j} + \mu_j^{UB} - \mu_j^{LB} = 0 \qquad \forall j \in rxn - \{bio\\
(*)&\quad v_{biomass} = \Sigma UB_j y_j \mu_j^{UB} - \Sigma LB_j y_j \mu_j^{LB} \qquad \forall j \in rxn\\
&\quad \sum(1-y_j) = \kappa\\
&y_j \in \{0,1\},\ v_j \in \mathbb{R},\ j \in rxn,\ i \in met,\ \lambda_i \in \mathbb{R},\ \mu_j \in \mathbb{R}^+
\end{aligned}
$$

$$
(*)\quad
\begin{aligned}
&v_{biomass} = \sum UB_j \alpha^{UB} - \sum LB_j \alpha^{LB}\\
&\alpha^{UB} \leq M y_j\\
&\alpha^{UB} \geq 0;\ \alpha^{UB} \geq \mu_j^{UB} - M(1-y_j)\\
&\alpha^{LB} \leq M y_j\\
&\alpha^{LB} \geq 0;\ \alpha^{LB} \geq \mu_j^{LB} - M(1-y_j)
\end{aligned}
$$

Photoautotrophic growth

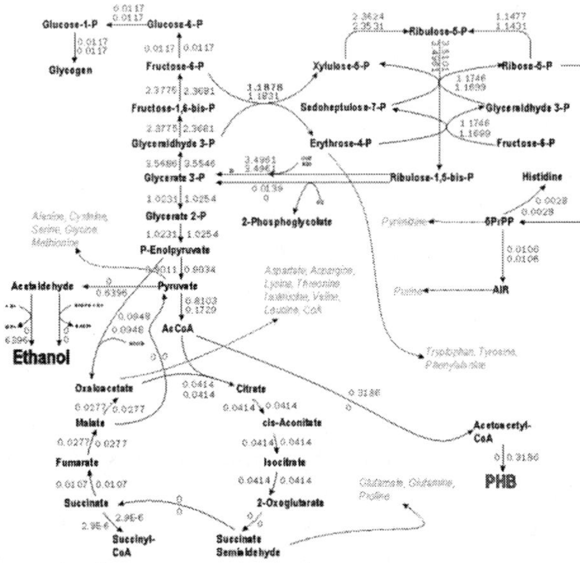

LP (FBA): Max Biomass Production

MILP: Max Eth Prod

Mixotrophic growth

Fluxes [mmol gDW⁻¹ h⁻¹]		Case A	Case B
Biomass		0.033	0.042
Ethanol		0.697	1.094
RubisCo	Biomass	3.496	3.565
	Ethanol	3.496	3.565
Photons	Biomass	60	60
	Ethanol	60	36.645
Glutamate	Biomass	0	0.372
	Ethanol	0	0.372
No_3^-	Biomass	0,3	0
	Ethanol	0,3	0
Co_2	Biomass	3.7	3.7
	Ethanol	3.7	3.7
P_i	Biomass	0.024	0.03
	Ethanol	0.024	0.03

Case A: Photoautotrophic

Case B: L-Glutamate in culture

	Biomass [mmol gDW⁻¹ h⁻¹]	Ethanol [mmol gDW⁻¹ h⁻¹]
L-Arginine	0.041405	0.730936
L-Glutamate	0.041405	1.094434
L-Glutamine	0.041405	0.862182
L-Histidine	0.034412	0.687907

Biomass & Ethanol Prod: Different Aminoacids

Conclusions

- Photoautotrophic ethanol production from *Synecocystis* sp
- Bilevel optimization problem to maximize ethanol production (max biomass)
- Reformulation to single level MILP
- Glutamate in the culture medium improves ethanol prod in 36%

Current work

- Biomass production involves PHB production

Simultaneous Optimization of Heat Exchanger Network Synthesis and Heat-exchanger Design Using Genetic/Simulated Annealing Algorithm

Wu Xiao[a], *Kaifeng Wang*[a], *Xiaobin Jiang*[a], *Xiangcun Li*[a], *Xuemei Wu*[a], *Gaohong He*[a,*], *Mahmoud M. El-Halwagi*[b,c]

[a] *State Key Laboratory of Fine Chemicals, Dalian University of Technology, Dalian, Liaoning, China*
[b] *Chemical Engineering Department, Texas A&M University, College Station, United States*
[c] *Chemical and Materials Engineering Department, King Abdulaziz University, Jeddah, Saudi Arabia*

Introduction

Although many research results about heat exchanger network synthesis (HENS) have been published in the open literatures in the last 50 years, numerous assumptions for simplifying the heat exchanger network, involving the neglect of the impact of heat exchanger fouling and pumping duties, as well as the assumption of constant stream heat-transfer coefficients were still applied. The assumptions will lead to a local or infeasible result,[1] or make the result failed to be applied in the industrial condition.

Recently, great efforts have been made to develop a more-detailed heat exchanger network design to be ensured that the studied results are consistent with industrial conditions.[2, 3] Sun and Luo[4] proposed multi-pass heat exchanger network synthesis considering both parallel and countercurrent flows, and the number of tubes and shells was optimized to minimize the total annual cost using pinch analysis. Ravagnani et al.[5] addressed the issue of detailed design of shell-and-tube heat exchanger in HENS, they pointed out that the actual and the expected area were quite different without detailed design of heat exchangers, and identification and loop breaking algorithm were used to optimize the heat exchanger network. Allen et al.[6] used pinch analysis for the design of heat exchanger network, and a genetic algorithm was used to optimize each heat exchanger of the network.

Mizutani et al.[7] proposed a methodology that the detailed heat-exchanger design model was combined with the network superstructure model developed by Yee and Grossman[8], and a logic-based outer approximation algorithm was used to solve the mathematical model. Similarly, Ravagnani and Caballero[9] considered pressure drop and fouling, as well as the mechanical aspects of the heat exchanger, and the HENS model and the detailed heat-exchanger model were optimized with a decomposition method. However, heuristic rules were used in the decomposition algorithm when terminate condition was determined, which made the results easily get trapped in local solutions.

In this paper, an approach is presented for the simultaneous optimization design of heat exchanger network and each of its heat exchangers, and the heat-transfer coefficients are the relevant parameters of HENS and detailed heat-exchanger design, the design of heat exchanger refers to the standards of the TEMA to guarantee a feasibility solution, the effects of fouling and pressure drop are considered for a more-detailed HENS, and Genetic/Simulated Annealing algorithm (GA/SA) is used to solve the problem.

Problem formulation

The objective is to obtain a lowest annual cost of the heat exchanger network with detailed heat-exchanger design, including the capital cost of heat exchangers, the operating cost of pumps, and the energy cost of utilities. Thus, the problem is to search for the operational conditions of each feasible match in heat exchanger network and the corresponding heat-exchanger design parameters. The given data for cold and hot streams are supply and target temperatures, flow rates, as well as the physical properties (thermal conductivity, heat capacity, viscosity, density and fouling factor), they also conclude cold and hot utility temperatures, and corresponding area and operating cost parameters.

To simplify the mathematical model, some assumptions are made as follows:

(1) Only E-type shell-and-tube heat exchangers without change of phase for HENS.

(2) The detailed design of heat exchangers for cold and hot utilities is not considered.

(3) The overall heat-transfer coefficients for process streams and utilities are assumed as constant parameters.

(4) Thermal and hydraulic design is the main consideration, without the detailed mechanical design.

Mathematics model

A simultaneous optimization model is proposed for HENS and detailed design of corresponding heat exchangers. The superstructure of the network developed by Yee and Grossmann[8] is used to generate possible network configurations, and non-isothermal mixing is considered for HENS in figure 1, simultaneously, the design variables of heat exchangers are assigned within a certain range. To make the heat exchanger network and its heat exchangers be designed at once, the heat exchangers should meet the requirements of HENS, as well as the operational and geometric constraints. The number of tubes is calculated as a continuous variable, so that the area required for each match of the HENS is approximately equal to the actual area of the corresponding heat exchanger, and rounding measures should be done for the number of tubes at the end of the algorithm.

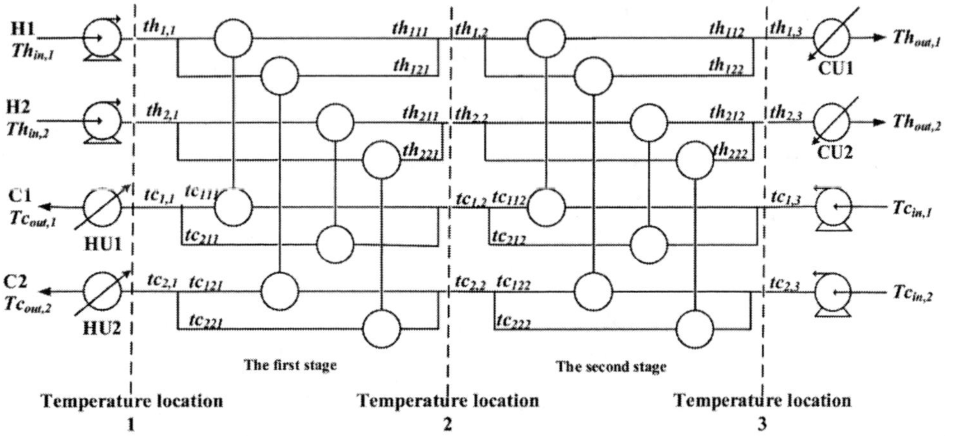

Figure 1. The superstructure of the network considering non-isothermal mixing.

The objective function

The objective function is to minimize the total annual cost (min TAC).

$$\min \mathrm{TAC} = C_{cap} + C_{pump} + C_{uti} \tag{1}$$

Constraints for the model

The area required for each match of the HENS:

$$A_r = \frac{q}{K \cdot Ft \cdot \Delta T} \tag{2}$$

The actual area of the corresponding heat exchanger:

$$A_a = \pi \cdot dout \cdot lt \cdot nt \cdot np \tag{3}$$

$$\left| A_r - A_a \right| < \varepsilon \tag{4}$$

Where the parameter ε is a really tiny amount.

Heat and mass balance in each stage considering non-isothermal mixing:

$$q_{i,j,k} = (th_{i,k} - th_{i,j,k}) \cdot fh_{i,j,k} \tag{5}$$

$$q_{i,j,k} = (tc_{i,j,k} - tc_{j,k+1}) \cdot fc_{i,j,k} \tag{6}$$

$$fh_i = \sum_{i \in N_h} fh_{i,j,k} \tag{7}$$

$$fc_i = \sum_{j \in N_c} fc_{i,j,k} \tag{8}$$

Heat balance for each stream in the HENS:

$$\sum_{k \in N_s} \sum_{i \in N_h} q_{i,j,k} + Q_{cu} = (th_{in,i} - th_{out,i}) \cdot fh_i \tag{9}$$

$$\sum_{k \in N_s} \sum_{j \in N_c} q_{i,j,k} + Q_{hu} = (tc_{out,j} - tc_{in,j}) \cdot fc_j \tag{10}$$

Pressure drop constraints for each stream in the HENS:

$$\Delta P_i = \sum_{k \in N_s} \Delta P_{i,k} \tag{11}$$

$$\Delta P_j = \sum_{k \in N_s} \Delta P_{j,k} \tag{12}$$

Stream splitting is considered in the superstructure, and the pressure drop of the match in parallel for each stage is referred to the method used by Zhu and Nie[10], maximum pressure drop of stream branches is taken for the stream in each stage.

$$\Delta P_{i,k} - \Delta P_{i,j,k} \geq 0 \tag{13}$$

$$\Delta P_{j,k} - \Delta P_{j,i,k} \geq 0 \tag{14}$$

$$\Delta P_{\min} \leq \Delta P_{i,k} \leq \Delta P_{\max} \tag{15}$$

$$\Delta P_{\min} \leq \Delta P_{j,k} \leq \Delta P_{\max} \tag{16}$$

Geometric constraints for heat exchangers

(1) Tube internal and external diameters are based on the standards of the TEMA.

(2) The number of baffles:

$$nb_{\min} \leq nb \leq nb_{\max} \tag{17}$$

(3) Tube arrangement pattern: square and parallel triangular.

(4) Four head types are considered when the inside shell diameter is calculated: Fixed tubesheet head (M), Outside packed floating head (P), Split-ring floating head (S), Pull-through floating head (T).

(5) The number of tube passes: 1, 2, 4, 6, and 8.

Bell-Delaware method is used to calculate the pressure drop and the heat-transfer coefficients for the shell-side fluid, and the Sieder-Tate correlations are used for tube-side heat-transfer coefficients, and a more detailed description about Bell-Delaware method can be referred to in the literatures[23, 24].

Application within the GA/SA

Premature convergence is a drawback for genetic algorithm (GA) so that it is possible to get trapped in local solutions, and simulated annealing algorithm (SA) can guarantee a global optimum solution by asymptotic convergence. GA/SA was presented by Yu et al.[11], the advantages that the GA/SA have good robustness and fast calculation speed make it possible to search for a global optimum, and the algorithm was successfully applied for HENS.[12] In the paper, GA/SA was used to solve the MINLP model. The real-value coding is used in the algorithm, and 11 design variables are the genes in the GA/SA, which can be denoted as ($q_{i,j,k}$, $fh_{i,j,k}$, $fc_{i,j,k}$, $din_{i,j,k}$, $dout_{i,j,k}$, $nb_{i,j,k}$, $nt_{i,j,k}$, $fa_{i,j,k}$, $ht_{i,j,k}$, $lt_{i,j,k}$, $np_{i,j,k}$).

Case and discussions

The case taken from Mizutani et al[7] is used to demonstrate the application of the model and the algorithm. The data, which include two cold streams, two hot streams, one cold utility and one hot utility, are shown in table 1, the parameters of the area cost and the pumping cost are set as follows: $a = 1000$, $b = 60$, $c = 0.6$, $\alpha = 0.7$.

Table 1. Data for case 1.

Stream	m (kg/s)	T_{in} (K)	T_{out} (K)
H1	8.15	368	348
H2	81.5	353	348
C1	16.3	303	363
C2	20.4	333	343
CW		300	320
S		500	500

The physical properties of the process streams are shown in Table 2, and they are assumed as constant parameters which are temperature independent. The overall heat-transfer coefficients for process streams and utilities are assumed as 444 W/ ($m^2 \cdot K$) and the minimum temperature difference in the heat exchanger network is ΔT_{min} equal to 10 K. The capital cost of each heat exchanger is estimated using the following equation, area cost= $a + b \cdot A^c$ \$/yr. The operation cost of the pumps are calculated using an equation, pumping cost= $\alpha \cdot (\Delta P_t \cdot m_t / \rho_t + \Delta P_s \cdot m_s / \rho_s)$. The energy cost of cold utilities is taken as 6 \$/kW·yr and 60 \$/kW·yr is for hot utilities.

The network obtained with the simultaneous optimization of HENS and heat-exchanger design is shown in Figure 2. The detailed designs of two heat exchangers are given in Table 3, and Table 4 presents detailed comparisons of results obtained in this work with those presented respectively by Mizutani[7] and Serna[15].

Table 2. Physical properties for process streams of case 1.

Item	**Value**	**Item**	**Value**
μ (kg/m·s)	2.4×10^{-4}	κ (W/m·K)	0.114
ρ (kg/m^2)	634	r_d (m^2·K/W)	1.7×10^{-4}
Cp (J/kg·K)	2454		

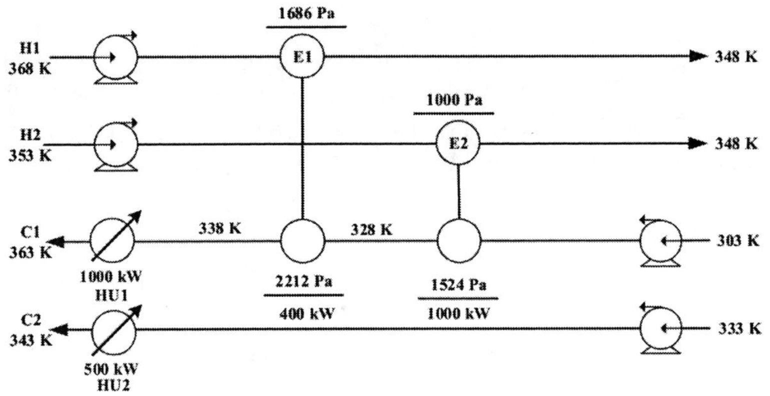

Figure 2. Optimal network for case 1.

Table 3. Detailed designs of heat exchangers for case 1

Parameter	E1	E2	Parameter	E1	E2
din (mm)	8.11	17.27	*fc* (kW/K)	40	40
dout (mm)	9.53	19.05	*q* (kW)	400	1000
nb	3	3	*Ft*	1	1
nt	536	653	h_t (W/m^2·K)	1803.03	1431.87
fa (hot fluid allocation)	shell	tubes	h_s (W/m^2·K)	1540.75	1180.89
ht	M	M	K (W/m^2·K)	593.15	501.66
ta	square	square	ΔP_t (Pa)	2212	1000
lt	1.7	1.5	ΔP_s (Pa)	1686	1524
np	1	1	Area (m^2)	27.34	58.58
fh (kW/K)	20	200			

Table 4. Results comparison for case 1.

Item	Mizutani et al.[7]	Serna et al.[15]	this work
Utility cost ($/yr)	90000	90037.26	90000
Capital cost for exchangers ($/yr)	5608	5500.60	5624.56
Total pumping cost ($/yr)	244	268.55	172.43
Total annual cost ($/yr)	95852	95806.42	95796.99

Utility cost obtained in this work is same with that reported by Mizutani, which is 0.04% lower than that acquired by Serna. As discussed by Serna, utility cost is accounted for the dominating cost, and we can notice that total annual cost obtained in this work is 95796.99, which is only 0.06% and 0.01% lower than that reported by Mizutani and Serna. The operational conditions and the geometric parameters of each heat exchanger are optimized simultaneously, the allocation of heat load and heat capacity flow rate of streams for the heat exchanger network influence the choice of heat-exchanger geometric parameters, at the same time, the differences in heat exchanger geometric parameters lead to different overall heat-transfer coefficients and different pressure drop, and then the capital cost for exchangers and pumping cost can be balanced. The optimized geometric parameters are used in the heat exchanger network, which make the capital cost for exchangers obtained in this work 0.29% and 2.2% higher than that obtained by Mizutan and Serna respectively, and the algorithm used in this work decreases the pumping cost by 41.51% and 55.74% as compared with Mizutani and Serna.

The simultaneous optimization models were solved with C++ based on GA/SA, the results were obtained in a personal computer with AMD Phenom FX-5000, 2.20GHz. The computing time of the case 1 was 10 minutes.

Conclusions

This paper has presented an approach that simultaneously optimizes heat exchanger network synthesis and detailed design of heat exchangers. Inconstant overall heat-transfer coefficients are the interconnecting parameters, which combine heat exchanger network synthesis and detailed design of each heat exchanger. GA/SA is used to solve the simultaneous optimization problem. An example is used to illustrate the availability of the proposed model and algorithm. The results showed the method can be better to balance the capital cost for exchangers and pumping cost, as well as utility cost, and therefore by searching global optimal solution it can offer options on obtaining main structure parameters of each heat exchanger with minimum total annual cost.

Acknowledge

We grateful thank the financial support from the Project of China Scholarship Council (201506060258); National Natural Science Foundation of China (21206014, 21306017); Support Project of the China Petroleum and Chemical Corporation (X514001); Program for Changjiang Scholars (T2012049); Fundamental Research Funds for the Central Universities (DUT16TD19)

Formula Symbols

A = heat exchanger area	q = heat load
A_a = actual area of the corresponding heat exchanger	r_d = deposit resistance
A_r = area required for each match of the HENS	S= split-ring floating head
a = area cost constant	T= pull-through floating head
b = area cost constant	TAC= the total annual cost
C_{cap} = the capital cost of heat exchangers	Ti = iterative times
C_{pump} = the operating cost of pumps	tc_{in} = inlet temperature of cold stream
C_{uti} = the energy cost of utilities	tc_{out} = inlet temperature of cold stream
Cp = fluid heat capacity	$th_{i,k}$ = temperature of hot stream
c = area cost constant	th_{in} = inlet temperature of hot stream
din = tube internal diameters	th_{out} = outlet temperature of hot stream
$dout$ = tube external diameters	
fa = fluid allocation	Greek Symbols
fc = heat capacity flow rate of cold stream	ΔP = pressure drop
fh = heat capacity flow rate of hot stream	ΔT = log-mean temperature difference
Ft = temperature correction factor	α = power cost constant
h = stream heat-transfer coefficient	ρ = density
ht = head type	μ = fluid viscosity
K = overall heat-transfer coefficient	κ = thermal conductivity
lt = tube length	
M= fixed tubesheet head	Subscripts
m = stream flow rate	cu = cold utility
nb = number of baffles	hu = hot utility
nt = number of tubes	i = hot stream
nt_l = lower limitations on the number of tubes	j = cold stream
nt_u = upper limitations on the number of tubes	k = index for stage
np = number of tube passes	max= maximum value
ta = tube arrangement pattern	min= minimum value
P= outside packed floating head	s = shell side
Q_{cu} = heat load of cold utility	t =tube side
Q_{hu} = heat load of hot utility	

References

1. Ponce-Ortega, J. M.; Serna-Gonzalez, M. (2007); Jimenez-Gutierrez, A., Heat exchanger network synthesis including detailed heat exchanger design using genetic algorithms. *Industrial & Engineering Chemistry Research*, 46, (25), 8767-8780.
2. Klemeš, J. J.; Varbanov, P. S.; Kravanja, Z. (2013), Recent developments in process integration. *Chemical Engineering Research and Design*, 91, (10), 2037-2053.
3. Jin, Z.; Qiwu, D.; Minshan, L. (2008), Heat exchanger network synthesis with detailed heat exchanger design. *Chemical engineering & technology*, 31, (7), 1046-1050.

4. Sun, L.; Luo, X.; Zhao, Y. (2015), Synthesis of multipass heat exchanger network with the optimal number of shells and tubes based on pinch technology. *Chemical Engineering Research and Design*, 93, 185-193.
5. Ravagnani, M.; Da Silva, A.; Andrade, A. (2003), Detailed equipment design in heat exchanger networks synthesis and optimisation. *Applied Thermal Engineering*, 23, (2), 141-151.
6. Allen, B.; Savard-Goguen, M.; Gosselin, L. (2009), Optimizing heat exchanger networks with genetic algorithms for designing each heat exchanger including condensers. *Applied Thermal Engineering*, 29, (16), 3437-3444.
7. Mizutani, F. T.; Pessoa, F. L.; Queiroz, E. M.; Hauan, S.; Grossmann, I. E. (2003), Mathematical programming model for heat-exchanger network synthesis including detailed heat-exchanger designs. 2. Network synthesis. *Industrial & engineering chemistry research*, 42, (17), 4019-4027.
8. Yee, T. F.; Grossmann, I. E. (1990), Simultaneous optimization models for heat integration—II. Heat exchanger network synthesis. *Computers & Chemical Engineering*, 14, (10), 1165-1184.
9. Ravagnani, M.; Caballero, J. A. (2007), Optimal heat exchanger network synthesis with the detailed heat transfer equipment design. *Computers & Chemical Engineering*, 31, (11), 1432-1448.
10. Zhu, X.; Nie, X. (2002), Pressure drop considerations for heat exchanger network grassroots design. *Computers & chemical engineering*, 26, (12), 1661-1676.
11. Mizutani, F. T.; Pessoa, F. L.; Queiroz, E. M.; Hauan, S.; Grossmann, I. E. (2003), Mathematical programming model for heat-exchanger network synthesis including detailed heat-exchanger designs. 1. Shell-and-tube heat-exchanger design. *Industrial & engineering chemistry research*, 42, (17), 4009-4018.
12. Serna-González, M.; Ponce-Ortega, J. M.; Castro-Montoya, A. J.; Jiménez-Gutiérrez, A. (2007), Feasible design space for shell-and-tube heat exchangers using the Bell-Delaware method. *Industrial & engineering chemistry research*, 46, (1), 143-155.
13. Yu, H.; Fang, H.; Yao, P.; Yuan, Y. (2000), A combined genetic algorithm/simulated annealing algorithm for large scale system energy integration. *Computers & Chemical Engineering*, 24, (8), 2023-2035.
14. Wei, G.; Yao, P.; Luo, X.; Roetzed, W. (2004), Study on multi-stream heat exchanger network synthesis with parallel genetic/simulated annealing algorithm. *Chinese Journal of Chemical Engineering*, 12, (1), 66-77.
15. Serna-González, M.; Ponce-Ortega, J. M. (2011), Total cost target for heat exchanger networks considering simultaneously pumping power and area effects. *Applied Thermal Engineering*, 31, (11), 1964-1975.

472490 Production of Bioethanol from Lignocellulosic Biomass in Mexico: Evaluation on Technical Potential and the Analysis of Two Biorefinery Configurations

Abstract:

As residual crop, lignocellulosic biomass is an important resource for biofuel production. Mexico occupies the third place in terms of cropland in Latinamerica (Valdez-Vazquez et al., 2010). Agricultural residues, in particular, become an important source for biofuel production.

The agricultural residue approach avoids high capital and operating costs as it could be considering other possibilities. In Mexico, the most important cultivars are wheat, maize, sorghum and barley.

In the first part of the work, an historical data analysis through the last ten years (2005-2014) of the agricultural residue production in Mexico was carried out. The amount of residues per cultivar per year was estimated and the theoretical, sustainable and technical potentials were calculated per year and as total through the ten years. As a result, the most productive residues regions are identified and, therefore, the specific locations for bioethanol production are suggested.

The development of new engineering processes and appropriate technologies represent a wide field of opportunities. Sánchez et al. (2013) reported that 500 ton/day is the minimal plant capacity to guarantee profitability for producing bioethanol in a biorefinery from agroresidues.

Thus, in the second part of this study the real production of residual crops in terms of the yield (residue-to-product ratio, total cultivated area and high heating value) was determined. Moreover, the evaluation of biomass processing (e.g. bioethanol production) ought to be revised with the central objective of minimize the energy consumption.

In this context, we present the analysis of several residual crops using two biorefiney configurations: conventional biochemical platform (CBP) (Sánchez et al., 2013) that use fungal polysaccharides hydrolase enzymes and consolidated bioprocess (CB) (Lynd et al., 2005) that use bacteria that possess an enzyme to hydrolyze biomass. Both processes transform the resulting sugars in bioethanol; nevertheless, CBP needs to produce enzymes in a dedicated reactor and CB produces enzyme in the same reactor involving. Finally, a thermodynamic analysis in exergy terms has been developed for both schemes in order to compare the efficiency of energy use (Marmolejo-Correa and Molina, 2015; Marmolejo-Correa et al., 2016).

Keywords:

Lynd, L.R., van Zyl, W.H., McBride, J.E., Laser, M. (2005). Consolidated Bioprocessing of Cellulosic Biomass: An Update. Current Opinion in Biotechnology, 16, 577–583.

Marmolejo-Correa, D., Molina, C. (2015). Conceptual Design of Bio-Fuels Production by Using Second Law Analyses. In the AIChE Annual Meeting. Salt Lake City, USA.

Marmolejo-Correa, D., Hernández-Hoyos, M., Molina, C., Sánchez, A., Cárdenas-Guerra, J.C. (2016). Advances on Exergy Targeting of a Biofuel Plant. In 29th International Conference on Efficiency, Cost, Optimisation, Simulation and Environmental Impact of Energy Systems (Accepted). Portorož, Slovenia.

Efficiency of 2G Enzymatic Ethanol Production. Fuel, 113, 165-179.

Valdez-Vazquez, I., Acevedo-Benítez, J.A., Hernández-Santiago, C. (2010). Distribution and Potential of Bioenergy Resources from Agricultural Activities in Mexico. Renewable and Sustainable Energy Reviews, 14, 2147-2153.

472888 Dynamic Optimization Model for a Heat and Power Plant

Abstract:

Industrial facilities are always seeking the optimal process configuration. Whether it is introduction of new process equipment or the load management on existing ones, optimization guides the decision making process. In this work, we are considering an existing facility with a goal to minimize an objective function (e.g. economic, environmental, etc.). This objective function is subject to seasonally varying process demands and restrictive operational constraints with non-linearity, often, adding a degree of difficulty to the problem. This case study focuses on an existing heat and power plant that generates steam and electricity to meet the process demands of a crude processing facility. The existence of a global minima will be contemplated, and a dynamic model will be developed, which will optimize the selection of steam and power generators and the loads on which they operate based on changing process demands.

473014 Selection of the Optimal Operating Conditions in Fired Heaters Based on Two Phase CFD Simulations

Abstract:

Fired heaters are widely used in the chemical industry. Helical coils configuration is preferred over others due to the high heat transfer area that can be accommodated in a relative small volume.

The effective gas temperature is usually calculated assuming complete flue gas mixing (Lobo-Evans, 1939) in the firebox. In other words, they ignored the existence of a longitudinal and transverse temperature gradient effects. A numerical method is usually used in order to determine the amount of heat in the radiant section and the effective gas temperature (Chapra Steven, 2005). For those cases where this assumption is not valid (e.g. high temperature heaters with tall narrow fireboxes), the radiant section is divided into zones for the energy balance calculations.

One of the most typical issues faced in this type of equipment is the short lifetime if the operating conditions are not adequate for the two-phase flow developed inside coils. In this work, a CFD study is carried out in order to provide the best configuration of the combustion chamber, specially closed to the location where the nucleate boiling point occurs. For this evaluation, different diameters of piping in the radiation zone will be studied. In addition, different insulating materials, usually placed in the inner wall of the vessel, will be used in order to minimize the radiation over the typical affected zones.

Chapra Steven C, Applied Numerical Methods with MATLAB for Engineers and Scientists, McGraw-Hill Companies, Inc, 2005. 1st edition.

W.E. Lobo, J.E. Evans, "Heat transfer in the radiant section of petroleum heaters", Trans AIChE , 743â€"778. 1939

462690 Monte-Carlo-Simulation-Based Optimization Methods for Copolymerization Process

Abstract:

Molecular weight distribution (MWD) and chemical composition distribution (CCD) are two most important microstructural indices for many copolymers [1]. They can provide detailed information about the end-use properties of the polymer. One method to obtain the microstructural information of polymer molecules, which is difficult to determine via traditional equation-based methods, is through Monte Carlo (MC) simulation [1]. This method is essentially "math-free" and do not need to solve complex differential-algebraic equations. However, the MC method could be very slow due to the fact that a large number of chains must be simulated to generate accurate results. On the other hand, when it is applied to optimization, the MC method results in the challenge that accurate derivatives are not available due to the stochastic nature of the MC simulation.

Therefore, an efficient derivative-free optimization method using trust region concepts is developed in this project. Along with the trust region method, a series of local surrogate models are systematically constructed to approximate the original system described by the embedded MC simulation [2]. Moreover, during the implementation of this trust region method, the balance between the computational time cost and the accuracy of the MC simulation should be well dealt with, as a larger molecular chain number can improve the simulation accuracy, but at the same time slow down the process. Under the framework of parallel MC method [3], this study also proposes a strategy with adaptive molecular chain numbers for MC simulations to reduce the computational time cost. On the basis of the stochastic feature in MC simulation, the violations of the microstructural quality constraints and the simulation errors of specific simulated chains are taken into account to determine the appropriate number of simulated chains at each iteration. Finally, an example of a ternary polymerization process is presented to illustrate the efficiency and convergence of the proposed methods.

Keywords: Monte Carlo simulation, Chemical composition distribution, Optimization

Reference:

[1] Soares, J. B. P., McKenna, T. F. L. (2012). Polyolefin Reaction Engineering. Weinheim: Wiley-vch Verlag GmbH & Co. KGaA

[2] Eason, J. P., Biegler, L. T. (2015). Reduced model trust region methods for embedding complex simulations in optimization problems. *Computer Aided Chemical Engineering*, 37, 773-778.

[3] Weng, J., Chen, X., Yao, Z., Biegler, L. T. (2015). Parallel Monte Carlo Simulation of Molecular Weight Distribution and Chemical Composition Distribution for Copolymerization on a Graphics Processing Unit Platform. *Macromolecular Theory & Simulations*, 24(5), 521–536.

468902 Dynamic Discrepancy Reduced Modeling: Overview and Applications

Abstract:

Dynamic discrepancy reduced modeling [1]-[2] is a "grey box" method of reduced order modeling, uncertainty quantification and propagation. The fundamental idea is a dynamic network reduction, which obtains a system of drastically reduced order using a projection of a high-dimensionality system onto a low-dimensionality space. The reduced models retain the first principles structure of the high-order system. Variability lost in the order reduction is replaced by application of Gaussian process (GP) stochastic functions within the reduced model form. The reduced models acquire information from the high-fidelity antecedent and/or experimental data through Bayesian calibration [3]-[4], a process which results in a stochastic version of the reduced model form that replicates the behavior of the high-fidelity / experimental system in the training set while providing an estimate of the uncertainty encountered due to model interpolation or extrapolation through a space of time-dependent functional inputs.

The method will be presented in general mathematical terms, to include a rigorous definition of extrapolation. Examples will then be discussed in steam reformation of methane along with a pair of examples in carbon capture. Emphasis will be placed on an application of the method to a solid sorbent carbon capture system, utilizing a sophisticated reaction-diffusion model of the sorbent [5]. A dynamic discrepancy version of the chemical adsorption model was implemented and calibrated to experimental data. The reduced model was then incorporated into a process-scale model of an adsorber.

[1] K.S. Bhat, D.S. Mebane, C.B. Storlie and P. Mahapatra, "Upscaling uncertainty with dynamic discrepancy for a multi-scale carbon capture system," arXiv:1411.2578, 2014.

[2] D.C. Miller, M. Syamlal, D.S. Mebane, C.B. Storlie, D. Bhattacharyya, N.V. Sahinidis, D. Agarwal, C. Tong, S.E. Zitney, A. Sarkar, S. Sundaresan, E. Ryan, D. Engel and C. Dale, "Carbon capture simulation initiative: A case study in multiscale modeling and new challenges," Ann. Rev. Chem. Biomol. Eng. 5 (2014) 301.

[3] D.S. Mebane, K.S. Bhat, J.D. Kress, D.J. Fauth, M.L. Gray, A. Lee and D.C. Miller, "Bayesian calibration of thermodynamic models for the uptake of CO2 in supported amine sorbents using ab initio priors," Phys. Chem. Chem. Phys. 15 (2013) 4355.

[4] J. Kalyanaraman, Y. Fan, Y. Labreche, R.P. Lively, Y. Kawajiri and M.J. Realff, "Bayesian estimation of parametric uncertainties, quantification and reduction using optimal design of experiments for CO2 adsorption on amine sorbents," Comput. Chem. Eng. 81 (2015) 376.

[5] D.S. Mebane, J.D. Kress, C.B. Storlie, D.J. Fauth, M.L. Gray and K. Li, "Transport, zwitterions and the role of water for CO2 adsorption in mesoporous silica-supported amine sorbents," J. Phys. Chem. C 117 (2013) 26617.

466166 A Heterogeneous High Performance Computing Implementation of Thin Film Growth Simulation

Abstract:

Thin film growth processes are one of the most important types of nanofabrication processes existing many industrial applications such as semiconductor manufacturing, solar cell fabrication, and nanomaterial coating, etc. There are quite a number of methods to simulate various kind of thin film growth, among which, the kinetic Monte Carlo (kMC) method turned out to be one very promising technique for its high efficiency in computation, high fidelity of process description, and high flexibility to accommodate complex process mechanisms.

However, because of the serial execution nature of kMC algorithm, it is quite challenging to leverage the high performance computing (HPC) power available nowadays in kMC computation since most HPC capability come from parallelization. Furthermore, as HPC architecture moving from homogeneous (all CPU computation) to heterogeneous (CPU + GPU or CPU + MIC/Intel Many Integrated Core Architecture[1]), it has become even less straight forward to efficiently implement kMC algorithm on HPC.

In this work, we carefully investigated the implementation of kMC algorithm on a heterogeneous computing platform which consists of two Xeon CPUs and four Xeon Phi many integrated core coprocessors. Our parallel algorithm follows the synchronous method proposed by Martínez et al.[2] of which time synchronization is perfect but computation is semi rigorous with boundary conflicts. We simulate a thin film growth process using a solid-on-solid lattice model with surface adsorption, migration, and desorption. We study the performance of the algorithm in terms of speedup factor, boundary error in different Xeon Phi execution mode (offload, native and symmetrical), as well as its relationship to simulation parameters such as lattice size, microscopic event activation energy level, etc.. Recommendations on implementation strategy is provided finally together with methods to minimize boundary conflicts.

[1] R. Rezaur, Intel Xeon Phi Architecture and Tools, Apress, 2013.

[2] E. MartÃnez, J. Marian, M.H. Kalos, J.M. Perlado, Synchronous parallel kinetic Monte Carlo for

continuum diffusion-reaction systems, Journal of Computational Physics, 2008, 227, pp3804â€"3823

460467 Multiscale Computational Fluid Dynamics: Methodology and Application to Film Microstructure Control in PECVD

Abstract:

Plasma enhanced chemical vapor deposition (PECVD) remains a popular technique for the manufacture of silicon thin films used in the microelectronics and photovoltaic industries [1]-[2]. Over the past decade, there have been significant strides in the multiscale modeling of the silicon deposition process in an effort to better understand and improve product manufacturing [3]-[4]. Nonetheless, these models typically rely on simplified Navier-Stokes and continuity equations and therefore cannot accurately predict the composition and dynamics of the deposition species of interest throughout the PECVD reactor gas-phase. Integrating computational fluid dynamics modeling with microscopic simulation techniques within a parallel computation environment is an important research challenge.

Motivated by the above considerations, a novel, multiscale modeling strategy is developed which allows for the concurrent simulation of both the macroscopic gas phase and the microscopic surface interactions that contribute to thin film growth, and is applied to a common manufacturing problem of drift in the thin film product thickness caused by fouling during the conditioning phase of reactor operation [5]. In a previous publication, a detailed kinetic Monte Carlo (kMC) algorithm was developed which has been shown to accurately reproduce thin film growth rates and surface morphologies at four distinct locations across the wafer surface [3]. In this work, we propose the addition of a computational fluid dynamics (CFD) model that can account for the complex geometry of the PECVD reactor allowing for the prediction of accurate plasma flow fields and chemistry which are required within the microscopic kMC domain. Furthermore, parallel computation is introduced through the use of the message passing interface (MPI) framework, which has made possible the simultaneous simulation of numerous reactor zones with far increased lattice dimensions. Specifically, the parallel-plate PECVD reactor of interest in this work is divided into ten discrete, radial zones, each of which interacts with both of their neighboring zones, as well as the process gas (i.e., plasma) flowing over the wafer surface. Although open-loop simulations reveal significant variability in the film thickness across the wafer surface and between successive batch deposition sequences [6], the multiscale model developed in this work demonstrates that through the use of a carefully designed exponentially weighted moving average (EWMA) algorithm, the offset in the product thickness can be reduced to <1% within 10 batches of reactor operation. Additionally, the run-to-run control strategy is shown to recover the desired product thickness (i.e., return to the thin film thickness set-point of 300 nm) in the presence of disturbances in the plasma quality caused by fluctuations in the radio frequency (RF) power and feed gas composition.

[1] Reif R. Plasma enhanced chemical vapor deposition of thin films for microelectronics processing. Noyes Park Ridge. 1990.

[2] Shah A, Torres P, Tscharner R, Wyrsch N, Keppner H. Photovoltaic technology: the case for thin-film solar cells. Science. 1999;285:692-698.

[3] Crose M, Kwon JSI, Nayhouse M, Ni D, Christofides PD. Multiscale modeling and operation of PECVD of thin film solar cells. Chemical Engineering Science. 2015;136:50-61.

[4] Christofides PD and A. Armaou. Control and optimization of multiscale process systems. Computers and Chemical Engineering. 2006;30:1670-1686.

[5] Gabriel O, Kirner S, Klick M, Stannowski B, Schlatmann R. Plasma monitoring and PECVD process control in thin film silicon-based solar cell manufacturing. EPJ Photovoltaics. 2014;55202:1â€"9.

[6] Crose M, Kwon JSI, Tran A, Christofides PD. Multiscale modeling and run-to-run control of PECVD of thin film solar cells. Renewable Energy. 2016; submitted.

468669 Fault Tolerant Computing through Machine Learning

Abstract:

In modern, massively parallel scientific computation, domain decomposition approaches lead to different segments of a domain, and different subfields/equations solved for in each segment, being computed on different processors.
If a processor fails during a "computation era", before information is exchanged between nodes, one is faced with a serious problem about if and how the computation can proceed.

In many cases, the different fields that these processors compute are all functions of some intrinsic lower-dimensional coarse variables (e.g., time during the computation, long-wavelength features of the solution).
If the computational algorithms share some such common information, we can use machine learning, and in particular diffusion maps, a nonlinear manifold learning algorithm, to ``register" the computational data in the coarse space and to ``fill in", to the best of our ability, data that are missing or corrupted because of a processor failure.

This allows us to learn functional relationships between aspects of the data fields that are not common across processors, effectively fusing the data sets.

We demonstrate our approach on two illustrative PDE systems with various spatiotemporal patterns of missing data.

The approach meshes well with equation-free computation schemes, in particular with patch dynamics; beyond helping to partially restore corrupted or missing data, it can help determine the size of the computational domain over which simulations need not be performed, and can help determine processor redundancy for different anticipated failure patterns.

This is joint work with Prof. G. Karniadakis and Dr. Seungjoon Lee at Brown University.

469990 An Information Entropy Based Consistency Index for Evaluating the Performance of Variable Selection Methods

Abstract:

An information entropy based consistency index for evaluating the performance of variable selection methods

Q. Peter He

Department of Chemical Engineering, Auburn University, Auburn, AL 36849, USA

Data-driven soft sensors have been widely used in both academic research and industrial applications for predicting hard-to-measure variables or replacing physical sensors to reduce cost. It has been shown that the performance of these data-driven soft sensors could be greatly improved by selecting only the vital variables that strongly affect the primary variables, rather than using all the available process variables.

In the past few decades, many different variable selection approaches have been reported for various applications with different soft sensor modeling methods. In order to evaluate the performance of different variable selection methods, several performance indices have been proposed in the literature. The most common ones are the average mean absolute percentage error (MAPE), coefficient of determination (R2), and geometric mean of selection sensitivity and specificity (G). Among them, only G directly measures the accuracy of variable selection results, while MAPE and R2 indirectly measure the effects of variable selection through the prediction performance of a soft sensor, such as PLS. However, when the information on the true relevant variables is not available, which is the case for most industrial applications, selection sensitivity and specificity (therefore G) cannot be obtained and there is no direct metric existing for variable selecting in the literature.

We recently reported an entropy based variable selection index to access the variable selection performance, which does not require the ground truth of variable relevance ADDIN \s <Colwiz><citation><biblioref linkend="f20f9a604c9f9c2" citekey="wang2015comparison" /></citation></Colwiz> [1]. The index evaluates the consistency of the variable selection performance, and is termed consistency index. It was shown that the consistency index describe the variable selection performance well for both simulated (with ground truth) and industrial (without ground truth) cases studies. However, the consistency index does not fully agree to a common expectation: the consistency is the lowest when a variable is being selected for a model 50% of the time. Instead, the minimum (the lowest consistency) occurs at probability of 0.3679 (see Figure 1); in addition, the consistency index cannot make use of the ground truth even when it is available.

To address these limitations, in this work we propose a modified consistency index based on information entropy. It has a symmetric response curve as shown in Figure 2, and can be applied to all cases – no matter the information on the true relevant variable is available or not. Simulated and industrial cases studies are provided to compare the performance of the proposed index to the existing indices.

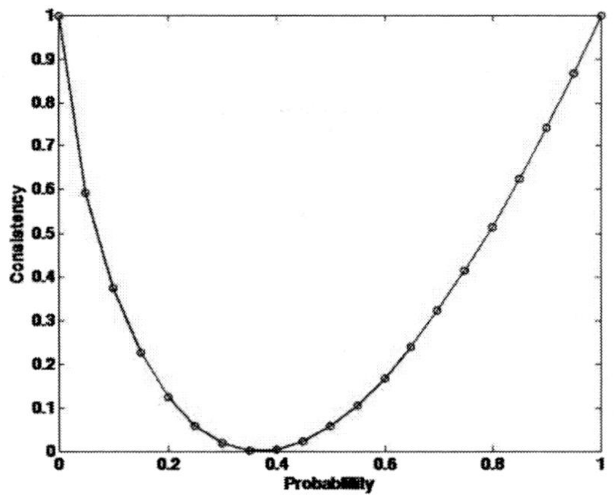

Figure 1. Probability vs. consistency based on [1]

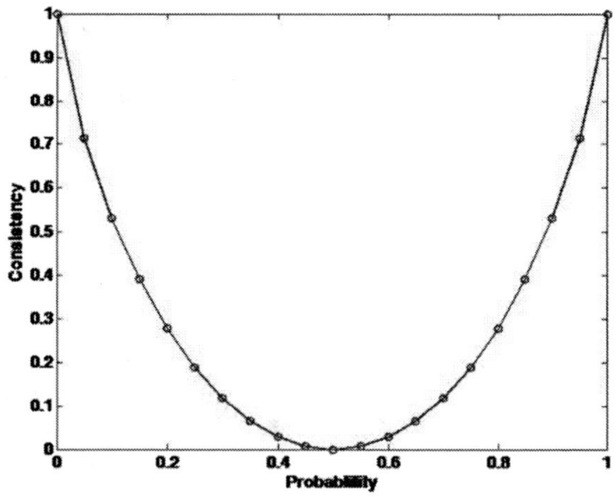

Figure 2. Probability vs. consistency based on the proposed index

References:

ADDIN \s <CWZ.BIB></CWZ.BIB>

[1] Z. X. Wang, Q. P. He, and J. Wang, "Comparison of variable selection methods for PLS-based soft sensor modeling," *J. Process. Control.*, vol. 26, pp. 56–72, 2015.

457352 Parallel Solution of Parameter Estimation Problem for Polymer Models Using Multiple Grade Transition Curves

Abstract:

Parallel solution of parameter estimation problem for polymer models using multiple grade transition curves

Francisco Trespalacios and Thomas Badgwell

ExxonMobil Research & Engineering Company

The dynamic modelling of grade transitions in polymer production typically requires the solution of a large scale differential-algebraic equation (DAE) system. These models involve many parameters of different types, including kinetic constants, material properties, diffusion coefficients, etc. Not only can these parameters differ from laboratory estimates, they can also change with time. For this reason, efficient parameter estimation is essential if such models are to be used in industrial polymerization applications.

One of the approaches for the parameter estimation problem is to solve a least-squares problem. In this problem, the objective is to find the value of the desired parameters that minimizes the difference between the predicted output and real data (or its variant errors-in-variables (EVM) problem, in which both the input and the output are considered as variables). With this approach, solving the parameter estimation problem requires the solution of an optimization problem with DAE constraints. For example, if one data set is available for a transition between two products, the parameter estimation problem solves a least-squares problem subject to the DAE system that defines that grade transition. If there are multiple data sets of multiple transitions, and even multiple models, the DAE system that constraints the optimization problem quickly grows in size.

In the parameter estimation problem, each scenario (which represents a data set and its corresponding DAE system) has a set of local variables and parameters. The only links between different scenarios are the parameters being estimated. Therefore, while the parameter estimation problem can increase to intractable sizes with many scenarios, the problem is sparse. In order to solve such problems, it is necessary to exploit the problem structure.

Two main approaches are typically used to solve this type of problem: the sequential approach and the simultaneous approach. In the sequential approach an optimizer updates the parameters and passes them to a DAE solver, solving each scenario independently. In the simultaneous approach, the parameter optimization and solution of DAE systems is solved together. This is achieved by a full discretization of the differential equations, yielding a large scale NLP problem. Developing solution frameworks for the sequential approach is relatively simple, but it requires solving each DAE system multiple times. The second approach requires the solution of a much larger but sparse problem. The work by Zavala et al. [1] shows the advantages for the simultaneous approach, especially for large scale systems.

In this work, motivated by a real application for Unipol polymerization reactors, we extend the approach suggested by Zavala et al. [1] in two ways. First, we consider a heterogeneous structure in which we allow different reactor models for each sub-problem. Second, we consider much larger problems with several data sets and multiple polymer grade transitions. We solve the problem with multiple cores (one for each scenario), using PIPS-NLP [2]. PIPS-NLP allows the modeler to specify first stage (global parameters) and second stage (local) variables. The solver parallelizes the linear algebra, allowing considerable speed-ups in the large scale dynamic optimization problem.

The application we present in this work is copolymerization of propylene and ethylene, using a bubbling fluidize bed reactor [3,4]. We present the performance of the parallelized solution method for the parameter estimation of this problem. The results show that the problem scales well when using multiple processors. The results also show that it is possible to solve the problem in reasonable time, even with multiple data sets of multiple transitions.

[1]. Zavala, Victor M., Carl D. Laird, and Lorenz T. Biegler. "Interior-point decomposition approaches for parallel solution of large-scale nonlinear parameter estimation problems." Chemical Engineering Science 63.19 (2008): 4834-4845.

[2]. Chiang, Naiyuan, Cosmin G. Petra, and Victor M. Zavala. "Structured nonconvex optimization of large-scale energy systems using PIPS-NLP." Power Systems Computation Conference (PSCC), 2014. IEEE, 2014.

[3]. Choi, Kyu-Yong, and W. Harmon Ray. "The dynamic behaviour of fluidized bed reactors for solid catalysed gas phase olefin polymerization." Chemical Engineering Science 40.12 (1985): 2261-2279.

[4] McAuley, K. B., and J. F. MacGregor. "Optimal grade transitions in a gas phase polyethylene reactor." AIChE Journal 38.10 (1992): 1564-1576.

467610 Design of Multi-Actor Distributed Processing Systems:a Game-Theoretical Approach

Abstract:

The manufacturing of a product could be the result of a value chain involving several processing plants distributed across distinct owners who form a network that exchange materials and/or energy among the involved actors. An example of such networks is given by the petrochemical value-chain which consists of refineries, petrochemical crackers, intermediates processing facilities and manufacturers of consumer products. Another example is given by the design of biorefineries, where conversion of biomass into intermediate platform chemicals and upgrading of these into other products will most probably take place at different corporations.

In our previous work, we proposed a novel strategy for the design of distributed manufacturing networks in which, suppliers and consumers linked through different intermediate products (mass or energy) sought maximization of their own profit. Under this assumption, we formulated a multiobjective optimization problem and proved that a 2-level Lagrangian approach could be used to obtain the optimal solution for the overall network (amount of mass/energy to be exchanged as well as transfer prices) through the optimal designs of the individual actors that participate in the network. This approach, assumed the existence of a centralized coordinator that collects the information from all the actors and thus implicitly knows in detail the form and parametric values of their objective functions and internal constraints.

In this presentation we propose and develop a game-theoretical framework and specific methodologies, which, without a centralized coordinator or share of internal information, allows for the optimal design of distributed processing systems. Within this framework the conditions of optimality established through the previously mentioned 2-level Lagrangian approach are satisfied through a negotiation process between suppliers and consumers which leads to a Nash equilibrium point. As this equilibrium point generally leads to an unequal distribution of the benefits between the actors, a Nash bargaining problem is analyzed to reach to an equitable distribution. The presentation will also discuss the use of penalty-term approaches to extend the framework to problems for which the underlying convexity assumptions of the 2-level Lagrangian approach may not be possible to ascertain. A series of case studies that illustrate the application of the proposed ideas to processing networks of various structures will be presented.

This work was funded by the Cooperative Agreement between the Masdar Institute of Science and Technology (Masdar Institute), Abu Dhabi, UAE and the Massachusetts Institute of Technology (MIT), Cambridge, MA, USA - Reference 02/MI/MI/CP/11/07633/GEN/G/00 for work under the Second Five Year Agreement.

458454 Optimization of Reverse Water-Gas Shift Chemical Looping for Continuous Production of Syngas from CO_2

Abstract:

Syngas is widely used in industry as a building block for the production of many bulk chemicals and chemical intermediates. The production method of syngas strongly influences its composition, most importantly the H_2/CO-ratio. Consequently, the syngas composition determines for which application it is suitable. Some of the most important uses of low H_2/CO syngas include the Monsanto process for acetic acid production and hydroformylation processes. High H_2/CO syngas may be used for alcohol synthesis and methanation. Today, syngas is produced almost exclusively from fossil fuels by steam reforming of natural gas or gasification of coal, depending on the desired H_2/CO-ratio. In light of the current and future global problems associated with the rising CO_2 levels in the atmosphere, traditional processes based on fossil fuels must be substituted to minimize the anthropogenic CO_2 production.

One possible way to shift towards a more sustainable future is to use CO_2 as the main carbon source instead of fossil fuels. In the context of sustainable syngas production, the reverse water-gas shift (RWGS) reaction is of high importance, as it converts CO_2 to CO with the help of process heat and hydrogen. The traditional RWGS reaction can be improved by introducing a metal oxide as an oxygen carrier, splitting the original reaction in a reduction and oxidation reaction on the metal oxide. This cyclic two-step process is referred to as reverse water-gas shift chemical looping (RWGS-CL). It has been shown that by using this approach, efficiency improvements can be obtained because the two-step operation yields partially separated gas streams and thus simplifies downstream gas separation [1]. This is especially true for syngas production with low H_2/CO-ratio. Furthermore, unwanted side reactions can be eliminated by separating the reactants spatially or temporally and favorable thermodynamics can be obtained by choosing suitable oxygen carrier materials. Due to the cyclic switching between two process steps, RWGS-CL is more complex than the traditional RWGS process. A steady state operation in the traditional sense is not possible. However, a cyclic steady state (CSS) is achievable after prolonged cycling. Therefore, simulation and optimization techniques are used to understand the dynamic behavior and to improve the process.

Different reactor designs and configurations for the RWGS-CL process are examined and evaluated. For a fixed bed configuration, a 1D model for the two-step RWGS-CL process is derived based on [2] with kinetic information from [3]. Process simulation is used to compute the cyclic steady state of the process for different flow regimes. On top of the simulation, a multi-objective optimization problem is formulated and solved to maximize the gas conversion and the utilization of the oxygen carrier material. Since both objectives are conflicting they lead to Pareto optimal solutions. The trade-off between gas conversion and the utilization of oxygen carrier material is investigated. The influence of the decision variables and the flow regime on the optimal solution is analyzed and discussed. Constraints on the decision variables are formulated such that continuous production of CO can be achieved with the RWGS-CL process. The simulation and optimization results are compared to those of the traditional RWGS process.

References:

[1] Wenzel, M., Rihko-Struckmann, L. and Sundmacher, K. (2016), Thermodynamic Analysis and Optimization of RWGS Processes for Solar Syngas Production from CO_2. Submitted to AIChE Journal.

[2] Heidebrecht, P., Hertel, C. and Sundmacher, K. (2008), Conceptual Analysis of a Cyclic Water Gas Shift Reactor. International Journal of Chemical Reactor Engineering, 6

[3] Wenzel, M., Dharanipragada, N. V. R. A., Galvita, V. V., Poelman, H., Marin, G. B., Rihko-Struckmann, L. and Sundmacher, K. (2016), CO Synthesis From CO_2 via Reverse Water-Gas Shift Reaction Performed in a Chemical Looping Mode: Kinetics on Modified Iron Oxide. Submitted to Journal of CO_2 Utilization.

463689 Model-Based Comparison of Novel Hollow-Fibre Sorbent Adsorber to Packed Bed Process for Dilute Gas Streams

Abstract:

Adsorption provides a relatively less energy intensive alternative to capture dilute streams from gaseous mixtures. In certain applications such as removal of CO_2 from air, recovery of valuable VOCs from exhaust streams, separating streams with almost identical thermodynamic properties (such as separation of ethylene/ethane); adsorption is often the only route to making the process commercially viable. Process improvement for the recovery of dilute gases has become an important area of research in the recent past. This may be attributed to the unprecedented rise in atmospheric concentration of CO_2 (>400 ppm for the first time in 4 million years[1]) and, the implementation of the Maximum Achievable Control Technology (MACT) provisions of the Clean Air Act, which make the reporting and control of VOC emissions mandatory. Capture of dilute streams also finds applications in upgrading processes for methane, inevitably extracted as a co-stream from natural gas reservoirs.

Traditional processes for adsorption involve use of TSA, PSA or VSA in the packed bed configuration. To improve on the current process, a novel hollow fibre configuration for adsorbents was proposed by Lively et. al. using 13X as a â €œproof of conceptâ€ zeolite[1]. The key advantage of this configuration is the substantial reduction in pressure drop relative to the packed beds. This is of primary importance when dealing with dilute streams where, additional pressure drop may involve significant capital investment. Additionally, the presence of the impermeable lumen layer allows the use of cooling agents without being limited by its adsorption properties, unlike the situation in a packed bed. This is a significant advantage for dilute streams, where TSA is a more economical process alternative compared to PSA. The fast heat transfer in the hollow fibre configuration shortens the cycle time, therefore increasing the throughput and reducing the quantity of adsorbent. One possible drawback of the hollow-fibre configuration maybe the parasitic heat load of the fibre material itself during the desorption step, which can be mitigated by heat integration.

This talk addresses the comparison of an adsorption-desorption cycle (TSA) of a dilute gas stream in both the previously mentioned configurations, packed bed and hollow fibre, using mathematical models. Coupled PDEs are obtained from heat and mass balance. The form of the equation is dictated by the system geometry. Mass transfer models appropriate for each configuration are also used, along with isotherm equations which are configuration independent. The commercial, dynamic process modelling software gPROMS is used to solve the system of equations. The models are used to compare the pressure drop of a dilute gas stream in both configurations, during adsorption. The separation performance is compared for the two configurations taking into account the pressure drop, which is dictated by the gas velocity. Finally, the parasitic heating load of the fibre material is estimated.

[1] 'Modeling of rapid temperature swing adsorption using hollow fiber sorbents', 2014, Rezaei, F, Subramanian, S, Kalyanaraman, J, Lively, R P, Kawajiri, Y, Realff, M J ,*Chemical Engineering Science*, **113**, 62-76

471719 Profitability and Risk in Conceptual Plant Design:Dealing with Key Financial Parameters Rigorously and Simultaneously

Abstract:

Profitability and Risk in Conceptual Plant Design:

Dealing with Key Financial Parameters Rigorously and Simultaneously

Duncan A. Mellichamp

Professor Emeritus

Department of Chemical Engineering

University of California, Santa Barbara, CA USA 93106

e-mail: dmell@engineering.ucsb.edu **Phone**: 805-569-9858 **Fax**: 805-893-4731

Abstract

Many companies impose a minimum *IRR* ("hurdle rate") to justify a "go-forward decision" with a new project, e.g., the *IRR* must be at least 25% to be deemed profitable enough. Presumably, this hurdle rate is chosen sufficiently higher than the sum of their ongoing *Enterprise Rate, ER*(the average year-over-year after-tax return on investment to the organization from all continuing operations) plus a "risk cushion" to cover uncertainties that statistically arise, causing loss of profitability or even outright failure of some projects.

Most users of discounted cash flow methods understand that the *ER*, when used as *Discount Rate* (*DR*) in net present value calculations imposes an appropriate "opportunity cost" on a potential project —i.e., foregone return that presumably could be obtained from alternative investment in ongoing internal operations. But few users understand just how problematic *IRR* is because the calculated value (i.e., *Discount Rate*) seldom has real physical meaning. Nor can the portion of the *DR* that is greater than *ER* be interpreted as "risk compensation." The key problem is that *IRR* is defined without parameters, so nothing is strictly associated either with "profit" or with "risk." Furthermore, *IRR*is a non-linear function, making any allocation of profit and risk difficult to extrapolate mathematically to different circumstances.In 2013 the author proposed a new index of profitability, $NPV_{\%}$, (*NPV*, evaluated with *DR* = *ER*) normalized by the *Total Capital Investment* and annualized by the *Project Lifetime*).* This metric was shown to be linearly related to ROI_{BT}, thus substantially represents "inverse risk" in the same way that *Pay-Out Time* approximates the length of time a capital investment will be at risk (time to recover the original investment via cash flows). Thus, $NPV_{\%}$ can serve as an inverse measure of business risk while its antecedent, *NPV*, provides a direct estimate of profitability. Further, since $NPV_{\%}$ includes at least one more parameter than *IRR*, it allows for an independent choice of the *Discount Rate*. Because *IRR* and $NPV_{\%}$ are both directly related to the ROI_{BT} of the proposed project, one can fix ROI_{BT}, calculate $NPV_{\%}$ as a function of *ER*, then use a manifold projection (onto the ROI_{BT}, *IRR* axis) to establish equivalence.

Several key outcomes are discussed: (1) Company policy ought to be to adjust the hurdle rate to optimize profitability (*NPV*) and the risk surrogate (inverse of $NPV_{\%}$) while updating *IRR* [as inferred function of $NPV_{\%}(ER)$] *whenever* experience with profits (internal *Enterprise Rate*) changes significantly. (2) A better approach might be to abandon the usual pairing <*NPV* (with *DR* = *ER*), *IRR*> altogether and adopt <*NPV*, $NPV_{\%}$> or a related evaluation metric. Ongoing research shows that such a choice can make evaluating project profitability particularly easy, allowing any reasonable optimization strategy to employ 1, 2, or 3 independent variables--e.g., profitability, risk, and/or capital investment--*simultaneously*.

---

* Mellichamp, D.A., *Computers and Chemical Engineering* **48** (2013) 251– 263.

459874 Characterization and Optimal Site Matching of Wind Turbine: Effects on the Economics of Synthetic Methane Production

Abstract:

Wind can easily provide the energy that mankind needs. However, harvesting energy from wind is challenging. There are only a few regions where it is economically interesting to use it. Furthermore, the variability in the wind velocity and the particular design of the turbine determine the turbine model to use for a particular site location based on its characteristic power curve. Instead of just producing power, with the challenge of its availability, it is possible to store wind energy in the form of chemicals. Davis and Martín (2014a&b) used solar and wind energy to produce methane from renewable energy and CO_2 not only to store both sources of energy in a more handy form, but also to capture and reuse CO_2 as a carbon source. However, the design of such processes must consider seasonal wind variability and uncertainty (Halemane and Grossmann (1983), Grossmann and Sargent (1978)). Thus, the selection of the appropriate turbine type for a specific site location requires accurate modeling of the power curve to address its operation under variable wind velocity conditions. Previous studies on turbine site –matching are based on the capacity factor, using piecewise approximations of the power curve to obtain analytical solutions Jowder (2009) El-Shiny (2010). A more systematic study and power curve modeling is required.

In this work, we evaluate the optimal site matching of wind farms developing a generic single equation model to characterize the power curve of a number of current commercial onshore and offshore wind turbines. Next, mathematical MINLP formulations are developed that allow considering the velocity distribution on a monthly basis, including wind variability using Weibull distribution. We apply the formulation to select the appropriate turbine type and number for the production of synthetic methane. This allows evaluating the effect of the site on the wind turbine selection and the investment and production costs of synthetic methane from hydrolytic hydrogen and CO_2. The chemical facility was developed in a previous work Davis and Martín (2014a&b). In that work it was shown that 94% of the investment cost is due to the wind farm. Thus, we can identify advantages in selecting the proper place and turbine in terms of economic savings for producing an easy to use and store/transport fuel.

The optimal selection of turbine and place result in competitive prices for synthetic methane, below 2€/MMBTU, as long as the appropriate turbine and allocation are selected. By using the formulation to select the allocation of such a facility in Spain, the best place selected corresponds to the one in Davis and Martín (2014a) paper, Cádiz. However, the use of the appropriate turbine design resulted in savings of 42 MM€ compared to the results in Davis and Martín (2014a) and a production cost 25% lower. Therefore, the formulation presented in this work provides a useful tool for the selection of turbines and allocations. Finally, the selection under uncertainty results in the fact that, although in general the cost is expected to be higher, since a more robust solution is expected, sometimes lower production prices are found due to the fact that the more detailed wind distribution fits better with the power curve than just an average velocity.

References

Archer CL., 2004. The Santa Cruz Eddy and U.S. wind power, Ph.D. Thesis, 190 pp., Stanford University,

Stanford, 1 April 2004.

Davis, W., Martín, M ., 2014a. Optimal year-round operation for methane production from CO_2 and Water using wind energy. Energy 69, 497-505

Davis, W., Martín, M., 2014b. Optimal year-round operation for methane production from CO_2 and Water using wind and/or Solar energy. J. Cleaner Prod. 80, 252-261.

El-Shimy, M. Optimal site matching of wind turbine generator: Case study of the Gulf of Suez region in Egypt. Renew. Energ. 35 (8), 1870-1878

Grossmann, I. E., Sargent, R.W.H., 1978. Optimum Design of Chemical Plants with Uncertain Parameters, *AIChE J.*, 24 (6), 1021-1028

Halemane, K.P.,Grossmann, I. E. , 1983. Optimal process design under uncertainty *AIChE, J.*, 29 (3), 425-433

470289 Multi-Objective Optimization of Biorefinery and Uncertainty Analysis

Abstract:

Abstract:

The renewable energy industry is growing continuously, and it has several challenges and opportunity for research community. Biomass is a potential source of energy for future as it can be converted to transportation fuels, chemicals and electricity using bio-chemical and thermo-chemical conversion routes. Bio-alcohols production *via* hydrolysis and fermentation can only use cellulosic part, leaving the lignin portion as waste. On the contrary, thermo-chemical processes such as gasification and pyrolysis can use both cellulose and lignin. Gasification produces syngas which can be converted into liquid fuels, chemical and electricity. Fischerâ''Tropsch (FT) crude, methanol (MeOH) and dimethyl ether (DME) can be synthesised by syngas. Further, solid oxide fuel cell with gas turbine (SOFC-GT) has shown higher thermodynamic performance for converting syngas into electricity and particularly considering the biomass conversion integration. Decision on the best pathways highly depends on bioproducts and biofuels market prices, which are highly uncertain. Further, investment cost and price and availability of raw materials are also uncertain.

The biorefinery superstructure for this study includes following processes: gasification, SOFC-GT, FT, MeOH and DME. Design and operating conditions of this biorefinery superstructure is optimized for profit and annualized capital cost, simultaneously, using stochastic multi-objective optimization method. The final selection of a solution from the obtained Pareto-optimal front depends on its sensitivity to the uncertain parameters, such as feed and product prices, plant life and operating time. Engineers are mainly interested in selecting one or few robust solutions which are less sensitive to the uncertain parameters, and so the uncertainty analysis of the obtained non-dominated solutions may help in identifying robust solutions. In this study, effect of several uncertain operating and market parameters is studied on the performance of selected biorefinery superstructure. The uncertainty analysis is able to identify most promising

biorefinery configurations, based on the profit as main decision crieriton. The results of this study will be presented at 2016 AIChE Annual Meeting.

Keywords:

Multi-objective Optimization, Gasification, Fischer-Tropsch Process, Methanol, Dimethyl Ether, Solid Oxide Fuel Cell, Gas Turbine, Uncertainty Analysis.

References:

1. Caliandro P., Tock L., Ensinas A.V, Maréchal F., Thermo-economic optimization of a solid oxide fuel cell - gas turbine system fuelled with gasified lignocellulosic biomass. Energy Conversion and Management 2014; 85: 764-773.
2. Facchinetti E., Favrat D., Maréchal F., Design and optimization of an innovative solid oxide fuel cell - gas turbine hybrid cycle for small scale distributed generation. Fuel Cells 2014; 14(4): 595-606.
3. Gassner M., Maréchal F., Methodology for the optimal thermo-economic, multi-objective design of thermochemical fuel production from biomass. Computers & Chemical Engineering 2009; 33(3): 769-781.
4. Tock L., Maréchal F., Decision support for ranking Pareto optimal process designs under uncertain market conditions. Computers & Chemical Engineering 2015; 83:165-175.
5. Tock L., Gassner M., Maréchal F., Thermochemical production of liquid fuels from biomass: Thermo-economic modeling, process design and process integration analysis. Biomass and Bioenergy 2010; 34(12): 1838-1854.

467946 Development of a Semantically-Enabled Model-Sharing Biorefinery Platform

Abstract:

The production of chemicals and fuels based on biomass has been steadily increased, as a consequence of the environmental issues that have arisen by fossil fuels in addition to the steady decrease of their reserves. There is a huge potential in capitalizing fragmented work around the world on biorefineries and biorenewables, since this would assist and accelerate the progress in both research and industrial communities. The objective of this work is to integrate biomass pathways, models and tools for biorefineries and create a repository that contains information on these existing biomass pathways and models. The extremely large number of possible pathways as well as the diversity of data consist important challenges that need to be overcome.

An ontology engineering approach will be presented that emulates the types of data as they are available in selected projects, mainly RENESENG[1] but also others. The aim is to develop an intelligent environment able to scale-up to also include and organize resources outside the current project. Ontology engineering has recently attracted a lot of attention in chemical engineering community for managing, searching and enriching knowledge[2,3]. Advantages of ontology engineering include the classification of large amount of information, the classification of complex relations, as well as the generation of new information by the use of reasoners. The services of the repository will include (a) access to all data included in the tool, (b) support for high-throughput search and analysis (c) integration with physical property models and data models, (d) synthesis technologies, and (e) uploading of new data of a variety of scales and types from the user. Having a user-friendly interface, the repository targets users of a wide range; from students to engineers in SMEs or international companies. The state of progress of the development of the repository will be presented and example operations will be demonstrated.

References

1. www.reneseng.com
2. Weisner et al., Computers and Chemical Engineering 35 (2011) 692-708
3. Trokanas, et al., Computers and Chemical Engineering 66 (2014) 259 - 268

459820 Optimal Production of Power in a Combined Cycle from Manure Based Biogas

Abstract:

Most countries have powerful agricultural systems to provide for food. As a result, large volumes of residues are generated. In particular, cattle and pigs farms generate vast amounts of residues (manures). Apart from the difficulty of dealing with such quantities of waste, the composition is also dangerous. Anaerobic Digestion (AD) provides the technology not only to dispose it but also to generate further value. The production of biogas through AD offers significant advantages over other forms of bioenergy production. It has been deemed as one of the most energy-efficient and environmentally beneficial technologies for bioenergy production [1]. Furthermore, biogas generation can drastically reduce greenhouse gases compared to fossil fuels by utilization of locally available resources. Compared to other fossil fuels, methane produces fewer atmospheric pollutants and generates less carbon dioxide per unit energy; as methane is comparatively a clean fuel, the trend is towards its increased use for appliances, vehicles, industrial applications and power generation [2]. Finally, the digestate represents an improved soil conditioner which can substitute mineral fertilizer [3]. However, most of the studies on power production from biogas limit to the use of Stirling engines or basic CHP plant analysis [4-6]

In this work, the production of power using a combined cycle gas turbine/steam turbine for wast heat recovery, which operates with biogas as fuel, is evaluated. The process begins with the production of biogas from pig and /or cattle slurry manure(s) using AD. Afterwards, the gas is cleaned up to remove humidity, hydrogen sulphide, carbon dioxide and ammonia. The cleaned gas (biomethane) is then used in a Brayton cycle (gas turbine) to produce energy. The flue gas that exits the Brayton cycle is typically at high temperature and it is further utilized to produce steam that generates power in a regenerative Rankine cycle (steam turbine). Two alternative steam production schemes are evaluated: either splitting the flue gas to have high temperature gas for the reheating step of the steam or sequential heating up. The model is formulated as a Mixer Integer Nonlinear Programming (MINLP) solved in GAMS® for the optimal production

of power and topology selection.

For a typical production capacity of manure in farms in Spain, 2.6 MW are produced. The hot flue gas is used in sequence for producing superheated steam and next to reheat up the stream after the first expansion in the steam turbine. While typical efficiencies of biogas combustion engines are from 25-40%, the system presented in this work reaches 65% efficiency from the biogas to electricity. The investment for the plant turns out to be 26M€ and the production cost of the electricity is 0.35€/kWh before including the credit from the conditioned digestate, that could be sold as fertilizer. The electricity cost goes down to 0.15€/kWh considering a reasonable credit from the digestate, whose composition depends on the feedstock processed in the facility.

References

[1] Fehrenbach H, Giegerich J, Reinhardt G, Schmitz J, Sayer U, Gretz M, et al. Criteria for a Sustainable Use of Bioenergy on a Global Scale. Germany: Federal Environment Agency; 2008; 245 (prepared by the Institute for Energy and Environmental Research (IFEU), Heidelberg).

[2] Chynoweth DP, Owens JM, Legrand R. Renewable methane from anaerobic digestion of biomass. Renew Energy 2001;22(1):1–8.

[3] Weiland P. Biogas production: current state and perspectives. Appl Microbiol Biotechnol 2010;85(4):849–6

[4] Kang, J. Y., Kang, D.W., Kim, T S., Hur, K.B. (2014) Economic evaluation of biogas and natural gas co-firing in gas turbine combined heat and power systems Appl.Thermal Eng. 2014; 70: 723-731

[5] Japaraju, P., Rintala, J. 17 - Generation of heat and power from biogas for stationary applications: boilers, gas engines and turbines, combined heat and power (CHP) plants and fuel cells 404-427, In The Biogas Hand book. 2013.

[6] Kang JY, Kang DW, Kim TS, Hur KB (2014) Comparative economic analysis of gas turbine-based power generation and combined heat and power systems using biogas fuel Energy 2014; 67: 309-318

464293 A State-Space Formulation for Autocovariance-Based Plant-Model Mismatch Estimation in Model Predictive Control

Abstract:

Model predictive control (MPC) has become the preferred strategy for handling complex, multivariate control problems in the process industries. MPC uses a model of the plant dynamics to predict response over a control horizon, and provides an optimal sequence of control moves at each time step. MPC is considered a mature technology, with thousands of implementations currently in place, benefiting from several decades of technological developments from both academic and industrial contributors [1]. In recent years, state-space model forms of MPC have gained traction. This is due to their ability to handle the full range of process dynamics, such as integrating and unstable systems, which are not in general dealt with by other model forms, such as the more traditionally used (and still predominant in applications) convolution models. State-space models also offer more flexibility in accounting for unmeasured disturbances in the state estimation step [2].

Although MPC confers benefits over, e.g., multi-loop approaches for many multivariable systems, as with any model-based method it is limited in performance by the model quality, i.e. how well the true process dynamics are represented by the model. Frequent system re-identification could in principle be implemented, but this tends to be cost prohibitive in practice; alternative methods are needed to ensure MPC performs as expected. As such, in there has recently been increasing interest in the problem of monitoring and assessing MPC controller performance. More general techniques for controller performance monitoring and assessment have been applied to MPC. Examples include data-driven methods such as Principal Component Analysis (PCA) and Partial Least Squares (PLS)-based fault detection [3], as well as benchmarking against performance indices, e.g. Linear Quadratic Gaussian (LQG) and Minimum Variance Controllers (MVC) [4],[5]. While these methods are often successful in indicating degraded control quality, they are often limited in their ability to provide useful model diagnosis. The development of techniques for quantifying MPC plant-model mismatch remains thus an area of active research.

Several approaches have been proposed, e.g., making use of partial correlation analysis [6] and frequency domain analysis with setpoint excitation [7]. In our previous work, we introduced an approach for estimating plant-model mismatch from closed loop operating data by means of an explicit output autocovariance-mismatch relation [8,9]. We demonstrated its effectiveness for MPC in with models in both convolution and transfer function form.

In this presentation, we extend our approach to the increasingly important class of MPCs which use state-space form models. We develop an autocovariance-mismatch relation, which explicitly links the mismatch in each of the state-space model parameters to the autocovariance of the process outputs. We then pose mismatch estimation as an optimization problem, wherein we seek to minimize (in the least squares sense) the discrepancy between the predicted output autocovariance and that computed from the actual data. We present a multivariable case study demonstrating the use of this approach, and we discuss some computational and statistical features of the method.

References

[1] S. J. Qin. Statistical process monitoring: Basics and beond. Journal of Chemometrics, 17:480–502, 2003.

[2] Mark L. Darby and Michael Nikolaou. MPC: Current practice and challenges. Control Engineering Practice, 20(4):328–342, 2012.

[3] A. AlGhazzawi and B. Lennox. Monitoring a complex refining process using multivariate statistics. Control Engineering Practice, 16:294–307, 2008.

[4] J. Yu and S. J. Qin. Statistical MIMO controller performance monitoring. Part I: Data-driven covariance benchmark. J. Proc. Contr., 18:277–296, 2008.

[5] T. J. Harris. Assessment of closed loop performance. Can. J. Chem. Eng., 67:856–861, 1989.

[6] Abhijit S. Badwe, Ravindra D. Gudi, Rohit S. Patwardhan, Sirish L. Shah, and Sachin C. Patwardhan. Detection of model-plant mismatch in MPC applications. Journal of Process Control, 19(8):1305–1313, 2009.

[7] Sehej Kaw, Arun K Tangirala, and Alireza Karimi. Improved methodology and set-point design for diagnosis of model-plant mismatch in control loops using plant-model ratio. Journal of Process Control, 24:1720–1732, 2014.

[8] S. Wang and M. Baldea. Autocovariance-based MPC model mismatch estimation for SISO systems. In Proceedings of 54th IEEE Conference on Decision and Control, pages 3032–3037, Osaka, Japan, 2015.

[9] S. Wang, J. Simkoff, M. Baldea, L. Chiang, I. Castillo, R. Bindlish, and D. Stanley. Data-driven plant-model mismatch quantification in input- constrained linear mpc. In Proceedings of The 11th IFAC Symposium on Dynamics and Control of Process Systems, Trondheim, Norway, 2016.

457017 Development of an Integrated Framework for Stochastic Model Predictive Control with Moving Horizon Estimation

Abstract:

In any practical application of Model Predictive Control (MPC), there is always a certain degree of mismatch between the predictions of the controller model and the plant actual dynamics. To address this challenge, robust control methodologies are a possible alternative, but they are often over conservative [1]. Stochastic Model Predictive Control (SMPC) approaches search for a compromise solution between conservativeness and performance [2], [3]. The central idea of SMPC is to consider probabilistic characteristics of the disturbances in the controller design, thus relaxing some constraints so that they only have to be met at certain probabilistic levels. To enable the industrial implementation of SMPC, a state estimation layer, which is often disregarded, should be included [4]. For this reason, this work integrates a Moving Horizon Estimation (MHE) algorithm into the SMPC framework, bridging the gap between recent academic breakthroughs and tangible industrial needs.

Previous contributions on output feedback SMPC considered observers with fixed gain [5], time-varying gain [6] and based on Kalman filter [7]. The integration between MHE and robust MPC was performed by deriving an error bound when an unconstrained observer is applied to linear systems with bounded disturbances, then proposing a tube controller formulation [8]. However, these previous studies did not consider the case of MHE coupled with SMPC, which is the main novelty of this work. Specifically, here we extend the work in [9] to the case of output feedback through the use of an MHE scheme [10]. The main contribution in the proposed SMPC formulation corresponds to a procedure to handle probabilistic constraints despite of the estimation error. The proposed approach considers that the exact value of the current state is unknown, but lies inside a set centered at the estimated state and shaped according to an error set calculated similarly as in [8]. The size of this set is used to tighten the probabilistic constraints, as

in the tube-based control strategies. The implementation of such constraints is translated in a deterministic form based on the cumulative density function of the disturbances.

In this presentation, we show the theoretical results on recursive feasibility and closed-loop stability guarantees in the integrated setup of SMPC with MHE. In particular, the stability analysis follows the general idea of standard MPC that uses a feasible shifted solution to prove the convergence of the control cost. Therefore, the estimated and the real state are driven towards an invariant set around the set point. The resulting controller is less conservative than the purely robust counterpart due to the relaxation of constraints. Simulation examples of the output-feedback controller show that the derived error bounds are accurate and illustrate the convergence to neighborhoods of the set point.

References

[1] L. Magni, D. Pala, and R. Scattolini, "Stochastic model predictive control of constrained linear systems with additive uncertainty," in *Proceedings of the European Control Conference 2009*, Budapest, Hungary, pp. 2235–2240, 2009.

[2] P. D. Couchman, M. Cannon, and B. Kouvaritakis, "Stochastic MPC with inequality stability constraints," *Automatica*, vol. 42, no. 12, pp. 2169–2174, 2006.

[3] D. Van Hessem and O. Bosgra, "Stochastic closed-loop model predictive control of continuous nonlinear chemical processes," *Journal of Process Control*, vol. 16, no. 3, pp. 225–241, 2006.

[4] J. Yan and R. R. Bitmead, "Incorporating state estimation into model predictive control and its application to network traffic control," *Automatica*, vol. 41, no. 4, pp. 595–604, 2005.

[5] M. Cannon, Q. Cheng, B. Kouvaritakis, and S. V. Raković, "Stochastic tube MPC with state estimation," *Automatica*, vol. 48, no. 3, pp. 536–541, 2012.

[6] M. Farina, L. Giulioni, L. Magni, and R. Scattolini, "An approach to output-feedback MPC of stochastic linear discrete-time systems," *Automatica*, vol. 55, pp. 140–149, 2015.

[7] P. Hokayem, E. Cinquemani, D. Chatterjee, F. Ramponi, and J. Lygeros, "Stochastic receding horizon control with output feedback and bounded controls," *Automatica*, vol. 48, no. 1, pp. 77–88, 2012.

[8] D. Sui, L. Feng, and M. Hovd, "Robust Output Feedback Model Predictive Control for Linear Systems via Moving Horizon Estimation," in *2008 American Control Conference*, Seattle, Washington, pp. 453–458, 2008.

[9] B. F. Santoro and D. Odloak, "Stochastic Model Predictive Control for Output Tracking with Bounded Control Inputs," in *2014 AIChE Annual Meeting*, Atlanta, USA, 2014.

[10] F. V Lima and J. B. Rawlings, "Nonlinear stochastic modeling to improve state estimation in process monitoring and control," *AIChE Journal*, vol. 57, no. 4, pp. 996–1007, 2011.

466954 Nonlinear Model Predictive Control with Explicit Performance Specification

Abstract:

The operation of chemical plants is hindered by numerous challenges such as inherent nonlinearity, stochastic disturbances, complex variable interactions and process constraints. One well established control method that is effective in handling these challenges is model predictive control (MPC), where an open-loop optimal control problem is solved at each sampling instance over a finite time horizon, subject to the dynamic response of the plant and constraints. The abundant studies and various industrial applications of MPC have also contributed towards better understanding of the rigorous stability and performance properties of MPC.

One form of contribution has involved the development of Lyapunov-based MPC, which can explicitly characterize the region from where stability of the closed-loop system is guaranteed in the presence of input and state constraints [1, 2]. This approach is also capable of handling uncertainty through the stochastic characterization of disturbances and robust control formulations [3]. On one hand, robust control approaches counteract the worst-case scenario for the realization of the uncertainty. On the contrary, offset-free MPC augments the state variables with fictitious states that estimate and counteract the uncertainty. The prevailing economic incentive in operating chemical processes has fostered the recent development of economic MPC formulations where the controller determines the set-point internally to satisfy the prescribed economic objective, while a rigorous analysis ensures that stability is preserved.

Regardless of the nature of the predictive controller used, relatively fewer studies have explored the explicit characterization of the desired closed-loop behavior in the MPC formulation. The explicit specification of the control performance has been addressed for linear, single-input single-output (SISO) systems in the frequency domain using internal model control (IMC) [5], where the controller is designed to achieve a pre-specified desired closed-loop transfer function. However, it is not always guaranteed that the transfer function of the IMC-based control law will be consistent with the proportional-integral-derivative (PID) control structure. Moreover, the IMC approach does not generalize readily for multi-input multi-output (MIMO) systems with constraints. In an effort to handle MIMO systems, the funnel control approach [6] is proposed where the time-varying output error feedback controller forces the tracking error to be within a bounded prescribed function. Although the funnel control approach is capable of handling nonlinear MIMO systems, the method does not explicitly consider input constraints. To address this inherent limitation, recent contributions proposed a framework that incorporates explicit performance specification for MIMO systems while cognizant of input constraints [7], which is then implemented in conjunction with offset-free model predictive control. The work considers linear time invariant (LTI) systems that are invertible (i.e. the inputs can be explicitly computed). The proposed explicit performance specification based offset-free MPC, using a linearized plant model, was implemented on a nonlinear chemical process. The previous work, however, does not explicitly handle the issue of lack of full state measurements, constraints on the rate of change of the manipulated input, and process nonlinearity.

Motivated by the above considerations, in this work we generalize the explicit performance specification based control design to handle the output feedback problem, constraints on the rate of change of the manipulated input, and process nonlinearity. The proposed framework consists of a bi-layer optimization scheme. In the first tier, the optimal trajectory is computed for the given performance specification in the form of a prescribed function. Then, in the second layer, a nonlinear model predictive control scheme computes the optimal inputs so as to closely follow the trajectory as given by the first tier. Results are first presented for the output feedback control for a linear systems, then the handling of rate constraints is illustrated for a nonlinear CSTR example, and finally, the approach is implemented on a reactor separator system.

[1] Mahmood, M. and Mhaskar, P., 2014. Constrained control Lyapunov function based model predictive control design. International Journal of Robust and Nonlinear Control, 24(2), pp.374-388.

[2] Mhaskar, P., El-Farra, N.H. and Christofides, P.D., 2006. Stabilization of nonlinear systems with state and control constraints using Lyapunov-based predictive control. Systems & Control Letters, 55(8), pp.650-659.

[3] Mahmood, M. and Mhaskar, P., 2012. Lyapunov-based model predictive control of stochastic nonlinear systems. Automatica, 48(9), pp.2271-2276.

[4] Wallace, M., Das, B., Mhaskar, P., House, J. and Salsbury, T., 2012. Offset-free model predictive control of a vapor compression cycle. Journal of Process Control, 22(7), pp.1374-1386.

[5] Garcia, C.E. and Morari, M., 1982. Internal model control. A unifying review and some new results. Industrial & Engineering Chemistry Process Design and Development, 21(2), pp.308-323.

[6] Ilchmann, A., Ryan, E.P. and Sangwin, C.J., 2002. Tracking with prescribed transient behaviour. ESAIM: Control, Optimisation and Calculus of Variations, 7, pp.471-493.

[7] Wallace, M., Pon Kumar, S.S. and Mhaskar, P., 2016. Offset-Free Model Predictive Control with Explicit Performance Specification. Industrial & Engineering Chemistry Research.

465654 Optimization-Based Event-Triggered Predictive Control of Process Systems with Control and Communication Constraints

Abstract:

The design and analysis of networked control systems have been the subject of significant research interest over the last two decades, both in the industrial and academic circles (e.g., see [1]-[3] for surveys of results in this area). The operational flexibility and substantial economic savings realized through networked control architectures have been a major driving force behind the increased reliance in industrial practice on sensor and control systems that are accessed over shared communication networks. The integration of wireless sensor networks in process control systems, for example, is an appealing goal that promises to expand the capabilities of existing control technology beyond what is possible with wired devices alone, and is regarded as a key enabling mechanism for the transition towards smart plant operations [4]. At the same time, the various fundamental challenges introduced by this technology from a control point of view have motivated numerous research studies. One of the key challenges is the development of resource-aware control methods that can systematically balance the desired stability and performance requirements against the intrinsic constraints on the sensing, computation and communication resources of the networked devices as well as the occasional unreliability of the communication medium due to interference in the field or environmental impact. On the one hand, maximizing control system performance often requires frequent sensor-controller communication and increased levels of network resource utilization. Reducing communication costs, on the other hand, favors keeping the sensing and communication levels to a minimum which can lead to reduced control system performance.

A common approach for dealing with communication constraints in networked control systems is the use of event-triggered control strategies. In these strategies, sensor-controller communication over the network is typically suspended or restored in response to certain events which are tied to some specified closed-loop stability and/or performance thresholds. The aim is to keep network utilization to a minimum while guaranteeing minimal levels of control performance. The approach typically starts with the design of a model-based stabilizing feedback controller and ends with the derivation of a constant or state-dependent stability threshold which is used as a trigger for sensor-controller communication. The implementation generally involves monitoring the model estimation error during periods of communication suspension and updating the model states in the event that the stability threshold is breached. This approach is appealing in the context of sensor/actuator networks where reducing network utilization can also reduce the energy expenditures of battery powered wireless devices, and has been widely studied in recent years (e.g., see [5]–[8] for some results in this area).

The appealing features of this approach notwithstanding, a close examination of existing event-triggered control strategies reveals a general lack of explicit control performance optimization and constraint handling capabilities in the controller design methodology. In the context of balancing the intrinsic tradeoff between control performance and network resource utilization, there is an apparent increased emphasis on reducing network utilization, often at the expense of control system performance. In addition to the lack of performance optimization considerations, practical implementation considerations such as the presence of constraints on the state and manipulated variables, which are commonly encountered in practical applications, are also ignored in the design formulation. These considerations not only limit the achievable control quality but can also lead to closed-loop instability if not explicitly accounted for at the controller design stage.

Motivated by these considerations, we present in this work an optimization-based event-triggered control methodology for systems subject to state, control and sensor-controller communication constraints. The objective is to optimize the control system performance while minimizing communication costs and network utilization at the same time. To this end, a model-based state-feedback controller is initially designed and used to obtain a state-dependent stability threshold which is used as the basis for event-triggered communication. An explicit characterization of the threshold -- in terms of the model and controller design parameters -- is obtained via Lyapunov techniques and used to provide an explicit link between the communication cost and the choice of controller design parameters. A finite-horizon optimal control problem in which the objective function includes explicit penalties on the states, the control action and the extent of sensor-controller communication is then formulated subject to the appropriate state, control and stability constraints. The penalty placed on the sensor-controller communication in the objective function is derived directly from the communication-triggering threshold. The optimization problem is solved in a receding horizon fashion to obtain the optimal controller design parameters over the horizon. Unlike conventional model predictive control formulations, however, the optimization problem is re-solved and the model states are updated only at times when the communication threshold is breached. To this end, the model estimation error is monitored at each sampling time, and communication is restored when the stability threshold is violated. In addition to reducing sensor-controller communication costs and optimizing system performance, the resulting reduction in the computational load associated with the proposed on-line optimization scheme is another appealing feature, especially in the context of nonlinear systems. Finally, the developed methodology is tested and illustrated using a simulated chemical process example.

References:

[1] J. P. Hespanha, P. Naghshtabrizi, and Y. Xu, "A survey of recent results in networked control systems," *Proc. of IEEE*, vol. 95, pp. 138–162, 2007.

[2] R. A. Gupta, "Networked control systems: Overview and and research trends," *IEEE Trans. Ind. Electron.*, vol. 57, pp. 2527–2535, 2010.

[3] K.-Y. You and L.-H. Xie, "Survey of recent progress in networked control systems," *Acta Automatica Sinica*, vol. 39, pp. 101–117, 2013.

[4] P. D. Christofides, J. Davis, N. H. El-Farra, D. Clark, K. Harris, and J. Gipson, "Smart plant operations: Vision, progress and challenges," *AIChE Journal*, vol. 53, no. 11, pp. 2734–2741, 2007.

[5] M. Mazo and P. Tabuada, "Decentralized event-triggered control over wireless sensor/actuator

networks," *IEEE Transactions on Automatic Control*, vol. 56, no. 10, pp. 2456–2461, 2011.

[6] X. Wang and M. D. Lemmon, "Event-triggering in distributed networked control systems," *IEEE Transactions on Automatic Control*, vol. 56, no. 3, pp. 586–601, 2011.

[7] E. Garcia and P. J. Antsaklis, "Model-based event-triggered control for systems with quantization and time-varying network delays," *IEEE Transactions on Automatic Control*, vol. 58, no. 2, pp. 422–434, 2013.

[8] D. Xue and N. H. El-Farra, "Networked event-triggered control of spatially distributed processes using a dual-mode communication strategy," *Proceedings of 54th IEEE Conference on Decision and Control*, pp. 1899-1904, 2015.

456977 Output-Feedback Predictive Control for Stochastic Nonlinear Systems

Abstract:

Model predictive control (MPC) is widely used for control of complex systems since it is systematically able to handle multivariable dynamics and constraints on the system states and inputs. Robustness to modeling uncertainties and system disturbances is a desired aspect in a MPC algorithm. While the receding-horizon implementation imparts a certain degree of robustness to MPC, nominal MPC formulations cannot systematically account for the effect of stochastic uncertainties in the system [1]. Stochastic MPC (SMPC) with chance constraints provides a systematic framework to address this issue by explicitly incorporating the probabilistic descriptions of uncertainties in the optimal control problem [2]. This work presents an output-feedback stochastic MPC approach for nonlinear systems that are subject to probabilistic modeling uncertainties and system disturbances. The output-feedback approach accounts for the fact that in most engineering systems all the system states cannot be regularly measured, and there is a need for reconciliation of system states from the available noisy measurements.

In this work, the generalized polynomial chaos (gPC) framework [3] is used for computationally efficient propagation of the uncertain initial conditions and parametric uncertainties in the system model through the nonlinear system dynamics. The main challenge in using the gPC framework is to enable efficient propagation of the stochastic system disturbances since this requires a large number of basis functions in the polynomial chaos (PC) expansions for describing the independent time-varying system disturbances. To overcome this limitation, this work uses a two-step approach [4] for propagation of the parametric uncertainties and additive disturbances in the space of the PC expansion coefficients under the assumption that these coefficients can be adequately described by a Gaussian distribution. The output-feedback approach is implemented by using a gPC-based histogram filter [5], which is a Bayesian state and parameter estimation algorithm for estimating the posterior probability density functions of the states and uncertain parameters using the available measurements. The constraints on the states are reformulated as

a joint chance constraint (JCC), which is satisfied in with a certain probability. A tractable implementation of the JCC is obtained by using a sample-based approach [6].

The proposed SMPC approach is implemented for control of the thermal effects of an atmospheric pressure plasma jet (APPJ) simulation case [7]. The system model consists of seven states (three of carrier gas temperature, three of carrier gas composition and one of target surface temperature) of which four are measured and three are unmeasured. The control objective is to regulate the target surface temperature. Chance constraints are imposed on the carrier gas temperature and surface temperature. The simulation results indicate that closed-loop control of the plasma device enables achieving the desired target surface temperature in the presence of system disturbances and modeling uncertainties, while maintaining the constraint satisfaction above the pre-specified probability level.

References:

1. Mayne, D. Q. Model predictive control: Recent developments and future promise. *Automatica,* 50:2967—2986, 2014.
2. Mesbah, A. Stochastic model predictive control: An overview and perspectives for future research. *IEEE Control Systems Magazine*, accepted, 2016.
3. Xiu, D. and Karniadakis, G. E. The Wiener-Askey polynomial chaos for stochastic differential equations. *SIAM Journal of Scientific Computation*, 24:614—644, 2002.
4. Konda, U., Singla, P., Singh, T. and Scott, P. D. State uncertainty propagation in the presence of parametric uncertainty and additive white noise. *Journal of Dynamic Systems, Measurement and Control, Transactions of the ASME*, 133(5):051009-1—051009-10, 2011.
5. Bavdekar, V. A. and Mesbah, A. A polynomial chaos-based nonlinear Bayesian approach for estimating state and parameter probability density functions. *In Proceedings of the American Control Conference*, accepted, Boston, 2016.
6. Alamo T., Tempo, R. and Luque, A. On the sample complexity of probabilistic analysis and design methods. In *Perspectives in Mathematical System Theory, Control, and Signal Processing*, pages 39—50. Springer—Verlag, Berlin Heidelberg, 2010.
7. Gidon, A., Graves, D. B. and Mesbah, A. Model predictive control of thermal effects of an atmospheric pressure plasma jet for biomedical applications. *In Proceedings of the American Control Conference*, Accepted, Boston, 2016.

451185 Strategies Towards the Robust Multi-Parametric Control of Continuous-Time Systems

Abstract:

Strategies towards the robust multi-parametric control of continuous-time systems

Muxin Sun[a,b], Mario E. Villanueva[b] Benoît Chachuat[a], Efstratios N. Pistikopoulos[b]

[a]Centre for Process Systems Engineering, Department of Chemical Engineering, Imperial College London, London, UK

[b]Artie McFerrin Department of Chemical Engineering, Texas A&M University, College Station, TX, USA.

Keywords: multi-parametric programming,NCO-tracking,robust optimal control

The need to improve performance and reduce economic costs of industrial processes has led to an increased interest in real-time model-based optimal control strategies. These strategies need to be able to deal with fast dynamics and hard process constraints, such as operational restrictions and safety requirements. Moreover, the presence of uncertainty in the form of process disturbances or model mismatch can lead to suboptimal performance and constraint violation. Model predictive control (MPC) [1] computes a feedback control law by repeatedly solving an optimization problem on a receding horizon, which is often a rather computationally demanding task and might cause serious delays and/or suboptimal performance.

In the multi-parametric programming paradigm [1], the optimization problems are solved off-line, which results in an a priori explicit mapping of the solutions as a function of measurable quantities, thus avoiding the need for repeatedly solving optimization problems online. The case of continuous-time dynamic systems calls for the solution of multi-parametric infinite-dimensional programs, namely multi-parametric dynamic optimization (mp-DO). Another approach for reducing the on-line computationally burden is to enforce the necessary conditions for optimality (NCO). In NCO tracking [2,3], feedback control laws are derived in order to track the NCOs along each arc of the nominal optimal solution, effectively converting an optimal control problem into a self-optimizing feedback control problem. However, NCO tracking often relies on the assumption that the switching structure of the optimal solution remains unchanged, which may not be the case when the process disturbances or the model mismatch are significant.

The implementation of classical MPC controllers is based on a certainty-equivalence principle, whereby the system is optimized as if neither process disturbances nor model mismatch were present. Due to its inherent ability to reject such disturbances, these controllers exhibit a certain degree of robustness. However, certain constraints may become violated when large disturbances occur, since this is not taken into account when optimizing system trajectories. One way of dealing with this problem involves solving a robust counterpart of the optimal control problem [4], i.e. to guarantee feasibility of the system assuming the worst-case realization of the uncertainty.

This paper explores strategies for robust multi-parametric control of continuous-time systems. First, we present a methodology for the development of so-called mp-NCO-tracking controllers (see Figure 1), which unifies mp-DO and NCO-tracking for linear dynamic systems [5]. From an NCO-tracking perspective, the use of mp-DO provides a means for relaxing the constant switching structure assumption by constructing critical regions that correspond to a unique switching structure of the optimal control trajectories; whereas from a mp-DO perspective, NCO-tracking leads to a reduction in the number of critical regions compared to the use of discretization methods. Next, we present a robust-counterpart formulation of the mp-DO problem in order for the mp-NCO-tracking controllers to be robust towards given process disturbances or model mismatch that are not measured on-line. This robustification involves backing-off the terminal and path constraints based on rigorous set-propagation techniques [6]. The methodology is illustrated with numerical case studies drawn from chemical engineering, including the control of an FCC unit as shown in Figure 2.

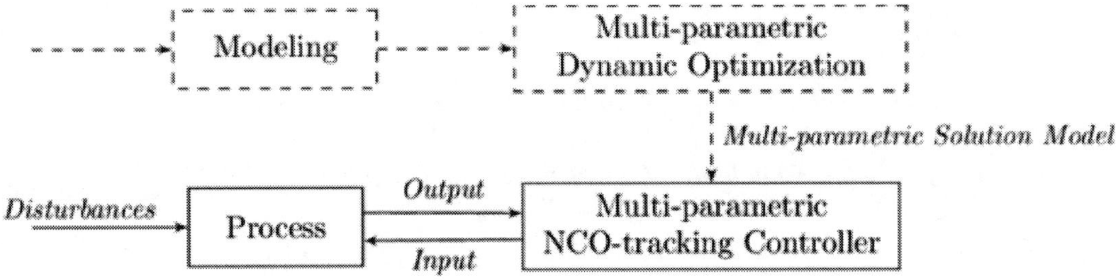

Figure 1: Framework of multi-parametric NCO-tracking

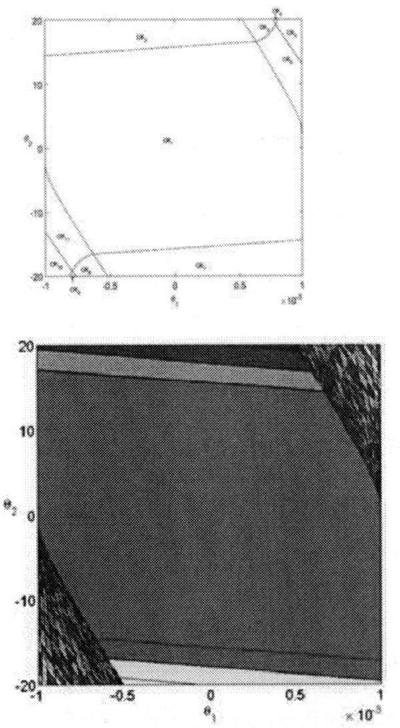

(a) mp-DO (b) discretized mp-QP

Figure 2: Critical regions in the controllers of an FCC unit

References

1. E. N. Pistikopoulos. From multi-parametric programming theory to MPC-on-a-chip multi-scale systems applications. *Computer and Chemical Engineering*, 47:57-66, 2012.

2. J. V. Kadam, M. Schlegel, B. Srinivasan, D. Bonvin and W. Marquardt. Dynamic optimization in the presence of uncertainty: From off-line nominal solution to measurement-based implementation. *Journal of Process Control*, 17:389-398, 2007.

3. D. Bonvin and B. Srinivasan. On the role of the necessary conditions of optimality in structuring dynamic real-time optimization schemes. *Computers and Chemical Engineering*, 51:172-180, 2013.

4. B. Houska, F. Logist, J. V. Impe and M. Diehl. Robust optimization of nonlinear dynamic systems with application to a jacketed tubular reactor. *Journal of Process Control*, 22(6):1152-1160, 2012.

5. M. Sun. , B. Chachuat and E. N. Pistikopoulos, Design of multi-parametric NCO tracking controllers for linear dynamic systems. *Computer and Chemical Engineering*, in press (DOI: 10.1016/j.compchemeng.2016.04.038).

6. B. Chachuat, B. Houska, R. Paulen, N. Peric, J. Rajyaguru, M.E. Villanueva, "Set-theoretic approaches in analysis, estimation and control of nonlinear systems," *IFAC-PapersOnLine,* 48(8):981-995, 2015.

466299 On the Moment-Based Robust MPC Formulations

Abstract:

The control technique commonly used in process industry for constrained multivariable optimal control is the model predictive control (MPC) [1]. The performance of these controllers depends to a large extent on the quality/validity of process models ([2]). In the mean time, more and more rigorous process models that cover wider operating conditions have become available. A valid hypothesis is then the use of this type of models for MPC. The practice, on the other hand, is conflicting since the model based estimation and control algorithms require relatively simple explicit models for efficient and robust calculations in their algorithms. This practical drawback has led to research directions in the area of model simplifications ([3] and references therein) or multiple model MPC approaches ([4]). In addition to the plant-model mismatch as a result of inaccurate identification of model parameters and its dynamics, such simplifications introduce uncertain behavior between the observations induced from the true process and the model predictions. In order to limit the effect of uncertainty on the closed loop performance, robust control techniques are required.

In the last couple of decades, different MPC algorithms have been introduced to achieve the desired control objectives while reducing the effect of uncertainty ([5]). We can broadly classify these algorithms into two classes; i) the worst case based techniques (WC-MPC) ([6]), including min-max based, tube based or input-to-state stability based arguments, where under any effect of predefined maximum level of uncertainty, the process variables should not violate the specifications; or ii) chance based MPC algorithms ([7]), including quantile based or scenario based formulations, where the specifications are softened, meaning that the constraints are allowed to be violated, but the chance of being off the specification is kept small, thus violations are rare events. Both of these formulations have their own drawbacks. The WC-MPC implementations are known to be pessimistic, resulting in deterioration of the actual closed loop performance compared to the nominal case where the uncertainties are realistically described with the nominal model. Furthermore the formulation of the problem tends to be overly complex for large-scale process models. The chance-based formulation reduces the conservatism inherent to WC-MPC by discarding the severe-but-rare events out of consideration. This reduces the pessimism of WC-MPC however the computational problems should be expected to occur, due to the tedious numerical integrations over the process trajectories. It is straightforward to see that the stochastic approach to the robustified MPC problem is more powerful than the WC-MPC. The ability to discard rare and severe realizations, by a tuning parameter, leads to balancing the resulting performance and constraint satisfaction tradeoff.

In this contribution we make a first attempt to apply stochastic MPC approach with a robustfication step different than the worst case or chance based MPC formulations. Instead of worst case realization or the probability distribution function of state, we consider the statistics, (finite order) moments, of the state predictions to calculate the control actions. We introduce Mean-MPC (M-MPC) and Mean-Variance MPC (MV-MPC), which consider the expectations and variances of the state trajectories and their functions, cost and constraints. By making use of variance of state predictions, one can effectively back-off the operating conditions, as a function of the process dynamics and the uncertainty model. We consider different types of uncertainty models and show parallelism between moment based MPC and aforementioned MPC techniques.

We distinguish multiple observations from the M-MPC and the MV-MPC results. Under several technical assumptions we show that the M-MPC is equivalent with nominal MPC. This explains the inherent robustness aspects of the nominal MPC, commented in other publications, e.g., [8]. In case of the MV-MPC formulation we observe that when the cost and constraint functions are mapped to robustified equivalents, the structure of these functions (linear, quadratic etc.) and the uncertainty model effect the resulting MPC problem. For additive disturbance case with quadratic cost and linear constraint functions, the case in this contribution, the resulting MPC problem consists of a cost function weighting both the process variables' current/predicted values and the spread of the state trajectories. The cost term relating to the variance is growing monotonically with the prediction horizon, which in turn allows us to establish stability without terminal cost function, decreasing the effect of artificial terminal cost. Furthermore, the robustified constraints are improving the chance of satisfying the constraints, by making use of the covariance matrix of the predicted states, hence the process dynamics, explicitly.

One another important aspect of the moment based MPC is the computational benefits. Through the use of moment operators, one obtains a MPC problem that is explicit in the optimization variables, meaning that no maximization or integration step is required. Hence for processes with medium to large scale dynamics, the moment based MPC formulation remains computationally tractable. This property leads to adjustable performance vs. robustness specifications, allowing the practitioners to implement moment based MPC for process systems.

Acknowledgement:

This work has been done within the project 'Improved Process Operation via Rigorous Simulation Models (IMPROVISE)' in the Institute for Sustainable Process Technology (ISPT)

References:

[1] – Bauer, M. and Craig, I.K., 2008. Economic assessment of advanced process control–a survey and framework. *Journal of Process Control*, *18*(1), pp.2-18.

[2] – Qin, S.J. and Badgwell, T.A., 2003. A survey of industrial model predictive control technology. *Control Engineering Practice*, *11*(7), pp.733-764.

[3] – Hovland, S., Willcox, K. and Gravdahl, J.T., 2006, December. MPC for large-scale systems via model reduction and multiparametric quadratic programming. In *45th IEEE Conference on Decision and Control* (pp. 3418-3423).

[4] – Özkan, L., Kothare, M.V. and Georgakis, C., 2000. Model predictive control of nonlinear systems using piecewise linear models. *Computers & Chemical Engineering*, *24*(2), pp.793-799.

[5] – Goodwin, G.C., Kong, H., Mirzaeva, G. and Seron, M.M., 2014. Robust model predictive control:

reflections and opportunities. *Journal of Control and Decision*, *1*(2), pp.115-148.

[6] – Kothare, M.V., Balakrishnan, V. and Morari, M., 1996. Robust constrained model predictive control using linear matrix inequalities. *Automatica*, *32*(10), pp.1361-1379.

[7] – Schwarm, A.T. and Nikolaou, M., 1999. Chance constrained model predictive control. *AIChE Journal*, *45*(8), pp.1743-1752.

[8] – Mayne, D.Q., 2014. Model predictive control: Recent developments and future promise. *Automatica*, *50*(12), pp.2967-2986.

470645 Robust Optimization Approximation of Chance Constrained Model Predictive Control

Abstract:

Uncertainty is an important issue to be considered in the control of process systems. Uncertainties in the system include parameter uncertainty, system disturbance and measurement noise. Conventional model predictive control (MPC) can have severe problems if the uncertainty is not properly dealt with. The problems include: severe constraint violation, instability of the system, unrealistic results, etc. In stochastic MPC, chance constraints are introduced to handle the uncertainty. A probability level of constraint satisfaction is defined by the user to seek a trade-off between robustness and optimality.

There are generally two approaches to handle the chance constraints: analytical methods and sampling methods. Typical analytical methods include inverse cumulative distribution function method, ellipsoidal method, and risk allocation method. Most of the aforementioned analytical methods require that the uncertainty distribution being Gaussian. Inverse cumulative distribution function method only applies under the condition of individual chance constraints. For one-dimensional Gaussian distribution, the inverse cumulative distribution function can be calculated directly. The resulting deterministic constraints are then equivalent to the chance constraints [1]. For joint chance constraints, the problem is much more complicated as there is no closed form of cumulative distribution function for multidimensional Gaussian distribution. But closed form exists if the integration area is an ellipsoid. However, the ellipsoidal method leads to very conservative results [2]. Risk allocation method is based on the Boole's inequality: the probability that at least one of the events happens is not greater than the sum of the probabilities of the individual events. So the joint chance constraint can be decomposed into individual chance constraints, which are much easier to solve [3]. Risk allocation can also be improved by optimizing the risk allocation to each individual chance constraint [4]. Sampling method has the advantage that it is not restricted to certain types of distribution. It includes mixed integer programming based sample average approximation approach and scenario optimization approach. However, those methods are computationally expensive [5,6] and scenario approach may lead to very conservative result [7].

In this work, we propose robust optimization approximation method to handle both the individual and the joint chance constraint in MPC problem. By introduction of the uncertainty set, the individual and joint chance constrained problem can be converted to tractable robust optimization problem. Various types of uncertainty set can be introduced to deal with different types of distribution. Box and polyhedral types result in linear constraints and ellipsoidal type results in second order cone constraints. Robust optimization approximation method preserves the simplicity of analytical methods and it is not restricted to certain type of probability distribution. Compared with other classical methods in the case study, robust optimization method leads to good performance in numerical case studies. The proposed method is also applied to the steam assisted gravity drainage (SAGD) process in the oil sands industry. Recursive data-driven modeling methods are applied to build models for the process. Chance constrained MPC is applied and the technique is tested with a reservoir simulator: Petroleum Experts. The robust optimization approximation based chance constrained MPC leads to satisfactory results.

Reference:

[1] Schwarm, Alexander T., and Michael Nikolaou. "Chance constrained model predictive control." *AIChE Journal* 45.8 (1999): 1743-1752.

[2] Van Hessem, D. H., C. W. Scherer, and O. H. Bosgra. "LMI-based closed-loop economic optimization of stochastic process operation under state and input constraints." *Decision and Control, 2001. Proceedings of the 40th IEEE Conference on*. Vol. 5. IEEE, 2001.

[3] Nemirovski, Arkadi, and Alexander Shapiro. "Convex approximations of chance constrained programs." *SIAM Journal on Optimization* 17.4 (2006): 969-996.

[4] Ono, Masahiro, and Brian C. Williams. "Iterative risk allocation: A new approach to robust model predictive control with a joint chance constraint."*Decision and Control, 2008. CDC 2008. 47th IEEE Conference on*. IEEE, 2008.

[5] Ruszczyński, Andrzej. "Probabilistic programming with discrete distributions and precedence constrained knapsack polyhedra." *Mathematical Programming* 93.2 (2002): 195-215.

[6] Blackmore, Lars. "A probabilistic particle control approach to optimal, robust predictive control." *Proceedings of the AIAA Guidance, Navigation and Control Conference*. No. 10. 2006.

[7] Calafiore, Giuseppe C., and Marco C. Campi. "The scenario approach to robust control design." *IEEE Transactions on Automatic Control* 51.5 (2006): 742-753.

463609 Recent Advances in the BARON Project

Abstract:

The BARON project for the global optimization of NLPs and MINLPs began in the early 1990s [1]. BARON has been at the forefront of the development of global optimization technology, pioneering techniques including the use of bounds tightening techniques [1, 2], polyhedral relaxations [2,3,4,5], convexity detection and use of multi-term relaxations [6,7], and multi-constraint relaxations [8,9].

In this talk, we (1) provide a brief account of key software engineering and algorithmic developments in the history of the software, (2) discuss recent advances, including the development of a systematic approach for diagnosing the cause of infeasibilities in MINLPs, acceleration via exploitation of optimality conditions, and learning approaches that result in dynamic solution strategies for solving difficult problems, and (3) present extensive computational results on a collection of benchmarks.

References cited

1. Ryoo, H. S. and N. V. Sahinidis, Global optimization of nonconvex NLPs and MINLPs with applications in process design, Computers & Chemical Engineering, 19, 551-566, 1995.
2. Tawarmalani, M. and N. V. Sahinidis, Global optimization of mixed-integer nonlinear programs: A theoretical and computational study, Mathematical Programming, Ser. A, 99, 563-591, 2004.
3. M. Tawarmalani and N. V. Sahinidis, A polyhedral branch-and-cut approach to global optimization. Mathematical Programming, 103:225-249, 2005.
4. Khajavirad, A. and N. V. Sahinidis, Convex envelopes generated from finitely many compact convex sets, Mathematical Programming, 137, 371-408, 2013.
5. Khajavirad, A. and N. V. Sahinidis, Convex envelopes of products of convex and component-wise concave functions, Journal of Global Optimization, 52, 391-409, 2012.
6. Bao, X., N. V. Sahinidis, and M, Tawarmalani, Multiterm polyhedral relaxations for nonconvex, quadratically-constrained quadratic programs, Optimization Methods and Software, 24, 485-504, 2009.
7. Bao, X., A. Khajavirad, N. V. Sahinidis, and M. Tawarmalani, Global optimization of nonconvex problems with multilinear intermediates, Mathematical Programming Computation, 7, 1-37, 2015
8. Zorn, K. and N. V. Sahinidis, Computational experience with applications of bilinear cutting planes, Industrial and Engineering Chemistry Research, 52:7514-7525, 2013.
9. Zorn, K. and N. V. Sahinidis, Global optimization of general non-convex problems with intermediate bilinear substructures, Optimization Methods and Software, 29:442-462, 2014

464673 Using Functional Programming to Recognize Named Structure in an Optimization Problem: Application to Pooling

Abstract:

The pooling problem is a nonconvex nonlinear optimization problem with applications including [1]: crude oil scheduling [2], water networks [3], natural gas production [4], fixed-charge transportation with product blending [5], hybrid energy systems [6], and multi-period blend scheduling [7]. It is possible to integrate additional complexity into the pooling problem, e.g., allowing mutable topological decisions [8] or nonlinear blending rules [9]. A wide variety of pooling variants with generic process networks applications can be found in MINLPLib.

In solving process networks optimization problems there is a common theme: it is much easier to solve large-scale instantiations of the standard, archetypal pooling problem than it is to solve variants including mutable topology, nonlinear blending, or temporal aspects. For example, a recent primal heuristic performs consistently well on the order of 10k variables and constraints [10], but the approach exploits the standard pooling network structure and does not apply to pooling variants.

Previous work in mixed-integer linear optimization (MILP) has found that using network structure can significantly help generate strong cutting planes [11]. Automatically identifying these embedded networks in large-scale optimization models is NP-hard [12], but there exist several polynomial time approximation algorithms to find good networks [13]. State-of-the-art MILP solver software, such as CPLEX uses preprocessing heuristics to automatically find these network patterns. More recent work has considered detecting more complex structures such as multi-commodity flow [14].

This manuscript proposes automatically recognizing pooling structure within a mixed integer nonlinear optimization problem (MINLP). The pooling structure inside of a generic process optimization problem is a subset of the entire problem, so specialized, pooling-specific, cutting planes will also be valid bounds for the entire process networks problem. We also show that primal heuristic solutions to the standard pooling problem would be a good starting point for primal heuristics for the entire optimization problem.

Identifying pooling problem structure hinges on pattern matching. Patterns are defined by which variables and coefficients are expected in a constraint and by constraint bounds. The implementation is in F#, a strongly typed functional programming language targeting the .NET runtime environment. Pattern matching, one of the most distinctive F# features, has many uses, from decomposing data to control flow. The core concept is defining how data is expected to look and acting accordingly.

We tested the implementation on 3 sets of input OSiL files: the 70 large-scale standard pooling Dey and Gupte [10] examples, the 16 standard and extended Misener et al. [15] examples, and the 1342 MINLPLib test cases. For each test set, we read in a flat optimization problem and try to produce a pooling network. The implementation successfully deduces the original network structure for all 70 Dey and Gupte [10] examples, i.e., large-scale, standard pooling problems with up to 40 input, 30 pool, and 50 output nodes. Running our implementation on the 1342 test cases in MINLPLib2, 3 produce complete pooling problems and 78 more produce a pooling-like network. We also show that, after detecting the pooling problem in the flat MINLP, we can apply a good heuristic approach to get a good approximation solution.

References:

[1] R Misener and CA Floudas. Advances for the pooling problem: Modeling, global optimization, and computational studies. Appl. Comput. Math., 8(1):3 – 22, 2009.

[2] J Li, A Li, IA Karimi, and R Srinivasan. Improving the robustness and efficiency of crude scheduling algorithms. AIChE J., 53(10):2659–2680, 2007.

[3] B Galan and IE Grossmann. Optimal design of distributed wastewater treatment networks. Ind. Eng. Chem. Res., 37(10):4036 – 4048, 1998.

[4] X Li, E Armagan, A Tomasgard, and PI Barton. Stochastic pooling problem for natural gas production network design and operation under uncertainty. AIChE J., 57(8):2120–2135, 2011.

[5] DJ Papageorgiou, A Toriello, GL Nemhauser, and MWP Savelsbergh. Fixed-charge transportation with product blending. Transport. Sci., 46(2):281–295, 2012.

[6] RC Baliban, JA Elia, R Misener, and CA Floudas. Global optimization of a MINLP process synthesis model for thermochemical based conversion of hybrid coal, biomass, and natural gas to liquid fuels. Comput. Chem. Eng., 42:64 – 86, 2012.

[7] SP Kolodziej, IE Grossmann, KC Furman, and NW Sawaya. A discretization-based approach for the optimization of the multiperiod blend scheduling problem. Comput. Chem. Eng., 53:122 – 142, 2013.

[8] R Misener and CA Floudas. Global optimization of large-scale pooling problems: Quadratically constrained MINLP models. Ind. Eng. Chem. Res., 49(11):5424 – 5438, 2010.

[9] R Misener, CE Gounaris, and CA Floudas. Mathematical modeling and global optimization of large-scale extended pooling problems with the (EPA) complex emissions constraints. Comput. Chem. Eng., 34(9):1432 – 1456, 2010.

[10] SS Dey and A Gupte. Analysis of MILP techniques for the pooling problem. Oper. Res., 63(2):412–427, 2015.

[11] TJ Van Roy and LA Wolsey. Solving mixed integer programming problems using automatic reformulation. Oper. Res., 35(1):45 – 57, 1987.

[12] GG Brown and WG Wright. Automatic identification of embedded network rows in large-scale optimization models. Math. Program., 29(1):41 – 56, 1984.

[13] RE Bixby and R Fourer. Finding embedded network rows in linear programs I. extraction heuristics. Manag. Sci., 34(3):342–376, 1988.

[14] T Achterberg and C Raack. The MCF-separator: Detecting and exploiting multi-commodity flow structures in MIPs. Math. Program. Comput., 2(2):125–165, 2010.

[15] R Misener, JP Thompson, and CA Floudas. APOGEE: Global optimization of standard, generalized, and extended pooling problems via linear and logarithmic partitioning schemes. Comput. Chem. Eng., 35(5):876–892, 2011.

466366 Tools and Workflows in the Design of Urban Energy Systems

Abstract:

This paper describes a tool for the interactive design of urban energy systems. The tool incorporates an optimisation model that can be used to select the resources, conversion processes and transport networks required to satisfy a spatially distributed and time varying pattern of demands across an urban area. The optimisation model can be formulated with specific goals and constraints such as minimising annualised costs while satisfying emission constraints. The optimisation model uses a temporal discretisation with two levels of time slices: major time periods for investment decisions, and minor time periods to represent the variation of demands within the major periods. The tool uses a Resource Technology Network to represent the potential technological choices for satisfying the demands. The tool provides a graphical interface for selecting the available conversion processes, resources and network links, and for specifying their operational parameters. This information is retrieved from a database implemented as a Protégé ontology and can be customised to reflect local conditions. The spatial grid and associated demands are specified as CSV files. Optimisation scenarios are constructed and submitted to a remote server. The results of the optimisation are retrieved and displayed within the tool. Resource flows and locations of conversion processes are visualised on a 2d representation of the spatial grid. Key performance indicators and a summary of results are displayed within tables and charts.

The tool described above is intended for use in interactive model development with an integrated database and results visualisation. The database and visualisation framework are implemented in Java, and the GAMS modelling language is used to formulate and solve the optimisation model. We are also developing an alternative implementation of the optimisation model written in the Pyomo modelling language. This is intended for use in integrated workflows where it could be connected with agent based models for estimating energy demands, and with other tools such as semantic models of urban areas. The Pyomo implementation could also be more convenient for use with tools for sensitivity analysis where the model has to be solved repeatedly.

467692 Plasmo: Platform for Scalable Modeling and Optimization

Abstract:

PLASMO (Platform for Scalable Modeling and Optimization) is a Julia-based open-source software platform that facilities the construction, solution, instantiation, and management of large-scale optimization problems on high-performance computers.

PLASMO is comprised of four core components: C1) Workflows: Enable fast and consistent instantiation and management of concurrent simulations generated from pre-specified model templates. C2) Modeling: Enables scalable and compact representations of applications by interfacing different modeling components. C3) Data and Analysis: Provides scalable and extensible templates to specify and communicate large-scale input model data from multiple sources in distributed environments. C4) Solver: Provides scalable and extensible solution templates that are compatible with the applications.

This presentation will focus on the modeling component of PLASMO. We discuss how to use the platform to target problems in multi-scale/hierarchical networks, integrated planning/scheduling/control, and stochastic programming formulations. For instance, PLASMO can be used to express hierarchical networks as trees where each node represents a component or a network itself. Each node has its own model, objective function, and connectivity of its children. In this way, PLASMO allows highly compact and scalable object-oriented expressions of the problem. PLASMO is already interfaced with serial solvers like IPOPT and parallel solvers like PIPS-NLP and DSP. When PIPS-NLP is used, PLASMO enables not only parallel model instantiation but also parallel evaluation of the Hessian, Jacobian, and objective/constraint functions.

469176 Analytics Framework and Infrastructure - Integrating Big-Data Cloud Technologies with First-Principle Modeling

Abstract:

In many industries sources of data are geographically distributed. Oil well pumps, electric motor drives, and heavy construction equipment are examples of such sources. These units require advanced and robust analytics capabilities to maximize their efficient utilization and minimize costs associated with unplanned downtime and uncertainty. While embedded control solutions manage the safe operation of the asset, analytics technologies can perform fault-diagnostics, operational predictions, and performance modeling for a battery of assets.

This paper presents the analytics framework and infrastructure built to integrate big-data cloud technologies with traditional process system engineering tools such as first-principle modeling, deterministic optimization techniques for scheduling, and process control. The engines include

- Proprietary model identification system which identifies the basic first-principles behaviors such as mass and energy balances, along with time-depended performance metrics such as fouling and physical performance degradation
- Scheduling platform developed for production scheduling of continuous processes with special emphasis on utility systems along with interactive interfaces for visual construction of optimization problems

We also present the deployment strategies for such engines which range from in-chassis deployment tightly coupled with the controller to deployment as cloud services for extreme scalability. The use of cloud facilities such as high throughput, multi-source data ingestion, stream processing of data, and elastic web applications is demonstrated through case-studies.

471420 Recent Developments in Pyomo

Abstract:

Computational tools for modeling mathematical programs are widely used within both academia and industry. Available commercial and open-source modeling packages support generic modeling by separating modeling constructs from instance data through concepts like sets, parameters, and parameterized constraints. However, their model representations are limited to constructs that directly correspond to established solver inputs. In general, this implies that mathematical programs must be expressed as either linear or nonlinear algebraic models; that is, a list of variables, an algebraic objective expression, and a list of algebraic constraints.

In contrast, the Pyomo environment is designed to be an open Python-based environment for developing and exploring complex, optimization-based analytical strategies. This includes support for extending the modeling environment to include non-algebraic modeling constructs, developing automated model transformation routines, and automating problem decomposition and interrogation.

In this presentation we highlight recent developments in the Pyomo project, including new language constructs for supporting modeling with units, preliminary support for native automatic differentiation, and new strategies for accelerating the solution of block-decomposable models.

Thermodynamic Consistency-Based Validation Approach for Equation of State Methods in Process Simulators

Seiya Hirohama, Prasad Narasimhan, Francisco Brana-Mulero,
Zeynep Nevin Gerek Ince, and David Bluck,
Schneider Electric Software LLE

Abstract

A software validation approach is proposed for phase equilibrium calculation module which uses equations of state (EOS). The approach is based on thermodynamic consistency through the entire process of the derivation and the programing. Test of the proposed approach was conducted for cases of Soave-Redlich-Kwong EOS combined with Huron-Vidal Mixing rule[1)] (SRK-HV) and Perturbed Statistical Associating Fluid Theory [2)] (PC-SAFT) EOS. The results indicated that the approach is enough sensitive to detect error, and eventually contribute a lot to ensuring the software quality.

1. Objectives

Development of physical property estimation module on process simulators requires proper validation approaches to confirm that the entire process of derivation and programing of a number of equations has been correctly performed. Conventional approaches, such as comparison between analytical solutions and numerical solutions, can be applicable to each step in the process. However, recent equations of state method involve huge numbers of complex equations and steps in the process, and eventually require increased efforts for validation. So, this work seeks for a new validation approach which can check consistency between input and output of the EOS methods as schematically indicated in Fig.1 to detect any error in the entire process.

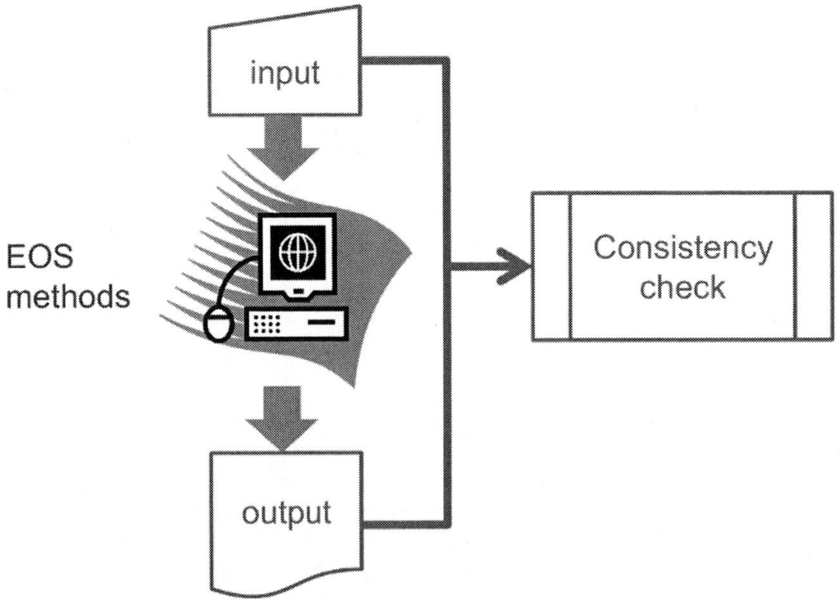

Fig.1 Concept of consistency check for phase equilibrium calculation

2. Methodology

Extensive variables such as Gibbs energy show 1st order homogeneousness[3)]. For example residual Gibbs energy at constant temperature and pressure can be expressed by the following equation using its composition derivatives.

$$G_{res,T,P} = \sum_i n_i \left(\frac{\partial G_{res,T,P}}{\partial n_i}\right)_{T,P,nj\neq i} \quad ...(1)$$

Applying similar concept to excess Gibbs energy yields the well-known Gibbs-Duhem equation for activity coefficient, which has been applied to variation of experimental data. In this work, we discuss validation approach based on Eq.(1) for EOS methods in process simulators.

3.Proposed Approach

Many of recent EOSs are described with residual Helmholtz energy at constant temperature and volume ($A_{res,T,V}$). Even if a EOS is defined by a traditional form of Eq.(2) that is explicit for pressure, mostly the form can be converted to that for $A_{res,T,V}$ by Eq(3). Thermodynamic properties can be derived from $A_{res,T,V}$.

$$P = f(n, V, x_i) \quad ...(2)$$

$$A_{res,T,V} = \int_{\infty}^{V} \left(\frac{nRT}{V} - \mathrm{P}\right) dV \quad ...(3)$$

Prediction of physical properties often gets through density calculation and fugacity coefficients using the density. This work proposed a validation approach which checked that the following two procedures gave exactly the same value for residual Gibbs energy at constant temperature and pressure ($G_{res,T,P}$).

1)Direct calculation from $A_{res,T,V}$

The following equation was directly obtained from the definition of $G_{res,T,P}$ and $A_{res,T,V}$.

$$\frac{G_{res,T,P}}{nRT} = \frac{[G-G_{ig}]_{T,P}}{nRT} = \frac{[G-G_{ig}]_{T,V}}{nRT} - ln(Z)$$

$$= \frac{[A-A_{ig}]_{T,V}}{nRT} + z - 1 - ln(Z) \quad ...(4)$$

2)Calculation from fugacity coefficient.

Just substituting fugacity coefficient into Eq.(1) yielded;

$$\frac{G_{res,T,P}}{nRT} = \sum_i x_i ln(\phi_i) \quad ...(5)$$

The dimensionless value of ($G_{res,T,P}/nRT$) by Eq.(4) must match that of Eq.(5), if all the derivation and programing of fugacity coefficient and density from EOS is perfect, and if the convergence tolerance for density is enough small.

4. Example Case

The following two tests were conducted. Eventually the test results indicated that the above validation approach was enough sensitive to detect errors.

1)SRK-HV

For the purpose of testing the approach, a standalone program for predicting vapor-liquid equilibrium (VLE) using SRK-HV was created to confirm the ability of the above validation approach. VLE was calculated as indicated Fig.2. When all the derivation and the programing had been perfectly done, ($G_{res,T,P}/nRT$) calculated by Eq.(4) perfectly matched that of Eq.(5) in a precision limit of computation. The deviation was less than a value typically like $1.0\text{x}10^{-15}$[-] (depending on calculation environment).

When we have intentionally bought an error of 0.01% in compressibility factor in the liquid phase, significant deviation in a magnitude of $1.0\text{x}10^{-4}$[-](depending on calculation environment) was found between the values of ($G_{res,T,P}/nRT$) as shown in Fig.3, even when the calculated VLE stays almost the same as shown in Fig.2 after the introduction of the error.

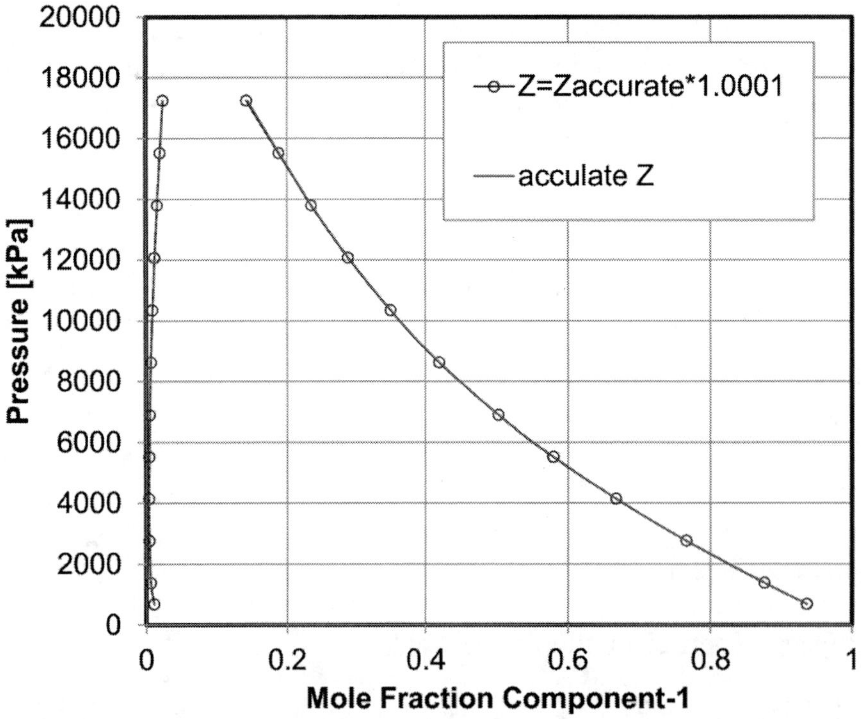

Fig. 2 Vapor-liquid Equilibrium for methane-heptane system at 233K

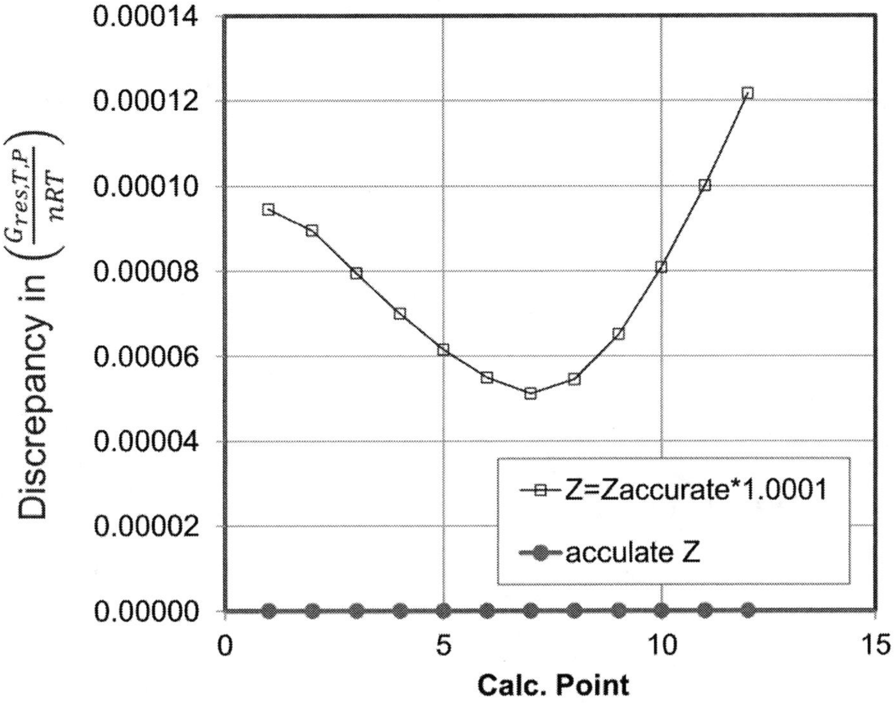

Fig. 3 Observed discrepancy for liquid phase in vapor-liquid equilibrium for methane-heptane system at 233K

2)PC-SAFT

Similar test was conducted for PC-SAFT in calculating VLE of propane-butanoate system in temperature ranges of 270K-370K. When all the derivation and the programing had been perfectly done, ($G_{res,T,P}/nRT$) calculated by Eq.(4) matched that of Eq.(5). The deviation was less than a value typically like $1.0\text{x}10^{-13}$ [-] (depending on calculation environment).

When we have intentionally bought an error of 0.01% in liquid density, significant deviation in a magnitude of $2.0\text{x}10^{-4}$[-](depending on calculation environment) was found between the values of ($G_{res,T,P}/nRT$), even when the calculated VLE stays almost the same as shown after the introduction of the error.

5. Conclusions

We have established a software validation approach based on thermodynamic consistency to confirm that the entire process of the derivation and the programing was perfectly conducted in prediction modules for phase equilibrium. The approach turned out to be enough sensitive to detect error, and eventually contribute a lot to ensuring the software quality.

[Nomenclature]

(Alphabet)

A: Helmholtz energy
G: Gibbs energy, n: quantity
n: Total molar number
R: Gas constant
T: Temperature
V: volume
x: mole fraction
Z: Compressibility factor
Ø: Fugacity coefficient

(Subscript)

i: component identification number,
ig: ideal gas,
res: residual

[Literature Cited]

1) Marie-Jose Huron and Jean Vidal; Fluid Phase Equil.3,255-271(1979)
2) Gross,J. et al. ; Ind. Eng. Chem. Res., 40 (4), 1244–1260 (2001)
3) Reid, C.E.; "Chemical Thermodynamics", McGraw-Hill, New York (1990)

466874 Managing Trade-Offs Between Food, Renewable Energy and Ecosystem Services

Abstract:

Competition between renewable energy and food production, driven by increasing food and energy demands, creates incentives to increase agricultural productivity. In the short term, productivity can be increased through the use of intensive farming management practices that use extensive tillage and large quantities of chemical fertilizers. Such practices have been shown to contribute to ecological degradation of farmland via an increase in soil erosion and a decrease in soil carbon, among other impacts. The surrounding watershed is also affected negatively by the nitrate and phosphorus-contaminated runoff from fertilized fields, which degrades the water quality and contributes to the growth of harmful cyanobacteria and other microorganisms. In the long term, this ecological degradation results in reduced agricultural productivity, such that increasing amounts of chemical fertilizers and other technological inputs are required. Farming for maximum productivity without accounting for the ecological impacts thus creates a vicious cycle, in which productivity is maintained by inflicting ecological harm.

Experimental studies have demonstrated that conservation farming practices, with reduced tillage and in some cases reduced amounts of fertilizers, insecticides and herbicides, can mitigate the ecological damage of conventional agriculture. Farming with reduced or no tillage helps to increase and maintain the carbon content in soil, providing an offset for the emissions produced in the agricultural life cycle, and reduces the rate at which soil erodes. Reducing chemical fertilizers or partially replacing them with soil amendments such as agricultural residues reduces the concentration of nitrates and phosphates in contaminated runoff and cuts down on wasted fertilizer as more of the applied fertilizer is taken up by crops. Conservation practices offer myriad ecological benefits but do not necessarily offer similar productivity benefits: the rate at which biomass is produced can be decreased under conservation farming, creating trade-offs between ecological preservation, agricultural production, and farm economics. With demand for both food and energy likely to increase for the foreseeable future, the need for food and bioenergy crops must be balanced against the need for ecological preservation.

The objective of this work is to design a system for the co-production of food and renewable energy with respect to three conflicting criteria: food productivity, energy productivity and net ecosystem service (ESS) supply. Ecosystem services are the goods (food, fuel, materials) and services (climate regulation, water quality regulation, erosion control) provided by nature that are necessary to support human activities, including agriculture. Here we define net ESS supply as the difference between ESS supply, provided by nature, and ESS demand, required by human activity. Climate regulation is a simple example. The net climate regulation supply is the quantity of greenhouse gas emissions generated by human activities, less the quantity of emissions that can be sequestered or otherwise removed from the atmosphere by ecological processes such as biomass growth and sequestration in soil. Net ESS supply provides a quantitative way to measure ecological degradation: a negative net supply means that the demand for ESS is greater than can be supplied by nature, and ecological degradation is the likely consequence. In this work, we will identify options for food and renewable energy co-production that have a net ESS supply greater than or equal to zero, for key services.

The system we consider is comprised of two stages: land use and biomass-to-energy conversion. Land use includes several options for crops both grain and cellulosic, rotation systems and farming management practices, as well as the production of wind and/or solar electricity. Land can also be used to create additional ESS supply through reforestation; however, land used for ESS supply cannot also provide food or energy, thus there is competition between the land uses. Options for biomass-to-energy conversions cover a range of liquid fuels, both ethanol and hydrocarbon, as well as electricity. Grain crops can be sold directly as food products or converted to liquid fuels; the system thus has a variety of configurations in which both food and renewable energy is produced from the same farmland. The land use and biomass-to-energy conversions are modeled at the value chain and equipment scales, respectively, of the process to planet (P2P) multi-scale modeling framework. For this system, the value chain scale may be thought of as a regional scale. The third and largest scale of the P2P framework, the economy scale, is used to capture the life cycle of all inputs to the land use and biomass-to-energy conversion stages. Net supplies of two ESSs are quantified as decision criteria: climate regulation, quantified with carbon dioxide emissions and carbon sequestration, and air quality regulation, quantified with nitrogen oxide emissions and removal. Carbon dioxide emissions (climate regulation demand) is quantified at all scales, equipment, regional and life cycle. Carbon sequestration (climate regulation supply) is quantified at the regional scale only, as the land use options are the only components of the system that provide a variable supply of ESS. Air quality regulation is quantified only at the smaller two scales, with supply again provided only by the land use options and demand created by both the land use options and the biomass-to-energy conversion options. Air quality regulation is not considered at the life cycle scale because the spatial presence of criteria air pollutants and its effects, are either local or regional in nature depending on the pollutant molecule.

To analyze trade-offs between food, energy and ecosystem services, the co-production system is optimized under several constraints. Food and energy production are constrained relative to the maximum production possible, to ensure that all optimal system designs are truly co-production systems. Where possible, net ESS supply is also constrained to be greater than or equal to zero; this is not always achievable for air quality regulation, but net climate regulation supply can reach above zero for many system configurations. Within these constraints, the system is optimized for maximum food production, maximum energy production, and maximum net ESS supply for both services. With the design space thus defined, Pareto curves are found between food production and net ESS supply, and between energy production and net ESS supply, in order to identify sets of compromise system designs. Preliminary results indicate that maximizing either food or energy productivity leads to the worst possible performance for net ESS supply, as expected. However, by either sacrificing some amount of productivity or by using land for both food and energy production and ESS supply, designs can combine adequate net ESS supply with relatively high food and energy productivity.

461799 Efficient Solar Thermal Hydrogen, Electricity and Fresh Water Coproduction Process Synthesis

Abstract:

Global population growth, increase in per capita GDP and change in consumption habits create increasing pressure on food, energy and water (FEW) supply. For the past century, with lower FEW needs and abundant fossil resources, energy, chemical, fertilizer and even fresh water supply via desalination have been predominantly dependent on limited fossil resources. However, concerns regarding limited fossil fuel reserves and increasing greenhouse gas (GHG) emissions from fossil fuels accelerate advancements and implementation of renewable energy conversion processes. Among these alternatives, solar energy conversion processes are particularly prominent because of sunâ€™s tremendous potential [1]. Yet intermittencies and land availability constraints for solar energy collection are the grand challenges for solar energy conversion technologies and demand high conversion efficiency and synergistically integrated energy storage solutions [2-3]. Recently, it has been shown that hydricity concept, synergistic solar thermal hydrogen and electricity coproduction, presents a potential solution for continuous and efficient power supply and also an exciting opportunity to envision and create a sustainable economy to meet all the human needsâ€"namely, food, chemicals, transportation, heating, and electricity [1].

Here, we introduce an integration strategy for the coproduction of fresh water, hydrogen and electricity from concentrated solar thermal energy and present potential impacts of the proposed system on FEW nexus. The integration of solar thermal power production and solar thermal hydrogen production techniques reduces the exergy losses associated with each one of the systems and thus provide an efficient solution. Solar conversion processes are evaluated based on the process *exergy efficiency* that refers to the fraction of incident solar exergy that is directly recovered as the net exergy output, which is defined as the sum of electricity and the hydrogen exergy output.

The proposed efficient solar trigeneration process presents: i) a continuous power and fresh water supply solution that

can achieve high solar conversion efficiencies when combined with the proper hydrogen power cycle and ii) numerous opportunities in chemical industry, transportation sector and in particular in food production via continuous supply of fresh water for irrigation and fertilizer production from coproduced hydrogen. Coproduced products can supply all FEW needs especially in the case of production of fertilizers from the coproduced hydrogen.

[1] Gençer E, Mallapragada DS, Marechal F, Tawarmalani M, Agrawal R. Round-the-clock power supply and a sustainable economy via synergistic integration of solar thermal power and hydrogen processes, Proceedings of the National Academy of Sciences (PNAS), 112(52), 15821-15826, 2015.

[2] Gençer E, Al-musleh E, Mallapragada D, Agrawal R. Uninterrupted Renewable Power through Chemical Storage Cycles. Current Opinion in Chemical Engineering, 5, 29-36, 2014.

[3] Gençer E, Agrawal R. A commentary on the US policies for efficient large scale renewable energy storage systems: Focus on carbon storage cycles, Energy Policy, 88, 477-484, 2016.

465212 Food-Energy-Water Nexus: Modeling Energy and GHG Emissions of Water Embodied in U.S. Domestic Food Transfers

Abstract:

Global population growth, climate change, and increasing urbanization are pressurizing already constrained essential food, energy, and water resources. Food, energy and water (FEW) systems are deeply interwoven, interconnected, and necessitate an integrated management approach. Central to understanding the complexity of the FEW nexus is quantitatively mapping and understanding their complex interdependencies. Existing studies have focused mostly on a single or two dimensions of the FEW nexus. There is a critical need for a holistic systems-level approach for studying FEW systems in a joint manner as opposed to studying them in isolation to avoid unintended consequences and encourage optimal outcomes. We integrate concepts of network theory with life cycle information to develop an understanding of the domestic FEW nexus for the United States.

Using publicly available disparate datasets, we develop a weighted and directed network model of interstate food transfers for the United States. The interstate food transfer model is translated into networks of embedded irrigation water, energy, and greenhouse gas (GHG) emission flows utilizing information on water footprints, irrigation datasets, and life cycle inventory databases. Specifically, we focus on four food commodity groups: livestock, cereal grains, meat, and milled grains. We utilize network theory tools and metrics to understand the structure, robustness, and environmental sustainability of these networks. Preliminary results indicate that over 600 million tons of food commodities were transferred across the U.S. along with 230 billion m^3 of virtual irrigation water. Additionally, 450 billion MegaJoules (MJs) of primary energy and 30 billion kg of CO_2-equivalent emissions were embedded in virtual irrigation water transferred across the U.S. It is observed that livestock and meat contributed only 13% by mass but accounted for 60% of embodied water, energy, and GHG transfers compared to grain based commodities. From a network perspective, the unweighted food transfer network is well-connected with majority of states participating in high volume of trade. However, the weighted network structure reveals that the majority of the food flow and embodied resources and emissions are controlled by a few states. Furthermore, meat and milled grain networks are denser compared to livestock and cereal grains networks. A robustness analysis highlights that despite the presence of vulnerable key states, the network is fairly robust to both random and targeted disruptions. These results reveal that while the domestic food transfer network is robust against extreme disruptions on specific nodes, the dense food transfer patterns of resource intensive products have important implications for sustainability of the food network.

465536 Water Footprint of Hydrotreated Renewable Jet Fuel Produced through Rapeseed Rotation with Wheat and Other Crops in North Dakota

Abstract:

Hydrotreated Renewable Jet (HRJ) fuel from oilseed plants has increasingly become important for the aviation sector to address energy security and climate change mitigation. Rapeseed is considered a promising HRJ source for its high quality oil content and the potentially attractive agro-economic benefit to replace the fallow period in wheat/fallow rotations. However, concerns have been raised about the viability of this important change in land use, within which water supplies is one important concern as large scale biofuels production may place pressure on fresh water supplies and water quality. Despite these achievements on analyzing the water impact of production of selected biofuel crops, investigation of rapeseed production for biofuel has been limited. This study develops assessment informed by inputs from multiple models to evaluate the environmental impacts of water utilization and water quality of the different production options for HRJ production from rapeseed cultivation in North Dakota. The biogeochemical based EPIC model is incorporated to simulate the crop growth that influences the hydrological cycle. Output of the land-use management changes from the EPIC is used to run the blue water footprint analysis using the LCA software SimaPro 8.0. Evapotranspiration data from the EPIC model is incorporated to calculate green water footprint during cultivation. The UNESCO Grey Water Footprint Accounting Tier 1 Guideline is followed to estimate the grey water footprint caused by nutrient leaching and runoff. In addition, inputs for multi-modal transportation logistics were obtained from the Alternative Fuel Transportation Optimization Tool (AFTOT) models in collaboration with the US Department of Transportation Volpe National Transportation Systems Center. Co-product allocation methods were employed in scenario analyses, such as energy allocation and market value allocation, therefore producing a robust analysis of water footprint of the rapeseed HRJ life cycle with compiled life cycle inventory.

Our presentation details the methods, assumptions and initial water footprint results for different scenarios of importance in North Dakota, one of the 10 US states studied in this project, when considering the rapeseed HRJ life cycle. Preliminary results indicate that HRJ fuel derived from rapeseed in rotation with wheat has a much larger water footprint when comparing to the conventional jet fuel. In addition, although it is usually neglected, the indirect blue water footprint caused by the energy usage and the material inputs during HRJ production pathway is worth considering due to its important impact on the overall water footprint. Therefore, policy makers should consider highly the water factor when making decisions on transition to large share of biofuels.

468516 NexSym – a Local Nexus Simulation System

Abstract:

Food, energy and water are essential needs to sustain human life. The complex interactions and interdependencies among the infrastructures and decision making processes involved in the provisioning of such needs have recently generated great debate under the umbrella of the Food-Energy-Water (FEW) nexus (also referred to as Water-Energy-Food nexus). As society becomes increasingly aware that resources available to satisfy the needs of a growing population are finite, there is a call to look at the interactions and start creating more efficient solutions to manage the wide range of nexus scenarios, especially under conditions of climate change, increased urbanisation and waste generation. Furthermore, there is awareness of the importance of maintaining ecosystems to achieve sustainable production while expanding land use and using it more intensively to obtain more food, energy and clean water. The nexus may manifest in unique ways in different localities. Similarly, the global conditions will affect each location differently. Therefore, it is urgent to develop holistic analytical tools that could inform decision making for managing the interdependencies of the FEW nexus at the local scale. This is one of the ways chemical engineers are able to contribute with solutions by applying their skills of modelling processes and managing constraints associated with an interconnected system.

In this work, we present the **Local Nexus Simulation System (NexSym)** that integrates models for technological process components involved in the production of food, energy and clean water as well as waste treatment. Furthermore, the techno-ecological interactions can be simulated by integrating models for ecosystem processes, including agricultural production. The tool then allows the user to build a "flowsheet" of the local system with the component of interest (e.g. bioenergy production) and their interacting local system components (e.g. biomass cultivation, water provisioning). This enables examination of the whole picture of FEW nexus interactions at the local scale, which can then be investigated within a global context. Options for managing the nexus from an engineering perspective can then be devised and further evaluated under scenarios of interest to the user. The **NexSym** is expected to have wide application by engineers working across the nexus system components, planners for urban or rural development, policy makers and decision makers in general. The tool will be illustrated with sample case studies featuring interactions between food, energy, water provisioning and waste treatment to demonstrate the usefulness of such a tool in understanding, managing and creating sustainable solutions to the nexus.

472086 Gibbsian Game Theory for Tragedy of the Commons Problems in Food-Energy-Water Sustainability

Abstract:

Sustainability problems in the food-energy-water nexus often involve "tragedy of the commons" (TOC) problems. TOC challenges are often modeled in game theory using n-player Prisoner's Dilemma (NPD) games and other games. However, classical solutions to game theory problems – here called "Nashian games" – have a number of shortcomings, and do not always conform to experimental data. In this talk we treat strategic games as a set of chemical reactions, and we treat choices as types of reactive molecules. Thus, this model provides new solutions to classical games. By using the usual Gibbsian approach to chemical reaction equilibrium, we are able to employ entropic considerations and minimize a type of Gibbs free energy to find game theory equilibria, overcoming the usual shortcomings in game theory. We solve classical games, including the 2-player PD game and the NPD game, using the "Gibbsian game theory" approach, which accounts for "entropic choices". We then show how treating strategic games as a set of chemical reactions can be used to represent decision situations more accurately than typical games, and we solve for the equilibrium of sets of reactions. Then, we apply Gibbsian solutions to NPD games to problems in sustainability. By quantifying concepts like "trust", "power", and "corruption", we explore opportunities for nudging outcomes of contested decisions in sustainability decisions. When overlapping considerations occur, as in the food-energy-water nexus, the NPD games reveal fundamental political conditions that increase the probabilities for when genuine human solutions can advance. In this talk we describe Gibbsian game solutions for NPD problems, and show how they apply to sustainability problems in the food-energy-water nexus.

461657 Water and Energy Systems in Sustainable City Development: Agent-Based Modelling and Resource Technology Optimization

Abstract:

The grand challenges facing human societies are all closely interconnected with the sustainable provisioning of energy, water and material resources for constantly growing and developing populations, as well as the subsequent processing and management of wastes and pollutions [1]. Urban water and energy systems are expanding and require attention and careful planning to minimize their economic and environmental impact both from an operational and investment perspective. However, there is a lack of a sufficiently systematic approach to address these impacts and achieve the optimal strategies either in the context of research or real-world decision making [2].

Facing the challenges of energy and water systems in urban development, this work describes a decision-making platform combining a comprehensive database, agent-based simulation modelling and resource technology network optimization. The current regional demographics, land-use, infrastructure and economic information is used as input for the initial state of the focused urban system incorporating energy, water and other related resources. Detailed spatio-temporal demand data, obtained by simulating a synthetic population using agent-based models, is further used to plan capacity utilization and expansion by supply-side matching on a cost optimal basis, eventually aiming to explore the optimal design and operational strategies for residential, commercial, industrial, and other sectors in urban energy and water systems, and especially their nexus. Moreover, long-term socio-economic scenarios are addressed in the urban system development through a real-time feedback loop.

The outputs depict, with fine spatial and temporal resolutions, an overall energy-water landscape of the studied urban region (e.g., Greater Accra Metropolitan Area in Ghana as a prototype [3]). The material and energy balance of supply and demand from both macro and micro perspectives are obtained, and further used to propose environmental friendly and cost effective sustainable city investment and development strategies. Several representative use cases are built

under different scenarios to maximize utilization of resources, minimize net waste generation, and improve social welfare to engage local stakeholders. This work is to become a core component of the *resilience.io* platform as an open-source, data-driven, integrated systematic tool gathering social, environmental and economic data to inform urban planning, investment and policy-making for city-regions globally.

References

[1] Bringezu, S. and Bleischwitz, R. eds., 2009. *Sustainable resource management: global trends, visions and policies*. Greenleaf Publishing.

[2] Keirstead, J. and Shah, N., 2013. *Urban energy systems: an integrated approach*. Routledge.

[3] Koppelaar, R., Sule, M.N., Kis, Z., Mensah, F., Wang, X., Triantafyllidis, C., van Dam, K.H., Shah, N., 2016. Urban water and sanitation flows in the Greater Accra Metropolitan Area. *Urban Water Journal* (submitted).

464359 Visualization and Analysis of Periodic Process Data

Abstract:

The operation of periodic processes differs significantly from that of continuous processes and batch processes: the well-defined operating state is replaced by a limit cycle, which can in principle repeat an infinite number of times. Typical examples include adsorption-based separation systems (e.g., pressure or temperature swing adsorption); however, the periodic operation of, e.g., chemical reactors and biological systems 1, 2 has also been explored. Research efforts have also been expended on fast simulation of periodic processes3.

While the design, operation and simulation of periodic processes have received some attention in the literature, periodic process monitoring remains an open question. In this presentation, we discuss a novel geometric method for data-driven monitoring of periodic processes. Our work is an extension of the time-explicit Kiviat (radial) plot visualization and fault detection frameworks that were previously developed for both continuous processes4 as well as batch processes5. In this framework, each sample data point of the multivariate time-series collected from process operations is represented in a radial plot. These plots are stacked vertically in the order they were acquired, resulting in a time-explicit representation of the multivariate time series. The geometric properties of this setup allow for the process state at a given time to be represented as a single point, the centroid of the corresponding plot in radial coordinates.

To extend these ideas to periodic processes, we propose a two-step monitoring and fault detection algorithm. In the first step, we conduct inter-cycle fault detection by defining a centroid for each cycle and identifying problematic cycles. Based on the principles developed for fault detection in continuous processes, this is done by using several normal operating cycles to obtain a confidence ellipse for the centroids of cycles represented in radial coordinates â€“ any cycles whose centroids falls outside of this region is deemed to be a problematic or faulty cycle.

After the problematic cycles are determined, it is often desirable to determine the timing of a fault within a cycle. Thus, in the second step, we conduct intra-cycle fault detection. In this application, we build x confidence ellipses for normal operating cycles that have a period of x. The samples in the problematic cycle are then compared on a sample-by-sample basis with these confidence ellipses to identify when the deviations begin.

We also introduce ancillary mechanisms for determining the period of operation of the process and for defining the normal operating state, and discuss a strategy for online implementation. We demonstrate our methodology on two simulation case studies â€“ a CSTR simulation with oscillatory set points, and a pressure-swing adsorption (PSA) system.

References

Sowa, S. W., Baldea, M., & Contreras, L. M. (2014). Optimizing metabolite production using periodic oscillations. PLoS Comput Biol, 10(6), e1003658.

Sterman, L. E., & Ydstie, B. E. (1990). The steady-state process with periodic perturbations. Chemical Engineering Science, 45(3), 721-736.

Pattison, R.C., Schmal, P., & Pantelides, C. (2015). Efficient Computation of Cyclic Steady States in Periodic Adsorption Processes Using the JFNK Method. 2015 AIChE Annual Meeting, 431146

Wang, R. C., Edgar, T. F., Baldea, M., Nixon, M., Wojsznis, W., & Dunia, R. (2015). Process fault detection using timeâ€explicit Kiviat diagrams. AIChE Journal, 61(12), 4277-4293

Wang, R., Baldea, M., & Edgar, T.F. (2015). Visualization and Data-Driven Monitoring of Batch Processes. 2015 AIChE Annual Meeting, 418567

470311 Iot-Enabled Cybermanufacturing: Challenges and Possibilities

Abstract:

Although there are many names used to describe next generation manufacturing systems, such as cybermanufacturing and smart/advanced manufacturing, the essence of these is the application of increasingly powerful and low-cost computation and networked information-based technologies in manufacturing enterprises[1].

One potential enabler for cybermanufacturing is the Internet of Things (IoT), and there is a general consensus that factories and plants connected to the Internet are more efficient, productive and smarter than their non-connected counterparts[2], [3].

Industrial IoT devices are sensors/actuators, computers with wireless networks, and, most importantly, systems that are small and easy to embed. IoT devices offer the opportunity to instrument systems with massive numbers of sensors. With the huge amount of data and the programmability of IoT devices, comes the opportunity to shape the data received, to address local redundancy of information, and to improve both the accuracy and precision of measurements locally and across a distributed parameter system such as a reactor. However, many challenges remain in the areas of hardware such as sensor/actuator/network reliability, sensor accuracy; software such as protocols, data management, cyber security; and data analytics such as how to address the 4V's of Big Data[4].

In this work, we test IoT temperature sensors on a batch system with the goal of better understanding and control of the system. Complete description of the batch system setup, including IoT hardware and software, will be given. Results obtained based on the batch system, challenges encountered and ways to address them, and potential application of IoT devices in chemical processes will be discussed.

References:

[1] S. C. of the A. M. P. 2. 0. (AMP2.0), "Report to the president: accelerating U.S. advanced manufacturing." 2014.

[2] P. C. Evans and M. Annunziata, "Industrial internet: Pushing the boundaries of minds and machines," *Gen. Electr.*, 2012.

[3] F. Shrouf, J. Ordieres, and G. Miragliotta, "Smart factories in Industry 4.0: A review of the concept and of energy management approached in production based on the Internet of Things paradigm," in *Industrial Engineering and Engineering Management (IEEM), 2014 IEEE International Conference on*, 2014, pp. 697–701.

[4] S. J. Qin, "Process data analytics in the era of big data," *AIChE J.*, vol. 60, pp. 3092–3100, 2014.

463600 An Optimization-Based Approach for Learning Simple Parametric Surrogate Models

Abstract:

Data obtained through simulations or experiments are routinely used to inform process decisions and build system models. In order to describe behavior in complex systems with non-linear behavior, non-parametric methods, such as artificial neural networks or support vector machines are routinely used to build accurate system models. However, these models often suffer from overfitting. Moreover, their non-convex functional forms make them difficult to interpret and incorporate directly into algebraic optimization algorithms. In order to develop simple, yet accurate algebraic models we have recently developed the ALAMO methodology to learn models from exogenous data [1]. ALAMO performs a number of nonlinear transformations of input variables to populate a regression basis set.

In this paper, we present a systematic computational study of several fitness metrics that can be used in an optimization-based subset selection methodology to identify an optimal subset of regression variables. These metrics include Mallows' Cp, Akaike's information criterion, and Bayesian information criterion amongst others. The resulting models consist of a linear combination of nonlinear transformations of input variables, and their simple algebraic form can help provide insight on the system at hand. We complement these exact optimization algorithms with fast heuristics and describe their computational performance in ALAMO. Moreover, we present a systematic comparison between ALAMO's optimization-based approach to model fitting from data with a number of other parametric model building methods, including the lasso implementation in Matlab [2] and R's leaps routine [3].

References

[1] Cozad, A., N. V. Sahinidis, and D. C. Miller, Automatic learning of algebraic models for optimization, AIChE Journal, **60**, 2211-2227, 2014.

[2] http://www.mathworks.com/help/stats/lasso.html

[3] https://cran.r-project.org/web/packages/leaps/leaps.pdf

463221 M.R.Q.P.: Prediction of Final Batch Quality Using a Multi-Resolution Framework

Abstract:

Introduction

Global competition, legislation, and consumer demands impose ever stricter targets for efficiency, safety, profitability, and reduction of environmental impact on process industry. In this context, accurate and quick assessment of product quality is a critical task as it speeds up production by reducing the delay incurred by quality analysis, enabling either quick response to deviating product quality or faster product release.

The issue of quality assessment is most pronounced for batch processes, which are widely used for the production of goods with a high added value such as pharmaceuticals, specialty chemicals, etc. [1] Batch processes are non-stationary by design, commonly owing to physical and/or (bio)chemical limitations on the processes. Online measurement of instantaneous quality is difficult or even impossible [2], as quality of a batch can often only be sensibly defined after batch completion. Because all batch are independent from each other, quality must also be assessed after completion of each run. As a result, slow quality assessment can represent a substantial delay in production as it must be performed after. This is in large contrast with continuous processes, where process and quality variations are much smaller and quality measurements must not be carried out as often.

Mathematical models provide a cheap and fast alternative for quality estimation compared to lab analysis [3]. For batch processes, models based on Multiway Partial Least Squares (MPLS)[2] are the most popular [4]. These PLS models use profiles of online measurements to predict the final quality. Typically, the selected sensors are included in the PLS model at their highest available sampling rate, which is determined during commisioning of the control system and not guaranteed to be optimal for quality prediction as different variables carry information on different time scales [5]. Although PLS models are capable of efficiently dealing with measurement noise as well as the data redundancy present in batch data [6,7], a priori removal of uncorrelated and/or noisy model inputs yields better prediction accuracy [8].

This work presents a Multi-Resolution Quality Prediction (MRQP) framework for batch quality estimation. Based on the insight that batch data inherently contains structured correlation (namely the time evolution of each sensor), MRQP searches for the optimal resolution for each variable's measurement profile with respect to batch-end quality.

Methodology

The presented Multi-Resolution Quality Prediction (MRQP) framework for batch quality first employs a multi-resolution wavelet decomposition [9] to obtain a set of approximations at different resolution of each variable's measurement profile.

Next, an adapted (robust) version of Sequential Forward Floating Selection (SFFS)[10] is used to select the set of variable/resolution combinations that yields the best prediction of the batch quality. In this step, each different resolution of each variable is treated as a separate candidate input for an MPLS model. To improve the robustness of optimal variable/resolution selection, multiple crossvalidations are performed to assess model performance, and variability over these repetitions is exploited.

This combination of simultaneous optimal variable and resolution selection results in a multi-resolution MPLS batch quality prediction model that is theoretically guaranteed to perform at least as well as the single-resolution MPLS prediction model (i.e., the model employing all sensors at the highest sample rate). The MRQP approach yields a model with improved accuracy and stability.

Case Studies

The proposed MRQP is tested in three case studies.

The first case study consists of a simulated batch process for growth of *Saccharomyces cerevisiae* [11]. When comparing the performance of the "minimally optimal" models (i.e. models for which the addition of an extra variable/resolution combination to the input set does not improve the crossvalidation performance singificantly), the multi-resolution model achieves a median crossvalidation error 3.2% lower than the single-resolution model. When the overall lowest crossvalidation error is compared, the multi-resolution models outperforms the single-resolution model by 3.5%. In both situations, model complexity was similar.

The second case study is a simulated fermentation process for the production of penicillin [12,13], a widely used benchmark for Statistical Process Control. Here, the median error of the minimally optimal multi-resolution model is 2.2% lower than that of the corresponding single-resolution model. The overall lowest median crossvalidation error for the MRQP model is 1.9% than for the single-resolution approach. In both situations, the better MRSP performance requires more latent variables in the MPLS model, but the dimensionality of the input matrix is substantially lower (4 vs. 2048, and 5 vs. 3072), making the multi-resolution model much easier to interpret.

While the first two case studies were simulated, the final case study is taken from an industrial polymerization process [7] to demonstrate the practical applicability of the MRQP. The minimally optimal MRQP model outperforms its single-resolution equivalent by 7.6%, but requires a slightly larger input space dimensionality. The overall best crossvalidation performance of the multi-resolution model is 5.2% better than that of the best single-resolution MPLS model. In this situation, the multi-resolution model requires fewer latent variables and a lower input space dimensionality than the single-resolution model.

In all case studies, it was observed that less significant variables are typically included in the model at coarser resolutions. This has the benefit that the dimensionality of the input space is only increased a little, reducing the potential for spurious correlations to influence the model. Hence, the MRQP framework results in models that are inherently more robust towards the selection of too many predictors.

Conclusions

This work introduced Multi-Resolution Quality Prediction (MRQP) for batch processes. Instead of employing all available sensors at their maximal available sample rate, MRQP exploits the presence of structured correlation in the batch data, namely between subsequent time points.

Hereto, a multi-resolution framework is first used to compute approximations of varying resolution for each variable, followed by an optimal input selection approach to identify the optimal set of variable/resolution combinations.

The MRQP model for quality prediction is theoretically guaranteed to be at least as accurate as the single-resolution model. In three case studies, however, performance improvements of 1.9%-7.6% were observed, often in combination with more parsimonious model structures. In addition, multi-resolution models were found to be more robust with respect to the selection of too many predictors. The presented MRQP framework is readily extendable towards multi-phase processes, as well as towards other regression techniques such as multiway, dynamic, or kernel models [14,15,16].

In conclusion, MRQP for multi-resolution batch-end quality prediction offers significant benefits over traditional single-resolution approaches.

Acknowledgements

Work supported in part by Project PFV/10/002 (OPTEC Optimization in Engineering Center) of the Research Council of the KU Leuven, Project IAP VII/19 (DYSCO Dynamical Systems, Control and Optimization) of the Belgian Program on Interuniversity Poles of Attraction initiated by the Belgian Federal Science Policy Office, and the SCORES4CHEM knowledge platform (www.scores4chem.be). Marco S. Reis acknowledges financial support through project PTDC/QEQ-EPS/1323/2014 co-financed by the Portuguese FCT and European Union's FEDER through the program "COMPETE 2020". The authors assume all scientific responsibility.

References

1. S. Stubbs, J. Zhang, J. Morris (2013) Multiway Interval Partial Least Squares for batch process performance monitoring. *Ind Eng Chem Res* 52:12399-12407
2. P. Nomikos, J.F. MacGregor (1995) Multi-way partial least squares in monitoring batch processes. *Chemom Intell Lab Syst* 30:97-108
3. S.J. Qin, Y. Zheng (2013) Quality-relevant and process-relevant fault monitoring with concurrent Projection to Latent Structures. *AIChE J* 59(2):496-504
4. N. Lu, F. Gao (2005) Stage-based online quality control for batch processes. *Ind Eng Chem Res* 45:2272-2280
5. M.S. Reis, P.M. Saraiva (2006) Multiscale statistical process control using wavelet packets. *AIChE J* 54(9):2366-2378
6. R. Bro (1996) Multiway calibration. Multilinear PLS. *J Chemometr* 10:47-61
7. G. Gins, B. Pluymers, I.Y. Smets, J. Espinosa, J.F.M. Van Impe (2011) Prediction of batch-end quality for an industrial polymerization process. *LNCS* 6870:314-328
8. J. Trygg, S. Wold (2002) Orthogonal projection to latent structures (O-PLS). *J Chemometr* 16:119-128
9. I. Daubechies (1992) *Ten lectures on wavelets*. SIAM, Philadelphia (PA, USA)
10. P. Pudil, J. Novovicová, J. Kittler (1994) Floating search methods in feature selection. *Pattern Recogn Lett* 15:1119-1125
11. F. Lei, M. Rotbøll, S.B. Jørgensen (2001) A biochemically structured model for Saccaromyces cerevisiae. *J Biotechnol* 88:205-221
12. G. Birol, C. Ündey, A. Cinar (2002) A modular simulation package for fed-batch fermentation: Penicillin production. *Comput Chem Eng* 26:1152-1565
13. J. Van Impe, G. Gins (2015) An extensive reference dataset for fault detection and identification in batch processes. *Chemomter Intell Lab Syst* 148:20-31
14. D.J. Louwerse, A.K. Smilde (2000) Multivariate statistical process control of batch processes based on three-way models. *Chem Eng Sci* 55(7):1225-1235
15. J. Chen, K.-C. Liu (2002) On-line batch process monitoring using dynamic PCA and dynamic PLS models. *Chem Eng Sci* 75:63-75
16. R. Rosipal, L.J. Trejo (2001) Kernel Partial Least Squares regression in Reproducing Kernel Hilbert Space. *J Mach Learn Res* 2:97-123

462425 Process Knowledge Discovery and Selecting Number of Non-Zero Loadings in Sparse Principal Component Analysis

Abstract:

Process knowledge discovery and selecting number of non-zero loadings in sparse principal component analysis

Shriram Gajjar*, Murat Kulahci**, Ahmet Palazoglu*

*University of California, Davis, CA 95616, USA

(Tel: 530-752-8774; e-mail: anpalazoglu@ucdavis.edu).

**Technical University of Denmark, Lyngby, Denmark and Luleå University of Technology, Luleå, Sweden

(e-mail:muku@dtu.dk)

Background: Smart production technologies that are implemented today have dramatically intensified data generation and collection through networked information-based technologies throughout the chemical industry and other manufacturing enterprises. The data generation and collection are so fast-paced that humans have to rely on computers for consuming as well as processing the data. This, in turn, leads to an ever increasing pace for the development of algorithms and methods to improve process performance and facilitate process monitoring. The algorithms and methods should, at first, be able to extract significant information from the large datasets. Second, they should provide accurate means to reduce process variability and boost performance. Third, they should allow discovery of the underlying process dynamics that can substantially improve decision-making. Finally, steps can then be taken to move towards recommending preemptive actions (preventive decisions that are made before a failure occurs or is even observed).

Prior Work: Researchers have used principal component analysis (PCA) to capture meaningful information in a reduced dimensional space. PCA-based monitoring methods are among the most widely used multivariate statistical methods (Cinar et al., 2007). Using PCA for dimension reduction has one specific drawback where each principal component (PC) is a linear combination of all *m* variables and the loadings are typically nonzero. Such nonzero loadings (NZL) make it difficult to interpret the derived PCs and may confound subsequent analyses. To address this challenge, Zou et al. (2006) proposed sparse principal component analysis (SPCA) in which sparse loadings are obtained by imposing the lasso (elastic net) constraints on the coefficients (i.e., loadings) of the PCA model. SPCA essentially is the result of an optimization of the trade-off between variance captured by PCs and the sparsity imposed on PCs. It allows the user to control the sparsity of the loadings and improve the ability to identify the important variables.

Preliminary results: One of the challenges in using SPCA is in deciding the penalty parameters or choosing the number of non-zero variables/loadings (NNZL). We propose three approaches *viz.* exhaustive selection, forwards selection and sensitivity analysis that simplify the process of selecting penalty parameters and provide a more intuitive solution for understanding the physical meaning of variables monitored in chemical processes. In the exhaustive search approach one goes through all possible combinations of NZL in each PC, then chooses a solution that meets the required criteria. The downside of this approach is that it is computationally intensive and, in scenarios with large number of NZL, it is impractical and even infeasible to go through all combinations. In the forward selection approach we impose constraints on each SPC and a lower limit on the total variance captured by the sparse principal components (SPCs) is also imposed. By doing so, the search space for the optimum number of NNZL for each SPC is drastically reduced. Sensitivity analysis is a systematic review of the NNZL on SPCs. In this approach, the NNZL on a PC is varied keeping all other aspects constant. The goal is to determine if the NNZL on a PC can be made sparser without losing information. Thus, the traditional PCA can be altered in such a way that the obtained loadings would have a clear interpretation without significant loss of information extracted in each PC in terms of explained variance. Such an approach would also assist in the application of PCA in process surveillance as better understanding of the impact of PC loadings can clearly facilitate process monitoring, i.e., fault detection and diagnosis. Furthermore, we discuss the advantages of SPCA for process knowledge discovery with a synthetic example and the Tennessee Eastman benchmark process. The paper will highlight the substantially improved performance of process fault detection and diagnosis strategies using SPCA when compared with traditional approaches.

References

Cinar, A., Palazoglu, A. and Kayihan, F. (2007) 'Multivariate Statistical Monitoring Techniques', *Chemical Process Performance Evaluation Chemical Industries: CRC Press*, pp. 37-71.

Zou, H., Hastie, T. and Tibshirani, R. (2006) 'Sparse Principal Component Analysis', *Journal of Computational and Graphical Statistics*, 15(2), pp. 265-286.

473188 Deep Reinforcement Learning Approach for Process Control

Abstract:

In the recent years, there has been significant progress in the fields of computer vision and natural language processing that followed the success of deep learning [1]. Human level control has been attained in games [2] and physical tasks [3] by combining deep learning with reinforcement learning resulting in 'Deep Q Network' [2]. In the process industry, Model Predictive Control (MPC) has been found to be an effective control strategy. However, application of MPC on nonlinear stochastic systems can be computationally demanding and may require estimation of hidden states in a complex system. The performance of MPC depends significantly on the quality of model and hence any Model Plant Mismatch (MPM) would be detrimental to the performance. In this work, we use deep learning and reinforcement learning for controlling process variables. We present an actor-critic based approach to deterministic policy gradient reinforcement learning algorithm for control. During the training phase of the learning algorithm, we consider the standard reinforcement learning setup, where an agent (controller) interacts with

an environment (process) through control actions and receives a reward in discrete time steps. The training is done with full state observation. Deep neural networks serve as function approximators and are used to learn the control policies. Once trained, the learned network acquires a policy that maps system output to control actions and it can be used to control the plant without full state observations. Since online optimization is not required as in MPC our algorithm is computationally less intensive during online operation phase. We evaluated our approach on Single Input Single Output Systems and Multiple Input Multiple Output Systems with varying set points and initial conditions and compared it with MPC.

References:

[1] Krizhevsky, Alex, Sutskever, Ilya, and Hinton, Geoffrey E. Imagenet classification with deep convolutional neural networks. In Advances in neural information processing systems, pp. 1097â€"1105, 2012.

[2] Mnih, Volodymyr, Kavukcuoglu, Koray, Silver, David, Rusu, Andrei A, Veness, Joel, Bellemare, Marc G, Graves, Alex, Riedmiller, Martin, Fidjeland, Andreas K, Ostrovski, Georg, et al. Human-level control through deep reinforcement learning. Nature, 518(7540):529â€"533, 2015.

[3] Timothy P. Lillicrap , Jonathan J. Hunt , Alexander Pritzel, Nicolas Heess, Tom Erez, Yuvalassa, David Silver & Daan Wierstra, Continuous control with deep reinforcement learning, International Conference on Learning Representations, 2016.

452397 The Distribution of Online News Evaluated By Chemical Engineering and Process Control Tools

Abstract:

Many decisions we make in our personal and professional life are influenced by news articles in the mass media. Throughout the last years our access to news has significantly changed. The constant and immediate access to new articles facilitated by the global connection has greatly accelerated the dissipation of information. Triggered by the need of understanding the flow of information we decided to utilize traditional chemical engineering tools to describe the dissipation rate of news from an unique event through the mass media to the final reader.(1)

As data source we firstly focused on news triggered by scientific events (publication of new articles in scientific journals), as the flow of such news can be monitored precisely by a free-to-use online tool (Altmetrics). Data was aggregated from the online tool and a linear system identification method was applied to identify the main characteristics of the processes involved. The data is discussed in terms of time scale analysis, but also in terms of preferential attachment (i.e. aggregation) phenomena.

In a second step the derived characteristics of scientific news were compared with data on political news today and in the past (prior to the internet). In this comparison the metrics of time constants and gains were highly beneficial and resulted in a quantitative insight into describing the flow of information.

We believe that the discussed approach is not only of educational value, but that traditional chemical engineering tools (e.g. linear system response, aggregation and diffusion phenomena) are highly valuable in identifying and quantifying processes in big data.

(1) RN Grass, WJ Stark, AIChE J. 62, 1104 (2016).

472562 Identifying Circadian Drug Targets for Maintained Oscillatory Precision

Abstract:

The mammalian circadian clock is a complex dynamic system comprised of interlocked transcription-translation feedback loops. Approximately 10% of all mammalian transcripts are under circadian regulation, highlighting the importance of this system. Disruptions such as shift work, high-fat diets, or aging have been shown to reduce the amplitude of circadian oscillation, and these diminished amplitudes have been associated with a higher incidence of metabolic disorders such as diabetes and obesity. There has therefore been significant recent interest in identifying pharmaceutical modulators of the clock to support robust circadian function. The cellular circadian oscillator is subject to intrinsic molecular fluctuations caused by the low copy number of biomolecules, resulting in cycle-to-cycle variability in period length for cells in the absence of synchronizing cues. Recent work has shown that potential therapeutic treatments affect the precision of the clock, in addition to the period, phase, or amplitude. Furthermore, the vast majority of clock hits within a recent genome-wide siRNA screen resulted in either decreased oscillatory amplitudes or decreased clock precision, indicating that identifying suitable drug targets in the circadian network is a significant challenge.

Rational design of circadian modulators is desirable; however, computational screening for changes in oscillator precision generally requires discrete stochastic simulation of hundreds or thousands of trajectories and then fitting a phase diffusion coefficient for each possible drug target. This approach is prohibitively computationally expensive. Here, we present an efficient analytical method for determining the precision of the circadian oscillator subject to intrinsic molecular noise. Briefly, we construct a model-reduced oscillator with relaxation dynamics governed by the Floquet multipliers of the system. By calculating the phase sensitivity to perturbation and approximating the noise distribution, we are able to analytically calculate a phase diffusion coefficient. We demonstrate agreement between this method and full stochastic simulation. Finally, we apply this method to a computational model of the circadian clock, and identify potential therapeutic targets which allow manipulation of clock period without diminishing precision.

469181 A Systems-Biology Approach to Investigate the Antimicrobial Activity of Oleuropein

Abstract:

In recent years, there is a considerable interest regarding the use of natural antimicrobial agents as food additives to inhibit the growth of food pathogens. Oleuropein and its hydrolysis products, including 3,4-dihydroxy-phenylethanol (hydroxytyrosol), elenolic acid and aglycon [1], are olive phenolic compounds that have antimicrobial effects to a variety of bacteria and fungi such as *Staphylococcus aureus*, *Pseudomonas aeruginosa*, *Escherichia coli* and *Candida albicans* [2-3]. The proposed antimicrobial mechanism of oleuropein is due to its surface activity at the bacterial cell membrane: molecular interactions between oleuropein and the phosphate group might modify the phospholipid/water interface properties [4]. In particular, oleuropein was proven to cause leakage of intracellular components such as potassium, phosphate and glutamate from bacteria along with the degradation of the cell membrane[5]. Additionally, oleuropein showed the ability to enhance nitric oxide production in macrophage cells, thus to fight against endotoxins and facilitate the elimination of pathogens[6]. From this, we hypothesize that intracellular metabolic reactions associated with nitric oxide should have larger fluxes for the antimicrobial function.

In this study, a genome-scale modeling approach which can help predict systems-level changes of intracellular metabolisms was adopted to investigate the effect of oleuropein on *S. aureus*, a common food-borne pathogen. Genome-scale modeling approaches, which incorporate genomic genes and metabolites in a metabolic reaction network, have been proved of excellent performance in predicting the systems-level changes of intracellular metabolisms in organisms under specific nutrient environment or treated by specific stimuli. Among all the pathogens that are inhibited by oleuropein, *S. aureus* has been widely studied experimentally [1], [7] as well as by modeling [8]–[10]. In this work, the model developed by Becker et al.[8] was first extended to incorporate the nitric oxide reaction into the intracellular metabolism based on experimental research. The stress that oleuropein imposes on *S. aureus* was mimicked by regulating the export of potassium, phosphate and glutamate out of the cell. With decreased phosphate uptake rate and increased glutamate secretion rate, *S. aureus* growth fits experimental results [1] in the presence of oleuropein at various concentrations. In addition, the reactions and enzymes that are highly related to oleuropein stress are identified by flux balance analysis. This study provides the first genome-scale model for oleuropein metabolism study, and the results indicates that genome-scale modeling is a promising method for revealing the intracellular metabolisms of oleuropein biological properties and thus help to find the food and pharmaceutical applications of oleuropein.

References:

[1] C. C. Tassou and G. J. E. Nychas, "Inhibition of Staphylococcus aureus by Olive Phenolics in Broth and in a Model Food System," *J. Food Prot.*, vol. 57, no. JANUARY, p. 5, 1994.

[2] H. P. Fleming, W. M. Walter Jr., and J. L. Etchells, "Antimicrobial properties of oleuropein and products of its hydrolysis from green olives," *Appl Microbiol*, vol. 26, no. 5, pp. 777–782, 1973.

[3] A. P. Pereira, I. C. F. R. Ferreira, F. Marcelino, P. Valentão, P. B. Andrade, R. Seabra, L. Estevinho, A. Bento, and J. A. Pereira, "Phenolic compounds and antimicrobial activity of olive (Olea europaea L. Cv. Cobrançosa) leaves.," *Molecules*, vol. 12, no. 5, pp. 1153–1162, 2007.

[4] N. Caturla, L. Pérez-Fons, A. Estepa, and V. Micol, "Differential effects of oleuropein, a biophenol from Olea europaea, on anionic and zwiterionic phospholipid model membranes," *Chem. Phys. Lipids*, vol. 137, no. 1–2, pp. 2–17, 2005.

[5] B. Juven, Y. Henis, and B. Jacoby, "Studies on the Mechanism of the Antimicrobial Action of Oleuropein*," *J. Appl. Bacteriol.*, vol. 35, no. 4, pp. 559–567, 1972.

[6] F. Durlu-özkaya and M. T. Özkaya, "Oleuropein using as an Additive for Feed and Products used for Humans," *Food Process. Technol.*, vol. 2, no. 3, pp. 1–7, 2011.

[7] J. Casas-Sanchez, A. M. Alsina, M. K. Herrlein, and C. Mestres, "Interaction between the antibacterial compound, oleuropein, and model membranes," *Colloid Polym. Sci.*, vol. 285, no. 12, pp. 1351–1360, 2007.

[8] S. A. Becker and B. Ø. Palsson, "Genome-scale reconstruction of the metabolic network in Staphylococcus aureus N315: an initial draft to the two-dimensional annotation," *BMC Microbiol.*, vol. 5, no. 1, p. 8, 2005.

[9] M. Heinemann, A. Kümmel, R. Ruinatscha, and S. Panke, "In silico genome-scale reconstruction and validation of the Staphylococcus aureus metabolic network," *Biotechnol. Bioeng.*, vol. 92, no. 7, pp. 850–864, 2005.

[10] D. S. Lee, H. Burd, J. Liu, E. Almaas, O. Wiest, A. L. Barabasi, Z. N. Oltvai, and V. Kapatral, "Comparative genome-scale metabolic reconstruction and flux balance analysis of multiple Staphylococcus aureus genomes identify novel antimicrobial drug targets," *J. Bacteriol.*, vol. 191, no. 12, pp. 4015–4024, 2009.

472684 VEGFR1 Mediates Cell Migration through Activation of PI3K and PLC

Abstract:

VEGFR1 mediates cell migration through activation of PI3K and PLC

Introduction:

VEGF-VEGFR regulate neovascularization; despite this, efforts to control this signaling axis towards directing neovascularization have yet to be achieved [1]. This challenge underlies a continuing need to increase our understanding of signal-to-response. Indeed, there is significantly less knowledge of the VEGFR1 role, which is often described as a decoy receptor due to its high VEGF affinity but low kinase activity [2], relative to the strongly pro-angiogenic VEGFR2. However, our recent computational models have identified that high-VEGFR1 levels may negate the effect of a VEGF inhibitor [3]. Thus, to understand the VEGFR1 signaling role, we develop computational models of the initial intracellular signaling mechanisms connected to VEGFR1: tyrosine site-specific VEGFR1-adapter interactions. Furthermore, we examine how cooperativity in adapter binding to specific VEGFR1 tyrosine sites, and adapter-adapter interactions, direct VEGFR1 signaling. We introduce a novel weighing scheme that correlates adapter-VEGFR1 binding to a cell response (e.g., proliferation, migration, or ubiquitination). We validate our computational predictions using wound healing assays, in vitro.

Materials and Methods:

Computational methods: We develop and compare four endothelial VEGFR1-adapter interaction models in MATLAB

Simbiology: each increases in complexity to gauge how each mechanistic component affects the physiologic response (Fig. 1A). The four models are as follows: Model 1, simplest: adapters bind to one C-terminal Tyr site; Model 2: Model 1 + adapter-adapter binding: adapter-adapter interactions can occur, with binding to one Tyr site; Model 3: Model 1 + Tyr site specificity: adapters bind to physiologically known Tyr sites; Model 4, most physiologically accurate: adapter-adapter interactions occur with binding to known Tyr sites. Therefore, the following comparisons provide insight into VEGFR1 signaling: Model 1 vs. Model 3 = site-specific binding, Model 1 vs. Model 2 = adapter cooperativity, and Model 1 vs. 4 = site-specific binding + cooperativity. We take a data-driven approach towards model development: our experimental qFlow (quantitative) cytometry provides membrane-receptor concentrations while literature mining provides adapter concentrations via western blots, kinetics via surface plasmon resonance, and protein-protein interactions via co-immunoprecipitation. We mine literature to determine how the cell response (migration, proliferation, or ubiquitination) changes when each adapter is inhibited. We quantify these changes in migration, proliferation, and ubiquitination, and assign a corresponding weight to each adapter and linearly correlate adapter contributions to cell response.

Experimental methods: Wound healing assays are performed using mouse RAW 264.7 macrophages: their high VEGFR1 levels (4,820 ± 112 VEGFR1/cell) relative to VEGFR2 (1,770 ± 132 VEGFR2/cell) allows us to focus on VEGF-VEGFR1 response. RAW macrophages are seeded into a 12-well plate and cultured as monolayer to ~90% confluence. The cells are serum starved overnight with DMEM supplemented with 0.5% FBS and 1% PS. The monolayer is scratched with a 100 μL pipette tip and washed once with PBS to remove floating cells. Wells are treated with 750 μL of the serum starved growth factor media, VEGF-A_{164} (50 ng/mL), 10 μM Wortmannin (Anti-PI3K, IC50 = 3 nM), 10 μM U73122 (Anti-PLC , IC50 = 1 μM), 10 μM Imatinib Mesylate (Anti-Abl, IC50 = 600 nM), or a combination of VEGF-A_{164} and an inhibitor. Images of the wounded cell monolayer are taken using a brightfield microsocope at 0 h and 24 h after scratching. All experiments are independently carried out in triplicate. Cell migration is quantified as the number of cells contained within the total gap area relative to the number of cells immediately after the scratch, using Image J.

Results and Discussion:

Adapter negative cooperativity decreases VEGFR1-mediated migration. We find that VEGFR1 regulates HUVEC migration: in Model 1, the integrated migratory response is 2.4-fold greater than the proliferation, and 4.6-fold greater than the ubiquitination, integrated responses. Examining how adapter cooperativity (Model 2) affects VEGFR1-mediated cell responses reveals two key findings. First, VEGFR1 still primarily promotes migration when accounting for adapter cooperativity, evidenced by the migration pathway having the greatest integrated response, 2.1-fold and 3.1-fold greater than proliferation and ubiquitination, respectively. Second, accounting for adapter cooperativity significantly alters the phosphorylation of multiple adapters. Specifically, the integrated response and phosphorylation amplitude of the adapter Nck decrease 88% and 86%, respectively, when adapter-adapter interactions occur (Model 2) compared to no adapter cooperativity (Model 1). Thus, cooperativity in adapter binding mechanistically preventing VEGFR1-mediated Nck activation, implying that Nck negatively cooperates with the other adapters.

VEGFR1 Tyr sites preferentially activate the PI3K and PLC adapters. Accounting for VEGFR1 tyrosine site specificity directs the preferentiality of adapter-VEGFR1 binding and subsequent adapter phosphorylation. Specifically, we find that accounting for VEGFR1 tyrosine sites (Model 3) increases PI3K and PLC signaling; we observe a 1.7-fold and 1.5-fold increase in PI3K and PLC signaling, respectfully, when accounting for VEGFR1 tyrosine sites (Model 3) compared to no tyrosine site specificity (Model 1). Thus, when accounting for specific VEGFR1 tyrosine sites (Model 3), we identify that VEGFR1 is structured to preferentially activate PI3K and PLC .

Tyr site specificity with adapter cooperativity amplifies VEGFR1 through increased PLC activation. By comparing VEGFR1 signaling in the simplest (Model 1) and most physiologically relevant (Model 4) cases, we find that VEGFR1. First, the cell response is unchanged: VEGFR1 signaling primarily promotes cell migration, followed by proliferation, with ubiquitination having the lowest activation. Second, both adapter cooperativity and binding specificity are necessary to enable signal amplification. We arrive at this conclusion because all integrated responses increase in Model 4 compared to Model 1: 6.5-fold for proliferation and 3.5-fold for both migration and degradation.

These increased cell responses do not occur with only Tyr site specificity (Model 3) or adapter-adapter interactions (Model 2). Furthermore, we find that signal amplification primarily results from the preferential activation of PLC ; the PLC integrated response is 2.5-fold greater than the next most activated adapter. We believe this high PLC activation is due to coordination between VEGFR1 Tyr sites and adapter cooperativity. Indeed, PLC has the highest number of available Tyr binding sites on VEGFR1, in addition to undergoing cooperative binding with

Wound healing assays validate VEGFR1-induced migration predictions. Our results indicate that PI3K and PLC are the two adapters primarily regulating cell migration within the VEGF-VEGFR1 signaling axis. To test the robustness of this prediction, we extend our model to macrophages, changing model parameters to match literature-minded adapter and receptor concentrations (e.g., [VEGFR1] >> [VEGFR2]). Again, we predict that PI3K and PLC regulate macrophage migration. Moreover, we rank-predict that PLC inhibition (max=58%/min=24%) > PI3K inhibition (max=51%/min=15%) >> Abl inhibition (< 1% decrease) (Fig. 1B). We validate these predictions to wound healing assays (Fig. 1C).

Conclusions:

Our findings further define VEGFR1 as an active controller in angiogenesis-rather than a mere decoy receptor. Moreover, we define the key proteins mediating this migration role, as PI3K and PLC . Since these pathways are linked to Ca^{2+}signaling [4], these models offer a path towards further understanding VEGFR1-migration via calcium signaling. Future studies can also apply this modeling methodology to understand and control VEGFR1-signaling in diseases where VEGFR1 concentrations are high, such as in breast cancer [3] and ischemia [5].

References:

[1] Eichmann A. *Curr Opin Cell Biol* (2012) 24:188. [2] Zhang Z. *Cell Death Differ* (2010) 17:499-512. [3] Weddell JC. *PLOS ONE* (2014) 9:e97271. [4] Ramch LE. *J Biol Chem* (1998) 273:23750. [5] Imoukhuede PI. *Am J Physiol Heart Circ Physiol* (2013) 304:H1085.

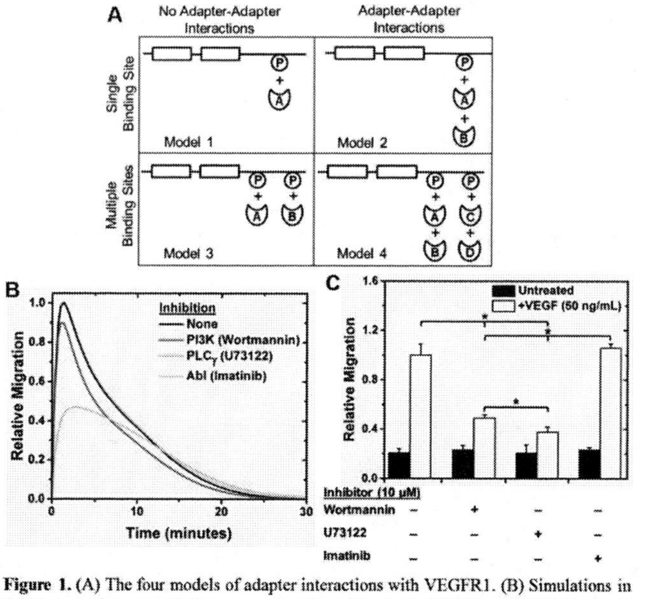

Figure 1. (A) The four models of adapter interactions with VEGFR1. (B) Simulations in RAW macrophages predicting mean migration with adapter inhibition. (C) RAW migration was experimentally quantified with wound healing assays.

469683 Developing a Stochastic Model of LPS-Induced TNF-α Production in Macrophages

Abstract:

Developing a stochastic model of LPS-induced TNF-α production in macrophages

Macrophages are immune cells that are present throughout the body and are involved in a variety of physiological responses. One of its most important roles is its regulation of innate and adaptive immune responses [1]. Toll-like receptors (TLR) embedded in the cell membrane mediate these responses by recognizing microbial components in the body. One such microbial component is LPS (lipopolysaccharide), which is a major constituent of bacterial membranes [2]. Binding of LPS to TLRs leads to the initiation of intracellular signaling pathways and downstream gene expression to rapidly eliminate the bacterial components. One of the major signaling pathways activated by TLR signaling is through the transcription factor NF-κB, which leads to the production of the pro-inflammatory cytokine TNF-α (tumor necrosis factor-α). TNF-α, in turn, initiates signaling through the TNF signaling pathway and leads to propagation of the inflammatory response to microbial components. Hence, LPS-induced signaling pathway is an important part of the early immune response.

Previous studies [3, 4, 5] have formulated population-level models to study the dynamics of signaling through the NF-κB pathway. In these models, the numerical model predicts population-level activation of NF-κB and downstream TNF-α gene expression. However, predictions from the population-level model could be misleading since the model masks the behaviors of individual cells. It has been shown that the effect of LPS stimulation can be highly stochastic, especially to low concentrations of LPS where only a fraction of cells respond to stimulation [6]. Therefore, we developed a stochastic model of LPS-induced NF-κB signaling and TNF-α production in macrophages. Specifically, in addition to gene activation and cellular membrane receptor activation, enzymes and proteins involved in the intermediate stages of LPS-induced NF-κB signaling were modeled as states. Then, the corresponding kinetic parameters for intracellular signaling pathway were identified via the sequential approach of model reduction and inverse modeling. Model simulations were validated and refined using experimental data on LPS stimulated macrophages. Specifically, the activation of NF-κB, TNF-α and IκB in RAW macrophages upon stimulation with different LPS doses was studied using flow cytometry and intracellular staining at the single-cell level. These models are expected to predict the diversity of cellular responses that cannot be captured by deterministic and population-based models. Furthermore, it was possible to simulate intercellular interactions with this stochastic model to analyze cell-to-cell communications - a step toward constructing the multicellular or tissue model. More importantly, the developed stochastic model was used to design a feedback control system that helps producing TNF-α at the desired rate through the autocrine regulation of the LPS-induced intracellular signaling pathway in macrophages.

References:

[1] Parameswaran N, Patial S. 2010. Tumor Necrosis Factor-α Signaling in Macrophages. Crit Rev Eukaryot Gene Expr. 20: 87-103.

[2] Takeuchi O, Hoshino K, Kawai T, Sanjo H, Takada H, Ogawa T, Takeda K, Akira S. 1999. Differential Roles of TLR2 and TLR4 in Recognition of Gram-Negative and Gram-Positive Bacterial Cell Wall Components. Immunity 11: 443-451.

[3] Choudhary S, Kalita M, Fang L, Patel K, Tian B, Zhao Y, Edeh C.B, Brasier A.R. 2013. Inducible TNF receptor associated factor-1 expression couples the canonical to non-canonical NF-κB pathway in TNF stimulation. Journal of Biological Chemistry 288:14612-14623.

[4] Lipniacki T, Paszek P, Brasier A.R, Luxon B, Kimmel M. 2004. Mathematical model of NF-κB regulatory module. Journal of Theoretical Biology 228: 195-215.

[5] Shih V, Kearns D, Basak S, Savinova O, Ghosh G, Hoffmann A. 2009. Kinetic control of negative feedback regulators of NF-κB/RelA determines their pathogen- and cytokine-receptor signaling specificity. Proceedings of the National Academy of Sciences 106: 9619-9624.

[6] James C, Moorman M, Carson B, Branda C, Lantz J, Manginell R, Martino A, Singh A. 2009. Nuclear translocation kinetics of NF-κB in macrophages challenged with pathogens in a microfluidic platform. Biomed Microdevices 11: 693-700.

468915 Parameter Estimation for Sparse Biological Data: Glucose-Dependence of Renin-Angiotensin System in Podocytes Cells during Diabetic Kidney Disease

Abstract:

Diabetic kidney disease (DKD) is the primary cause of end-stage renal failure. Hyperglycemia is known to initiate and exacerbate the pathophysiology of DKD. Podocytes are terminally differentiated glomerular epithelial cells that play a key role in maintaining the structure and function of glomerular filtration barrier in kidneys. Podocytes express a local renin-angiotensin system (RAS) that is altered in hyperglycemia. Studies have shown that angiotensin II (ANG II), which is a RAS peptide, is modulated in hyperglycemic conditions and triggers podocyte injury and apoptosis. The progression of DKD could be slowed by controlling the ANG II levels to prevent irreversible podocyte loss. However, experimental evidence for glucose-dose-dependency of ANG II is scarce in the literature. Hence, for better understanding of the underlying mechanism, we use mathematical modeling to describe the glucose-stimulated RAS signaling in podocytes that produces ANG II.

We have formulated a mathematical model to describe the glucose-sensitive reaction network that triggers the synthesis of ANG II. The local podocyte RAS signaling pathway is represented by a system of ordinary differential equations to track RAS peptides, enzymes, and receptors without explicit glucose-dependence. The system is assumed to be at steady state, thus the model reduces to a system of nonlinear equations. Previous experimental studies were used to estimate the unknown parameters and kinetic constants for the model. Due to sparse experimental data on RAS at sufficiently many glucose states, two different approaches were used to parameterize the model to add glucose-dependency through the parameters: (1) parameterizing the RAS model at normal and high glucose states where all the RAS components are known or estimated and approximating the parameters for intermediate glucose states where no RAS component information is available by considering glucose-dependent linear mathematical functions for each of the sensitive parameters, and (2) fitting the parameters for the RAS model with different proposed glucose-dependent mathematical functions to experimental data from the literature for ANG II vs. glucose [1]. In both the approaches, only the sensitive parameters were varied whereas the non-sensitive parameters were assumed to be independent of glucose.

The model was used to study the change in ANG II concentrations in hyperglycemic conditions with varying glucose levels. The results showed a rise in ANG II levels with increasing glucose concentrations consistent with experimental observations. Sensitivity analysis was conducted. The two parameter estimation approaches resulted in multiple candidate models that are reasonable for glucose-dependence of podocyte intracellular RAS.

Using this model, we were also able to discriminate between possible models for the glucose dose-dependency of ANG II production in podocytes. The key RAS-modulating biomarkers can be identified by studying their effect on ANG II levels. The model can also be used to study the effect of different combinations of various ANG II modulating therapies, which could be useful for drug development.

References: [1] Durvasula & Shankland. 2008. Am J Physiol-Renal Physiol, 294, F830-F839.

461205 A Method for Learning a Sparse Classification Model in the Presence of Missing Data

Abstract:

This work addresses two issues in building classification models: finding a subset of a large number of possible predictors and learning in the presence of missing data. Computational and interpretability concerns promote interest in learning sparse models. Modern dataset often have many measurements but few replicates which can lead to numerical issues. Furthermore, models with many parameters may not provide insight to the analyst. This setting is particularly common in biological or medical studies. Missing data can occur for a wide variety of reasons and is not atypical in the analysis of real datasets. This work focuses on supervised generative binary classification models, specifically linear discriminant analysis (LDA). The parameters are found using an expectation maximization [1] (EM) algorithm. EM allows both missing observations to be handled more naturally as well as the introduction of priors to promote sparsity. An algorithm that finds a sparse LDA model for datasets with and without missing data is presented.

To test the algorithm, a case study for the classification of two types of acute leukemia is presented. The dataset is gene expression data from a microarray. It is a public benchmark problem and has been widely studied [2]. Missing data is artificially added to be representative of missing data in microarrays [3]. The proposed approach is compared to the nearest shrunken centroids algorithm [4] and sparse linear discriminant analysis [5]. Missing data is handled with complete case analysis, mean imputation and k-nearest neighbor imputation, all common approaches in the field. The proposed approach outcompetes alternative methods.

[1] A. P. Dempster, N. M. Laird and D. B. Rubin. Maximum likelihood from incomplete data via the EM algorithm. Journal of the Royal Statistical Society. Series B Methodological, 39: 1-39, 1977.

[2] T. R. Golub, D. K. Slonin, P. Tamayo, C. Huard, M. Gaasenbeek, J. P. Mesirov, H. Coller, M. L. Loh, J. R. Downing, M. A. Caligiuri, C. D. Bloomfield, and E. S. Lander. Molecular classification of cancer: Class discovery and class prediction by gene expression monitoring. Science, 286: 531-537, 1999.

[3] O. Troyanskaya, M. Cantor, G. Sherlock, P. Brown, T. Hastie, R. Tibshirani, D. Botstein, and R. B. Altman. Missing value estimation methods for DNA microarrays. Bioinformatics, 17: 520-525, 2001.

[4] R. Tibshirani, T. Hastie, B. Narasimhan, G. Chu. Diagnosis of multiple cancer types by shrunken centroids of gene expression. PNAS, 99: 6567-6572, 2003.

[5] K. Sjöstrand, L. H. Clemmensen, R. Larsen, B. Ersbøll. SpaSM: A Matlab toolbox for sparse statistical modeling. Journal of Statistical Software, 2012.

456695 Optimal Experiment Design for Uncertain Biological Systems

Abstract:

Computational modeling has become an indispensable tool to aid discovery in biological systems, which are commonly characterized by highly nonlinear and uncertain behavior. The predictive power of the constructed model largely depends, among others, on the accuracy of the estimates of model parameters. Model-based Optimal Experiment Design (OED) techniques are being increasingly adopted for the purpose of parameter estimation in systems biology [6], as they allow for designing experiments that yield informative data for parameter estimation while accounting for the practical system considerations. However, classical OED approaches cannot account for the intrinsic heterogeneity and uncertainty of biological systems [4]. In addition, uncertainty in the estimates of model parameters cannot be accounted for. In OED for biological systems, the disregard of uncertainty originating from the intrinsic dynamics of the biological system under study as well as inaccuracy in the estimates of the model parameters can lead to severely suboptimal experimental designs [1].

This work considers OED for uncertain biological systems. Two different approaches to OED are comparatively investigated - OED based on deterministic descriptions of uncertainties and OED based on probabilistic descriptions of uncertainties. Deterministic experiment design with respect to all uncertainty realizations, including worst-case uncertainties, can potentially lead to overly conservative designs that underperform when maximizing information content of experiments. On the other hand, the probabilistic OED approach allows for taking into consideration the probability associated with each uncertainty realization. In this case, hence, the uncertainty realizations that have a very small probability of occurrence can be accounted for in the design of experiments, thus resulting in possibly less conservative designs. In the Probabilistic OED approach, the generalized polynomial chaos framework [11] is employed along with Galerkin projection [3] to propagate the uncertainty of the parameter estimates through the system dynamics, resulting in the full probability distribution of system states and outputs at each time point [9]. The deterministic OED approach is formulated through a max-min optimization problem [5]. The D-optimality criterion

[10], which corresponds to the determinant of the Fisher Information (FI) Matrix, is used as a quantitative measure for information content of experiments in both approaches.

The efficacy of both approaches is compared for the canonical Wnt (Wnt/Î²-catenin) signaling pathway, which is ubiquitous in mammalian cell systems and orchestrates critical processes such as cell growth and differentiation [2,7]. A previously reported compartmentalized stochastic model of the Wnt/Î²-catenin pathway is adopted in this work [8]. The OED approaches are intended to design an optimal Wnt stimulation profile (input), which aims at maximizing the information content of the response of the signaling pathway.

The use of a stochastic model that quantitatively accounts for all sources of uncertainty enables obtaining a more representative description of the naturally occurring behavior in a given cell population, thus resulting in a superior performance of experimental design. Stimulation of the signaling pathway with the optimal Wnt profiles show that the probabilistic OED approach yields more accurate parameter estimates, as compared to the deterministic OED. However, the higher accuracy obtained with the probabilistic approach comes at the expense of allowing for a pre-specified probability of constraint violation.

References:

[1] Asprey, S.P. and Macchietto, S. (2002). Designing robust optimal dynamic experiments. *Journal of Process Control*, 12, 545-556.

[2] Clevers, H. and Nusse, R. (2012) Wnt/Î²-Catenin signaling and disease. *Cell*, 149, 1192-1205.

[3] Ghanem, R.G. and Spanos, P.D. (1991) Stochastic Fine Elements: A Spectral Approach. Springer-Verlag, New York.

[4] Kaern, M., Elston, T.C., Blake, W.J., Collins, J.J (2005). Stochasticity in gene expression: From theories to phenotypes. *Nature Reviews Genetics*, 6 , 451- 463.

[5] Korkel, S., Kostina, E., Bock, H.G., Schloder, J.P. (2004). Numerical methods for optimal control problems in design of robust optimal experiments for nonlinear dynamic processes. *Optimization Methods and Software Journal*, 19, 327-338.

[6] Kreutz, C. and Timmer, J. (2009). Systems biology: Experimental design. *FEBS Journal*, 276, 923-942

[7] MacDonald, B., Tamai, K., and He, X. (2009). Wnt/beta-catenin signaling: Components, mechanisms, and diseases. *Developmental Cell*, 17, 9-26.

[8] Martin-Casas, M. and Mesbah, A. (2016). A Stochastic model for the canonical Wnt Signaling pathway. *Conference in Foundations of Systems Biology and Engineering*, Submitted.

[9] Mesbah, A. and Streif, S. (2015). A probabilistic approach to robust optimal experiment design with chance constraints. In *Proceedings of the 9th International Symposium on Advanced Control of Chemical Processes*, 100 -105.

[10] Pronzato, L. (2008). Optimal experiment design and related control problems. *Automatica*, 44, 303-325.

[11] Xiu, D. and Karniadakis, G. (2002). The Wiener-Askey polynomial chaos for stochastic differential equations. *SIAM Journal on Scientic Computing*, 24, 619-644.

464761 Data-Driven Parameter Reduction in Sloppy Models

Abstract:

The past decades have seen significant research progress towards developing reduced, low-dimensional descriptions of complex dynamic models, leading to simpler systems that are easier to interpret. However, the parallel field of reducing the number of parameters required to capture relevant system behavior is relatively nascent: the Manifold Boundary Approximation Method and active subspaces are two promising approaches to this problem of model â €œsloppinessâ€. In this talk we present advances in our own, data-based technique that is capable of discovering sloppy directions in a high-dimensional parameter space without recourse to the analytical equations that constitute the model. We illustrate its effectiveness in the context of biological networks, and show how these methods can be used to perform selective optimization over important parameter combinations, ignoring sloppy directions.

456011 Process Systems Engineering Beyond Chemical Plants: Signed Digraph As a Modeling Tool for Analyzing Systemic Risk in Financial Networks

Abstract:

Process systems engineering (PSE) has been an active and important research area in chemical engineering for more than half a century. It encompasses the control, optimization, and risk management of large-scale chemical engineering systems. Complex sociotechnical systems, such as modern financial systems, are characterized by the same interdependencies and a large number of units one often observe in a chemical plant. Systemic disasters, such as â €™08 financial crisis and â€™10 BP Deepwater Horizon oil spill, share alarming similarities. The commonalities naturally lead to the question: Can we use PSE to understand and, more ambitiously, design complex sociotechnical systems?

Human factors such as self-interest and cognitive limitation are critical in the course of a systemic event. They are also the biggest challenges of applying PSE to such systems. Without a central planner and being subject to the limited information, each individual (person or institution) tends to maximize its own utility and oversee the local stability only. This myopia might lead to system-wide failures as we have seen in the financial crisis and other systemic disasters.

In this talk, we apply the signed digraph (SDG) technique used in fault diagnosis to modeling systemic risk in a financial system. This approach helps us identify whether and how a causal pathway in the financial network could potentially lead to a systemic risk outcome. The framework is capable of recreating known runaway situations and discovering potential vulnerabilities. More importantly, since each causal relationship is derived from underlying economic principles and rational individual behaviors, one would expect global stability for the overall system. Quite

the contrary, we identified systemic runaway events such as fire sale and funding run from the diagnosis despite individual institutions being conservative. Actions that dampen risk on a local level can contribute positive feedback and cascades on the global level, leading to large-scale unintended consequences.

Automation plays an important role in such fault diagnosis. In the case study, one can perhaps manually identify and enumerate all the feedback loops. However, for a more realistic version of the network, where there are multiple interacting subsystems, it is virtually impossible to analyze such loops manually. The SDG framework can be automated to handle larger systems via artificial intelligence and optimization techniques. The end-goal of this framework is providing a systematic method to detect potential threats and design such complex systems.

464492 Multiperiod Inventory Pinch Algorithm for Integrated Planning and Scheduling of Oil Refineries

Abstract:

A petroleum refinery plant transforms raw crude oils into more valuable products that can be found almost anywhere such as liquid fuels, plastics, oils, solvents, asphalt, and many chemicals used in production processes across different industries. Oil refineries involve several intertwined complex physical and chemical processes, and they operate in a highly competitive and dynamic market under strict safety, environmental, and governmental regulations. For these reasons, planning and scheduling of refinery operations are very important tasks. Moreover, in order to obtain high quality solutions, integration of planning and scheduling decisions is necessary[1].

Commercial software found in industrial practice to compute production plans for a refinery employ multiperiod models and solve them using linear programming or successive linear programming techniques, e.g. Aspen PIMS, Haverly GRTMPS, and Honeywell RPMS. Nonlinear programming (NLP) algorithms can be found as well, e.g. Aspen PIMS AO. However, in order to avoid intractable problems, such software often use simplified process unit-operation models[2]. Refinery scheduling is usually carried out via interactive simulation (e.g. Aspen Petroleum Scheduler and Haverly H/Sched) with the production plan targets used as constraints. Mixed-integer linear programming (MILP) optimization techniques to solve the scheduling problem represent an opportunity for companies to improve their operations and profits; however, more efficient and reliable MILP models and solution algorithms are required to be developed in order to be applied in practice.

Regarding planning models, current trend is to increase their accuracy by considering their nonlinear nature and developing formulations that can be tractable[2,3,4]. For the scheduling problem, several MILP models have been published in the literature for the entire refinery or some of its subsystems[5,6,7]. Maravelias and

Sung[1] reviewed different strategies to integrate planning and scheduling levels. They discussed the importance of developing more computationally effective models and improving decomposition and iterative algorithms, as well as hybrid methods. Several researchers have formulated full-space models (i.e. simultaneous planning and scheduling), usually as mixed-integer nonlinear programming (MINLP) models, and proposed algorithms to solve them[8,9,10]. In general, these models can be solved directly only for a small number of time periods. Various bi-level decomposition algorithms for MILP and MINLP models have been proposed[11,12,13]. Castillo-Castillo and Mahalec[14] developed a three-level decomposition approach based on the inventory pinch concept to integrate planning and scheduling decisions in gasoline blending operations, which reduces the number of blend recipes and can handle nonlinear blending rules.

Based on the work by Castillo-Castillo and Mahalec[14], we present a multiperiod inventory pinch algorithm, denoted as "extended MPIP-C algorithm", to integrate planning and scheduling decisions of an oil-refinery (not only for the gasoline blending section). The original full-space problem is decomposed into three levels: 1) optimization of operating conditions, 2) derivation of an approximate schedule, and 3) computation of a detailed schedule. At the 1st level, a multiperiod NLP model is formulated, where periods are delineated initially by inventory pinch points of various product pools (e.g. gasoline and diesel). This reduces the number of periods and number of different blend recipes. Various formulations for the processing unit models can be utilized (e.g. fixed yield, multi-mode, or nonlinear). A global optimization algorithm is employed to solve the 1st level if such model is nonconvex. The 2nd level is solved via multiperiod MILP model, where periods are delineated by scheduler based on the demand and supply data, and the minimum time requirements to complete major tasks (i.e. blend runs, product tank service). The 3rd level uses a continuous-time MILP model to determine the exact times to carry out the necessary tasks to meet the targets imposed by the 2nd and 3rd level solutions. The 2nd and 3rd levels are MILP models since the nonlinear constraints associated with the operating conditions are solved at the 1st level, and the optimal conditions found at the 1st level are fixed at the other levels. The algorithm minimizes the total cost comprised by the crude oil cost, production cost, switching cost, and demurrage cost. Switching cost involves blend runs, delivery runs, product changeovers in the swing tanks, and mode transitions in the processing units. The algorithm eliminates infeasibilities found at the 2nd and 3rd levels by re-optimizing operating conditions at the 1st level.

For our case studies, the refinery system consists of a crude distillation unit, a catalytic reformer, a hydrocracker, a fluid catalytic cracking unit, five hydrotreaters, three blenders, and various mixers and splitters. We consider five products and eight quality properties under specification, as well as different demand and supply scenarios. Results show that the extended MPIP-C algorithm computes solutions close to the optimum and in relative shorter execution times than solving the full-space model. Most of the computational time is spent at the 1st and 3rd level models. Future work will be to test mathematical decomposition strategies to solve the 3rd level model (detailed scheduling) more efficiently.

References

1. Maravelias CT, Sung C. Integration of production planning and scheduling: Overview, challenges and opportunities. *Comput Chem Eng.* 2009; 33(12): 1919-1930.
2. Menezes BC, Kelly JD, Grossmann IE. Improved swing-cut modeling for planning and scheduling of oil-refinery distillation units. *Ind Eng Chem Res.* 2013; 52(51): 18324-18333.
3. Fu G, Sanchez Y, Mahalec V. Hybrid model for optimization of crude oil distillation units. *AIChE J.* 2016; 62: 1065-1078.
4. Li J, Boukouvala F, Xiao X, Floudas CA, Zhao B, Du G, Su X, Liu H. Data driven mathematical modeling and global optimization framework for entire petrochemical planning operations. *AIChE J.* 2016. DOI: 10.1002/aic.15220
5. Shah NK, Ierapetritou MG. Short term scheduling of a large scale oil refinery operations: Incorporating logistics details. *AIChE J.* 2011; 57(6): 1570-1584.
6. Shi L, Jiang Y, Wang L, Huang D. Refinery production scheduling involving operational transitions of mode switching under predictive control system. *Ind Eng Chem Res.* 2014; 53(19): 8155-8170.
7. Castillo-Castillo PA, Mahalec V. Improved continuous-time model for gasoline blend scheduling.

Comput Chem Eng. 2016; 84: 627-646.
8. Kopanos GM, Puigjaner L, Maravelias CT. Production planning and scheduling of parallel continuous processes with product families. *Ind Eng Chem Res*. 2010; 50(3): 1369-1378.
9. Yue D, You F. Planning and scheduling of flexible process networks under uncertainty with stochastic inventory: MINLP models and algorithm. *AIChE J*. 2013; 59(5): 1511-1532.
10. Gutiérrez-Limón MA, Flores-Tlacuahuac A, Grossmann IE. MINLP formulation for simultaneous planning, scheduling, and control of short-period single-unit processing systems. *Ind Eng Chem Res*. 2014; 53(38): 14679-14694.
11. Dogan ME, Grossmann IE. A decomposition method for the simultaneous planning and scheduling of single-stage continuous multiproduct plants. *Ind Eng Chem Res*. 2006; 45(1): 299-315.
12. Li Z, Ierapetritou MG. Integrated production planning and scheduling using a decomposition framework. *Chem Eng Sci*. 2009; 64(16): 3585-3597.
13. Chu Y, You F, Wassick JM, Agarwal A. Integrated planning and scheduling under production uncertainties: Bi-level model formulation and hybrid solution method. *Comput Chem Eng*. 2015; 72: 255-272.
14. Castillo-Castillo, PA, Mahalec V. Inventory pinch gasoline blend scheduling algorithm combining discrete-and continuous-time models. *Comput Chem Eng*. 2016; 84: 611-626.

464707 Improved Data-Driven Mathematical Modeling and Global Optimization Framework: An Application in Refinery Planning Operations

Abstract:

Improving the planning and scheduling operations is one of the major challenges in the petrochemical industry considering the tight competition, environmental regulations, and lower profit margins [1]. A recent work of our group developed a data-driven modeling and global optimization-based planning formulation for highly integrated refinery-petrochemical complexes [2]. In this work, an improved and more generalized framework for planning operations is applied to an actual refinery operated by Hyundai Oilbank and located in Korea.

The refinery planning problem has received considerable attention since the introduction of linear programming in 1950's. Planning is a topic of special interest for many process systems engineers and in this respect, nonlinear models and specialized algorithms have been proposed for refinery operation but global optimality was not guaranteed in any method [3]. Integration and interaction between refineries and chemical plants is another level of complexity that is addressed only by few researchers [2,4]. Although there are commercially available highly detailed mathematical models for simulation of processing units, such detailed models cannot be used in enterprise-wide operations due to computational expense. Data-driven modeling offers a promising way to obtain inexpensive nonlinear models [5,6], which can relate relevant inputs to relevant outputs to describe planning operations.

The first step of this work consists of organizing, analyzing, and processing the raw data provided by the industrial partner. In order to reduce the variability in the daily data, weekly averages are calculated and used. A detailed outlier analysis is carried out and missing points in the data set are imputed. Having decided on the functional relationship between inputs and outputs, data-based nonlinear models are developed for the processing units. The primary approach is to use quadratic, polynomial, signomial, exponential, and logarithmic models to predict product yields and properties for all of the production units. The parameters of the postulated models are globally optimized using the state-of-the-art, commercial deterministic global optimization solver ANTIGONE [7], based on a large data set of daily production and property measurements. All the models are created in a generalized and an automated fashion, which improved the applicability of the framework.

The second step of the work is creating a superstructure containing all possible connections and operating modes in the refinery. The resulting single-period planning model is a large-scale non-convex mixed integer nonlinear optimization model, which is solved to ε-global optimality using ANTIGONE [7]. Several case studies with candidate months of operation are chosen and the profit calculated by the formulation is compared with actual plant profit. All the outlier analysis, data processing, and planning are done with close collaboration with our industrial partner. Results of the case studies are provided to illustrate the efficiency of our proposed model and global optimization approach.

References:

1. CA Floudas, AM Niziolek, O Onel, LR Matthews, Multi-Scale Systems Engineering for Energy and the Environment: Challenges and Oppurtunities, AIChE Journal, 62 (3) (2016), pp. 602-623
2. J Lie, X Xiao, F Boukouvala, B Zhao, G Du, X Su, H Liu, CA Floudas, Data-Driven Mathematical Modeling and Global Optimization Framework for Entire Petrochemical Planning Operations, AIChE Journal, 2016, DOI: 10.1002/aic.15220
3. NK Shah, Z Li, MG Ierapetritou, Petroleum refining operations: key issues, advances, and opportunities. Industrial and Engineering Chemisty Research, 50 (3) (2011), pp. 1161-1170
4. K Al-Qahtani, A Elkamel, Multisite refinery and petrochemical network design: optimal integration and coordination, Industrial and Engineering Chemisty Research, 48 (2) (2009), pp. 814-826
5. P Guyonnet, FH Grant, MJ Bagajewicz, Integrated model for refinery planning, oil procuring, and product distribution, Industrial and Engineering Chemisty Research, 48 (1) (2009), pp. 463-482
6. SJ Qin, Process data analytics in the era of big data, AIChE Journal, 60 (9) (2014), pp.3092-3100
7. R Misener, CA Floudas, ANTIGONE: Algorithms for coNTinuous Integer Global Optimization of Nonlinear Equations, Journal of Global Optimization, 59 (2–3) (2014), pp. 503–526

466503 Evaluation of Smart Manufacturing (SM) Benefits in Industrial Steam-Methane Reformers (SMR)

Abstract:

Steam methane reformers (SMRs) provide the bulk of the hydrogen consumed in refineries and for production of important chemicals such as ammonia and methanol. Consequently, these are large scale systems, with modern hydrogen plants producing above 100 million standard cubic feet hydrogen per day (MMSCFD)[1]. In this context, energy efficient operation is an imperative necessity for long-term economic viability.

Mathematical modeling of SMRs becomes a crucial tool for achieving high energy efficiency. The models can be either high-fidelity (e.g., using computational fluid dynamics), data-driven (and typically low-order) empirical models, or a combination of both.[2,3] Furthermore, the development of mathematical models should be accompanied by online deployment. The latter must be supported by a smart manufacturing framework, that includes, i) capabilities for acquiring appropriate process data via systematically placed sensors, ii) adequate high-performance computational resources for just-in-time computations, and, iii) a user-friendly visualization interface for operator use. The model-based computations are themselves a sequence/workflow of calculations being executed in series or parallel. Seamless interactions between these different layers require an appropriate IT infrastructure that allows the intensive computations to be performed on cloud and intercommunications between various components of the workflow, and provides appropriate database management capabilities. Moreover, rapid adoption of such practices demand that such an IT infrastructure should be easy to use and deploy. These requirements constitute the natural and essential form of â€˜Smart manufacturingâ€™[4] (SM). In this work, we present an example application of SM concepts to an industrial scale steam methane reformer along with the benefits obtained.

The reforming process studied in this work takes place in a large scale, high-temperature SMR, where the endothermic

reforming reactions are carried out in hundreds of catalyst-filled tubes placed in a gas-fired furnace. The overall productivity (energy consumed per unit H_2 produced) of the plant is strongly dependent on how efficiently the SMR is operated, which further depends on the spatial temperature distribution inside the furnace, where a more uniform distribution paves the way for reduced plant-wide energy use. An integrated framework comprising of state-of-the-art infrared cameras, kepler-based workflow for Matlab and Ansys Fluent-based calculations, PI OSI-based database system, and Tableau-based visualization interface assisted by a novel open-source IT platform/wireframe was deployed for the furnace temperature homogenization leading to ~ 44% improvement in the furnace operation.[5] The cloud-based SM Platform provides on demand scalability and reusability of resources which opens the possibility of scaling IT infrastructure and deploying similar systems much faster with significantly lower costs and reduced staff resources at remote plants and facilities around the globe. Specifically, we demonstrate how the SM approach rendered itself further useful by allowing reusability of the integrated framework for quick deployment at other SMR facilities as well as other similar furnace-based manufacturing processes leading to reduction in deployment costs.

References:

[1]Latham, D. A., McAuley, K. B., Peppley, B. A., & Raybold, T. M. (2011). Mathematical modeling of an industrial steam-methane reformer for on-line deployment. *Fuel Process. Technol.*, 92 , 1574-1586.

[2]Kumar, A., Baldea, M., Edgar, T. F., & Ezekoye, O. A. (2015). Smart Manufacturing Approach for Efficient Operation of Industrial Steam-Methane Reformers. *Industrial & Engineering Chemistry Research, 54* (16), 4360-4370.

[3]Kumar, A., Baldea, M., Edgar, T. F. Real-time optimization of an industrial steam-methane reformer under distributed sensing. *Control Engineering Practice* [Submitted, Jan 2016].

[4]Korambath, P., Wang, J., Kumar, A., Hochstein, L., Schott, B., Graybill, R., Baldea, M., & Davis, J. (2014). Deploying Kepler Workflows as Services on a Cloud Infrastructure for Smart Manufacturing. *Procedia Computer Science*, 29 , 2254-2259.

[5]Korambath, P., Wang, J., Kumar, A., Davis, J., Graybill, R., Schott, B., & Baldea, M. (2016). A smart manufacturing use case: Furnace temperature balancing in steam methane reforming process via Kepler workflows. Procedia Computer Science [Accepted, April 2016].

468067 Development of a High Fidelity and Reduced Order Data-Driven Model of Hydrogen Plant

Abstract:

Hydrogen is one of the most indispensable chemical components extensively used in petroleum and chemical industries. An economic way to commercially produce hydrogen is to utilize steam methane reforming [1]. The process consists of a number of material and heat flows, resulting in an intricate network of material and energy. Natural gas (NG) and superheated steam are fed to a chemical reactor called reformer, which consists of catalyst tubes filled with nickel reforming catalyst. Here, majority of the hydrogen is produced. The gas out of the reformer is then processed through another reactor to further produce hydrogen. Finally, the hydrogen flow is purified in a pressure swing adsorber (PSA), where high purity hydrogen is produced. The reformer exit temperature is an important process variable for this process and is expected to be kept at a desired level by heating the reformer. This heat is provided by burning the off-gas from the PSA and NG fuel stream. A fan is used to supply air to the burners and another one draws the combustion products, termed as flue gas, out of the reformer box. This process presents various constraints and challenges during the start-up as well as nominal operation of the plant such as, the reformer exit temperature should be maintained at a desirable level. Further, the firebox pressure should not breach its lower and upper limits for safety. If the pressure is too low, the fire can be extinguished. If it is too high, it may impose safety hazards to facility and personnel [2].

A typical startup procedure consists of series of events that take place over a period of time based on a predefined recipe. Startup can broadly be categorized as cold startup and warm startup based on the initial state of the plant from which the plant is required to be driven to nominal operation. The focus of this work will be on cold startup where the plant is started from completely shutdown conditions. Typical steps involve ramping the natural gas fuel to reach a sufficient reformer exit temperature, circulation of natural gas feed and steam, regulation of nitrogen flow in the reactor tubes. The startup also involves making discrete decisions such as starting up the PSA, recirculating the tail gas from the PSA and ramping up to full capacity in several stages. Thus, the startup time is not fixed and varies based on the decisions taken at various stages. Existing startup recipes have been built heuristically from past successful startups, with the possibility of optimizing the startup procedure remaining unexplored.

One key recognition that we make is that due to the nature of startup/shutdown phase, they can be treated as a batch like operation with variable lengths, where the objective is to drive the process from one state to the other. For modeling and control of batch processes, a variety of a data-driven modeling techniques exist. One of the most widely used approach is partial least squares (PLS), which models the process in a projected latent space [3]. These models are essentially time-varying linear models, linearized around mean past trajectories, and therefore require the batches to be of same length, or to recognize an appropriate alignment variable. To account for these limitations, a multi-model approach was proposed in [4]. These models were based on the 'current measurements' of the process instead of the 'time'. These developments were followed by contributions in the area of integration of these data-driven models with the advanced control formulations [4-5]. More recently a subspace identification based batch control approach was proposed in [6] where a LTI state-space model of the batch process is estimated, and does not require the training batches to be of equal length. The existing formulations have, however, not been applied to startup/shutdown operations in general, and for standard operating procedure (SOP) synthesis/optimization problem for the hydrogen plant in particular.

Motivated by these considerations, in this work we address the problem of determining a data driven model for the startup of a hydrogen plant. To this end, first a high fidelity test bed model of the entire plant is developed in UniSim, capable of simulating the startup and shutdown phase, with appropriate adaptation of the plant SOP. Several simulated startups are performed to generate the training data. Then, an LTI data-driven model (higher order) of the process using subspace identification based method [6-7] is estimated. Simulation results are presented that illustrate the prediction capabilities of the identified model and its suitability for deployment in an optimization framework.

Note that the development of a good model that can handle variable batch times paves the way for application of optimization algorithms that can improve not just the process economics along the startup/shutdown but also the operating time of the startup/shutdown procedure. From an optimization standpoint, the problem is quite different compared to existing application of data driven models for control of lumped parameter (such as in [6]) or distributed parameter process [7], where the objective was to reach a certain quality requirement in a given, predetermined batch length. The optimization problem for the present application is an optimal control problem where the objective is to reach the desired terminal state in minimum time, while adhering to various constraints, and also making discrete decisions along the way. The data driven model developed in the present work will subsequently be utilized within such an optimization framework to synthesize the startup SOP.

[1] JL Lynn and BH Bory. Kirk-othmer encyclopedia of chemical technology, 2000.

[2] Miao Du, Prashant Mhaskar, Yu Zhu, and Jesus Flores-Cerrillo (2014). Safe-Parking of a Hydrogen Production Unit. Industrial & Engineering Chemistry Research, 53 (19), 8147-8154.

[3] J. Flores-Cerrillo, J. F. MacGregor, Control of particle size distributions in emulsion semibatch polymerization using mid-course correction policies, Industrial and Engineering Chemistry Research 41 (7) (2002) 1805-1814.

[4] S. Aumi, B. Corbett, P. Mhaskar, and T. Clarke-Pringle. Data-based modeling and control of nylon-6, 6 batch polymerization. IEEE Transactions on Control Systems Technology, 21(1):94-106, Jan 2013.

[5] S. Aumi, P. Mhaskar, Integrating Data-Based Modeling and Nonlinear Control Tools For Batch Process Control, AIChE Journal 58 (2012) 2105-2119.

[6] B. Corbett, P. Mhaskar, Subspace identification for data-driven modelling and quality control of batch processes, AIChE Journal (2016) n/a-n/a.

[7] Abhinav Garg, P. Mhaskar, Data-Driven Model Predictive Control of Crystal Size Distribution in a Seeded Batch Crystallization Process, Chemical Engineering Science (2016) Submitted.

472915 Computation and Analysis of Optimal Operating Strategies for Air Separation Units Using High Fidelity Dynamic Models

Abstract:

Manufacturing operations are frequently motivated and shaped by external environmental factors, such as raw material availability, demand level and government regulations. The traditional interpretation of such external variations as disturbances and the practice of decoupling a plant from its environment are not suitable for today's transient and competitive global market. This means that such plants need to switch from largely steady-state operations to more intentionally dynamic ones in response to the time-varying environment [1].

Cryogenic air separation units (ASUs) based on distillation technology are an example of such manufacturing plants. Due to the high degree of heat and material integration of the process, ASUs typically have limited agility and low flexibility, which make process transitions challenging. Notwithstanding, the current market environment creates a significant incentive for dynamic operation of ASUs. Although the raw material, namely ambient air, is free and plentiful, ASUs consume large amounts of electricity, with industrial gas producers in the U.S. consuming in excess of $ 1 billion dollars of electricity per year [2]. Since electricity price deregulation, energy providers introduced dynamic electricity pricing strategies, such as use-of-time and real-time policies. These changes in the utility market may further amplify variations on the demand side, as now the ASU's customers may also practice demand response operations.

Studies on the operations of ASUs have focused mainly on transition agility [3], short-term planning [4] and long-term scheduling [5]. It is demonstrated in several studies that when an ASU adjusts its production loads in response to the electricity price, improvements in the plant's economic performance can be achieved. However, most of the studies focused on electricity price induced operation changes. Except for the work done by Pattison and co-workers [6], most of the planning and scheduling studies do not include accurate representations of the process dynamics, despite its importance in determining the actual feasibility of required transitions. In these studies, one of the commonly practiced strategies is to use liquid inventory to buffer the effects of electricity price variations.

In this paper, we conduct a more comprehensive analysis on optimal operation strategies for air separation systems through a multi-tiered dynamic optimization framework. Here, the response of the integrated plant is captured using collocation-based reduced order dynamic models that are shown to be highly accurate. The overall objective is to improve the plant's economic performance under electricity price or demand variation. One of the operational issues assessed is the economic incentive for practicing a storage and evaporation strategy in response to time-varying electricity price/demand. The trade-off between profitability and feasibility (i.e. due to process dynamics) and the resulting selection of collection methods (i.e. collecting liquid product or liquefying over-produced gas product) and utilization mechanisms (i.e. vaporizing to meet gas demand or introducing as additional reflux) are systematically explored. Another issue investigated is the benefit of having cooperative supply chain subsystems. With information on the upcoming demand change, the air separation plant could potentially take pre-emptive action so that the demand of the customer can be satisfied, while also maximizing the profitability of the plant. All these optimization problems are solved using the commercially available software package, gPROMS 4.1.0.

References

[1]

T. Backx, O. Bosgra and W. Marquardt, "Towards intentional dynamics in supply chain conscious process operations," in *Foundations of Computer Aided Process Operations Conference*, 1998.

[2]

Z. Chen, M. Henson, P. Belanger and L. Megan, "Nonlinear model predictive control of high purity distillation columns for cryogenic air separation," *IEEE Transactions on Control Systems Technology,* vol. 18, no. 4, p. 811 –821, 2010.

[3]

Y. Cao, S. C.L.E., M. Baldea and S. Blouin, "Optimization-based assessment of design limitations to air separation plant agility in demand response scenarios," *Journal of Process Control,* vol. 33, pp. 37-48, 2015.

[4]

Y. Zhu, S. Legg and L. C.D., "A multiperiod nonlinear programming approach for operation of air separation plants with variable power pricing," *AIChE Journal,* vol. 57, no. 9, pp. 2421-2430, 2011.

[5]

S. Mitra, J. Pinto and I. Grossmann, "Optimal multi-scale capacity planning for power-intensive continuous processes under time-sensitive electricity prices and demand uncertainty. Part I: Modeling," *Computers & Chemical Engineering,* vol. 65, pp. 89-101, 2014.

[6]

R. Pattison, C. Touretzky, T. Johansson, I. Harjunkoski and M. Baldea, "Optimal Process Operations in Fast-Changing Energy Markets: Framework for Scheduling with Low-Order Dynamic Models and an Air Separation Application," *Industrial and Engineering Chemistry Research,* vol. 55, no. 16, pp. 4562-4584, 2016.

454846 Tailor-Made Solvent and Process Design for the Separation of Azeotropic Mixtures

Abstract:

Tailor-made Solvent and Process Design for the Separation of Azeotropic Mixtures

Marta Gonzalez Garcia[1], Ioulietta Paraskeva[1], Anjan Kumar Tula[1], Emmanouil Papadakis[1], Deenesh Kavi Babi[2], Rafiqul Gani[1]

[1]SPEED, Department of Chemical and Biochemical Engineering, Technical University of Denmark, Soltofts Plads 1, Building 229, DK-2800 Lyngby, Denmark

[2]Utility and Solvents, Department of Insulin Manufacturing 1, Novo Nordisk A/S, Hallas Allé DK-4400 Kalundborg, Denmark

A typical example of a process that does not require the addition of a solvent for achieving the desired separation of an azeotropic mixture is pressure swing distillation. An important condition for its operational feasibility is that the azeotrope must be pressure sensitive over a moderate change in pressure[1]. However, if the azeotrope does not satisfy this condition or the process is energy intensive and/or environmentally unfriendly then a promising alternative is a solvent – based separation process[2]. The solvent – based process configuration for a two phase system (vapor-liquid) consists of an extraction column followed by a solvent recovery column and the process design together with solvent selection is usually obtained through a trial and error approach (heuristics, mathematical programming, and/or hybrid). However, can the design be obtained without any trial and error? It is possible by first determining the process design for a desired (target) solvent and then finding a solvent that matches the desired solvent properties. This decomposition of the design problem is feasible because the process design is based on the properties of the system and does not need the identities of the chemicals involved. On the other hand, the selection problem needs the identities of the chemicals involved but not details of the process. Also, a unique solvent-azeotropic mixture cannot be obtained for all homogeneous azeotropic mixtures. This means that the azeotropic mixtures need to be classified first and for each class, the optimal solvent behavior and the corresponding *apriori* process design determined. Also, can the solvent-based separation problem further improved in terms of energy consumption for the same separation specification?

Assuming a desired solvent-azeotropic mixture behavior and a set of solvent properties, the design of the two column process can be determined, for example, in terms of product purities, number of stages, feed location, and reboiler-condenser duties without knowing the identity of the solvent. This means that having *apriori* designed the separation process, if a solvent that matches the targeted behavior and properties could be found, a simultaneous tailor-made solvent and process design could be very easily achieved. The conceptual basis for this method is that solvent properties have a direct and unique influence on the separation process behavior, for example, separation feasibility is defined by the distillation boundary (corresponding to the ternary system of solvent and the azeotropic mixture); energy consumption is influenced by solvent capacity and solvent pure component properties; waste generation and environmental impact is influenced by solvent selectivity and solvent environmental properties. By using the diving-force concept[3], it is ensured that for the *apriori* design, the extraction and recovery columns operate at the maximum driving force, which translates into the design of a more sustainable energy efficient process.

The following reverse design problem consisting of two sub-problems is formulated and solved: Sub-problem 1: Given ternary phase equilibrium data, driving force diagrams and properties of the chemical system, determine the optimal energy efficient design of the extraction and the solvent recovery columns. Sub-problem 2: Given, a homogenous azeotropic mixture and an *apriori* process design; find the best (optimal) solvent that satisfies the process design in the form of targeted solvent-azeotropic mixture behavior plus solvent properties such as normal boiling point, heat of vaporization, solubility of solute and many more. Note that matching these targets implicitly mean satisfying all the process constraints such as mass and energy balance, separation feasibility and energy demands. Sub-problem 3: Replace the solvent recovery column (SRC) with a liquid-liquid extraction column plus an evaporator (LL+E) by swapping the original solvent with a swap solvent. The swap solvent takes the solute from the original solvent in the liquid-liquid extraction column and the solute is removed by simply heating the solute-swap solvent mixture. The original solvent is recycled back to the original extraction column and the swap solvent is recycled to the liquid-liquid extraction column. The energy needed for the removal of the solute by evaporation is significantly lower than that required by the SRC using the original solvent.

The main features of the method are: classification of the homogeneous azeotropic mixtures, classification of the solvent-azeotropic mixture behavior; classification of the swap solvent mixture behavior, apriori process design based on driving force, solvent-azeotropic mixture behavior and swap solvent-original solvent-solute mixture behavior; finding solvents and swap solvents that match the design targets; and a database containing apriori process design from which designs of other problems can be extrapolated or interpolated. For each class of azeotropic mixture in the database, a list of pre-selected solvents is already available. Therefore, for a separation involving a known azeotrope, the optimal solvent (or swap solvent) and process design are obtained with minimal calculations. For an azeotropic mixture not available in the database, simply identifying the class of the azeotropic mixture, all other necessary details can be extrapolated or interpolated from the available data.

The presentation will highlight the concepts and the method for tailor-made solvent and swap solvent selection and process design, together with tools developed to implement and test this method. The method comprises of 4 –stages. In stage 1, the azeotropic mixture class is identified. For this stage, a database of 43 homogenous azeotropic mixtures that have an azeotropic composition between 30-70% mol is developed using available experimental data. In stage 2, the solvent target behavior is retrieved for the identified azeotropic mixture from a second database. For this stage, a classification of the solvent-azeotropic mixture behavior is developed using a graphical approach based on ternary plots and the position of the distillation boundaries. The distillation boundary of interest is drawn from the azeotropic point (located on the binary vertex of the two compounds that constitute the azeotrope with compound 1 not removed by the solvent) until it touches and overlaps the vertex of compound 2-solvent binary line. The classification is done between compound 2 and solvent, that is, high or low composition (more or less than

50%mol) of compound 2 in the azeotrope and high or low composition (more or less than 50%mol) of solvent in the distillation boundary overlap. For example, if a mixture has an azeotrope composition of 40%mol of compound 2 (low) and a distillation boundary overlap of 60% of solvent composition (high), it will be classified as Low-High. Therefore, the 43 azeotropic mixtures are classified into four groups: High-High, High-Low, Low-High and Low-Low. This behavior also requires that the solvent must not form azeotropes with either of the compounds and that the ternary system must be totally miscible. In stage 3, based on the selection from stages 1 & 2, the corresponding driving force based *apriori* design is retrieved from the database. For this stage, validated process designs by rigorous simulation for all classes of azeotropic mixtures and solvent-azeotropic behavior are determined and stored in a database. Also, additional information that helps to extrapolate from the stored data are provided. For example, plots of solvent flow rate vs composition of solvent in the compound 2-solvent mixture and solvent flow rate vs driving force. In stage 4, the solvent selection problem is solved by database search and/or standard solvent selection-design techniques (3). Final design is verified through rigorous simulation using as initial estimate, the retrieved *apriori* design data. For the swap-solvent selection, a target liquid-liquid phase boundary is defined together with properties such as melting points and boiling points, which must be much greater than the solute.

The method is generic and able to directly determine the best solvent, solvent flow rate, the driving force and the energy required for separation of the classes of homogeneous azeotropic mixtures available in the database. It allows safe extrapolations for azeotropic mixtures not covered in the database but that match the classified behavior. The method has been tested for a wide range of homogeneous azetropes and a collection of these results will be presented. Until the final rigorous simulation verification stage (for which also a good initial estimate is provided), the method hardly needs any calculations and is able to provide more sustainable and near optimal simultaneous selection of solvents and its corresponding optimal process design. The concept and method, can in principle, be applied for many other simultaneous product-process design problems. The savings is not only in energy and material but also time to operate and solve problems. References

1. Seader, J. D., Henley, Ernest J. y Roper, Keith. *Separation Process Principles.* s.l. : John Wiley, 2011.

2. Igor Mitrofanov, Sascha Sansonetti, Jens Abildskov, Gürkan Sin and Rafiqul Gani, 2012, "The Solvent Selection framework: solvents for organic synthesis, separation processes and ionic-liquids solvents", Computer-Aided Chemical Engineering , 30, 762-766

3. *R. Gani*, E. Bek-Pedersen, 2001, "A Simple New Algorithm for Distillation Column Design", *AIChE Journal*, 46 (6), 1271-1274

460397 A Target Upper Bound on Reaction Selectivity Via Feinberg's Cfstr Equivalence Principle

Abstract:

A Target Upper Bound on Reaction Selectivity via Feinberg's CFSTR Equivalence Principle

Mole balances are widely used in the design of chemical plants and processing equipment as well as in conducting techno-economic analyses of potential chemical business ventures [1]. They allow us to calculate unknown flowrates into and out of a system once an appropriate number of specifications are made. On a broader level, mole balances also serve to bound the attainable region of a reaction system. We should not, however, expect that the entire attainable region be accessible for a process on which capacity constraints are imposed. For example, certain areas of the attainable region may be accessible only by permitting extremely low single-pass reactor conversions to achieve high selectivities resulting in impractically large recycle flowrates. It is the underlying kinetics of a chemistry that determine the size and location of these inaccessible regions.

Feinberg and coworkers developed the Continuous Flow Stirred Tank Reaction (CFSTR) Equivalence Principal which allows one to decompose any arbitrary steady-state reactor separator system with total reaction volume "V" into a new system comprising "R+1" CFSTRs (where R is the number of linearly independent reactions) with the same total reaction volume and a perfect separations system [2, 3]. Using this methodology in conjunction with the kinetics of a system of interest, the attainable region given by the mole balances can be refined. Our work aims to further refine the attainable region by introducing flowrate capacity constraints on the CFSTR Equivalence Principle. The constraints on molar flowrates between the "R+1" CFSTRs and the separations system prevent large recycle streams and small reactor conversions. By optimizing this constrained system of CFSTRs, we can determine the maximum possible selectivity of several chemistries completely independent of reactor design given capacity constraints. We have investigated serial reaction networks and more complex reaction networks involving serial and parallel reactions. With the added constraints, the maximum possible selectivity for both types of reaction networks is less than 100%.

We have used this methodology to analyze the maximum possible selectivity of a selection of realistic chemistries and have compared the results to archetypal reactors (e.g., CFSTR, PFR). The results all support the hypothesis that this CFSTR reactor decomposition can be constrained and optimized to obtain a target upper bound on selectivity for chemistries completely independent of reactor-separator design.

References:

[1] Douglas, J. "Conceptual Design of Chemical Processes", McGraw-Hill, New York (1988).

[2] Feinberg, M.; Ellison, P. *Ind. Eng. Chem. Res.* **2001**, *40*, 3181-2194.

[3] Tang, Y.; Feinberg, M. *Ind. Eng. Chem. Res.* **2007**, *46*, 5624-5630.

460985 A Nonsmooth inside-out Algorithm for Robust Flash Calculations

Abstract:

Dependable algorithms for nonideal vapor-liquid equilibrium calculations are essential for effective process design, simulation and optimization. At present, the inside-out formulation[1] for flash calculations serves as the basis for many of the algorithms used by process simulation software due to its robustness with respect to initialization and inexpensive computational cost. However, the original inside-out algorithm will fail if the specified conditions lead to a single-phase result, as the solution is constrained to obey equilibrium relationships which are only valid in the two-phase region. While heuristic modifications of the algorithm have shown effectiveness in allowing process simulators to converge to single-phase solutions, [2] these provide no guarantee of success. Such behavior is particularly undesirable in a process simulation/optimization problem where flash calculations must be performed on streams where the phase regime at the solution is not known ***a-priori.*** To address these issues, we present a modification of the classical inside-out algorithm, which features an additional nonsmooth equation in the inner loop to relax equilibrium conditions when necessary and a modified scheme for updating the outer loop variables. Additional nonsmooth equations are also employed in the outer loop to avoid numerical issues and nonphysical solutions, such as those needed to identify the correct compressibility factor roots from cubic equations of state. These modifications guarantee existence of a solution that is physically correct, regardless of the number of phases present. The nonsmooth subproblems are efficiently solved with a linear-programming based Newton method[3] supplied with automatically calculated elements of generalized derivatives,[4] while the outer loop may be converged either by a similar nonsmooth Newton-type method or simply by successive substitution. Numerical results for simulations involving different flash types and property packages are shown, highlighting the capability of the algorithm to handle both two-phase and single-phase behavior robustly and efficiently. This flash module can also readily be included as a subroutine in more complex problems, since desirable properties of the generalized derivative elements and our nonsmooth framework allow objects which behave in the same way as classical derivatives to be passed around a flowsheet for use in Newton-type methods for simulation and optimization.

---

[1] J. F. Boston and H. I. Britt, "A radically different formulation and solution of the single-stage flash problem," ***Computers & Chemical Engineering***, Vol. 2, pp. 109-122, 1978.

[2] V. S. Parekh and P. M. Mathias, "Efficient flash calculations for chemical process design – extension of the Boston-Britt "Inside-out" flash algorithm to extreme conditions and new flash types," ***Computers & Chemical Engineering***, Vol. 22, No. 10, pp. 1371-1380, 1998.

[3] F. Facchinei, A. Fischer and M. Herrich, "An LP-Newton method: nonsmooth equations, KKT systems, and nonisolated solutions," ***Mathematical Programming***, Vol. 146, pp. 1-36, 2014.

[4] K. A. Khan and P. I. Barton, "A vector forward mode of automatic differentiation for generalized derivative evaluation." ***Optimization Methods & Software,*** Vol. 30, No. 6, pp. 1185-1212, 2015.

461804 Simultaneous Chemical Process Synthesis and Heat Integration with Unknown Cold/Hot Process Streams

Abstract:

In chemical process synthesis, the heat integration between cold and hot process streams reduces the external utility consumptions and leads to economic benefits. Sequential heat integration approaches rely on the assumption that temperatures and flow rates of process streams are known, which means that these methods can only be applied after stream conditions have been determined. On the other hand, simultaneous approaches can handle variable stream temperatures and flow rates, which means that the heat integration subsystem can be optimized along with the process and thus stream conditions. It has been shown that simultaneous approaches lead to better solutions compared to the sequential approaches [1]. While various simultaneous heat integration models have been proposed [1, 2], they require all the process streams to be classified as cold or hot streams (depending on whether the streams require heating/cooling) prior to solving the problem. However, in some cases there exists a subset of process streams that cannot be predetermined as cold or hot streams, especially in the context of superstructure optimization that involves the selection of alternative processing units. For example, the product stream from one of the two alternative reactors (A and B) is sent to one of the two alternative separation units (C and D) for purification. Units A, B, C and D might operate at different temperature ranges. Therefore, we might not be able to determine if the stream that connects the reactor and separator is a cold or hot stream before solving the problem.

Accordingly, in this work, we present a mixed-integer nonlinear programming (MINLP) optimization model for the simultaneous chemical process synthesis and heat integration with general, that is undetermined, process steams. The model extends the transshipment model proposed by Papoulias and Grossmann [3], to account for (1) streams that cannot be classified as hot or cold, and (2) variable stream temperatures and flow rates. The cold/hot stream â €œidentitiesâ€ are represented by classification binary variables which are (de)activated based on the relative stream inlet and outlet temperatures. Variables including stream temperatures and heat loads are disaggregated into two

mutually exclusive parts (cold and hot), and each part is (de)activated by the corresponding classification binary variable. The dynamic temperature intervals constructed from the variable process stream inlet temperatures are implicitly ordered from high to low temperatures and stream inlet/outlet temperatures are assigned to the correct temperature intervals. Stream heat loads at each temperature interval depend on the relative positions of stream inlet and outlet temperatures with respect to the temperatures at the interval boundaries. Preprocessing is used to reduce the size of the problem and speed up the solution process. For example, if a process stream can be predetermined as a cold stream, we can fix the corresponding classification binary variable to one, and deactivate the hot part of the disaggregated variables. It is important to note that the only nonlinearity in this model are the bilinear [flow rates]*[temperature difference] terms. Therefore, in the case that the stream flow rates or temperatures are fixed, the proposed model reduces to a MILP model; and if both of them are fixed, we recover the original transshipment model.

The proposed model is generalized to handle intermediate utilities at arbitrary temperatures. Further, the model is able to deal with both isothermal and non-isothermal phase changes. Should phase changes of pure components are present in a process stream, the original stream can be disaggregated into sub-streams and the heat integration model can be reformulated to include phase detection (i.e. decide at what phase this stream starts and ends).

The heat integration model can be integrated with a process superstructure model. To achieve this integration effectively, we discuss a series of modeling techniques that lead to simplified unit models and connectivity equations. For example, we show that the minimum number of temperature intervals needed for heat integration equals to the minimum number of process streams over all the feasible solutions, which is fewer than the total number of process streams in the original superstructure.

The proposed model is applied on several illustrative examples with variable flows and temperatures, in which some streams have unknown cold/hot classifications. Finally, we present a case study in which the proposed model is integrated with a realistic bioethanol production superstructure and is optimized based on an economic criterion [4].

References

1. Duran, M.A. and I.E. Grossmann, *Simultaneous-Optimization and Heat Integration of Chemical Processes.* Aiche Journal, 1986. **32**(1): p. 123-138

2. Navarro-Amorós, M.A., et al., *An alternative disjunctive optimization model for heat integration with variable temperatures.* Computers & Chemical Engineering, 2013. **56**: p. 12-26

3. Papoulias, S.A. and I.E. Grossmann, *A Structural Optimization Approach in Process Synthesis .2. Heat-Recovery Networks.* Computers & Chemical Engineering, 1983. 7(6): p. 707-721

4. Kong, L., et al., *A superstructure-based framework for simultaneous process synthesis, heat integration, and utility plant design.* Computers & Chemical Engineering, 2016 DOI: http://dx.doi.org/10.1016/j.compchemeng.2016.02.013.

461915 A Method for Minimization of Total Exergy Loss over the Complete Search Space of Regular Distillation Configurations

Abstract:

Exergy of a stream is its potential to do work when it is brought to the reference conditions via a reversible path. And it is a good metric to account for the temperature level at which heat is used or rejected in a distillation column. The thermodynamic efficiency of a distillation configuration can be derived through exergy analysis. Thus, performing exergy analysis, in conjunction with minimization of total heat duty, will help us identify distillation configurations with low energy requirement and high thermodynamic efficiency. Such configurations are particularly valuable for cryogenic as well as natural gas liquids (NGL) separations.

Here, we present a global optimization algorithm based on Underwood's equations that minimizes the total exergy loss of any regular-column distillation configuration (i.e. distillation sequence separating n components using n-1 columns) which separates any ideal or near-ideal multicomponent mixture. This objective function tends to favor configurations that use high temperature condensers and low temperature reboilers, relative to other configurations with similar vapor duties. For the first time, a global optimization based rank-list of the complete search space of regular-column configurations based on minimizing total exergy loss is synthesized. The identification of configurations that simultaneously have low vapor duty and high thermodynamic efficiency provides industrial practitioners with strong candidates for heat pump assisted multicomponent distillation.

462457 Shale Gas to Light Olefins: Global Optimization of an Integrated NGL Recovery, Steam Cracking, and Methane Conversion Superstructure

Abstract:

Due to the decreased prices of natural gas in recent years [1], production of "wet" gas that contains natural gas liquids (NGLs), such as ethane, propane, butane, and natural gasoline, is favorable to extract for commercial purposes [2]. In some wet shale plays, the combined ethane and propane composition can exceed 30% [3]. These recent trends motivate industrial efforts to build several new ethane crackers with a combined ethylene production capacity of 12.5 million tonnes/year in the United States [4]. Although the NGLs can be effectively used as petrochemical feedstocks toward light olefins, the optimal processing technology (catalytic and/or thermal), feed composition (mixed feed or a pure feed), and operating conditions are unknown.

Previous work has shown that light olefins including ethylene, propylene, and butene isomers can be produced from natural gas via reforming in a profitable manner out of a superstructure of process alternatives [5]. However, extraction of NGLs prior to methane conversion is an untapped opportunity for these refineries that can simultaneously maximize carbon conversion and avoid high reforming costs. This NGL recovery section is introduced to the existing superstructure of novel/competing process alternatives. A demethanizer column is utilized [6,7] to recover more than 83% of ethane and more than 99% of higher hydrocarbons in the NGL stream. The NGLs can be further separated and sent to steam cracking or catalytic dehydrogenation alternatives. Several steam cracking reactors incorporate a pure or mixture of hydrocarbon feed to produce olefins. The operating conditions and reactor topologies of the steam cracking alternatives are dynamically optimized using the well-established cracking kinetics in the literature [8,9]. Catalytic dehydrogenation process alternatives incorporate commercially available catalysts for propane and/or isobutane dehydrogenation. The methane rich gas from the demethanizer is converted via reforming or direct conversion processes introduced into the process superstructure [5].

The mathematical model of the overall process superstructure is solved using a novel branch and bound global optimization framework. The objective is to maximize the profit of light olefin production from the integrated NGL extraction, cracking, catalytic dehydrogenation, and methane conversion refinery. A technoeconomic analysis will be presented with major topological decisions across several case studies. Different natural gas compositions will be shown to present the topological trade-offs in the NGL recovery section. The net present values can be significantly improved (>10%) through the use of the integrated superstructure.

[1]: Floudas, C. A.; Niziolek, A. M.; Onel, O.; Matthews, L. R. Multi-scale systems engineering for energy and the environment: Challenges and opportunities. AIChE Journal 2016, 62 (3), 602-623.

[2]: Kopalek, M.; High value of liquids drives U.S. producers to target wet natural gas resources, 2014, Energy Information Administration

[3]: Hill, R. J.; Jarvie, D. M.; Zumberge, J.; Henry, M.; Pollastro, R. M.; Oil and gas geochemistry and petroleum systems of the Fort Worth Basin, 2007, AAPG Bulletin, 91(4), 445-473.

[4]: Chang, J.; New projects may raise US ethylene capacity by 52%, PE by 47%, 2014, ICIS Petrochemicals.

[5]: Onel, O.; Niziolek, A. M; Floudas, C. A Optimal Production of Light Olefins from Natural Gas via the Methanol Intermediate. Industrial & Engineering Chemistry Research 2016, 55 (11), 3043-3063.

[6]: Luyben, W. L.; NGL Demethanizer Control, 2013, Industrial and Engineering Chemistry Research, 52, 11626-11638.

[7]: Luyben, W. L.; Effect of Natural Gas Composition on the Design of Natural Gas Liquid Demethanizers, 2013, Industrial and Engineering Chemistry Research, 52, 6513-6516.

[8]: Sundaram, K. M.; Froment, G. F.; Modeling of thermal cracking kinetics – I: Thermal cracking of ethane, propane and their mixtures, 1977, Chemical Engineering Science, 32(6), 601-608.

[9]: Sundaram, K. M.; Froment, G. F.; Modeling of thermal cracking kinetics – II: Cracking of iso-butane, of n-butane, and of mixtures ethane-propane-n-butane, 1977, Chemical Engineering Science, 32(6), 609-617.

467281 An Algorithm for Integrated Molecular and Process Design: Physically-Driven Domain Reduction for Liquid-Liquid Extraction

Abstract:

Computer Aided Molecular and Process design (CAMPD) involves the simultaneous optimisation of processes and processing materials (such as mass separating agents, catalysts) [1]. In CAMPD one considers the minimisation of a process-wide objective subject to process design constraints and molecular feasibility constraints. Molecular-level (material) decisions are represented using discrete variables such as the number of groups of a given kind, whereas process variables can either be continuous or discrete. The resulting design problem is a mixed integer nonlinear problem (MINLP) which is difficult to solve for a number of reasons. The large molecular design space leads to combinatorial explosion [2]; the non-convexity of the domain can lead to poor solutions depending on the starting point; convergence failure can arise for some combinations of molecular and process variables. In particular, numerical difficulties can be encountered in the absence of a good initial guess to the highly nonlinear, and coupled, process and property models. Numerical issues can also occur if a combination of molecular and process inputs violates constraints of phase behaviour implicit in the model. For instance, an evaluation of an absorber model can fail if the stream at its liquid inlet is in the vapour state. Also, the feasible region with respect to these constraints is a function of the solvent: for example, the range of temperatures where a molecule is in the liquid state varies with the choice of solvent.

To address these challenges we have recently a developed a new algorithm for CAMPD of separation systems [3, 4]. The algorithm builds on the concept of embedding screening of molecules within an optimizer [5, 6]. Novel tests were introduced to screen both molecular *and* process variables, in the context of gas-liquid separation. In this approach, we first identify the process domain where the feed is stable. At each major iteration of an outer approximation (OA) algorithm [7], we find a reduced process domain for a solvent. The reduced process domain is an overestimation of the feasible region for that solvent. In a first test, the feasibility of solvent handling and storage is checked. Ina second test, the range of process variables over which the solvent and the feed can form a two-phase mixture is found. The range of process variables over which the required degree of purity of the treated stream may be attained is identified in a third test. If the solvent fails any of the tests it is eliminated. If the tests are feasible, the primal problem is solved using bounds identified by the tests and initial guesses within the reduced process domain. Information from the tests and the primal problem is used to construct the master problem for the OA algorithm.

The extension of this methodology to liquid-liquid extraction systems is presented here. The three tests, initially developed for absorption processes, are recast for the case of liquid-liquid extraction. The approach is illustrated with a case study of the separation of butanol from a fermentation broth. As the broth is dilute in butanol, separation of this mixture by distillation is highly energy intensive. Hence, an optimal solvent for extraction, and the corresponding optimal process variables, are designed using the algorithm proposed here. A group contribution equation of state, the group contribution version of the statistical associating fluid theory with a Mie potential, SAFT-Î³Ã Mie[8,9], is used in this work to predict the relevant thermodynamic properties. A systematic study of the performance of the algorithm is undertaken. The algorithm succeeds in avoiding expensive process evaluations for infeasible solvents and enhances convergence to the solution from multiple starting points.

1 Adjiman, C. S., Galindo, A., and Jackson, G. (2014). Proceedings of the 8th International Conference on Foundations of Computer-Aided Process Design FOCAPD 2014, edited by Eden M.R., Siirola J.D., Towler G.P., Computer Aided Chemical Engineering, Elsevier B.V.

2 Joback, K.G. and Stephanopoulos, G. (1995). In Advances in Chemical Engineering, 21: 257-311

3 Gopinath S, Galindo A, Jackson G, Adjiman CS (2016). Proceedings of the 26th European Symposium on Computer Aided Process Engineering, edited by Kravanja Z., Computer Aided Chemical Engineering. Elsevier B.V.

4 Gopinath S, Galindo A, Jackson G, Adjiman CS (2016). Computer-aided molecular and process design for absorption-desorption systems using physically-driven domain reduction (2016). Manuscript submitted

5 Buxton, A., Livingston, A. G., and Pistikopoulos, E. N. (1999). AIChE Journal, 45: 817-843

6 Giovanoglou, A., Barlatier, J., Adjiman, C. S., Pistikopoulos, E. N., and Cordiner, J. L. (2003). AIChE Journal, 49: 3095-3109

7 Duran, M. A. and Grossmann, I. E. (1986). Mathematical Programming, 36: 307-339

8 Papaioannou, V., Lafitte, T., Aveñdano, C., Adjiman, C. S., Jackson, G., Müller, E. A., and Galindo, A. (2014). The Journal of Chemical Physics, 140: 054107

9 Lafitte, T., Apostolakou, A., Avendaño, C., Galindo, A., Adjiman, C. S., Müller, E. A., and Jackson, G. (2013). The Journal of Chemical Physics, 139: 154504

469862 Optimisation-Based Design of a Heat-Integrated Crude Oil Distillation System Using Rigorous Simulation and Surrogate Models

Abstract:

In petroleum refining industry, the crude oil distillation system plays a key role in the overall process. The system comprises a complex heat-integrated distillation column and a heat recovery system where the crude oil feed is partially vaporised. The presence of large number of degrees of freedom (column structure and operating conditions) and the complex interactions existing between individual units makes the design of crude oil distillation unit one of the most challenging tasks within process system engineering. Due to the complex nature of the process, traditional design methods consider designing the complex column and the heat recovery system in separate steps. In this approach, however, interactions between the two subsystems are not taken into account. Some researchers have applied standard optimisation techniques to simultaneously design the crude oil distillation unit and the heat recovery system, but these approaches typically rely on shortcut column models, to avoid the convergence problems associated with rigorous column models. The use of these shortcut models can lead to large errors, as they might fail to predict the behaviour of the complex crude oil distillation unit accurately.

This paper introduce a new framework for the design of crude oil distillation system that integrates a rigorous tray-by-tray crude oil distillation system model implemented in a process simulator (Aspen HYSYS) with an optimiser coded in MatLab. The approach takes advantage of the physical and thermodynamic property models, crude oil characterisation models and column sizing models available in the process simulator.

The proposed approach is implemented in three steps. First, the column superstructure of [1] for optimal feed location, total number of trays and optimal operating conditions is adapted to build the column superstructure using a rigorous tray-by-tray model in Aspen HYSYS. The superstructure consists of conditional and unconditional trays in all column sections, which are represented using Murphree tray efficiencies. In the second step, the problem is formulated as an

MINLP with the number of trays in each column section defined as integer variable and the operating conditions as continuous variables. The heat recovery system is represented using Pinch analysis in order to determine the minimum utility requirements that will be used in the calculations. In the final step, an optimisation algorithm (genetic algorithm) is applied to optimise the system, thereby determining the optimal column configuration and operating conditions that minimise a given objective (e.g. total annualised cost).

To expedite the solution procedure, the proposed approach also explored the use of surrogate models based on neural networks that are constructed from a set of samples generated using the original simulation model.

The proposed approach has been applied to the design of a crude oil distillation system that separates crude oil into five products. Two design objectives of interest are considered namely, minimum total annualised cost and minimum energy cost. The results show that significant savings in total annualised cost and energy cost can be achieved using our approach without compromising the solution accuracy.

References

[1]Â Yeomans, H. & Grossmann, I. E. Optimal design of complex distillation columns using rigorous tray-by-tray disjunctive programming models. Ind. Eng. Chem. Res. 2000, 39 (11), 4326â''4335

465151 A Neutron Scattering Study of Ion-Conduction Mechanisms in Nanocomposite Polymer Electrolytes

Abstract:

Polymer Electrolytes are the key to light-weight and energy-dense lithium ion batteries but suffer from low room temperature ion conductivities. Spherical ceramic fillers are known to improve ion conductivity and mechanical properties but are still orders of magnitude away from commercialization. Understanding the ion conduction mechanism enables better design of the electrolyte system. Two important factors that govern ion conduction include electrolyte microstructure and polymer dynamics. While electron microscopy and small angle neutron scattering (SANS) provide structural information, quasi-elastic neutron scattering (QENS) examines polymer dynamics. QENS probes two processes: segmental dynamics of polymer chains and rotation of highly-conductive crystalline channels. While the former process remains unaffected by presence of particles, the latter becomes more restricted. Based on this information, polymer electrolytes were designed with high aspect ratio fillers to enable further stabilization of the highly conducting channels. This results in one-order of magnitude improvement in lithium ion-conduction over spherical fillers.

459477 Mechanical Properties of Graphene–Polymer Nanocomposites

Abstract:

The exceptional mechanical, electronic, and thermal properties of graphene render it a highly promising filler for polymermatrix nanocomposites. While preliminary experiments indeed show significant enhancement in the thermomechanical properties of graphenepolymer nanocomposites compared to those of the polymer matrix even at relatively low loading of filler, our fundamental understanding of structure–property relationships that govern such enhanced material response remains very limited.

Toward this end, in this presentation, we report results of moleculardynamics simulation studies on the mechanical behavior of polymer (high-density polyethylene) nanocomposites reinforced by graphene and fullerenes with the aim of elucidating the underlying mechanisms that govern the mechanical response of these composite materials. Using a united-atom-based model of the glassy polymer matrix, we show systematic trends in the enhancement of the mechanical stiffness of the polymer composite as a function of filler concentration, filler size, and matrix–filler interfacial interaction strength that is governed by dispersive forces. For comparison of different filler reinforcement effects in such polymermatrix nanocomposites, we also present systematic mechanical behavior studies of fullerene–polymer composites and show that, for fullerene fillers, the response is only weakly dependent on filler size with relatively high loadings required for a considerable improvement in the composite's stiffness. In contrast, graphene-polymer nanocomposites, with graphene circular flakes used as fillers, show an appreciable improvement in the stiffness reinforcement with increasing filler size. We explain the differences in the elastic response of (0D) fullerene and (2D) graphene reinforced polymer matrix composites through detailed atomic scale characterization of the nanocomposites in conjunction with the predictions of an analytical continuum-mechanics model for understanding the fillersize dependent response of the polymer nanocomposites. We also derive scaling relations for the dependence of the composite modulus for polymer-graphene nanocomposites as a function of the flake size at given filler concentration. Moreover, we report results of

our continuum mechanical analysis for polymer-matrix reinforcement on the basis of shear-lag modeling and show a parametric dependence of the axial and interfacial shear stress at the graphene-polymer interface on the graphene flake size. Based on our atomistic simulations and this shear-lag analysis, we identify the range of filler length scale over which classical shear-lag theory becomes most effective for describing the polymer-matrix reinforcement.

471794 Probing Multiscale Structure in Aligned Nanotube-Polymer Composites

Abstract:

Composites made from vertically aligned carbon nanotube (CNT) forests are important for a broad range of applications, such as for thermal management, high-flux membranes, protective fabrics, mechanical reinforcement, and energy dissipation. Despite progress, there remains a need for quantitative understanding of how the multiscale structure of CNT forests dictates both the fabrication of advanced composites and ultimately their performance. Drawing such quantitative relationships is challenging because measuring the morphological structure and order in CNT forests is difficult for a) multi-component systems, b) over large range of length scales, and c) without destroying the material.

To nondestructively investigate salient CNT forest characteristics, we built upon a suite of complementary soft and hard X-ray scattering techniques including polarized resonant soft X-ray scattering (R-SoXS â€“ micro/meso/nano), small- and wide-angle X-ray scattering (SAXS/WAXS â€“ nano/atomic) and X-ray attenuation. R-SoXS is a novel technique at the Advanced Light Source (ALS) ideally suited for composite characterization because it enables selective probing of the CNTs and their polymer matrix *via*scattering with chemical specificity by tuning of incident photon energy. Using our combined synchrotron techniques, we analyzed high-performance nanocomposite membranes fabricated by: 1) synthesizing self-aligned CNT forests by chemical vapor deposition; 2) infiltrating the void space between CNTs with a polymeric material; and 3) then etching open the ends of a large number of CNTs to expose their core volume to fluid transport.

In particular, we mapped structural characteristics in self-aligned CNT forests/composites with *chemical*, *spatial*, and *length-scale* resolution. We developed a unified analytical model to quantitatively describe the structural hierarchy of aligned CNTs and to extract key structural parameters via curve-fitting of our X-ray scattering data. From the bottom

up, our model parameters define the graphitic lattice and wall number (atomic), CNT diameter (nano), CNT bundling and spacing (meso), regular corrugations (micro), and number density (macro). Comparison with electron microscopy reveals good agreement for forests with a wide range of characteristics (e.g., 1-11 walls, 1.5-15 nm diameter, 10^{10}-10^{12} cm^{-2}density). With R-SoXS as our complementary chemical probe, we also analyzed the structure of the polymer matrix around the outside of the CNTs and ions inside the CNT nanochannels.

Finally, we show how the structural characteristics of our CNT forests inform the fabrication of large-area, flexible polymer composites toward their application as ultrabreathable and protective membranes.[1] We leveraged composite structural information to precisely quantify the fluid flow enhancement measured experimentally through the CNT nanochannels in our membranes. This is an exciting advancement because insufficient information regarding the CNT membrane structural parameters is thought to be the primary source of orders-of-magnitude discrepancies in the ultrafast fluid transport inside CNT nanochannels reported in the literature.[2] More precise understanding of the transport in these bulk materials may help elucidate open controversies regarding transport through individual CNT nanochannels.[3] While we focused here on fluidic properties, our methodology is broadly applicable for drawing quantitative, multiscale relationships between structure and other (mechanical, thermal, electrical, etc.) properties in nanocomposite systems.

This work is supported by the Defense Threat Reduction Agency (DTRA) D[MS]2 project under Contract No. BA12PHM123 and was performed under the auspices of U.S. Department of Energy by Lawrence Livermore National Laboratory under Contract DE-AC52-07NA27344.

1. Bui, N.; Meshot, E. R.; Kim, S.; Peña, J.; Gibson, P. W.; Wu, K. J.; Fornasiero, F., Ultrabreathable and Protective Membranes with Sub-5 nm Carbon Nanotube Pores. *Advanced Materials* **2016**.

2. Kim, S.; Fornasiero, F.; Park, H. G.; In, J. B.; Meshot, E.; Giraldo, G.; Stadermann, M.; Fireman, M.; Shan, J.; Grigoropoulos, C. P.; Bakajin, O., Fabrication of Flexible, Aligned Carbon Nanotube/Polymer Composite Membranes by in-Situ Polymerization. *Journal of Membrane Science* **2014**, *460*(0), 91-98.

3. Guo, S.; Meshot, E. R.; Kuykendall, T.; Cabrini, S.; Fornasiero, F., Nanofluidic Transport through Isolated Carbon Nanotube Channels: Advances, Controversies, and Challenges. *Advanced Materials* **2015,** *27* (38), 5726-5737.

460070 Thermal and Mechanical Property Predictions for Conjugated Polymers Using Atomistic Simulations

Abstract:

Paralleling successful application in catalyst design, drug discovery, and protein structure analysis, molecular simulations are emerging as reliable property-prediction tools for synthetic polymers as well. Simulations are not only useful for determining functional properties such as band gaps of п-conjugated polymers, but also for accessing microscopic information that is not easily available from experiments.

In this work, we used atomistic simulations to predict thermophysical and mechanical properties of two p-conjugated polymers, polyacetylene (PA) and poly(*para*-phenylene vinylene) (PPV) [1]. Although the electron transport and optical behaviors of these polymers are well-characterized, information on properties like elastic modulus and glass transition temperature are uncertain. Glass transition temperature, , the temperature above which the polymer softens considerably undergoing a large decrease in stiffness, and the elastic modulus of the polymer, are useful parameters in the design of devices such as flexible LED displays and solar panels that use conjugated polymers.

The literature values of the elastic modulus of PA and PPV at room temperature are scattered over a range spanning two to three orders of magnitude [1]. Similarly, differences in the reported values of the of PPV are greater than 100 °C. Polymer properties are generally quite dependent on their synthesis and processing history. The dispersity in the lengths of the polymer chains, the variations in polymer architecture caused by branching, chemical defects such as oxidation, the presence of plasticizers or other low molar mass impurities, and even the method of measurement can greatly alter the measured property values. These effects seem to be particularly important in the case of the two conjugated polymers investigated. Molecular simulations, on the other hand, allow one to probe a virtual ensemble of perfectly defect-free identical polymer molecules as a basis for comparison with measurements. Architectural variations and chemical defects can be introduced in a controlled manner, if their effects need to be

studied.

We used the variations of specific volume and the potential energy of intermolecular interactions with temperature to identify the glass transition temperature. Properties such as the coefficient of thermal expansion, the torsion angle distributions, and the mean square end-to-end distance, were calculated at different temperatures. п-Conjugation of the sp^2 hybridized carbon atoms in the PPV chain would be expected to impart planarity to the molecule, but a non-planar structure, consistent with previously reported observations in neutron diffraction experiments [2], was predicted.

Conventional mechanical characterization techniques such as tensile testing stretch the polymer specimen to different lengths and measure the force required to produce a given extension of the specimen. However, polymer chains in the sample can align or orient along the stretching direction. We used the technique of dynamic nanoindentation instead to measure polymer stiffness by applying relatively small oscillatory loads, of micronewton magnitude, on the polymer films using a diamond microindenter [3]. The experimental results were in good agreement with the MD predictions employing the static deformation method [4].

This study demonstrates that atomistic simulations can reliably complement existing tools for the prediction of the properties of solid polymers.

References:

1. Venkatanarayanan, Ramaswamy I., Sitaraman Krishnan, Arvind Sreeram, Philip A. Yuya, Nimitt G. Patel, Adama Tandia, and John B. McLaughlin (2016), "Simulated Dilatometry and Static Deformation Prediction of Glass Transition and Mechanical Properties of Polyacetylene and Poly(*para*-phenylene vinylene)," *Macromolecular Theory and Simulations*, 25(3), pp. 315–321.
2. Mao, G., J. E. Fischer, F. E. Karasz, and M. J. Winokur (1993), "Nonplanarity and Ring Torsion in Poly (*p* phenylene vinylene). A Neutron Diffraction Study," *The Journal of Chemical Physics*, 98(1), pp. 712–716.
3. Sreeram, Arvind, Nimitt G. Patel, Ramaswamy I. Venkatanarayanan, John B. McLauglin, Stephan J. DeLuca, Philip A. Yuya, and Sitaraman Krishnan, "Nanomechanical Properties of Poly(*para*-phenylene vinylene) Determined Using Quasi-static and Dynamic Nanoindentation," *Polymer Testing* (2014), 37, pp. 86–93.
4. Theodorou, Doros N. and Ulrich W. Suter (1986), "Atomistic Modeling of Mechanical Properties of Polymeric Glasses," *Macromolecules*, 19(1), pp. 139–154.

467782 Importance of Selecting Right Analytical Tools for Successful Product Development: Case Study of Single and Two Phase Polymerization Reactions

Abstract:

The importance of Chemical Reaction Engineering, which deals with chemical kinetics, transport processes, and subsequent reactor design, for product development cannot be overstated. Understanding the chemical kinetics and underlying transport phenomena is critical for successful reactor scale-up to enable commercial production. The success of a process and hence a product is inherently tied to laboratory reactors and concomitant process analytical techniques which assist in development of intrinsic kinetics devoid of mass and heat transfer limitations.

This talk will focus on two examples which highlight the importance of right analytical tools for process and product development. The first example will discuss capturing â€œintrinsicâ€ kinetics of a potentially hazardous exothermic chemistry. In particular, ethylene and propylene oxide based polyalkylene glycols are widely used in high performance formulated fluids and lubricants. The current alkoxylaton kinetics employed in their reactor design and operation limit optimization at industrial scale due to the need for empirical adjustments. Thus development of intrinsic kinetics, not affected by heat and mass transfer limitations, is necessary for improved process performance and for product optimization. This work addressed the challenge by development of an experimental methodology coupled with an in-situ analytical technique to accurately capture the true kinetics. Kinetic model development, regression of kinetic parameters and model validation for a commercial process will be presented.

The second example will address the role of analytical techniques in development of energy efficient processes and products. Single phase solution polymerization for making polyethylene resins renders itself aptly for efficient processing. In particular, energy intensive solvent recovery after polymerization results in a high energy foot print.

Selection of appropriate solvent to lower polymer solubility *during* polymerization to form two phases i.e. polymer phase and solvent phase provides an energy efficient alternative. The heavier polymer phase can be separated using a decantation step after polymerization to remove up to 50% of the solvent leading to lower solvent separation costs. Evaluation of this new processing route and development of analytical methodology to prevent risk of failure on final implementation will be discussed.

474341 Computer Aided Design of Ionic Liquids through a New Group Contribution Approach Based on Surface Charge Density and COSMO-SAC Predictions

Abstract:

The practical use of computer-aided ionic liquid design (CAILD) approach towards the design of solvents for separation processes, such as CO_2 absorption and liquid-liquid extraction, is greatly hindered by the very limited availability of UNIFAC interaction parameters that cover diverse ionic liquid building blocks (i.e. different cation cores, anions, and side chain groups) and other molecular groups (e.g. CO_2). The non-availability of interaction parameters can be linked to the limited thermodynamic data available for systems involving ionic liquids. COSMO-RS and COSMO-SAC methods provide a fully predictive approach for activity coefficients that rely on surface charge density obtained from quantum chemical calculations. However, these methods are not directly compatible with CAMD or CAILD approach as electronic structure calculations are possible only through *ab initio* quantum chemical programs such as Gaussian or Turbomole and they are specific for each ionic liquid as a whole. Further, these calculations are extremely time consuming and sometimes can take several hours for a single compound. We have recently developed a novel methodology that involves group contribution prediction of ionic liquid surface charge density (sigma profiles) and cavity volumes. This is further integrated with COSMO-SAC methodology to enable prediction of solution thermodynamics (activity coefficient) of systems involving ionic liquids. This approach is fully predictive in nature and does not require fitting group interaction parameters using experimental data or quantum chemical calculations. Further, this method is fully compatible with CAMD approach. This predictive thermodynamic model is integrated with optimization methods to reverse engineer novel ionic liquids. We demonstrate the utility of the model through case studies.

468269 Identification of Magnetic Nanoparticles to Eradicate Cancer Cells Using the Hyperthermia Approach a Computational Search Approach

Abstract:

In recent years, nanomaterials have received significant attention due to their unique properties. Magnetic nanoparticles (MNPs) in particular are of high importance due to their magnetic properties. These particles are being tested for various applications including cancer cell detection and targeted cell temperature increase.

Hyperthermia treatment of cancer cells has been one of the most promising ideas of last decade. The theory is to design materials which are very efficient in targeting the cancer cells. The designed particles are to have magnetic properties such that at certain magnetic field conditions the particle develops a higher temperature than the surrounding atmosphere/environment.

This project aims to investigate a way to model magnetic nanoparticle behavior in the presence of different magnitudes of magnetic fields. The aim is to model Brownian and Neel relaxation parameters to determine the heat generated by different MNPs to study which is best suited for hyperthermia treatment of cancer cells. For that, approaches were used to determine AC susceptibility of the out-of-phase particles which was then used in a 3-D model containing healthy and cancer cell. The 3-D model is used to determine temperature output for such materials in the presence of an optimal magnetic field.

Selected References:

1. Rudolf Hergt, Silvio Dutz, Robert Muller, Matthias Zeisberger, *Magnetic particle hyperthermia: nanoparticle magnetism and materias development for cancer treatment*, 2006, J. Phys.:Condens. Matter, Vol 18, S2919-S2934.
2. Shihu Wang and Elena E. Dormidontova, *Nanoparticle Design Optimization for Enhanced Targeting: Monte Carlo Simulations*, 2010, Biomacromlecules, Vol 11, 1785-1795
3. Andreas Jordan, Regina Scholz, Peter Wust, Horst Fahling, Ronald Felix, Magnetic fluid hyperthermia: *Cancer treatment with AC magnetic field induced excitation o biocompatible supermagnetic nanoparticles*, 1999, Journal of Magnetism and Magnetic Materials, Vol 201, 413-419
4. Stefan T Bromley, Iberio de P. R, Moreira, Konstantin M. Neyman, Francese Illas, *Approaching nanoscale oxides: models and theoretical methods*, 2009, Chem. Soc. Rev., Vol 38, 2657-2670.

467310 Machine Learning and Natural Language Processing for Pharmaceutical Product Engineering

Abstract:

Pharmaceutical product engineering is a “Big Data” discipline. It requires understanding of details of the drug chemistry during production and within the body, the manufacturing processes and conditions, and the pharmacokinetics of a disease – all data intensive. In fact, a typical New Drug Application (NDA) contains more than 100,000 pages of a variety of information. In this talk, we present a framework, called HOLMES, for the automatic extraction of knowledge from primary sources related to pharmaceutical product engineering. The information extracted is then stored in ontologies. These ontologies are a computer readable semantic knowledge representation used in artificial intelligence. We describe Machine Learning (ML) algorithms and Natural Language Processing (NLP) techniques that are used in HOLMES for Entity and Concept Recognition and Relation Extraction. We will discuss our progress on the creation of an entity-concept-and-relation databank (7968 entities and concepts, 1665 relations); the application of different ML algorithms for the purpose of joint Entity and Concept detection; and the development of a relation clustering algorithm using common feature sets.

467921 Algorithms for a Closed-Loop Artificial Pancreas: Challenges and Solutions Pertinent to Chemical Process Operations and Control

Abstract:

The development of a closed-loop artificial pancreas to regulate the blood glucose concentration of individuals with type 1 diabetes has been a focused area of research for over 50 years, with very rapid progress during the past decade. In this presentation we first provide an overview of the state of the art of diabetes technology, including self-monitoring blood glucose meters (SMBG), continuous glucose monitors (CGM) and insulin pumps, as well as the daily challenges faced by someone with type 1 diabetes. The major steps toward a closed-loop artificial pancreas include (i) monitoring and overnight alarms for hypoglycemia (low blood glucose), (ii) overnight low glucose suspend (LGS) systems to prevent hypoglycemia, and (iii) fully closed-loop systems that adjust insulin (and perhaps glucagon) to maintain desired blood glucose levels day and night.

The development of a closed-loop artificial pancreas involves the efforts of many individuals, oversight committees and regulatory agencies. In this talk we concisely review the clinical trial process, including Investigational Device Exemption (IDE) applications to the US Food and Drug Administration (FDA), oversight by Institutional Review Boards (IRB), Data and Safety Monitoring Boards (DSMB), and clinical trial registration. Typically, short-term (1-3 day) studies to verify safety and conducted in an inpatient setting (hospital) are followed by hotel- or camp-based studies that approximate the home environment but allow additional safety through monitoring and rapid response of nursing and engineering staff. Finally, based on inpatient and hotel/camp results, longer-term (weeks, months) outpatient (at-home) studies are performed.

While we have been involved in many of the major steps towards an artificial pancreas, we focus on a multiple model probabilistic predictive control (MMPPC) strategy for a fully automated closed-loop artificial pancreas that does not require meal announcement (e.g. without feedforward control). We point out lessons learned and engineering and clinical protocol changes as we moved from simulation to inpatient and hotel/camp, and finally outpatient (at home) studies.

One of the major challenges in developing a fully closed-loop artificial pancreas is that subjects will neglect to provide a meal-related insulin bolus (feedforward control), which often results in a period of post-meal hyperglycemia (high blood glucose). Our MMPPC approach detects when a meal is being consumed, estimates the meal size, and begins to more aggressively inject more insulin. Included is knowledge of typical eating patterns; for example, if someone has just consumed a large meal they are unlikely to eat again for several hours. At each time step the controller delivers insulin to assure a low probability that an individual will become hypoglycemic (low blood glucose). Similarly, when a subject is likely to be sleeping (which can be inferred for the orientation of an accelerometer), the insulin delivery is less aggressive, but again targeting a low probability of hypoglycemia. The algorithm also accounts for the increased insulin sensitivity and reduced need for insulin during exercise.

A summary glucose plot for ten subjects (over 300 subject-hours) in an in-clinic study is shown in Figure 1. The overnight performance of the closed-loop strategy is clearly significantly improved compared to open-loop control. In the presentation we will also include results of 15 subjects in a hotel trial for 3 days and 2 nights. Further, we discuss studies currently being conducted in an outpatient (at-home) trial of 20 subjects, each undergoing two weeks of closed-loop control as well as two weeks of open-loop control, for a direct comparison of performance.

So what have we learned that can be generalized and applied to chemical process operations and control? The most direct application is to batch and semi-batch processes, where the dynamic behavior and objectives change during the course of the batch. Individuals have schedules/activities that change from day-to-day, imposing challenges that are similar to multi-product chemical plants. The major concern with hypoglycemia is analogous to chemical product purity and related specifications, where there can be severe economic penalties when these specifications are violated. The long time-scale pharmacodynamic action of insulin makes it important to keep track of "insulin on board" (IOB), since insulin injected two hours ago continues to have an effect on glucose for the next three hours or so; this is most analogous to an exothermic semi-batch reactor, where overfeeding can cause excess reactant to accumulate, possibly resulting in a large exotherm. Continuous glucose monitors used for the feedback signal create what is essentially an inferential control (state estimation) problem, since at least two reference glucose (fingerstick) measurements per day are required to keep the sensor in calibration; this is analogous to the combined use of continuous process measurements along with infrequent product composition measurements that have some level of uncertainty.

There are also process development analogies – for example, initial clinical trials are conducted in a hospital environment (inpatient) on a limited number of subjects, analogous to laboratory scale chemical process development. Like process development, lessons learned and data collected during the initial inpatient studies allow model development and a tailoring of the protocols for the next clinical trial stages. Similar to pilot plant testing where multiple scale reactors are studied, hotel and camp studies involve "scale-up" where many subjects, with different characteristics and behaviors, can be studied simultaneously as a transition between R&D and use at-home. Initial at-home studies in a sense represent the final pilot plant scale, with shorter-term (two week) studies followed by much larger trials of up to six months. A number of artificial pancreas groups throughout the world are now at the larger clinical trial stage, which will be followed by extensive trials for commercially available devices.

470369 A Robust Hybrid Model Predictive Control Framework for Hill Curve Model-Based Systems

Abstract:

The Hill curve, introduced by A.V. Hill to describe the equilibrium relationship between oxygen tension and the saturation of haemoglobin [1], is an empirical relation exhibiting an S shape characteristic used widely as part of diverse mathematical models in different fields. Examples include: (i) biosciences as a descriptive-deterministic model [2], (ii) in physiology, to describe nonlinear and saturable mechanisms as the renal uptake of aminoglycosides by proximal tubular cells [3], the tubuloglomerular feedback of the glomerular filtration in the kidney [3, 4], and the effect of ligand binding on the conductance of voltage-dependent ion channels [5], (iii) in pharmacology, on the rational basis of the receptor occupancy theory [6] as a model for describing the drug concentrationâ€"effect relationship [7], for the treatment of many diseases.

In this work we present a general framework for modelling and advanced control of systems involving the Hill curve as part of their mathematical model. The developed framework features (i) a piece wise affine modelling representation of the Hill curve [8], (ii) a hybrid mixed integer optimisation formulation leading to a hybrid model predictive control (hMPC) formulation [9], (iii) a state-of-the art multiparametric mixed integer quadratic programming (mp-MIQP) solution step for the derivation of explicit controllers [10] and (iv) advanced robust control and estimation strategies addressing issues related to variability and uncertainties. The framework will be demonstrated with its application to two representative biomedical systems (i) the intravenous anaesthesia process [11] where robust advanced control strategies are presented for the administration of Propofol to control the level of the depth of anaestheisa for a set of 12 real patients data in the induction and maintenance phases; and (ii) the acute myeloid leukemia process [12] where robust personalised treatment schedules for the delivery of chemotherapy are derived and performed for 6 real patient data.

References

1. Hill, A.V., *The possible effects of the aggregation of the molecules of haemoglobin on its dissociation curves.* Physiol, 1910. **40**.

2. Michaelis, L. and M.L. Menten, *Die kinetic der Invertinwirkung.* Biochem. Z., 1913. **61**: p. 530-546.

3. Rougier, F., et al., *Aminoglycoside nephrotoxicity: Modeling, simulation, and control.* Antimicrobial Agents and Chemotherapy, 2003. **47**(3): p. 1010-1016.

4. Schnermann, J., et al., *Tubuloglomerular feedback control of renal vascular resistance*, in *Handbook of renal physiology*, O.U. Press, Editor. 1992: Oxford, UK. p. 1675â€"1705.

5. Haynes, L.W., A.R. Kay, and K.W. Yau, *Single cyclic GMP-activated channel activity in excised patches of rod outer segment membrane.* Nature, 1986. **321**(6065): p. 66-70.

6. Wagner, J.G., *Kinetics of pharmacologic response I. Proposed relationships between response and drug concentration in the intact animal and man.* Journal of Theoretical Biology, 1968. **20**(2): p. 173-201.

7. Csajka, C. and D. Verotta, *Pharmacokinetic-pharmacodynamic modelling: History and perspectives.* Journal of Pharmacokinetics and Pharmacodynamics, 2006. **33**(3): p. 227-279.

8. Nascu, I., R. Oberdieck, and E. Pistikopoulos, *An explicit Hybrid Model Predictive Control Strategy for Intravenous Anaesthesia.* 9th IFAC Symposium on Biological and Medical Systems, 2015. **in press**.

9. Bemporad, A. and M. Morari, *Control of systems integrating logic, dynamics, and constraints.* Automatica, 1999. **35**(3): p. 407-427.

10. Oberdieck, R. and E. Pistikopoulos, *Explicit hybrid model-predictive control: The exact solution.* Automatica, 2014. **submitted**.

11. NaÅŸcu, I., et al., *Advanced model-based control studies for the induction and maintenance of intravenous anaesthesia.* IEEE Transactions on Biomedical Engineering, 2015. **62**(3): p. 832-841.

12. Pefani, E., et al., *Chemotherapy Drug Scheduling for the Induction Treatment of patients with Acute Myeloid Leukemia.* IEEE Trans Biomed Eng, 2014.

457996 Stochastic Nonlinear Model Predictive Control Using Adaptive Polynomial Chaos: Application to an Atmospheric Pressure Plasma Jet

Abstract:

Stochastic Nonlinear Model Predictive Control Using Adaptive Polynomial Chaos: Application to an Atmospheric Pressure Plasma Jet

Model predictive control (MPC) is a control strategy that can effectively handle competing control objectives for multivariable systems in the presence of constraints [1,2]. Key to MPC is a dynamic system model to predict future behavior so that optimal control action can be computed. Standard MPC considers deterministic system models, whereas in reality probabilistic uncertainties in model parameters, disturbance, and measurement noise are ubiquitous. Stochastic model predictive control (SMPC) provides a framework for systematically incorporating the probabilistic description of system uncertainties into the optimal control problem and enabling constraint handling in a probabilistic sense (e.g., see [3] and the references therein). However, efficient propagation of probabilistic uncertainties through system dynamics poses a key challenge in SMPC, particularly for nonlinear systems.

Stochastic nonlinear MPC (SNMPC) approaches using generalized polynomial chaos (gPC) for uncertainty propagation have been reported in [4, 5, 6]. gPC is a framework that approximates stochastic states and parameters as a weighted sum of orthogonal polynomials [7,8]. The orthogonality property of the polynomials allows for efficient on-line propagation of the statistics of state distributions or, alternatively, efficient construction of the state distributions. One major shortcoming of the present nonlinear SMPC approaches employing gPC is, however, the inability to account for time-varying stochastic disturbances. With time-varying stochastic disturbances, the disturbance realization (and possibly its underlying distribution) may change in time. Considering each disturbance realization as a stochastic variable in gPC results in intractable formulations for real-time control since the number of terms in the gPC expansion increases factorially with the number of stochastic variables. Uncertainty propagation approaches that can handle time-varying stochastic disturbances in the context of polynomial chaos remains an open problem.

This work presents an SNMPC approach based on adaptive polynomial chaos (aPC) framework [9] for systems with time-invariant and time-varying probabilistic uncertainties. The distinguishing characteristic of aPC is that instead of selecting one set of orthogonal polynomial basis functions for all times, the basis functions are dynamically adapted at each sampling time based on the moments of the states. A primary reason that gPC cannot handle time-varying probabilistic uncertainties is due to the fact that constant orthogonal polynomial basis functions cannot accurately describe the distribution of stochastic states over long prediction horizons and when the states have been corrupted by several realizations of time-varying uncertainties. By updating the polynomial basis functions to account for the changes in the state distribution, aPC can effectively handle time-varying additive noise regardless of its distribution. This aPC-based SNMPC approach has been demonstrated for high-performance control of an atmospheric pressure plasma jet [10]. Simulation results show that the SNMPC approach is able to successfully achieve the control objectives while satisfying the state constraints in the presence of stochastic disturbances.

References

[1] M. Morari and J. H. Lee, "Model predictive control: Past, present and future," *Computers & Chemical Engineering*, vol. 23, pp. 667-682, 1999.

[2] D. Q. Mayne, "Model predictive control: Recent developments and future promise," *Automatica*, vol. 50, pp. 2967-2986, 2014.

[3] A. Mesbah, "Stochastic model predictive control: An overview and perspectives for future research," *IEEE Control Systems Magazine, In Press,* 2016.

[4] A. Mesbah, S. Streif, R. Findeisen, and R. D. Braatz, "Stochastic nonlinear model predictive control with probabilistic constraints," in *Proceedings of the American Control Conference*, pp. 2413–2419, Portland, Oregon, 2014.

[5] L. Fagiano and M. Khammash, "Nonlinear stochastic model predictive control via regularized polynomial chaos expansions," in *Proceedings of the 51st IEEE Conference on Decision and Control*, pp. 142–147, Maui, 2012.

[6] V. A. Bavdekar and A. Mesbah, "Stochastic nonlinear model predictive control with joint chance constraints," in *Proceedings of the 10th IFAC Symposium on Nonlinear Control Systems*, accepted, Monterey, 2016.

[7] N. Wiener, "The homogeneous chaos," *American Journal of Mathematics*, vol. 60, pp. 897–936, 1938.

[8] D. Xiu and G. E. Karniadakis, "The Wiener-Askey polynomial chaos for stochastic differential equations," *SIAM Journal of Scientific Computation*, vol. 24, pp. 619–644, 2002.

[9]. S. Oladyshkin and W. Nowak, "Data-driven uncertainty quantification using the arbitrary polynomial chaos expansion," *Reliability Engineering & System Safety*, vol. 106, pp. 179-190, 2012.

[10] D. Gidon, D. B. Graves, and A. Mesbah, "Model predictive control of thermal effects of an atmospheric pressure plasma jet for biomedical applications," in *Proceedings of the American Control Conference*, accepted, Boston, 2016.

470597 Proportional State-Feedback Controller Design Using MPC Structure and Carleman Approximation Method

Abstract:

Model Predictive Control (MPC) has gained widespread recognition due to its ability to account directly for constraints. The cost is that all the advanced Model Predictive Controllers need to solve a nonlinear optimization problem at each sampling time. Thus, MPC is usually a computationally expensive control approach and this issue may cause a delay in the input signal which potentially deteriorates the controller's performance and may cause closed-loop system instability. Thus, MPC designs are not always practical, especially for large scale or fast evolving processes. Some papers utilize Carleman approximation method to represent nonlinear dynamic systems in a bilinear form. This approximation not only provides the analytical solution of the system, but also facilitates sensitivity derivations of the objective function with respect to decision variables [1-4]. This sensitivity information significantly accelerates the online computations. However, these articles do not discuss the stability of the resulting closed-loop system.

Motivated by the above discussion, this work tunes a proportional state feedback controller employing the nonlinear MPC structure combined with the Carleman approximation method. Therefore, the analytical solution and sensitivity of the objective function with respect to the proportional gain vector are available. This proposed state feedback control in the MPC structure, gives a smoother control law rather than a traditional nonlinear MPC which generates a piecewise constant input. Also, the analytical calculation of the Hessian matrix in addition to the gradient vector, in this work, reduces the online computation efforts even more.

The proposed controller would inherit the same nominal stability properties of the ideal nonlinear MPC, when there exists no Carleman approximation error. However, Carleman approximation method, with a finite dimension, always has a dynamic error which might endanger the stability of the closed-loop system. This work discusses the conditions required to guarantee the input-to-state stability in the presence of the Carleman approximation.

References:

[1] N. Hashemian and A. Armaou, "Simulation, model-reduction and state estimation of a two-component coagulation process," AIChE J., 2016, 62, pp 1557-1567, DOI:10.1002/aic.15146.

[2] N. Hashemian and A. Armaou, "Fast Moving Horizon Estimation of nonlinear processes via Carleman linearization,"Proceedings of the American Control Conference, pp. 3379 - 3385, Chicago, IL, USA, 2015,DOI: 10.1109/ACC.2015.7171854.

[3] Y. Fang and A. Armaou, "Carleman approximation based Quasi-analytic Model Predictive Control for Nonlinear Systems, " AIChE J., 2016, Accepted Author Manuscript, DOI:10.1002/aic.15298.

[4] Y. Fang and A. Armaou, "Nonlinear Model Predictive Control Using a Bilinear Carleman linearization-based Formulation for Chemical Processes," Proceedings of the American Control Conference, pp. 5629 - 5634, Chicago, IL, USA, 2015, DOI: 10.1109/ACC.2015.7172221.

469384 Constrained Control Lyapunov Functions: Design and Application for Nonlinear Systems

Abstract:

Constrained Control Lyapunov Function: Design and application for nonlinear systems
Systems exhibiting nonlinearity, constraints, and uncertainty are ubiquitous in practice, from chemical reactors to flight control systems. In many applications, particularly those systems with input constraints, control designs are sought that guarantee stabilization to the origin from the largest possible set of initial conditions. This set has been termed the null controllable region (NCR) [1, 2].

There currently do not exist results that identify the NCR of a general nonlinear system or give controls laws for stabilization everywhere within the NCR. Motivated by this, in this work we consider the problems of constructing the NCR for control-affine input-constrained nonlinear systems and determining a control law which guarantees stabilization from everywhere within the NCR.

It is known that the NCR is described by time-optimal trajectories of the reverse-time system [7]. We use the fact that reachable points in reverse-time are stabilizing for the nominal system. Our construction reinforces the special relationship that time-optimal controls have with a system's reachable set: that they steer to and then traverse the boundary of the NCR [5].
In [2], it was shown that the NCR for linear systems is covered by certain bang-bang trajectories of the reverse-time system. Our construction of the NCR for nonlinear systems can be seen as a generalization of this result. In that work, the boundary of the NCR is formed by a collection of alternating bang-bang arcs with arbitrary switching times, whereas in this paper, the switching times of the reverse-time bang-bang arcs are indicated by the evolution of a certain switching function™ ODE [6].
The notion of Constrained Control Lyapunov Functions (CCLF) was recently defined as Lyapunov functions designed so that the control laws which guarantee their decay are admissible and stabilizing for the constrained system with a region of attraction equal to the NCR [3]. Clearly, control designs using CLFs that are not CCLFs only guarantee stabilization within subsets of the NCR [4].

Notice that that decay of the trajectory into successive sub-NCRs, corresponding to the same dynamics but successively smaller input constraints, is sufficient to result in stabilization to the origin. Thus, by recognizing that such a collection of sub-shells of the NCR is a CCLF, we implement a control law that guarantees stabilization from everywhere in the NCR. We employ this CCLF in a model predictive controller that constrains the control action to those which force the trajectory into successive sub-shells of the NCR. This result will be illustrated with an example.

References
[1] T. Hu, Z. L. and Qiu, L. (2001). Stabilization of exponentially unstable linear systems with saturating actuators. IEEE Transactions on Automatic Control, 46(6), 973-979.

[2] T. Hu, Z. L. and Qiu, L. (2002). An explicit description of null controllable regions of linear systems with saturating actuators. Systems & Control Letters, 47, 65-78.

[3] Mahmood, M. and Mhaskar, P. (2014). Constrained control Lyapunov function based model predictive control design. Int. J. Robust Nonlinear Control, 24, 374-388.

[4] Mahmood, M. and Mhaskar, P. (2008). Enhanced stability regions for model predictive control of nonlinear process systems. Proceedings of the American Control Conference,
1133-1138.

[5] Lewis, A. (2006). The Maximum Principle of Pontryagin in Control and in Optimal Control. Department of Mathematics and Statistics, Queen's University, Kingston, Canada.

[6] Shattler, H. and Ledzewicz, U. (2012). Geometric Optimal Control. Springer, New York.

[7] Sussmann, H. (1987). The structure of time optimal trajectories for single input systems in the plane. SIAM Journal of Control and Optimization, 25(2), 433-465.

460282 Input-Output Paring Accounting for Both Structure and Strength in Coupling

Abstract:

Within process control, the decentralized and distributed control frameworks have attracted great attention due to the merits in terms of structural flexibility and fault tolerance. Input-output pairing is one fundamental step in both decentralized and distributed control system designs in order to achieve desired control performance. Input-output pairing has been an active research topic for a few decades and different approaches have been proposed. The relative grain array (RGA) analysis that was used to measure the strength of interaction between inputs and outputs is a widely-used tool [1]. However, RGA may give pairing results that are not consistent with the process physical topology, which is not favorable for control system designs. The relative degree analysis adopted in the evaluation of control configurations is another promising input-output pairing approach [2]. It provides valuable information on how directly a manipulated input can affect an output variable from a structural point of view. While the relative degree analysis provides a measure of physical closeness of inputs and outputs, it neglects entirely the strength of interaction. Consequently, it may also fail to work in different situations.

Motivated by above observations, in this work, we aim at developing an input-output pairing approach that takes into account both structural and strength information in coupling. Specifically, the relative degree analysis is taken advantage of to measure the physical closeness of input-output pairs and relative sensitivity is used to explore the strength of interaction progressively with respect to the relative degree. The proposed measure is the so-called relative sensitivity array (RSA) between inputs and outputs. Detailed analysis reveals the relationship between the gain matrix in the RGA analysis and the sensitivity matrix defined in the RSA, which implies that the RSA can be interpreted as truncated RGA with respect to a specific relative degree. To account for more general input-output paring problems, an iterative algorithm for the RSA-based method as the implementation guideline is proposed. Since the RSA is essentially an analog of the RGA, many existing pairing rules that were originally developed for RGA can be adopted in the RSA-based approach as the pairing rules.

The proposed RSA-based method is compared with RGA- and relative degree-based approaches in the input-output pairing of several chemical process examples. From the results, it can be seen that (a) the pairs formed by the proposed RSA-based approach are more consistent with the physical topologies of the studied processes; (b) results obtained by only considering the coupling strength as in RGA could be different at different operating points which is not desirable from a practice point of view; (c) results obtained by RGA-based methods may not consistent with the physical topologies of the processes which may increase the information coupling of the entire control system; (d) the relative degree analysis may even fail to give a complete pairing recommendation since coupling strength is ignored. Furthermore, it is demonstrated that the proposed approach is able to handle input-output pairing tasks for systems with larger scales.

[1] Bristol, E. On a New Measure of Interaction for Multivariable Process Control. IEEE Transactions on Automatic Control, 11:133-134, 1966.

[2] Kravaris, C. and Soroush, M. Synthesis of Multivariable Nonlinear Controllers by Input/output Linearization. AIChE Journal, 36:249-264, 1990.

458936 Elucidating and Handling Valve-Induced Nonlinearity in Industrial Feedback Control Loops

Abstract:

In much of the process control literature, valves are treated as having either very fast dynamics such that the flowrate out of the valve is equal to the valve output set-point that is the control signal to the valve, or the valve dynamics are approximated with a simple linear model (e.g., a first-order process model [1]). However, the flowrate out of the valve may be related nonlinearly to the control signal received by the valve, through static nonlinearities between the valve position and outlet flow rate (e.g., an equal percentage valve characteristic [1]) or dynamic effects such as stiction and deadzone [2]-[3]. Nonlinearities in process control loops are known to cause closed-loop performance or even stability issues; for example, the valve stiction-induced nonlinear dynamic behavior is well-known for causing control loop oscillations [2], and as a result, a number of methods for improving process control by accounting for valve stiction have been investigated [4]-[5]. In general, accounting for valve nonlinearities in process control design can be expected to positively impact the closed-loop performance obtained.

In this work, we clarify the effects that static and dynamic nonlinearities can have on a closed-loop system under classical control methods and under model predictive control (MPC), both of which are control methods widely used in process industries. Using the insight of the results obtained, we discuss methods for improving industrial control-loop performance when such negative effects occur. For example, when control loop oscillations occur in a loop containing a controller with an integral term (e.g., proportional-integral control) and a sticky valve, we propose an anti-windup-inspired method for reducing the windup-like effect caused by stiction on the integrator. To improve set-point tracking when MPC is used in the presence of valve stiction, we propose a set of modifications of the MPC ranging from the inclusion of suitable, computationally-efficient models of the valve dynamics within the MPC to the incorporation of constraints on the variation of the control action within two successive MPC sampling times. We provide theoretical results establishing conditions under which our methods are guaranteed to work and demonstrate their applicability and performance using several chemical process examples.

[1] Coughanowr DR, LeBlanc SE. Process Systems Analysis and Control, 3rd ed. Boston, MA: McGraw-Hill, 2009.

[2] Brásio ASR, Romanenko A, Fernandes NCP. Modeling, detection and quantification, and compensation of stiction in control loops: The state of the art. Industrial & Engineering Chemistry Research. 2014;53:15020-15040.

[3] Choudhury MAAS, Thornhill NF, Shah SL. Modelling valve stiction. Control Engineering Practice. 2005;13:641-658.

[4] Srinivasan R, Rengaswamy R. Approaches for efficient stiction compensation in process control valves. Computers & Chemical Engineering. 2008;32:218-229.

[5] Hägglund T. A friction compensator for pneumatic control valves. Journal of Process Control. 2002;12:897-904.

470331 Design and Implementation of a Biologically-Inspired Optimal Control Strategy (BIO-CS) for Advanced Energy Systems

Abstract:

Biologically-inspired strategies or biomimetics are the human-made designs that imitate nature. In the last few decades, this area of research has provided ideas for the solutions of various engineering problems. For example, the behavior of natural groups such as ants, bees and swarms has previously inspired the development of control approaches in other engineering fields. In these natural groups, self-organization and cooperation between individual members by following simple rules of interaction can result in a wide range of optimal patterns [1]. In this work, a biomimetic optimal control strategy is proposed for chemical process control. This strategy is inspired by the ant's rule of pursuit idea combined with optimal control concepts. In this novel approach, entitled Biologically-Inspired Optimal Control Strategy (BIO-CS), the cooperation of multiple agents (or ants) enables the generation of optimal control trajectories between the current state and the desired setpoint. In particular, starting from an initially feasible trajectory for the leader agent, each follower agent improves its path towards the setpoint by employing optimal control laws. As the number of agent progresses, the trajectories converge to an optimal solution.

To demonstrate the effectiveness of the proposed approach, two chemical and energy systems are addressed for the application of BIO-CS. The first implementation case study corresponds to the setpoint tracking and disturbance rejection associated with a fermentation process for bioethanol production. The challenges in this fermentation system consist of nonlinearities present in the dynamic process model in addition to steady-state multiplicity. The BIO-CS implementation results for this fermentation process show smooth and offset free output trajectories with improved performance when compared to other classical and optimal control approaches [2, 3, 4]. The second application of BIO-CS addressed is the Acid Gas Removal (AGR) unit of an Integrated Gasification Combined Cycle (IGCC) process. In this case, an

innovative framework is proposed for the closed-loop implementation of BIO-CS. For this implementation, an IGCC-AGR process model in DYNSIM (software used for dynamic simulations of chemical processes) is employed. As the first step, a subsystem from the IGCC-AGR process simulation is selected to define the control loops for the BIO-CS implementation. Then, a simplified dynamic model for use by the controller is derived employing system identification techniques. Specifically, an autoregressive model with exogenous inputs (ARX) is a developed from input/output data of the IGCC-AGR DYNSIM plant. The optimal control trajectories computed by BIO-CS are implemented online for the selected subsystem considering two different scenarios: (i) BIO-CS controller model and the actual plant model are identical; and (ii) implementation for the actual IGCC-AGR sub-process simulation in DYNSIM. This case presents additional challenges to the controller associated with plant-model mismatch. For the communication between the BIO-CS controller designed in MATLAB and the DYNSIM plant, the MATLAB-DYNSIM link developed at West Virginia University is employed. The optimal control laws computed by the BIO-CS are transmitted as setpoint trajectories for the PID controllers already in place in the DYNSIM plant. The implementation results demonstrate the potential of the BIO-CS for online implementation for processes with different challenges, including nonlinearities, high dimensionality, steady-state multiplicity and plant-model mismatch.

References:

1. Hristu-Varsakelis D. and Shao C., "A bio-inspired pursuit strategy for optimal control with partially constrained final state". *Automatica* **2007**;43:1265-1273.
2. Lima F. V., Li S., Mirlekar G. V., Sridhar L. N. and Ruiz-Mercado G. J., "Modeling and advanced control for sustainable process systems". Sustainability in the Analysis, Synthesis and Design of Chemical Engineering Processes, G. Ruiz-Mercado and H. Cabezas (eds.), Elsevier, **2016**.
3. Mirlekar G. V., Gebreslassie B. H., Diwekar U. M. and Lima F. V., "Design and implementation of a biomimetic control strategy for chemical processes based on efficient ant colony optimization." Presented at AIChE Annual Meeting, Salt Lake City, Utah, November **2015**.
4. Li S., Mirlekar G. V., Ruiz-Mercado G. J., and Lima F. V., "Development of Chemical Process Design and Control for Sustainability". Submitted for publication, **2016**.

464702 The Mesoscopic Behavior of Stochastic SchloG Model

Abstract:

An important concept for describing chemical reaction networks is the thermodynamic limit, where the number of molecules and the size of the system are asymptotically large. Systems in the thermodynamic (macroscopic) limit can be modeled with a deterministic modeling formalism; away from it, at the microscopic limit, a stochastic approach is more suitable [1]. However, due to numerical and computational difficulties, little is known for the behavior of networks in the mesoscopic area, where the system is sizable yet still under the influence of thermal noise.

We attempt to provide useful information about the mesoscopic behavior of chemical reaction networks with the help of ZI-closure scheme method [2]. This is a numerical solution of chemical master equations, which utilizes the maximization of the system entropy. For our analysis, we study the Schlögl model reaction network, a simple single component system that can exhibit bistability.

The closure scheme makes it possible to investigate the behavior of the system for a wide variety of volumes. Notably, the stochastic mesoscopic behavior only partially matches the deterministic solution; in the region of kinetic parameter values that the deterministic system exhibits two attractors, the stochastic system may retrieve only one of them in the mesoscopic limit. The stochastic formalism distinguishes between the attractors of the system and recovers the more stable one, whereas the deterministic system cannot distinguish between the two attractors. An investigation of the passage time of the system gives a better insight into this inability of the deterministic model.

A look at the dependence of the system entropy on the kinetic constants provides an important insight into the mesoscopic system behavior. The entropy of the system has a global maximum with respect to the kinetic constants of the system. The analysis also reveals a first order phase transition and a critical point of the system with respect the kinetic parameters. Results suggest the coexistence of two, appropriately normalized delta-like functions at the

thermodynamic limit of this critical point. Such behavior is also supported by a basic stability analysis [3]. To our knowledge this is the first time such an analysis is presented for the transition between stochastic and deterministic chemical reaction models.

1. Constantino PH, Vlysidis M, Smadbeck P, Kaznessis YN, " Modeling stochasticity in biochemical reaction networks", J Phys D: Appl Phys 2016 Feb 1; 49(9): 093001 10.1088/0022-3727/49/9/093001
2. Smadbeck P, Kaznessis YN, "A Closure Scheme for Chemical Master Equations", Proc Natl Acad Sci U S A. 2013 Aug 27; 110(35): 14261-5 10.1073/pnas.1306481110
3. Smadbeck P, Kaznessis YN, "On a theory of stability for nonlinear stochastic chemical reaction networks", J Chem Phys 2015 May 8; 142: 184101 10.1063/1.4919834

468537 Reaction Network Analysis for Thin-Film Deposition Processes: Physical Interpretation of Reaction Invariants

Abstract:

Modeling the dynamics of atomic layer deposition processes is challenging because of the nonlinear behavior of these systems, their multiple and widely-ranging timescales, and by the relative lack of validated reaction kinetics information. Those data that do exist are typically derived from quantum chemical computations or experimental examinations of reaction sequences that define only a portion of the complete ALD reaction cycle.

In this talk, we will describe our efforts to develop mathematical methods that address the numerical challenges of simulating dynamic Atomic Layer Deposition (ALD) and Chemical Vapor Deposition (CVD) processes while providing a rational path to creating well-posed models of these deposition processes. Our model reduction approach is based on a two-step procedure where in the first, the chemical species surface balance dynamic equations are factored to decouple the (nonlinear) reaction rates, eliminating redundant dynamic modes. The second phase further reduces the dynamic dimension when species relatively minor in concentration can be identified. The overall technique extracts physically significant reaction invariants [1-3] and points to potential model structural problems if they exist. An alumina ALD process is used for an example consisting of nineteen reactions and twenty-three surface and gas-phase species. Using our approach, the model is reduced by nineteen modes to a four-dimensional dynamic system without any knowledge of the reaction rate values. Results are interpreted in the context of identifying rate limiting steps and validating the predicted dynamic behavior.

[1] Asbjornsen, O.A., Reaction invariants in the control of continuous chemical reactors. Chemical Engineering Science, 27(4), 709-717 (1972).

[2] Rodrigues, D., S. Srinivasan, J. Billeter, and D. Bonvin, Variant and invariant states for chemical reaction systems.

Computers Chemical Engineering, 73, 23-33 (2015).

[3] Zhao, Z., J. M. Wassick, J. Ferrio, and B. E. Ydstie, Reaction variants and invariants based observer and controller design for CSTRs, DYCOPS 2016 (2016).

473173 Inferring Gene Regulatory Networks from Single Cell Expression Data

Abstract:

Recent advances in cell profiling technology, such as RNA-sequencing and real-time PCR (polymerase chain reaction), have provided researchers with the ability to obtain expression data of a large set of genes at single cell resolution. These revolutionary tools produce single cell data, which we could use to answer important scientific questions that are previously not possible using population-averaged measurements (Sandberg 2013). For example, by looking at single cell expression data, we could address the functional role of cell-to-cell variability arising from gene expression stochastic dynamics, in cell lineage decision-making during physiological differentiation process. However, new computational tools are also needed to take advantage of information contained in single cell data, which existing algorithms were not originally designed for.

In this work, we focused on the inference of gene regulatory network (GRN) from single cell expression data. More specifically, we considered time-stamped cross-sectional expression datasets, consistent with time series measurements taken using Fluidigm Biomark©platform. Recently, several algorithms have been published for such GRN inference based on Boolean networks (Chen et al. 2014; Moignard et al. 2015), stochastic modelling (Teles et al. 2013), gene co-expression/correlation (Kouno et al. 2013; Moignard et al. 2013; Pina et al. 2015), and nonlinear ordinary differential equation models (Ocone et al. 2015). But, the direct application of these algorithms to time-stamped cross-sectional datasets face a few challenges due to, for example, the requirement of dense time course data and high computational complexity that scales exponentially with the size of the network.

Here, we developed a novel method for inferring the GRN structure, called Sparse Network Inference For Single cell data (SNIFS). SNIFS produces a directed graph model of the GRN by analyzing the time evolution of the distribution of single cell gene expression levels. Briefly, the algorithm begins with the computation of the changes in single cell transcriptional expression distribution over time for each gene. By employing the Kolmogorov-Smirnov (KS) distribution distances (Massey 1951) between two subsequent time points, the GRN inference involves solving a linear regression problem of the type $\mathbf{y}=\mathbf{X}\boldsymbol{\alpha}$. More specifically, the KS distance of a gene at each time step $\mathbf{y}$ is modelled as a linear function of the KS distances of all other genes at a previous time step $\mathbf{X}$. SNIFS then uses the elastic-net regularization (Zou and Hastie 2005) to find the optimal (sparse) solution $\boldsymbol{\alpha}$ by solving the following penalized least square optimization problem:

$$\min ||\mathbf{y}-\mathbf{X}\boldsymbol{\alpha}||_2^2 + \lambda(m||\boldsymbol{\alpha}||_1 + (1-m)||\boldsymbol{\alpha}||_2) \text{ subject to } \alpha_j \geq 0.$$

Note that by setting m to 1 or to 0 turns the elastic net regularization into Lasso or Tikhonov (ridge regression) regularization, respectively. In the implementation of SNIFS, we used GLMNET (r (Friedman et al. 2010) to solve for the optimal $\boldsymbol{\alpha}$.

We evaluated the performance of SNIFS by inferring 10- and 20-gene random subnetworks of *E. coli* and yeast GRNs using *in silico*time-stamped cross-sectional single cell expression datasets. Given the structure of the GRN, we generated single cell expression data by simulating a stochastic differential equation (SDE) model: (Pinna et al. 2010)

$$dx_j = V(\beta \Pi(1+a_{ij}x_i/(x_i+1)) - \theta x_j) + \sigma x_j dW(t)$$

where x_j represents the mRNA level of gene j, $a_{i,j}$ describes the regulation of the expression of gene j by gene i, β denotes the basal transcriptional rate, q is the mRNA degradation rate constant, and σ and V are scaling parameters. The variable $dW(t)$ describes the random Wiener process, which accounted for intrinsic stochastic dynamics of the gene expression (Wilkinson 2009). We set a_{ij} to 1 for activation, to −1 for repression, and to 0 otherwise. For the main datasets in the case study, we further set the parameters to the following: V=30, β =1, q=0.2, and σ=0.1. In total, we generated single cell data for 8 equally-spaced time points between t = 0.1 and t = 2.

We assessed the accuracy of the GRN predictions by computing the area under the receiver operating characteristics (AUROC) and the precision recall (AUPR) curves. We compared the GRNs predicted by SNIFS with those predicted using the population-averaged expression data by TSNI (Time Series Network Inference) (Bansal et al. 2006), and using a tree-based ensemble regression method called GENIE3 (GEne Network Inference with Ensemble of trees) (Huynh-Thu et al. 2010). The averaged AUROC and AUPR values in Table 1 indicated that for any $m$values, SNIFS could significantly outperform the predictions of TSNI and GENIE3. This result demonstrated the advantage of considering information contained in the single cell distributional data for the purpose of GRN inference, as done in SNIFS.

	Table 1. Evaluation of GRN Inference using TSNI, GENIE3, and SNIFS											
	10-GENE NETWORK						**20-GENE NETWORK**					
	AUROC			**AUPR**			**AUROC**			**AUPR**		
m	TSNI	*GENIE3*	SNIFS	TSNI	*GENIE3*	SNIFS	TSNI	*GENIE3*	SNIFS	TSNI	*GENIE3*	SNIFS
0												

(Ridge)	0.41	0.48	0.75	0.10	0.14	0.31	0.41	0.50	0.63	0.06	0.07	0.15
0.1	0.41	0.48	0.76	0.10	0.14	0.31	0.41	0.50	0.68	0.06	0.07	0.19
0.2	0.41	0.48	0.73	0.10	0.14	0.29	0.41	0.50	0.66	0.06	0.07	0.20
0.3	0.41	0.48	0.70	0.10	0.14	0.28	0.41	0.50	0.66	0.06	0.07	0.21
0.4	0.41	0.48	0.67	0.10	0.14	0.27	0.41	0.50	0.65	0.06	0.07	0.22
0.5	0.41	0.48	0.65	0.10	0.14	0.25	0.41	0.50	0.64	0.06	0.07	0.22
0.6	0.41	0.48	0.63	0.10	0.14	0.25	0.41	0.50	0.64	0.06	0.07	0.23
0.7	0.41	0.48	0.61	0.10	0.14	0.25	0.41	0.50	0.63	0.06	0.07	0.23
0.8	0.41	0.48	0.61	0.10	0.14	0.26	0.41	0.50	0.62	0.06	0.07	0.23
0.9	0.41	0.48	0.60	0.10	0.14	0.25	0.41	0.50	0.61	0.06	0.07	0.23
1 (Lasso)	0.41	0.48	0.58	0.10	0.14	0.25	0.41	0.50	0.60	0.06	0.07	0.24

REFERENCES

Bansal, M., Gatta, G. Della and di Bernardo, D. (2006). Inference of gene regulatory networks and compound mode of action from time course gene expression profiles. *Bioinformatics*, 22(7), pp.815–822.

Chen, H. et al. (2014). Single-cell transcriptional analysis to uncover regulatory circuits driving cell fate decisions in early mouse development. *Bioinformatics*, 31(7), pp.1060–1066.

Friedman, J., Hastie, T. and Tibshirani, R. (2010). Regularization Paths for Generalized Linear Models via Coordinate Descent. *Journal of statistical software*, 33(1), pp.1–22.

Huynh-Thu, V.A. et al. (2010). Inferring regulatory networks from expression data using tree-based methods. *PloS one*, 5(9), p.e12776.

Kouno, T. et al. (2013). Temporal dynamics and transcriptional control using single-cell gene expression analysis. *Genome biology*, 14(10), p.R118.

Massey, F.J. (1951). The Kolmogorov-Smirnov Test for Goodness of Fit. *Journal of the American Statistical Association*, 46(253), pp.68 – 78.

Moignard, V. et al. (2013). Characterization of transcriptional networks in blood stem and progenitor cells using high-throughput single-cell gene expression analysis. *Nature cell biology*, 15(4), pp.363–72.

Moignard, V. et al. (2015). Decoding the regulatory network of early blood development from single-cell gene expression measurements. *Nature Biotechnology*, advance on(3).

Ocone, a. et al. (2015). Reconstructing gene regulatory dynamics from high-dimensional single-cell snapshot data. *Bioinformatics*, 31(12), pp.i89–i96.

Pina, C. et al. (2015). Single-Cell Network Analysis Identifies DDIT3 as a Nodal Lineage Regulator in Hematopoiesis. *Cell reports*, 11(10), pp.1503–10.

Pinna, A., Soranzo, N. and de la Fuente, A. (2010). From knockouts to networks: establishing direct cause-effect relationships through graph analysis. *PloS one*, 5(10), p.e12912.

Sandberg, R. (2013). Entering the era of single-cell transcriptomics in biology and medicine. *Nature Methods*, 11(1), pp.22–24.

Teles, J. et al. (2013). Transcriptional regulation of lineage commitment--a stochastic model of cell fate decisions. *PLoS computational biology*, 9(8), p.e1003197.

Wilkinson, D.J. (2009). Stochastic modelling for quantitative description of heterogeneous biological systems. *Nature reviews. Genetics*, 10(2), pp.122–33.

Zou, H. and Hastie, T. (2005). Regularization and variable selection via the elastic net. *Journal of the Royal Statistical Society: Series B (Statistical Methodology)*, 67(2), pp.301–320.

469142 Modeling for Complex Interactive Behaviors Among *Enterococcus Faecalis* Donor and Recipients in Biofilms

Abstract:

The pathogen, *Enterococcus faecalis,* is a leading cause of hospital acquired infections. This species utilizes the mechanism of conjugation to transfer antibiotic resistance from plasmid-harboring antibiotic-resistant donors to plasmid-free antibiotic-sensitive recipients. Specifically, the plasmid **pCF10** carried in the donors provides cells with the ability to resist to antibiotics. The two types of signaling molecules which regulate the conjugative transfer of this plasmid are inducer **cCF10** and inhibitor **iCF10**. Peptide **cCF10** is produced by recipients. In the presence of both **cCF10** and donors in the environment, the peptides enter donor cells, leading to an induction of conjugation. Inhibitor **iCF10**, on the other hand, is produced by donors and functions to suppress the conjugation. Both types of signaling molecules, which are produced by cells to the environment, can then transfer to the donor cells. Depending on the domination of either **cCF10** or **iCF10**, donors will have different responses: promoting the conjugation when the recipient cells are abundant and restraining the process when the recipient concentration is low.

We have formulated a mathematical stochastic model that captures the dynamic interaction and responses of both recipients and donors in the entire population. Both types of cells were embedded inside the biofilm and were influenced by both extracellular **cCF10** and **iCF10**. The model also attempts to capture the plasmid spreading effect through which some recipients are converted into new donors. The biofilm is formulated as a 2D model and is composed of both donors and recipients. This model accounts for the following effects: 1) molecular interaction between intracellular quantities prior to conjugation, 2) mechanistic genetic regulation of the conjugations and the corresponding outcomes, and 3) population level, signal peptides and donor/recipient population in the biofilm and recipients. We anticipate that the initial locations of donors and recipients and the availability of signaling molecules within the biofilm can affect the plasmid transfer between cells. A higher donor concentration means a higher number of available plasmids to transfer but also leads to more production of **iCF10**, suppressing the conjugation to avoid any wasteful plasmid transfer. This model helps to evaluate these competitive effects in different scenarios and thus provides a full set of conditions at which some factors can be more dominant than others and has the potential to fully capture all the different effects from both donors and recipients. Understanding the balancing between different signaling mechanisms in this system may shed light on new ways of controlling the spread of antibiotic resistance.

470922 Elucidating the Meaning of Alternative Optimal Solutions in Flux Balance Analysis

Abstract:

The genotype-phenotype relationship of an organism is fundamental to biology. Predicting different phenotypes based on the sequenced genome is one of the main goals for genome-scale metabolic model development. These models provide a holistic view of the organism's metabolism. One common approach for analyzing metabolic models is flux balance analysis (FBA) which uses linear program to optimize an objective function, which in most applications is biomass growth. A stoichiometric matrix consisting of metabolites as rows and reactions as column is created based on the reactions included in the model. The elements of the matrix consist of the stoichiometric coefficient of a metabolite if it is involved in the reaction. The rows (metabolites) represent the linear equations that define the model, and the columns (reactions) represent the variables. For FBA, a carbon source and oxygen is provided to the model, and a flux distribution that provides the optimal amount of biomass growth is given by the linear program solver. As is common in metabolic network models, there are more reactions than metabolites giving an underdetermined system of linear equations. Since the system is underdetermined, a unique optimal solution is not guaranteed. The presence of alternative optimal solutions may lead to mischaracterization and/or undetected phenotypes.

Last year, we presented a systems identification enhanced phenotype phase plane (SID-PhPP) analysis to help characterize different phenotypes predicted by a model, how different pathways interact with each other for a given phenotype, and how such interactions differ from different phenotypes. We utilized SID-PhPP on an Escherichia coli (E. coli) core model which is a central carbon metabolic network. Traditional PhPP identified four phenotypes for the core model. By applying SID-PhPP, we discovered that one of the phenotypes is actually two distinct phenotypes undetectable by the traditional PhPP analysis. It is undetectable because the shadow prices for oxygen and carbon are the same in both phenotypes; however, SID-PhPP elucidated the presence of different active reaction sets in each phase confirming that the phases are unique phenotypes.

In this work, we further investigated the two "sub-phases", P3' and P3", using a uniform random sampling algorithm, optGpSampler developed by Megchelenbrink et al., which also revealed the presence of alternative solutions. To better understand the alternative optimal solutions existing in P3', we applied our SID-PhPP to analyze all alternative solutions corresponding to the same carbon and oxygen uptake rates. Our analyses showed that the alternative solutions correspond to different linear combinations of two extreme reaction pathways that produce the same amount of NADPH. In addition, we developed a systems approach for calculating shadow price of each metabolite through utilizing the loadings obtained by SID-PhPP. The obtained shadow prices agree with the shadow prices obtained from the traditional PhPP method. But SID-PhPP based shadow price provides further insight into how the metabolite is utilized by the network.

472966 The Impact of Resource Constraints on the Reverse Engineering of Biological Pathways

Abstract:

Every element within a gene network requires chemical species supplied by the cellular environment to enable biological processes. In many cases, these chemical species, or resources, act in an indiscriminate manner and are consequently shared within the network. This global coupling of the network elements through these resources can have measurable effects on the network as it operates. Here we employ an engineered gene circuit, which was introduced in a recent paper by the authors (Kang et al. 2016), and report additional data, analysis, and interpretation. Using this synthetic network as a benchmark, we find that network inference results can to be influenced by changes in the resources allocated to each node. Specifically, we show that applying Modular Response Analysis to this network leads to the inference of a nontrivial "ghost" regulation edge which was not explicitly engineered into the network and which is, in fact, not immediately apparent from experimental measurements. Various mathematical models including a coupled system of ODEs, a multiclass queueing network, and a graph theory consideration are explored in closed-form. Taking resource availability into account during reverse engineering allows closer approximation of the complex cellular environment and potentially uncovers network operation constraints to be considered for improved synthetic circuit design.

470390 Interaction Network-Based Antiviral Protein Discovery

Abstract:

Drug discovery faces several challenges. Most drug attrition occurs in Phase III during human-trial clinical testing[1]. While recent advances in drug design has led to a reduction in attrition due to drug safety[2], there has been no significant improvement in drug efficacy[3]. "Whole-cell" networks, such as protein-protein interaction or protein phosphorylation networks, may provide reasonable models of disease, but the challenge remains in isolating the components associated with disease. Here, we demonstrate a method of combining human protein-protein interaction (PPI) data with virus-host protein interaction data to improve influenza virus antiviral protein (i.e. drug target) discovery. We show that influenza virus proteins interact with human proteins that are in network positions significant for information flow. We then isolated a subnetwork of the human PPI network enriched for virus replication processes. Selecting proteins based on their network topology, we performed an siRNA screen to determine if network position offers an advantage in drug target prioritization. We found that the subnetwork is highly enriched for antiviral proteins – even more enriched than the set of host proteins previously found to directly interact with virus proteins. Moreover, we demonstrate that network positions can be used to further improve candidate prioritization. Lastly, we aim to discuss the implications of this study in its relation to the current state of host response models (specifically RNA virus sensing models) and which pathways should be given priority in order to develop more accurate, physiologically relevant models of the host response.

1. Arrowsmith, J. A decade of change. *Nat. Rev. Drug Discov.* **11,**17–8 (2012).

2. Bowes, J., Brown, A. J., Hamon, J., Jarolimek, W., Sridhar, A., Waldron, G. & Whitebread, S. Reducing safety-related drug attrition: the use of in vitro pharmacological profiling. *Nat. Rev. Drug Discov.* **11,**909–22 (2012).

3. Arrowsmith, J. Trial watch: Phase II failures: 2008-2010. *Nat. Rev. Drug Discov.* **10,** 328–329 (2011).

Complex Traveling Wave Phenomena:
Coupling of Phosphorylation Cascades and Extracellular Diffusion in Cellular Populations

Michal Přibyl, Vladislav Nevoral, Igor Schreiber

Department of Chemical Engineering
University of Chemistry and Technology, Prague

pribylm@vscht.cz

Department of Chemical Engineering, UCT Prague

Outline

- Motivation and goals
- Ligand induced ligand release
  - Autocrine signaling
- Description of reaction-transport processes
  - Ligand release, transport and binding in extracellular domain
  - Switch-like behavior of intracellular processes
  - The Huang-Ferrell model of MAPK cascades in intracellular domain
- Results
  - Reduction of complexity of the kinase cascade
  - Feedback loop based on the simplified reaction network
  - Bifurcation analysis of the reduced system
  - Signal transmission studies
- Conclusions

Department of Chemical Engineering, UCT Prague

Motivation and goals

- Cell self-organization is typically controlled by autocrine/paracrine signaling mediated by extracellular ligands
- Interplay between ligand transport and complex intracellular processes
- To understand developmental processes on molecular level
- To find as simple as possible but robust and physiologically meaningful representation of autocrine signaling

Gut epithelium

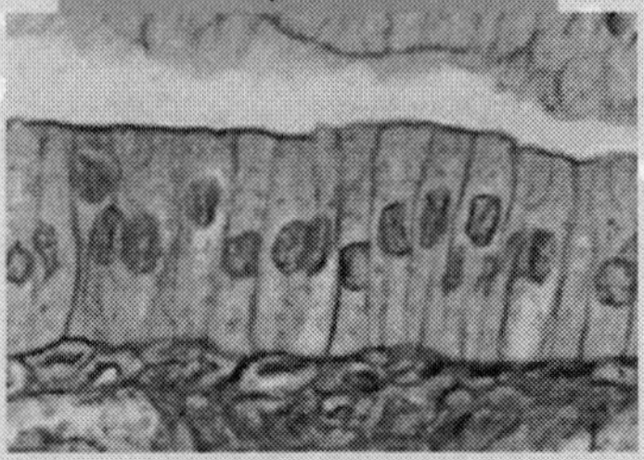

Coleman, Technion

Follicular epithelium – *Drosophila melanogaster*

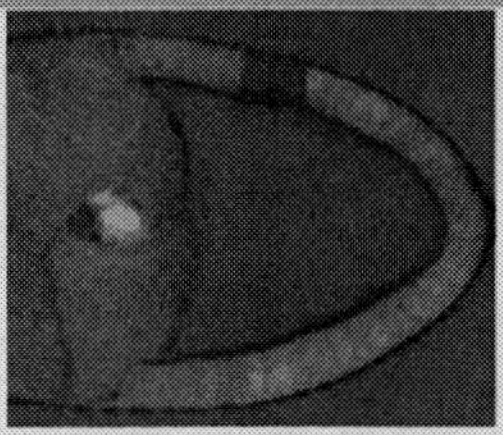

Baum, *Nature Cell Biol.* 3, 2001, 883.

Vascular system morphogenesis

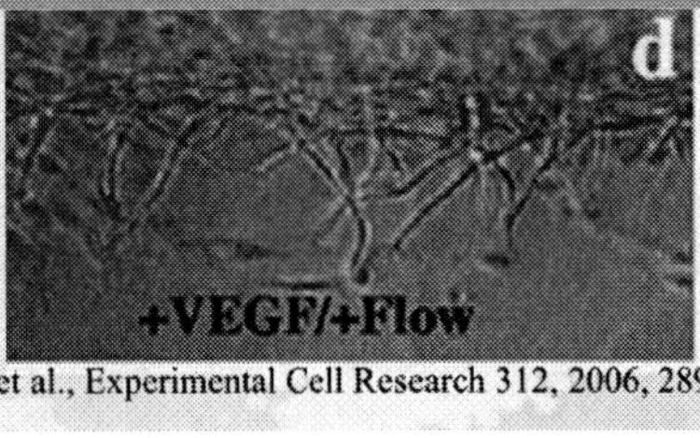

Semino et al., Experimental Cell Research 312, 2006, 289– 298

Department of Chemical Engineering, UCT Prague

Ligand induced ligand release

- Phases
  - Ligand release
  - Ligand transport
  - Ligand-receptor binding
  - Intracellular processing
- Effects
  - Proliferation
  - Differentiation
  - Apoptosis
  - Movement
  - Stable regime
  - …

Autocrine signaling

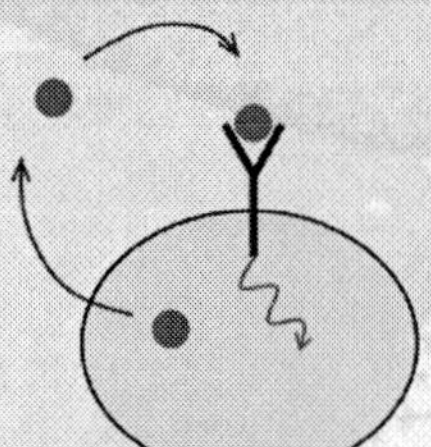

Paracrine signaling

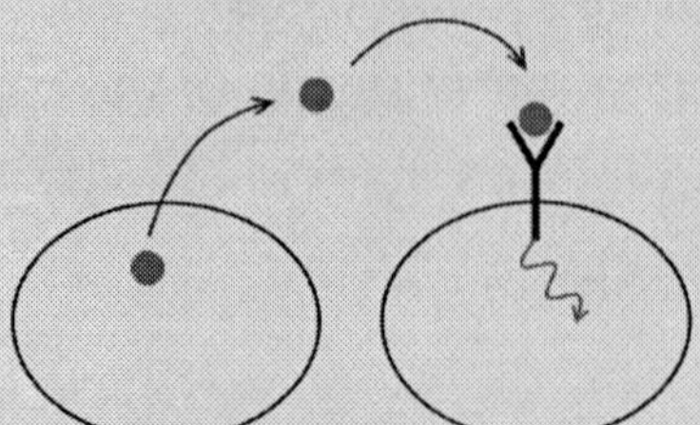

Department of Chemical Engineering, UCT Prague

Autocrine loop

Examples of reaction-transport processes

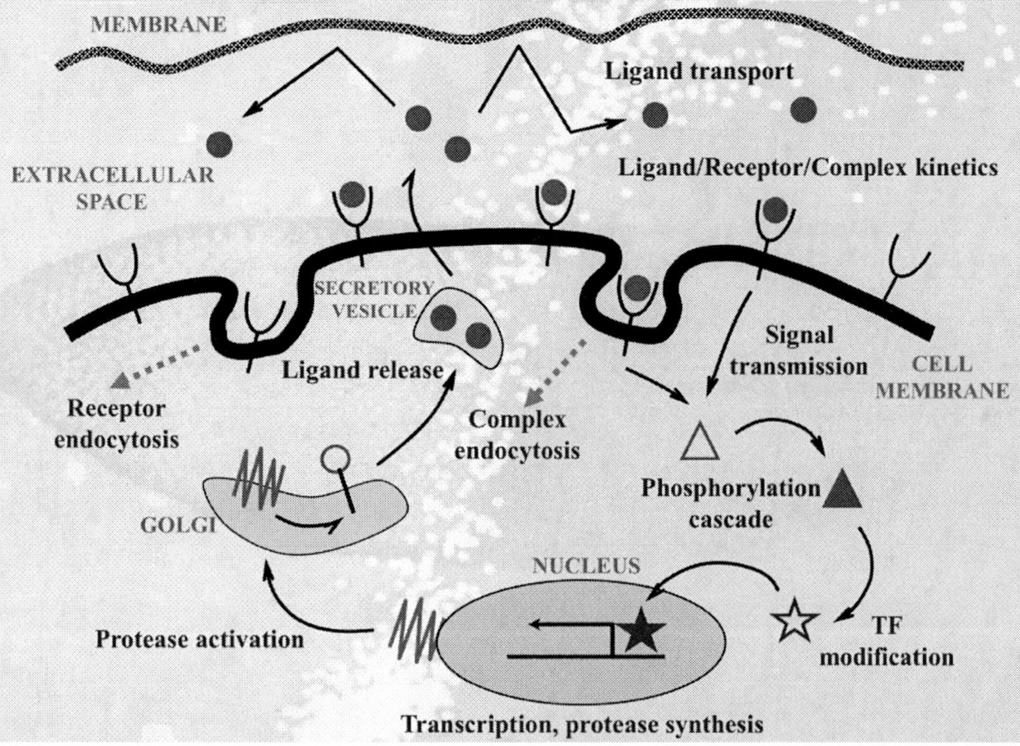

Department of Chemical Engineering, UCT Prague

Extracellular domain – reaction-transport processes

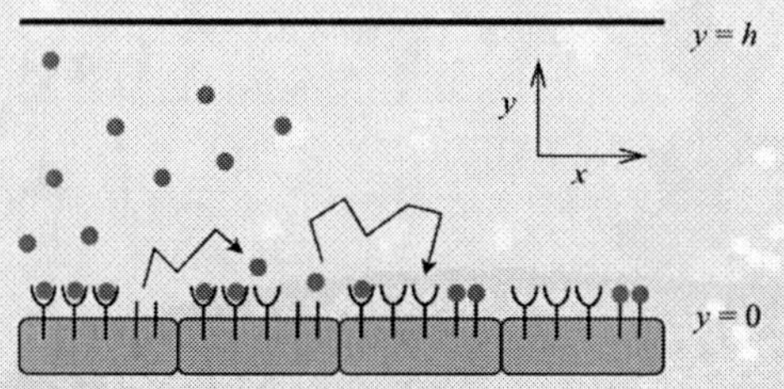

Přibyl, *Biophys. J. 84, 2003, 883.*
Nebyla, *Biophys. J. 102, 2012, 990.*

$$\frac{\partial S}{\partial t} = D\left(\frac{\partial^2 S}{\partial x^2} + \frac{\partial^2 S}{\partial y^2}\right)$$

$$\left(D\frac{\partial S}{\partial y} - k_{on} R S\right)\bigg|_{y=0} = -k_{off} C - g_r P \qquad \frac{\partial S}{\partial y}\bigg|_{Y=h} = 0$$

$$\frac{\partial R}{\partial t} = -k_{on} R \bar{S} + k_{off} C - k_{er} R + Q_r$$

$$\frac{\partial C}{\partial t} = k_{on} R \bar{S} - \left(k_{off} + k_{ec}\right) C$$

Coupling of intracellular and extracellular domains

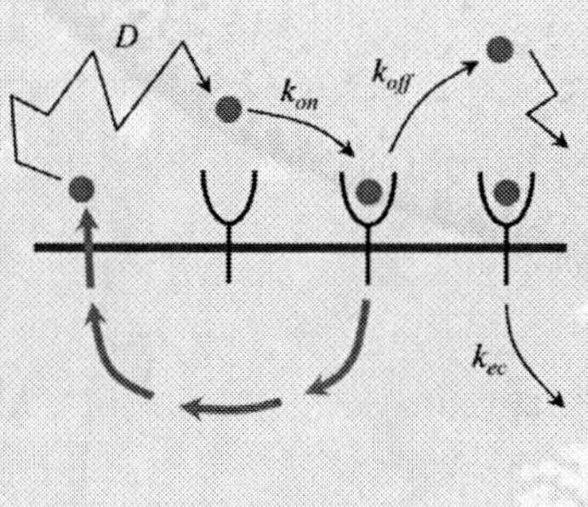

Variables

S – ligand concentration
P – activity of ligand releasing protease
R – surface concentration of receptors
C – surface concentration of ligand-receptor complex

Reakční a transportní procesy

Ligand diffusion and convection, ligand release, ligand-receptor complex formation, receptor renewal and endocytosis, complex endocytosis

Department of Chemical Engineering, UCT Prague

Tyrosin kinase receptors and intracellular processes

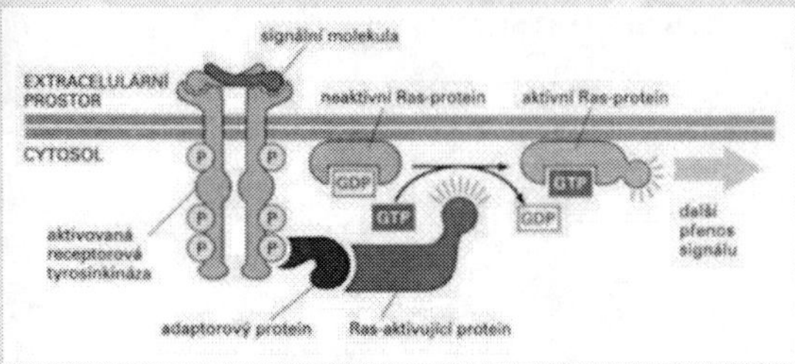

Alberts et al., Molecular Biology of the Cell

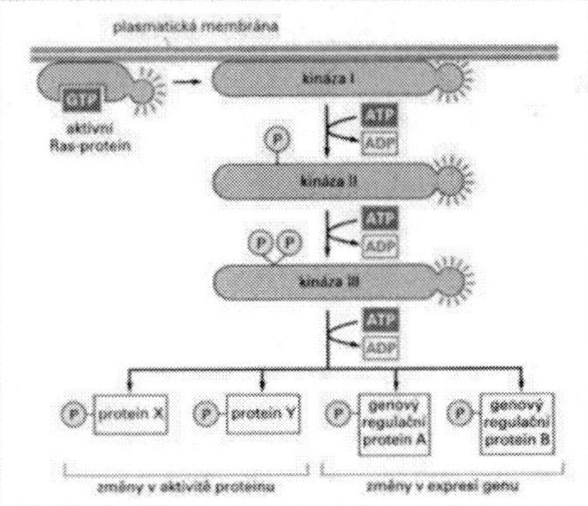

- Ligand-receptor binding
- Receptor dimerization
- Phosphorylation of intracellular tyrosin kinase domains
- Activation of membrane-bound Ras protein and other proteins
- Activation of MAPK cascade
- Modification of target proteins (cytosolic proteins, transcription factors etc.)

Department of Chemical Engineering, UCT Prague

Switch-like behavior of intracellular processes

Protease activity depends on protase denaturation and formation due to ligand-receptor signaling

$$\frac{\partial P}{\partial t} = -k_p P + g_p \sigma(C - C_T),$$

Graphical representation of steady states

Hill kinetics: $\sigma(c) = c^n/(k^n + c^n)$

Step-like function: $\sigma(c) = H(c - c_0)$

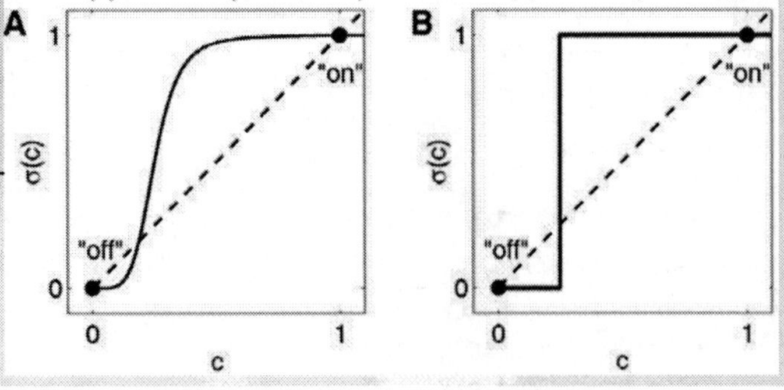

- Coupling between the activation of intracellular cascade and ligand release is provided by a simple sigmoidal or step-like function
- Sigmoidal function is characterized be the degree of cooperation (*n*-the power in the Hill kinetics)
- High sensitivity of protease activity to complex concentration at certain threshold value
- Two stable steady states typically coexist – “on” and “off” states
- Autocatalysis and signal propagation – switching from the “off” to “on” state

Přibyl, *Biophys. J. 84, 2003, 883.*

Department of Chemical Engineering, UCT Prague

Signal transmission in cell layers is mediated by moving concentration front of ligand

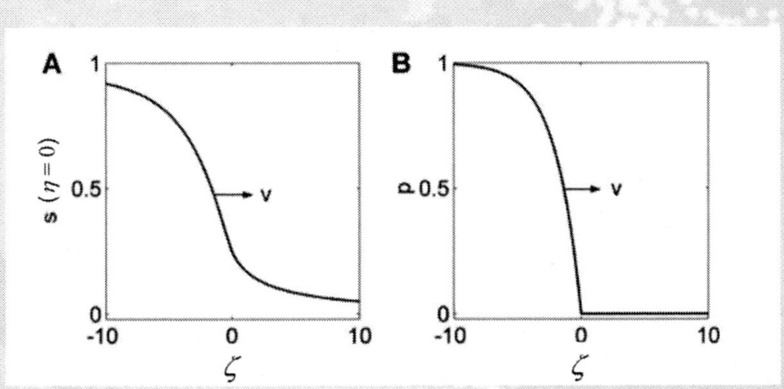

Intracellular processes represented by the step-like function

$$\sigma(c) = H(c - c_0)$$

Přibyl, *Biophys. J. 84, 2003, 883.*
Nebyla, *Biophys. J. 102, 2012, 990.*

Implicit relationship for propagation velocity v was derived

$$c_0 = \sum_{n=1}^{\infty} \frac{4\alpha\lambda_n^2 \sin\lambda_n \cos\lambda_n / (\lambda_n + \sin\lambda_n \cos\lambda_n)}{\alpha(\tau_s v^2 + 1)(4\lambda_n^2 + v^2\alpha^2\tau_s^2) + v[2\lambda_n^2 + \alpha^2\tau_s(\tau_s v^2 + 1)](4\lambda_n^2 + v^2\alpha^2\tau_s^2)^{\frac{1}{2}}}$$

where $\lambda_n \tan(\lambda_n) = \alpha\beta \qquad n = 1, 2, 3, \ldots, +\infty$

Department of Chemical Engineering, UCT Prague

Huang-Ferrell of MAPK cascades

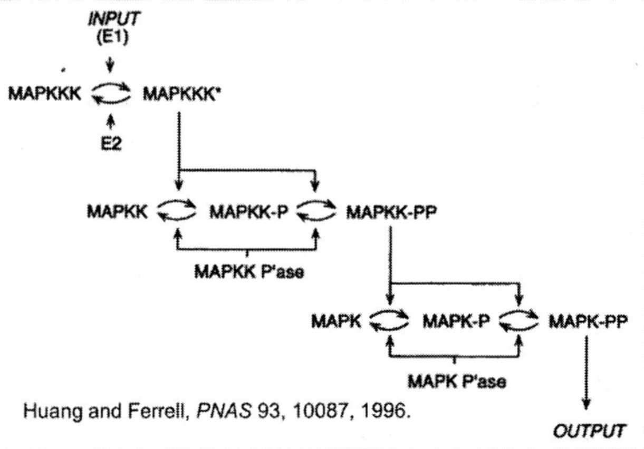

Huang and Ferrell, *PNAS* 93, 10087, 1996.

Table 1. Frequencies of the I/O Map Classes for the Subsystems within the MAPK Cascade

System	D_p [a]	D_v [b]	Single-Valued (Percent)	Oscillatory (Percent)	Hysteretic (Percent)
1 + 2 + 3 [c]	36	15	81.99 ± 0.65	9.13 ± 0.49	8.88 ± 0.48
2 + 3	28	12	73.35 ± 0.75	7.52 ± 0.45	19.13 ± 0.67
1 + 2	22	9	93.65 ± 0.41	1.48 ± 0.20	4.87 ± 0.36
1	8	3	100	0	0
2	14	6	83.88 ± 0.62	0	16.12 ± 0.62
3	14	6	85.75 ± 0.59	0	14.25 ± 0.59

- Key part of EGFR related autocrine communication
- Provides bistability, which is necessary for chemical signal propagation
- Provides periodic behavior
- Complex reaction network
- Task – to find a subnetwork that exhibits qualitatively similar behavior
- Only two stages of MAPK cascade are necessary to provide bistabilities and oscillations

Qiao et al., PLoS Comput Biol 3, e184, 2007.

Department of Chemical Engineering, UCT Prague

Huang-Ferrell of MAPK cascades

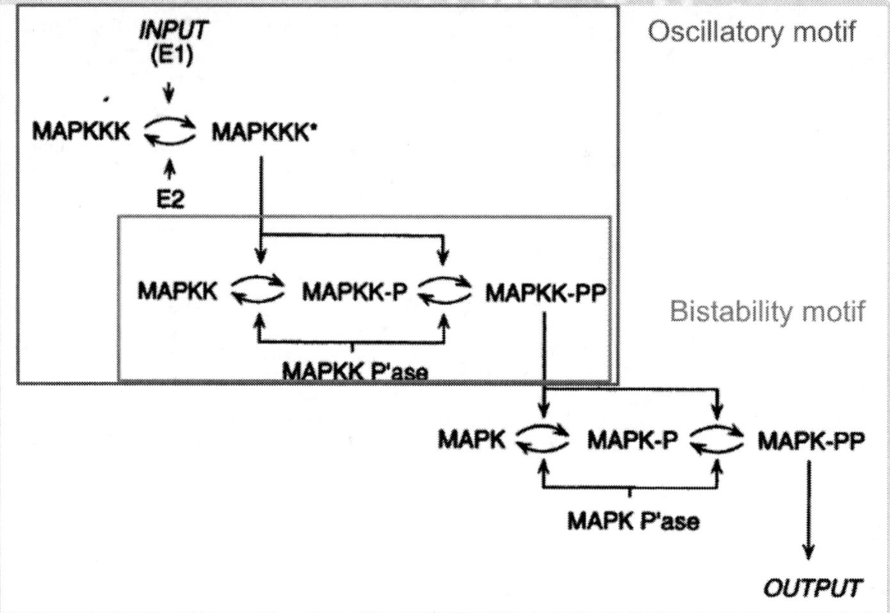

Huang and Ferrell, *PNAS* 93, 10087, 1996.

Department of Chemical Engineering, UCT Prague

Minimal subnetwork with bistability

- Stoichiometric network analysis
- Identification of sources of instabilities
- Simplifications
  - no dissociation of enzyme-substrate complex
  - Protease C is in pool conditions
  - quasi-steady state approximations for enzyme-substrate complexes

Symbol	Chemical species
A	MAPKKK*
B	MAPKK
B_1	MAPKKP
B_2	MAPKKPP
C	MAPKKP'ase
AB	MAPKKK*MAPKK
AB_1	MAPKKK*MAPKKP
CB_1	MAPKKP'aseMAPKKP
CB_2	MAPKKP'aseMAPKKPP

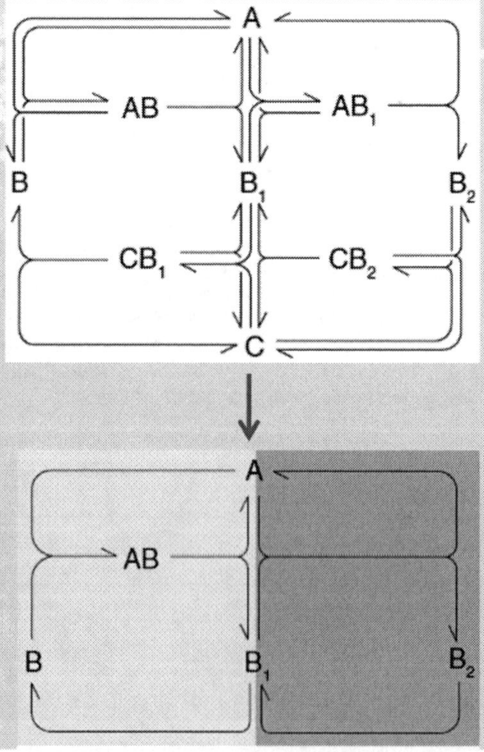

Department of Chemical Engineering, UCT Prague

Feedback loop based on the simplified reaction network

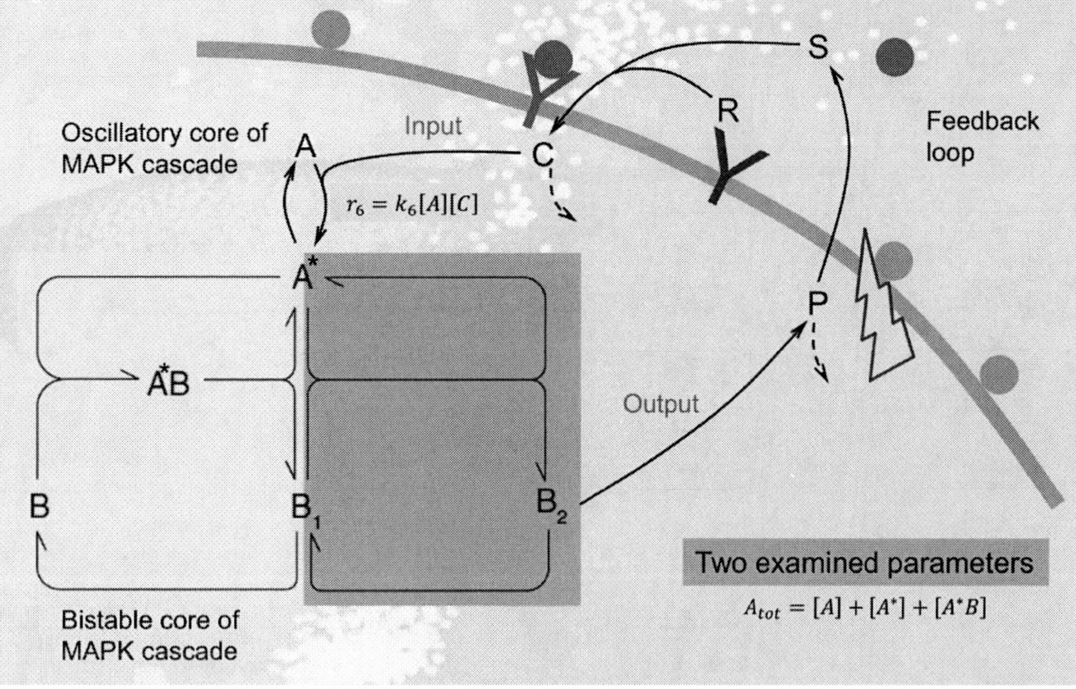

Department of Chemical Engineering, UCT Prague

Bifurcation analysis

High values of these parameters promote autocatalytic process – ligand induced ligand release

- A_{tot} – total concentration of MAPKKK kinase
- k_6 – kinetic constant of MAPKKK activation

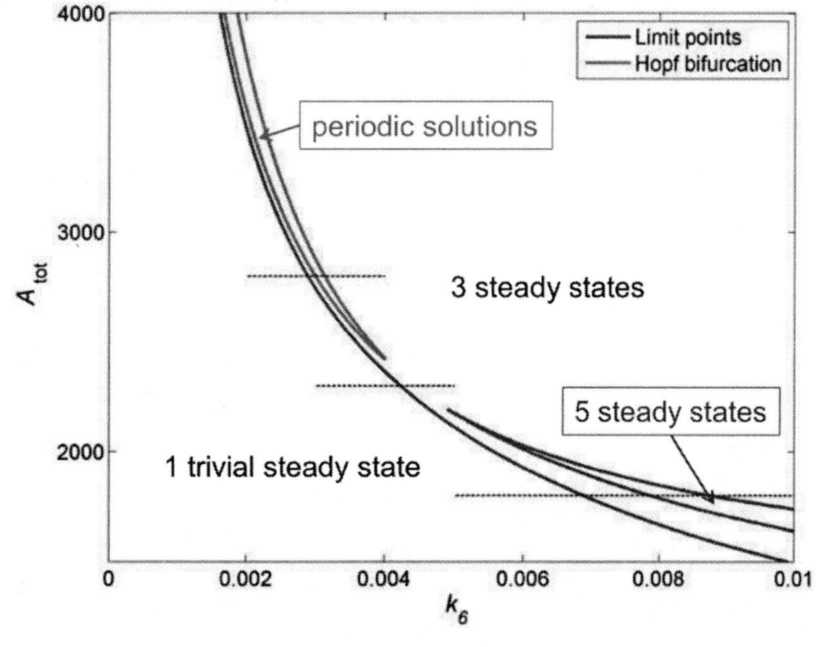

Department of Chemical Engineering, UCT Prague

Bifurcation analysis

A_{tot} = 2300 nmol m^{-3}

- No signal propagates in spatially distributed system.
- Ligand concentration smoothly connects two steady states.

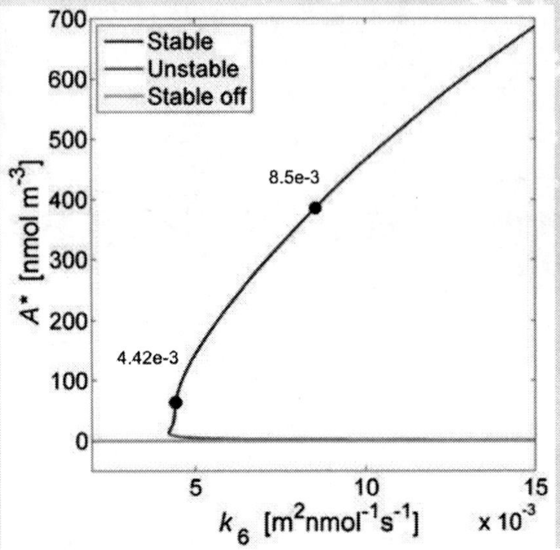

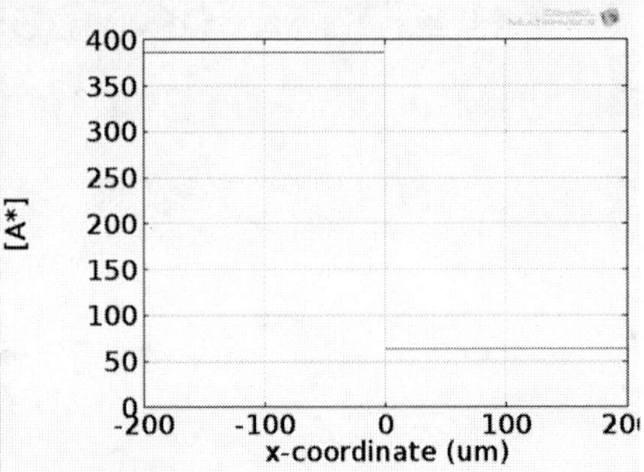

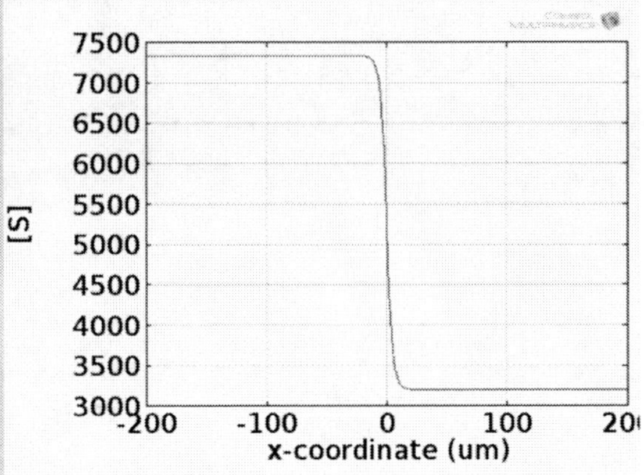

Department of Chemical Engineering, UCT Prague

Bifurcation analysis

A_{tot} = 1800 nmol m^{-3}

- Jump from the middle to highest steady state.
- $v_P \approx 0.0207\ \mu\text{m min}^{-1}$

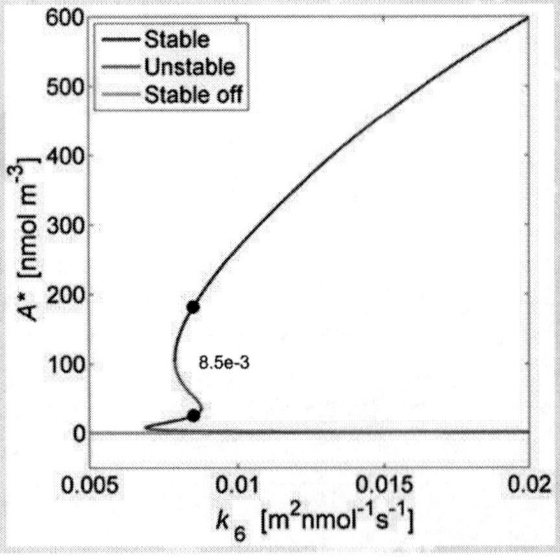

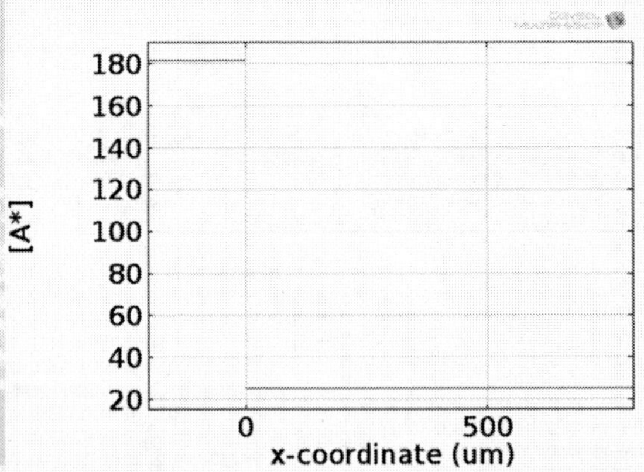

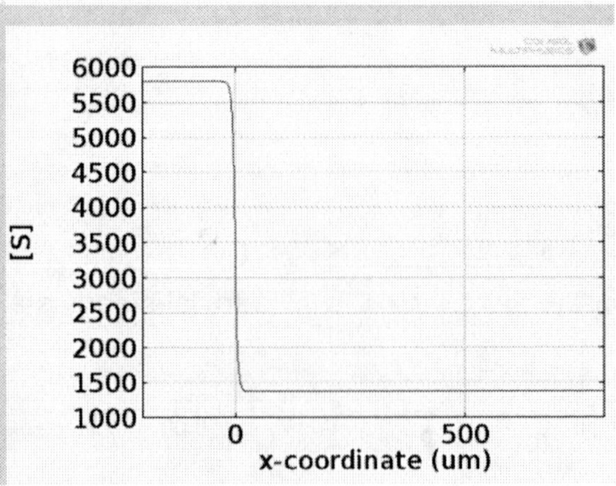

Department of Chemical Engineering, UCT Prague

Bifurcation analysis

A_{tot} = 1800 nmol m^{-3}

- Jump from the trivial to middle steady state.
- $v_P \approx 0.032\ \mu\text{m min}^{-1}$

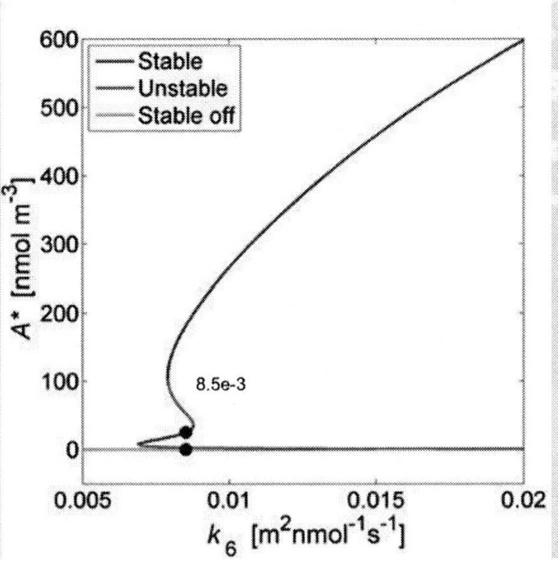

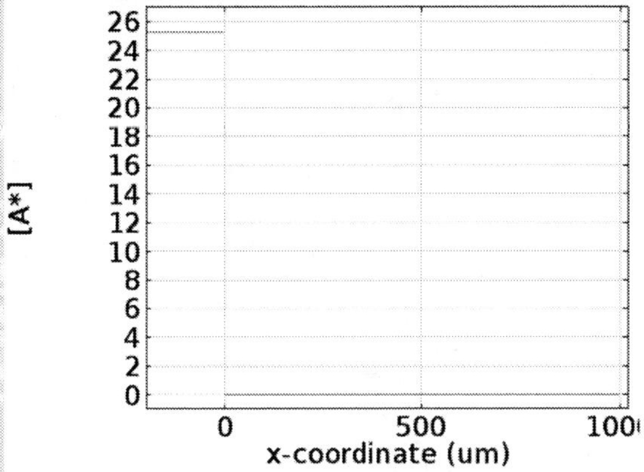

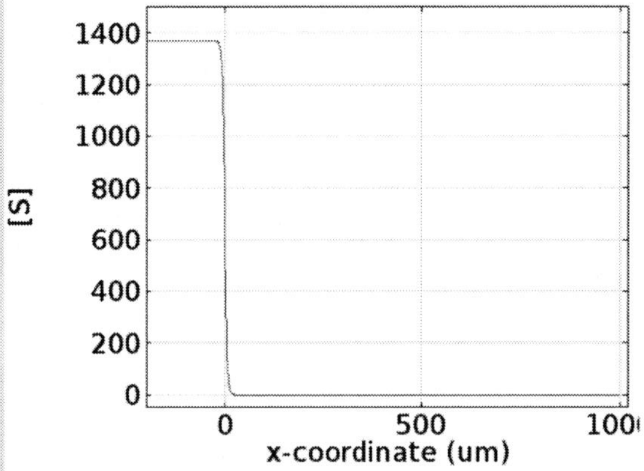

Department of Chemical Engineering, UCT Prague

Bifurcation analysis

A_{tot} = 1800 nmol m^{-3}

- Jump from the trivial to the highest steady state.
- Emergence of the middle state.

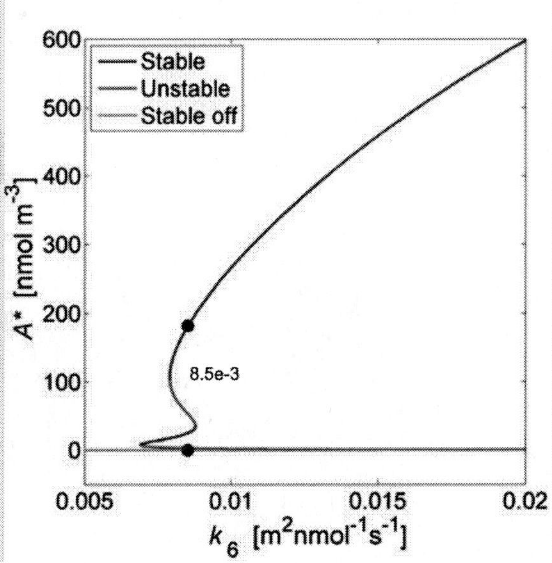

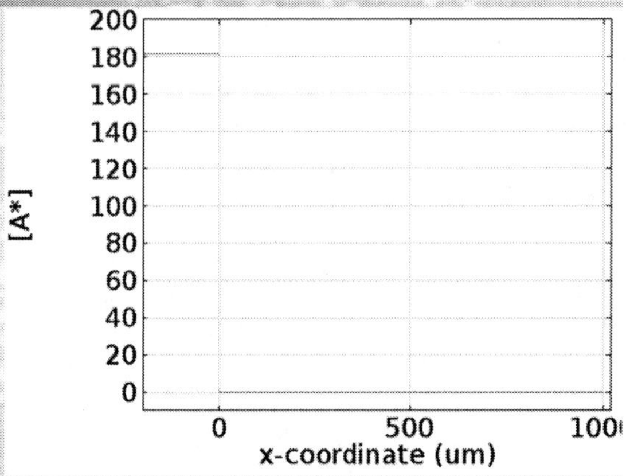

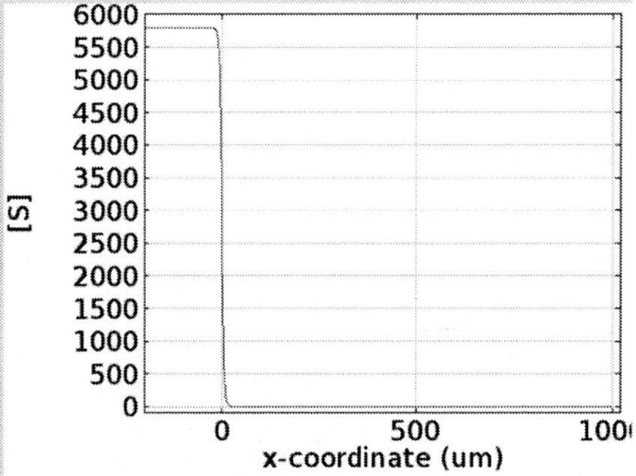

Department of Chemical Engineering, UCT Prague

Bifurcation analysis

A_{tot} = 2800 nmol m^{-3}

- Narrow region of oscillations
- Initially damped oscillations going to stable periodic regime when in contact with cell population in a steady state

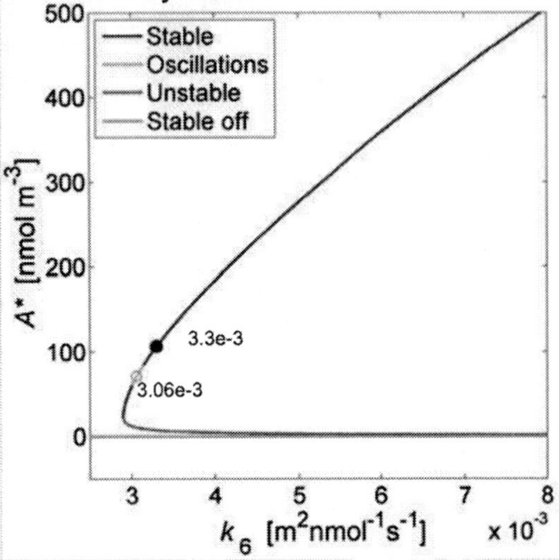

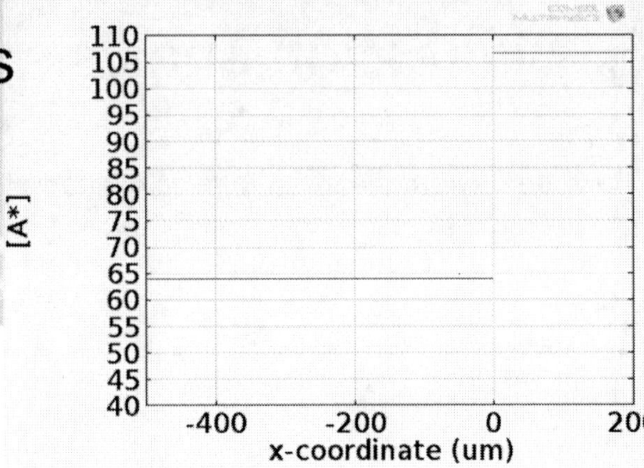

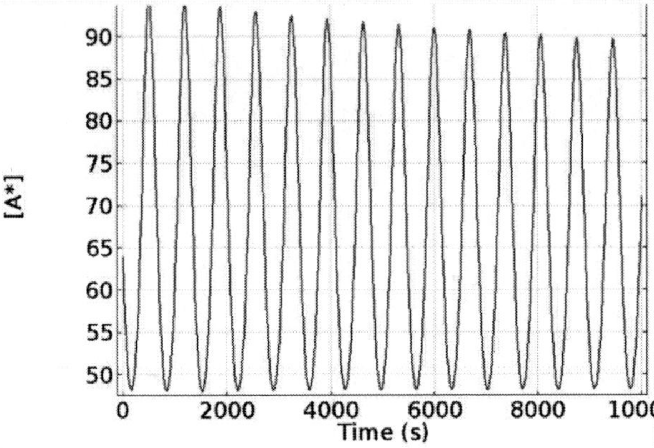

Department of Chemical Engineering, UCT Prague

Bifurcation analysis

A_{tot} = 2800 nmol m^{-3}

- Jump from the trivial steady state to stable periodic regime
- $v_P \approx 0.015\ \mu m\ min^{-1}$

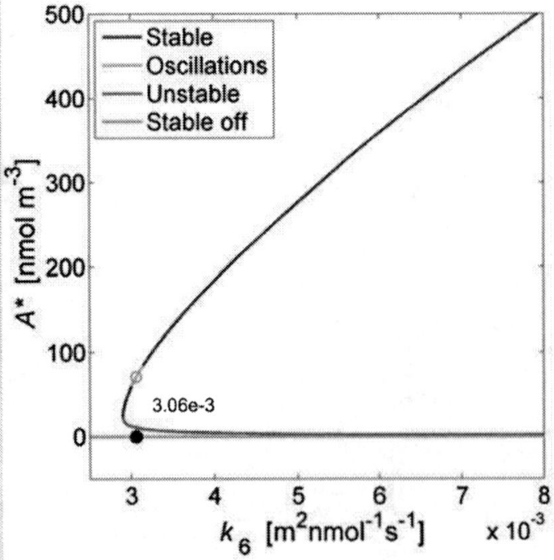

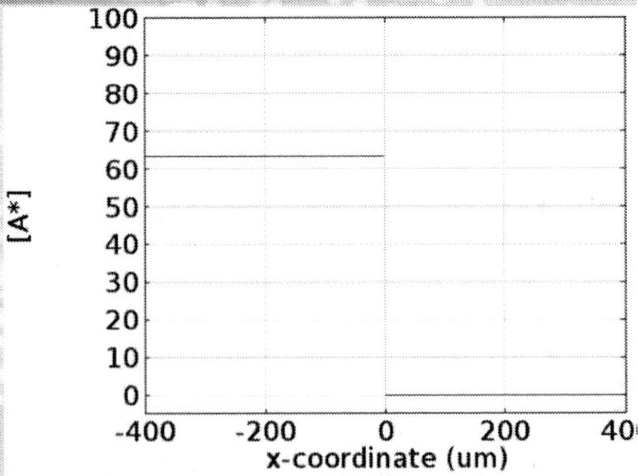

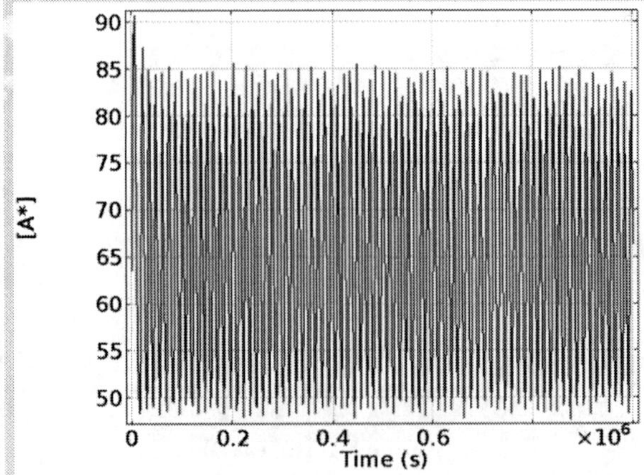

Department of Chemical Engineering, UCT Prague

Conclusions

- Minimal reaction subnetworks responsible for bistability and oscillations in MAPK cascade were identified
- Mathematical model of positive feedback loop based on the subnetwork and ligand-receptor interaction was developed
- Bifurcation analysis of the coupled model revealed:
  - Region with coexistence of two or three stable steady states
  - Region with coexistence oscillations and stable steady state
  - Region with one trivial steady state (no ligand, no complex, no phospohorylated kinase forms)
- Dynamical analysis od spatially distributed system revealed:
  - Possible coexistence of stable spatial structures
  - Existence of propagating concentration fronts in multistability regions (signal transmission) always leading to switch to a regime with higher enzyme activity
  - Observed spatial propagation of oscillatory regime coupled with the formation of complex wave structures

The authors thank for the support by the grant of the Ministry of Education, Youth and Sports of the Czech Republic, grant no. LH14009.

Department of Chemical Engineering, UCT Prague

Thank you for your attention

453300 Global Optimization of Crude Oil Scheduling with Discrete and Continuous-Time Formulations

Abstract:

Optimization of scheduling operations in refineries has been recognized as a relatively inexpensive and fast way to improve key performance indices [1]. A distinctive feature of refinery scheduling problems is the blending of different types of crudes or liquid fuels for meeting quality specifications on process streams or customer orders. These are challenging problems since they involve modeling tanks of varying composition with non-convex bilinear terms, as well as complicating logistic constraints.

In this work, we focus on crude oil scheduling [2-8], which can be classified as a non-convex, mixed-integer nonlinear programming (MINLP) problem, belonging to one of the most complex fields in optimization [9]. Recent developments have occurred at the level of the: (i) mathematical formulation, with tighter source-based formulations for the closely related multi-period pooling problem of refined petroleum products [10,11]; (ii) global optimization algorithm, with efficient piecewise relaxation techniques [12,13]. The latter also include major performance improvements in the most recent versions of commercial solvers BARON [14] and GloMIQO [15].

In our previous work [8], a source-based continuous-time formulation was used underneath a rather complex Resource-Task Network (RTN) process model. The main novelty of this work is to present a simpler formulation by modelling logistic constraints through Generalized Disjunctive Programming (GDP) [9,16-17]. Furthermore, we also present its discrete-time counterpart. It should be highlighted that the discrete-time representation concept was part of the seminal paper [2] but has lost relevance in the last decade or so (when applied to this particular problem). The main motivation for revisiting discrete-time is that the inventory cost term in one of the two alternative objective functions considered becomes linear compared to continuous-time.

Results over a set of test problems from the literature show indeed that the discrete-time formulation is better when minimizing total cost, despite requiring an order of magnitude larger number of time slots. It was particularly surprising to find out that nonlinear scheduling problems with roughly 100 time points can be tackled to global optimality. On the other hand, the continuous-time approach is preferable when maximizing the gross margin.

The research study also involved the comparison of different global optimization algorithms. A two-stage MILP-NLP strategy relying on the standard McCormick relaxation [18] on the first iteration and on normalized multiparametric disaggregation [13] to close the gap, was shown competitive to commercial solvers BARON 16.3.4 and GloMIQO 2.3 for the objective of cost minimization. More specifically, the McCormick relaxation unexpectedly resulted in zero MILP relaxation gap when using a discrete-time formulation; (ii) normalized multiparametric disaggregation was slightly better when relying on continuous-time.

Acknowledgments: Financial support from Fundação para a Ciência e Tecnologia (FCT) through the Investigador FCT 2013 program and project UID/MAT/04561/2013.

References:

[1] Kelly JD, Mann JL. Crude oil blend scheduling optimization: an application with multimillion dollar benefits-Part 1. Hydrocarbon Processing 2003; 82(6): 47.

[2] Lee H, Pinto JM, Grossman; IE, Park S. Mixed-integer linear programming model for refinery short-term scheduling of crude oil unloading with inventory management. Ind. Eng. Chem. Res. 1996; 35: 1630.

[3] Jia, Z, Ierapetritou M, Kelly JD. Refinery Short-term Scheduling Using Continuous Time Formulation: Crude-Oil Operations. Ind. Eng. Chem. Res. 2003; 42: 3085.

[4] Li J, Li W, Karimi IA, Srinivasan R. Improving the Robustness and Efficiency of Crude Scheduling Algorithms. AIChE J. 2007; 53: 2659-2680.

[5] Mouret S, Grossmann IE, Pestiaux P. A novel priority-slot based continuous-time formulation for crude-oil scheduling problems. Ind. Eng. Chem. Res. 2009; 48: 8515.

[6] Li J, Misener R, Floudas CA. Continuous-Time Modeling and Global Optimization Approach for Scheduling of Crude Oil Operations. AIChE J. 2012; 58: 205.

[7] Yadav S, Shaik MA. Short-Term Scheduling of Refinery Crude Oil Operations. Ind. Eng. Chem. Res. 2012; 51: 9287.

[8] Castro PM, Grossmann IE. Global Optimal Scheduling of Crude Oil Blending Operations with RTN Continuous-time and Multiparametric Disaggregation. Ind. Eng. Chem. Res. 2014; 53, 15127-15145.

[9] Trespalacios F, Grossmann IE. Review of Mixed-Integer Nonlinear and Generalized Disjunctive Programming Methods. Chem. Ing. Tech. 2014; 86(7): 991–1012.

[10] Castro PM. New MINLP Formulation for the Multiperiod Pooling Problem. AIChE J. 2015b; 61(11): 3728-3738.

[11] Lotero I, Trespalacios F, Grossmann IE, Papageorgiou DJ, Cheon M-S. An MILP-MINLP decomposition method for the global optimization of a source based model of the multiperiod blending problem. Comput. Chem. Eng. 2016; 87: 13-35.

[12] Kolodziej S, Castro PM, Grossmann IE. Global optimization of bilinear programs with a multiparametric disaggregation technique. Journal of Global Optimization 2013; 57: 1039-1063.

[13] Castro PM. Normalized multiparametric disaggregation: an efficient relaxation for mixed-integer bilinear problems. Journal of Global Optimization 2016; 64: 765-784.

[14] Tawarmalani M, Sahinidis NV. A polyhedral branch-and-cut approach to global optimization. Mathematical Programming 2005; 103(2): 225-249.

[15] Misener R, Floudas CA. GloMIQO: Global mixed-integer quadratic optimizer. Journal of Global Optimization 2013; 57: 3-50.

[16] Balas E. Disjunctive Programming and a Hierarchy of Relaxations for Discrete Optimization Problems. SIAM J. Algebraic Discrete Methods. 1985; 6(3): 466-486.

[17] Raman R, Grossmann IE. Modeling and computational techniques for logic based integer programming. Comput. Chem. Eng. 1994; 18: 563-578.

[18] McCormick GP. Computability of global solutions to factorable nonconvex programs. Part I. Convex underestimating problems. Mathematical Programming 1976; 10, 147-175.

461735 Optimal Scheduling of a Renewable Ammonia Plant with Wind Generation Under Variable Electricity Pricing

Abstract:

The increasing penetration of renewable power into the overall energy landscape presents several challenges for power system operators. In particular, it has been well documented that as the amount of renewable sources in the power market increases, the ramp rates of fossil fuel generation sources and the risk of overgeneration of power also increases. Variable electricity pricing mitigates these challenges by providing economic incentives to use power when the risk of overgeneration is high, and curtail power usage when renewable supply is low and low-efficiency fossil fuel generators would need to be used. Chemical plants typically have sufficient process flexibility to be able to take advantage of such variable pricing. Indeed, previous research has shown that cost savings can be achieved by utilizing a variable pricing structure in batch-continuous plant scheduling [1] and by integrating a plant containing combined heat and power with the grid [2]. Since the industrial sector accounts for 26% of electricity usage, chemical plants which account for variable pricing can help the integration of additional renewable energy sources.

Recently, a first of its kind renewable ammonia plant has been developed in Morris, MN, which uses wind energy to electrolyze water, separate air, and produce ammonia in a Haber-Bosch reactor. We have shown previously that such a plant can be cost-competitive in the ammonia supply chain when scaled up to 140,000 t/y [3]. However, it is envisioned that renewable plants be implemented on a smaller, more distributed scale to promote local and sustainable fertilizer production for farming. To help achieve this goal, it is hypothesized that the economics of the renewable ammonia plant can be improved by implementing a scheduling program that takes advantage of variable electricity pricing both through the chemical process flexibility and through the integration of the wind turbine with the grid.

This work develops a framework for the optimal scheduling for plants which contain both chemical processes and renewable power generation under variable electricity pricing. The scheduling problem is formulated as a mixed integer program (MIP) that minimizes the operating costs of the system over some time horizon by determining set points for the chemical processes and grid interaction such that product demands are met. Such a framework will be useful for the industrial sector to play its part in being amenable to increasing renewable power penetration. As a case study, this framework will then be applied to the renewable ammonia system. In this study, we consider hourly scheduling over a 24 hour rolling horizon. The scheduling optimization problem takes advantage of the variable pricing, producing extra hydrogen and nitrogen when electricity prices are low and selling excess wind energy when prices are high. As such, we establish that the economic feasibility of smaller scale renewable ammonia plants is improved when considering variable power pricing.

[1] P. M. Castro, I. Harjunkoski, and I. E. Grossmann, "New continuous-time scheduling formulation for continuous plants under variable electricity cost," Ind. Eng. Chem. Res., vol. 48, pp. 6701-6714, 2009.

[2] S. Mitra, L. Sun, and I. E. Grossmann, "Optimal scheduling of industrial combined heat and power plants under time-sensitive electricity prices," Energy, vol. 54, pp. 194-211, 2013.

[3] A. Allman and P. Daoutidis, "Ammonia supply chains: a new framework for renewable generation with a case study for Minnesota," Comput. Aided Process Des., Accepted, 2016.

462308 Discrete-Time Mixed-Integer Programming Models for Scheduling in Sequential Environments

Abstract:

Discrete-time Mixed-Integer Programming Models for

Scheduling in Sequential Environments

Hojae Lee, Andres F. Merchan, and Christos T. Maravelias

Department of Chemical and Biological Engineering, University of Wisconsin-Madison

1415 Engineering Dr., Madison, WI 53706, USA

Most mixed-integer programming (MIP) models that have been developed to address problems in sequential production environments (e.g., multi-stage processes) adopt a continuous modeling of time. The major disadvantage of these models is that accommodating common process features such as shared renewable resources and time-varying resources leads to nonlinearities in the model or introduction of additional variables and constraints. On the other hand, discrete-time representation offers a natural framework to model such type of features without significant increase in the complexity of the model. Moreover, a previous study has shown that discrete-time formulations are not only more effective in terms of computational time, but also in finding better solutions[1].

In this work, we study four discrete-time MIP models for short-term scheduling in multi-stage facilities with non-uniform parallel units. First, we revisit the Resource-Task Network (RTN) based model originally proposed by Castro and Grossmann[2], where resource balance constraints are enforced for both batches and units. Then, we develop a State-Task Network (STN) based model, in which clique constraints are used to enforce unit utilization, in place of the unit balance constraints in the previous model. Next, we propose two models inspired from the time-aggregated and disaggregated formulations for the Resource-Constrained Project Scheduling Problem (RCPSP)[3,4]. The batch precedence constraints in these models are inspired from the precedence constraints written for activities and their successors in the RCPSP formulations.

Furthermore, we propose modeling extensions to account for a more complicated production environment, namely multi-purpose facilities. In the multi-purpose facilities, each product goes through a series of product-specific stages, which consist of multi-purpose units that may belong to different stages for different products[5]. To explicitly model product-specific routing and the multi-purpose units, we introduce new sets to denote product-specific stages for each batch and units that belong to a specific stage of a batch, respectively.

We also propose extensions to model intermediate batch storage for both production environments. Since we model the storage of each intact batch, we only consider vessel compatibility, neglecting vessel capacity. For RTN- and STN-based models, we implicitly model the storage of intermediate batches through vessel and batch balance constraints. In the RCPSP inspired models, additional flow variables are introduced to explicitly model batch flow in and out of the vessels.

We provide illustrative examples to showcase the advantages of the proposed discrete-time formulations in modeling limited and time-varying resources. Furthermore, through computational study, we compare the four models with continuous-time models to show the effectiveness of the proposed formulation in terms of computational efficiency.

Reference

1. Sundaramoorthy A, Maravelias CT. Computational Study of Network-Based Mixed-Integer Programming Approaches for Chemical Production Scheduling. *Industrial & Engineering Chemistry Research.* May 4 2011;50(9):5023-5040.

2. Castro PM, Grossmann IE. New continuous-time MILP model for the short-term scheduling of multistage batch plants. *Industrial & Engineering Chemistry Research.* Nov 2005;44(24):9175-9190.

3. Pritsker AAB, Waiters LJ, Wolfe PM. Multiproject Scheduling with Limited Resources: A Zero-One Programming Approach. *Management Science.* 1969;16(1):93-108.

4. Christofides N, Alvarez-Valdes R, Tamarit JM. Project Scheduling with Resource Constraints â€" A Branch and Bound Approach. *European Journal of Operational Research.* Jun 1987;29(3):262-273.

5. Maravelias CT. General framework and modeling approach classification for chemical production scheduling. *Aiche Journal.* Jun 2012;58(6):1812-1828.

458814 Emission Considered Scheduling for Crude Unloading, Transferring, and Processing

Abstract:

Petroleum refinery is a leading segment of the entire chemical process industry. It uses crude oils to produce various fuels such as gasoline, aviation kerosene, diesel, heavy fuel oil, and chemical raw materials such as naphtha and benzene. Because of the increasingly strict economic competitions and environmental regulations, refineries in one hand are eager to improve production solutions to leverage profitability margins in nowadays volatile market; on the other hand, they have to pursue cost-effective pollution prevention technologies to comply with more stringent environmental mandates, such as air emission requirements. Air emissions from refineries are significant and emission reduction in refineries is a very challenging task. It involves a large-scale complex manufacturing system, which consists of over 17 major facilities with more than 85 main operation units, and thousands of process streams. In such a large industrial system, air emissions could spatially scatter at many reactors, furnaces, and flaring facilities; and temporally occurs at any time. Thus, cost-effective emission-reduction strategies should neither depend on end-of-pipe approaches, nor focus on solutions for single units or subsystems. The major thrusts should focus on the comprehensively study of material, energy, and information exchanges within the entire manufacturing system, so as to identify the best solutions to the entire plant. Therefore, to optimally balance emission source generations and utilizations among the entire system might be the smartest way to reduce refinery emissions.

A concept of profitable emission reduction (PER) is proposed in this research, which has merits of economically attractive, environmentally benign, and technologically viable for air emission reductions in petroleum refineries. To identify PER strategies for refinery plants, a production scheduling model covers both front-end and refinery has to be involved. The key is to optimize the overall material and energy flows to make the plant net profit maximum, meanwhile to ensure emission source generations and utilizations to be smartly balanced. Other manufacturing constraints, such as operation specifications and inventory limits, should also be satisfied as well. Scheduling of front-end and refinery crude oil operations is an important part of petroleum supply-chain management.

In this research, a new methodology framework and a new general scheduling model have been developed for emission-considered crude unloading, transferring, and processing (ECUTP) system to achieve profitable emission reduction. The scope of the scheduling problem includes crude oil unloading from vessels to storage tanks at onshore berths, transfer of crude from these tanks to charging tanks, charging crude distillation units and further processing of crude inside the refinery, which includes crude distillation, cracking, coking, reforming, hydrotreating, product blending and component recovery. It also couples characterization and quantification of major air emissions from refineries, such as CO_2, volatile organic compounds (VOC), nitrogen oxides (NO_X), and particulate matters (PM). A general refinery process with more emphasis on air emission characterization and quantification will be introduced. The scheduling model is a large-scale mixed integer nonlinear programming problem (MINLP). The efficacy of the PER concept and the developed methodology have been demonstrated by different case studies.

461970 A Game Theoretic Framework for Strategic Production Planning

Abstract:

Strategic planning allocates future production targets which link plant scheduling procedures to high level organizational decisions. In most literature strategic planning is driven by two common objectives: minimization of the costs associated with meeting fixed production objectives, or maximizing the profitability of a process according to a set of fixed product prices [1]. From a strategic perspective neither of these objectives addresses the potential for changes in market conditions nor the role of competition in organizational planning. The solutions obtained by these types of approaches represent cost effective plans and may yield reasonable production targets that maximize profitability, but such results may not represent competitive strategies, and the value reported by the model may not be realized, or be achievable, as a result.

All industries face competition and need to make decisions accordingly in order to avoid losses in profits or market shares. The study of competitive decision making processes falls into the field of game theory which has developed methods to quantify competitive behaviour. Our research examines methods of incorporating game theoretic descriptions of competition into strategic production planning, which connects organizational decision making to process models allowing competitive decisions to be made within the operating space of existing process equipment. The production targets emerging from such strategic planning methods represent competitive strategic planning results.

We present a framework for game theoretic strategic production planning based on the potential game formulation of a Cournot oligopoly model of market prices which incorporates the planning efforts of multiple organizations competing in different markets [2][3][4]. The framework is illustrated using strategic production planning models of crude oil refineries. In this framework competing organizations each produce multiple products which are sold in local and export markets. Product prices in those markets react according to the total supply of each product received from all suppliers. Refiners are forced to set production targets which take into account not only market conditions but also the behaviour of their competitors such that the resulting strategic production plans form Nash equilibria.

We demonstrate the use of the framework in strategic production planning and extend its use to market structure planning and process expansion planning. As Nash equilibria, the strategic results obtained are robust to changes in competitor behaviour and represent optimal responses to both market conditions and competitor behaviour taking into account the operating costs and limitations of the underlying process model. The resulting strategic plans have a property not shared by plans generated using cost minimization or profit maximization approaches; they are rational according to the description of competitive market forces, and thus avoid production plans which under fixed price analysis may be considered optimal but are unachievable in a game theoretic perspective.

References

[1] Papageorgiou, L., "*Supply chain optimization for the process industries: Advances and opportunities,*" *Comput. Chem. Eng.*, vol. 33, 2009.

[2] Monderer, D., Shapley, L., "*Potential Games,*" *Game. Econ. Behav.*, vol. 14, 1996.

[3] Slade, M., "*The Fictitious Payoff-Function: Two Applications to Dynamic Games,*" *Ann. Econ. Stat.*, 1989.

[4] Slade, M., "*What does an oligopoly maximize?*" *J. Ind. Econ.*, vol. 42, 1994.

462427 Solution Methods for Discrete-Time Formulations for Scheduling in Multi-Stage Facilities

Abstract:

Solution Methods for Discrete-time Formulations for

Scheduling in Multi-stage Facilities

Andres F. Merchan, Hojae Lee and Christos T. Maravelias

Department of Chemical and Biological Engineering, University of Wisconsin-Madison

1415 Engineering Dr., Madison, WI 53706, USA

Most research on scheduling in multi-stage facilities focuses on models based on continuous representation of time[1]. However, discrete-time mixed-integer programming (MIP) models have some distinct advantages, such as their extendibility. For example, they can be readily used to accommodate shared renewable resources (e.g. utilities, manpower), time-varying resource cost and availability, and linear modeling of inventories and resource utilization costs[2]. At the same time, discrete-time scheduling formulations have some disadvantages as well. Since the decision variables are defined at each time point, the length of the time horizon and/or irregular time related data (e.g. processing times, release/due dates) have a high impact on the model size. Also, a predefined coarse time grid may lead to poor solution quality in some cases[3]. Thus, studies on solution methods for discrete-time formulations for scheduling in sequential environments is necessary.

In this work, we propose four different solution methods for discrete-time models for scheduling in multi-stage facilities. First, we develop tightening constraints based on fixed time windows. Using the earliest start and latest finish times of a unit, we develop a tightening constraint which enforces that the total processing time of all the batches assigned to that unit does not exceed its time window. Second, we propose a different tightening method based on variable time windows. Here, earliest start time and shortest tail of each batch are treated as variables, and are used to formulate constraints which limit the total time for a batch to be completed by its available time window. Then, we propose a simple reformulation by introducing a new discrete variable that denotes the total number of batches assigned to a unit. This reformulation leads to efficient bound improvements in the branch-and-bound algorithm, since branching on these variables facilitates pruning of multiple assignments at once. Finally, inspired by the fact that production schedules in sequential environment are highly influenced by the schedule of the bottleneck stage, we develop a priority-based branching method that enables us to put higher importance on the decision variables related to these stages and/or the corresponding units.

Furthermore, we develop a solution refinement method which refines an approximate solution obtained by a coarse discretization by resolving for the correct timing of process events. The main idea behind this method is that realistic timing of events can be recalculated using a continuous-time model[4] by just solving a Linear Program (LP) with fixed assignments and sequencing obtained from the discrete-time models. To increase the probability of obtaining the optimal solution, we automatically generate a pool of solutions while solving the discrete-time model, in order to collect multiple candidates for the refinement process.

Finally, we perform an extensive computational study with more than 3000 runs to show the effectiveness of the solution methods proposed. We test the solution methods on different discrete-time models and different objective functions and we study various combinations of the proposed solution methods for each model and objectives. We show that our solution methods lead up to three order of magnitude improvement in the computational time thus allowing us to solve problems that were thought to be intractable.

1. Harjunkoski I, Maravelias CT, Bongers P, et al. Scope for industrial applications of production scheduling models and solution methods. *Computers & Chemical Engineering.* Mar 2014;62:161-193.

2. Velez S, Maravelias CT. Advances in Mixed-Integer Programming Methods for Chemical Production Scheduling. In: Prausnitz JM, Doherty MF, Segalman RA, eds. *Annual Review of Chemical and Biomolecular Engineering, Vol 5.* Vol 5. Palo Alto: Annual Reviews; 2014:97-121.

3. Floudas CA, Lin XX. Continuous-time versus discrete-time approaches for scheduling of chemical processes: a review. *Computers & Chemical Engineering.* Oct 2004;28(11):2109-2129.

4. Mendez CA, Henning GP, Cerda J. An MILP continuous-time approach to short-term scheduling of resource-constrained multistage flowshop batch facilities. *Computers & Chemical Engineering.* May 2001;25(4-6):701-711.

463235 Production Planning and Scheduling Integration through Multiparametric Bilevel Mixed-Integer Optimization

Abstract:

Planning and scheduling optimization problems with seasonal demand variability can be often expressed within a hierarchical structure, where optimal decisions at an aggregate upper level (planning) provide constraints for the detailed decision making (scheduling) at a lower level, typically posed as bilevel optimization problems [1-4]. Since discrete decisions are involved most likely at both levels, the resulting formulations typically correspond to bilevel mixed-integer linear programming problems (B-MILP). The solution of these problems is very challenging, and typically requires the use of global optimization techniques, even for the derivation of approximate solutions [5, 6].

In this work, we propose a novel algorithm for the exact and global solution of bilevel planning/scheduling problems. The algorithm is based on multiparametric programming theory and our past work on bilevel programing with continuous variables [7, 8]. The main idea of our approach is to treat the scheduling problem as a multi-parametric mixed-integer linear programming problem (mp-MILP), in which the production targets (optimization variables of the upper level problem) are considered as the parameters for the lower level scheduling problem. The resulting exact parametric solutions are then substituted into the upper level planning problem, which can be solved as a set of single-level deterministic mixed-integer programming problems. The proposed algorithm was applied and tested on a range of planning and scheduling example problems from the literature [2, 3].

References: [1] You, F., Grossmann, I.E. Design of responsive supply chains under demand uncertainty (2008) Computers and Chemical Engineering, 32 (12), pp. 3090-3111. [2] Dogan, M.E., Grossmann, I.E. A decomposition method for the simultaneous planning and scheduling of single-stage continuous multiproduct plants (2006) Industrial and Engineering Chemistry Research, 45 (1), pp. 299-315.

[3] Li, Z., Ierapetritou, M.G. Integrated production planning and scheduling using a decomposition framework (2009) Chemical Engineering Science, 64 (16), pp. 3585-3597.

[4] Ryu, J.-H., Dua, V., Pistikopoulos, E.N. A bilevel programming framework for enterprise-wide process networks under uncertainty (2004) Computers and Chemical Engineering, 28 (6-7), pp. 1121-1129.

[5] Mitsos, A. Global solution of nonlinear mixed-integer bilevel programs (2010) Journal of Global Optimization, 47 (4), pp. 557-582.Global Optimization, 47 (4), pp. 557-582.

[6] Gümüş, Z.H., Floudas, C.A. Global optimization of mixed-integer bilevel programming problems (2005) Computational Management Science, 2 (3), pp. 181-212.

[7] Faísca, N.P., Saraiva, P.M., Rustem, B., Pistikopoulos, E.N. A multi-parametric programming approach for multilevel hierarchical and decentralized optimisation problems (2009) Computational Management Science, 6 (4), pp. 377-397. [8] Faisca, N.P., Dua, V., Rustem, B., Saraiva, P.M., Pistikopoulos, E.N. Parametric global optimisation for bilevel programming (2007) Journal of Global Optimization, 38 (4), pp. 609-623.

466374 Scheduling and Internet of Things - Vision or Reality?

Abstract:

In the last decade there has been an increasing discussion around the topic internet of things (IoT). In fact the term was first coined by MIT in the late 90s and refers a world where objects are connected, monitored and optimized through either wired, wireless or hybrid systems (Zhou et al., 2015). As a consequence there are academic-industrial activities world-wide e.g. Smart Manufacturing in the US and Industry 4.0 driven by Germany. The term cyber-physical system, originating from the area of mechatronics, refers to a system composed of physical entities such as mechanisms controlled or monitored by computer-based algorithms. Even if the first actual developments have been done for discrete manufacturing, this opens up lots of interesting application opportunities also within chemical processes.

From a supply chain management perspective, IoT should enable automated decision making reducing the need of human intervention. Connectivity is one of the key issues here since it is predicted that there will be around 50 billion connected devices already in 2020. Much attention has been paid on mainly three elements: Big Data, Cloud computing and Cyber Security. These are natural areas of research since increasing connectivity also multiplies the available data, of which a large part is unstructured (Babiceanu and Seker, 2016), the data processing is by itself a challenge and the value or outcome of it is unclear. Cloud computing offers means for hosting and analyzing larger set of data collected from various sources and in this context the security of data becomes an issue - also within an intranet. Thus, data is one of the key enablers of new applications and it is expected to improve the enterprise, manufacturing and business decision support. It is expected that having more information about the processes improves their resilience, i.e. the capability to cope with complex events and respond in acceptable time. This following the vision of providing physical components awareness of the context in which they operate to enable more independent decision making in a collaborative manner. Even though the IoT-focus is also moving towards mobility (5G) etc., the realization of "smart space", where users can easily enjoy a variety of services at any time and any place, is still in its infancy (Shen et al., 2015).

The fact that developments have been lead by the discrete manufacturing does not mean that it has less potential within the process industry. Maybe even the opposite is true but the more complex physical processes and established and installed solutions evidently slows down the realization of the IoT scheme into its full scale. Furthermore, before agreeing to a mainly SW-driven change, process industries should also be able to see the expected value. Successful use cases are still few and therefore the IoT-activity is often seen as a technology push. While some opportunities may be realizable in short-term, it can be expected that in long term the well-established automation pyramid (Engell and Harjunkoski, 2012) will dissolve. This has a severe impact on current process control and scheduling technologies as the hierarchical structure is replaced by an any-to-any communication increasing the opportunities for collaboration and especially through better tracking make it possible to adapt the schedules on-the-fly. Having faster and more flexible communication channels across business systems also enables totally new business models and new vendors to enter the market.

Nevertheless, the basic function of and need for production scheduling will certainly remain, as it is important to be able to coordinate and manage a network of physical processes in an economical, safe and energy-efficient way. The IoT development makes the role of scheduling even more important part of an intelligent manufacturing environment (Li et al., 2015). The vision of being fully able to track production, use historic and current data to predict possible incidents and re-schedule the production including maintenance operations in order to avoid incidents before they actually occur and ensure that the production is always done in the most optimal way will one day be technically realizable. Before this can become reality, there are still many obstacles related to communication standards, cross-interactions and conflict handling but also in the modeling and solution of the resulting problems. This presentation aims at catalyzing the discussion on the real obstacles and needs to bring the IoT-vision into reality. Some of the key questions are: Do we need IoT in the first place and does it bring added value and are the conventional approaches sufficient to meet the upcoming challenges?

Zhou, L., Chong, A.Y.L., Ngai, E.W.T. (2015). Supply chain management in the era of the internet of things (Editorial), International Journal of Production Economics, 159, pp. 1-3

Babiceanu, R.F., Seker, R. (2016). Big Data and virtualization for manufacturing cyber-physical systems: A survey of the current status and future outlook, Comput. Industry, http://dx.doi.org/10.1016/j.compind.2016.02.004

Engell S., Harjunkoski, I. (2012), Optimal Operation: Scheduling, Advanced Control and their Integration, Computers and Chemical Engineering, 47, pp. 121-133

Shen, F., Liu, L., Palesi, M. (2015). Introduction to the special issue on "Emerging research in Internet of Things" (Editorial). Computers and Electrical Engineering 44 (2015), pp. 104â€"106

Li, G., Ge, Y., Wang, Y., Hu, S. (2015). Job Scheduling Management Research of Intelligent Manufacturing Based on the Internet of Things, Chemical Engineering Transactions, 46, pp. 829-834

471164 Nature in Engineering: Expanding the Engineering Design Space By Including Ecosystem Goods and Services

Abstract:

Currently, most engineering design methods ignore the role played by nature in supporting engineering systems. Analogous to economic systems, ecological systems also include limits on the stocks of ecosystems goods and services. These can include limits on the availability of ecological resources like fossil fuels, minerals, biomass, fiber and feed, as well as limits on the provisioning of ecosystem services like carbon sequestration, water provisioning, nutrient cycling etc. Ignoring these limits while designing systems have resulted in several unintended consequences, resulting in ecological degradation. For instance, switching farming practices to more intensive agricultural habits to increase crop yield has resulted in harmful algal blooms in Lake Erie due to fertilizer runoff from farms.

In the engineering context, the main focus of conventional design has been to maximize profitability while simultaneously minimizing environmental impacts. Accounting for the role of nature in engineering has been confined to ideas like Biomimicry [1] and Eco-technology [2] such as the use of wetlands for water treatment. However, these methods lack proper integration with concepts of engineering design and such methods do not necessarily identify possible synergies and trade-offs that exist between engineered systems and nature. In addition, all these approaches also neglect the ecological carrying capacity, thus resulting in an increase in disparity between the demand for ecological goods and services and their supply from nature.

The Techno-Ecological Synergy (TES) framework [3] was developed to account for the carrying capacity of ecosystems by quantifying the demand and supply of ecosystem services (ES). Resource inputs for operating engineering systems creates a demand for ecological goods while the emissions, waste and wastewater generated by engineering systems creates a demand for ecosystem services. The supply of ecosystem goods and services is based on the capacity of an ecological system, in the vicinity of an engineering system to provide the necessary ES demanded by the process. The sustainability index for this framework is based on the difference in the supply and demand for ES.

This work integrates the TES framework with concepts of engineering design, resulting in the design of integrated Techno-Ecological systems. The TES design framework accounts for both demand and supply of ecosystem services while designing systems and the objective is to maximize profitability and minimize the disparity between the demand and supply of ES. Thus, the problem focuses on reducing both demand and increasing supply by preservation and restoration of ecological systems, in contrast to conventional approaches of merely minimizing demand.

The presentation conveys the significance of including the role of nature in supporting engineering design. This is indicated by demonstrating that including nature in the design problem expands the solution space of conventional engineering design resulting in the identification of plausible "win-win" scenarios. Design problems are formulated as multi-objective optimization problems, with the objective of maximizing the economic value of the system and maximizing the sustainability index. Trade-offs between these objectives are identified by scalarizing multiple objectives using the epsilon constraint methods. Tradeoff designs are obtained for different ecosystem services, depending on the type of service relevant to the system of interest and the spatial presence of ecological systems.

Four diverse case studies are discussed to demonstrate the significance of including nature in engineering. The first study describes the design of a residential system accounting for ecosystems like trees, lawn and a vegetable garden. While designing a residential systems, technological, ecological and behavioural variables are considered in the design. The objectives are to minimize both the cost compared to basecase and carbon emission. The problem is formulated as a simulation-based Mixed Integer Linear Program (MILP) using EnergyPlus as the simulation engine.

The second study discusses the design of a biosolids network flow system along with the design of ecological systems that supply carbon sequestration ecosystem services, like timberland, forest lands and geological sequestration. Superstructure of possible biosolid flow pathways along with possible pathways for carbon sequestration are identified, and optimal designs of these coupled Techno-Ecological networks are determined by minimizing the network cost and minimizing the difference between carbon sequestration service demand and supply. The entire problem is formulated and solved as a MILP in the Eco-Flow tool. In addition, feedback from ecosystems in terms of biomass from timberland is considered as a feedstock for supplying energy to the wastewater treatment sites.

Third study describes the benefits of including ecosystems in a process engineering design context whereby ecological systems are included in engineering design analogous to unit operations. An integrated design of a biofuel system with a wetland ecosystem is discussed wherein the system is design to maximize provisioning of ecosystem services like water provisioning service and water quality regulation service, while maximizing the net present value.The problem is formulated as an Non Linear Program implemented in GAMS.

The last study focuses on designing agricultural landscapes for producing energy and food, while accounting for services provided by agro-ecosystems. These services include carbon sequestration, water provisioning, and others. The trade-off between monetary objectives and costs are evaluated for various biofuel production alternatives.

Non-inferior pareto frontiers are generated for each ecosystem service for all case studies. Results indicate that including ecological variables in the design results in possible solutions that do not exist in the techno-centric solution space of conventional engineering methods. In addition to this, accounting feedback flows from ecosystems in terms of ecological goods shifts the Pareto frontier to a new range, revealing innovative and novel design solutions. These new solutions obtained from including ecological systems are economically superior compared to conventional solutions; they also have environmental benefits in terms of reduced overshoot.

References

[1] Benyus, Janine M. Biomimicry. New York: William Morrow, 1997.

[2] Straškraba, M. "Ecotechnology as a new means for environmental management." Ecological Engineering 2.4 (1993): 311-331.

[3] Bakshi, Bhavik R., Guy Ziv, and Michael D. Lepech. "Techno-ecological synergy: A framework for sustainable engineering." Environmental science & technology 49.3 (2015): 1752-1760.

466829 Heat-Integrated Work Exchange Network Design

Abstract:

Energy loss in the manufacturing sectors is very high (~50% of the total consumption in 2010, according to a DOE report). While thermal energy efficiency has been greatly improved through adopting effective heat integration technologies over the past decade, the improvement of mechanical energy efficiency is yet to be fully investigated. Mechanical energy, often characterized by process pressure change as the driving force, is more difficult to recover; this is especially true as the pressurization/depressurization involves temperature change, which makes the process non-isothermal. Thus, effective mechanical energy recovery should be coupled with heat integration. Mechanical energy recovery can be realized using work exchangers that are currently used in industrial scale for desalination. The concept and fundamentals of this type of unit, as well as a thermodynamic analysis method for identifying the maximum amount of recoverable mechanical energy from a process system under isothermal condition has been investigated by us previously.

In this paper, a thermodynamic analysis on mechanical energy recovery under non-isothermal condition is described, and a process synthesis methodology for designing a heat-integrated work exchange network is introduced. The methodological efficacy is illustrated by two examples, where both mechanical and thermal energy efficiencies as well as economic feasibility analysis are provided. Furthermore, certain key dynamic performance of the system is analyzed, which is valuable for developing operational strategies for the integrated system.

Reference:

Energetics Incorporated, Manufacturing Energy and Carbon Footprint, Prepared for the Industrial Technologies Program (ITP), DOE, 2010.

468486 Discovery and Assessment of Integrated Waste Biorefinery Paths with Conventional Industry Using an Ontology Engineering Approach

Abstract:

The environmental impacts caused by the extensive use of decreasing fossil fuels has brought tremendous attention towards valorisation of renewable and untapped resources such as urban and industrial wastes. The need to reduce the use of virgin resources gives rise in research areas such as waste Industrial Symbiosis (IS), waste biorefineries, and circular economies. The chemistries or else known as â€œsynthesis pathsâ€, able to extract materials from wastes can be considered abundant.

Unwanted residues are multicomponent materials, containing valuable platform chemicals and numerous added-value specialties. If recovered, they can become standalone feedstock or be also integrated with alike feedstocks to be processed by existing industrial facilities. Such sustainable, retrofitting practice will enlarge process capacity, benefiting from economies of scale, improve LCA performance and cost-effectiveness originating from wasteâ€™s off-market prices.

To achieve such task, a) to appropriately integrate waste resources with virgin feedstocks and b) to identify suitable alternative synthesis paths, producing a diverse gamut of bio-based products, requires the combined use of two well-established research fields: ontology engineering and mathematical optimization. Therefore, it is firstly required a large well-structured database to organize knowledge such as waste properties and valorising technologies characteristics, thus achieving interoperability of information across various manufacturing organizations and processing facilities. Secondly, a screening-level decision support platform is necessary to holistically assess waste synthesis paths.

This work provides a framework that merges an ontological based platform with an optimization process systems tool to first organize and manage tacit knowledge such as chemistries, process models, paths, technology specifications, feedstock characteristics etc. for knowledge sharing across diverse users, and secondly to discover the optimum synthesis path according to a set of environmental, economic, and technical objectives. The semantic algorithm is used to establish layers of interactions and translate process models into networks that are further fathomed using optimization and superstructure technology.

468645 Ionic Liquid Design and Sustainable Process Simulation for Decarbonization of Shale Gas and Gas Separations

Abstract:

Ionic liquids (ILs) have been receiving increasing attention as a potential decarbonization solvent as well as solvent for gas separations, which are usually energy intensive. However, the enormous number of potential ILs that can be synthesized makes it a challenging task to search for the best IL as a decarbonization solvent, for example, for CO2 removal from methane. Also, for gas separations, for example, gas purification processes, ILs can play a major role by replacing the extractive distillation column with a liquid-liquid extractive column, while the recovery column is replaced by an evaporator, thereby saving a lot of energy. The main question is what properties should the ionic liquids and solute (gas)-IL mixtures have to design the optimal separation process? If the IL properties can be matched, then the target process design would be obtained.

In this work, a systematic method is proposed to screen suitable ILs based on a collection of ILs (generated by Computer Aided Molecular Design and collected data); using the COSMO-RS (Conductor-like screening model for real solvents) model and available data in the database, a model for absorption mechanism as well as process simulation-design tools. Several IL properties related to gas absorption, such as the Henryâ€™s constant, the viscosity, gas-solvent solubility data as well as toxicity of ILs are considered for evaluation of targeted decarbonization process and gas separation processes. Furthermore, process simulation is performed to evaluate the new IL-based decarbonization and gas separation technologies. Considering CO2 solubility, CO2/CH4 selectivity, gas solubility, gas/gas selectivities, viscosity and toxicity of ILs, several IL such as [Bmim][NTf2] has been found as a potential solvent from among more than 5000 ILs in the database. Based on reliable experimental data (available for only a fraction of the ILs present in the database), a rigorous thermodynamic model is established by comparing the simulation results with available experimental data. Different process flowsheet options obtained from targeted solvent behavior have been investigated to determine the best IL-based flowsheet. For the decarbonization process, compared

with the well-known MDEA (methyldiethanolamine) process for CO2 capture, the single-stage and multi-stage process alternatives with IL reduce the total energy consumption by 42.8%, 66.04%, respectively. For selected gas separation, the potential for energy reduction is even higher â€“ a reduction of more than 70% can be achieved. Net carbon emission as well as eco efficiency indicators are also evaluated to highlight the more sustainable and innovative aspects of the process designs.

455407 Developing Rapid Life Cycle Inventories for Chemical Processes

Abstract:

A product life cycle assessment (LCA) presents an opportunity to consider environmental effects over multiple categories for a system that includes the supply chain and product network (i.e., from cradle to grave). LCA benefits from the flexibility of being able to consider any boundaries for the system. However, this flexibility also means that each system studied can be different. Thus there is a need for rapid methods for developing life cycle inventories. Efforts at the U.S. Environmental Protection Agency Office of Research and Development are focusing on rapid inventory development for processes involved in a product LCA, including top-down and bottom-up methods. The top-down method uses Agency databases of releases to determine an average chemical process inventory. The bottom-up method uses process design methods, available process information, and emission factors to develop an inventory. For this latter method, a step-by-step procedure will be discussed. This presentation will explore the rapid inventory methods, focusing on acetic acid as a case study.

The views expressed in this presentation are those of the authors and do not necessarily reflect the views or policies of the U.S. Environmental Protection Agency.

472750 Sustainable Integration of Refineries and Biorefineries with Forest Plantations to Reduce Emissions

Abstract:

Fossil fuel production typically generates large amounts of greenhouse gas emissions, which are associated to the global warming problem. It is important to note that small increments in the global temperature involve drastic consequences in the world population. For this reason, some authors have stated that it is necessary to limit the greenhouse gas emissions from human activities. Furthermore, several authors have suggested alternatives to address this problem. For instance, biomass has been identified as a high potential renewable energy source for production of biofuels, chemicals and other products using several processing technologies. However, to consider the large scale production of biofuels from biomass, the entire supply chain must be analysed from the economic, environmental and social points of view. It is worth noting that the most of the papers published in the literature present several alternatives for the reduction of the overall environmental impacts during the energy production; however, did not consider a scheme to integrate two or more ways to reduce the global emissions in the supply chain. Therefore, this paper presents a mathematical programming model for the optimal planning for the production of fuels through refineries and biorefineries integrated with the forest plantations focused on decreasing the total greenhouse gas emissions, maximizing the net profit of the system and maximizing the generation of jobs simultaneously. This way, the proposed approach takes into account two ways for the reduction of greenhouse gas emissions, which is given by the production of biofuels in the biorefineries and the forest plantations to capture the CO_2. Additionally, the present approach considers several important issues such as the location of facilities, availability of raw materials, demand of products, inventory of the greenhouse gas emission, as well as the generated jobs for the different activities involved in the supply chain. It is important to mention that the problem is defined by a set of possible oil reservoirs distributed for the extraction of oil to be sent to refineries. Also, the supply chain includes a set of possible harvesting sites to produce different raw materials to be sent to biorefineries. In addition, there is the possibility to produce fuels from oil in the existing refineries, as well as installing necessary refineries to satisfy the fuel demand. On the other hand, it considers the possibility to install new biorefineries to produce biofuels and the option to use existing biorefineries. Furthermore, the proposed model incorporates pollution trading concepts to reduce emissions through a set of forest plantations with different capacities and sizes. It should be noted that the fuel demand, which can be satisfied via fossil fuels or biofuels, is received from several distribution centers. Therefore, the problem consists in determining the topology of the supply chain, the flowrates of products and raw materials, capacities of the refineries, biorefineries and forest plantations in order to obtain the best environmental, social and economic benefits.

The model is formulated as a mixed integer linear programming problem, which involves three objective functions in order to perform the sustainable planning of the production process of fuels in Mexico considering reduction of greenhouse gas emissions, mainly CO_2. In general way, the proposed mathematical formulation is composed of constraints for mass balances, capacity limitations, distribution, carbon emissions and capture as well as objective function equations. The economic objective function considers the maximization of the total profit, which is the sum of the profit from refineries, biorefineries and forest plantations. The profit of refineries and biorefineries includes the revenue from the sale of products, and all the involved costs such as the cost of raw materials for refineries and biorefineries, the operating costs, the capital costs for implementation of new refineries and biorefineries, the transportation costs for raw materials from harvesting places to processing plants, and the transportation costs for products. It also considers a cost for the emissions; for instance, the refineries should pay the forest plantations to reduce their emissions. In case of forest plantations, the profit considers the revenue for the payment that refineries or biorefineries have for their emissions minus the costs of their implementation. The environmental objective is to minimize the total CO_2 emissions, which considers the sum of total emissions from refineries and biorefineries minus the emissions captured by the forest plantations. The total CO_2 emissions from refineries include the ones produced in all the involved activities, including oil extraction, oil refining, transportation of products as well as the emissions obtained from burning petroleum fuels. Whereas the emissions of the biorefineries include the emissions for each of the activities involved in biorefineries such as emissions of CO_2 for biomass transportation, production in biorefineries, emissions produced by the transportation of bioproducts from biorefineries to distribution centers, emissions for the use of bioproducts and the captured emissions of CO_2 by the biomass growth. Furthermore, the captured CO_2 emissions by the eco-industries are given by a unitary factor multiplied by the total number of trees for the forest plantations. Additionally, the model considers a social objective function, which is the maximization of the total generated jobs for the implementation of the proposed scheme. This objective function takes into account the jobs generated in refineries, biorefineries and forest plantations. The jobs of refineries include the ones for processing and installing only in the new refineries, it is important to note that the transportation jobs are not considered because they are given in the transportation network for the existing refineries. On the other hand, the jobs of biorefineries include the ones for biomass plantations, biomass transportation, installing and processing of new biorefineries. With respect to the forest plantations, the considered jobs correspond to the needed people to take care of the plantations.

The results have shown that the relationship between the environmental and the social objectives is inversely proportional, while the behaviour of the economic objective with respect to the environmental objective is different. Also, the results illustrate that petroleum refineries are more economically attractive than the biorefineries, although the biorefineries can satisfy the fuel demand with a minor environmental cost. Additionally, based on the results, it is possible to conclude that the forest plantations can be important sinks for the CO_2 emissions from the fuel production process, which can be afforded by the refineries.

455577 A Hybrid Framework for Process Synthesis-Design: Application to Biorefineries

Abstract:

As energy demand and emissions increase steadily, the need for more sustainable processes is becoming evident. To this end, innovation in synthesis and design of processing routes needs to take into account raw materials availability, existence of appropriate processing technologies and products that have demand and use. When more than one alternative (technology or process interval) exists for one or more processing steps, a processing network results from which the optimal processing route needs to be determined.

The synthesis and design of processing networks for any chemical process in general and biorefineries in particular is an industry-relevant problem. Mathematical programming techniques, where the model equations representing all alternatives present in the network are included as constraints within an optimization problem, are commonly applied to solve this type of synthesis-design problems. The determination of the optimal processing route from a network (superstructure) of alternatives therefore involves decision-making through continuous variables (such as mass flow-rates and separation factors) and discrete variables (such as selection of an alternative in a processing step). The size of the problem depends on the number of alternatives included within the network, while, the complexity of the problem depends on the type of the models and decision variables used to represent the process intervals. Within the context of a biorefinery, considering a network involving a maximum of NS processing steps, NI process intervals (alternative) organized in processing steps, NF different biomass (raw materials), NP different products, and, allowing for processing step bypass and material flow recycle, would result in a mixed integer nonlinear programming problem. The size of the problem depends on the values of NS, NI, NF, NP and NC (number of chemicals). Note also that the models for each process interval also include model parameters that would need to be specified before any problem can be solved.

The objective of a hybrid framework for biorefinery synthesis-design is to provide the means to study the feasibility of establishing biorefineries as a function of biomass availability, location and characteristics, product types (fuels and/or chemicals), available technologies (use of different types of conversion technologies) and many more. The requirements for such a framework are: availability of appropriate numerical solvers for MILP and MINLP problems; a tool to generate the network model equations and to formulate the problem specific mathematical programming problems; and a database of information-knowledge-data, such as, different types and sources of biomass, different types of products, different types of known conversion technologies, processing routes and prices. This hybrid framework is based on a previous work (Quaglia et al. 2012) and has been extended in order to accommodate a wider range of process synthesis-design problems that can be formulated and solved. Key additions to the framework are a data management system that allows the generation of the superstructure (network) based on the biorefinery case study under investigation and retrieval of location-dependent data. A data structure has been designed and implemented in a knowledge base, which contains all necessary information for a range of problem formulations and their solution. The knowledge base currently contains 11 types of biomass, from 10 different locations, 102 intervals representing processing technologies and 9 products. It is a dynamic knowledge base that is ever-growing as more data are being collected and more problems are being solved. Moreover, once data is stored in it, superstructures are generated systematically including only the relevant alternatives for a specific problem definition with respect to a specified location. In addition, a user-friendly interface has been developed covering several of the steps (Bertran et al. 2016).

The main features and application of the hybrid framework are illustrated through the formulation and solution of a multi-feedstock multi-product problem. Six biomass-based raw materials are considered, yielding various products, in eight geographical locations; this leads to a superstructure with 9 processing steps (NS) and 35 processing intervals (NI). The solution is a (set of) raw material(s), a (set of) products, a location and the process configuration. Furthermore, several scenarios, such as minimum waste, maximum profit and different product demands, derived from the main problem have been investigated and the location dependency highlighted. A second case study involving different biomass to produce liquid fuel products is also briefly highlighted, which involves 2 feedstocks (NF), 2 products (NP), 13 processing steps (NS) and 34 processing intervals (NI), which leads to 79118 equations. Note that although the hybrid framework was originally designed for chemical flowsheet synthesis-design and extended for biorefineries, it is generic and has also proven useful for other processes such as carbon dioxide utilization and waste-water treatment networks.

References

Bertran M., Frauzem R, Zhang L, Gani R. A generic methodology for superstructure optimization of different processing networks. In: Kravanja Z, editor. Proc 26th Eur Symp Comput Aided Process Eng â€" ESCAPE 26. PortoroÅ¾, Slovenia: Elsevier B.V.; 2016.

Quaglia A, Sarup B, Sin G, Gani R. Integrated business and engineering framework for synthesis and design of enterprise-wide processing networks. Computers & Chemical Engineering, 2012, 38:213â€"23.

461880 A New Heat Integration Model for Simultaneous Utility and Total Heat Exchanger Area Targeting

Abstract:

The heat integration between cold and hot process streams is beneficial because it minimizes the utility consumption and increases energy efficiency. There are two approaches to this problem, sequential/decomposition and simultaneous. In the sequential approaches, heat integration is performed after the chemical process has been designed (all the process stream conditions are determined), and it is performed in three sequential steps to determine (1) the minimum utilities for the entire system, (2) minimum number of heat exchangers, and (3) total heat exchanger area and total cost. The major limitations of the sequential approaches are that they neglect (1) the interactions between the chemical process and heat integration subsystems, and (2) the trade-off among utility consumption, number of exchangers, and area. To address the first limitation, simultaneous approaches have been proposed [1, 2], where process synthesis and utility targeting (i.e. minimize total unity consumption under variable stream conditions) are considered simultaneously. However, these approaches do not take into account the tradeoff between energy cost and capital investment (e.g. heat exchanger area) in the heat integration subsystem. Another approach that addresses both limitations is the formulation of rigorous mixed-integer nonlinear programming (MINLP) superstructure heat exchanger network synthesis (HENS) model in which the utility consumption, heat exchanger area, and network configurations are simultaneously optimized [3]. However, this approach can lead to suboptimal solutions and it is computationally expensive when adopted for simultaneous process synthesis and HEN optimization.

To address the aforementioned challenge, we propose a MINLP optimization model for simultaneous utility and area targeting. The model can handle process streams with variable temperatures and flow rates so that it can be integrated with process synthesis models. The model is based on a mathematical representation of the balanced composite curves (BCCs) in the pinch analysis. The cold and hot composite curves (including cold and hot utilities) are each constructed from the cumulative enthalpies from low to

high temperatures using implicit orderings. The BCCs are then constructed by "bringing" the two composite curves together but allowing the heat recovery approach temperature (HRAT) to vary. The BCCs are divided vertically into enthalpy intervals wherever the curves have slope changes, which corresponds to a process stream starting or finishing along the BCCs. Therefore, the total number of intervals needed is twice the amount of process streams. It is important to note that if any one of the two BCCs has a slope change (a "kink" in the graphic representation), we calculate the temperature of the other BCC at the same enthalpy. In each interval, it is assumed that the hot streams and cold streams transfer heat vertically. Using the boundary temperatures and enthalpies at each interval, we can calculate the mean temperature difference and the approximate heat exchanger area needed at each interval. Finally, the objective is to minimize total area cost and utility

The nonlinearities in this model come from either the bilinear [flow rates]*[temperature difference] terms, or the terms in the area cost calculation that appear in the objective function. If either the stream flow rates or temperatures are fixed, the feasible space will be defined strictly by linear constraints, and the only nonlinearities will be in the objective function. In the case that flow rates, temperatures, and HRAT are all fixed, the model reduces to the problem table algorithm in the pinch analysis.

The proposed model can be generalized to consider intermediate utilities. Further, we extend the model so that it handles process streams with unknown cold/hot classifications. In other words, the model can deal with streams that cannot be determined as cold/hot streams prior to solving the optimization problem, which is in particular useful when the proposed model is integrated with a chemical process superstructure model that involves the (de)activation of units.

We apply the proposed model on several examples that involve process streams with variable flow rates and temperatures. Finally, we illustrate its applicability by integrating it with a rigorous biofuel production superstructure model.

References

1. Duran, M.A. and I.E. Grossmann, *Simultaneous-Optimization and Heat Integration of Chemical Processes.* Aiche Journal, 1986. **32**(1): p. 123-138

2. Grossmann, I.E., H. Yeomans, and Z. Kravanja, *A rigorous disjunctive optimization model for simultaneous flowsheet optimization and heat integration.* Computers & chemical engineering, 1998. **22**: p. S157-S164

3. Yee, T.F. and I.E. Grossmann, *Simultaneous optimization models for heat integration—II. Heat exchanger network synthesis.* Computers & Chemical Engineering, 1990. **14**(10): p. 1165-1184

459837 Optimal Composition of Biogas for Methanol Production Via Dry Reforming

Abstract:

Biogas is an interesting fuel that is produced from anaerobic digestion of organic wastes. Its composition is mainly methane, with a certain amount of carbon dioxide and other species in smaller quantities like H_2S or NH_3 [1]. Therefore, it can be considered as a source of methane. Methane or natural gas has been used as energy source or as fuel, but it has also been used as a raw material for the production of syngas. There are a number of alternatives to process methane. We can use steam to reform methane, an endothermic process. Alternatively, oxygen can be used to partially oxidize the methane. Autoreforming combines both, steam reforming and partial oxidation, so that the process is adiabatic. Lately, another technology has been added to the portfolio, dry reforming using carbon dioxide. It is also an endothermic process with a lower yield to hydrogen. The hydrogen to CO ratio varies widely depending on the method used. This ratio is key for the proper production of different species. For instance, the production of methanol requires a H_2:CO ratio of 2, the production of FT-liquids suggests values from 1 to 2 depending on the length of the hydrocarbon chain and the operating conditions at the reactor. The production of ethanol and DME needs a H_2:CO ratio of 1 [2,3].

Biogas already has CO_2 within its composition. However, the syngas produced is expected to show a low ratio H_2:CO. To tune this ratio we can follow different alternatives. Biogas composition is variable and can be adjusted. Furthermore, traditionally dry reforming of methane has been combined with other reforming modes [4-6].

In this paper, we present a mathematical optimization approach for the optimal dry reforming of biogas for the production of methanol. The raw biogas is cleaned up before reforming. We allow for the use of steam and variable biogas composition to increase the hydrogen content of the syngas. Part of the biogas is used to provide energy for the process. Next, the unreacted hydrocarbons and CO_2 are removed. Subsequently, its composition may be adjusted (using either water gas shift reaction or pressure swift adsorption if there is an excess of hydrogen). Finally, methanol is synthesized. The problem is formulated as an NLP with simultaneous heat integration for the optimal biogas composition and methanol production. Two objective functions are considered: a simplified production cost and we proposed an environmental one based on carbon footprint considering the contributions to CO_2 emissions due to the water usage as cooling agent, raw material or steam production and CO_2 consumption and release across the process.

Biogas is expected to have around 50-52% of CH_4 and 45-47% of CO_2, depending on the objective. The production cost of methanol is $1.7/gal, for a plant size that uses 10% of the potential biogas to be produced in Madrid (Spain), with an investment of $46 MM. The production cost is becoming competitive with the current price of methanol ($0.9/gal) with the advantage of its sustainability. The results obtained for both optimizations show a negative carbon footprint, creating a net capture of CO_2 while producing a valuable intermediate for chemicals or fuels.

References.

[1] Japaraju, P., Rintala, J. 17 - Generation of heat and power from biogas for stationary applications: boilers, gas engines and turbines, combined heat and power (CHP) plants and fuel cells 404-427, In The Biogas Hand book 2013

[2] Martín, M., Grossmann, I.E., On the systematic synthesis of sustainable biorefineries Ind. Eng. Chem. Res. **2013,** 52 (9), 3044-3064

[3] Peral, E., Martín, M. Optimal production of DME from switchgrass based syngas via direct synthesis. Ind. Eng. Chem. Res., **2015**, 54, 7464-7475 [4] Lim, Y.; Lee, C.J:; Su Jeong, Y.; Song, I-H.; Lee, C.J.; Han, C. Optimal Design and Decision for Combined Steam Reforming Process with Dry Methane Reforming to Reuse CO_2 as a Raw Material *Ind. Eng. Chem. Res.* **2012**, 51, 4982–4989

[5] Vernon, P. D. F.; Green, M. L. H.; Cheetham, A. K.; Ashcroft, A. T. Partial oxidation of methane to synthesis gas, and carbon dioxide as an oxidising agent for methane conversion. *Catal. Today* **1992**, 13 (2–3), 417–426.

[6] Song, C.; Pan, W. Tri-reforming of methane: a novel concept for catalytic production of industrially useful synthesis gas with desired H2/CO ratios. *Catal. Today.* **2004**, 98 (4), 463–484.

462416 Production of Benzene, Toluene, and Xylenes from Natural Gas Via Methanol: A Process Synthesis and Global Optimization Approach

Abstract:

Efficient utilization of domestic resources, such as natural gas, is crucial toward meeting U.S. energy independence targets and addressing energy security, energy affordability, and the generation of lower-carbon energy. [1] As a result of the developments within the shale gas industry, recent efforts have focused on converting natural gas into more valuable products. [2,3] A subset of these valuable products are aromatics, which represent approximately one-third of the market for commodity petrochemicals. [4] Between 2005 and 2008, the global demand for benzene, para-xylene, ortho-xylene, and meta-xylene was approximately 40, 26, 6, and 0.4 million metric tons per year, respectively. [4] Benzene is a precursor to over 250 different products, including styrene and phenol. Para-xylene, the most valuable xylene isomer, is converted into polyethylene terephthalate (PET) fibers, resins and films. [5] Opportunities exist for widespread penetration of natural gas based chemical refineries, albeit with challenges associated with which types of chemicals to produce and which technologies to invest in. In order to address these challenges, we propose novel methods of producing aromatics using a combined process synthesis and global optimization approach. [6]

We present a deterministic global optimization based process synthesis framework to perform a comprehensive technoeconomic and environmental assessment of a natural gas based aromatics refinery. [6] Several direct and indirect natural gas conversion technologies, including autothermal reforming, steam reforming, and partial oxidation to methanol, are investigated. Numerous novel, commercial, and/or competing technologies are modeled within the framework, including methanol-to-aromatics, toluene alkylation with methanol, selective toluene disproportionation, toluene disproportionation and transalkylation with heavy aromatics, para-xylene separation via adsorptive separation or crystallization, isomerization of xylenes, and dehydrocyclodimerization of liquefied petroleum gas, among others. To address the economic barriers often associated with the utilization of alternative feedstocks, we develop a novel branch-and-bound global optimization algorithm that is capable of determining the optimal technologies to produce aromatics from natural gas at the highest profit. [6,7] Our approach provides an adequate baseline for comparing competing technologies and identifying bottlenecks for natural gas conversion based technologies. Several key aspects of the algorithm are discussed, multiple case studies are presented to investigate the effect of refinery capacity and product output, and the key topological decisions are described.

[1] Floudas, C. A.; Niziolek, A. M.; Onel, O.; Matthews, L. R. Multi-Scale Systems Engineering for Energy and the Environment: Challenges and Opportunities. AIChE Journal 2016, 62, 602-623.

[2] Mokrani, T.; Scurrell, M. Gas conversion to liquid fuels and chemicals: the methanol route-catalysis and processes development. Catalysis Reviews 2009, 51:1-145

[3] Baliban, R. C.; Elia, J. A.; Floudas, C. A. Novel natural gas to liquids (GTL) technologies: Process synthesis and global optimization strategies. AIChE Journal 2013, 59, 505â€"531.

[4] de Klerk, A. Fischer-Tropsch Refining; Wiley-VCH Verlag & Co. KgaA: Weinheim 2011

[5] Niziolek, A. M.; Onel, O.; Elia, J. A.; Baliban R. C.; Floudas, C. A. Coproduction of Liquid Transportation Fuels and C6_C8 Aromatics from Biomass and Natural Gas. AIChE Journal 2015, 61, 831-856.

[6] Niziolek, A. M.; Onel, O.; Floudas, C. A. Production of Benzene, Toluene, and Xylenes from Natural Gas via Methanol: Process Synthesis and Global Optimization. AIChE Journal 2016, 62 (5), 1531 â€" 1556.

[7] Baliban, R. C.; Elia, J. A.; Misener, R.; Floudas, C. A. Global Optimization of a MINLP Process Synthesis Model for Thermochemical Based Conversion of Hybrid Coal, Biomass, and Natural Gas to Liquid Fuels. Computers and Chemical Engineering 2012, 42, 64-86.

452419 Extensive Sensitivity Analysis and Stochastic Global Optimization for Renewable Energy Businesses Under Operational Level Uncertainties

Abstract:

The current energy supply around the world and the US is highly dependent on fossil fuels. Even though in the past year oil price dropped from its previous attractive peak, cheap oil prices should not be a temptation for stepping away from renewables. The characteristic volatility and future uncertainty of petroleum should encourage governments, investors, shareholders and decision-makers to start preparing for the long farewell mankind has to say to fossil fuels. The modern portfolio of available renewables includes second generation biofuels produced from biomass which appears to be a sustainable solution for mobile energy services giving also an innovative usage of organic waste. However, market volatility, technological uncertainty, and limited experience in renewable energy processes can prevent the development of this burgeoning industry.

By processing biomass in biochemical or thermochemical pathways, renewable energy facilities are able to produce fuels, value-added chemicals, heat and bioelectricity. First and second generation biorefineries are already operating worldwide, and process systems engineering has contributed over the past few years in addressing the complexity of modelling and decision making to optimize and make cost-effective renewable energy projects. Usually, the optimization of renewable energy businesses considers a deterministic design approach in which all the model parameters are assumed to be known from literature. However, during conceptual design there is lack of information and reliability, especially in novel or recently developed processes. When ideality supposed, external factors that might affect the process behavior and the project profitability are mistreated. Underestimating these limitations may lead into non-optimal designs and generate extra expenses or difficulties during startup and operation. Technological risk management should be considered during decision making since it permits to select optimal conditions when uncertainties are taken into account. Therefore, the global optimization problem should be addressed considering risk management at technological level in initial stages. It will allow managers to have a conservative and reliable method

for making and supporting critical decisions. To identify significant uncertainty sources in the process, a sensitivity analysis of the parameters is required.

Our proposed framework is composed of different sections. First, a global sensitivity analysis based on Sobolâ€™s method is performed to identify the significant sources of uncertainty in the kinetic parameters involved in the biological reactions of biofuels and value-added chemicals production. The method is computationally efficient since it averages the results of the evaluated functions and increases the original size of the sample. The first and total sensitivity indices are obtained by calculating the total variance and the individual variances generated when each parameter changes separately. One of the main characteristics of the current work is to explore the efficiency of two different approaches for selecting significant uncertain parameters: 1) evaluating all parameters simultaneously, 2) evaluating parameters independently per product pathway. Then, the identified uncertain parameters are utilized to optimize the profitability of the plant under uncertainty. A stochastic global optimization technique (*ParLMSRBF-R*), similar to a metaheuristic method, is used to maximize the cash flow after tax of a renewable energy business while considering uncertainty and risk management. In this iterative optimization approach the decision variables are selected intelligently with the help of a Radial Basis Function as the response surface or surrogate model. Finally, the optimal points obtained are compared statistically with optimal values from other methods, e.g. Monte-Carlo based optimization. To evaluate the efficacy of this novel decision-making framework, a hypothetical multiproduct lignocellulosic biorefinery is used as a case study.

This optimization framework appears to be more effective than other optimization procedures, e.g. Monte-Carlo based optimization. The algorithm searches for the optimal operating conditions using a fitted and continuously updating Radial Basis Function without losing the clarity of the actual mechanism of the process. The algorithm selects candidate points to be evaluated near the current best solution neighbourhood achieving exploration and exploitation in the solution domain. In terms of computational effort, the optimization algorithm employed requires less expensive function evaluations to reach global optima. In fact, *ParLMSRBF-R* achieves optima with less function evaluations in comparison to Monte Carlo-based optimization approach. In addition, the objective value improves in contrast to the best value achieved with Monte-Carlo based optimization.

464830 Toward a Unified Method for Process Design, Integration and Intensification

Abstract:

Until now, process synthesis and integration methods have focused on classical unit operations, where units such as separators, reactors and storage tanks are integrated but they are not merged to perform multiple tasks *in situ*. Due to the historically standalone development of conversion, separation and storage technologies, excess energy is consumed in isolated units. In this work, we propose an original approach for process design that leverages on the construction of flowsheets using building blocks, where each block performs specific operation (e.g., separation, conversion, storage, etc.) via enabling materials (e.g., membrane, catalyst, adsorbent, etc.). An assembly of blocks of the same type obtains a classical unit, while an assembly of blocks with different types results in a hybrid or intensified unit. Our building block-based flowsheet superstructure not only contains all alternatives typically considered for process synthesis and integration, but it also systematically identifies and incorporates all plausible intensification alternatives at the flowsheet level. To this end, our proposed method can potentially lead to a unified approach for process synthesis, integration and intensification without exhaustive enumeration. The overall problem is formulated as a single mixed-integer nonlinear optimization (MINLP) problem which can be solved using commercial solvers. While the mass and energy flows between the blocks are modeled using continuous variables, the selection of the operations and their enabling materials are discrete decisions. The objective is to synthesize a process with intensified units by minimizing or maximizing a process metric given the feed and product specifications, feed and product prices, material properties and bounds on flow rates. We demonstrate the applicability of the proposed method for the innovative design of several emerging technologies for clean and sustainable energy, such as carbon capture and conversion, natural gas utilization, and multicomponent gas separation. We also demonstrate that the simultaneous synthesis and intensification approach leads to substantially smaller, cleaner, safer, and more energy-efficient designs.

469198 Work and Heat Exchange Networks - from Thermodynamic Insight to Optimization Based Design Procedures

Abstract:

Design of Work and Heat Exchange Networks (WHENs) represents an emerging area of research, and it involves three distinct activities that briefly can be described as: (1) Heat Integration, (2) Work Integration, and (3) Pressure Integration. The first is a classical Process Synthesis problem trying to recover as much heat as possible from hot process streams to heat up cold process streams while selecting an optimal set of hot and cold utilities to provide additional heating and cooling. The second involves the use of work (power) from expansion processes to supply work (power) to compression processes. This work integration can be done on the same shaft (e.g. the use of so-called companders for the extraction of heavier hydrocarbons in natural gas liquefaction plants) or separate shafts where electricity is produced and consumed. The third involves the utilization of additional heating and cooling resulting from changing the pressure of process streams, and this has been the focus of research in our group for the last 10 years [1].

The sparse and relatively recent literature on WHEN design can be grouped into publications on Work Exchange Networks (WENs, e.g. [2]), Work and Heat Exchange Networks (WHENs, e.g. [3]) and Work, Heat and Mass Exchange Networks (e.g. [4]). The extended problem definition for WHEN synthesis that has been used in our research group is:

Given a set of process streams with supply and target states (temperature and pressure), as well as utilities for power, heating and cooling; design a network of heat exchangers, expanders and compressors in such a way that exergy consumption is minimized or exergy production is maximized.

At this stage of methodology development, exergy is used in the objective function since energies of different qualities (thermal and mechanical) are involved. Later, when moving from a thermodynamic approach to an optimization-based approach, a more appropriate objective function involving total annual cost will be used.

The motivation for including compression and expansion processes in the traditional heat recovery problem is that considerable additional energy savings can be obtained. The main issue is to sacrifice modest amounts of mechanical energy to make significant savings in thermal energy, and/or increase work production by efficient utilization of low temperature thermal energy. As such, the research on WHENs has strong relations to heat pumping [5] and refrigeration cycles. It is also closely related to the research on the so-called self-heat recuperation technology [6].

In developing a design methodology for WHENs, the Appropriate Placement concept from Pinch Analysis applied to pressure changing equipment (primarily compressors and expanders) has been a key issue. New challenges compared to traditional Heat Exchanger Network (HEN) design are: (1) The thermodynamic path from supply to target state is unknown (changing both temperature and pressure), (2) Pinch points and utility demand may (often will) change, so that no explicit targeting is possible, and (3) stream identity (hot/cold) may (often will) temporarily change.

A set of 4 theorems have been proposed and proven using thermodynamics and mathematics for expanders above ambient temperature [7], with minor modifications for compressors above ambient as well as expanders and compressors integrated below ambient. Thus, 4 scenarios with a total of 16 theorems are handled with considerable symmetry between the 4 cases. Simultaneous compression and expansion has also been handled with the introduction of a new theorem that deals with the integration sequence for compression and expansion [8]. In line with earlier findings [9], Pinch Expansion and Compression (i.e. expander and compressor inlet temperature should be equal to the Pinch temperature) will often result in significant exergy savings.

An early version of a design procedure for WHENs, based on the Appropriate Placement concept with detailed guidelines from the above mentioned theorems, uses the Grand Composite Curve (GCC) to quantify the maximum amount of heating/cooling that can be accommodated from compression/expansion. The procedure is time consuming even for very small cases and a number of simplifying assumptions. In addition, using the GCC with its modified temperatures ignored the different features for hot and cold streams as well as the relative position of supply and target temperatures to the pinch temperature. As a result, minor (in most cases) errors in estimation of utility consumption were introduced.

To overcome problems related to prohibitive time consumption and inaccuracies in utility estimation due to ignoring the identity of streams (hot/cold), work has been initiated to develop new superstructures with corresponding mathematical models of varying complexity (from LP to MINLP) that can be used in an optimization framework [10]. In fact, early applications of such optimization models have revealed new insight on the WHEN synthesis problem. One example is that Pinch Compression and Expansion should start at the Pinch temperature (hot or cold) that corresponds to the identity (hot or cold) of the stream segment prior to compression or expansion, and not to the original identity of the stream. Two classes of optimization models were developed, i.e. with and without utilizing the insight provided by the theorems. The latter class was used as a final validation of the theorems, while the former class offered advantages in model complexity and thus reduced time to solve the optimization problem.

In conclusion, based on several case applications using the original thermodynamic approach as well as the optimization framework under development, the original hypothesis that Pinch Compression and Expansion represent optimal placement of pressure changing equipment is valid fully or partly in the majority of cases. However, there are cases where expansion/compression at hot/cold utility temperature or ambient temperature provides better solutions. No other temperatures result in WHENs with lower exergy consumption (or higher exergy production).

Preliminary studies on industrial applications such as CO_2 capture processes [11] show that considerable energy savings can be achieved following the new design procedures. An interesting feature of this research is that Exergy Analysis has been moved from post-design evaluation to conceptual stage design.

References

[1] A. Aspelund, D.O. Berstad and T. Gundersen, An Extended Pinch Analysis and Design Procedure utilizing

Pressure based Exergy for Subambient Cooling, *Applied Thermal Engineering*, 27, 2633-2649, 2007.

[2] M.S. Razib, M.M.F. Hasan and I.A. Karimi, Preliminary Synthesis of Work Exchange Networks, *Comput. & Chem. Engng.*, 36, 262-277, 2012.

[3] V.C. Onishi, M.A.S.S. Ravagnani and J.A. Caballero, Simultaneous Synthesis of Heat Exchanger Networks with Pressure Recovery, *AIChE Journal*, 60, 893-908, 2014.

[4] R. Dong, Y. Yu and Z. Zhang, Simultaneous Optimization of integrated Heat, Mass and Pressure Exchange Network using Exergoeconomic Method, *Applied Energy*, 136, 1098-1109, 2014.

[5] C. Fu and T. Gundersen, A Novel Sensible Heat Pump Scheme for Industrial Heat Recovery, *Ind. Eng. Chem. Res.*, 55, 967-977, 2016.

[6] Y. Kansha, N. Tsuru, K. Sato and A. Tsutsumi, A Self-Heat Recuperation Technology for Energy Saving in Chemical Processes, *Ind. Eng. Chem. Res.*, 48, 7682-7686, 2009.

[7] C. Fu and T. Gundersen, Integrating Expanders into Heat Exchanger Networks above Ambient Temperature, *AIChE Journal*, 61, 3404-3422, 2015.

[8] C. Fu and T. Gundersen, Correct Integration of Compressors and Expanders in above Ambient Heat Exchanger Networks, *Energy*, in review, 2016.

[9] T. Gundersen, D.O. Berstad and A. Aspelund, Extending Pinch Analysis and Process Integration into Pressure and Fluid Phase Considerations, *Chem. Engng. Trans.*, 18, 33-38, 2009.

[10] C. Fu, P. Maurstad Uv, B. Nygreen and T. Gundersen, Compression and Expansion at the right Pinch Temperature, *Chem. Engng. Trans.*, 51, 2016.

[11] C. Fu and T. Gundersen, Heat and Work Integration: Fundamental Insights and Applications to Carbon Dioxide Capture Processes, *Energy Conversion and Management*, 2016, DOI: 10.1016/j.enconman.2016.04.108.

470562 Analysis, Synthesis and Optimization of Multiple-Effect Evaporation Systems Using Mathematical Programming

Abstract:

Evaporation processes are used within the process industries in order to produce concentrated products by evaporating part of water from different feeds-diluted water solutions. Concentrated products can represent final products (fruit and vegetable juices) or intermediate products in cases that crystalized (salt, sugar) or dried (milk powder) final products should be produced. Large amounts of steam and cooling water are consumed in these processes. In order to reduce energy and water consumption within evaporation processes different systems can be applied, namely, multiple-effect evaporation, vapor recompression (thermal and mechanical) or their combinations. Additionally, these processes can be integrated with other process subsystems in order to achieve improved energy and water integration.

To address these issues different computer-aided tools have been proposed. However, most studies have focused on analysis and simulation of evaporation processes. Some of the initial studies [1, 2] considered the synthesis of evaporation processes in order to develop tools for computing the minimum utility use for a multiple-effect evaporation system, which was heat-integrated with process hot and cold streams. These studies were based on a modified grand composite curve and heat-path diagram. Also, the focus of the recent works have been on multiple-effect evaporation systems [3] and their energy integration with the background processes in order to minimize the energy consumption within the overall system [4]. These studies have motivated us to further expand research in this direction, by applying mathematical programming approach for the analysis of existing and the design of new evaporation systems as well as their heat integration with other process subsystems or process streams.

The main goal of this paper is to develop models based on mathematical programming that can be applied for the analysis, synthesis and optimization of multiple-effect evaporation systems. The proposed models will be developed in General Algebraic Modeling System (GAMS). The developed models will enable

examination of different scenarios of multiple-effect evaporation in order to address the analysis of existing, retrofit and/or design new evaporation process. Within the proposed framework, a network consisting of a multiple-effect evaporation system and heat exchanger network will be investigated in order to achieve the improved heat integration within the overall system. Two strategies will be considered to achieve this task, namely, sequential and simultaneous. The developed models will be tested on several examples, and also applied to different feed streams. New results are expected to be obtained within this field.

Keywords: multiple-effect evaporation, analysis, synthesis, optimization, mathematical programming.

Acknowledgment

The authors are grateful to the Swiss National Science Foundation (SNSF) and the Swiss Agency for Development and Cooperation (SDC) for providing financial support within the SCOPES 2013 2016 (Scientific Co-operation between Eastern Europe and Switzerland) joint research project (CAPE EWWR: IZ73Z0_152622/1).

References:

[1] Hillenbrand JJB, Westerberg AW. The synthesis of multiple-effect evaporator systems using minimum utility insights—I. A cascaded heat representation. Computers & Chemical Engineering. 1988;12:611.

[2] Westerberg AW, Hillenbrand JJB. The synthesis of multiple-effect evaporator systems using minimum utility insights—II. liquid flowpattern selection. Computers & Chemical Engineering. 1988;12:625.

[3] Khanam S, Mohanty B. Energy reduction schemes for multiple effect evaporator systems. Applied Energy. 2010;87:1102.

[4] Sharan P, Bandyopadhyay S. Energy Integration of Multiple Effect Evaporators with Background Process and Appropriate Temperature Selection. Industrial & Engineering Chemistry Research. 2016;55:1630.

473748 Process to Planet Framework for Sustainable Design: Systematic Approach for Developing a Multiscale Model and for Multiobjective Optimization

Abstract:

Over the last decade, sustainable process design (SPD) research has utilized life cycle assessment(LCA) methods to incorporate an environmental dimension in process systems engineering(PSE). Conventional SPD involves a â €œbottom up approachâ€ of modeling the system of interest at a smaller scale with detailed process specific fundamental engineering models, while the upstream life cycle of the product is modeled at a larger scale using process based LCA. System boundary selection is a significant challenge for process LCA due toÂ infinite nature of the upstream life cycle network for any process [1]. Process based LCA can lead to significant truncation errors in life cycle resource consumption and emissions, leaving a wide margin for shifting of impacts across narrow system boundaries and leading to illusive optimal design solutions that may in fact be inferior due to incomplete life cycle information [2]. Hybrid LCA(HLCA) methods aim to address the boundary selection problem by combining process level life cycle inventories with environmentally extended input-output analysis(EEIO), which allows for a significant part of the product life cycle to be included in the assessment. The Process to Planet (P2P) framework [3] was developed by introducing the concept of integrated hybrid LCA [4] to SPD. Detailed engineering models of the primary process being designed and other key processes in the supply chain are integrated within a macroeconomic system represented by economic input-output (EIO) model. This ensures solution of process design problems while reducing chances of environmental impacts being neglected due to incomplete life cycle inventories. While bridging the knowledge gap between PSE and LCA, the P2P framework not only improves upon the life cycle assessment component of SPD, but also addresses a major weakness in Hybrid LCA. HLCA is limited to the impact assessment stage of life cycle assessment. Very few studies have dealt with the fourth and final stage of LCA, improvement analysis. This situation is lamentable because inventory scope and impact analysis provide little benefit if not followed by recommendations and decisions that should be adopted to reduce environmental burden. The P2P framework can be visualized as an extension of HLCA method where it enables the incorporation of design variables through process design and allows optimization of the system to improve its environmental footprint. It was successfully demonstrated by application to a corn based ethanol production plant and highly insightful results were obtained that proved the efficacy of the framework [2].

This presentation proposes extensions and modifications that address some of the inadequacies that are inherent in the original P2P framework. While strictly applied to P2P, these extensions can be extrapolated to answer challenges that exist in broad areas of literature, like HLCA. The selection of significant processes that are modeled at the process based life cycle scale and the insignificant processes that are studied at the economy scale is arbitrary. In other words, delineation of the boundary between the different scales of the P2P framework was a subjective task. Previously, this was done based on data availability, accuracy and detail requirements and the complexity of the overall problem which is proportional to the solving time [5,6]. The problem was first mentioned and some methods were proposed for addressing it in a paper by Suh [7]. Lenzen proposed Structural Path Analysis (SPA) involving the decomposition of the Leontief Inverse for addressing this problem [8]. Treloar also used this method to study energy use of individual residential buildings in Australia [9,10]. Lave et al. [11] and Lenzen [12] used SPA to quantify the truncation errors in process-based LCA. A path exchange method using SPA was also developed as an alternative to the method of combining process LCA and IO LCA within integrated hybrid LCA as proposed by Suh. We intend to address this knowledge gap by proposing a path analysis algorithm where the P2P framework superstructure model generation is guided by a hierarchical structural path decomposition method which determines the coverage of the different scales. It is one of the first applications of SPA for solving the boundary selection problem involving the distribution of different upstream processes between multiple scales.

A large number of studies have dealt with multi objective optimization in process design and energy systems involving environmental and economic impacts. A comprehensive list of such literature can be found in Diwekar et al.[13] Reviewing these articles, it becomes apparent that it is necessary to simultaneously work towards both economic profitability and environmental sustainability of the system. Previously, the P2P framework focused only on the environmental dimension of process design. In this work, we propose a multi objective optimization problem, involving the trade-off between an economic objective and an environmental objective function. The novelty in this addendum is that it makes the P2P framework more fruitful for designing any generic process while exercising control over both the economic and environmental scope.

Preliminary results have provided insightful views on the application of SPA algorithm for model generation as well as exposed existing deficiencies to work on. While SPA is successful in mapping the upstream networks of the process to be designed and calculate their contributions to the entire life cycle, the quantification of the quality of the generated model still needs to be addressed. For example, considering two scenarios where an upstream sector is modeled at the IO model scale and process LCA scale respectively, we need to develop parameters that can quantitatively compare between the two cases and rank them objectively. As of now, subjective decisions are being used to build the model, but we intend to develop an algorithm based on error analysis or uncertainty information to determine the model quality. Multiobjective optimization solutions demonstrated that, highly erroneous optimal solutions and counter intuitive results are obtained from practicing conventional SPD approach rather than P2P. In addition to this, it was observed that there is a 59% chance of choosing an inferior solution over the optimal solution when solving the design problem using conventional SPD.

References

[1] Sangwon Suh. Functions, commodities and environmental impacts in an ecological economic model. 48:451-467, 2004.

[2] Rebecca J Hanes and Bhavik R Bakshi. Sustainable process design by the process to planet framework. AIChE Journal, 61(10):3320-3331, 2015.

[3] Rebecca J Hanes and Bhavik R Bakshi. Process to planet: A multiscale modeling framework toward sustainable engineering. AIChE Journal, 61(10):3332-3352, 2015.

[4] Sangwon Suh and Shinichiro Nakamura. Five years in the area of input-output and hybrid lca. The international journal of life cycle assessment, 12(6):351-352, 2007.

[5] H Hondo and S Sakai. Preliminary life cycle inventory analysis (pre-lci) using an economic input-output table. In The Fourth International Conference on EcoBalance, Tsukuba, Japan, pages 181-4, 2000.

[6] G Norris. Selecting and evaluating boundaries for life cycle assessment using economic input-output analysis. In Proceedings from the SETAC 17th Annual Meeting, Washington, DC, pages 17-21, 1996.

[7] Sangwon Suh, Manfred Lenzen, Graham J Treloar, Hiroki Hondo, Arpad Horvath, Gjalt Huppes, Olivier Jolliet, Uwe Klann, Wolfram Krewitt, Yuichi Moriguchi, et al. System boundary selection in life-cycle inventories using hybrid approaches. Environmental Science & Technology, 38(3):657-664, 2004.

[8] Manfred Lenzen. A guide for compiling inventories in hybrid life-cycle assessments: some australian results. Journal of Cleaner Production, 10(6):545-572, 2002.

[9] Graham J Treloar. Extracting embodied energy paths from input-output tables: to-wards an input-output-based hybrid energy analysis method. Economic Systems Re-search, 9(4):375-391, 1997.

[10] GJ Treloar, PED Love, OO Faniran, and U Iyer-Raniga. A hybrid life cycle assessment method for construction. Construction Management & Economics, 18(1):5-9, 2000.

[11] Lester B Lave. Using input-output analysis to estimate economy-wide discharges. Environmental Science & Technology, 29(9):420A-426A, 1995.

[12] Manfred Lenzen. Errors in conventional and input-output based lifecycle inventories. Journal of Industrial Ecology, 4(4):127-148, 2000.

[13] Urmila Diwekar and Yogendra Shastri. Design for environment: a state-of-the-art re-view. Clean Technologies and Environmental Policy, 13(2):227-240, 2011.

462158 Application of a Shale Environmental Footprint Optimization Tool to Enhance Operational Excellence– a Case Study of Selecting Process Fuels with Lower Cost and Less Potential for Impacts to Power Chevron Operations

Abstract:

Unconventional resources represent significant business opportunities; however, their development is subject to much public and environmental scrutiny. Currently there are over 400 local bans and 2 state bans on hydraulic fracturing. Many countries besides the U.S. have placed moratoriums on hydraulic fracturing. The challenge and opportunity that the oil and gas industry is facing is: how to unlock the potential in unconventionals, achieve better economic and operational performance, while still fulfilling our commitment to communities and the environment.

The Shale Environmental Footprint Optimization Tool is a tool developed at Chevron to enable the integration of environmental considerations in process design and decision making related to shale development. It simulates shale operations at the play level, where material and energy flows of in-scope processes are aggregated together and converted to quantitative environmental indicators (i.e. energy consumption, water consumption, and air emissions). It is built with a commercial life cycle assessment (LCA) software (GaBi® Professional) that contains >4000 datasets, and adopts international standards (ISO 14040-10444). The tool also allows sensitivity and uncertainty analysis to better assess and interpret results under varying reservoir and operational conditions.

This tool has been used by business units to better characterize their operations, assess potential environmental risks associated with different decision alternatives, and understand the trade-offs. It also provides them a more holistic view across the value chain, and help them pinpoint major contributors to the total footprint. These insights help business partners prioritize R&D efforts, make informed decisions on mitigation options, and ultimately lead to reduced environmental impact. Externally, the tool can be used address stakeholder concerns and for benchmarking our performance.

In the presentation, we'll show how this tool has been used to assess real world situations (using field data and business scenarios) and generate information to support a critical decision – evaluating and identifying the preferred alternative fuel to power drilling rigs. Optimizing the selection of process fuels could result in significant environmental, social, and economic benefits. One of the scenarios we'll include is to switch from LNG to gas produced in the field for power generation, which may reduce truck traffic on the roads. Reducing trucks is beneficial to safety (MVC, avoid backing/maneuvering incident on well pad), cost (hauling and road repair), land disturbance (less roads and processing units), and emissions (air pollutants & spills). In some locations, it also means improved reliability and fewer burdens on business continuity.

GHG Life Cycle Assessment for the United Arab Emirates Electricity Sector Combining Optimization and Simulation Tools

Paper ID #: 466062

*Alberto Betancourt-Torcat, Mohammed Alkatheri, Ali Almansoori**

Department of Chemical Engineering, The Petroleum Institute, Sas Al Nakhl Campus, Abu Dhabi, P.O. Box 2533, United Arab Emirates.

*Corresponding Author's e-mail: aalmansoori@pi.ac.ae

Keywords: Greenhouse Gas, Life cycle assessment, Electricity sector, Economics, Optimization, Simulation.

Abstract

The United Arab Emirates (UAE) relies heavily on natural gas for power generation, and the demand for gas in a country with one of the highest per capita power consumption rates in the world position the electricity sector as the leading CO_2 emission source. Although the country is very rich in energy resources it is currently facing challenges to meet its domestic gas demand. Thus, the country has been forced to develop its large, but highly sour (high-sulfur content) gas reservoirs. Given the gas high-sulfur content, the sweetening process is particularly very energy and greenhouse gas (GHG) emission intensive compared with conventional gas resources. Additionally, the volume of sweet gas yield is low due to the high proportion of acid gases in the feed gas (e.g., over 35% of H_2S and CO_2 combined). This work quantifies the well-to-grid (WTG) GHG emissions per kilowatt hour (kWh) of domestic gas-based electricity generated in the UAE using mathematical models, simulation tools, and engineering principles. WTG emissions include emissions from sour gas extracting, sour gas processing, sweet gas product transportation, electricity generation, and electricity transmission in the UAE. Two types of gas resources were examined: conventional and ultra-sour gas. The results show that the WTG GHG emissions of domestic gas-based power generation in UAE ranges between 444 to 501 g CO_2 eq./kWh; placing the country in the low-mid range of values compared with the literature 413-587 g CO_2 eq./kWh.

Introduction

Over the last decade, the rapid economic and demographic growth of the United Arab Emirates (UAE) has pushed the country's electricity grid to its limits. Most of the UAE's generation capacity is natural gas-based [1]. Although natural gas is considered to be a relatively clean primary energy source, it is still a fossil fuel source that generates significant greenhouse gas (GHG) emissions. The UAE electricity demand reached a domestic record of 105 billion kWh in 2013, placing the country as one of the highest per capita power consumers in the world [2]. Nuclear and renewable power capacities are currently being added to increase the country's energy mix and reduce GHG emissions [3]. As a result of the steadily growing electricity demand, the UAE plans to increase natural gas production to meet the domestic demand. The majority of the growth will come from the country's largely sour (high-sulfur content) gas reservoirs. The UAE's gas reservoirs have a relatively high sulfur content, which makes it corrosive and difficult to process [4]. The process is particularly very energy intensive and environmentally challenging. Thus, the process is highly GHG-emission intensive, as well vast amounts of sulfur are generated as by-product [5].

There is currently a need to assess the environmental footprints of the UAE electricity sector in terms of its GHG emission life cycle. This could allow benchmarking the UAE's electricity sector against other nations, while finding suitable niches for GHG mitigation strategies. Moreover, since over 98% of the electricity generated in the UAE is gas-based, an analysis on the UAE's natural gas life cycle represents a key factor on determining the overall environmental impact of the electricity sector. The present analysis considers a well-to-grid (WTG) GHG lifecycle assessment; which includes the upstream and downstream operations. The upstream operations includes all activities before fuel is combusted at power plants. Such activities comprise: sour natural gas extraction, sour gas processing, and sweet gas transport to the power plants gate. On the other hand, the downstream operations include the electricity generation and transmission. For the present study, fundamental engineering principles were used to mathematically estimate process specific energy consumption and related GHG emissions; namely for sour gas extraction, sweet gas product transportation, and electricity transmission. Additionally, ProMax® v3.2 [6] was used to simulate the sour gas processing, while performing a parametric analysis to improve the performance of these processes. Also, an optimization mathematical model developed in the General Algebraic Modeling System (GAMS®) [3] was used to illustrate the operation of the UAE power sector, and obtain the average GHG emissions of the country's gas-based power fleet.

Methodology

In the UAE the well-to-grid life cycle starts with the extraction of natural gas from the reservoirs and ends at the electricity transmission. Natural gas is extracted using wells and sent to processing plants where: water, carbon dioxide, sulfur, and heavier hydrocarbons are removed. Then, the resulted sweet gas product is transported through pipelines to the power plants to generate electricity, and transmitted through the electricity grid. The UAE's WTG GHG emissions life cycle can be divided into two inventories: upstream and downstream. The upstream inventory includes all activities before fuel is combusted at a power plant. Such activities comprise: sour natural gas extraction, sour gas processing, and sweet gas product transportation to the power plants gate. On the other hand, the downstream inventory includes electricity generation and transmission. In this study, the GHG emissions associated to gas extraction, sweet gas product transportation, and electricity transmission were estimated based on typical average emission values found in the literature [7]. On the other hand, the GHG emissions generated by the natural gas processing and combustion in power plants were calculated using ProMax® v3.2 (simulation software) [6] and a mathematical model developed in GAMS® (modeling system) [3], respectively. There are two main types of natural gas resources in UAE: 1) ultra/highly-sour natural gas, and 2) conventional natural gas. Therefore, for the present analysis the WTG GHG emissions life cycle was estimated taking into account both natural gas resources. Typical compositions for both types of gas resources are shown in Table 1.

Table 1. Typical compositions for ultra-sour and conventional natural gas resources in UAE.

Main Components	Ultra-sour gas [4]	Conventional gas [8]
H_2S	22	5
CO_2	10	4
C1	54	76
C2	4	8
C3+	5	3
H_2O	5	4

The first stage in the WTG GHG emissions life cycle is the natural gas extraction using a gas well. This operation encompasses various processes such as: rig construction, drilling, well completion, and gas production. Energy is consumed during the drilling process generating GHG emissions. The emissions of natural gas during the extraction stage are considered episodic emissions because they are not part of the daily, steady state of the well operations. For instance, it was assumed that around 20 g CO_2 eq./kWh is emitted to the atmosphere in gas extraction (e.g., mainly point source and fugitive losses) [7]. Moreover, gas production starts after the well is completed. Gas from the wellhead is sent to surface processing, where it is sweetened, dehydrated, and compressed for pipeline transmission. The raw gas from the wellhead is first sent to a three phase gravity separator to separate the gas, light liquids (e.g., mostly sour hydrocarbon liquids C3+), and heavy liquids (mostly sour water). Since no fuel is consumed during this process, it is not considered a significant source of GHG emissions.

After the phase separator, the gas is treated in a gas sweetening unit. In this work, di-glycol amine (DGA) is used as the gas sweetening solvent for ultra-sour natural gas; while methyl-di-ethanol amine (MDEA) is used as sweetening solvent for conventional gas. The units that consume energy in the gas sweetening process are: feed pre-coolers, amine reboilers, circulation pumps, solvent coolers, and acid gas condensers. The amine re-boiler is assumed to be powered by low pressure steam (LPS) (i.e. saturated steam at 3.5 barg), the circulation pump and solvent air cooler are powered by electricity, the feed pre-cooler and acid gas condenser employ cooling water. Later, the wet sweet gas is sent to the dehydration unit; whereas the acid gases (i.e. H_2S and CO_2) are sent to the sulfur recovery unit. In the dehydration unit, water is removed from the sweet natural gas since it promotes corrosion. Tri-ethylene glycol (TEG) is used as liquid desiccant to dehydrate the sweet gas. The glycol reboiler, sweet gas pre-cooler, and pumps are the major energy-consuming units in this process. In the dehydration unit, high pressure steam (HPS) (i.e., saturated steam at 41 barg) is used to power the glycol reboiler. Moreover, electricity is used to drive the pumps, whereas cooling water is used to pre-cool the wet sweet gas. The dry sweet gas from the dehydration unit can be combusted at an onsite power plant to produce electricity. Sometimes the dry sweet gas from the dehydration unit is sent to a natural gas liquid recovery unit to recover ethane and ethylene. However, for simplicity the natural gas liquid recovery unit was not considered in this study.

The liquid hydrocarbons are treated in a liquid sweetening unit using DGA as solvent; while the sour water is treated using a sour water stripper. The sole purpose of both aforementioned treatment processes is the removal of sour components from the hydrocarbon liquids and water. The liquid hydrocarbons sweetening process is similar to the gas sweetening process; thus, involving equivalent energy consuming units. Conversely, mainly two types of units consume energy in the sour water stripper: the water stripper reboiler and pumps.

All acid gases effluent from the gas sweetening unit, liquid sweetening unit, and sour water stripper are collected and sent to the sulfur recovery units (SRU). In these units, the H_2S in the acid gas stream is converted into elemental sulfur. A variety of configurations are available for sulfur recovery. In this study, it was used the "straight-through" Claus process. This process consists of an initial combustion step at high temperature (e.g., above 980 °C), followed by a series of reactions (carried out in a Claus Bed) and condensation steps. Generally two or three such steps are used. Steam is generated in the combustion step; thereby, contributing to meet the steam requirements of the process. The units consuming energy in sulfur recovery are condensers and re-heaters. Condensers and re-heaters are powered by cooling water and HPS, respectively.

The ProMax® v3.2 simulation software was used to calculate all forms of energy required to process conventional and ultra–sour natural gas. However, fundamental engineering principles were employed to calculate the associated GHG emissions. Furthermore, the amount of fuel gas (i.e., dry sweet natural gas) required to produce process steam (e.g., high and low pressure steam needed to process natural gas) was also estimated using the simulator. It is assumed that the fuel gas first combines with air, and then is combusted in a burner. The heat generated from the combustion process is used to boil water and generate high and low pressure steam. From this analysis, it was found that the average emission per slandered of fuel gas is around 66.7 g CO_2 eq./scf. This emission rate is in agreement with values reported by the US Environmental Protection Agency (EPA) [9].

The electricity requirements for the cooling water loop can be estimated using the cooling water flow rate. This can be done calculating the power requirements in pumps and cooling tower fans. For instance, the power requirements in a cooling water pump with a volumetric flow rate (V (m^3/s)) and pressure drop (ΔP (kPa)) can be calculated as follows:

$$Power = \frac{1}{\text{efficiency}} * V * \Delta P \quad (1)$$

For the cooling water loop, it was assumed an overall efficiency of 75% [10], the pressure drop around the water loop was estimated to be 266.7 kPa [10]. Accordingly, the power requirements can be calculated as follows [11]:

$$Fan\ Power = \left(V\frac{m^3}{s}\right)\left(\frac{60s}{min}\right)\left(\frac{264.172\text{gallon}}{m^3}\right)\left(\frac{0.5ft^2}{\frac{gallon}{min}}\right)\left(\frac{0.041\text{hp}}{\text{ft}^2}\right)\left(\frac{0.7457\text{kw}}{\text{hp}}\right) \quad (2)$$

The emissions of transporting the sweet gas product from process facilities to power plants was assumed to be approximately 30 g CO_2 eq./kWh according to information reported in the literature [7]. Furthermore, a mathematical model developed in GAMS on the UAE power infrastructure [3] was used to calculate the average GHG emissions generated by the gas-based power plants fleet in the country. The simulation results show that in average 370 g CO_2 eq. are generated per kWh of power produced. This value was used to calculate the GHG emissions associated to the electricity requirements involved in the analyzed life cycle. Finally, the GHG emissions associated to electricity transmission was also taken from the literature (i.e., 3.4 g CO_2 eq./kWh) [7].

Results and Discussions

The ProMax® v3.2 simulator was used to calculate the utilities (e.g., high/low pressure steam, cooling water, and electricity) consumed in the processing of conventional and ultra–sour natural gas. Table 2 shows the utility requirements for both types of gas resources per standard cubic feet (scf) of dry sweet gas produced.

Table 2. Utility requirements for conventional and ultra–sour natural processing.

Utility Type	Units	Ultra-sour gas	Conventional gas
Electricity	(Wh/scf)	20.8	4.0
High Pressure Steam (HPS)	(g/scf)	6.1	1.2
Low Pressure Steam (LPS)	(g/scf)	41.5	11.0
Low Pressure Steam (LPS) generated from SRU	(g/scf)	23.7	3.7
Net Required Low Pressure Steam (LPS)	(g/scf)	17.8	7.3
Cooling water	(liter/scf)	1.0	0.3

As shown in the table, the processing of ultra-sour gas requires much more energy than conventional gas per unit of sweet gas yielded. For example, the energy requirements to process ultra-sour gas are roughly four times higher than that for conventional gas. This can be explained by the fact that as the amount of H_2S in the feed gas increases so does the absorption plat utilities demand. This because more acid gases must be removed by the absorber leading to higher recirculating solvent flow rates. Therefore, the power consumed to cool and pump the regenerated solvent back into the absorber, as well as the steam used to regenerate the alkanolamines increases. Both energy forms (i.e., steam and electricity) account for the majority of the utilities consumed in the absorption plant. Likewise, more sulfur must be processed by the sulfur recovery units. Moreover, the sweet gas product recovery rate is much lower for the ultra-sour gas feed. Table 3 shows the recovery percentage for ultra-sour and conventional gas resources.

Table 3. Recovery percentages for ultra-sour and conventional gas.

Feed type	Ultra-sour gas	Conventional gas
Recovery rate	60 %	90 %

As shown in the table, almost 40 % of the ultra-sour gas feed volume is loss due to the high acid gases content; which are removed to obtain the sweet gas product. Conversely, for conventional gas only 10% of the volume is loss. Consequently, more energy is required per unit of sweet gas produced from ultra-sour gas. The electricity requirements for the cooling water loop were estimated using Equations (1)-(2). Accordingly, the GHG emissions corresponding to the electricity requirements were estimated using the gas power generation emission factor. Moreover, the process steam requirements associated to sour gas processing were obtained simulating the gas sweetening plants with ProMax® v3.2 [6]. The GHG emissions values related to HPS and LPS usage, and electricity for processing the natural gas are reported in Table 4. These values are given per standard cubic feet (scf) of dry sweet gas produced.

Table 4. GHG emissions related to conventional and ultra–sour gas processing.

Type of Utility Consumed	Units	Ultra-sour gas	Conventional gas
Electricity	(g CO_2 eq./scf)	7.7	1.5
High Pressure Steam (HPS)	(g CO_2 eq./scf)	0.7	0.4
Light Pressure Steam (LPS)	(g CO_2 eq./scf)	2.7	1.1
Total GHG emissions	(g CO_2 eq./scf)	11.1	3.0

From the table it can be clearly noticed that the GHG emissions associated with ultra-sour gas processing are much higher than those for conventional gas. Additionally, Table 4 shows that more than 50% of the GHG emissions are due to electricity consumption. Using the GAMS model [3], the average amount of natural gas needed to produce one-kilowatt hour of electricity in UAE was estimated. This value was found to be approximately 6.97 scf of gas/kWh. Additionally, all GHG emissions values from natural gas processing were also expressed in terms of per unit of kWh of electricity produced. The GHG emissions involved in natural gas extraction and final product transportation were estimated using factors reported in the literature [7]. The WTG GHG emissions life cycle for UAE domestic gas-based power generation can be estimated as the summation of the GHG emissions from: gas extraction, sour gas processing, sweet gas transportation, electricity generation (i.e., combustion of natural gas in power plants), and electricity transmission. Table 5 shows the single components and total GHG emissions life cycle for electricity produced from natural gas for both type of gas resources.

Table 5. Single emissions and total GHG emissions life cycle for domestic gas-based power generation in the UAE.

Life Cycle Stage	Ultra-sour gas (g CO_2 eq./kWh)	Conventional gas (g CO_2 eq./kWh)
Gas Extraction	20	20
Gas Processing	77.6	20.6
Product Transportation	30	30
Gas Combustion (power plants)	370	370
Electricity Transmission	3.4	3.4
Total GHG	501	444

As shown in the table, ultra-sour gas involves higher GHG emissions than conventional gas; which results from greater energy requirements. It is worth pointing out that most of the GHG emissions originate from the combustion of the sweet natural gas product in the power plants. Additionally, the WTG GHG life cycle analysis shows that for ultra-sour gas the emissions are approximately 13% greater than those involved using conventional gas. Figure 1 shows the percentage share of each stage/activity in the GHG emissions life cycle.

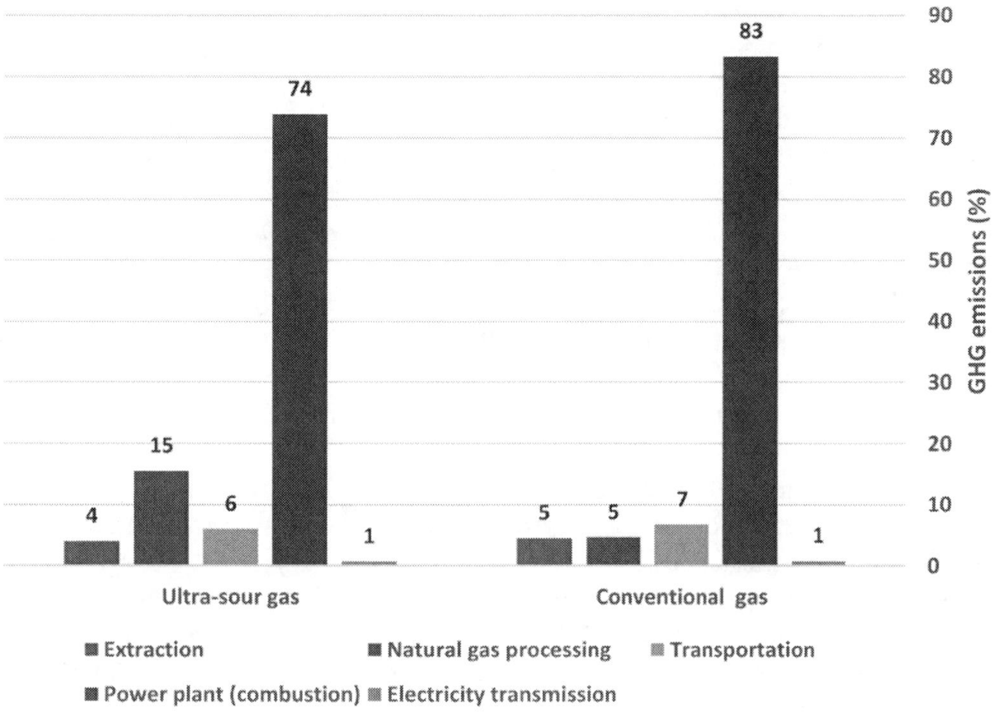

Figure 1. Percentage share per stage/activity in the GHG emissions life cycle for ultra–sour and conventional gas.

As shown in the Figure, sour gas processing and electricity generation (at the power plants) account for more than 88% of the total GHG emissions in the life cycle. Moreover, electricity generation is clearly the leading GHG emission source in the analyzed life cycle. Nevertheless, the GHG emissions generated by the electricity generation operations are rather low compared with other countries. This since the majority of the UAE's current power fleet is composed of modern gas-fired combined cycle plants; which are comparatively more efficient than other gas-based plants. Also, the total WTG GHG emissions of domestic gas-based power generation in UAE places the country in the low range (for conventional gas) and middle range (for ultra-sour gas) of values reported in the literature (e.g., 413-587 g CO_2 eq./kWh) for similar systems worldwide [7, 12, 13].

Conclusion

In this study, the well-to-grid (WTG) GHG emissions lifecycle for domestic gas-base power generation in the UAE is analyzed. The analysis results show that gas-based power generation using ultra-sour gas is more emission intensive than using conventional gas as the power plants' feed. This because much more energy is consumed in ultra-gas processing. Additionally, the yield of sweet gas product is much lower when processing ultra-sour gas compared with conventional gas. Regarding the total emissions of the life cycle, sour gas processing and electricity production account for almost 90 % of the total GHG emissions. Electricity generation is clearly the leading GHG emission source in the analyzed WTG lifecycle.

References

[1] A. Betancourt-Torcat and A. Almansoori, "Design multiperiod optimization model for the electricity sector under uncertainty – A case study of the Emirate of Abu Dhabi," *Energy Conversion and Management,* vol. 100, pp. 177-190, 8// 2015.

[2] Energy Information Administration, "United Arab Emirates " U.S. Energy Information Administration2015.

[3] A. Almansoori and A. Betancourt-Torcat, "Design optimization model for the integration of renewable and nuclear energy in the United Arab Emirates' power system," *Applied Energy,* vol. 148, pp. 234-251, 6/15/ 2015.

[4] Al Hosn Gas. (2015, 22/5/2016). *Abu Dhabi Gas Development Company Ltd. (Al Hosn Gas).* Available: : http://www.alhosngas.com/.

[5] J. Slagle, "Highly Sour Amine Sweetening Process Study for Conditions Common in the Middle East," presented at the The Abu Dhabi International Petroleum Exhibition and Conference, Abu Dhabi, UAE, 2013.

[6] Bryan Research & Engineering Inc., "ProMax," ed. Bryan, Texas, U.S.: Bryan Research & Engineering, 2013.

[7] T. Skone, J. Marriot, J. Littlefield, and B. A. Hamilton, "Life Cycle Assessment of Natural Gas Extraction, Delivery and Electricity Production," National Energy Technology Laboratory, United States of America2012.

[8] P. Pal, A. AbuKashabeh, S. Al-Asheh, and F. Banat, "Role of aqueous methyldiethanolamine (MDEA) as solvent in natural gas sweetening unit and process contaminants with probable reaction pathway," *Journal of Natural Gas Science and Engineering,* vol. 24, pp. 124-131, 5// 2015.

[9] Environmental Protection Agency, "U.S. annual non-baseload CO2 output emission rate," US Environmental Protection Agency, Washington, DC, US2015.

[10] R. Turton, R. C. Bailie, W. B. Whiting, and J. A. Shaeiwitz, *Analysis, Synthesis and Design of Chemical Process*, 3rd ed. Boston, MA, U.S.: Prentice Hall PTR International, 2008.

[11] R. H. Perry and C. H. Chilton, *Chemical Engineers' Handbook*, 5 ed. New York, US: McGraw-Hill, 1973.

[12] L. S. Abrahams, C. Samaras, W. M. Griffin, and H. S. Matthews, "Life Cycle Greenhouse Gas Emissions From U.S. Liquefied Natural Gas Exports: Implications for End Uses," *Environmental Science & Technology,* vol. 49, pp. 3237-3245, 2015/03/03 2015.

[13] World Nuclear Association, "Comparison of Lifecycle Greenhouse Gas Emissions of Various Electricity Generation Sources," World Nuclear Association
London, UK2011.

468547 Data Envelopment Analysis Coupled with Thermodynamic and Life Cycle Assessment Metrics for Solvent Screening: Application to CO_2 Capture

Abstract:

The growing trend towards more sustainable products and processes calls for advanced decision-making tools for the assessment and optimisation of the economic, environmental and social impact of chemicals and chemical processes. The pressure for more sustainable processes is particularly strong in the chemical industry, in which thousands of products are developed and launched at a very fast pace, raising the question (before scaling them up) of whether they are more sustainable than the existing ones or not.

Particularly, in green chemistry applications, engineers and scientists are confronted with the task of screening chemicals (e.g. mainly solvents, additives, etc.) according to sustainability principles. In practice, it is quite hard to find a single chemical/process performing best in the whole range of sustainability metrics, as inherent trade-offs typically arise between conflictive thermodynamic, environmental and safety metrics. There has been a substantial amount of research in the field of multi-criteria assessment of chemical processes.[1-3] In contrast, the problem of screening chemicals according to several criteria has received much less attention to date. In the absence of any rigorous and systematic screening method, practitioners assess them by defining, in either an implicit or explicit manner, subjective weights that are attached to every category of interest (e.g. viscosity, density, solubility, etc.). These weights enable the calculation of aggregated indicators on the basis of which the chemicals can be ranked.[3-5]While it is straightforward to implement, this approach is highly sensitive to the weights defined and provides in turn no insight into the underlying multi-criteria problem.

In this work we propose a new methodology to screen and select chemicals according to the extent to which they adhere to sustainability principles that is based on Data Envelopment Analysis (DEA), a technique developed in economics for efficiency assessment. Essentially, DEA is a non-parametric analytical tool for quantifying the relative efficiency of a set of decision making units (DMUs) that makes use of linear programming techniques.[6,7]DEA serves two purposes: i) it identifies, among a set of DMUs - each producing multiple outputs from multiple inputs - the efficient ones (i.e. those with the best ratio desired outputs to required inputs); and ii) it establishes efficient targets for the inefficient units that if achieved would make them efficient. DEA has recently found many applications in environmental studies, but to the best of our knowledge has never been applied to assess chemicals in green chemistry applications. In essence, we combine here DEA with a set of thermodynamic and life cycle assessment indicators so as to identify the chemicals that perform best according to technical (economic), environmental and social aspects simultaneously.

The capabilities of the DEA-based approach in the context of green chemistry are demonstrated through the screening of 125 conventional amine-based solvents for CO2 capture according to 10 performance indicators. Our approach eliminates 36% of the solvents for being inefficient, establishing quantitative targets and clear guidelines on how to improve them. The DEA models are capable of identifying the practical downsides of amines as main sources of inefficiency through the calculation of clear improvement targets for properties such as vapour pressure, acute toxicity and several life cycle impacts. The final aim of the DEA-based analysis is to facilitate the selection of more sustainable chemicals in the transition towards a more sustainable chemical industry.

References

1. C. Capello, U. Fischer and K. Hungerbühler, *Green Chem.*, 2007, **9**, 927.
2. G. Wernet, S. Papadokonstantakis, S. Hellweg and K. Hungerbühler, *Green Chem.*, 2009, **11**, 1826.
3. S. Hellweg, U. Fischer, M. Scheringer and K. Hungerbühler, *Green Chem.*, 2004, **6**, 418.
4. M. Tobiszewski, S. Tsakovski, V. Simeonov, J. Namieśnik and F. Pena-Pereira, *Green Chem.*, **17(10)**, 4773-4785.
5. A. I. Papadopoulos, S. Badr, A. Chremos, E. Forte, T. Zarogiannis, P. Seferlis, S. Papadokonstantakis, C. S. Adjiman, A. Galindo and G. Jackson, *Chem. Eng. Trans.*, 2014, **39**, 211–216.
6. W. D. Cook and L. M. Seiford, *Eur. J. Oper. Res.*, 2009, **192**, 1–17.
7. P. Zhou, B. W. Ang and K. L. Poh, *Eur. J. Oper. Res.*, 2008, **189**, 1–18.

470934 Using Multiobjective Optimization and Life Cycle Assessment for the Design of More Sustainable National and International Energy Systems

Abstract:

Using multiobjective optimization and life cycle assessment for the design of more sustainable national and international energy systems

Sabio, N[a]*., Volkart, K[b]., Densing, M[b]., Strachan, N[a].

[a] UCL Energy Institute, University College London, 14 Upper Woburn Place, WC1H 0NN

[b] Paul Scherrer Institut, Energy Economics Group, 5232 Villigen, Switzerland

Nowadays, an increasing energy demand driven by population and economic growth, and the need to reduce green-house gas emissions are the two major axis along which energy systems need to evolve. Not in vain, the design of sustainable energy systems has also been classified as a major challenge in the area of process systems engineering [1,2].

Energy systems models were developed after the first oil crisis using the methods of linear programming for informing policy makers concerned with the importance of energy systems planning and design at the time [3]. These models are still nowadays the main decision-support tools used for informing energy policy.

Although they were initially conceived for analysing energy security and costs, climate regulation has turned environmental factors in the major focus of energy systems analysis. For instance, after the UK committed to reduce its greenhouse gas emissions by 80% in 2050 [4], the Committee of Climate Change produced a set of carbon budgets based on energy systems analysis that outline the pathways and expected changes in the energy system needed to achieve the target [5].

Conflicting objectives lie at the heart of any design problem. Multi-objective optimization (MOO) frameworks were pervasive in PSE since the early days, the combination of multi-objective optimization frameworks combined and life-cycle analysis [6] has gained more recent attention [1,7,8]. Nevertheless, to the extent of our knowledge, this framework has not been applied to energy systems models.

One reason for this might be that multi-criteria analysis is not seen as a popular method in energy systems modelling community [9]. Although the advantages of multi-objective optimization have already been more widely recognised when applied to distributed generation problems involving a variety of stakeholders [10]. However, the challenges for the implementation of life cycle assessment in a whole energy systems model has been subject of more extensive attention in the literature [11].

In the present work we propose a framework for simultaneously optimizing several life cycle environmental impact indicators and the economic performance in energy systems models. For that purpose we use the classical ε-constraint deterministic multi-objective method that a side from being suitable for convex Pareto-sets, it allows to exploit the problem structure showing the different objective spaces without the need to articulate a priori preferences (i.e., weights), and avoiding the numerical problems arising from normalization.

Our results evaluate the effect of simultaneously optimizing several life-cycle environmental impact metrics against the more traditional total system cost optimisation with CO_2 emissions constrained for a newly developed national energy systems model of the UK (UKTM-UCL) [12] and for an international multi-regional energy systems model (GMM) [13].

***Corresponding author: n.sabio@ucl.ac.uk**

References

[1] Grossmann, I.E., Guillén-Gosálbez, G. (2010) Scope for the application of mathematical programming techniques in the synthesis and planning of sustainable processes, ***Computers and chemical engineering***, 34: 1365-1376

[2] Grossmann, I.E.(2012) Advances in mathematical programming models for enterprise-wide optimization, ***Computers and Chemical Engineering,*** 47: 2-18

[3] Stefan Pfenninger, Adam Hawkes, James Keirstead, Energy systems modeling for twenty-first century energy challenges, Renewable and Sustainable Energy Reviews, Volume 33, May 2014, Pages 74-86

[4] Acts of Parliament. ***Climate Change Act 2008***. The Stationary Office Limited, UK

[5] The Committee of Climate Change (CCC) The 5th Carbon Budget: The next step towards a low carbon economy, November 2015

[6] Guinee JB, Goree M, Heijungs R, Huppes G, Kleijn R, de Koning A, van Oers L, Wegener Sleeswijk A, Suh S, Udo de Haes HA, de Bruijn H, van Duin R, Huijbregts MAJ. Handbook on Life Cycle Assessment. Operational Guide to the ISO Standards. Dordrecht: Kluwer Academic Publishers, 2002.

[7] Azapagic A, Clift R. (1999) The application of life cycle assessment to process optimisation. ***Computers and Chemical Engineering***, 10:1509–1526.

[8] N Sabio, C Pozo, G Guillén Gosálbez, L Jiménez, R Karuppiah, Vasudevan, V., Sawaya, N., Farrell, J.T. Multiobjective optimization under uncertainty of the economic and life cycle environmental performance of industrial processes, ***AIChE Journal,*** 60 (6): 2098-2121

[9] Hall, M.H. L., Buckley, A.R. (2016) A review of energy systems models in the UK: Prevalent usage and

categorisation. ***Applied Energy,*** 169: 607-628

[10] Alarcon-Rodriguez, A., Ault, G., Galloway, S. (2010) Multi-objective planning of distributed energy resources: A review of the state-of-the-art. ***Renewable and Sustainable Energy Reviews,*** 14 (5): 1353-1366

[11] Volkart K. (2014).Development of a New Methodology for the Integration of LCA and Energy-Economic System Modelling, ***LCA XIV***, San Francisco, USA

[12]UCL Energy Institute Models Website, Available at: https://www.ucl.ac.uk/energy-models/models/uktm-ucl [10 May 2016]

[13] Turton H., Panos E., Densing M., Volkart K. (2013). Global Multi-regional MARKAL (GMM) model update: Disaggregation to 15 regions and 2010 recalibration, PSI Bericht 13-03

472120 Life Cycle Assessment and Multiobjective Optimization in a Natural Gas Based Petrochemical Complex

Abstract:

In this work, we address the optimization of a natural gas based petrochemical complex using economic and environmental criteria. The mathematical model includes linear and non-linear models for single plants, taking into account main operating variables, intermittent deliveries and inventory variable profiles. The environmental objective is measured with the global warming potential (GWP) and ReCiPe metrics according to the life cycle assessment procedures using indicators at two levels: eighteen midpoint indicators and three endpoint indicators realised in SimaPro (Pre-Consultants, 2015).The resulting mixed integer nonlinear programming (MINLP) models are implemented in GAMS. The bi-criteria MINLP model is solved with the Îµ-constraint method. Mid-point indicators are presented for current conditions. Numerical results show that the petrochemical complex can satisfy product demands while improving the environmental performance by decreasing greenhouse gases emissions in 33 %, from 1,042 to 695.33 kt CO_2-eq/year in the case of GWP minimization.

464636 Uncertainty Considerations for Surrogate Functions for Constrained Grey-Box Optimization

Abstract:

Derivative-free optimization methods have attracted a lot of attention recently, due to the increasing interest in multiscale expensive simulations which cannot be directly optimized, as well as the abundance of data from systems which cannot be described by analytic first-principle models [1-2]. We have previously presented our framework, AlgoRithms for Global Optimization of coNstrAined grey-box compUTational problems (ARGONAUT) [3-4], which identifies an appropriate surrogate representation for all of the unknown correlations of the constraints set and the objective, and subsequently combines all of the surrogate equations with any known equations, in order to form a non-convex non-linear programming (NLP) problem. The hybrid NLP is iteratively solved to global optimality in order to identify new samples and update the surrogate models. ARGONAUT addresses many elements ranging from optimal sampling, optimal sampling reduction, model identification, bound refinement, variable selection to global optimization. Insofar, we have shown that this framework exhibits competitive performance over a set of high-dimensional constrained benchmark problems [3] and a case study for the optimization of pressure swing adsorption simulation [4]. In this work, we will show improvements by (a) incorporating the effects of uncertainty introduced by sampling and surrogate model identification and (b) fully parallelizing our computational implementation to optimize efficiency.

One of the main components of any model-based derivative-free or black-box optimization method, is the identification of good surrogate approximations. This remains to be one of the major challenges of optimization without analytical forms or accurate derivatives, since the uncertainty caused by the fitted parameters of the surrogate models, as well as the mismatch between the underlying unknown function and the fitted model, highly affect the performance and reliability of any algorithm. Specifically, it has been observed that sampling plays a major role in the identification of the optimal parameters, since given a different set of samples, the optimal parameters of a surrogate model can be different. Moreover, the actual form of the input-output model is unknown, magnifying the observed variability in the surrogate model parameters given a different set of samples. It is important to develop surrogate models which provide good search directions, within which the surrogate models will be refined. Thus, the main challenge becomes the identification of promising refined search spaces, when the surrogate models cannot be entirely trusted as exact representations of the unknown functions.

In this work, we investigate the effects of surrogate parameter uncertainty caused by two main sources: (a) sampling of the search space and (b) selection of the appropriate surrogate function. We show that through the use of concepts from robust counterpart optimization [5-6], we can formulate a 'best-case' constrained grey-box representation, as well as a 'worst-case' constrained grey-box representation, which can serve as probabilistic lower and upper bounds on the actual optimum, respectively. We will show this concept through several examples using a set of different surrogate function types (i.e., linear, polynomial, kriging and radial basis functions). We will also discuss how these uncertainty-based bounds are incorporated within our constrained grey-box optimization framework to improve the reliability by mitigating the variability caused by the initial sampling stage. We will show results of our improved ARGONAUT framework which incorporates all of the above, as well as improved variable and term selection methods. Finally, we show that through efficient parallelization of the sampling and model identification stages, we can solve problems with large dimensionality and total number of constraints with reduced computational running time.

References:

1. Boukouvala, F., Misener, R., Floudas, C.A. (2016). Global Optimization Advances in Mixed-Integer Nonlinear Programming, MINLP, and Constrained Derivative-Free Optimization, CDFO. European Journal of Operational Research, 252(3), 701â€"727

2. Floudas, C.A., Niziolek, A.M., Onel, O., Matthews, L.R. (2016). Multi-scale Systems Engineering for Energy and the Environment: Challenges and Opportunities. AIChE J, 62(3), 602-623

3. Boukouvala, F., Hasan, M.M.F., Floudas, C.A. (2016). Global optimization of general constrained grey-box models: new method and its application to constrained PDEs for pressure swing adsorption, Journal of Global Optimization, DOI: 10.1007/s10898-015-0376-2

4. Boukouvala, F., Floudas, C.A. (2016). ARGONAUT: Algorithms for Global Optimization of Constrained Grey-Box Computational Problems. Optimization Letters,
DOI: 10.1007/s11590-016-1028-2

5. Guzman, Y.A., Matthews, L.R., Floudas, C.A. (2016). New *a priori* and *a posteriori* probabilistic bounds for robust counterpart optimization: I. Unknown probability distributions. Computers & Chemical Engineering, 84, 568-598

6. Li, Z., Tang, Q., Floudas C.A. (2012). A Comparative Theoretical and Computational Study on Robust Counterpart Optimization: II. Probabilistic Guarantees on Constraint Satisfaction, Ind.&Eng. Chem. Res., 51, 6769-6788

464725 Optimization of Constrained and Multidimensional Black-Box Problems Using Convex Hull Approximation and Single-Dimension Surrogate Model

Abstract:

Many important optimization problems in science and engineering are black-box, i.e., the analytical forms of the objective function and constraints are hidden or unknown. Identifying unknown feasible region and optimization of high-dimensional problems are the major challenges in black-box optimization. Evaluating black-box functions and their derivatives can be difficult, computationally expensive and unreliable, which limits the use of gradient-based optimization. One approach in derivative-free optimization is to replace expensive black-box functions with inexpensive surrogates through sampling and parameter estimation [1–6]. However, the accuracy of the surrogate models depends on the sampling set, surrogate function type, parameter estimation method, etc. These limitations often restrict the use of surrogate models to low-dimensional problems and few constraints, as the model size, complexity and parameters increase with the number of variables and constraints.

In the first part of the work, we present a data-driven approximation of the feasible region of constrained black-box problems without using surrogate models for the constraints. Data-driven construction of convex region has been recently proposed [7]. In this work, we approximate a feasible region (convex or nonconvex, continuous or disjoint) as the region described by the convex hull of feasible samples and their neighboring points, while subtracting the infeasible samples and their neighboring points from the convex hull. We represent each sample and its neighboring points using a Euclidean ball with the sample as its center and with the radius proportional to the minimum constraint satisfaction (if the sample is feasible) or violation (if the sample is infeasible). We solve a linear optimization (LP) model to select the largest Euclidean balls while avoiding any intersection between the feasible and infeasible balls. This increases the accuracy of our convex hull-based approximation as we obtain more samples. The design of experiments

(DoE) for new samples is now posed as a packing problem, which is a quadratically constrained program (QCP) and is solved to optimality using a global solver. Unlike surrogate model-based approaches, where we need to postulate a surrogate model and estimate the parameters for each constraint, our method exploits a single global parameter related to the size of the Euclidean balls, irrespective of the number of constraints.

In the second part of the work, we propose a novel approach to address high-dimensional black-box problems. Instead of constructing multivariate models, we project the original *n*-dimensional problem as a separate single-dimensional problem. Specifically, we consider a problem of minimizing $F(t)$, which geometrically can be interpreted as the projection of the original objective function $f(x_i)$, $i = 1, ..., n$ taken in the t-space following certain conditions imposed on the projection and the relations between x_i and t. We obtain $F(t)$ such that the minimum of $F(t)$ also corresponds to the solution of the original problem. The overall problem can be now decomposed into the following: fit a single-dimension surrogate model for $F(t)$, minimize $F(t)$ to approximately find t^*, find x_i^* corresponding to t^*, and check critically measure for x_i^*. This procedure is repeated iteratively in a trust region framework [8] to converge to an optimum t. The idea is extended to constrained problems by relating the objective function and constraint violations with $t$and applying filter technique [9] to converge to local optima.

In the derivative-free optimization paradigm, these developments can be promising because of the use of (i) a global parameter to approximate the feasible region irrespective of the number of constraints, and (ii) a single-dimension surrogate model to approximate the objective function irrespective of the number of variables. The applicability of the methods described above will be demonstrated on both benchmark black-box problems and chemical engineering applications.

References

[1] Rios, L. M.; Sahinidis, N.V. Derivative-free optimization: a review of algorithms and comparison of software implementations. *Journal of Global Optimization*, 56(3):1247–1293, 2013.

[2] Boukouvala, F.; Misener, R.; Floudas, C. A. Global optimization advances in Mixed-Integer Nonlinear Programming, MINLP, and Constrained Derivative-Free Optimization, CDFO. *European Journal of Operational Research*, 252(3):701–727, 2016.

[3] Boukouvala, F.; Hasan, M. M. F.; Floudas, C. A. Global optimization of general constrained grey-box models: new method and its application to constrained PDEs for pressure swing adsorption. *Journal of Global Optimization*, 2015. DOI 10.1007/s10898-015-0376-2.

[4] Biegler, L.T., Lang, Y-D, Lin, W-J. Multi-scale optimization for process systems engineering. *Computers & Chemical Engineering*, 60(10): 17-30, 2014.

[5] Boukouvala, F.; Ierapetritou, M. Surrogate-based optimization of expensive flowsheet modeling for continuous pharmaceutical manufacturing. *Journal of Pharmaceutical Innovation*, 8(2):131:145, 2013.

[6] Caballero, J. A.; Grossmann, I. E. An algorithm for the use of surrogate models in modular flowsheet optimization. *AIChE Journal*, 54(10):2633–2650, 2008.

[7] Zhang, Q.; Grossmann, I. E.; Sundaramoorthy, A.; Pinto, J. M. Data-driven construction of convex region Surrogate models. *Optimization and Engineering*, 2015. DOI 10.1007/s11081-015-9288-8

[8] Conn, A. R.; Scheinberg, K.; Vicente, L. N. Global convergence of general derivative free trust-region algorithms to first-and second-order critical points. *SIAM Journal on Optimization*, 20(1):387–415, 2009.

[9] Fletcher, R.; Gould, N.I.M.; Leyffer, S.; Toint, P. L.; Wachter, A. Global convergence of a trust-region SQP-filter algorithm for general nonlinear programming. *SIAM Journal on Optimization*, 13(3):635–659, 2002.

465463 Nonlinear Robust Optimization for Process Design and Operations

Abstract:

Robust optimization has become an active research area for addressing uncertainty in optimization problems and significant progress has been made in the past. However, most of the past works are limited to linear optimization (or linear in uncertain parameters) problems [1-3]. On the other hand, many realistic problems are formulated as nonlinear optimization problems and only very few work has studied the corresponding robust optimization technique [4]. To address this issue, we propose a novel robust optimization framework that can be used to address general nonlinear problems those are commonly seen in process design and process operations optimization. The major techniques used in the proposed method include local linearization with respect to the uncertain parameters, multiple local approximation and local affinely adjustable decision rule. The problems to be addressed can be classified as the following categories based on the complexity level of the nonlinear problems.

In the first case, the uncertainty is only involved in the inequality constraints. This is the simplest case of nonlinear optimization problems and it is often seen in process design or operations optimization problem with only static decisions. Here the objective is to find a robust decision that is feasible to all the possible uncertainty realizations in an uncertainty set. Linearization can be directly applied to the nonlinear inequality constraint and robust counterpart optimization formulation is then applied. In the second case, the nonlinear optimization problem involves design variables and state variables coupled by equality constraints, and inequality constraints are enforced for some state variables. In such optimization problems, there will be equality constraints containing uncertain parameters and the traditional robust optimization method cannot be directly applied. However, the state variables can be determined after the design variables as well as the uncertain parameters are fixed. Using the implicit function theorem, the state variables can be replaced by a function of uncertain parameters and design variables. Then the robust optimization formulation can be applied through linearization of the inequality constraints. Third, the nonlinear problem involves design variables, operation variables and state variables. For this type of problem, the operation variables can be adjusted based on the realization of the uncertainty. Correspondingly, a local affinely adjustable decision rule is adopted for the operation variables (i.e., an affine function of the uncertain parameter). The decision rule will be applied in the nonlinear problem and the problem can be reduced to the second case.

For all the above problems, the linearization and robust optimization formulation will be enforced around different samples of the uncertain parameters. The problem is solved by an iterative algorithm. Samples are randomly generated in each iteration, and the optimal solution obtained will be tested to find a new sample point that violates the constraints. The new sample violating the constraints is added to the robust optimization model, and the algorithm stops until no new violating sample is found. Application of the proposed method in various process design and process operations examples will be investigated to verify the efficacy of the nonlinear robust optimization algorithm.

Reference:

[1] Ben-Tal, A., Nemirovski, A. (1999). Robust solutions of uncertain linear programs. Operations research letters, 25(1), 1-13.

[2] Li, Z., Ding, R., Floudas, C.A. (2011). A comparative theoretical and computational study on robust counterpart optimization: I. Robust linear optimization and robust mixed integer linear optimization. Industrial & engineering chemistry research, 50(18), 10567-10603.

[3] Bertsimas D., Sim M. (2004). The Price of Robustness. Operations Research, 52(1):35–53.

[4] Zhang, Y. (2007). General robust-optimization formulation for nonlinear programming. Journal of optimization theory and applications, 132(1), 111-124.

470250 Differentiable McCormick Relaxations for Global Optimization

Abstract:

Several deterministic methods for nonconvex optimization require lower bounding information that is typically obtained by minimizing convex underestimators of the objective function on appropriate subdomains. McCormick's relaxation scheme [1] is an established, efficient method for generating and computing appropriate convex underestimators of factorable functions, and was recently generalized and improved by Tsoukalas and Mitsos [2]. However, both McCormick's original scheme and the variant in [2] can produce underestimators that are nonsmooth, particularly when the relaxed functions employ products of nontrivial terms. This nonsmoothness necessitates the use of nonsmooth optimization solvers to evaluate lower bounds; such solvers exhibit poorer convergence properties and performance than their smooth counterparts.

This presentation describes conditions under which Tsoukalas and Mitsos's relaxation scheme is guaranteed to produce continuously differentiable relaxations, and shows how these conditions may be satisfied for any factorable function by modifying the treatment of products and other multivariate intrinsic functions. The result is a continuously differentiable variant of McCormick's relaxations that retain the various computational benefits of the established schemes: the new relaxations may be generated and evaluated automatically, cheaply, and accurately, and converge rapidly to the relaxed function as the considered domain is reduced in size. Gradients may be evaluated using standard automatic differentiation techniques, removing the need for dedicated subgradient propagation methods. The new relaxations remain differentiable even if the relaxed function is nonsmooth. Moreover, since the relaxations are generated in closed form, they may be employed in established methods for generating convex underestimators for implicit functions and the solutions of parametric ordinary differential equations.

This presentation focuses on the construction of the differentiable relaxations themselves; a case study in optimization of an offshore LNG production employing these relaxations is the topic of a separate presentation. An implementation

in C++ based on the library MC++ [3] is described.

References:

[1]: G.P. McCormick, Computability of global solutions to factorable nonconvex programs: Part I - Convex underestimating problems. Math. Program., 10 (1976), pp. 147-175.

[2]: A. Tsoukalas and A. Mitsos, Multivariate McCormick relaxations. J. Glob. Optim., 59 (2014), pp. 633-662.

[3]: B. Chachuat, MC++: A toolkit for bounding factorable functions, https://projects.coin-or.org/MCpp

461096 Sensitivity Analysis of Uncertain Dynamic Systems Using Set-Valued Integration with Application to Complete-Search Optimization

Abstract:

Set-valued integration methods for uncertain dynamic systems find applications in many research areas, including reachability and invariance analysis, set-membership estimation, global optimization, and robust optimization and control [1,2]. Existing methods can be classified as either continuous-time or discretized approaches. The former considers an auxiliary set of differential equations, which propagate a parameterization describing an enclosure of the system's trajectories, pointwise in time. The latter relies on a predictor-validation approach, which constructs parameterized tubes using time-series expansion and determines time intervals whereby these tubes yield a guaranteed enclosure of the trajectories. Both methods can propagate a variety of tube parameterizations describing either convex enclosures (e.g., interval boxes, ellipsoids or zonotopes) or nonconvex enclosures (e.g., image sets of multivariate polynomials); see [3,4].

The main contribution of this paper is an extension of existing set-valued integrators to enable sensitivity analysis. Given an uncertain dynamic system in the form of a nonlinear parametric ODE,
$\mathrm{d}x/\mathrm{d}t = f(t,x(t),p)$,
with initial condition $x(0)=h(p)$ and uncertain parameters *p*â^^*P*âŠ,IR^n, we consider the problem to compute bounds on the gradients of state-dependent functions such as Î¦(*p*):=Ï†($x(T),p$), with $T>0$. Our focus is on continuous-time set-valued integration, and we consider both forward and reverse (adjoint) sensitivity analysis. A derivation of auxiliary ODEs describing enclosures of the sensitivity or adjoint trajectories of the parametric ODE is presented, and we discuss efficient numerical integration procedures for these ODEs based on SUNDIALS-CVODES [5] in order to exploit their underlying structures.

An application of this new sensitivity/adjoint bounding capability for parametric ODEs is presented in the context of

parameter estimation in dynamic systems using experimental data. Numerical experiments are carried out by introducing cuts derived from the first-order optimality conditions in order to tighten the relaxations and reduce clustering in branch-and-bound search for global optimization.

References:

1. B. Chachuat, B. Houska, R. Paulen, N. Peric, J. Rajyaguru, M.E. Villanueva, "Set-theoretic approaches in analysis, estimation and control of nonlinear systems," IFAC-PapersOnLine 48(8):981-995, 2015.
2. S. Streif, K.K. Kim, P. Rumschinski, M. Kishida, D.E. Shen, R. Findeisen, and R.D. Braatz, "Robustness analysis, prediction and estimation for uncertain biochemical networks," Journal of Process Control 42:14-34, 2016.
3. M.E. Villanueva, B. Houska, B. Chachuat, "Unified framework for the propagation of continuous-time enclosures for parametric nonlinear ODEs," Journal of Global Optimization 62(3):575-613, 2015.
4. B. Houska, M.E. Villanueva, and B. Chachuat, "Stable set-valued integration of nonlinear dynamic systems using affine set-parameterizations," SIAM Journal on Numerical Analysis 53(5):2307-2328, 2015.
5. A.C. Hindmarsh, P.N. Brown, K.E. Grant, S.L. Lee, R. Serban, D.E. Shumaker, and C.S. Woodward, "SUNDIALS: Suite of Nonlinear and Differential/Algebraic Equation Solvers," ACM Transactions on Mathematical Software, 31(3), pp. 363-396, 2005.

461446 A Novel Deterministic Global Optimization Algorithm and Its Application to Oil Refinery Planning

Abstract:

A petroleum refinery separates crude oils into various fractions with different quality properties (e.g. density, sulfur content, octane number, etc.). Such fractions are then transformed into intermediate products by chemical processes and they are either sold, used as raw materials for another process, or used as blend components for liquid fuels (e.g. gasoline, kerosene, and diesel).

One of the main issues regarding production planning of an oil refinery is the nonlinear nature of the processes involved. By assuming the processes to be linear, inaccuracy of the resulting model can make the solution suboptimal or unsuitable for real life applications. Current trend is to develop nonlinear planning models that are more accurate but that can be solved in a reasonable time. Li et al.[1] developed nonlinear models for the crude distillation unit (CDU) and the fluidize-bed catalytic cracker and integrated them into the refinery planning problem. Alhajri et al.[2] proposed a swing cut method for the CDU based on a polynomial formulation and fractional polynomials for the processing unit models. Alattas et al.[3] presented a nonlinear model for the CDU based on the fractionation index of a distillation column. Menezes et al.[4] formulated a swing-cut method that accounts for the difference in quality of the light and heavy splits of a swing cut fraction. Fu et al.[5] developed a hybrid nonlinear model for the CDU that predicts the true boiling point (TBP) curves of the products via partial least squares model with feed TBP curve and column operating conditions as inputs. Li et al.[6] described a least-squares methodology to use data-based nonlinear models for the processing units. If the planning model is nonconvex, then global optimization techniques are required to avoid computing a local solution far from the true optimum.

Deterministic global optimization algorithms compute feasible solutions as well as estimates of the best possible solution. These estimates are computed by relaxing the original nonconvex problem into a convex one. The main idea in deterministic global optimization algorithms is to reduce the difference between the best feasible solution and the estimate of the best possible solution to less than a non-zero tolerance value ε. To improve the quality of the estimates of the global optimum, the tightness of the relaxed model is usually increased by employing a spatial branch-and-bound strategy[7], addition of cutting planes[8], reducing the domain of the variables[9,10], and/or by increasing the number of partitions in the case of piecewise relaxation techniques[11]. Feasible solutions are usually computed using information from the convex model solution.

Bilinear terms (i.e. the product of two variables) are very common in planning models; for example, in material balances (mass or volume times concentration) and blending equations (mass or volume times a quality property). For bilinear terms, McCormick envelopes[12] provide the tightest linear relaxation. Piecewise linear relaxation techniques discretize the range of one of the variables involved in a bilinear term and then construct convex envelopes for each partition. Piecewise linear relaxations are tighter than simple linear ones; however, they introduce binary variables (making the relaxed model a mixed-integer linear program or MILP) and thus a trade-off between quality of the relaxed model and the computational effort required to solve it to optimality. Piecewise McCormick relaxation (PMCR) is a well-known technique of this type; it constructs McCormick envelopes for each partition. Normalized multiparametric discretization technique (NMDT) is a method recently developed[11] and it requires fewer number of binary variables than PMCR for the same discretization level.

We propose a deterministic global optimization algorithm for mixed-integer nonlinear programming (MINLP) problems, where nonlinearities are due to quadratic and bilinear terms. It relies on discretizing the nonlinear terms dynamically using either PMCR or NMDT. The resulting MILP model is solved using CPLEX and several feasible solutions are stored in CPLEX's solution pool and employed as starting points for a local nonlinear solver (e.g. CONOPT). These nonlinear models are solved in parallel. Then, the estimate of the global solution and the best feasible solution are updated. If the relative difference between these two (i.e. the optimality gap) is smaller than ε, then the algorithm stops; otherwise, it continues by reducing the range of the variables or increasing the number of partitions for the next iteration. If there is an improvement in the best feasible solution, then the domain of the variables involved in nonlinear terms is reduced using an optimality-based bound tightening (OBBT) method. This OBBT method consists in solving two optimization problems for each variable: a maximization and a minimization of the range of the variable subject to the MILP relaxation constraints. Parallelization of this step is required to avoid long execution times.

The algorithm was tested on several examples of a refinery planning model considering different demand scenarios and number of time periods. The refinery system comprises a crude distillation unit, a catalytic reformer, a hydrocracker, a fluid catalytic cracking unit, 5 hydrotreaters, 3 blenders, 5 products, 19 blend components, 8 quality properties, and a planning horizon of 7 days. Results show that the algorithm outperforms commercial solver BARON[13] and performs at least as well as commercial solver ANTIGONE[14] when handling large-scale multiperiod problems. For large-scale problems, increasing the number of partitions leads to larger MILP models, which are more accurate but cannot be solved to optimality. The major improvements in these examples are observed after applying OBBT method and solving the MILP model with the same number of partitions. Future work will be to improve efficiency of OBBT method in order to reduce execution times. Finally, the proposed algorithm can be applied to other problems by selecting and including the required convex relaxations for other types of nonlinear terms.

References

1. Li W, Hui CW, Li A. Integrating CDU, FCC and product blending models into refinery planning. *Comput Chem Eng*. 2005; 29(9): 2010-2028.
2. Alhajri I, Elkamel A, Albahri T, Douglas PL. A nonlinear programming model for refinery planning and optimisation with rigorous process models and product quality specifications. *Int J Oil Gas Coal Technol*. 2008; 1(3): 283-307.
3. Alattas AM, Grossmann IE, Palou-Rivera I. Refinery production planning: Multiperiod MINLP with nonlinear CDU model. *Ind Eng Chem Res*. 2012; 51(39): 12852-12861.

4. Menezes BC, Kelly JD, Grossmann IE. Improved swing-cut modeling for planning and scheduling of oil-refinery distillation units. *Ind Eng Chem Res.* 2013; 52(51): 18324-18333.
5. Fu G, Sanchez Y, Mahalec V. Hybrid model for optimization of crude oil distillation units. *AIChE J.* 2016; 62: 1065-1078.
6. Li J, Boukouvala F, Xiao X, Floudas CA, Zhao B, Du G, Su X, Liu H. Data driven mathematical modeling and global optimization framework for entire petrochemical planning operations. *AIChE J.* 2016. DOI: 10.1002/aic.15220
7. Quesada I, Grossmann IE. Global optimization of bilinear process networks with multicomponent flows. *Comput Chem Eng.* 1995; 19(12): 1219-1242.
8. Karuppiah R, Furman KC, Grossmann IE. Global optimization for scheduling refinery crude oil operations. *Comput Chem Eng.* 2008; 32(11): 2745-2766.
9. Faria DC, Bagajewicz MJ. A new approach for global optimization of a class of MINLP problems with applications to water management and pooling problems. *AIChE J.* 2012; 58: 2320-2335.
10. Yang Y, Barton PI. Refinery optimization Integrated with a nonlinear crude distillation unit model. *IFAC-PapersOnLine.* 2015; 48(8): 205-210.
11. Castro PM. Normalized multiparametric disaggregation: An efficient relaxation for mixed-integer bilinear problems. *J Glob Optim.* 2016; 64(4): 765-784.
12. McCormick GP. Computability of global solutions to factorable nonconvex programs: Part I—Convex underestimating problems. *Math Program.* 1976; 10(1): 147-175.
13. Tawarmalani M, Sahinidis NV. A polyhedral branch-and-cut approach to global optimization. *Math Program.* 2005; 103(2): 225-249.
14. Misener R, Floudas CA. ANTIGONE: Algorithms for continuous/integer global optimization of nonlinear equations. *J Glob Optim.* 2014; 59(2-3): 503-526.

460541 The Cluster Problem in Constrained Global Optimization

Abstract:

A key issue in continuous global optimization is the so-called cluster problem where a large number of boxes may be visited in the vicinity of a global optimizer [1–3]. While it is well-known that at least second-order Hausdorff convergence of the scheme of relaxations is usually required to eliminate the cluster problem for unconstrained optimization, a similar result for constrained optimization was unknown prior to this work. This work analyzes the cluster problem for constrained optimization based on a recently proposed definition of convergence order of bounding schemes for constrained problems. Conditions under which first-order and second-order convergent bounding schemes are sufficient to mitigate the cluster problem are provided, based on suitable assumptions.

[1] K. Du and R. B. Kearfott, "The cluster problem in multivariate global optimization," Journal of Global Optimization, vol. 5, no. 3, pp. 253–265, 1994.
[2] A. Neumaier, "Complete search in continuous global optimization and constraint satisfaction," Acta Numerica, vol. 13, pp. 271–369, 2004.
[3] A. Wechsung, S. D. Schaber, and P. I. Barton, "The cluster problem revisited," Journal of Global Optimization, vol. 58, no. 3, pp. 429–438, 2014.

457965 A Framework for Multi-Stakeholder Decision-Making and Conflict Resolution

Abstract:

The design and control of engineering systems is a complex decision-making process in which multiple conflicting objectives of social, economic, and environmental nature must be taken into account. Moreover, many decisions-making processes involve multiple stakeholders who disagree on how to prioritize the conflicting objectives. This is especially true in the field of sustainability, where hundreds of metrics may be considered and prioritized differently by a diverse population of stakeholders, such as different communities, regulatory agencies and companies. For even a moderate number of dimensions, however, exhaustively characterizing the trade-offs between objectives (i.e., computing the entire Pareto set) is computationally intractable.

We use the conditional-value-at-risk (CVaR) metric to create a general multi-stakeholder decision-making framework. We observe that stakeholder dissatisfactions (distance to their individual ideal solutions) can be interpreted as random variables. We thus shape the dissatisfaction distribution and find an optimal compromise solution by solving a CVaR minimization problem parameterized in the probability level. This enables us to generalize multi-stakeholder settings previously proposed in the literature that minimize average and worst-case dissatisfactions. We use the concept of the CVaR norm to give a geometric interpretation to this problem and use the properties of this norm to prove that the CVaR minimization problem yields Pareto optimal solutions for any choice of the probability level. We discuss a broad range of potential applications of the framework. We demonstrate the framework in a biowaste processing facility location case study, where we seek compromise solutions (facility locations) that balance stakeholder priorities on transportation, safety, water quality, and capital costs.

Reference:

Dowling, A. W., Ruiz-Mercado, G., Zavala, V. M. (2016), A framework for multi-stakeholder decision-making and conflict resolution, *Computers & Chemical Engineering*, in press.

461824 Systematic Process Design Strategies for Efficient and Synergistic Integration of Solar Thermal Hydrogen, Electricity and Fresh Water Production Processes

Abstract:

There is an ever-increasing demand for food, energy and water (FEW) due to population growth and change in consumption habits. For the past century, energy, chemicals, fertilizers and even fresh water via desalination or water treatment have been supplied by using fossil fuel resources. However, with decaying fossil fuel reserves and increasing greenhouse gas (GHG) emissions, the grand challenge is to sustainably meet FEW needs without compromising the current level of civilization. Development and implementation of renewable energy conversion processes by systems approach are essential to ensure a smooth transition to a sustainable economy. Among these alternatives, solar energy is particularly prominent because of its tremendous potential [1]. High conversion efficiency and integrated energy storage systems are indispensable to overcome intermittent and dilute nature of harnessing solar energy [2-3]. Recently proposed hydricity concept, synergistic solar thermal hydrogen and electricity coproduction, presents a potential solution for continuous and efficient power supply and also an exciting opportunity to envision and create a sustainable economy to meet all the human needsâ€"namely, food, chemicals, transportation, heating, and electricity [1].

Here, we introduce an integrated process design strategy based on the hydricity concept for the coproduction of fresh water, hydrogen and electricity from concentrated solar thermal energy. Process simulations for the proposed integration are performed in an integrated Matlab and Aspen Plus platform [4]. The operating conditions of various units and topological structure of the integrated process are determined via sensitivity analysis and optimization in Matlab. Detailed process simulations of thermal desalination processes are validated by comparison with performance data from the current operating desalination plants. The resultant desalination processes are further integrated with

solar thermal electricity and hydrogen processes to coproduce electricity, hydrogen and fresh water. Standalone and integrated solar conversion processes are evaluated based on the process energy efficiency and *exergy efficiency*that refers to the fraction of incident solar exergy that is directly recovered as the net exergy output, which is defined as the sum of electricity and the hydrogen exergy output. We have identified exergetically inefficient steps and implemented new process designs to improve the thermal efficiency of standalone solar desalination processes.

The integration of solar thermal power generation, solar thermal hydrogen production and thermal desalination techniques reduces exergy losses associated with each one of the standalone processes and thus provide a synergistic strategy. The implementation of the proposed process synthesis strategy results in efficient solar thermal process designs, which coproduce i) baseload power, ii) fresh water, and iii) hydrogen. Depending on the targeted end-use, hydrogen can be used as an energy career, transportation fuel, feedstock for the production of chemicals, fuels or fertilizers for food production. Coproduced products can supply all FEW needs unveiling a sustainability roadmap.

[1] Gençer E, Mallapragada DS, Marechal F, Tawarmalani M, Agrawal R. Round-the-clock power supply and a sustainable economy via synergistic integration of solar thermal power and hydrogen processes, Proceedings of the National Academy of Sciences (PNAS), 112(52), 15821-15826, 2015.

[2] Gençer E, Al-musleh E, Mallapragada D, Agrawal R. Uninterrupted Renewable Power through Chemical Storage Cycles. Current Opinion in Chemical Engineering, 5, 29-36, 2014.

[3] Gençer E, Agrawal R. A commentary on the US policies for efficient large scale renewable energy storage systems: Focus on carbon storage cycles, Energy Policy, 88, 477-484, 2016.

[4] Gençer E, Tawarmalani M, Agrawal R, Integrated solar thermal hydrogen and power coproduction process for continuous power supply and production of chemicals, Computer Aided Chemical Engineering, 37, 2291-2296, 2015.

472761 Ideas As a Process Intensification Tool with Application to Natural Gas Reforming Based Hydrogen Production

Abstract:

In this work, process intensification is identified as a process synthesis activity aiming at significant improvements over traditional process designs. The infinite-dimensional state-space (IDEAS) conceptual framework is put forward as a systematic tool for process intensification, which can identify fundamental limitations to the attainable performance by networks of technologies under consideration. The employed approach allows for a broad search of the state-space for process intensification opportunities, even by low dimensional IDEAS linear programming approximations that always retain design feasibility. The proposed process intensification methodology is applied to a case study of natural gas reforming based hydrogen production. The obtained results identify intensified process designs featuring hot/cold utility costs that in certain cases can be lower by more than an order of magnitude to those of traditional designs.

464734 Optimal Design of Integrated Upgrading Plant and Utility System for the Oil Sands Industry

Abstract:

Profitability in the oil sands industry historically has been quite unstable and it is significantly affected by the fluctuating crude oil price. Specially, the margin between bitumen production costs and the synthetic crude oil market price is insignificant these days. However, despite the unstable market and low profitability, oil sands production is predicted to be tripled by the year 2020, when taking into account existing projects, projects under development, and announced investments. Hence, it is imperative to reduce the operating costs for both existing and developing upgrading plants to allow for a more sustainable industry with a larger profit margin.

Limited number of works studied the optimization of oil sands upgrading and utility system. Ordorica-Garcia et al. [1] provided a model for the estimation of the energy requirements associated with producing crude oil products from oil sands according to the current commercially used production schemes. In their later work, Ordorica-Garcia et al. [2] presented a model for optimizing energy production for oil sands operations to reach the minimum of the total annual cost of energy consumptions in the oil sands industry. The model determined the optimal combinations of power and hydrogen plants to meet the given energy demands. Betancourt-Torcat et al. [3] incorporated the two previous studies into a more complex model simultaneously searching for the optimal set of oil production schemes and the corresponding energy infrastructures that meet the total production demands. The proposed superstructure involves multiple extraction and upgrading technologies and it is useful for production costs estimation, evaluating future production schemes and energy demand scenarios, and identifying the key parameters that affect the oil sands industry. In a more recent work [4], an energy optimization model for the oil sands upgrading operations was presented to reach the minimum downstream operating costs at the upgrading part in the oil sands industry, and the proposed superstructure of this work only focused on upgrading plants.

In this work, optimal design of integrated upgrading plants and utility systems in the oil sands industry is investigated. For the upgrading plant part, several state-of-the-art applicable upgrading schemes are considered, which include atmospheric and vacuum distillation, LC-fining, delayed coking and hydrotreatment. For the utility system part, a detailed superstructure model is proposed in this work. Compared to previous works [1-4] where the utility systems were just taken account as energy producers and very rough approximations were used for their operational and economic features, this work considers detailed features including process capacity, heating rate, capital cost, and operation and maintenance cost. A more complex polygeneration superstructure is developed in this study for the utility systems which can produce power, hydrogen and steam of varying quality. It includes gasifier (to produce syngas from coke or residue withdrawing from the upgrading plants), syngas cleaning, air separation unit, gas turbine, heat recovery steam generator, gas-fired boiler, and steam turbine. Results of various case studies demonstrate that the proposed work can efficiently address the optimal design of the integrated upgrading plant and utility system, which can lead to significant cost reduction of the oil sands industry operations.

References:

[1] G. Ordorica-Garcia, E. Croiset, P. Douglas, A. Elkamel, and M. Gupta, "Modeling the Energy Demands and Greenhouse Gas Emissions of the Canadian Oil Sands Industry," *Energy & Fuels*, vol. 21, pp. 2098-2111, 2007.

[2] G. Ordorica-Garcia, A. Elkamel, P. L. Douglas, E. Croiset, and M. Gupta, "Energy optimization model with CO2-emission constraints for the Canadian oil sands industry," *Energy and Fuels*, vol. 22, pp. 2660-2670, 2008.

[3] A. Betancourt-Torcat, G. Gutierrez, A. Elkamel, and L. Ricardez-Sandoval, "Integrated Energy Optimization Model for Oil Sands Operations," *Industrial & Engineering Chemistry Research*, vol. 50, pp. 12641-12663, 2011.

[4] J. Charry-Sanchez, A. Betancourt-Torcat, and L. Ricardez-Sandoval, "An optimization energy model for the upgrading processes of Canadian unconventional oil," *Energy*, vol. 68, pp. 629-643, 2014.

452326 Design and Operation of a 10 MWe Supercritical CO_2 Recompression Brayton Power Cycle

Abstract:

Supercritical carbon dioxide (sCO_2) Brayton power cycles are gaining increasing attention as an attractive alternative to conventional Brayton and Rankine cycles using steam. Brayton cycles with sCO_2 as the working fluid have the potential for more compact design, higher thermal efficiencies, and lower cost of electricity. In view of these promising advantages, the U.S. Department of Energy under its Supercritical Transformational Electric Power Program is in the process of designing, with the intent to build, a nominal 10 MWe (net) sCO_2 Brayton cycle test facility to accelerate scale-up, development, and commercial deployment. In this work, steady-state and dynamic process models of a 10 MWe indirect recuperated sCO_2 recompression Brayton cycle have been developed to analyze cycle design and operation for use with power generation applications. The process models are capable of predicting the nominal design point, off-design, and part-load performance.

In this presentation, the 10 MWe sCO_2 recompression Brayton process model, assumptions, and design parameters will be described and the key results for the nominal steady-state operating point will be presented. The additional specifications and methods applied to provide increased model fidelity for generating the rigorous pressure-driven dynamic model will also be discussed, including the use of performance curves for the turbomachinery and flow-dependent pressure drop correlations, volume specifications, and heat capacity calculations for the heat exchangers. The dynamic model predicts time-dependent profiles of key outputs such as power generated and thermal efficiency as a function of compressor inlet temperature and pressure, turbine inlet temperature, bypass recompression fraction, and other key variables for the purpose of designing effective control strategies.

From the dynamic perspective, this presentation will highlight the transient responses of the sCO_2 cycle to reductions in heat input, representing typical off-design and part-load scenarios arising from integration with a load-following energy plant. Based on the open-loop cycle results without controllers, several operational strategies will be analyzed for maximizing cycle efficiency by maintaining turbine inlet temperature during heat input turndown subject to process constraints.

464793 Integrated Thermochemical Process for Optimal Co-Production of Liquid Fuels and Chemicals

Abstract:

Although the current crude oil price is quite low, it is still necessary to develop reliable biorenewable conversion technologies from a long-term perspective due to diminishing petroleum reserves and increasing market requirements. As a very versatile energy resource, biomass can not only be converted to liquid transportation fuels for vehicular use but can also be transformed to aromatics, olefins, lubricants, industrial chemicals, and many other products currently derived from petroleum or natural gas. Indeed, co-producing these value-added chemicals in an integrated biorefinery can incentivize the de-risking of the â€œfront-endâ€ processes that are necessary for liquid fuels production.

An integrated thermochemical conversion based biorefinery plant for coproducing premium quality liquid transportation fuels, olefins, and aromatics is systematically designed and synthesized in this presentation. The central elements contain gasification, fast pyrolysis, hydrocarbon upgrading, methanol synthesis, methanol-to-olefins, methanol-to-propylene, and methanol-to-aromatics. In general, the pretreated biomass is directed to the gasifier for syngas production and to the fast pyrolyzer for crude bio-oil production, respectively. The major portion of the hot syngas from the gasifier is used to provide heat for the fast pyrolyzer and to promote the fluidization. The crude bio-oil is converted to liquid fuels through sequential hydrotreating and downstream separation. The fluidizing gas along with the syngas from the fast pyrolyzer will finally serve as the feedstocks for methanol production. The produced methanol will be converted to olefins and aromatics through alternative routes. In the developed process superstructure, which considers multiple catalysts and reactor types for certain processing units, is formulated as a generalized disjunctive programming model. To reduce the computational complexity of optimization model, a symbolic regression based approach is proposed for generating surrogate models to describe those complex processing units. The bilinear, trilinear, quadrilinear, and concave terms are reformulated for enhancing the optimization model to be solved to global optimality. Water, power and heat integration studies are simultaneously carried out to reduce the utility cost. Trade-offs between process economics and hydrogen source selection are made. The top adjustable factors on overall economic performance are identified through ranking relevant candidates. Through this analysis, the designed integrated process exhibits high flexibility to cope with the uncertainties on market demands and feedstock sources.

465824 Design and Optimization of Integrated Carbon Capture and Conversion with Natural Gas to Produce Syngas

Abstract:

Direct chemical recycling of CO_2 to fuels and chemicals can significantly reduce fossil fuels-related greenhouse gas emission, and create great economic opportunities for energy and chemical sectors worldwide. As much as 2 Gt/yr of fuels and 200 Mt/yr of chemicals can be produced from CO_2 utilization [1]. In this work, we develop and optimize a novel process which takes power plant flue gas and cheap methane sources (e.g., renewable biogas, landfill gas, coalbed methane, CO_2-contaminated natural gas, shale gas) as raw materials and produces syngas (CO and H_2) - the universal precursor for hydrocarbon-based fuels and chemicals. The process is integrated such that it can simultaneously separate CO_2 from power plant flue gas as a source of carbon, and convert CO_2 with methane as a source of hydrogen to produce syngas. The process also utilizes renewable energy such as concentrated solar energy to partially supply the process energy demand.

Specifically, we combine a CO_2-selective zeolite [2] and a dry-reforming catalyst in a sequential configuration to capture and store CO_2 from dilute flue gas, and convert it to syngas using renewables. A special instance of this is when both the zeolite and catalyst are packed inside a single column in a layered configuration. The process is dynamic and operates in a cyclic manner, where each cycle consists of at least two steps. In the first step, the zeolite section separates and intermittently stores CO_2 from flue gas at ambient temperature. In the second step, the zeolite section automatically supplies CO_2 and methane toward the catalyst section for conversion, due to desorption triggered by solar heating and/or the introduction of methane-rich feed. The high temperature for conversion is partially achieved by using concentrated solar energy. The simultaneous desorption and conversion of CO_2 completely eliminates a separate regeneration step, which is typically the most energy-intensive operation in current capture technologies.

The integration of CO_2 adsorption, intermittent storage, desorption using reactor feed, and conversion with methane is largely controlled by the trade-offs between these competing phenomena. Hence, a predictive model of the overall process has been difficult to create – largely because of the time-dependent, distributed and cyclic nature of the process and the complex interaction between the adsorption and conversion sections. In this work, we use a nonlinear, algebraic and partial differential equation (NAPDE)-based model that captures the equilibrium, transport and kinetic behavior of the gases for the dynamic adsorption, desorption and conversion operations. The optimization of the NAPDE model is also challenging. To this end, we have recently developed and applied a data-driven platform for the simulation-based optimization of complex process models, which treats the NAPDE-based model as a black-box system and use it to generate the required data for optimization. We optimize the process for different feed conditions and syngas specifications (H_2:CO ratio). Once optimized, the integrated technology shows significant promise for reducing the cost of CO_2 capture by eliminating the need for a material regeneration step, and reducing the cost of CO_2 conversion by storing solar energy as chemical energy while mitigating GHG emission.

References:

[1] Quadrelli, E. A.; Centi, G.; Duplan, J.-L.; Perathoner, S. Carbon Dioxide Recycling: Emerging Large-Scale Technologies with Industrial Potential. *ChemSusChem.*2011, 4:1194–1215.

[2] Hasan, M. M. F.; First, E. L.; Floudas, C. A. Cost-effective CO_2 capture based on *in silico* screening of zeolites and process optimization. *Physical Chemistry Chemical Physics* 2013, 15 (40):17601–17618.

464736 Optimal Design and Operation of a Semi-Closed Oxy-Combustion Combined Cycle Power Plant

Abstract:

Optimal design and operation of a semi-closed oxy-combustion combined cycle power plant

Holger Teichgraeber[1] Adam Brandt[1]

[1]Department of Energy Resources Engineering, Stanford University, Stanford, California 94305-2220, USA

As gas turbines provide flexibility to energy systems today, in a low carbon future energy system, natural-gas-based power systems with CO_2 capture may take over that role. Hence, we analyse the design and operations of an oxyfuel-based CCS system that is designed to operate flexibly on timescales of order one hour. Such a system can provide an intermediate level of supply flexibility beyond that provided by short response batteries. Such a system could also fill gaps in seasons of low renewable energy availability. (e.g., winter).

Specifically, we present optimal design and hour-to-hour operation of a flexible semi-closed oxy-combustion combined cycle power system. Our method includes co-optimization of the required air separation unit used to produce O_2. Computational optimization is used to maximize net present value, including both operational and design optimization. Operating profit is increased through running the air separation unit at full capacity at times of low electricity prices, storing oxygen and running the oxyfuel-combustion combined cycle at full load in times of high electricity prices. This avoids energy-intensive air separation when the economic penalty is the greatest.

The system consists of a cryogenic air separation unit, oxygen storage facilities, a natural gas combustion turbine, a heat-recovery steam generator, a steam turbine and a CO_2 compression unit. In this integrated system, all components are represented in a modular fashion using energy and mass balances. The air separation unit is modelled for operation at full capacity and partial load based on industrial data. The gas turbine's, steam turbine's and CO_2 compression unit's partial load operation are modelled based on existing literature. In order to model the oxyfuel-combustion gas turbine, the partial load gas turbine model is expanded with pure oxygen combustion and a recirculation fluid cycle. The heat recovery steam generator model is based on previous work [1]. System configuration and the sizing of various components have a substantial impact on operations and system economics.

The system in this work is able to operate flexibly and its overall operation including time-varying electricity prices is optimized. It also includes the air separation unit and an oxygen storage facility within the system boundary. Amongst all system components, the air separation unit is the least flexible and gaseous or liquid oxygen storage would be beneficial in order to provide fast ramping in case of changing electricity prices.

We maximize net present value (NPV) which is a function of design and operational decision variables. A similar optimization approach has been taken and described in previous work [2]. The design optimization is formulated as a mixed integer nonlinear programming problem. It uses the operational optimization as an inner loop in order to find the maximum NPV. This problem is solved using an algorithm that consists of Particle Swarm Optimization (PSO) and Mesh Adaptive Direct Search (MADS). The optimal operation maximizes annual profit using a Sparse Nonlinear OPTimizer (SNOPT) algorithm with multiple starting points due to the nonconvexity of the problem.

[1] Kang, Charles A., Adam R. Brandt, and Louis J. Durlofsky. "Optimizing heat integration in a flexible coal–natural gas power station with CO 2 capture." *International Journal of Greenhouse Gas Control* 31 (2014): 138-152.

[2] Brodrick, Philip G., Charles A. Kang, Adam R. Brandt, and Louis J. Durlofsky. "Optimization of carbon-capture-enabled coal-gas-solar power generation." *Energy* 79 (2015): 149-162.

462461 Natural Gas to Liquids, Olefins, and Aromatics Under Uncertainty in Feedstock and Product Prices

Abstract:

Although natural gas to liquid transportation fuels processes have been investigated for several decades, they were considered economically unfavorable due to high feedstock costs. [1,2] Recent studies have demonstrated the profitability of such refineries can be significantly increased by coproducing valuable chemicals, such as olefins and aromatics, in tandem with liquid fuels. [3,4] Even more recently, new studies have investigated the potential of stand-alone natural gas based chemical refineries, where the main products output are mainly olefins and aromatics, and assessed the technoeconomic and environmental impacts of such processes. [5,6]

The economics of such natural gas facilities is intricately dependent on the complexity of the refinery configuration and the type, quantity, and quality of products produced (i.e., the product slate). The large chemical space available for the manufacture of liquid fuels and chemicals has been expanded even further in recent years due to novel and emerging technologies that can produce liquid fuels, olefins, and aromatics. [3-6] The types of technologies to utilize and the types of products to produce is a tremendously difficult challenge to address, one that is further complicated with the uncertainty surrounding the costs and prices of feedstocks and products, respectively. In this talk, we address these challenges through a robust optimization framework that is incorporated in a large-scale process superstructure that contains several alternatives for the production of fuels, olefins, and aromatics from natural gas. [6-10] We introduce an important production parameter that allows us to investigate the economic effects of natural gas to liquids, olefins, and aromatics refineries. The resulting large-scale, nonconvex, mixed-integer, nonlinear optimization (MINLP) model is solved using our novel global optimization algorithm. [11]

Our comprehensive optimization-based process synthesis framework contains several direct and indirect natural gas conversion technologies and numerous hydrocarbon production alternatives, including Fischer-Tropsch refining and methanol synthesis; each of which is compared to determine the optimal processing pathway. [2-6] Multiple commercial and novel technologies for the production of the olefins and aromatics, such as methanol-to-olefins and toluene alkylation with methanol, are rigorously modeled. [5,6] The key production parameter is introduced as a function of the important products output from the refinery, namely, gasoline, diesel, kerosene, aromatics, and olefins. Robust counterparts are formulated in order to address the uncertainty associated with feedstock and product prices.

Several case studies are presented to investigate the effect of plant capacity and production ratios on the overall profit of the refinery and incorporate uncertainty in the optimal solutions obtained. A discussion on the robust refinery designs that incorporate the effect of uncertainty will follow with emphasis on the major topological decisions as a function of the production parameter. The economic and environmental tradeoffs of the refineries are presented.

[1] Floudas, C. A.; Niziolek, A. M.; Onel, O.; Matthews, L. R. Multi-Scale Systems Engineering for Energy and the Environment: Challenges and Opportunities. AIChE Journal 2016, 62, 602-623.

[2] Baliban, R. C.; Elia, J. A.; Floudas, C. A. Novel natural gas to liquids (GTL) technologies: Process synthesis and global optimization strategies. AIChE Journal 2013, 59, 505–531.

[3] Onel, O.; Niziolek, A. M.; Elia, J. A.; Baliban, R. C.; Floudas, C. A. Biomass and natural gas to liquid transportation fuels and olefins (BGTL+C2_C4): Process Synthesis and Global Optimization. Industrial & Engineering Chemistry Research 2014, 54, 359-385.

[4] Niziolek, A. M.; Onel, O.; Elia, J. A.; Baliban R. C.; Floudas, C. A. Coproduction of Liquid Transportation Fuels and C6_C8 Aromatics from Biomass and Natural Gas. AIChE Journal 2015, 61, 831-856.

[5] Onel, O.; Niziolek, A. M.; Floudas, C. A. Optimal Production of Light Olefins from Natural Gas via the Methanol Intermediate. Industrial & Engineering Chemistry Research 2016, 55, 3043-3063.

[6] Niziolek, A. M.; Onel, O.; Floudas, C. A. Production of Benzene, Toluene, and Xylenes from Natural Gas via Methanol: Process Synthesis and Global Optimization. AIChE Journal 2016, 62 (5), 1531 – 1556.

[7] Li, Z.; Ding, R.; Floudas, C. A. A comparative theoretical and computational study on robust counterpart optimization: I. Robust linear optimization and robust mixed integer linear optimization. Industrial & Engineering Chemistry Research 2011, 50, 10567–10603.

[8] Li, Z.; Tang, Q.; Floudas, C. A. A comparative theoretical and computational study on robust counterpart optimization: II. probabilistic guarantees on constraint satisfaction. Industrial & Engineering Chemistry Research 2012, 51, 6769–6788.

[9] Li, Z.; Floudas, C. A. A comparative theoretical and computational study on robust counterpart optimization: III. improving the quality of robust solutions. Industrial & Engineering Chemistry Research 2014, 33, 13112–13124.

[10] Guzman, Y. A; Matthews, L. R; Floudas, C. A New a priori and a posteriori probabilistic bounds for robust counterpart optimization: I. Unknown probability distributions. Computers & Chemical Engineering 2016, 84, 568-598.

[11] Baliban, R. C.; Elia, J. A.; Misener, R.; Floudas, C. A. Global Optimization of a MINLP Process Synthesis Model for Thermochemical Based Conversion of Hybrid Coal, Biomass, and Natural Gas to Liquid Fuels. Computers and Chemical Engineering 2012, 42, 64-86.

466909 Improving Industrial Polyethylene Production Via Data Analysis

Abstract:

Plant historians accumulate extensive data about plant operation and performance. Hidden in these data are possible solutions to chronical problems such as, polymer fouling, production bottlenecks, and product quality variations. Process engineers routinely analyze historian data in search of solutions, but the sheer amount of data and the complexity of industrial plants conspire to keep these solutions hidden from engineers using conventional data-analysis tools. Small data sets miss recurrent root causes. Plant disturbances obscure steady-state trends. Instrument errors propagate through data manipulations.

Here we give examples where data mining and rigorous statically analyses have uncovered simple solutions to real plant problems. We take advantage of specialized software to handle large amounts of data, covering a comprehensive set of plant variables and several years of plant operation. In addition, we place special emphasis on data reconciliation: we use noise reduction techniques and outlier analysis to refine data points before mining them for root causes. This approach has been successful in identifying simple solutions that could have been identified by some other means, but simply werenâ€™t.

470643 Parsimonious Modeling Approaches for Batch Process Analysis

Abstract:

Batch processes play a central role in semiconductors, chemical & petrochemical, food & beverage, and many other industries. These processes generate large amounts of data, which can be used to conduct process analysis & improvement, monitoring & abnormality management, and process control & optimization. However, the difficulties of dealing with data collected from batch processes are well-known. They arise as consequence of their non-stationarity, multi-stage nature and heterogeneity of information sources (raw materials, process, quality labs, etc.). This complexity has prompted researchers to develop methodologies with enough flexibility to handle the challenges raised by batch data. Frequently, these tools evolved from existing methodologies by preprocessing and reshaping the raw data structure (unfolding, synchronization, scaling) to a convenient format, and then adapting the algorithms to the usual conditions of batch processes (e.g., imputation of future observations for on-line use). Examples include the well-known and successful 2-way methodologies such as multi-way PCA [1] and PLS [2], 3-way methods [3] such as Tucker 3 and PARAFAC, and adaptive methods such as hierarchical PCA [4]. More recently, particular aspects not covered by the previous approaches were addressed, such as non-Gaussian distributions [5] and nonlinearities [6]. In most of these approaches, the computational complexity is proportional to the average length of the batch samples, implying that they do not scale well in certain industrial batch processes, namely those with long runs (e.g., fermentations) or with high sampling rates (e.g., microelectronics). Furthermore, these methodologies critically depend on successful batch alignment (or synchronization), which is another specialized task that imports complexity to the solutions. A thorough analysis of the technical literature confirms an overrepresentation of contributions on the high computational/methodological complexity side, whereas approaches with lower computational complexity levels remain vastly unexplored. Examples of possible lower complexity approaches include statistical pattern analysis [7] (SPA) and some versions of dynamic methods such as batch dynamic PCA [8] (BDPCA) and auto-regressive PCA[9] (ARPCA), depending on the preprocessing methodology employed.

In this article, we present a contribution with a lower characteristic complexity (i.e., with a lower number of parameters to be estimated from process data, simpler model structures and more straightforward procedures) and illustrate its application in several case studies, where it is compared with current methodologies. This approach describes the trajectory of process variables by a small number of features that capture the essence of their time evolution. These features are then used for process analysis, diagnosis or prediction, according to the application goal. Models build from profile features are more parsimonious and do not require preliminary synchronization and complex preprocessing tasks. On the other hand, some information is inevitably lost when using features instead of the more detailed information (the measurements).

This framework was applied to three cases studies covering two simulated scenarios and an industrial dataset. The results obtained indicate that implementing a lower complexity framework do not necessarily imply a reduction in performance. On the contrary, the feature-oriented models often outperform conventional methods due to their parsimonious and robust nature. Furthermore, in exploratory frameworks, features can be an effective alternative to uncover process disturbances since they are often related with general characteristics of the trajectory of the process variables, which can be more easily interpreted.

References

1. Nomikos, P. and J.F. MacGregor, *Monitoring batch processes using multiway principal component analysis.* AIChE Journal, 1994. **40**(8): p. 1361-1375.

2. Nomikos, P. and J.F. MacGregor, *Multi-way partial least squares in monitoring batch processes.* Chemometrics and intelligent laboratory systems, 1995. **30**(1): p. 97-108.

3. Louwerse, D. and A. Smilde, *Multivariate statistical process control of batch processes based on three-way models.* Chemical Engineering Science, 2000. **55**(7): p. 1225-1235.

4. Rännar, S., J.F. MacGregor, and S. Wold, *Adaptive batch monitoring using hierarchical PCA.* Chemometrics and intelligent laboratory systems, 1998. **41**(1): p. 73-81.

5. Yoo, C.K., et al., *On-line monitoring of batch processes using multiway independent component analysis.* Chemometrics and Intelligent Laboratory Systems, 2004. **71**(2): p. 151-163.

6. Zhao, C.H., F.R. Gao, and F.L. Wang, *Nonlinear batch process monitoring using phase-based kernel independent component analysis-principal component analysis.* Industrial & Engineering Chemistry Research, 2009. **48**: p. 9163-9174.

7. He, Q.P. and J. Wang, *Statistics pattern analysis: A new process monitoring framework and its application to semiconductor batch processes.* AIChE Journal, 2011. **57**(1): p. 107-121.

8. Chen, J. and K.-C. Liu, *On-line batch process monitoring using dynamic PCA and dynamic PLS models.* Chemical Engineering Science, 2002. **57**(1): p. 63-75.

9. Choi, S.W., J. Morris, and I.-B. Lee, *Dynamic model-based batch process monitoring.* Chemical Engineering Science, 2008. **63**(3): p. 622-636.

464413 Nonlinear SVM-Based Feature Selection for Fault Detection and Diagnosis of Continuous Processes

Abstract:

Dimensionality reduction is a key task in most data-driven applications, in areas such as multi-scale systems engineering [1], where vast amounts of data must be reduced to an essential subset that is used to provide actionable insights. Advances in machine learning and pattern recognition techniques, such as support vector machines (SVMs) [2-6], have allowed the use of very high dimensional feature sets for tasks such as classification and regression; however, it is often desired to reduce a high-dimensional feature space to a minimal set of features in order to gain practical insight, reduce overhead of data collection, and to improve accuracy. For the case of linear SVMs models, vector weights can be used to rank the contribution of features [7]. However, for nonlinear SVMs, feature ranking is nontrivial since the kernel trick obfuscates vector weights. Accurate feature ranking for nonlinear SVM models is highly desired, since nonlinear SVMs models are often necessary to describe the complexities of many systems of interest. In order to rank features of nonlinear SVM models, we have derived novel kernel-dependent SVM feature rank criteria [8] derived from sensitivity analysis of the dual SVM objective. Additionally, we have developed a platform for support vector-based feature selection that is powered by the newly derived criteria, and utilizes existing and novel greedy algorithms to rank features.

Process monitoring of continuous processes represents a key potential application of these theoretical and methodological advances in nonlinear support vector-based feature selection. Fault detection and diagnosis are specific aspects of process monitoring that can be addressed by using SVM classification and will benefit from the developed platform for feature selection. To this end, specific two-class SVM models are trained to detect known faults, while one-class support vector data descriptors are used to characterize normal operations. In these models, the manipulated and measured variables of the process compose our input feature space, and instances of normal and faulty operation are used as training samples for SVM models. The developed platform for support vector-based feature selection is then used to improve the accuracy of fault detection models, as well as to perform fault diagnosis. We present results for the Tennessee Eastman [9,10] process as a case study and compare our approach to existing approaches for fault detection and diagnosis.

[1]. Floudas, C. A., Niziolek, A. M., Onel, O., & Matthews, L. R. (2016). Multiâ€Scale Systems Engineering for Energy and the Environment: Challenges and Opportunities. AIChE Journal, 62(3), 602-623.

[2]. Vapnik, V. (1995). The Nature of Statistical Learning Theory. Springer.

[3]. Cortes, C., & Vapnik, V. (1995). Support-vector networks. Machine Learning, 20(3), 273-297.

[4]. SchÃ¶lkopf, B., Smola, A. J., Williamson, R. C., & Bartlett, P. L. (2000). New support vector algorithms. Neural Computation, 12(5), 1207-1245.

[5]. Schölkopf, B., Platt, J. C., Shawe-Taylor, J., Smola, A. J., & Williamson, R. C. (2001). Estimating the support of a high-dimensional distribution. Neural Computation, 13(7), 1443-1471.

[6]. Tax, D. M., & Duin, R. P. (2004). Support vector data description. Machine Learning, 54(1), 45-66.

[7]. Guyon, I., Weston, J., Barnhill, S., & Vapnik, V. (2002). Gene selection for cancer classification using support vector machines. Machine learning,46(1-3), 389-422.

[8]. Kieslich, C. A., Tamamis, P., Guzman, Y. A., Onel, M., & Floudas, C. A. (2016). Highly Accurate Structure-Based Prediction of HIV-1 Coreceptor Usage Suggests Intermolecular Interactions Driving Tropism. PloS one,11(2), e0148974.

[9]. Downs, J. J., & Vogel, E. F. (1993). A plant-wide industrial process control problem. Computers & chemical engineering, 17(3), 245-255.

[10]. Chiang, L. H., Braatz, R. D., & Russell, E. L. (2001). Fault Detection and Diagnosis in Industrial Systems. Springer Science & Business Media.

469249 Data Management and Integration for Continuous Pharmaceutical Manufacturing

Abstract:

As the pharmaceutical industry seeks more efficient methods for the production of higher value therapeutics, the associated data analysis, data visualization, and predictive modeling require dependable data origination, management, transfer and integration. As a result, the management and integration of data in a consistent, organized, reliable manner is a big challenge for the pharmaceutical industry.

The S88 recipe model, an international standard for describing process, has been adapted in this study to deliver a well-defined data structure that will improve the data communication inside the system architecture. This work has been divided into two parts due to the difference requirements between laboratory-based analytical measurements and the pilot-plant continuous pharmaceutical process. In the laboratory platform, recipes have been developed for a sub-set of material property tests that could be performed on analytical instrument (e.g. FT4 for flow). Drupal, an open source content management system, is implemented on an Amazon web service for data transfer between the analytical devices eventually a data management platform. A recipe module for Drupal is developed for recipe management, in which users could create, import, and modify recipes. Scientists can access recipe through Drupalâ€™s webpage interface and perform experiments following recipe steps. Experiment data can be recorded by manually input or automatically parsing data file on the backend of server. This system works like a recipe based electronic laboratory notebook.

In the pilot-plant, process data is generated by unit operation equipments and implemented process analytical technology (PAT) instruments [1]. A process control system (e.g. DeltaV (Emerson)) collects the data from equipments and a PAT data management tool (e.g. synTQ (Optimal Industrial Automation)). The PAT data management collects data from inline/online measurement system [2]. The recipe for the whole continuous process is implemented in DeltaV. Data in DeltaV is collected according to the recipe and is transferred to a data storage hub (PI system (OSI Soft)) in the same structure. The event frame (EF) feature from PI system allows the possibility to create individual recipe based on continuous data feeding. From PI system, the data is sent to online data storage box and cloud system. From the box/cloud, the data can be access at different physical company sites, can be analyzed and applied for different applications. It is the first attempt to apply S88, a batch control standard, to continuous pharmaceutical manufacturing.

All the detailed information of the lab-based experiment and process manufacturing, including equipments, samples and parameters are documented in the recipe. Recipes containing data can be exported from this system to be shared and transferred. After detaching the data from recipes, a reliable and consistent data source is provided for data visualization and process modeling. Another feature is the two dimensional barcode labels that are used in this strategy. Every ingredient and equipment of the analytical experiment or manufacturing process will have a unique barcode, which can be used to identify the item and trace all the information related. This enforces material traceability which is an important requirement in the overall QbD initiative.

References:

1. **Singh, R.,** Sahay, A., Fernando Muzzio, Ierapetritou, M., Ramachandran, R. (**2014**). Systematic framework for onsite design and implementation of the control system in continuous tablet manufacturing process. Computers & Chemical Engineering Journal, 66, 186-200.
2. **Singh, R.,** Román-Ospino, A. D., Romañach, R. J., Ierapetritou, M., Ramachandran, R. **(2015)**. Real time monitoring of powder blend bulk density for coupled feed-forward/feed-back control of a continuous direct compaction tablet manufacturing process. International Journal of Pharmaceutics, 495, 612-625.

461160 Sipat 5.0 the Next Generation of PAT Data Management

Abstract:

The mind shift from batch production to continuous manufacturing has been made in the pharmaceutical industry for a couple of years, but the IT infrastructure to cover the analyzer integration, data management and advanced process control was put under pressure by this new paradigm. The current generations of PAT data management platforms are bridging the gap when it comes to full robust and reliable integration of analyzers, high throughput dataflows, and a further and deeper integration in the process automation layer.

SIPAT 5.0 is the latest generation of Siemens PAT data management system. A series of inevitable functionalities will be demonstrated for different applications.

Optimized time alignment is critical to combine the right pieces of information. In this presentation we will dig a bit deeper in time alignment applied on hot melt extrusion lines with dynamic line speeds, and how this can be translated into optimized time alignment. It will be explained how PAT software can deal with dynamic data validity ranges to cover always the correct pieces of information in the decision cycle. This will emphasize the importance of a very close integration between the automation layer and the PAT data management on top.

The user interface is a key aspect in smoothening the integration of PAT data management into continuous manufacturing. UX or user experience is the new trend to optimize the user interfaces for your operators into the IT platforms. Logic user-focused workflows are guiding the operator through the different necessary steps of operation. SIPAT 5.0 is the new reference in configuring PAT data management strategies.

Enabling Data Integration in Pharmaceutical Digital Supply Chains Using Ontologies

Nikolaos Trokanas, Jagjit Singh Srai

Centre for International Manufacturing, Institute for Manufacturing (IfM), University of Cambridge, 17 Charles Babbage Road, Cambridge, CB3 0FS, UK

Pharmaceutical supply chains are becoming progressively complex and increasingly vital for all actors of the supply chain [1]. From suppliers and pharmaceutical companies to healthcare providers and patients, the benefits stemming from an integrated supply chain are numerous [2].

The ability to query all different systems involved in a supply chain network using a common ontology model would enable integration and also facilitate the utilisation of many diverse sources of information while improving visibility and traceability of information [3]. This approach would enable a single point access for querying all the systems while offering a common vocabulary therefore eliminating the heterogeneity between different sources [4].

The unceasing digitisation of firm's operations has profoundly affected supply chains, leading to increased information flows, therefore creating an urgent need for efficient and holistic data integration across the supply chain.

In this work, we propose an ontological approach towards enabling data integration across different stages and levels of the supply chain (REF _Ref450114776 \h * MERGEFORMAT Figure 1 08D0C9EA79F9BACE118C8200AA004BA90B02000000080000000E0000005F005200650066003400350030003100310034003700370036000000
).

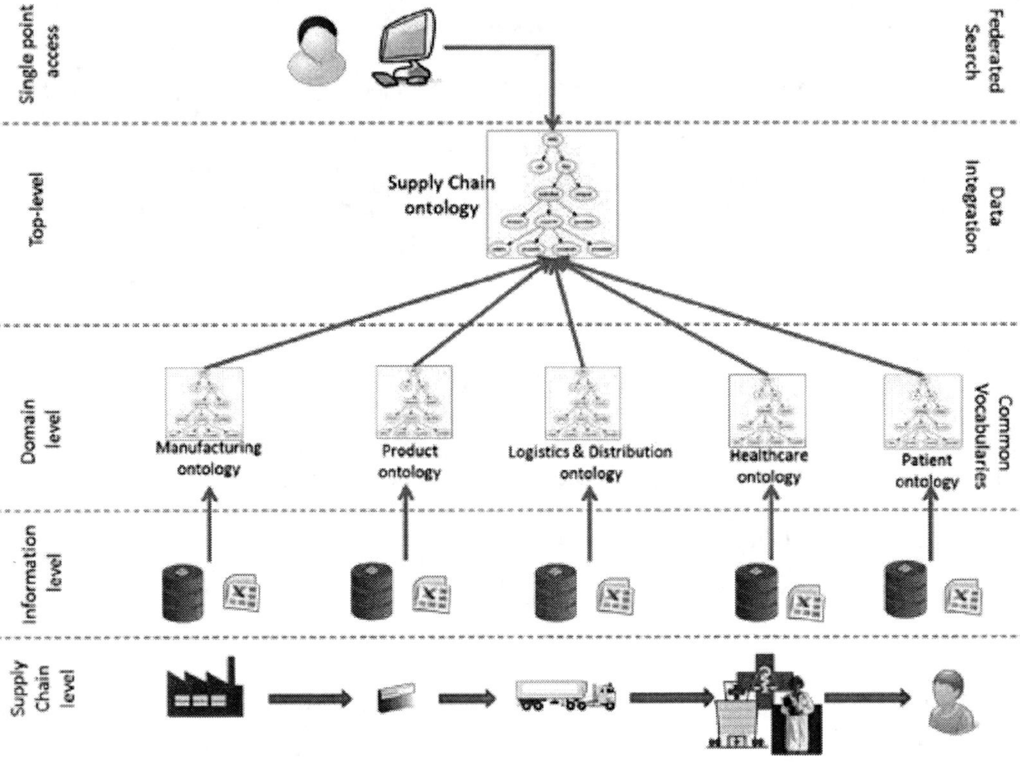

Figure SEQ Figure * ARABIC 1 Semantic Integration Architecture

The proposed ontological architecture consists of four levels:

i. Supply chain level: the physical supply chain enhanced by the use of digital technologies such as RFID, smart technologies, tracking devices etc.

ii. Information level: the legacy systems currently used across the pharmaceutical supply chain including databases, ERPs etc.

iii. Domain level: the level where different domains are represented using ontologies (local), offering common vocabularies hence eliminating any heterogeneity

iv. Top-level: the level that integrates the local ontologies from the domain level, offering a single point query that enhances visibility of the supply chain

This work is part of *REMEDIES* (*RE*-configuring *MEDI*cines *E*nd-to-end *S*upply) project and the proposed architecture is validated via a number of case studies in the pharmaceutical and healthcare industry.

References

[1] Rui T. Sousa, Songsong Liu, Lazaros G. Papageorgiou, Nilay Shah, Global supply chain planning for pharmaceuticals, Chemical Engineering Research and Design, Volume 89, Issue 11, November 2011, Pages 2396-2409, ISSN 0263-8762, http://dx.doi.org/10.1016/j.cherd.2011.04.005.

[2] Kefah Hjaila, José M. Laínez-Aguirre, Miguel Zamarripa, Luis Puigjaner, Antonio Espuña, Optimal integration of third-parties in a coordinated supply chain management environment, Computers & Chemical Engineering, Volume 86, 4 March 2016, Pages 48-61, ISSN 0098-1354, http://dx.doi.org/10.1016/j.compchemeng.2015.12.002.

[3] Edrisi Muñoz, Elisabet Capón-García, José M. Laínez-Aguirre, Antonio Espuña, Luis Puigjaner, Knowledge Management to Support the Integration of Scheduling and Supply Chain Planning using Lagrangean Decomposition, In: Krist V. Gernaey, Jakob K. Huusom and Rafiqul Gani, Editor(s), Computer Aided Chemical Engineering, Elsevier, 2015, Volume 37, Pages 989-994, ISSN 1570-7946, ISBN 9780444634290, http://dx.doi.org/10.1016/B978-0-444-63577-8.50010-3.

[4] Stef Lemmens, Catherine Decouttere, Nico Vandaele, Mauro Bernuzzi, A review of integrated supply chain network design models: Key issues for vaccine supply chains, Chemical Engineering Research and Design, Volume 109, May 2016, Pages 366-384, ISSN 0263-8762, http://dx.doi.org/10.1016/j.cherd.2016.02.015.

462278 An Ontology-Driven Knowledge Management Framework for Public Health

Abstract:

Global public health events such as the 2014 Ebola outbreak raises concerns about Emerging Infectious Diseases (EIDs) preparedness and the effectiveness of the response across the world. EIDs preparedness and response require public health coordination and communication, which face the challenge of knowledge sharing and reuse. While many studies have been done on public health surveillance or meta-knowledge analysis, limited progress has been made in knowledge management of public health documents. The complexity of public health knowledge is the major barrier. Similar information management problem has been studied in chemical engineering and biological science for years. Engineers and scientists address this problem by designing an ontology that facilitates knowledge management.

In this work, we propose a systems engineering inspired Ontology-driven Knowledge Management approach. This approach decomposes public health knowledge into concepts and relations and stores them in a function-based manner in an ontology. Semantic rules enable knowledge retrieval with respect to usersâ€™ queries. This novel approach takes a systems engineering perspective to manage public health knowledge. The function-based knowledge representation reduces the hierarchical complexity as well as the heterogeneity of public health knowledge. A West Nile Virus (WNV) outbreak scenario has been studied, and the reasoning results are compared with lessons learned from 1999 WNV outbreak in New York City. The comparison demonstrates the capability of this approach.

458111 Economic Nonlinear Model Predictive Control of an Integrated Solid-Sorbent Carbon Capture System

Abstract:

Post-combustion carbon capture is a practical technique to remove CO2 emission from coal-fired power plants. A major bottleneck for the commercial application of the carbon capture system is the high energy consumption required for plant operation such as CO2 separation and solvent/sorbent regeneration. The operation of the plant also needs to satisfy environmental regulations such as maintaining a certain CO2 capture fraction, under frequent disturbances resulting from the fast load changes in the power plants. Therefore, advanced control strategies are required to enhance the economic efficiency and control performance of the carbon capture system. In this study, we develop economic nonlinear model predictive control (NMPC) strategies for a solid-sorbent carbon capture system to reduce the operational cost while satisfying process constraints in the face of process disturbances. The computational cost of economic NMPC problem is greatly reduced by using an efficient NMPC formulation and a dynamic reduced model. A fast NMPC algorithm is then utilized to enable the online control of the carbon capture system.

The major components of the solid-based carbon capture system are bubbling fluidized bed (BFB) adsorber and regenerator, which are described by a one-dimensional, three-region, non-isothermal and spatially distributed model. The reactor's hydrodynamic behavior is described by partial differential and algebraic equations (PDAEs), constructed from mass and heat conservation and hydrodynamic correlations. The complete description of the BFB model can be found in [1, 2]. After spatial discretization, the BFB model becomes a highly nonlinear and large scale differential and algebraic equation (DAE) system which consists of 2000 differential equations and over 10000 algebraic equations. It is difficult to directly apply this rigorous model for time-critical applications such as NMPC. Therefore, we developed dynamic reduced models to reduce computational cost while maintaining accurate prediction capacity of the rigorous model.

Physics-based model reduction of the rigorous BFB model was conducted for both spatial and temporal aspects [3]. Orthogonal collocation on finite elements (OCFE) was applied to discretize the partial differential equations in space. Since OCFE is high order, it requires fewer discretization points to achieve the accuracy of the finite difference method. Further model reduction is due to unevenly distributed finite elements in space. For the temporal reduction, eigenvalue analysis was used to assess overall time scale differences in the rigorous BFB model. Then a quasi-steady state approximation replaced differential equations with algebraic states. In addition, a null space projection method eliminated reversible reactions for CO2 adsorption and led to a simplified kinetic model. The resulting reduced system therefore has essentially the same general behavior, but becomes much less stiff.

To maintain high CO2 capture under unsteady operation we develop a dynamic optimization strategy that discretizes the differential-algebraic equations in time and solves the corresponding nonlinear programming (NLP) problem. Moreover, we extend this approach to on-line optimization through NMPC. For a sufficiently long predictive horizon, the corresponding NLP problem raises a computational challenge, which we address through input and state blocking which significantly reduces the size of NLP subproblems for economic NMPC and maintains Lyapunov stability [4]. Based on the process dynamics, these nonuniform discretizations for state and control profiles also capture multiple time scales of the system and lead to control performance that is essentially the same as using uniform grids.

In this talk, we compare this approach with traditional setpoint-tracking NMPC, and we demonstrate that economic NMPC achieves significant reduction in the operational cost of the integrated carbon capture system, by directly considering process economics in the objective function. With reduced model and blocking strategies, we obtain a much smaller optimization problem and achieve one order of magnitude reduction in the solution time for economic NMPC problems. Moreover, advanced step NMPC (asNMPC) [5] is then applied to take these optimization calculations off-line in order to enable fast online control of the integrated carbon capture system. In addition, we apply the robust problem reformulation in [6] and relax inequality constraints with l1 penalty terms, with sufficiently large penalty parameters, to guarantee stability and improve NMPC robustness under process disturbance/uncertainty.

References

[1] Lee, Andrew, and David C. Miller. "A one-dimensional (1-d) three-region model for a bubbling fluidized-bed adsorber." Industrial & Engineering Chemistry Research 52.1 (2012): 469-484.

[2] Modekurti, Srinivasarao, Debangsu Bhattacharyya, and Stephen E. Zitney. "Dynamic Modeling and Control Studies of a Two-Stage Bubbling Fluidized Bed Adsorber-Reactor for Solid–Sorbent CO2 Capture." Industrial & Engineering Chemistry Research 52.30 (2013): 10250-10260.

[3] Yu, Mingzhao, David C. Miller, and Lorenz T. Biegler. "Dynamic reduced order models for simulating bubbling fluidized bed adsorbers." Industrial & Engineering Chemistry Research 54.27 (2015): 6959-6974.

[4] Yu, Mingzhao and Lorenz T. Biegler. A Stable and Robust NMPC Strategy with Reduced Models and Nonuniform Grids, the 11th IFAC Symposium on Dynamics and Control of Process Systems, 2016

[5] Zavala, Victor M., and Lorenz T. Biegler. "The advanced-step NMPC controller: Optimality, stability and robustness." Automatica 45.1 (2009): 86-93.

[6] Yang, Xue, Devin W. Griffith, and Lorenz T. Biegler. "Nonlinear Programming Properties for Stable and Robust NMPC." IFAC-PapersOnLine 48.23 (2015): 388-397.

469055 State Estimation of Energy Integrated Systems with Time-Scale Multiplicity

Abstract:

Energy integration requires dynamic interactions between different process systems units that give rise to challenges associated with the process operation. In particular, process integration results in models with time-scale multiplicity that are characterized by stiff systems of differential equations [1, 2]. For example, networks with large energy recycle, when compared to input/output energy flows, result in a system in which the individual units operate in a fast time scale as opposed to the whole network that operates in a slow one [3]. In the past, control strategies to address time-scale multiplicity have been studied and applied to several process systems [1-3]. However, state estimation approaches for systems with time-scale multiplicity are scarce [4]. In this presentation, we introduce a state estimation framework for energy integrated systems with multiple time scales. The developed framework relies on the time-scale separation of the system, in which state estimators are implemented to match the system dynamics associated with each of the time scales. For example, recursive and optimization-based estimators are investigated for the fast and slow time scales, respectively.

To demonstrate the effectiveness of the proposed framework, two case studies with time-scale multiplicity are addressed: (i) a chemical reactor system [5] as a motivating example. In particular, this example illustrates the challenges on the implementation of previously proposed state estimation techniques for low-dimensional systems with time-scale multiplicity; and (ii) an energy-integrated system that corresponds to a reactor-feed effluent heat exchanger (FEHE) network [2]. In addition to the presence of multiple time scales, the FEHE system is also high dimensional with more than 2,000 state variables, which introduces additional challenges to the proposed approach. To address these challenges, a framework with multiple extended Kalman filters (EKFs) for different time scales will be initially presented. Then, an implementation strategy with extended EKFs for the fast time scale and moving horizon estimators (MHEs) [6,7] for the slow time scale will be discussed.

References

[1] S. S. Jogwar and P. Daoutidis. Dynamics and control of vapor recompression distillation. J. Proc. Cont., 19(10):1737–1750, 2009.

[2] S. S. Jogwar, M. Baldea, and P. Daoutidis. Dynamics and control of process networks with large energy recycle. Ind. Eng. Chem. Res., 48(13):6087–6097, 2009.

[3] S. S. Jogwar, M. Baldea, and P. Daoutidis. Tight energy integration: Dynamic impact and control advantages. Comput. Chem. Eng., 34(9):1457–1466, 2010.

[4] S. C. Patwardhan, S. Narasimhan, P. Jagadeesan, B. Gopaluni, and S. L. Shah. Nonlinear Bayesian state estimation: A review of recent developments. Control Eng. Prac., 20(10): 933–953, 2012.

[5] C. Pantea, A. Gupta, J. B. Rawlings, and G. Craciun. The QSSA in chemical kinetics: As taught and as practiced. In N. Jonoska and M. Saito, editors, Discrete and Topological Models in Molecular Biology. Springer, 2013.

[6] F. V. Lima and J. B. Rawlings. Nonlinear stochastic modeling to improve state estimation in process monitoring and control. AIChE J., 57(4):996–1007, 2011.

[7] F. D. Rincón, G.C. Le Roux, F.V. Lima. The autocovariance least-squares method for batch processes: Application to experimental chemical systems. Ind. Eng. Chem. Res., 53(46):18005-15, 2014.

462447 Biomass-Based Production of Benzene, Toluene, and Xylenes Via Methanol: Process Synthesis and Deterministic Global Optimization

Abstract:

The pursuit towards an environmentally sustainable energy economy requires the development of economically competitive renewable processes, as discussed in a recent perspective article on multi-scale systems engineering for energy and the environment. [1] Of particular interest are biomass-based processes, which have the potential to significantly reduce well-to-wheel greenhouse gas (GHG) emissions. [2] Recent emphasis from academia, industry, and government agencies has solely focused on the production of electricity and liquid transportation fuels from biomass â€" such energy processes using single and hybrid feedstocks are reviewed in [3]. It is remarkable that very few studies in the literature have investigated the technical and economic feasibility of producing high-value aromatics from biomass. [4-6] Aromatics represent a significant portion (one-third) of the market for commodity petrochemicals, and a large portion (approximately 70 percent) of the worldâ€™s benzene, toluene, and xylenes supply comes from petroleum naphtha. [6,7] Biomass-based aromatics production additionally offsets import requirements for petrochemicals, crude oil, or naphtha, thus aiding in energy independence objectives set by the U.S. government. [2] Therefore, we propose novel methods of producing aromatics from biomass via methanol.

In this work, we present a comprehensive process synthesis superstructure consisting of several novel, commercial, and/or competing technologies that produce aromatics from biomass via methanol. [2] The optimal processes will be determined using our novel branch-and-bound global optimization framework. [2,9] We quantify the effect that biomass type has on the overall profit of the refinery by investigating forest residues, agricultural residues, and perennial crops as potential feedstocks. [2] Multiple case studies that investigate the effect of biomass moisture content, refinery capacity, and product distribution are discussed. The major topological decisions, as well as the overall profit of the biomass-based aromatics refinery, are presented.

[1] Floudas, C. A.; Niziolek, A. M.; Onel, O.; Matthews, L. R. Multi-Scale Systems Engineering for Energy and the Environment: Challenges and Opportunities. AIChE Journal 2016, 62, 602-623.

[2] Niziolek, A. M.; Onel, O.; Guzman, Y. A.; Floudas, C. A. Biomass-Based Production of Benzene, Toluene, and Xylenes via Methanol: Process Synthesis and Deterministic Global Optimization. Energy & Fuels 2016, DOI: 10.1021/acs.energyfuels.6b00619.

[3] Floudas, C. A.; Elia, J. A.; Baliban, R. C. Hybrid and Single Feedstock Energy Processes for Liquid Transportation Fuels. Computers and Chemical Engineering 2012, 41, 24-51.

[4] Lin, Z.; Ierapetritou, M.; Nikolakis, V. Aromatics from Lignocellulosic Biomass: Economic Analysis of the Production of p-Xylene from 5-Hydroxymethylfurfural. AIChE Journal 2013, 59, 2079-2087.

[5] Lin, Z.; Nikolakis, V.; Ierapetritou, M. Alternative approaches for p-xylene production from starch: Techno-economic analysis. Industrial & Engineering Chemistry Research 2014, 53, 10688-10699.

[6] Niziolek, A. M.; Onel, O.; Elia, J. A.; Baliban R. C.; Floudas, C. A. Coproduction of Liquid Transportation Fuels and C6_C8 Aromatics from Biomass and Natural Gas. AIChE Journal 2015, 61, 831-856.

[7] de Klerk, A. Fischer-Tropsch Refining; Wiley-VCH Verlag & Co. KgaA: Weinheim 2011.

[8] Meyers, R. A. Handbook of petroleum refining processes, Volume 3; McGraw-Hill New York, 2014.

[9] Niziolek, A. M.; Onel, O.; Floudas, C. A. Production of Benzene, Toluene, and Xylenes from Natural Gas via Methanol: Process Synthesis and Global Optimization. AIChE Journal 2016, 62 (5), 1531 â€" 1556.

464629 A Supervisory Control Framework for Fault-Tolerant Dispatch of Distributed Energy Resources

Abstract:

The convergence of competition in the power industry with the arrival of environmentally friendly micro-turbines, fuel cells, photovoltaics, small wind turbines, and other advanced distributed power technologies has sparked strong interest in distributed power generation [1]. This convergence of policy and technology promises to transform the electric power system from one relying primarily on central generation to one in which Distributed Energy Resources (DERs) provide most of the power needed. The resulting major improvement in power reliability, quality and efficiency; and the greater flexibility to respond to changing energy needs could save billions of dollars now lost each year because of power disruptions. These considerations have motivated a significant and growing body of research work on the modeling, analysis and control of DERs (e.g., see [2]-[6] for some results and references).

An important challenge with the evolution from central to distributed generation -- especially in light of the increasing number and diversity of DERs spread over the grid -- is the development of integrated monitoring, control and coordination strategies that would ensure optimal dispatch of the energy resources to provide highly reliable services under various disturbance and failure scenarios. This is significant given the fact that the distributed power market is driven by the need for reliable high-quality power, and the substantial impact that local disruptions in power flow can have when the DERs are integrated to support grid operations. In this context, process control and monitoring are critical tools for ensuring that the supply of DER networks matches demand whilst ensuring the stable and sustainable operation of each DER unit. However, the stability of DER networks and how they perform in the presence of local unit faults and failures has not been rigorously assessed. The need for these networks to function in the presence of faults is critical as to be a viable option for electricity generation, and these DER networks must be able to perform at the same standards of quality, stability, and robustness as current electrical generation. Meeting this challenge calls for a hierarchical monitoring and fault-tolerant control architecture where distributed monitoring and fault-tolerant control systems that perform timely fault identification and control system reconfiguration at the local level are integrated with a higher-level supervisor that optimally coordinates power generation in the event that local fault rectification is not possible. This hierarchy enables the timely mitigation of faults with supervisory oversight.

In prior work, fault-tolerant control of DERs has been studied in the context of a small-scale network of solid oxide fuel cells, where the focus has been on the detection and recovery from faults at the local level without supervisory oversight [7]. DER health monitoring was conducted through use of an alarm threshold on a properly designed observer-based output residual. Exploiting inherent actuator redundancy in the component subsystems, a number of stabilizing controller configurations were designed, and a methodology for active switching between the configurations in the event of a threshold breach was developed. The results were subsequently extended in [8] to include explicit fault estimation and performance-based fault accommodation capabilities in the local fault-tolerant control system design. In both studies, the system under consideration was only looked at from the local level with no supervisory communication. This negated the need for any logic or control at anything other than the local level which would be necessary in the event of a severe local unit fault in order to bring the local unit to an adequate safe parking state or initiate a safe unit shutdown. When all locally feasible corrective measures cannot rectify the local fault, an intervention by the supervisor becomes necessary. The local fault diagnosis information in this case must be communicated to the supervisor to be able to reconfigure the power load distribution for the entire network and compensate for the loss in power supply caused by the local failures.

The objective of this contribution is to develop an optimization-based supervisory control framework for fault-tolerant dispatch of DERs at the network level, which builds on the local fault identification and accommodation strategies developed earlier. The proposed framework extends the previous results in two important directions. One direction is incorporating logic on the local unit side to communicate unit fault scenarios at the local level with the supervisory controller. This would be in the case of a unit being hindered to the point where operation at a level deemed necessary by the supervisor is no longer a feasible option. The other direction is incorporating a supervisory controller to optimally allocate load to each unit on the network with logic in place to handle fault situations at the local level. The supervisory controller is designed using model predictive control (MPC) techniques which allow optimization of resource dispatch in the control action while incorporating various operational constraints on the local units. In the event of a local unit fault, the supervisor would communicate with the local unit to determine a sustainable operating condition, whether that be complete unit shut down or some diminished safe parking state. At that time, the supervisor would need to reallocate the load deficit caused by the faulted unit in an optimal way across the remaining units of the system. To this end, a finite-horizon constrained optimization problem is solved on-line to determine the optimal future power references for the functional DERs. Only the first part of the reference trajectories are transmitted to the local control systems, and the optimization is repeated when new measurements are sent to the supervisor. The receding horizon implementation strategy provides robustness against intrinsic uncertainties in the power generation capacity of certain DERs (e.g., renewables subject to environmental variations and intermittencies), as well as the uncertainty associated with long-term future power demand. The proposed framework is illustrated using a multi-source hybrid renewable DER system consisting of a photovoltaic energy conversion system, a wind energy conversion system, a fuel cell, an electrolyzer and a battery bank.

References:

[1] R. H. Lasseter, "Microgrids and distributed generation", *Journal of Energy Engineering*, **133**, 144–149, 2007.

[2] M. Deshmukh and S. Deshmukh, "Modeling of hybrid renewable energy systems", *Renewable and Sustainable Energy Reviews*, **12**, 235 –249, 2008.

[3] S. Barsali, M. Ceraolo, P. Pelacchi, and D. Poli, "Control techniques of dispersed generators to improve the continuity of electricity supply," *Proceedings of IEEE Power Engineering Society Winter Meeting*, pp. 789–794, 2002.

[4] M. Marei, E. El-Saadany, and M. Salama, "A novel control algorithm for the DG interface to mitigate

power quality problems," *IEEE Transactions on Power Delivery*, **19**, 1384-1392, 2004.

[5] Y. Sun, S. Ghantasala and N. H. El-Farra, ``Networked Control of Distributed Energy Resources: Application to Solid Oxide Fuel Cells," *Industrial Engineering and Chemistry Research*, **48**, 9590-9602, 2009.

[6] W. Qi, J. Liu, X. Chen, and P. D. Christofides, "Supervisory predictive control of stand-alone wind/solar energy generation systems," *IEEE Transactions on Control Systems Technology*, **19**, 199 – 207, 2011.

[7] Y. Sun, S. Ghantasala and N. H. El-Farra, "Monitoring and fault-tolerant control of distributed power generation: Application to solid oxide fuel cells," *Proceedings of American Control Conference*, pp.448-453, 2010.

[8] J. T. Allen and N. H. El-Farra, "A model-based framework for fault estimation and accommodation applied to distributed energy resources," *Renewable Energy*, in press.

464284 Grid-Relevant Demand Response Modeling of Chemical Processes

Abstract:

Manufacturing industries constitute a significant consumer of electricity ranging from about thirty three percent in developed countries[1] to twelve percent in developing countries[2]. In light of the fact that the world jointly seeks ways to improve energy efficiency, mitigate climate change, and optimize the efficiency of the grid[3]; it is beneficial that manufacturing industries, especially chemical industries, contribute to meeting this need. Several chemical processes have been identified as potential candidates owing to their high electricity consumption; including cement, pulp and paper, aluminum, and air separation. These industries can make modifications to their production schedules by increasing production during off-peak periods (i.e., times when total grid load and time-of-use prices are low) and storing excess product to satisfy customer demand during peak times when production rates are lowered. This way, the chemical industry can serve as a grid-level "storage battery."

While intuitive, this strategy – referred to as demand response (DR) – requires a shift in the established operating paradigm of chemical plants which is predicated on running at steady state as much as possible. In contrast, synchronizing chemical process operations with the state of the power grid calls for frequent changes in production rate and/or product grade. From an operational point of view, this requires that production scheduling decisions become not only aware of the steady state characteristics of the process (i.e. what products can be made and at what cost) but also of the process dynamics (i.e. how long it takes to quickly switch between products, and the transient levels of critical operating constraints).

In our previous work[4] we focused precisely on developing such a scheduling framework. Specifically, we proposed the use scale-bridging models (SBMs) to represent the process dynamics in the DR scheduling calculation. SBMs are a set of low-order representations of the closed-loop dynamic behavior of a process that is relevant to scheduling calculations (including, e.g., the evolution production rate and product quality following changes in production targets). We showed that the use of SBMs presents distinct advantages compared to embedding a detailed process model in scheduling calculations, namely, leading to significant reductions in computation time and providing an incentive for real-time implementation.

Increased participation of industrial players in DR initiatives is expected to have an impact on the operation of the electric grid and, as a consequence, developing representations of DR entities that can be incorporated in power system models is highly desirable. Most existing efforts in this direction focus, however, on thermal loads[5,6]. As a consequence, in this work we propose a new representation of the dynamics of DR operations for chemical processes that is relevant from the grid perspective. Starting with the premise that the aforementioned SBMs are not amenable to sharing with grid operators due to, e.g., business confidentiality concerns related to explicitly representing the performance of a process, we propose a new representation of process dynamics as a grid-level battery. We define a specific set of states (e.g., state of charge) and constraints (e.g., rate of charge) that are meaningful both from the point of view of plant operations and from the point of view of the grid. We demonstrate that such models can be expressed in terms of linear equations (with both continuous and binary variables), and are as such readily embeddable in the existing frameworks for modeling power systems. Finally, we illustrate our results with a case study.

Reference

1. Levin, K., Cashore, B., Bernstein, S., & Auld, G. (2008). Worldwide Trends in Energy Use and Efficiency: Key Insights from IEA Indicator Analysis.
2. Prasad V.S.N. T (2009). Nigeria's Electricity Sector- Electricity and Gas Pricing Barriers. International Association for Energy Economics Journal, 1st Quarter, Pg. 29 – 34.
3. G8 Gleaneagles (2005). Gleneagles Plan of Action: Climate Change, Clean Energy and Sustainable Development. Available online http://www.mofa.go.jp/policy/economy/summit/2005/ccc_a.pdf
4. R. C. Pattison, C. R. Touretzky, T. Johansson, I. Harjunkoski, and M. Baldea. Optimal process operations in fast-changing electricity markets: Framework for scheduling with low-order dynamic models and an air separation application. Ind. Eng. Chem. Res., 55(16):4562–4584, 2016.
5. L. Goransson, J. Goop, T. Unger, M. Odenberger, and F. Johnsson. Linkages between demand-side management and congestion in the European electricity transmission system. Energy, 69:860–872,2014.
6. A. Zerrahn and W. P. Schill. On the representation of demand-side management in power system models. Energy, 84:840–845, 2015.

468436 Model-Guided Optimization of Polymer-Electrolyte Dye Sensitized Solar Cells

Abstract:

Dye sensitized solar cells (DSSCs) have several appealing attributes. Their components and fabrication are inexpensive, are easy to fabricate, have tunable optical properties such as color and transparency, and can be integrated in unique ways into buildings and apparels. A DSSC is composed of three main components: (1) a photosensitive dye which converts solar light into electrons, (2) a mesoporous TiO_2 photoanode which provides a large surface area for dye adsorption and a pathway for electron transport, and (3) a liquid electrolyte comprised of a I^-/I^-_3 redox couple which is used to regenerate the dye and acts as a hole transport medium. These components have been studied over the past two decades but the conversion efficiency of the cells has increased only by ~2% to 13%. Most of the research to design and optimize DSSCs during this timeframe has been based on scientific intuition and trial-and-error experimentation, a slow and inefficient process. However, there have been a few computational/theoretical optimization studies. *First-principles-model-guided* design and optimization can lead to improved design and operation of DSSCs through systematically calculating optimal design specification and operating conditions.

First-principles modeling combined with experimental knowledge can provides a deeper understanding into DSSC material properties and processes than experimental knowledge alone. For example, Mathew et al. [1] used quantum chemical calculations to design a dye molecule that led to a record DSSC efficiency. This work showed that a rational, model-guided design of DSSC materials can improve the DSSC performance substantially. In our previous work, we used first-principles macroscopic mathematical models to theoretically and experimentally investigate the effects of polymer-electrolyte chemistry on the performance of polymer-electrolyte DSSCs [2]. We determined how the polymer chemistry affects DSSC interfacial processes and subsequent affected device performance. Polymer electrolytes can reduce interfacial recombination and allow for addressing the disadvantages of liquid electrolytes such as being prone to leakage and evaporation which hinder DSSC applications.

In this work, we first conduct DSSC optimization based on a first-principles DSSC mathematical model to calculate optimal design specifications for a polymer-electrolyte DSSC and then fabricate and test the optimal cell to validate the optimality of the cell.Â Optimal values of DSSC design variables such as photoanode thickness, dye absorption coefficient, iodide and triiodide diffusion coefficient, irradiation intensity, and polymer chemistry are will be calculated. The mathematical model includes Bulter-Volmer kinetics along with time-dependent continuity and transport equations for all charged speciesâ€"including electrons, iodide, triiodide, and lithium (the cation)â€" to describe the cell and to predict cell conversion efficiency and current-voltage behavior. The optimal design specifications that maximize the conversion efficiency are calculated by solving a constrained nonlinear optimization problem using a Nelder-Mead algorithm. Optimization validation will be conducted with two sets of DSSC experiments: (1) a DSSC is fabricated based on the predicted optimal design specifications, and the conversion efficiency of the fabricated cell is compared to the conversion prediction predicted by the model. (2) Cells with design specifications that are perturbed values of the theoretically-calculated optimal design specifications are fabricated and tested for conversion efficiency to investigate the true optimality of the design specifications.

[1] S. Mathew, A. Yella, P. Gao, R. Humphry-Baker, F. E. CurchodBasile, N. Ashari-Astani, I. Tavernelli, U. Rothlisberger, K. NazeeruddinMd, M. Grätzel, Nat Chem 2014, 6, 242.

[2] Y. Y. Smolin, S. Nejati, M. Bavarian, D. Lee, K. K. S. Lau, M. Soroush, Journal of Power Sources 2015, 274, 156.

466207 A Model-Based Feedback Optimization Method for Solar Dish Facets Alignment

Abstract:

The solar dish is a point-focusing concentrator with a very high concentration ratio ranging from hundreds to thousands, it is one of the most critical components of a solar thermal power generation system which governs the overall energy captured. However, large scale solar dishes are hard to assemble because they typically consist of hundreds of mirror facets, which reduce the complexity for manufacturing but greatly increase the difficulty to assemble, specially to align hundreds of facets to optimal positions. As a result, the alignment optimization of large scale dishes typically take months to years, which substantially contributes to the high cost of dish-type solar thermal power generation systems.

This work attempts to provide a systematic model-based feedback optimization method to align mirror facets instead of trial-and-error or empirical methods which has long information cycle and inefficient use of data obtained through alignment. The proposed method uses as projected light flux density data in front of the focus point to guide the alignment of the mirror facet in real-time without any special light patterns or complex equipment. The method first build a rigid body model of the entire solar dish assembly as if it is perfectly aligned, then based on the model predicted flux density distribution off focus and observed flux density distribution to identify a series of off-focus hot spots sorted by size. Then for each hot spot, flux density distributions at different distances are used to locate the misaligned mirror facet(s) primarily responsible for the hot spot and compute adjustments needed, adjustments on the targeted facet(s) is then adjust accordingly to remove the hot spot. The procedures are then repeated until all hot spots are resolved, which ensure near perfect alignment of the entire solar dish assembly. A three layer solar dish consists of inner, middle, and outer ring is investigated as an example, simulation showed successful alignment of solar dish facets to parabolic shape following this method and experimental study also confirm the effectiveness of the proposed strategy under realistic gravity induced shape deform.

458861 Model Predictive Control of Solar Thermal System with Borehole Seasonal Storage

Abstract:

This work addresses the model predictive controller design for a complex solar boreal-thermal storage system. A solar thermal power plant is used for heating dis- trict houses with bore- hole seasonal energy storage. The modelling of the overall system is inspired by the Drake Landing Solar Community in Okotoks, Alberta, Canada [1]. In this work, the discrete model of the integrated energy system is ob- tained by using energy preserving Cayley-Tustin discretization [2]. As the energy output from the solar thermal plant with borehole seasonal storage varies, the control system maintains the system achieving a desired thermal comfort level and energy savings. The model predictive control is designed by utilizing standard constrained optimization obtained control law which leads to quadratic regulator design account- ing for input or/and state/output constraints [3]. Finally, the controller performance is assessed by a numerical simulation with consideration of various possible scenarios. In addition, the proposed model development and constrained optimization regula- tion successful without approximations can account for the long range behaviour and variability in environmental and/or economic conditions associated with the overall operational costs of entire solar thermal community.

Highlights

1. Solar Thermal System with Borehole Seasonal Storage

2. Cayley-Tustin Discretization

3. Model Predictive Control

References

1. [1] B. Sibbitt, D. McClenahan, R. Djebbar, J. Thornton, B. Wong, J. Carriere, J. Kokko, The performance of a high solar fraction seasonal storage district heat- ing system–five years of operation, Energy Procedia 30 (2012) 856–865.
2. [2] V. Havu, J. Malinen, The Cayley transform as a time discretization scheme, Nu- merical Functional Analysis and Optimization 28 (7-8) (2007) 825–851.
3. [3] K. R. Muske, J. B. Rawlings, Model predictive control with linear models, AIChE Journal 39 (2) (1993) 262–287.

454885 Sustainability Assessment and Life Cycle Inventories of Bio-Based Chemicals Using Performance Indicators

Abstract:

The implementation in practice of sustainable development by society, government, and industry has demonstrated that sustainable solutions possess the potential for obtaining economic benefits without damaging the environment and human health. However, it is necessary to evaluate and quantify the effects of these new practices in the environment. Therefore, a more comprehensive description of environmental, social, and economic effects when implementing sustainable practices, e.g., for the design of chemical manufacturing processes, is needed. This contribution describes the GREENSCOPE (Gauging Reaction Effectiveness for the Environmental Sustainability of Chemistries with a multi-Objective Process Evaluator) tool as a support for decision makers when developing, designing, and evaluating sustainable processes and products. In addition, a life cycle inventory for chemical synthesis routes and processes is obtained as one of the main features of GREENSCOPE. A life cycle inventory of a chemical manufacturing stage is an important step in the development of life cycle assessments, which determine environmental, economic, and social effects throughout a productâ€™s life cycle stages. The GREENSCOPE sustainability evaluation is performed by employing environment, economics, energy, and efficiency process indicators. A case study demonstration using GREENSCOPE during a bottom-up design of bio-based chemicals will describe which indicators have shown improvement and simultaneously find additional opportunities for enhancements. Finally, this contribution will present the GREENSCOPE tool as an approach to achieving life cycle assessments of bio-based chemical products during their manufacturing life cycle stages.

470414 Modeling for the Life Cycle Assessment of Dry Reforming Technology â€“ Challenges and Lesson Learnt

Abstract:

The dry reforming reaction offers several advantages in fixation of carbon dioxide, a greenhouse gas. The process however, is dependent on choosing the right conditions and catalysts, which primarily govern the formation of syngas and byproducts. The quality of syngas produced also has a major role in the selection of subsequent products, where there may be the need for introducing hydrogen sources for hydrocarbons.

The choice of a system boundary and accounting for this process in terms of carbon emissions becomes very crucial, as the sources of carbon on one hand supplies mass, while the reactions can be endothermic, where external heat needs to be supplied and requires a source of energy. If fossil sources of energy are used, the net carbon balance tends to be negative. In this paper, the authors explore the possibilities of conventional and non-conventional energy sources to supplement the dry reforming process, and for targeted products in modeling the life cycle stage of CO_2 utilization via the dry reforming reaction.

466883 Performance Analysis of Chemical-Looping Fixed Bed Reactors Integrated in Combined Cycle Power Plants

Abstract:

Chemical-looping combustion (CLC) offers a promising method for combining electricity generation with in-situ CO_2 capture.[1,2] A metallic oxygen carrier is used to oxidize a hydrocarbon fuel in CLC, thus fuel and air are never mixed, and a pure stream of CO_2 can be separated after condensing water vapor. Thus, CLC has the potential to achieve higher thermal efficiency than most other CO_2 capture technologies. In this work, fixed bed CLC reactor configurations are studied, because their potential for seamless integration with combined cycle (CC) power plants and the capability to operate at high pressures.[3,4] Extensive analysis showed that the a key issue in the design of a fixed bed CLC reactor is the operating strategy that controls the temperature variations inside the reactor bed, especially during the exothermic oxidation step.[5–9] The dynamic nature of fixed bed reactors, as well as the many kinetically controlled reactions, necessitate the use of dynamic modeling to evaluate power plant performance, efficiency, stability and feasibility under transient operation.

This work presents a dynamic model for an integrated CLC-CC power plant and transient analyses of the integrated plant performance. A tool chain is presented for modeling the integrated CLC-based combined cycle power plant, coupling different equation-oriented platforms. Optimization of high-pressure fixed bed CLC reactor is performed to improve the efficiency of the downstream CC by maximizing the utilization of the gas turbine of the power plant. Optimal operating strategies are explored for the control of the temperature variations inside the reactor bed. The resulting time-varying gas stream exiting the CLC reactor is used by the gas turbine and steam cycle of a dynamic combined cycle power plant model. The power plant model is first validated against steady-state data from an existing natural gas-fueled, combined cycle power plant.[10] Then the combustion chamber of this plant is substituted by a CLC island operating optimally for improving the gas turbine performance. The transient variations of the integrated plant in terms of power, temperature, and pressure profiles are presented.

The CLC-CC power plant was estimated to operate at an average efficiency of 48.2%. This efficiency was accomplished by optimizing operational strategy for fixed bed reactors and the utilization of the exhaust streams of oxidation and heat removal. Dynamic simulation of the integrated CLC-CC plant showed that the combined cycle is only slightly affected by the batch nature of operation of the fixed bed CLC reactors. The CLC-CC plant was estimated to generate a relatively flat power output, especially from its gas turbine. Future work will focus on plant simulation under part-load operation and simulation of the integrated CLC-CC plant subject to time-varying loads. The developed simulation environment paves the way for conducting real-time optimization of power cycles with high-efficiency CO_2capture technologies.

Acknowledgements

This material is based upon work supported by the National Science Foundation under Grant No. 1054718.

References

1. Kvamsdal, H. M., Jordal, K. & Bolland, O. A quantitative comparison of gas turbine cycles with CO2 capture. Energy 32, 10–24 (2007).

2. Chen, C., Han, L. & Bollas, G. M. Dynamic simulation of fixed bed chemical looping combustion reactors integrated in combined cycle power plants. Energy Technol. (accepted) (2016).

3. Han, L. & Bollas, G. M. Chemical-looping combustion in a reverse-flow fixed bed reactor. Energy 102, 669–81 (2016).

4. Zhou, Z., Han, L. & Bollas, G. M. Overview of Chemical-Looping Reduction in Fixed Bed and Fluidized Bed Reactors Focused on Oxygen Carrier Utilization and Reactor Efficiency. Aerosol Air Qual. Res. 14, 559–571 (2014).

5. Spallina, V. et al. Reactor design and operation strategies for a large-scale packed-bed CLC power plant with coal syngas. Int. J. Greenh. Gas Control 36, 34–50 (2015).

6. Noorman, S. & van Sint Annaland, M. Packed Bed Reactor Technology for Chemical-Looping Combustion. Ind. Eng. Chem. Res. 46, 4212–4220 (2007).

7. Fernández, J. R. & Abanades, J. C. Conceptual design of a Ni-based chemical looping combustion process using fixed-beds. Appl. Energy 135, 309–319 (2014).

8. Spallina, V. et al. Investigation of heat management for CLC of syngas in packed bed reactors. Chem. Eng. J. 225, 174–191 (2013).

9. Hamers, H. P., Gallucci, F., Cobden, P. D., Kimball, E. & Van Sint Annaland, M. A novel reactor configuration for packed bed chemical-looping combustion of syngas. Int. J. Greenh. Gas Control 16, 1–12 (2013).

10. Kehlhofer, Rolf; Rukes, Bert; Hannemann, Frank; Stirnimann, F. Combined-Cycle Gas and Steam Turbine Power Plants. 165–209 (2011).

472811 Chemical Looping Carbon Neutral and Carbon Negative Schemes for a Gas to Liquid Fuel Facility Thermodynamic, Techno-Economic and Experimental Analysis

Abstract:

Carbon Capture Utilization and Sequestration (CCUS) is also a grand challenge for modern chemical engineers. Many new technologies strive to implement either CO_2 capture or CO_2 utilization, but the two processes are rarely thought of as mutually inclusive. In order to have a meaningful impact on the mitigation of CO_2 emissions it is paramount that CO_2 capture and utilization or sequestration technologies are developed in conjunction with each other. On a commercial scale, a fossil fuel process involving unprecedented simultaneous capture and utilization of CO_2 can be transformational and disruptive in conventional CO_2utilization and carbon capture markets. The OSU carbon neutral and carbon negative schemes presented using the methane to syngas process for liquid fuels production as a model example.

The methane to syngas (MTS) process developed at The Ohio State University (OSU) can efficiently convert methane to liquid transportation fuels and high value chemicals by producing a syngas of high purity with a desired H_2:CO molar ratio in a single step from natural gas while avoiding the use of catalysts, steam, and molecular oxygen produced from an air separation unit. This unique feature of the MTS process can lead to a dramatic reduction of the capital cost required for, as an example, the gas to liquid (GTL) plant to as low as $65,100/(bbl/day) from $86,200/(bbl/day) required for conventional auto-thermal reforming (ATR) process. Such a substantial reduction in the capital cost, confirmed by a third party engineering firm WorleyParsons, is due to a >60% reduction in capital investment for the syngas production unit by the MTS process, as compared to ATR. It, thus, allows the MTS process to remain economically competitive even with low West Texas Intermediate crude oil prices (~$35/bbl for a natural

gas cost of $3/MMBTU). This presentation initially presents will focus on presenting a detailed thermodynamic sensitivity coupled with a comprehensive techno-economic analysis (TEA) comparison to a conventional auto-thermal reforming technology for producing 50,000 bpd of liquid fuels from natural gas. This will be followed by the presentation of the design approach and philosophy used for developing a process performance model of the MTS technology and integrating it into a reference GTL facility, performing necessary heat and mass balance calculations, developing equipment configurations and lists, assessing capital and operating costs, and performing an initial and final economic evaluation to assess the costs and benefits of the MTS technology. Finally, the effect of application of the carbon negative and carbon neutral concept schemes to the developed TEA will be presented in terms of percentage improvement in carbon efficiency for liquid fuel production, comparison to the base MTS case TEA, variation of technology parameters which are important in terms of being de-risked and experimental validation of the carbon neutral and carbon negative schemes will be discussed

467227 Techno-Economic Analysis of Metal-Organic Framework Materials for Onboard, Light-Duty Vehicle Hydrogen and Natural Gas Storage

Abstract:

Metal-organic frameworks (MOFs) are a class of porous materials comprised of inorganic units bridged by coordinating organic linkers. Due to their high surface area and tunable binding energies, MOFs are promising materials for onboard, light-duty, automotive natural gas and hydrogen storage. A manufacturing cost analysis was conducted to understand the cost drivers for synthesis of Ni_2(dobdc), Mg_2(dobdc), Zn_4O(bdc)$_3$, and Cu_3(btc)$_2$ at automotive industry relevant production volumes (up to 2.5 Mkg/year) based on an engineering scale-up of demonstrated laboratory synthesis procedures. Conventional lab-scale synthesis of MOFs is carried out in excess organic solvents such as DMF, at elevated temperatures (~80°C-120°C), and are optimized for product purity rather than for yield. A number of extrapolations were assumed based on demonstrated high volume chemical manufacturing techniques and recent improvements in lab-scale synthesis. Results will be presented covering baseline costs using demonstrated solvothermal synthesis methods in organic solvents, aqueous synthesis as an alternative to the baseline synthesis in organic solvents, and liquid assisted grinding as a proposed alternative method to reduce solvent. Projected costs will be compared against U.S. Department of Energy targets for onboard hydrogen and natural gas storage, and sensitivity analyses will be presented to understand cost drivers. The results are expected to help identify a path towards low-cost, high volume MOF synthesis and to help guide the MOF research community towards avenues of investigation that lead to cost savings.

dobdc4 = 2,5-dioxido-1,4-benzenedicarboxylate

bdc2â€" = 1,4-benzenedicarboxylate

btc3â€" = 1,3,5-benzenetricarboxylate

461769 Global Optimization Algorithm for Miqcps Featuring Spatial Branch-and-Bound and Multiparametric Disaggregation

Abstract:

Mixed-integer quadratic constrained problems (MIQCPs) can be found in Process Systems Engineering when dealing with design and operational problems. Examples involve water networks [1-2], blending in petroleum refineries [3-9] and hydroelectric power plants [10]. Due to their non-convex nature, MIQCPs are often very difficult to solve to global optimality and sometimes it is already challenging to find a feasible solution. It is thus not surprising that our community has focused on the development of commercial global optimization solvers such as BARON [11] and ANTIGONE [12], which have improved dramatically over the years and can now effectively tackle small to medium sized problems.

A key element for global optimization is the computation of a tight lower bound. Three different strategies can essentially be used to achieve this goal: (i) find an equivalent formulation that differs for the better in the quality of the relaxation (perhaps the pooling problem is the most well-known [13]); (ii) rely on piecewise relaxation approaches [14-20], of which multiparametric disaggregation [19-20] has the advantage of adding a number of binary variables that scales logarithmically with the number of partitions (N) that is translated into a significantly better computational performance for $N \geq 10$; (iii) instead of reducing the variables domain simultaneously, do it one by one, an iterative procedure known as spatial branch-and-bound (SB&B) that can converge rather slowly.

The main novelty of this work is to integrate SB&B with the powerful mixed-integer linear programming relaxation derived from normalized multiparametric disaggregation [20]. Experiments are conducted after selecting one variable to discretize in every bilinear term and using two alternative settings for the number of partitions: N=10 and N=100. Optimality based bound tightening (OBBT) with the standard McCormick relaxation [21] is also performed at every node of the tree, after bisecting the domain of the variable contributing to the largest discrepancy, so as to compute tight lower and upper bounds for all bilinearly appearing variables.

Through the solution of a set of 43 NLP and MINLP benchmark problems from the literature, we show that the proposed global optimization algorithm achieves a better performance in terms of optimality gap compared to BARON 14.0.3 and GloMIQO 2.3 [22], being N=10 the best alternative. This is an indication that both piecewise relaxation approaches and OBBT should be used to a greater extent. On the other hand, GloMIQO led to the lowest computational time, benefiting from the faster solution time in the easiest problems, while BARON had some difficulties in the process network problems but excelled with the MINLP multiperiod blending problems. The latter result is an indication that we should explore branching on the binary variables in future work.

Acknowledgments: Financial support from Fundação para a Ciência e Tecnologia (FCT) through the Investigador FCT 2013 program and project UID/MAT/04561/2013.

References:

[1] D.C. Faria, M.J. Bagajewicz, *Novel bound contracting procedure for global optimization of bilinear MINLP problems with applications to water management problems*. Comput. Chem. Eng. 35 (2011), pp. 446–55.

[2] J.P. Teles, P.M. Castro, H.A. Matos, *Global optimization of water networks design using multiparametric disaggregation*. Comput. Chem. Eng. 40 (2012), pp. 132–147.

[3] H. Lee, J.M. Pinto, I.E. Grossmann, S. Park, *Mixed-integer linear programming model for refinery short-term scheduling of crude oil unloading with inventory management*. Ind. Eng. Chem. Res. 35 (1996), pp. 1630–41.

[4] Jia Z., Ierapetritou, M., Kelly, J.D.: Refinery short-term scheduling using continuous time formulation: crude-oil operations. Ind Eng Chem Res **42**, 3085–97 (2003).

[5] P.M. Castro, I.E. Grossmann, *Global optimal scheduling of crude oil blending operations with RTN Continuous-time and Multiparametric Disaggregation*. Ind. Eng. Chem. Res. 53 (2014), pp. 15127-15145.

[6] C.A. Meyer, C.A. Floudas, *Global optimization of a combinatorially complex generalized pooling problem*. AIChE J. 52(2) (2006), pp. 1027-1037.

[7] S.P. Kolodziej, I.E. Grossmann, K.C. Furman, N.W. Sawaya, *A discretization-based approach for the optimization of the multiperiod blend scheduling problem*. Comput. Chem. Eng. 53 (2013), pp. 122–142.

[8] P. Castro. New MINLP Formulation for the Multiperiod Pooling Problem. AIChE J. 61 (2015), pp. 3728-3738.

[9] I. Lotero, F. Trespalacios, M.-S. Cheon, D.J. Papageorgiou, I.E. Grossmann, *An MILP- MINLP decomposition method for the global optimization of a source based model of the multiperiod blending problem*. Comput. Chem. Eng. 87 (2016), pp. 13-35.

[10] P.M. Castro and I.E. Grossmann, *Optimality-based bound contraction with multiparametric disaggregation for the global optimization of mixed-integer bilinear problems*. J. Global Optim. 59 (2014), pp. 277-306.

[11] M. Tawarmalani, N.V. Sahinidis, *A polyhedral branch-and-cut approach to global optimization*. Math Program. 103(2) (2005), pp. 225–49.

[12] R. Misener, C.A. Floudas, *ANTIGONE: Algorithms for coNTinuous / Integer Global Optimization of Nonlinear Equations*. J. Glob. Optim. 59 (2014), pp. 503-526.

[13] N. Boland, T. Kalinowski, F. Rigterink, *New multi-commodity flow formulations for the pooling problem*. J. Global Optim. DOI 10.1007/s10898-016-0404-x.

[14] M.L. Bergamini, P. Aguirre, I.E. Grossmann, *Logic-based outer approximation for globally optimal synthesis of process networks*. Comput. Chem. Eng. 29 (2005), pp. 1914–1933.

[15] D.N. Wicaksono, I.A. Karimi, *Piecewise MILP under- and overestimators for global optimization of bilinear programs.* AIChE J. 54 (2008), pp. 991-1008.

[16] C.E. Gounaris, R. Misener, C.A. Floudas, *Computational comparison of piecewise-linear relaxations for pooling problems.* Ind. Eng. Chem. Res. 48 (2009), pp. 5742-5766.

[17] M.M.F. Hasan, I.A. Karimi, *Piecewise linear relaxation of bilinear programs using bivariate partitioning.* AIChE J. 56 (2010), pp. 1880-1893.

[18] P.M. Castro, *Tightening piecewise McCormick relaxations for bilinear problems*. Comput. Chem. Eng. 72 (2015), pp. 300-311.

[19] S. Kolodziej, P.M. Castro and I.E. Grossmann, *Global Optimization of Bilinear Programs with a Multiparametric Disaggregation Technique*. J. Global Optim. 57 (2013), pp. 1039-1063.

[20] P.M. Castro, *Normalized multiparametric disaggregation: an efficient relaxation for mixed-integer bilinear problems.* J. Global Optim. Doi: 10.1007/s10898-015-0342-z.

[21] G.P. McCormick, *Computability of global solutions to factorable nonconvex programs. Part I. Convex underestimating problems.* Math. Program. 10 (1976), pp. 147-175.

[22] R. Misener, C.A. Floudas, *GloMIQO: Global mixed-integer quadratic optimizer.* J. Global Optim. 53 (2013), pp. 3–50.

462360 Multi-Parametric Quadratic Programming: Past, Present and Future

Abstract:

In multi-parametric programming (mp-P), an optimization problem is solved for a range and as a function of certain parameters [1]. Due to its wide variety of applications, there has been a significant interest within the research community to solve mp-P problems efficiently. The most common class of mp-P problems are thereby multi-parametric quadratic programming (mp-QP) problems, as they arise in areas such as explicit model predictive control [2] of discrete-time linear systems and bilevel programming [3]. In this work, we examine the current state-of-the-art for mp-QP theory and algorithms. In particular, we draw upon a historical description of the evolution of the theory and algorithms related to mp-QP problems in order to highlight the current state-of-the-art and the research direction forwards with respect to mp-QP problems. The first comprehensive solution approach for mp-QP problems was presented in 2002 [2], where the geometrical interpretation of the parameter space was used to move from one region, where an active set is optimal, to another. However, as this approach introduces a large number of artificial cuts, several researchers have proposed modifications, including a variable step-size approach [4], inferring the active set of the adjacent critical regions from the irredundant constraints [5] and a combination thereof [6]. Each region of optimality, called critical region, is uniquely identified by its active set. Thus, soon after [2] appeared, researchers suggested the exhaustive enumeration of all active sets in order to solve mp-QP problems [7]. However, this combinatorial approach only became computationally tractable in 2011, when a branch-and-bound algorithm was suggested, which used the fact that if a candidate active set is infeasible, so is its superset [8]. The efficiency of the approach was refined in [9] by considering symmetry elements.

Although both approaches solve the same problem, so far no attempt has been made to unify the properties underpinning these algorithms. In this work, we discuss such a unified strategy by proving that the solution to a mp-QP problem is given by a connected graph. Each node is thereby an active set and a connection between two nodes is generated based on geometrical arguments which indicate adjacency.

Combined with the branch-and-bound approach in [8], this novel approach limits the number of candidate active sets to be considered. In the case of primal and dual degeneracy, it can be proven that there exists a single graph which solves the problem, resulting in a set of disjoint critical regions. This new development, part of our POP toolbox, is demonstrated through a unique computational study, where the computational abilities of state-of-the-art geometrical, combinatorial and connected-graph algorithm implementations are shown and compared. In addition, these results are compared with the latest version of the MPT toolbox [10], a software tool which solves mp-QP problems by reformulating them as multi-parametric linear complementarity problems. Based on these developments, we investigate the future research directions for mp-QP algorithms. In particular, we highlight the ability to apply parallelization strategies, the limitation of storage requirements and the extension of these results to general multi-parametric non-linear programming.

References

[1] Pistikopoulos, E. N.; Diangelakis, N. A.; Oberdieck, R.; Papathanasiou, M. M.; Nascu, I.; Sun, M. (2015) PAROC - an Integrated Framework and Software Platform for the Optimization and Advanced Model-Based Control of Process Systems. Chemical Engineering Science, 136, 115 – 138.

[2] Bemporad, A.; Morari, M.; Dua, V.; Pistikopoulos, E. N. (2002) The explicit linear quadratic regulator for constrained systems. Automatica, 38(1), 3 – 20.

[3] Faisca, N. P.; Dua, V.; Rustem, B.; Saraiva, P. M.; Pistikopoulos, E. N. (2007) Parametric global optimisation for bilevel programming. Journal of Global Optimization, 38(4), 609 – 623.

[4] Baotic, M. (2002) An Efficient Algorithm for Multiparametric Quadratic Programming. Technical Report, ETH Zurich.

[5] Tondel, P.; Johansen, T. A.; Bemporad, A. (2003) An algorithm for multi-parametric quadratic programming and explicit MPC solutions. Automatica, 39(3), 489 – 497.

[6] Spjotvold, J.; Kerrigan, E. C.; Jones, C. N.; Tondel, P.; Johansen, T. A. (2006) On the facet-to-facet property of solutions to convex parametric quadratic programs. Automatica, 42(12), 2209 – 2214.

[7] Seron, M. M.; Goodwin, G. C.; de Dona, J. A. (2002) Finitely parameterised implementation of receding horizon control for constrained linear systems. Proceedings of the American Control Conference, vol. 6, 4481 – 4486.

[8] Gupta, A.; Bhartiya, S.; Nataraj, P. S. V. (2011) A novel approach to multiparametric quadratic programming. Automatica, 47(9), 2112 – 2117.

[9] Feller, C.; Johansen, T. A.; Olaru, S. (2013) An improved algorithm for combinatorial multi-parametric quadratic programming. Automatica, 49(5), 1370 – 1376.

[10] Herceg, M.; Jones, C. N.; Kvasnica, M.; Morari, M. (2015) Enumeration-based approach to solving parametric linear complementarity problems. Automatica, 62, 243 – 248.

463428 Multiparametric Programming Based Algorithms for Bilevel Mixed-Integer Linear and Quadratic Programming Problems

Abstract:

Optimization problems involving two decision makers at two different decision levels are referred to as bilevel programming problems: the first decision maker (upper level; leader) is solving an optimization problem which includes in its constraint set another optimization problem solved by the second decision maker (lower level; follower). Bilevel programming problems are very challenging to solve, even in the linear case. For classes of problems where the lower level problem also involves discrete variables, this complexity is further increased, typically requiring global optimization methods for its solution. Solution approaches for mixed integer bilevel problems with discrete variables in both levels mainly include reformulation approaches [1, 2], branch and bound techniques [3] or genetic algorithms [4], all of which result in approximate solutions.

In this work, we present novel algorithms for the exact and global solution of two classes of bilevel programming problems, namely (i) bilevel mixed-integer linear programming problems (B-MILP) and (ii) bilevel mixed-integer quadratic programming problems (B-MIQP) containing both integer and continuous variables at both optimization levels. Based on multi-parametric theory and our earlier results [5, 6], the main idea is to recast the lower level problem as a multi-parametric programming problem, in which the optimization variables of the upper level problem are considered as parameters for the lower level. The resulting exact parametric solutions are then substituted into the upper level problem, which can be solved as a set of single-level deterministic mixed-integer programming problems. Extensions to problems including right-hand-side uncertainty on both lower and upper levels are also discussed.

References: [1] Mitsos, A. Global solution of nonlinear mixed-integer bilevel programs (2010) Journal of Global Optimization, 47 (4), pp. 557-582.

[2] Saharidis, G.K., Ierapetritou, M.G. Resolution method for mixed integer bi-level linear problems based on decomposition technique (2009) Journal of Global Optimization, 44 (1), pp. 29-51.

[3] Gümüş, Z.H., Floudas, C.A. Global optimization of mixed-integer bilevel programming problems (2005) Computational Management Science, 2 (3), pp. 181-212.

[4] Nishizaki, I., Sakawa, M. Computational methods through genetic algorithms for obtaining Stackelberg solutions to two-level integer programming problems(2005) Cybernetics and Systems, 36 (6), pp. 565-579.

[5] Faísca, N.P., Saraiva, P.M., Rustem, B., Pistikopoulos, E.N. A multi-parametric programming approach for multilevel hierarchical and decentralized optimisation problems (2009) Computational Management Science, 6 (4), pp. 377-397.

[6] Faisca, N.P., Dua, V., Rustem, B., Saraiva, P.M., Pistikopoulos, E.N. Parametric global optimisation for bilevel programming (2007) Journal of Global Optimization, 38 (4), pp. 609-623.

460182 Deterministic Global Optimization of Large-Scale Pooling Problems Via Topological Branch-and-Bound

Abstract:

The pooling problem is an industrially-relevant nonconvex nonlinear optimization problem minimizing cost on a feed-forward network of input nodes, intermediate storage or pool nodes, and output nodes [1]. The difficulty of the pooling problem arises from the nonconvex bilinear terms representing linear blending in the pools. These bilinear terms, which track quality across the network, typically multiply flow rates by concentrations or flow fractions.

This work develops an algorithm that solves specialized pooling problem instances to global optimality and integrates the algorithm into a branch-and-bound framework for more generic instances. Beginning with simple sub-problems, we parameterize with respect to pool concentrations and find embedded polyhedral properties that generate disjunctive cuts for more meaningful test cases. These disjunctive cuts correspond to distinct network topologies and motivate a novel branch-and-bound algorithm based on the set of active network flows.

Each sub-restriction of the standard pooling problem can be represented I-L-J-K where I is the number of input feed nodes, L is the number of intermediate storage pool nodes, J is the number of output nodes, and K is the number of tracked qualities in the network. For example, case I-1-1-1 has an arbitrary number of input nodes, 1 pool node, 1 output node, and 1 tracked quality.

For restriction I-1-1-1 with no limits on feed availability or pool capacity, this work shows that at most 3 input nodes are active and analytically calculates the optimal solution. Extending to restriction I-1-J-1, we develop a polynomial-time algorithm based on solving a univariate polynomial. This polynomial-time algorithm generalizes and extends recent work analyzing computational complexity of the pooling problem [2—4] since the analysis additionally integrates source-to-output streams and bounds on the network parameters. An interesting result is that subproblem I-1-J-1 preserves similar topological sparsity to I-1-1-1; there are at most J + 2 active input nodes for problems with no limits on feed availability or pool capacity. We also analyze several other limiting cases and thereby develop solution strategies for small problems that can be translated into effectively solving larger problems.

Transitioning from the polynomially solvable to the NP-hard instances, we design a topologically based branch-and-bound algorithm that branches on active network topologies and reduces the search space by eliminating infeasible topologies. Optimal flows are now simple to calculate for a fixed topology, so the topological branch-and-bound algorithm can avoid branching on the continuous variables in the pooling problem. We demonstrate the effectiveness of this method on a large number of problems including test cases in MINLPLib2 and the Dey and Gupte [5] test set. Further, we motivate how our results can extend from the pooling problem to any bilinear optimization problem based on process network topology.

This work is important for several reasons. First, we enable the solution of large scale pooling problems. Second, we are able to show why and how discretization is so useful for the pooling problem. Beginning in 2006, a significant body of work has shown that discretization from a nonconvex quadratically-constrained quadratic program (QCQP) into a mixed-integer linear program (MILP) is useful for generating rigorous lower bounds [6–8] and primal heuristic solutions [5] for process networks problems with bilinear terms. Our parametric characterization of the pooling problem, which uncovers embedded polyhedrality and disjunctions, show why this strategy has been working to our advantage and motivates how to choose disjunctions for new problems.

[1] R Misener and CA Floudas. Advances for the pooling problem: Modeling, global optimization, and computational studies. Appl. Comput. Math., 8(1):3 – 22, 2009.

[2] D Haugland. The computational complexity of the pooling problem. J. Glob. Optim., 64(2), 199–215, 2015.

[3] N Boland, T Kalinowski, and F Rigterink. A polynomially solvable case of the pooling problem. J. Glob. Optim., DOI: 10.1007/s10898-016-0432-6, 2016.

[4] D Haugland and EMT Hendrix. Pooling Problems with Polynomial-Time Algorithms. J. Optim. Theory Appl., DOI: 10.1007/s10957-016-0890-5, 2016.

[5] SS Dey and A Gupte. Analysis of MILP techniques for the pooling problem. Oper. Res., 63(2):412–427, 2015.

[6] CA Meyer and CA Floudas. Global optimization of a combinatorially complex generalized pooling problem. AIChE J., 52(3):1027 – 1037, 2006.

[7] R Karuppiah and IE Grossmann. Global optimization for the synthesis of integrated water systems in chemical processes. Comput. Chem. Eng., 30:650 – 673, 2006.

[8] R Misener, JP Thompson, and CA Floudas. APOGEE: Global optimization of standard, generalized, and extended pooling problems via linear and logarithmic partitioning schemes. Comput. Chem. Eng., 35(5):876–892, 2011.

460618 Solving Mpccs with IPOPT

Abstract:

Mathematical Programs with Complementarity Constraints (MPCCs) are used to model certain types of disjunctions without introducing binary variables. Mathematically, bi-level optimization, piecewise functions, and equilibria all can be reformulated as MPCCs. In process optimization, complementarities allow for models containing disappearance of phases, flow reversal, safety valve options, and some other discrete events within an NLP framework. However, MPCCs pose well-known difficulties, particularly because they violate constraint qualifications (CQs), such as the Linear Independence Constraint Qualification and the Mangasarian Fromovitz Constraint Qualification, at all feasible points. Without a CQ, the classical necessary optimality conditions do not apply so the optimum may or may not be a KKT point. Therefore MPCCs are challenging for most NLP solvers, which are based on the KKT conditions.

Nevertheless, nonlinear programming reformulation strategies may be used to find strongly stationary MPCC solutions, which are characterized by solutions to relaxed nonlinear programming (RNLP) problems that satisfy LICQ. Such reformulations include i) inequality relaxations of the complementarity constraints, ii) replacing complementarity constraints by smoothed NCP functions and iii) embedding complementarity terms as exact penalties. Inequality relaxations solve a sequence of more and more ill-posed NLPs. Similarly, the NCP function reformulation is extremely nonlinear near the solution leading to numerical difficulty. The exact penalty reformulation requires a priori knowledge of a suitable penalty parameter to be effective.

This talk discusses our experiences in extending the well-known primal-dual barrier solver, IPOPT, which is designed for large scale NLP problems, to deal with the solution of MPCCs. First, we propose two automatic penalty adjustment approaches in IPOPT. The first approach is similar to that from Leyffer, Lopez-Calva and Nocedal (2007), in which the complementary error is checked and the penalty parameter is potentially adjusted after each barrier problem is converged. The second approach only adjusts the penalty term after the penalty NLP converges. Next, we propose an

improvement to the equality reformulation of MPCCs. When a complementary constraint is biactive, the equality constraint leads to a singular row in the Jacobian. For this case, our previous work on structured regularization can detect dependent equalities automatically and numerically eliminate them at each iteration. Therefore, we propose an altered equality reformulation of MPCCs using structured regularization. If only one part of a complementary constraint is active, we relax the corresponding inequality constraint to effectively remove this constraint. As a result, the problem is well posed at all points and the modified Newton step leads to fast convergence.

We evaluate the proposed approaches on 120 problems from the MacMPEC library, other examples that do not have strongly stationary solutions, and several larger MPCCs derived on engineering systems. The latter examples are also scalable to variable problem sizes. For all of these case, we compare the efficiency of the new approaches are compared to previous NLP reformulation strategies.

460660 Feasibility Pump for Solving Convex MINLP Problems with Dicopt

Abstract:

Feasibility pump for solving convex MINLP problems with DICOPT

David E. Bernal, S. Vigerske, F. Trespalacios and I.E. Grossmann

DICOPT (short for Discrete Continuous Optimizer) is an MINLP solver based on the outer-approximation method[1,2]. This solver combines the algorithms of outer-approximation, equality relaxation and augmented penalty with other existing NLP and MIP solvers [3]. The outer approximation and equality relaxation algorithm proceeds by alternatingly solving NLP subproblems with fixed discrete variables (in the first iteration is the relaxed MINLP) and an outer-approximation MILP master problem that generates cuts to bound the feasible region. The criteria used by DICOPT for establishing optimality in convex MINLP is a crossover, which is obtained as soon as the bound defined by the objective of the last MIP master problem is worse than the best NLP solution found [3].

For some problems, DICOPT has difficulty in finding a first feasible solution. The main reason for this is that by default, DICOPT does not include constraint linearization in infeasible solutions, only no-good cuts. Furthermore, even if the linearizations are included for infeasible solution the issue persists in some problems. In order to allow DICOPT to find an initial feasible solution of convex MINLP problems, the implementation of a feasibility pump [4] is incorporated into DICOPT.

The feasibility pump is similar to the outer-approximation, but its objective is to find feasible solutions rather than optimal ones. The main idea of this algorithm is to decompose the original mathematical programming problem in two parts: integer feasibility and constraint feasibility. By solving an MIP subproblem its solution is an integer feasible solution, which may violate the constraints; and by solving a continuous relaxation of the original MINLP (NLP subproblem) the solution is constraint feasible but might not be integral. By minimizing in successive iterations the distance between these two types of solutions it is expected to achieve a solution that is both constraint and integral feasible.

The feasibility pump can be used as a standalone solver for convex MINLP problems. This is achieved by iteratively applying the method, while including a bound to the objective function. This bound is obtained by the best known solution and an epsilon improvement. The drawback of this algorithm is that may require many iterations, since each time the objective function is restricted to improve only by epsilon.

In this work, the feasibility pump is applied in DICOPT before the outer-approximation method. In the feasibility pump, large improvements in the objective function are enforced at each iteration. After the method finishes, all the cuts and the best known solution are passed to the outer-approximation method to find and prove optimality. The described algorithm has been implemented in DICOPT and is available in GAMS 24.5. We present computational results of the new method with over 50 instances of convex MINLP problems, and show that it outperforms the previous version of DICOPT.

References

1. Duran, Marco A., and Ignacio E. Grossmann. "An outer-approximation algorithm for a class of mixed-integer nonlinear programs." Mathematical programming 36.3 (1986): 307-339.

2. Fletcher, Roger, and Sven Leyffer. "Solving mixed integer nonlinear programs by outer approximation." Mathematical programming 66.1-3 (1994): 327-349.

3. G. Kocis and I. Grossmann, "Computational experience with DICOPT solving MINLP problems in process systems engineering," Computers and Chemical Engineering 13.3 (1989): 307-315.

4. P. Bonami, G. Cornuéjols, A. Lodi and F. Margot, "A feasibility pump for mixed integer nonlinear programs," Mathematical Programming, vol. 119, pp. 331-352, 2009.

460661 Rigorous Surrogate-Based Optimization Strategies That Integrate Glass Box/Black Box Process Models

Abstract:

Modern nonlinear programming solvers can efficiently handle very large scale optimization problems when accurate derivative information is available. However, black box or derivative free modeling components are often unavoidable in practice when the modeled phenomena may cross length and time scales. This work is motivated by examples in chemical process optimization where most unit operations have well-known equation oriented representations, but some portion of the model (e.g. a complex reactor model) may only be available with an external function call.

Several approaches have been proposed to handle this situation. Perhaps the easiest method to implement is to consider the entire system as a black box and apply black box optimization in the space of independent variables. Inequality constraints can be handled using a penalty function method. In a flowsheeting context, this could mean repeatedly simulating the flowsheet in a sequential modular manner at various operating points. Unfortunately, converging the equality constraints (converging the flowsheet) requires many calls to the black box function for each function call used by the derivative free solver. Another approach to this problem is the use of surrogate models (i.e. reduced models). A surrogate model is an equation oriented approximation of the black box that allows traditional derivative based optimization to be applied directly. However, optimization tends to exploit approximation errors in the surrogate model leading to inaccurate solutions and repeated rebuilding of the surrogate model. However, even if the surrogate model is perfectly accurate at the solution, this only guarantees that the original problem is feasible. Because optimality conditions require gradient information, a higher degree of accuracy is required.

In this work, we consider the general problem of hybrid glass box/black box optimization, with focus on guaranteeing a sufficient accuracy level in the surrogate model to bound optimality error at a desired level. We first propose an algorithm that combines ideas from SQP filter methods and derivative free trust region methods to solve this class of problems. The black box portion of the model is replaced by a sequence of surrogate models in trust region subproblems. By carefully managing surrogate model construction, the algorithm is guaranteed to converge to optimal solutions of the original, rigorous problem. We then examine how the hybrid structure of the problem can be exploited to reduce the number of updates to the surrogate model. Moreover, with error bounds for a given surrogate model, it is sometimes possible to propagate this error through to optimality error of the surrogate model optimization problem. This information can be used to derive effective termination criteria for the trust region algorithm, or even to inform surrogate model construction by suggesting desired accuracy bounds. These concepts are demonstrated on optimization benchmarks as well as illustrative examples for process optimization.

464425 Capacity Planning of Industrial Gas Plants with Rational Markets Under Demand Uncertainty

Abstract:

Capacity planning under uncertainty is a crucial problem for industrial gases companies due to the significant capital requirements. With dynamic market conditions, it is extremely important for these companies to develop capacity expansion plans based on realistic models. Insufficient capacity may lead to loss of market share and excessive investments in capacity may have a detrimental impact on their economic performance. Uncertainty in future demands and competition among producers to satisfy them, increase the complexity of the capacity planning problem. However, the role of uncertainty and rational decision-makers is fundamental for the success of capacity expansion projects.

Capacity planning for industrial gas producers is a challenging task because expansion plans are highly dependent on market behavior and the future demands [1]. A suboptimal expansion strategy may lead to unnecessary investments or loss of market share. To mitigate the risk associated with uncertain demands and to account for market preferences in the face of competition, a stochastic bi-level MILP optimization model is proposed [4].

The stochastic bilevel MILP optimization model involves an upper level that maximizes the expected Net Present Value (NPV) of an industrial producer, and a lower level that minimizes the expected cost paid by the markets over different demand scenarios. The upper level determines capacity planning decisions, which are modeled with discrete variables; whereas the lower level is a Linear Program (LP) which assigns market demands to competing producers. Our formulation finds a Stackelberg equilibrium between the leading producer and the markets. The producer makes the first move by deciding on the expansion plan, considering the rational responses of the markets in each scenario; then, the markets respond according to their cost minimization criterion. The strong Stackelberg equilibrium is found from the solution of the bilevel optimization problem. In order to solve the capacity planning problem, we reformulate it as a single-level MILP optimization problem. This reformulation leverages the strong duality property of the lower-level LP [2,3]. Uncertainty is considered in scenarios describing different realizations of demand. In this way, the upper level of the bi-level program maximizes the expected NPV obtained from the capacity expansion plan, while the lower level minimizes the cost paid by the markets in different scenarios. The model also accommodates the possibility to distribute capacity among different products and has the flexibility to constrain their production ratios.

The proposed capacity planning model is first implemented on a small illustrative example. The results demonstrate the advantages over its equivalent deterministic formulation. In particular, in the face of large uncertainties in demands, the proposed stochastic model is more conservative in the investments but resulting in a higher expected NPV compared to implementing the deterministic model under conditions of uncertainly. An effective domain reduction strategy developed to solve large-scale problems is demonstrated with an industrial size example. A stochastic bi-level model for capacity planning with non-uniform time periods is also proposed to better handle demand forecasting and to reduce the problem size of the resulting MILP.

[1] Eppen, G.D., R.K. Martin, L.Schrage, 1989. A scenario approach to capacity planning. Operations Research 37 (4), 517â€"527.

[2] Garcia-Herreros, P., Misra, P., Arslan, E., Mehta, S., Grossmann, I. E., 2015. A duality-based approach for bilevel optimization of capacity expansion. In: Krist V. Gernaey, J. K. H., Gani, R. (Eds.), 12th International Symposium on Process Systems Engineering and 25th European Symposium on Computer Aided Process Engineering. Vol. 37 of Computer Aided Chemical Engineering. pp. 2021â€"2026.

[3] Garcia-Herreros, P; Zhang, L., Misra, P., Arslan, E., Mehta, S., Grossmann, I.E., 2015. Mixed-integer bilevel optimization for capacity planning with rational markets. Computers & Chemical Engineering 86, pp. 33-47.

[4] Ryu, H., V. Dua, E.N. Pistikopoulos, 2004. A bilevel programming framework for enterprise-wide process networks under uncertainty. Computers & Chemical Engineering 28, 1121â€"1129

458643 Implementing Robust Vapor-Liquid Equilibrium Calculations in Nonsmooth Multi-Stream Heat Exchanger Models

Abstract:

Optimization of the liquefaction process for natural gas has proven to be a challenging task in particular for two reasons. First, the refrigeration cycles introduce computational loops that cause convergence problems. Second, the refrigerants are mostly multicomponent mixtures with strong interactions in the search for optimal heating profiles to match the cooling profile of the natural gas. In addition, the driving forces in the heat exchangers are very tight, asking for rigorous modeling to make sure the simulation mimics the true behavior of the process.

Both gradient based methods and stochastic search methods have been used to try to optimize processes for Liquefied Natural Gas (LNG) production, where commercial simulators are used to provide the function values in the optimization, however, with limited success beyond the most simple liquefaction processes. Our research groups have been involved in the development of a new model for these processes with the inherent features required to be able to utilize the latest developments in the field of Global Optimization [1].

As an important part of this new approach to LNG optimization, Barton and coworkers at MIT [2] has developed a new non-smooth multi-stream heat exchanger model accounting for phase changes in the process. The model was successfully used to simulate the Poly Refrigerant Integrated Cycle Operations (PRICO) process for liquefied natural gas production. However, it suffered from convergence difficulties when studying more complex single mixed refrigerant (SMR) processes. The primary challenge was flash calculations, which frequently failed even for initial guesses close to the solution. These problems will of course increase even further when moving to Dual Mixed Refrigerant (DMR) processes and the Mixed Fluid Cascade (MFC) process.

In this project, an alternative model structure is studied, replacing the equation-oriented framework with a hybrid solution. Rather than converging the vapor-liquid equilibrium calculations simultaneously with the other model equations, they are included as nested subroutines and solved sequentially.
Nonsmooth Newton-type methods and the implicit function theorem for lexicographically smooth functions are used for computing analytical derivatives in each subroutine [3, 4]. In addition, more robust flash calculations are attempted by developing a nonsmooth version of the "inside-out" algorithm by Boston and Britt [5] that can be incorporated in the existing model framework. The same SMR processes are then simulated in order to investigate whether the alternative framework is more robust.

[1] Floudas, C.A. and Gounaris, C.E., A review of recent advances in global optimization, *Journal of Global Optimization*, vol. 45(1), pp. 3-38, 2009.

[2] Watson, H.A.J., Khan, K.A. and Barton, P.I., Multistream heat exchanger modeling and design, *AIChE Jl.*, vol. 61(10), pp. 3390-3403, 2015.

[3] Facchinei, F., and Pang, J.S., *Finite-dimensional variational inequalities and complementarity problems*, vol. 2. Springer Science & Business Media, 2007.

[4] Khan, K. A., and Barton, P.I., Generalized derivatives for hybrid systems, *Submitted*, 2015.

[5] Boston, J.F. and Britt, H.I., A radically different formulation and solution of the single-stage flash problem, *Computers & Chemical Engineering*, vol. 2(2), pp. 109-122,

465812 Shared and Practical Approach to Conserve Utilities in Eco-Industrial Parks

Abstract:

Eco-Industrial Park (EIP) is a community of manufacturing and service facilities in close proximity. These enterprises collaborate with each other for improving the social, environmental, and economic performance of the whole complex in a sustainable manner by utilizing each other's resources efficiently, and sharing facilities, services, and utilities (Boix et al., 2015; Côté & Cohen-Rosenthal, 1998; Lowe et al., 1996). Examples of such successful industrial symbiosis are Jurong Island (Singapore), Kalundborg (Denmark), and Rotterdam (Netherlands). EIPs such as Jurong Island in Singapore are interested in assessing the potential for inter-plant heat integration among its inhabitants. The aim of such an enterprise wide heat exchanger network system is to exploit the synergistic heating/cooling needs and reduce total utility costs. However, there are many practical challenges in the implementation, construction, and operation of such an enterprise wide network (Chew et al., 2013) that need to be addressed while synthesizing inter-plant heat exchange network (HEN).

(1) The inter-plant heat exchange network synthesis (HENS) methodology should involve all the capital and operating expenses. Since, the exchanger and pipeline costs are comparable in case of long distance transport, inter-plant piping and transport costs cannot be ignored in configuring an EIP-wide HEN.

(2) The independent and diverse profit-making enterprises should be willing to participate and collaborate in such an EIP-wide HEN. Hence, there should be attractive economic benefits and fairness guaranteed to motivate the enterprises for such collaboration.

(3) The exchangers in an EIP-wide direct heat integration must have real physical locations. Furthermore, if a plant hosts a heat exchanger, it must bear the added responsibilities of maintaining and operating that exchanger, and guaranteeing the safety, security, and confidentiality of any guest streams.

(4) Operational disturbances in one plant shouldn't propagate to other plants, as the plants should be able to operate independently.

In this work, we present a novel strategy for EIP-wide HENS that addresses most of the issues highlighted above. We propose a central, shared, mutually agreed location for the HEN. The participating enterprises can appoint a third party logistics provider to manage and operate such a facility. We have considered pumping costs in the operating costs and the pipeline, pump and heat exchanger costs in the capital investment. By optimizing the HEN location, we can minimize the transport and pipeline costs. In our proposed model, we allowed the cooling (heating) of a hot (cold) stream both before entering and after leaving the HEN. To give operational flexibility and control to the parent company, existing in-plant utility heaters/coolers will be retained. Such an arrangement offers several advantages in terms of flexibility, control, cost, safety, security, confidentiality, accountability, operation, and management. We then propose a more practical and rational cost-benefit sharing arrangement. The contribution of an enterprise to capital expenses (CAPEX) and operating expenses (OPEX) is proportional to the utility cost savings it enjoys through the shared HEN. This ensures fairness among the enterprises as each enterprise enjoys an identical rate of return.

The EIP-wide HEN must trade off capital and operating costs versus utility savings. Therefore, we develop a mixed-integer non-linear programming (MINLP) model to maximize the net present value (NPV) of an EIP-wide HEN. The superstructure-based formulations allow sub-streams and non-isothermal mixing in a stage (Huang & Karimi, 2013; Yee & Grossmann, 1990). Since we wish to maximize the net present value of the shared HEN, discounted flows and rate of return has been included in the cost equation. Our model allows arbitrary cost expressions for exchangers with both fixed and variable components and cost multipliers for material and installation considerations. The single HEN location allows the full flexibility of multiple heat exchanges among streams from different plants. We have also improved the efficiency of formulation by adding additional constraints for eliminating redundant or complicated network designs. We have limited the number of sub-streams in a stage, avoided repeat identical matches, and included lower limits on exchanger areas and temperature drops to avoid exchangers with small-duty or small-area. If the HEN looks profitable (NPV $>$ 0), each enterprise will enjoy annual savings. In return, it must share the various costs of the HEN in proportion to its annual savings.

Our proposed model for the case study in Hiete et al. (2012) offers an NPV of \$5.71 million with a utility savings of \$1.74 million/a for the EIP. The piping and pumping costs are major costs in this shared HEN, which restates their importance in EIP-integration. Enterprise E1 contributes 67.4 % to the CAPEX and OPEX, as it enjoys the most savings. In the game theory based formulation in Cheng et al. (2014), enterprise E2 trades streams to E1 and E3, but pays nothing for the exchangers. In contrast, our optimal HEN apportions costs on a more rational and practical basis of utility savings. Since E1 gains the most from the EIP-integration with a utility cost savings of \$126 456/a (\$29 411 /a for E2 and \$71 183 /a) for E3, it contributes the largest share of the costs (E1:55.7 %, E2:13.0 % and E3:31.3 %). Like the case study of direct heat integration in Wang et al. (2015), our HEN also achieves a total utility savings of 95 %, which is the thermodynamic limit. Wang et al. (2015) did not clearly indicate the locations of their heat exchangers (or equivalently the streams being transported). E2 contributes 90.5 % of investment costs, as it supplies the cold streams that would otherwise require expensive hot utilities. Hence, solutions are fair, as companies saving expensive heating utilities will have to contribute more in the investment costs.

Heat integration has been employed over the years in several chemical processes, in particular refineries and petrochemical. As discussed, EIP-wide heat integration poses several unique economic, design, logistics, operations, and safety challenges. But, the utility costs for refineries and petrochemical plants can be huge every year, which is an incentive to reduce these costs in spite of the various challenges. In fact, modern EIPs like Jurong Island in Singapore and Jubail in Saudi Arabia are very much interested in assessing the potential for inter-plant heat integration. The methodology presented here paves the way for assessing and implementing heat integration for such "futuristic" systems.

Boix, M., Montastruc, L., Azzaro-Pantel, C., & Domenech, S. (2015). Optimization methods applied to the design of eco-industrial parks: a literature review. Journal of cleaner production, 87, 303-317.

Cheng, S.-L., Chang, C.-T., & Jiang, D. (2014). A game-theory based optimization strategy to configure inter-plant heat integration schemes. Chemical Engineering Science, 118, 60-73.

Chew, K. H., Klemeš, J. J., Wan Alwi, S. R., & Abdul Manan, Z. (2013). Industrial implementation issues of Total Site Heat Integration. Applied thermal engineering, 61, 17-25.

Côté, R. P., & Cohen-Rosenthal, E. (1998). Designing eco-industrial parks: a synthesis of some experiences. Journal of cleaner production, 6, 181-188.

Hiete, M., Ludwig, J., & Schultmann, F. (2012). Intercompany Energy Integration. Journal of industrial Ecology, 16, 689-698.

Huang, K. F., & Karimi, I. A. (2013). Simultaneous synthesis approaches for cost-effective heat exchanger networks. Chemical Engineering Science, 98, 231-245.

Lowe, E. A., Moran, S. R., Holmes, D. B., & Martin, S. A. (1996). Fieldbook for the Development of Eco-Industrial Parks: Final Report: Indigo Development.

Wang, Y., Chang, C., & Feng, X. (2015). A systematic framework for multi-plants Heat Integration combining Direct and Indirect Heat Integration methods. Energy, 90, Part 1, 56-67.

Yee, T. F., & Grossmann, I. E. (1990). Simultaneous optimization models for heat integration--II. Heat exchanger network synthesis. Computers & chemical engineering, 14, 1165-1184.

462355 Optimal Design and Operation of Integrated Multi-Vector Energy Networks

Abstract:

Integrated multi-vector energy networks are becoming more important because we have a much greater variety of primary resources of energy such as wind, solar, biomass etc. Whereas existing energy systems are dominated by fossil fuels as the primary resource, with electricity and natural gas as the energy vectors, increasing the penetration of renewable energy will change the energy mix and other energy vectors, such as hydrogen, methanol, syngas, ammonia etc. may also play important roles. There are many different pathways to satisfy energy service demands through different energy vectors: for example, hydrogen, biofuels or electricity are considered promising alternatives to petroleum for the transport sector.

The aim of this work is to determine the best design and operation of integrated energy network to obtain the most value from limited available resources while minimising its negative socio-economic and environmental impacts. This endeavour requires the use of mathematical models to provide an accurate representation of the potential technologies and transport infrastructures that may become part of the network, as well as optimisation techniques to determine the most promising designs among the many possible alternatives. This involves selecting the right combinations of conversion and storage technologies, where to locate them, what their capacities should be and how to interconnect them via transport infrastructures. Moreover, the mathematical models need to be sufficiently detailed to account for all of the operational issues at different time scales (e.g. hours, days, seasons, years) and interactions between the different energy vectors. In this conference, we will present a general mixed integer linear programming model that can simultaneously determine the design and operation of any integrated energy networks comprising technologies for conversion and storage and infrastructures for transport of resources (both material and energy). The model takes into account the spatial and temporal distribution of system properties and simultaneously determines the spatial structure, i.e. the location, number, capacity of the technologies and the structure of the transport infrastructures that connect them, as well as the hourly operation of the entire network. It is a dynamic model in order to properly account for the intermittency of renewables and dynamics of energy storage. The model includes detailed and coherent storage inventory balances, thus it is able to track the inventory level of each resource in storage at every hour over the entire planning horizon (e.g. out to 2050).

This presentation will describe the model and discuss a decomposition method that enables this very large, and otherwise intractable, optimisation model to be solved. We will also present case studies wherein the design and operation of an integrated electricity-hydrogen-syngas-natural gas network are determined in order to maximise the value (or minimise the cost or environmental impacts) of meeting the demands for heat, electricity and transport fuels in Great Britain.

464617 Simultaneous Optimization of Design and Operation Strategies for CHP Systems

Abstract:

Combined heat and power (CHP) facilities are a very promising path to reducing CO_2emissions and increasing efficiency in the power generation sector, especially when combined with residential solar photovoltaic (PV) power generation. CHP facilities rely on natural gas, a cleaner fuel than coal, to generate electricity, as well as hot and chilled water to provide heating and cooling. The ability to supply these essential residential utilities in an efficient way on a medium to large scale opens the path for combining district cooling, heating and power generation, and suggests that CHP plants are an appealing choice as an integrated utility supplier for the neighborhood of the future. Yet, there are currently no CHP plants that serve exclusively residential neighborhoods, and published works exploring this possibility are scarce.

In our previous work [1], we demonstrated that a CHP plant with PV integration can in principle meet the demands of a residential neighborhood during the summer months. However, a CHP plant operating in island (i.e., grid-disconnected) mode must be optimally sized to maximize efficiency and to lower the capital and marginal costs, while producing enough electricity, heating, and cooling to meet the neighborhood demands at all times. A majority of the approaches described in literature for the optimal CHP plant sizing tend to select equipment by analyzing the various utility demands (i.e. electricity, steam, and chilled water) individually. Also, they only minimize the capital and O&M costs, overlooking the transitional costs of equipment turning on, by either not allowing the equipment to turn off or ignoring the monetary value of the wear and tear on the equipment. Lastly, many micro-grid papers assume that the main grid will be there to support the system at all times, if the CHP plant is not able to produce the needed energy.

In the paper, we describe a novel simultaneous optimization of design and operation strategies for a CHP plant as a utility producer for a residential neighborhood operating in island (i.e., grid-disconnected) mode, incorporating residential photovoltaics, implemented in a centralized fashion. In order to more accurately predict the operational costs of the CHP plant over a long period of time, the integrated design optimization is combined with an operational-oriented scheduling, forming a two-level problem. The first level consists of the presented integrated design and scheduling optimization, which selects equipment for a plant that must meet hourly residential neighborhood utility demands while minimizing the overall cost of the plant. The second level uses the optimal equipment sizes determined from the first level, and schedules the plant for a year, giving a better result of the operation economics of the CHP plant over a longer period of time. Utilizing data collected by Pecan Street Research Inc., a non-profit smart grid demonstration project headquartered at The University of Texas at Austin, a year of hourly residential heating, cooling, and electricity demand are analyzed and evaluated. These data are then used to create a time-resolved energy demand profile describing residential energy use, which is subsequently serves as a basis for optimizing the size and operation of the CHP plant so that all utility demands are met, while minimizing the capital and marginal costs, including transition penalties.

References:

[1] Ondeck, A.D., Baldea, M., Edgar, T.F. (2015). *Data-Driven Modeling and Optimal Operation of District-level Combined Heat and Power and Photovoltaic Power Generation System*. Applied Energy, 156, pp. 593-606.

462056 Exploiting Dynamic Flexibility to Enable Participation in Multi-Scale Electricity Markets

Abstract:

There is significant interest in re-designing the control architectures for a broad spectrum of energy and manufacturing systems to exploit time-varying electricity prices. For example, the Alcoa Point Comfort Power Plant, which is a utility plant that provides electricity and steam to the adjacent aluminum manufacturing facility, re-optimizes its operations every 15 minutes in response to electricity and natural gas price. These new energy-oriented architectures provide load flexibility to the power grid in exchange for monetary payments. Most academic literature on industrial demand response optimization focuses on scheduling and planning decision layers and considers coarse (e.g., 1 hour) time discretizations with multi-day planning horizons. Emphasis has been placed on optimal planning under time-vary electricity prices, although recent work considers providing reserve capacity and interruptible loads. This emphasis missing flexibility at fast timescales (e.g., minutes to seconds) that may be monetized through participating in real-time energy markets and/or providing regulation capacity and other ancillary services.

We develop a mathematical model for multi-scale electricity markets run by the California Independent System Operator (CAISO). The model incorporates prices from both the day-ahead (1 hour intervals) and real-time markets (5 â€“ 15 minute intervals), and considers both energy and ancillary service (regulation, reserves, etc.) sales. Market rules are expressed as linear algebraic expressions with continuous and optionally discrete variables. The models are general and applicable to both conventional thermal generators and storage devices (e.g., batteries).

Using the market model, we optimize the operation policy of a combined heat and power utility system using real price data for all of 2015. We observe participating in energy markets run at multi-times decreases net operating costs (fuel minus market proceeds) by 14 â€“ 54%, depending on fuel price, without degrading satisfaction of onsite steam and electricity demand. The savings increase if the onsite steam and/or electricity demands are made flexible.

Approximately one-quarter to one-half of the anticipated savings are from regulation capacity sales to the market. In CAISO, the automatic generator control (AGC) system sends new electricity production setpoints to generators providing regulation every 5 seconds that are within a specified flexibility band (units: MW_e). Capacity payments are based on the size of this band. Thus regulation offers a way to monetize the inherent flexibility in manufacturing/energy systems at faster timescales (seconds compared to electricity price variations that occur at slower timescales (minutes to hours). In practice exploiting this flexibility and realizing revenues from electricity markets requires sophisticated controls strategies that manage multiple time-scales and consider uncertainty in market prices.

468112 A Rolling Horizon Scenario-Based Approach for Smart House Management Under Uncertainty

Abstract:

The development of sustainable, efficient and environmental friendly energy supply chains has led to advances in the area of Energy Systems Engineering. In this context, the growing interest in the exploitation of renewable energy sources has led to an increasing use of microgrids, which are decentralized networks able to integrate several heat/power generation systems of low/medium capacity (Hodge et al., 2011; Zhang et al., 2013), with the main purpose of distributing the generated heat/energy locally so as to reduce energy losses and improve the responsiveness to energy demand fluctuations (Kopanos et al., 2013; Tong et al., 2015). These kind of networks usually involve the presence of renewable generators. So, it is imperative that the variations in the behaviour of natural energy sources requires the consideration of uncertainty in microgrids. This variability must be considered in order to ensure the operability and the reliability of the system, since the optimal solution obtained under deterministic conditions can be sub-optimal or even infeasible. Particularly, the operations management of microgrids can be affected by several types of uncertainty, such as energy production and energy demand variations.

This work addresses the presence of uncertainty through a discrete-time two-stage stochastic programming approach, which is combined with a rolling horizon approach (Kopanos et al., 2014) for the simultaneous management of energy supply and demand in microgrids. The proposed Mixed Integer Linear Programming (MILP) formulation contains inequality constraints corresponding to the available heat and energy generation technologies, as well as equality constraints corresponding to heat and energy balance equations that describe flows, production, storage and consumption levels. This mathematical formulation uses a scenario-based stochastic programming approach where the scenarios are associated to internal variations in the duration of the energy consumptions as well as in the overall heat demand. However, the high complexity related to the estimated weather forecast, makes the consideration of all possible external scenarios, computationally intensive. The variability in weather conditions may affect the availability and production capacity of renewable energy generators. Thereby, updating input data (i.e., wind profile, heat and energy demand) is needed to ensure the adequate quality in the obtained results. Hence, an MILP two-stage stochastic programming approach was incorporated within a rolling horizon scheme that periodically updates input data information as the uncertain input parameters are revealed or considered to be known with certainty. The main decisions to be made to maximise the profit of the microgrid includes the execution of consumptions, the amount of heat and energy to be produced or purchased, the heat and energy storage levels and the amount of energy to export to the power grid. The methodology proposed above was tested on a case study and promising results demonstrating the applicability of the methodology for decision making were obtained. This work also forms a basis for future work to address more complex problems.

References

Hodge, B.M., Huang, S., Siirola, J.D., Pekny, J.F. & Reklaitis, G.V. (2011). A multi-paradigm modeling framework for energy systems simulation and analysis. Computers & Chemical Engineering, 35, 1725-1737.

Kopanos, G.M., Georgiadis, M.C. & Pistikopoulos, E.N. (2013). Energy production planning of a network of micro combined heat and power generators. Applied Energy, 102, 1522-1534.

Kopanos, G.M. & Pistikopoulos, E.N. (2014). Reactive scheduling by a multiparametric programming rolling horizon framework: A case of a network of combined heat and power units. Industrial & Engineering Chemistry Research, 53, 4366-4386.

Tong, C.D., Palazoglu, A., El-Farra, N.H. & Yan, X.F. (2015). Energy demand management for process systems through production scheduling and control. AIChE Journal, 61, 3756-3769.

Zhang, D., Shah, N. & Papageorgiou, L.G. (2013). Efficient energy consumption and operation management in a smart building with microgrid. Energy Conversion and Management, 74, 209-222.

450742 Optimization Models for Shale Gas Well Refracture Treatments

Abstract:

Optimization Models for Shale Gas Well

Refracture Treatments

Markus G. Drouven[1], Diego C. Cafaro[2] and Ignacio E. Grossmann[3]

[1,3]Department of Chemical Engineering

Carnegie Mellon University

Pittsburgh, PA 15213, USA

[2]INTEC (UNL – CONICET)

Güemes 3450, 3000 Santa Fe, Argentina

[1]mdrouven@cmu.edu, [2]dcafaro@fiq.unl.edu.ar [3]grossmann@cmu.edu

Abstract

Shale gas wells are well-known for their characteristically steep decline curves. In fact, some shale wells produce more than half of their total estimated ultimate recovery (EUR) within the first year of operation. The initial production peak after well completion is caused by the sudden release of previously trapped hydrocarbons in the pores of the reservoir. The subsequent decline is ultimately driven by pressure depletion and the ultra-low permeability of the shale play. Upstream operators oftentimes struggle with the production decline curves. For one, they are contractually obligated to providing steady gas deliveries to midstream distributors over time – which is difficult to accomplish given the characteristic, dramatic production decline immediately after wells are turned in line. Moreover, as Cafaro and Grossmann [1] suggest and Drouven and Grossmann [2] confirm, operators need to maximize the utilization of production and gathering equipment – such as pipelines and compressors – in order to stay profitable. In reality, however, production and gathering equipment is usually sized based on the initially high production rates. This means that within a matter of months shale gas wells feed into oversized pipelines and compressor stations, and equipment utilization drops. Worse even, in order to satisfy contractual gas delivery agreements, operators see themselves forced to open up new wells continuously to honor their obligations, and hence the process is repeated over and over again.

Refracturing presents a promising strategy for addressing the characteristically steep decline rates of shale gas wells [3]. The core idea behind refracturing is to restimulate the reservoir such that it yields previously untapped hydrocarbons and improves the overall production profile of a well. Whether or not a refracture treatment will reinvigorate a shale gas well depends on a number of factors including the characteristics of the reservoir and the initial completion design. Historically, refracture treatments have been applied predominantly to shale gas wells suffering from low production rates due to known suboptimal initial stimulations and completions. However, Dozier et al. [4] argue that even wells with effective initial treatments have shown significant production improvements when restimulated after an initial period of production and partial reservoir depletion.

The problem addressed in this paper can be stated as follows. We assume that a candidate shale gas well has been identified for refracturing. For this well, a long-term production forecast as well as the production profile after additional refracture treatments at any point in time over the planning horizon is given. A gas price forecast along with expenses for drilling, fracturing, completions and refracturing operations are also available as problem data. The problem is to determine: (a) whether or not the well should be refractured, (b) how often the well should be refractured over its entire lifespan, and (c) when exactly the refracture treatments should be performed. The objective is to maximize either: (a) the estimated ultimate recovery (EUR) of the well, or (b) the net present value (NPV) of the well development project.

First, we present a continuous-time nonlinear programming (NLP) model to determine whether or not a shale gas well should be refractured, and when to schedule the refracture treatment. The proposed NLP model relies on the assumption that the well productivity profile follows a decreasing power function of time. To predict well performance prior to and after a refracture treatment, we propose an effective forecast function that mimics real-life curves. Next, we extend the proposed framework to allow for multiple refracture treatments and present a discrete-time mixed-integer linear programming (MILP). In an attempt to reduce solution times to a minimum, we compare three alternative formulations against each other (big-M formulation, disjunctive formulation using Standard and Compact Hull-Reformulations) and find that the disjunctive models yield the best computational performance. Finally, we apply the proposed MILP model to two case studies to demonstrate how refracturing can increase the expected recovery of a well and improve its profitability by several hundred thousand USD.

References

[1] Cafaro, D. C.; Grossmann, I. E. Strategic Planning, Design, and Development of the Shale Gas Supply Chain Network. *AIChE J.* 2014. doi: 10.1002/aic.14405.

[2] Drouven, M. G.; Grossmann, I. E. Multi-Period Planning, Design and Strategic Models for Long-Term Quality-Sensitive Shale Gas Development. *AIChE J.* 2016 (accepted for publication).

[3] Jacobs T. Renewing Mature Shale Wells Through Refracturing. *SPE News*. 2014, May Issue.

[4] Dozier, G., Elbel, J., Fielder, E., Hoover, R., Lemp, S., Reeves, S., Siebrits, E., Wisler, D., Wolhart, S., Refracturing works. *Oilfield Review*. 2003; 10(8):38-53.

460724 A Composite-Curve-Based Biomass Procurement Planning Approach

Abstract:

A Composite-Curve-Based Biomass Procurement Planning Approach

Wenzhao Wu, Daniel Kurniawan, WenBo Zhu, Christos T. Maravelias*

Dept. of Chemical and Biological Engineering, University of Wisconsin-Madison, Madison, WI 53706

The production of fuels and chemicals from biomass has received considerable attention recently due to environmental concerns[1]. Since the production of biofuels involves relatively expensive feedstock and energy-intensive biomass transportation, any biomass-to-fuels strategy should include an efficient, both in terms of cost and environmental impact, biomass (feedstock) supply chain. Unlike fossil fuels, biomass, as a low-energy density resource, is sparsely distributed. The efficient biomass transportation thus requires biomass procurement planning methods[2]. In many studies, the farms are treated as points without shape or area[3]. This is a reasonable assumption when the transportation distance is so large that the shape and size of the farms can be neglected. In this case, the transportation problem is modeled as a point-to-point (farm-to-refinery) problem. However, the shape and size of the farms cannot be neglected when the refinery is close to the farm, which means that the size of the farm is not significantly smaller than the transportation distances, which in turn means that the error in approximating the real transportation distance with the distance between the center of the farm and the bio-refinery can be quite large. In this case, transportation should be treated as a region-to-point problem. To this end, we discuss a novel approach to biomass procurement planning on a region-to-point basis.

In terms of transportation, we propose a region-to-point modeling approach based on mathematical integration (in a polar coordinate system) over the sourcing region that has unique characteristics such as shape, location, and productivity. The transportation cost is correlated with the amount of biomass ("mass") procured from each farm. The final result is a function $C = f(M)$, where M is the mass, and C is the transportation cost. In other words, the proposed methods generate a function that returns the total cost of procuring M mass, which is graphically represented by a "procurement curve". Both algebraic and numerical solution methods are discussed and demonstrated with examples.

In terms of system-level procurement planning, we develop a composite-curve-based approach that incorporates the regional transportation modeling method, and aims at identifying the biomass procurement plan that minimizes the total procurement cost (including biomass purchasing, harvesting and transportation). The specific steps for the generation of the composite curve using the individual procurement curves, as well as insights into the procurement planning problem are discussed. An analogy to the cold/hot composite curves in the pinch design method for heat exchanger networks[4] is also presented. A case study involving 12 farms (which are approximated as polytopes) and one refinery is presented (see Figure 1A). The corresponding composite curve is shown in Figure 2, and the final procurement strategy is graphically represented in Figure 1B.

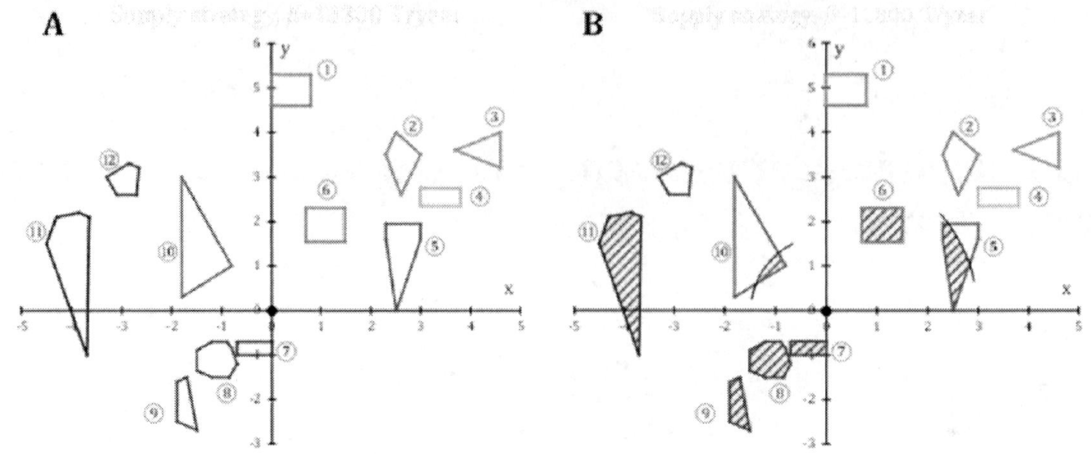

Figure 1. (**A**) Map of farms (the polytopes) surrounding a bio-refinery (the black dot at the origin) for the case study; (**B**) procurement strategy (represented by the green dashed areas) for a 11800 T/year demand. The farm numbers are labeled accordingly.

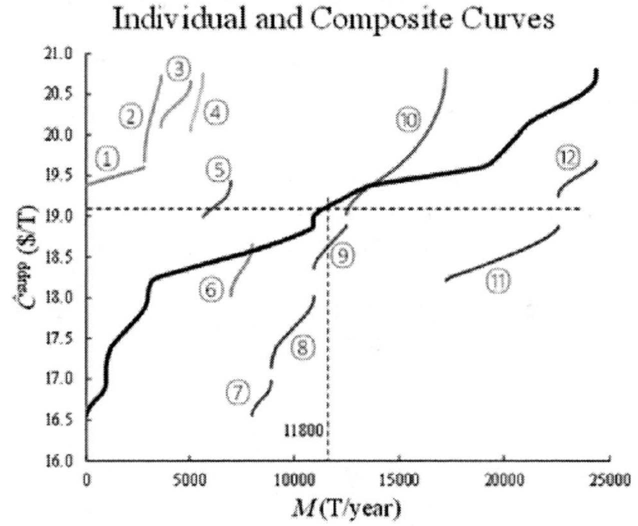

Figure 2. Individual procurement curves and the composite curve for the case study. The composite curve is marked thick. The demand of 11800 T/year and the corresponding per-mass supply cost ($\hat{C}^{supp}$) on the y-axis are marked with dashed lines. The farm numbers are labeled accordingly.

The proposed composite-curve-based method allows us to more accurately calculate transportation distance, and thus transportation costs and GHG emissions due to transportation. It also provides some key insights into the design of biofuel supply chains. In addition, the methods proposed in this work are integrated with mathematical programming to address complicated problems involving, for example, multiple feedstocks, multiple refineries, and multiple periods. We can solve such problems by either generating multiple composite curves, or directly incorporating the $C = f(M)$

function for each farm into a general supply chain optimization model.

References

[1] DOE Bioenergy Technologies Office, 2014. Multi-year program plan, Washington DC, USA: DOE.

[2] DOE/EERE, 2013c. Feedstock supply and logistics: biomass as a commodity, US: Department of Energy, Office of Energy Efficiency & Renewable Energy.

[3] You, F., Tao, L., Graziano, D. & Snyder, S. W., 2012. Optimal Design of Sustainable Cellulosic Biofuel Supply Chains: Multiobjective Optimization Coupled with Life Cycle Assessment and Input-Output Analysis. AIChE Journal, Volume 58, pp. 1157-1180.

[4] Linnhoff, B. & Hindmarsh, E., 1983. The pinch design method for heat exchanger networks. Chemical Engineering Science, 38(5), pp. 745-763.

459309 CFD Modeling and Computation for an Industrial Steam Methane Reforming Furnace

Abstract:

Hydrogen is one of the most important raw materials for petroleum refineries that convert crude oil into a variety of products with higher economic value, and its unavailability can limit the production rates of petroleum products. Hydrogen is produced on commercial scale most commonly and economically by the steam methane reforming (SMR) process. Specifically, the steam reforming process is an overall endothermic process in which raw natural gas, e.g., methane, reacts with high-pressure, high-temperature steam (superheated steam) in the presence of a nickel-based catalyst to produce hydrogen, carbon dioxide and carbon monoxide. Due to large production rates of hydrogen (e.g., the largest North American plant in 2004 could produce up to 300, 000 Nm^3/h of hydrogen [1]), the cost of raw materials can be on the order of millions of dollars [1]. Therefore, a small improvement in the process efficiency results in a great gain in profit margin of a plant. However, determining the optimal process parameters may be a time-consuming or risky process if performed experimentally, because it may require increasing the outer reforming tube wall temperature to improve efficiency, which can pose issues for the reactor material [1] [2]. To determine optimal processing parameters without experimental techniques, computational fluid dynamics (CFD) modeling is attractive because it can precisely capture all geometry characteristics of a given reformer through computer-aided design software, which in turn allows CFD models to generate simulation results that can be expected to serve as reasonable substitutes for experimental data, and the parametric effect can be visualized and quantified through CFD simulations.

This work develops a computational fluid dynamics (CFD) model of an industrial-scale steam methane reformer comprised of 336 industrial-scale reforming reactors, 96 industrial-scale burners and 8 industrial-scale flue gas tunnels. The industrial-scale reformer CFD model is developed by analyzing well-established physical phenomena, i.e., the transport of momentum, material, energy and turbulence, and chemical reactions, i.e., combustion and the SMR process, that take place inside the steam methane reformer.

Specifically, the $P - 1$ radiation model, standard $k -$ turbulence model, compressible ideal gas equation of state and finite rate/eddy dissipation (FR/ED) turbulence-chemistry interaction model are adopted to simulate the macroscopic and microscopic events in the reformer. The simulation results generated by the industrial-scale reformer CFD model are verified to be in agreement with the available typical plant data, and also data reported in literature. Furthermore, the CFD model of a reformer can provide insights into the system which cannot be captured by experimental data generated by on-site parametric study or by solution of a complete reformer mathematical model (e.g., the species distributions inside the combustion chamber.) Through this work, a converging strategy that allows one to quickly acquire the converged solution of complex CFD models is developed.

[1] Latham D. Masters Thesis: Mathematical Modeling of an Industrial Steam Methane Reformer. Queen's University, 2008.

[2] Pantoleontos G, Kikkinides ES, Georgiadis MC. A heterogeneous dynamic model for the simulation and optimisation of the steam methane reforming reactor. International Journal of Hydrogen Energy. 2012;37:16346-16358.

[3] Lao L, Aguirre A, Tran A, Wu Z, Durand H, Christofides PD. CFD modeling and control of a steam methane reforming reactor. Chemical Engineering Science. 2016;148:78-92.

[4] Aguirre A, Tran A, Lao L, Durand H, Crose M, Christofides PD. CFD Modeling of a Pilot-Scale Steam Methane Reforming Furnace. Advances in Energy Systems Engineering, Kopanos G, Liu P and Georgiadis M (Eds.), Springer, in press.

462415 Toward Optimal NGL Conversion to Olefins: Advances in Steam Cracking Optimization

Abstract:

Recent advances in the shale gas industry resulted in a significant increase in natural gas liquids (NGL) production [1]. NGL production increased by more than 100% since 2008 and is expected to grow by another 27% until 2040 [2,3]. In some wet shale plays, the combined ethane and propane composition can exceed 30% [4]. These recent trends motivate industrial efforts to build several new ethane crackers with a combined ethylene production capacity of 12.5 million tonnes/year in the United States [5]. Similar to ethane, other natural gas liquids can undergo steam cracking to produce olefins [6,7] with different underlying kinetics and mechanisms. This suggests that each hydrocarbon has a unique set of optimal operating conditions for the steam cracking process.

Although the steam cracking kinetics are well-established [6,7], several side reactions exist and the underlying reactor model contains a complex set of ordinary differential equations. In a stiff system such as this one, there is a need to develop a mathematical optimization framework to simultaneously address all trade-offs for optimal sizing operation. Therefore, the differential equations regarding mass, energy, and momentum balances are transformed using orthogonal collocation on finite elements [8]. In this work, the proposed reactor model also incorporates the kinetics of coke formation as well as decoking downtime. Finally, the reactor length is optimized to minimize the reactor volume and the associated investment costs. The resulting mixed integer nonlinear optimization model (MINLP) is solved to optimize the inlet flowrates, external heat flux, and reactor length (volume) to maximize the profit. The optimal reactor topologies and configurations are compared with the industrial cases, since the overall profit increased significantly. Finally, novel reactor designs are proposed that input a mixture of hydrocarbon feeds as opposed to a pure feed (ethane, propane, or butane). These optimal designs unravel the untapped opportunity of optimal allocation of NGLs as petrochemical feedstocks, while establishing a systematic framework for kinetic reactor design and optimization.

In order to illustrate the practical benefits of our systematic framework over a complete shale gas to olefins process, the optimal cracking designs are investigated in an integrated process superstructure [9]. Through the use of optimal steam cracking reactors that we developed, the net present values of these systems are significantly improved (>10%).

[1]: Floudas, C. A.; Niziolek, A. M.; Onel, O.; Matthews, L. R. Multi-scale systems engineering for energy and the environment: Challenges and opportunities. AIChE Journal 2016, 62 (3), 602-623.

[2]: Energy Information Administration, Annual Energy Outlook, 2015

[3]: Energy Information Administration, Petroleum and Other Liquids, 2015

[4]: Hill, R. J.; Jarvie, D. M.; Zumberge, J.; Henry, M.; Pollastro, R. M.; Oil and gas geochemistry and petroleum systems of the Fort Worth Basin, 2007, AAPG Bulletin, 91(4), 445-473.

[5]: Chang, J.; New projects may raise US ethylene capacity by 52%, PE by 47%, 2014, ICIS Petrochemicals.

[6]: Sundaram, K. M.; Froment, G. F.; Modeling of thermal cracking kinetics â€" I: Thermal cracking of ethane, propane and their mixtures, 1977, Chemical Engineering Science, 32(6), 601-608.

[7]: Sundaram, K. M.; Froment, G. F.; Modeling of thermal cracking kinetics â€" II: Cracking of iso-butane, of n-butane, and of mixtures ethane-propane-n-butane, 1977, Chemical Engineering Science, 32(6), 609-617.

[8]: Cuthrell, J. E.; Biegler, L. T.; On the optimization of differential-algebraic process systems, 1987, AIChE Journal, 33(8), 1257-1270.

[9]: Onel, O.; Niziolek, A. M; Floudas, C. A Optimal Production of Light Olefins from Natural Gas via the Methanol Intermediate. Industrial & Engineering Chemistry Research 2016, 55 (11), 3043-3063.

467043 Decarbonization of Electricity Grids: A Multi-Scale Challenge

Abstract:

Decarbonization of Electricity Grids: A multi-scale challenge

Sharifzadeh Mahdi[1] , Shah Nilay.

Centre for process Systems Engineering (CPSE), Chemical Engineering Department, South Kensington Campus, Imperial College London, SW7 2AZ, UK.

Abstract - Electricity power consumption in a country has a direct relation to the quality of life and economic growth. Nevertheless, about a quarter of global greenhouse gas emissions is associated with electricity and heat production, [1]. Therefore, decarbonisation of electricity grids is one of the major environmental impact pathways. The main methods for the desired decarbonization include carbon capture from exiting fossil-fired power generation stations and substituting fossil fuels with renewable or cleaner resources. This presentation covers a part of our recent research activities for design and operation of low carbon electricity grids.

In the first part of the presentation, we review our research for integrated design and operation of electricity grids. We proposed a comprehensive framework in which design and operation of electricity grid is considered simultaneously using stochastic mixed integer optimization programming. The formulated mathematical model is comprehensive and include conventional members of the electricity grid, the potentials for including renewable generators such as wind and solar energies, energy storage and the transmission network. The proposed method is demonstrated for the case of retrofitting the UK electricity grid to include 50% penetration of the renewables energies. Five years of historical data was applied for generating the representative scenarios regarding availability of wind and solar power and fluctuations in demand. The features of the interest include the configuration of the electricity grid after retrofit, investment in renewable power generators and the operational scheduling for power generation under various stochastic scenarios, [2].

Electricity grids are subject to high degrees of uncertainties. Examples of uncertainties in electricity grid include the stochastic behaviours of the solar and wind energies and fluctuation in electricity demand. The key observation is that future electricity grids will consist of a heterogeneous portfolio of energy conversion technologies and will be associated with a large degree of uncertainties. Such operational uncertainties can be accommodated using standby power generation capacitates, often provided by "fast-acting" grid members such as gas-fired power plants or by investing in energy storage. However, there is a much cheaper third option and that is to enhance the "smartness" of the grid. In this part of our research, we investigate the application of artificial intelligence (AI) for enhancing the predictability of electricity demand, in addition to wind speed and solar radiance. Various AI methods are incorporated and compared and the benefits of grid intelligence are quantified in economic terms, [3].

Currently, 80% of the household energy consumption is associated with heating and is supplied by fossil fuels. While electrification of residential heating systems can deliver an important environmental impact, electricity grids, unlike natural gas grids, are not resilient enough to accommodate the uncertainties in the heat demand. The third and last part of this presentation will study the potentials for decarbonisation of urban energy systems using combined heat and power systems integrated with carbon capture and storage. The feature of interest is the overall network flexibility and is demonstrated on the case of the urban energy system in London, [4].

Keywords

Smart grid, renewable energy, optimization, Big-data, Uncertainty, Combined Heat and Power (CHP) generation

References

[1] EPA, Global Emissions by Economic Sector, accessed online:

https://www3.epa.gov/climatechange/ghgemissions/global.html

[2] Sharifzadeh M, Lubiano H, Shah N. Optimal design and operation of integrated renewable energy systems. Under review.

[3] Sharifzadeh M, Sikinioti-Lock, A., Lubiano H, Shah N. The economic implications of enhance electricity grid smartness. In preparation.

[4] Sharifzadeh M, Shah N. Decarbonization of urbane energy using combined heat and energy cycles. In preparation.

1 Corresponding author, Email: mahdi@imperial.ac.uk , **Address:** Room C603, Roderic Hill Building, Centre for Process Systems Engineering (CPSE), Department of Chemical Engineering, Imperial College London, South Kensington, London SW7 2AZ, UK.

472604 Efficient Global Optimization for a Mixed AC-DC Power Distribution System

Abstract:

This presentation is concerned with global optimization for integrated design and operation of a mixed AC-DC power distribution system. The system under consideration is a benchmark electricity distribution system of medium office buildings. Frank and Rebennack [1] developed an integrated design and operation problem for a subpart of the system. The goal is to determine the network architecture (i.e., AC or DC) to which each branch or bus is assigned, in order to minimize the total energy supplied from the AC electric utility while satisfying physical restrictions, for a certain number of operating time periods. This problem is a large-scale nonconvex mixed-integer nonlinear programming (MINLP) problem. The integer variables are binary design decisions on network architectures. The nonconvex constraints come from the operational subproblems over the different time periods, each of which is essentially a mixed alternating current optimal power flow (ACOPF) and direct current optimal power flow (DCOPF) problem. While the DCOPF part of the problem is linear, the ACOPF part of the problem is nonconvex quadratic.

ACOPF problem plays a central role in optimal design and operation of electrical power networks, and the formal study of this problem can date back to 1962 [2]. The vast majority of ACOPF literature focuses on operational ACOPF problems that are nonconvex continuous optimization problems, and only a few of the existing methods (e.g., [3]) can guarantee finding a global optimal solution. This is because operational ACOPF problems usually need to be solved many times a day and none of the existing global optimization methods can solve the problem fast enough. However, global optimization seems to be more attractive for ACOPF design problems, as it is reasonable to make a design decision in a day or a week. Frank and Rebennack [1] successfully applied a decomposition-based global optimization method, called nonconvex generalized Benders decomposition (NGBD [4]), to some mixed AC-DC power distribution systems, and demonstrated the computational advantage of NGBD over state-of-the-art global optimization solvers.

The performance of NGBD is highly dependent on tightness of the convex lower bounding problem used. When the size of the distribution system becomes larger and the convex lower bounding problem becomes less tight, the performance of NGBD degrades quickly, as identified by Frank and Rebennack [1]. Here we propose to integrate domain reduction techniques into NGBD, in order to enhance the tightness of the convex lower bounding problem progressively during the decomposition procedure. Two types of domain reduction techniques are considered for NGBD. One is to solve customized convex bound contraction problems whose sizes are independent of the number of time periods. These bound contraction problems are also strengthened with valid linear cuts derived from the previously solved convex NGBD subproblems. The other is to perform range reduction calculations [5] for nonconvex NGBD subproblems, according to the primal and dual solutions of the previously solved convex NGBD subproblems. The domain reduction techniques not only can reduce the number of NGBD iterations, but also can speed up the solution of nonconvex NGBD subproblems. It will be shown in the case study that, the proposed method is significantly faster than the NGBD used by Frank and Rebennack [1] for the mixed AC-DC power distribution system.

In the case study, the convex lower bounding problem used for NGBD is obtained through linear relaxation, which is known to be weaker than a second-order cone programing (SOCP) relaxation or semi-definite programming (SDP) relaxation for ACOPF problems (e.g., [6]). Therefore, our method can be further improved via the use of a tighter nonlinear convex lower bounding problem.

References

[1] S. M. Frank and S. Rebennack, "Optimal design of mixed AC-DC distribution systems for commercial buildings: A nonconvex generalized Benders decomposition approach," *European Journal of Operational Research*, vol. 242, no. 3, pp. 710 – 729, 2015.

[2] J. Carpentier, "Contribution to the economic dispatch problem," Bulletin de la Socit Franaise des Electriciens, vol. 8, no. 3, p. 431447, 1962.

[3] C. Chen, A. Atamturk, and S. S. Oren, "Bound tightening for the alternating current optimal power flow problem," IEEE Transactions on Power Systems, vol. PP, no. 99, pp. 1–8, 2015.

[4] X. Li, A. Tomasgard, and P. I. Barton, "Nonconvex generalized Benders decomposition for stochastic separable mixed-integer nonlinear programs," *Journal of optimization theory and applications*, vol. 151, no. 3, pp. 425–454, 2011.

[5] Ryoo, H. S., Sahinidis, N. V., "A branch-and-reduce approach to global optimization," *Journal of Global Optimization*, vol. 8, no. 2, pp. 107-138, 1996.

[6] X. Bai, H. Wei, K. Fujisawa, and Y. Wang, "Semidefinite programming for optimal power flow problems," International Journal of Electrical Power and Energy Systems, vol. 30, no. 67, pp. 383–392, 2008.

469324 Data-Driven Modeling of Gas Leakage from Shale Natural Gas Wells

Abstract:

Availability of large amount of data recorded during natural gas well construction and completion stages can be put to use for studying underlying reasons for some of the critical problems which occur during the production phase. One such problem is natural gas leakage from the cemented annulus of the gas wells. The gas leakage may severely pollute the environment and underground water aquifers. Therefore, it is important to understand the reasons responsible for gas leaks and subsequently seek solutions to limit such leakage from wells drilled and completed in the future.

A number of factors may cause poor cementing, which may lead to development of cracks and micro-channels in the cement, providing a continuous path for gas to migrate from the production zone to the well head (Nelson, 2006). The extent of gas leakage from the cemented annulus is measured by a parameter called *Sustained Casing Pressure* (SCP) which is simply the pressure exerted by gas in the cemented annulus at the well-head. In a previous case study that we conducted on a set of gas wells data, we developed linear models to identify factors causing high SCP (Wehrens, 2011). Although the linear models provided a reasonable approximation of the underlying cause-and-effect relationships, the model accuracy was somewhat restricted by nonlinearities existing in the process (Hastie et al., 2009). In this paper we present a nonlinear modeling approach to study underlying causes of high gas leakage. A number of data-driven nonlinear modeling approaches, revolving around dimensionality reduction, as well as model specifics will be included in the final presentation.

References:

1. Hastie, T., Tibshirani, R., & Friedman, J. (2009). The Elements of Statistical Learning: Data Mining, Inference, and Prediction (2nd ed.).

2. Nelson, E. B. (2006). Well Cementing (2nd ed.).
3. Wehrens, R. (2011). Chemometrics with R: Multivariate Data Analysis in the Natural Sciences and Life Sciences. New York: Springer.

467393 The Simulation of an Industrial Wet Flue Gas Desulfurization Absorber

Abstract:

A comprehensive computational fluid dynamics (CFD) model was developed to study the process of desulfurization in the absorption tower of a wet flue gas desulfurization (WFGD) unit at a coal fired power station. This study was motivated by the need for South African coal-fired power stations to meet internationally aligned compliance levels w.r.t. SO_2 emissions, and the fact that current WFGD designs have been largely tailored for European power stations. Currently, the design of WFGD absorbers are based on empirical and semi-empirical correlations which address a limited set of design parameters; therefore, CFD modeling is considered to be essential for further enhancement of WFGD design to ensure compliance with the stringent environmental requirements w.r.t. SO_2 emissions. The main objective was to develop a comprehensive CFD model for an industrial (plant size) WFGD to provide an in-depth analyses of (i) the slurry injection dynamics and the droplet behavior with the inclusion of droplet properties such as size distribution, droplet distortion, wall-droplet interaction and evaporation, (ii) the hydrodynamics of the respective slurry and the gas phases to analyze the velocity distribution, the L/G ratio and temperature profiles, and (iii) the SO_2 absorption efficiency by the limestone slurry. The model was subsequently validated against plant test data of an operational WFGD absorber, with respect to the flue gas velocity, temperature and desulfurization efficiency.

The modelling of the injectors was successfully achieved with a statistical model based on an optimized number of parcel streams to ensure converged results. The detailed modelling of the injector assembly enabled accurate description of the gas and droplet behaviour with particular reference to the velocity fields, the L/G ratio, and the droplet size distribution. The gas and droplet velocities were found to be within the ranges suitable for controlling conditions to avoid carry-over of the smaller droplets in the outlet gas stream, and thus to ensure effective operation of the mist eliminator. The distortion of droplets was found to be very small with the largest effect occurring at the gas inlet to the tower. The inclusion of evaporation in the model proved to be important to accurately describe the temperature and moisture concentration profiles within the column. Consequently, the model predicted that saturation is achieved relatively close to the flue gas inlet, and that saturated conditions are prevalent over most of the column. The dependence of the desulfurization efficiency on the slurry pH as predicted by the model were found to correlate well with the plant test results.

463432 Modeling Chemotactic Bacterial Transport in Physically Homogeneous Groundwater Systems Containing Distributed Contaminant Sources

Abstract:

Groundwater pollution is of major concern as contaminants are continuously released into the subsurface via oil spills, chemical waste streams, and other means. Often times these contaminants become trapped within regions of the soil matrix that are characterized by low hydraulic conductivity making them difficult to remediate. In bioremediation strategies, chemotaxis â€" a phenomenon in which pollutant-degrading bacteria have the ability to detect chemical concentration gradients and swim toward the contaminant source â€" may enhance the transport of bacteria to sources of contamination and concomitantly lead to increased pollutant accessibility and biodegradation, even in regions with low water permeability. The influence of chemotaxis on bacterial transport in aquifers containing a distribution of contaminants with localized concentration gradients is not yet fully understood. This study aimed to utilize computational techniques to predict and better understand the migratory response of chemotactic bacteria to contaminant microniches within a model subsurface environment. Focus was placed on a sand-packed column containing randomly distributed pollutant sources. The bacterial population was introduced as a pulse input at the column inlet. A modified advection-dispersion equation containing an additional advection-like term to describe chemotaxis was used to model bacterial transport. The chemotactic velocity is a function of the chemotactic sensitivity coefficient, which describes the strength of chemotaxis, and the chemotactic receptor constant, which is the concentration at which the chemotactic velocity of the bacterial population is equal to one-half of its maximum value. Sensitivity analyses were performed on the chemotactic sensitivity coefficient and receptor constant to determine their influence on bacterial transport. It was equally important to predict the pollutant concentration profile within the porous media, via a standard advection-dispersion equation, because the chemotactic velocity is also a function of the pollutant concentration and concentration gradient. Our results showed that when chemotaxis was occurring, bacterial

transport in the direction of groundwater flow was retarded and the bacterial population was retained within the contaminated porous media for an extended period of time, compared to when chemotaxis was not occurring. Predictions from our numerical simulations were consistent with experimental observations from a sand column system analogous to our model set-up. Modeling chemotactic bacterial transport in porous media systems containing a distribution of contaminant sources is important for assessing the significance of chemotaxis in enhancing the retention of bacteria in sites of contamination.

472925 Analytical Calculation of Laminar Flow Reactor Effluent Concentrations without the Necessity of a CFD Approach

Abstract:

This work seeks to develop an analytical model capable of calculating effluent concentrations for segregated laminar flow reactors (SLFRs) based on the segregated flow model (SFM). The distinguishing element of the presented model is its applicability not only to isothermal, but also non-isothermal SLFRs. This essentially enables the accurate determination/calculation of effluent concentration for any type of laminar flow reactor, regardless of size or geometry. The sole requirement is the knowledge of the reactorâ€™s residence time distribution function (RTD), which is easily determined through simple pulse experiment. The accuracy of the developed model is demonstrated on two case studies featuring a Trambouze reaction scheme for both the isothermal and non-isothermal cases respectively. The obtained results are compared with outlet concentrations obtained from CFD simulations conducted for the same cases. The comparison clearly indicates the high accuracy of the analytical model consequently making a CFD approach superfluous.

458120 Active Fault Diagnosis for Stochastic Nonlinear Systems

Abstract:

Detection of process malfunction and faults is becoming increasingly challenging for high-performance technical systems. The increased level of dynamical complexity in these systems makes them prone to abnormalities such as sensor faults, actuator faults and disturbances. Fault detection and isolation (FDI) has become a crucial component of advanced process monitoring and control applications to meet stringent requirements of complex systems. FDI is intended to first detect whether a fault has occurred and then identify which component has failed [1][2]. Active fault diagnosis (FD) involves designing the system input signals such that the outputs pertaining to various fault scenarios can be differentiated. Active FD is particularly useful when the outputs corresponding to the nominal behavior and different faulty scenarios overlap or when the control system masks the effect of faults [3]. Thus, active FD enables detection and isolation of faults that may not otherwise be possible under nominal operation.

Motivated by the above considerations, several studies have considered active FD for uncertain systems. An active FD approach for closed-loop uncertain systems is developed in [4] based on the so-called fault signature matrix. In [5], an active FD problem is formulated via explicitly characterizing the nonlinear way that faults affect the process evolution through supervisory feedback control. In [6], an optimization-based approach has been proposed to robustify the filter with respect to signatures of dynamic nonlinearities in the presence of given disturbance patterns with known statistical information. The active FD problem has also been formulated in a probabilistic framework to deal with time-invariant system uncertainties using generalized polynomial chaos (gPC) [7]. gPC replaces the implicit relations between the uncertain system parameters and dynamic state variables with a series of orthogonal polynomials, whose statistical moments can be readily computed from the expansion coefficients.

This work presents a model-based active FD approach for stochastic, nonlinear systems subject to time-invariant probabilistic uncertainties (e.g., uncertainties in initial conditions and parameters) and time-varying system disturbances. The gPC framework [8][9] is adapted for efficient propagation of stochastic disturbances. Chance constraints are included in the active FD problem to account for the practical considerations and system constraints in a stochastic setting. The proposed approach is demonstrated for continuous Acetone-Butanol-Ethanol (ABE) fermentation [10] under various sources of probabilistic uncertainties and multiple fault scenarios. The goal is to enhance fault diagnosability by isolating nominal process operation and multiple faulty scenarios. The case study involves four uncertainties related to initial conditions and enzyme production rates as well as additive process noise, where the process is subject to an actuator fault that affects the dilution rate and to a process fault characterized by butanol inhibition. The simulation results show the promise of the proposed approach for active fault diagnosis under stochastic uncertainty.

References

[1] Nikoukhah, R. and Campbell, S. L. (2004), Auxiliary Signal Design for Failure Detection. Princeton University Press.

[2] Blanke, M., Kinnaert, M., Lunze, J. and Staroswiecki, M. (2003), Diagnosis and Fault-Tolerant Control. Springer, Berlin.

[3] Zhang, X. J. (1989), Auxiliary Design in Fault Detection and Diagnosis. Springer, Berlin.

[4] Niemann, H. (2006), "Active Fault Diagnosis in Closed-Loop Uncertain Systems". In Proceedings of the 6th IFAC Symposium on Fault Detection, Supervision and Safety of Technical Processes, pp. 587–592, Beijing.

[5] Du, M. and Mhaskar, P. (2013), "Active Fault Isolation of Nonlinear Process Systems". AIChE Journal ,vol. 59, no. 7, pp. 2435-2453.

[6] Esfahani, P. M. and Lygeros, J. (2016), "A Tractable Fault Detection and Isolation Approach for Nonlinear Systems With Probabilistic Performance". IEEE Transactions on Automatic Control, vol. 61, no. 3, pp. 633-647.

[7] Mesbah, A., Streif, S., Findeisen, R. and Braatz, R. D. (2014), "Active Fault Diagnosis for Nonlinear Systems with Probabilistic Uncertainties". In Proceedings of the IFAC World Congress. Cape Town, pp. 7079-7084.

[8] Xiu, D. and Karniadaki, G. E. (2003), "Modeling Uncertainty in Flow Simulations via Generalized Polynomial Chaos". Journal of Compurational Physics, vol. 187, pp. 137–167.

[9] Bavdekar, V. A. and Mesbah A. (2016), "Stochastic Nonlinear Model Predictive Control with Joint Chance Constraints." In Proceedings of the 10th IFAC Symposium on Nonlinear Control Systems, accepted, Monterey.

[10] Haus, S., Jabbari, S., Millat, T., Janssen, H., Fischer, R. J., Bahl, H., King, J. R., Wolkenhauer, O. (2011), "A Systems Biology Approach to Investigate the Effect Of pH-induced Gene Regulation on Solvent Production by Clostridium Acetobutylicum in Continuous Culture". BMC Systems Biology, vol.5, pp. 1-13.

Sparse nonlinear features based locally weighted kernel partial least squares for virtual sensing of nonlinear time-varying processes

Xinmin Zhang * **Manabu Kano** *

* *Department of Systems Science, Kyoto University, Kyoto 606-8501, Japan (e-mail: manabu@human.sys.i.kyoto-u.ac.jp).*

Abstract: This research introduces a novel locally weighted kernel PLS (LW-KPLS) regression approach based on sparse nonlinear features for tackling both strong nonlinearity and time-varying dynamics of industrial processes. Unlike the conventional locally weighted PLS (LW-PLS), the proposed method constructs a regression model in high-dimensional feature spaces by using sparse kernel regression factors (SKRFs), which describe nonlinear dependency between the query and training samples. By integrating the nonlinear features into the locally weighted regression framework, LW-KPLS not only can cope with the time-varying characteristics but also is more suitable for highly nonlinear processes. The prediction performance of the proposed LW-KPLS is compared to those of PLS, KPLS, and LW-PLS using two case studies. The application results have demonstrated that LW-KPLS shows superior prediction performance.

Keywords: Soft-sensor; virtual sensing; nonlinear time-varying processes; sparse kernel regression factors; kernelized Lasso; locally weighted kernel partial least squares

1. INTRODUCTION

In industrial processes, it is important to monitor and measure some important variables in real time. However, hardware analyzers have some practical problems such as time-consuming maintenance, aged deterioration, insufficient accuracy, long dead-time, and slow dynamics as clarified through the questionnaire survey in Japan (Kano and Fujiwara, 2013). To solve these problems, soft-sensors have been developed over the past decades. Among them, data-driven soft-sensors such as multiple linear regression (MLR), principal component regression (PCR), and partial least squares (PLS) regression are the most popular modeling approaches. However, these linear regression methods limit their applications to nonlinear industrial processes because they assume that process data are linear.

To handle the nonlinearity, kernel regression methods such as kernel PLS (KPLS) (Rosipal and Trejo, 2001) have been developed. KPLS can efficiently capture the nonlinearity through the introduction of the kernel function, and it has been successfully applied to many industrial processes. Furthermore, industrial processes are time-varying due to changes of process characteristics and operating conditions. Such changes cause accuracy deterioration of soft-sensors, and the model maintenance becomes the most important issue concerning soft-sensors in the process industry (Kano and Ogawa, 2010). To this end, just-in-time (JIT) modeling methods have been developed to cope with changes in process characteristics as well as nonlinearity. Locally weighted PLS (LW-PLS) is JIT modeling method that has successful industrial applications (Kim et al., 2013; Hirai and Kano, 2015). Even LW-PLS can achieve high prediction performance in particular applications, it may not function well for industrial processes with strong nonlinearity. The reason is that the weights used in the conventional LW-PLS usually do not adapt to changes in the strength of nonlinearity around a query, which may be caused by the time-varying nature of industrial processes. In addition, input features used for LW-PLS are still the original process variables, and the nonlinearity of the regression model only rests in the different treatment of samples, i.e., local weighting; therefore, LW-PLS does not always meet the demand of the prediction accuracy for highly nonlinear industrial processes.

In this work, a novel locally weighted kernel partial least squares (LW-KPLS) regression method based on sparse nonlinear features is proposed. Unlike the conventional LW-PLS, the proposed method uses sparse kernel regression factors (SKRFs) to weight the training samples. SKRFs are derived from a nonlinear kernelized Lasso model and describe nonlinear dependency between the query and training samples in the Hilbert feature space. By integrating the nonlinear features into the locally weighted regression framework, LW-KPLS not only is more suitable for highly nonlinear processes than the conventional LW-PLS but also can cope with time-varying characteristics.

This paper is organized as follows. The concepts of LW-PLS and kernelized Lasso are introduced in Section 2. Then, locally weighted kernel PLS based on sparse nonlinear features is presented in section 3. In section 4, the prediction performance of LW-KPLS is compared with PLS, KPLS, and LW-PLS using two case studies. The conclusions are given in section 5.

2. EXISTING METHODS

2.1 Locally weighted PLS (LW-PLS)

Assume $\boldsymbol{X} \in \Re^{N\times M}$ and $\boldsymbol{Y} \in \Re^{N\times L}$ are the historical input and output data, and whose n-th row are $\boldsymbol{x}_n^T$ and $\boldsymbol{y}_n^T$. N is the number of the samples, M and L denote the numbers of input and output variables, respectively.

In LW-PLS (Kim et al., 2011), when an output estimation is required for a query sample $\boldsymbol{x}_q$, the similarity ω_n between $\boldsymbol{x}_q$ and $\boldsymbol{x}_n$ is calculated, then a local PLS model is constructed by weighting samples with a similarity matrix $\boldsymbol{\Omega} \in \Re^{N\times N}$ defined by

$$\boldsymbol{\Omega} = \mathrm{diag}(\omega_1, \omega_2, \cdots, \omega_N) \tag{1}$$

where diag($\bullet$) represents a diagonal matrix.

The output estimate $\hat{\boldsymbol{y}}_q$ corresponding to the query sample $\boldsymbol{x}_q$ is calculated as follows.

1). Determine the number of latent variables R, and set $r = 1$.

2). Calculate the weight matrix $\boldsymbol{\Omega}$.

3). Calculate $\boldsymbol{X}_r$, $\boldsymbol{Y}_r$, and $\boldsymbol{x}_{q,r}$

$$\boldsymbol{X}_r = \boldsymbol{X} - \boldsymbol{1}_N[\bar{x}_1, \bar{x}_2, \cdots, \bar{x}_M] \tag{2}$$

$$\boldsymbol{Y}_r = \boldsymbol{Y} - \boldsymbol{1}_N[\bar{y}_1, \bar{y}_2, \cdots, \bar{y}_L] \tag{3}$$

$$\boldsymbol{x}_{q,r} = \boldsymbol{x}_q - [\bar{x}_1, \bar{x}_2, \cdots, \bar{x}_M]^T \tag{4}$$

$$\bar{x}_m = \frac{\sum_{n=1}^N \omega_n x_{nm}}{\sum_{n=1}^N \omega_n} \tag{5}$$

$$\bar{y}_l = \frac{\sum_{n=1}^N \omega_n y_{nl}}{\sum_{n=1}^N \omega_n} \tag{6}$$

where $\boldsymbol{1}_N \in \Re^N$ is a vector of ones.

4). Derive the r-th latent variable of $\boldsymbol{X}$

$$\boldsymbol{t}_r = \boldsymbol{X}_r \boldsymbol{w}_r \tag{7}$$

where $\boldsymbol{w}_r$ is the eigenvector of $\boldsymbol{X}_r^T \boldsymbol{\Omega} \boldsymbol{Y}_r \boldsymbol{Y}_r^T \boldsymbol{\Omega} \boldsymbol{X}_r$, which corresponds to the maximum eigenvalue.

5). Derive the r-th loading vector of $\boldsymbol{X}$ and the regression coefficient vector.

$$\boldsymbol{p}_r = \frac{\boldsymbol{X}_r^T \boldsymbol{\Omega} \boldsymbol{t}_r}{\boldsymbol{t}_r^T \boldsymbol{\Omega} \boldsymbol{t}_r} \tag{8}$$

$$\boldsymbol{q}_r = \frac{\boldsymbol{Y}_r^T \boldsymbol{\Omega} \boldsymbol{t}_r}{\boldsymbol{t}_r^T \boldsymbol{\Omega} \boldsymbol{t}_r} \tag{9}$$

6). Derive the r-th latent variable of $\boldsymbol{x}_q$.

$$t_{q,r} = \boldsymbol{x}_{q,r}^T \boldsymbol{w}_r \tag{10}$$

7). If $r = R$, finish the output estimate

$$\hat{\boldsymbol{y}}_q = [\bar{y}_1, \bar{y}_2, \cdots, \bar{y}_L]^T + \sum_{r=1}^R t_{q,r} \boldsymbol{q}_r. \tag{11}$$

Otherwise, set

$$\boldsymbol{X}_{r+1} = \boldsymbol{X}_r - \boldsymbol{t}_r \boldsymbol{p}_r^T \tag{12}$$

$$\boldsymbol{Y}_{r+1} = \boldsymbol{Y}_r - \boldsymbol{t}_r \boldsymbol{q}_r^T \tag{13}$$

$$\boldsymbol{x}_{q,r+1} = \boldsymbol{x}_{q,r} - t_{q,r} \boldsymbol{p}_r \tag{14}$$

8). Set $r = r + 1$, and go to step 4.

The estimation performance of LW-PLS depends on the similarity ω_n, which is defined on the basis of the weighted Euclidean distance as follows:

$$\omega_n = \exp\big(-\frac{\varphi d_n}{\sigma_d}\big) \tag{15}$$

$$\mathrm{d}_n = \sqrt{(\boldsymbol{x}_n - \boldsymbol{x}_q)^T \boldsymbol{\Theta} (\boldsymbol{x}_n - \boldsymbol{x}_q)} \tag{16}$$

$$\boldsymbol{\Theta} = \mathrm{diag}(\theta_1, \theta_2, \cdots, \theta_M) \tag{17}$$

where σ_d is a standard deviation of $\{d_n\}$, φ is a localization parameter, $\boldsymbol{\Theta} \in \Re^{M\times M}$ is a weighting matrix for Euclidean distance, and θ_m is a weight for each input variable. The estimation accuracy of LW-PLS is higher than or at least the same as that of PLS if the parameter φ is adjusted properly, because LW-PLS is equal to PLS when $\varphi = 0$.

2.2 Kernelized LASSO

LASSO (least absolute shrinkage and selection operator) regularizes ordinary least square regression with the L_1 norm and can lead to sparse solutions through shrinking the coefficients of the irrelevant or redundant features to zero (Roth, 2004). LASSO has succeed in selecting features for regression. To learn the nonlinear dependency between output and input variables, kernelized LASSO has been developed on the basis of some underlying basis functions (Gao et al., 2010). Given the input data $\boldsymbol{X} \in \Re^{N\times M}$ and the corresponding target vector $\boldsymbol{y} = [y_1, y_2, \cdots, y_N]^T \in \Re^N$, let $\boldsymbol{K} = \left[K(\boldsymbol{x}_i, \boldsymbol{x}_j)\right]_{i,j=1}^N \in \Re^{N\times N}$ be the kernel Gram matrix with the entries $K(\boldsymbol{x}_i, \boldsymbol{x}_j)$, which denotes the kernel function. The kernelized LASSO with the L_1 penalty minimizes the following objective function

$$\begin{aligned} L(\boldsymbol{\beta}) &= \arg\min_{\boldsymbol{\beta}} \frac{1}{2}\|\boldsymbol{y} - \boldsymbol{K}\boldsymbol{\beta}\|_2^2 + \lambda\|\boldsymbol{\beta}\|_1 \\ &= \arg\min_{\boldsymbol{\beta}} \frac{1}{2}\sum_{i=1}^N \Big(y_i - \sum_{j=1}^N \beta_j K(\boldsymbol{x}_i, \boldsymbol{x}_j)\Big)^2 + \lambda\sum_{j=1}^N |\beta_j| \end{aligned} \tag{18}$$

where $L(\boldsymbol{\beta})$ denotes the regularized empirical loss, $\boldsymbol{\beta} = [\beta_1, \beta_2, \cdots, \beta_N]^T \in \Re^N$ is the vector of the regression coefficients, λ is the regularization parameter, $\|\bullet\|_2$ and $\|\bullet\|_1$ denote the L_2 and L_1 norm of a vector, respectively. Since L_1 norm is used, some regression coefficients tend to be zero, and this characteristic can be used for feature selection.

3. LOCALLY WEIGHTED KERNEL PLS REGRESSION BASED ON SPARSE NONLINEAR FEATURES

In this section, we propose a novel locally weighted kernel PLS based on sparse nonlinear features. First, the original input variables are mapped into a high-dimension feature space. Second, sample selection based on kernelized LASSO is performed in this feature space, which gives sparse kernel regression factors to represent the importance of each sample to express the current query. Then sparse kernel regression factors are used to weight the training samples and construct a LW-KPLS model.

Given $\boldsymbol{X} \in \Re^{N\times M}$ and $\boldsymbol{Y} \in \Re^{N\times L}$ as the training input and output data, a nonlinear transformation of the original samples $\{\boldsymbol{x}_i\}_{i=1}^N \in \Re^M$ into a feature space F, i.e., mapping $\Phi : \boldsymbol{x}_i \in \Re^M \to \Phi(\boldsymbol{x}_i) \in F$, can be implemented by a kernel function. Let us define a kernel Gram matrix by

$$[\boldsymbol{K}]_{ij} = K(\boldsymbol{x}_i, \boldsymbol{x}_j) = \langle \Phi(\boldsymbol{x}_i), \Phi(\boldsymbol{x}_j) \rangle \quad (19)$$

where $K(\boldsymbol{x}_i, \boldsymbol{x}_j)$ denotes the kernel function. In this work, the Gaussian kernel function,

$$K(\boldsymbol{x}_i, \boldsymbol{x}_j) = \exp\Big(-\frac{\|\boldsymbol{x}_i - \boldsymbol{x}_j\|^2}{2\delta^2}\Big) \quad (20)$$

is selected due to its good generalization and smooth estimation ability.

When an output estimation is required for a query $\boldsymbol{x}_q \in \Re^M$, the test kernel vector $\boldsymbol{K}_q \in \Re^N$ is computed by

$$[\boldsymbol{K}_q]_j = K(\boldsymbol{x}_q, \boldsymbol{x}_j) = \langle \Phi(\boldsymbol{x}_q), \Phi(\boldsymbol{x}_j) \rangle \quad (21)$$

where $\boldsymbol{x}_j \in \Re^M (j = 1, 2, \cdots, N)$. Then, LW-KPLS solves the following optimization problem using the training kernel matrix $\boldsymbol{K}$ and test kernel vector $\boldsymbol{K}_q$.

$$L(\boldsymbol{\beta}_q) = \arg\min_{\boldsymbol{\beta}} \frac{1}{2}\|\boldsymbol{K}_q - \boldsymbol{K}\boldsymbol{\beta}\|_2^2 + \lambda\|\boldsymbol{\beta}\|_1$$

$$= \arg\min_{\boldsymbol{\beta}} \frac{1}{2}\sum_{i=1}^{N}\Big(K(\boldsymbol{x}_q, \boldsymbol{x}_i) - \sum_{j=1}^{N}\beta_j K(\boldsymbol{x}_i, \boldsymbol{x}_j)\Big)^2 + \lambda\sum_{j=1}^{N}|\beta_j| \quad (22)$$

Since L_1 penalty is used, a part of sparse kernel regression factors (SKRFs) $\boldsymbol{\beta}_q = [\beta_1, \beta_2, \cdots, \beta_N]^T$ will be equal to zero. Note that the SKRFs show the nonlinear dependency between the query and training samples in the Hilbert feature space. Samples corresponding to greater $|\beta_n|$ are more important to express the current query. Hence, the absolute value of SKRFs $|\boldsymbol{\beta}_q| = [|\beta_1|, |\beta_2|, \cdots, |\beta_N|]^T$ are useful as the similarity indexes to weight the training samples in the Hilbert feature space. In this way, the similarity matrix $\boldsymbol{\Omega}$ for the current query is determined.

$$\boldsymbol{\Omega} = \mathrm{diag}(\omega_1, \omega_2, \cdots, \omega_N) = \mathrm{diag}(|\boldsymbol{\beta}_q|) \quad (23)$$

The output estimation $\hat{\boldsymbol{y}}_q$ corresponding to the query $\boldsymbol{x}_q$ is derived through the following procedure.

1). Determine the number of latent variables R, and set $r = 1$.

2). Calculate $\boldsymbol{K}_r$, $\boldsymbol{Y}_r$, and $\boldsymbol{K}_{q,r}$

$$\boldsymbol{K}_r = \boldsymbol{K} - \boldsymbol{1}_N[\bar{K}_1, \bar{K}_2, \cdots, \bar{K}_N] \quad (24)$$

$$\boldsymbol{Y}_r = \boldsymbol{Y} - \boldsymbol{1}_N[\bar{y}_1, \bar{y}_2, \cdots, \bar{y}_L] \quad (25)$$

$$\boldsymbol{K}_{q,r} = \boldsymbol{K}_q - [\bar{K}_1, \bar{K}_2, \cdots, \bar{K}_N]^T \quad (26)$$

$$\bar{K}_j = \frac{\sum_{i=1}^N \omega_i K(\boldsymbol{x}_i, \boldsymbol{x}_j)}{\sum_{i=1}^N \omega_i} \quad (27)$$

$$\bar{y}_l = \frac{\sum_{i=1}^N \omega_i y_{il}}{\sum_{i=1}^N \omega_i} \quad (28)$$

where $\boldsymbol{1}_N \in \Re^N$ is a vector of ones.

3). Derive the r-th latent variable of $\boldsymbol{K}$

$$\boldsymbol{t}_r = \boldsymbol{K}_r \boldsymbol{w}_r \quad (29)$$

where $\boldsymbol{w}_r$ is the eigenvector of $\boldsymbol{K}_r^T \boldsymbol{\Omega} \boldsymbol{Y}_r \boldsymbol{Y}_r^T \boldsymbol{\Omega} \boldsymbol{K}_r$, which corresponds to the maximum eigenvalue.

4). Derive the r-th loading vector of $\boldsymbol{K}$ and the regression coefficient vector.

$$\boldsymbol{p}_r = \frac{\boldsymbol{K}_r^T \boldsymbol{\Omega} \boldsymbol{t}_r}{\boldsymbol{t}_r^T \boldsymbol{\Omega} \boldsymbol{t}_r} \quad (30)$$

$$\boldsymbol{q}_r = \frac{\boldsymbol{Y}_r^T \boldsymbol{\Omega} \boldsymbol{t}_r}{\boldsymbol{t}_r^T \boldsymbol{\Omega} \boldsymbol{t}_r} \quad (31)$$

5). Derive the r-th latent variable of $\boldsymbol{K}_q$.

$$t_{q,r} = \boldsymbol{K}_{q,r}^T \boldsymbol{w}_r \quad (32)$$

6). If $r = R$, finish the output estimate

$$\hat{\boldsymbol{y}}_q = [\bar{y}_1, \bar{y}_2, \cdots, \bar{y}_L]^T + \sum_{r=1}^{R} t_{q,r}\boldsymbol{q}_r. \quad (33)$$

Otherwise, set

$$\boldsymbol{K}_{r+1} = \boldsymbol{K}_r - \boldsymbol{t}_r \boldsymbol{p}_r^T \quad (34)$$

$$\boldsymbol{Y}_{r+1} = \boldsymbol{Y}_r - \boldsymbol{t}_r \boldsymbol{q}_r^T \quad (35)$$

$$\boldsymbol{K}_{q,r+1} = \boldsymbol{K}_{q,r} - t_{q,r}\boldsymbol{p}_r \quad (36)$$

7). Set $r = r + 1$, and go to step 3.

4. CASE STUDIES

In this section, the prediction performance of LW-KPLS is compared with PLS, KPLS, and LW-PLS using a nonlinear numerical example and a case study of the penicillin fermentation process.

4.1 Nonlinear numerical example

The nonlinear numerical example contains two operation modes (Zhang et al., 2008). For the first operation mode, process data is generated from the following equations:

$$\begin{aligned} x_1 &= t^2 - t + 1 + \varepsilon_1 \\ x_2 &= \sin(t) + \varepsilon_2 \\ x_3 &= t^3 + t + \varepsilon_3 \\ y &= x_1^2 + x_1 x_2 + 3\cos(x_3) + \varepsilon_4 \end{aligned} \quad (37)$$

where t is uniformly distributed between $[-2, 2]$ and $\varepsilon_i (i = 1, 2, 3, 4)$ are the noises with uniformly distributed between $[-0.1, 0.1]$. For the second operation mode, process data is generated from the following equations:

$$\begin{aligned} x_1 &= t^3 + t^2 + 5 + \varepsilon_1 \\ x_2 &= 0.6\cos(t) + t + \varepsilon_2 \\ x_3 &= -5t^2 + \varepsilon_3 \\ y &= 3x_1 + x_2 + x_3 + \varepsilon_4 \end{aligned} \quad (38)$$

where t is uniformly distributed in $[-1.5, 1.5]$ and $\varepsilon_i (i = 1, 2, 3, 4)$ are the uniform distributed noises over $[-0.1, 0.1]$.

To build the prediction models, 500 samples were generated in each operation mode. Among a total of 1000 samples, 400 samples were used for model training, 200 samples were used for parameter tuning, and the remaining 400 samples were used for model testing. The prediction ability was evaluated in terms of the root mean squared error (RMSE). The selected optimal parameters, RMSE, and RMSEP (RMSE of Prediction) of four methods are listed in Table 1. Here LV denotes the number of latent variables, δ is the Gaussian kernel parameter, and φ is the localization parameter. PLS gave poor prediction performance due to its intrinsic linearity. KPLS and LW-PLS outperform PLS in dealing with the nonlinearity and provided good prediction performance. LW-PLS was better than KPLS, because LW-PLS can also cope with time-varying characteristics. However, RMSEP of LW-PLS was larger than that of LW-KPLS. This is mainly because LW-PLS utilizes the original samples to build local linear regression models and uses the locally weighting strategy to describe the nonlinearity. Thus, LW-PLS does not always succeed to efficiently cope with strong nonlinearity. In contrast, LW-KPLS takes nonlinear characteristics into consideration from the weight selection to the regression model construction, thus it is suitable for highly nonlinear processes. Furthermore, by integrating the nonlinear features into locally weighted framework, LW-KPLS can also cope with the time-varying characteristics.

Table 1. Comparison of four regression methods in the numerical example.

Methods	LV	δ	φ	RMSE	RMSEP
PLS	2			0.5885	0.6277
KPLS	3	0.1		0.1116	0.1136
LW-PLS	2		20	0.0893	0.0898
LW-KPLS	3	0.5		0.0671	0.0581

4.2 Penicillin fermentation process

The penicillin fermentation process is a typical nonlinear time-varying process, which is often used as a benchmark to demonstrate the validity of soft-sensors (Zhang et al., 2015). The penicillin fermentation process consists of three phases: the pre-culture phase, the exponential growth phase, and the stationary phase. To build the prediction models, 1200 samples were generated with sampling interval of 1h from the PenSim v2.0 (Birol et al., 2002). One third of the data was used for model training, one third of the data was used for parameter tuning, and the remaining one third of the data was used for model testing.

The prediction results of PLS, KPLS, LW-PLS, and LW-KPLS are summarized in Table 2. PLS provided the worst prediction result with large RMSEP. Compared to PLS, KPLS and LW-PLS gave better prediction performance because they can cope with the nonlinearity. However, they still presented a relatively large RMSEP. This is because KPLS cannot deal with time-varying natures in the penicillin fermentation process and the locally weighting strategy of LW-PLS cannot always effectively deal with the nonlinearity. In contrast, LW-KPLS gave the smallest RMSEP among four methods owing to the following facts: 1) LW-KPLS integrates the nonlinear features into the locally weighted framework; this makes LW-KPLS more efficient in dealing with high nonlinearity; 2) Sparse kernel regression factors which describe nonlinear dependency between the query and training samples are used to construct LW-KPLS models. 3) LW-KPLS inherits the merits of locally weighted regression, and it can cope with changes in process characteristics.

5. CONCLUSIONS

In this work, we proposed a novel locally weighted kernel partial least squares based on sparse nonlinear features.

Table 2. Comparison of four regression methods in the penicillin fermentation process.

Methods	LV	δ	φ	RMSE	RMSEP
PLS	5			0.2915	0.2933
KPLS	5	50		0.1109	0.0901
LW-PLS	5		3	0.0656	0.0678
LW-KPLS	5	50		0.0569	0.0486

LW-KPLS integrates the nonlinear features into the locally weighted framework; this makes LW-KPLS more efficient in dealing with strong nonlinearity. Furthermore, sparse kernel regression factors which describe nonlinear dependency between the query and training samples are used to build LW-KPLS models. LW-KPLS can also cope with changes in process characteristics. Two case studies have demonstrated that LW-KPLS gives better prediction performance compared with PLS, KPLS, and LW-PLS.

REFERENCES

Birol, G., Ündey, C., and Cinar, A. (2002). A modular simulation package for fed-batch fermentation: penicillin production. *Computers and Chemical Engineering*, 26(11), 1553–1565.

Gao, J., Kwan, P.W., and Shi, D. (2010). Sparse kernel learning with lasso and bayesian inference algorithm. *Neural Networks*, 23(2), 257–264.

Hirai, T. and Kano, M. (2015). Adaptive virtual metrology design for semiconductor dry etching process through locally weighted partial least squares. *IEEE Transactions on Semiconductor Manufacturing*, 28(2), 137–144.

Kano, M. and Fujiwara, K. (2013). Virtual sensing technology in process industries: trends and challenges revealed by recent industrial applications. *Journal of chemical engineering of Japan*, 46(1), 1–17.

Kano, M. and Ogawa, M. (2010). The state of the art in chemical process control in japan: Good practice and questionnaire survey. *Journal of Process Control*, 20(9), 969–982.

Kim, S., Kano, M., Nakagawa, H., and Hasebe, S. (2011). Estimation of active pharmaceutical ingredients content using locally weighted partial least squares and statistical wavelength selection. *International journal of pharmaceutics*, 421(2), 269–274.

Kim, S., Okajima, R., Kano, M., and Hasebe, S. (2013). Development of soft-sensor using locally weighted pls with adaptive similarity measure. *Chemometrics and Intelligent Laboratory Systems*, 124, 43–49.

Rosipal, R. and Trejo, L.J. (2001). Kernel partial least squares regression in reproducing kernel hilbert space. *Journal of machine learning research*, 2(Dec), 97–123.

Roth, V. (2004). The generalized lasso. *IEEE transactions on neural networks*, 15(1), 16–28.

Zhang, X., Yan, W., and Shao, H. (2008). Nonlinear multivariate quality estimation and prediction based on kernel partial least squares. *Industrial and Engineering Chemistry Research*, 47(4), 1120–1131.

Zhang, X., Li, Y., and Kano, M. (2015). Quality prediction in complex batch processes with just-in-time learning model based on non-gaussian dissimilarity measure. *Industrial and Engineering Chemistry Research*, 54(31), 7694–7705.

463509 Fault Detection and Isolation and Optimal Parking for HVAC Systems

Abstract:

Fault detection and isolation and optimal parking for HVAC systems

Hadi Shahnazari, Craig McDonald, Prashant Mhaskar*, John House and Tim Salsbury

Government regulations and initiatives have placed a large emphasis on the reduction of energy consumption and increase in energy efficiency. Heating, ventilation, and air-conditioning (HVAC) systems are responsible for 40-50% of total building energy consumption. In general, 15 to 20% per annum of energy consumption can be reduced by efficient and optimal operation of buildings. A major factor contributing to this inefficiency are undiagnosed faults (such as stuck dampers) being compensated through wasted, energy intensive effort (higher static pressures). In US alone, the fault detection and isolation (FDI) and fault tolerant control methods are estimated to be capable of saving 10-40% of HVAC energy consumption (see e.g., [1]).

These realizations have motivated significant research effort on devising FDI frameworks for HVAC systems (see e.g., [1], [2], [3] and [4]). In [2], a fault detection tool is proposed that uses a set of expert rules derived from mass and energy balances to detect faults in air handling units (AHUs). A subset of the expert rules which correspond to that mode of operation are then evaluated to determine whether a fault exists. In [3], a principal component analysis (PCA) based approach is presented to detect single sensor faults in heating, ventilation and air conditioning (HVAC) systems and faults are isolated using joint angle plot, which compares the new fault vector with known ones in the library. In [4], a PCA approach is used to extract the correlation of measured variables in heating/cooling billing system and reduce the dimension of measured data. Square prediction error (SPE) statistic is used to detect sensor faults in the system. Then, sensor validity index (SVI) is employed to identify faulty sensors and a reconstruction algorithm is presented to recover the correct data of faulty sensor in accordance with the correlations among system variables. In [5], a combination of PCA and wavelet transform has been used for FDI of HVAC systems. To this end, at first wavelet transform has been used to remove effect of normal weather conditions changes from data. Then PCA has been used as a data driven methodology for fault detection and isolation. The existing results, however, consider only isolation of single fault scenarios or at best multiple sensor faults or multiple actuator faults, and do not consider simultaneous actuator and sensor faults.

There also exist results on fault tolerant control (FTC) of HVAC systems. In [6], the control design compensates for the fault effect as much as possible by switching between different control modes available in the air handling unit design. In [7] and [8], single sensor faults are diagnosed via estimation of healthy value of sensors using PCA method and handled by utilizing the healthy values of sensors in the closed loop, upon fault isolation. In [9], the fault tolerant control design is based on real time estimation of the fault magnitude, and determining MPC constraints (input constraints) based on those values. In this way, MPC design explicitly accounts for fault effect. These fault-tolerant control approaches, however, are all predicated on the idea of maintaining nominal operation, which might simply be impossible, or very expensive, in case of certain faults. Recently, safe-parking based approaches for fault-tolerant control have been proposed (see e.g., [10] and [11]) that park the process at an appropriate operating point, instead of trying to maintain nominal operation. These ideas, however, have not been applied to HVAC systems. In summary there is lack of results in simultaneous actuator and sensor fault detection and isolation and optimal parking of HVAC systems to minimize energy expenditure.

Motivated by the above considerations, in this work, we design and implement an integrated framework for fault diagnosis and fault handling in HVAC system. The results include simulations as well as application to a room at McMaster University (room 207 of Hamilton Hall). To this end, first, we identify a linear model for the room HVAC system using the existing data. Then we design a model based FDI framework based on the methodology proposed in [12]. The key idea is to exploit analytical redundancy in the system through state observer design. We consider subset of actuator and sensor faults and design observers that only use information of inputs and outputs that are not subject to fault. Then we generate residuals that are only sensitive to a subset of fault. The fault is detected if the corresponding residuals breach their thresholds and isolated using a bank of residuals and a logic rule. We also compare the result obtained using model based approaches for fault detection and isolation (FDI) with a principal component analysis (PCA) based approach (see e.g., [13]). Finally, to minimize energy expenditure, we design and implement an optimal parking approach that minimizes energy expenditure in the presence of a fault.

References

[1] Jeffrey Schein, Steven T Bushby, Natascha S Castro, and John M House. A rule-based fault detection method for air handling units. Energy and Buildings, 38(12): 1485-1492, 2006

[2] John M House, Hossein Vaezi-Nejad, and J Michael Whitcomb. An expert rule set for fault detection in air-handling units. Transactions-American Society of Heating Refrigerating and Air Conditioning Engineers, 107(1):858-874, 2001.

[3] Zhimin Du and Xinqiao Jin. Detection and diagnosis for sensor fault in HVAC systems. Energy Conversion and Management, 48(3):693-702, 2007.

[4] Youming Chen and Lili Lan. Fault detection, diagnosis and data recovery for a real building heating/cooling billing system. Energy Conversion and Management, 51(5):1015-1024, 2010.

[5] Shun Li and Jin Wen. A model-based fault detection and diagnostic methodology based on PCA method and wavelet transform. Energy and Buildings, 68:63-71, 2014.

[6] John E Seem. Integrated control and fault detection of HVAC equipment, May 1 2001. US Patent 6,223,544.

[7] Xiaoli Hao, Guoqiang Zhang, and Youming Chen. Fault-tolerant control and data recovery in HVAC monitoring system. Energy and buildings, 37(2):175-180, 2005.

[8] Xinqiao Jin and Zhimin Du. Fault tolerant control of outdoor air and AHU supply air temperature in VAV air conditioning systems using PCA method. Applied Thermal Engineering, 26(11):1226-1237, 2006.

[9] Sorin C Bengea, Pengfei Li, Soumik Sarkar, Sergey Vichik, Veronica Adetola, Keunmo Kang, Teems Lovett, Francesco Leonardi, and Anthony D Kelman. Fault-tolerant optimal control of a building HVAC system. Science and Technology for the Built Environment, 21(6):734-751, 2015.

[10] Rahul Gandhi and Prashant Mhaskar. Safe-parking of nonlinear process systems. Computers & Chemical Engineering, 32(9):2113-2122, 2008.

[11] Miao Du and Prashant Mhaskar. A safe-parking and safe-switching framework for fault tolerant control of switched nonlinear systems. International Journal of Control, 84(1):9-23, 2011.

[12] Miao Du, James Scott, and Prashant Mhaskar. Actuator and sensor fault isolation of nonlinear process systems. Chemical Engineering Science, 104:294-303, 2013.

[13] Seongkyu Yoon and John F MacGregor. Statistical and causal model-based approaches to fault detection and isolation. AIChE Journal, 46(9):1813-1824, 2000.

465693 Pipe-Flows Leak Detection, Size Estimation and Localization

Abstract:

Monitoring of pipelines integrity is important for fluid transportation since it is necessary in timely manner to detect malfunctioning, e.g. leaks. In particular, for the petroleum industry, leaks from pipelines may potentially result in environmental damage and pollution, as well as to the economic loss. These motivating factors with requirement from environmental authorities leads to the development of ecient leak detection systems. However, some leak detection methods are hardware-based, where the leak detection techniques rely on physical equipment being installed along the pipeline. In this work, we focus on software-based approaches that work for cases with limited instrumentation. In particular, the instrumentation is restricted to pressure and ow measurements at the pipeline inlet and outlet.

Some software-based leak detection methods utilize statistical analysis on measurements (black box), while others are on the basis of mathematic models from physical principles. The approach in this work belongs to the latter category. For more details about leak detection

techniques, we refer to a recent survey [1]. In this work, a dynamic model of the pipe described by a set of two coupled hyperbolic partial differential equations (PDEs) is utilized to realize leak detection system. In [2] and [3], a bank of observers are used to potentially detect multiple leaks. However, multiple simultaneous leaks may not likely happen. Therefore, we employ the idea from internal model control and adaptive control, addressing the size and location estimation of a single point leak based on a set of two coupled one dimensional first order hyperbolic PDEs. Moreover, the backstepping transformation is applied to simplify the design of the observer for the size estimation and the observer gains can be obtained in closed-form. Finally, simulations are provided to show the ability of the proposed system to detect, quantify and locate leaks.

[1] P.-S. Murvay and I. Silea (2012) A survey on gas leak detection and localization techniques, J. Loss Prevent. Process Ind., vol. 25, pp. 966--973.

[2] C. Verde (2001) Multi-leak detection and isolation in fluid pipelines, Control Eng. Pract., vol. 9, pp. 673--682.

[3] C. Verde (2004) Minimal order nonlinear observer for leak detection, J. Dynam. Syst., Measur., Control, vol. 126, pp. 467--472.

467436 Model-Based Fault Detection for Nonlinear Process Systems

Abstract:

Faults can result in fatal damages and economic loss if they are not handled properly. Safe operation of process systems depends upon efficient computational techniques to quickly and reliably detect the faults. Most of the model-based fault detection (FD) approaches are based on observer, parity relation or parameter estimation based techniques. The observer-based FD is widely used in fault detection approach. The basic idea behind this technique is to estimate the outputs of the system from the measurements by using Luenberger observer in a deterministic setting or Kalman filter in a stochastic setting. Then, the output estimation error is used as a residual for the detection of the faults. The parity relation technique uses the parity check of the consistency of parity equation to generate residual. A set of properly modified system equations is derived based on the measured signals from the process to decouple the residuals from the system states. The inconsistency in the parity relation indicates the presence of fault. On the other hand, the parameter estimation approach is based on the assumption that the faults are reflected in the physical system parameters. The parameters of the actual process are repeatedly estimated on-line and the results are compared with the reference model [1-4].

This work presents a model-based fault detection methodology for nonlinear process systems. The objective is to detect faults by estimating the model parameters online by minimising the error between model predictions and the observed plant data. Algorithms for parameter estimation of dynamic systems require integration of the ordinary differential equations (ODEs), which is performed using artificial neural network (ANN) in this work [5]. The parameters are estimated through an augmented optimization problem where the objective is to simultaneously solve the ODEs as well as estimate the parameters. The results are compared with the parameters of the observed plant data under fault-free situation. The estimated parameters should be close to observed parameters when no fault is present. Any substantial discrepancy indicates changes in the process and may be interpreted as a fault if the discrepancy goes above a certain threshold. One key issue in fault detection is to estimate the model parameters precisely (accuracy) and as fast as possible (speed). Accuracy is important to avoid false-positives, while speed ensures that the faults and their location are identified quick enough to be able to take corrective actions in a timely manner. It is shown that this issue can be addressed by using the proposed approach. Details of the proposed approach are illustrated for an isothermal continuous stirred tank reactor (CSTR) where it is assumed that the faults can be represented by changes in values of reaction rate constants. To demonstrate the proposed approach, different scenarios are presented: a fault-free situation and some faulty situations. The scenarios considered show the effectiveness of the proposed approach.

References

[1] V. Venkatasubramanian, R. Rengaswamy, and K. Yin, "A Review of Process Fault Detection and Diagnosis Part I : Quantitative Model-Based Methods," Comput. Chem. Eng., vol. 27, pp. 293–311, 2003.

[2] R. Isermann, "Model-Based Fault-Detection and Diagnosis - Status and Applications," Annu. Rev. Control, vol. 29, pp. 71–85, 2005.

[3] P. Mhaskar, A. Gani, N. H. El-farra, C. Mcfall, P. D. Christofides, and J. F. Davis, "Integrated Fault-Detection and Fault-Tolerant Control of Process Systems," Am. Inst. Chem. Eng., vol. 52, no. 6, pp. 2129–2148, 2006.

[4] I. Hwang, S. Kim, Y. Kim, and C. Seah, "A Survey of Fault Detection, Isolation, and Reconfiguration Methods," IEEE Trans. Control Syst. Technol., vol. 18, no. 3, pp. 636–653, 2010

[5] V. Dua and P. Dua, "A Simultaneous Approach for Parameter Estimation of a System of Ordinary Differential Equations, Using Artificial Neural Network Approximation," Ind. Eng. Chem. Res., vol. 51, no. 4, pp. 1809–1814, 2011.

465691 Fault Estimation and Performance-Based Accommodation in Multi-Rate Sampled-Data Process Systems

Abstract:

Fault handling is a critical component of modern-day process control systems. When left unchecked, faults in process elements such as measurement sensors and control actuators can lead to substantial loss of product quality and in some cases instabilities and safety hazards. The critical nature of this problem has led to a significant body of research work on the topic of fault-tolerant control (e.g., see [1]-[4]). Classically, this problem has been addressed in the context of a conventional feedback control setting where continuous measurements of the states are assumed to be available for the controller to utilize in driving the process to the desired operating point. However, with real-time operation practices this is generally not the case, and measurements of the state variables are limited in availability by the measurement sensors and/or the resources of the sensor-controller communication medium. Discretely sampled-data systems are commonly found in industrial processes as continuous measurement of system data often proves logistically or technologically prohibitive. In sampled-data systems, the stability of the closed-loop system is critically dependent on the rate at which the sampled measurements are taken.

Fault-tolerant control of sampled-data systems has been the focus of prior work [5] where a model-based framework was developed for fault detection and reconfiguration of a control system with sampled and delayed measurements. This was implemented in the context of control actuator faults. While reconfiguration of the control system is a viable option for fault handling, it is not always an ideal go to solution for fault handling in situations where the availability of component redundancy is either costly on intrinsically limited by process design considerations. The ability of the system to operate satisfactorily in the faulted control configuration is a worthwhile pursuit as was demonstrated in [6] where a stability-based fault accommodation strategy was developed. These results where subsequently generalized in [7] to include fault estimation capabilities through an optimization-based approach to aid in the implementation of the fault accommodation framework mentioned prior.

A key aspect of previous works is that all states in the system were assumed to be sampled at the same rate. In many practical situations, however, limitations on the measurement capabilities of different sensors may result in a significant gap between the sampling rates, and in such cases a synchronized sampling mechanism may not be the best choice [8]. Moreover, the importance of the measurement collected is another factor that can trigger the use of multi-rate sampling. It is reasonable to apply a fast sampling rate to the sensors placed at certain critical locations in the process (e.g., where frequent monitoring and tight control are required), while reducing the sampling rates of the other sensors in order to reduce cost and optimize energy resource consumption. A more robust framework for fault-tolerant control would therefore be to allow for the possibility of sampling each state at a different rate. However, to date there has not been any rigorous assessment or characterization of the stability properties of multi-rate sampled-data systems in the context of fault-tolerant control, nor has there been any thorough assessment of performance-based accommodation for multi-rate sampled systems. These are important gaps that the current work aims to address.

Motivated by these considerations, we present in this work a combined data-based and model-based framework for the detection and accommodation of control actuator faults in process systems with multi-rate sampled measurements. A key feature of this contribution is to explicitly incorporate performance based fault accommodation into the framework, which would allow for post accommodation performance to be characterized and determine if the accommodation measure is sufficient or if more drastic measures must be taken to ensure the desired quality of the process output. Initially, a model-based controller is designed and its closed-loop stability and performance properties are explicitly characterized in terms of the model and controller design parameters, the sensor sampling rates and the magnitudes of the faults. The model is used to compute the control action, and its states are updated at different times whenever the measurements become available from the sensors. A data-based moving-horizon parameter estimation scheme, which takes the available sampled data from the process, is developed and used to estimate the severity of the faults and to identify their locations. Owing to the different sampling rates of the states, the data set used for fault estimation contains missing data. To address this problem, suitable propagation models are incorporated in the fault diagnosis unit to provide estimates of the missing states based on the available measurements, which are then used in solving the optimization problem. The fault accommodation logic will then determine an appropriate response while using this estimated fault values to meet some baseline performance needed by the process, whilst ensuring stability. The fault accommodation logic is based on the parametrization of the closed-loop stability region and the performance metric obtained at the controller design stage as a function of the fault size, the controller design and model parameters as well as the frequency of each of the sampled measured states. Finally, the developed framework is illustrated using a simulated chemical process example.

References:

[1] M. Blanke, M. Kinnaert, J. Lunze, and M. Staroswiecki, *Diagnosis and Fault-Tolerant Control*. Berlin, Germany: Springer, 2003.

[2] R. Isermann, *Fault-Diagnosis Systems: An Introduction from Fault Detection to Fault Tolerance*. Berlin, Germany: Springer, 2005.

[3] Y. Zhang and J. Jiang, "Bibliographical Review on Reconfigurable Fault-Tolerant Control Systems," *Annu. Rev. in Contr.*, vol. 32, pp. 229–252, 2008.

[4] P. Mhaskar, J. Liu, and P. D. Christofides, *Fault-Tolerant Process Control: Methods and Applications*. London, England: Springer-Verlag, 2013.

[5] Y. Sun and N. H. El-Farra, "Model-Based Fault Detection and Fault-Tolerant Control of Process Systems with Sampled and Delayed Measurements," *Proceedings of 18th IFAC World Congress*, pp. 2749-2754, 2011.

[6] T. Napasindayao and N. H. El-Farra, "Fault Detection and Accommodation in Particulate Processes with Sampled and Delayed Measurements," *Ind. Eng. Chem. Res.*, 52, 12490–12499, 2013.

[7] T. Napasindayao and N. H. El-Farra, "Model-based Fault-Tolerant Control of Uncertain Particulate Processes: Integrating Fault Detection, Estimation and Accommodation," *Proceedings of 9th IFAC Symposium on Advanced Control of Chemical Processes*, pp. 872-877, 2015.

[8] J. Wang, T. Chen, and B. Huang, "Multi-rate Sampled-Data Systems: Computing Fast-Rate Models," *J. Proc. Contr.*, vol. 14, pp. 78–88, 2004.

469137 Multi-Rate Reduced-Order Observer Design with Application to Monitoring of Gas-Phase Polyethylene Reactors

Abstract:

State observers find many applications in the areas of process monitoring, feedback control, and fault detection. Many available observer design methods tackle the state estimation problem by using a continuous-time model with continuous outputs, and this works well under fast sampling rates [1-4]. In polymerization processes, however, it is inevitable to use measurements with large sampling periods (e.g., gel permeation chromatography, gas chromatography, melt index deviceâ€¦) in order to monitor product quality. This poses a significant challenge on observer design. To deal with the sampled measurements, one possible approach is to first develop a discrete-time description of the dynamics, with discretization time step equal to the sampling period. But this would not account for the inter-sample behavior and a part of the information is lost at the moment when the continuous-time model is being discretized.

A full-order observer based on an available continuous-time observer design, coupled with inter-sample output predictor, was proposed recently [5]. During the time period in between two consecutive output measurements, an inter-sample predictor is designed to obtain an estimate of the rate of change of the output, in order to be able to continuously apply a correction on the most recent measurement during this period. This predictor is re-initialized once the most recent measurement is available. It is proved in [5] that the sampled-data implementation inherits the stability, performance and robustness properties of the continuous-time observer as long as the sampling period is not too large.

The present paper is motivated by the same idea of inter-sample predictor, but the development will be in a reduced-order observer framework, to eliminate the redundancy of the full-order formulation, where the sampled states are estimated twice, in the observer and in the inter-sample predictor. In the proposed formulation, a continuous-time reduced-order observer will be coupled with an inter-sample predictor for the slow-sampled measurements. Moreover,

the present work will provide a generalization in the sense of handling multi-rate output measurements, allowing the possibility of both continuous and slow-sampled measurements to be processed. Stability analysis of the error dynamics of the overall multi-rate reduced-order sampled-data observer will be performed in the linear case. Explicit conditions that allow the estimation of the maximum allowable sampling period for stability will be derived by using Lyapunovâ€™s second method [6].

Moreover, in the present work, we apply the proposed observer design framework to a gas-phase polyethylene reactor to monitor the number of active catalyst sites. Polyethylene is the most popular synthetic commodity polymer. A large fraction of polyethylene is produced in gas-phase reactors using Ziegler-Natta catalysts [7].Â A dynamic process model that also includes the dynamics of the product properties (i.e., instantaneous/cumulative melt index and density) was derived in [8, 9]. Gas compositions in the bleed stream leaving from the top of the reactor are measured by on-line gas chromatography in a discrete manner. Cumulative melt index and density are measured off-line in a quality control lab, and these are directly related to molecular weight and copolymer composition respectively. The only continuous measurement is the temperature of the reactor. Using the proposed design framework of multi-rate reduced-order observer, we are able to obtain a continuous estimate of the unmeasured state, with a convergence rate that is comparable to the case of continuous measurements. In addition, by using the inter-sample predictor, we are able to estimate the outputs in between consecutive samples as well.

References:

[1] D. G. Luenberger, "Observing the state of a linear system," *IEEE Transactions on Military Electronics,* vol. 8, pp. 74-80, 1964.

[2] J. P. Gauthier, H. Hammouri, and S. Othman, "A Simple Observer for Nonlinear-Systems Applications to Bioreactors," *IEEE Transactions on Automatic Control,* vol. 37, pp. 875-880, 1992.

[3] J. Tsinias, "Further Results on the Observer Design Problem," *Systems & Control Letters,* vol. 14, pp. 411-418, 1990.

[4] N. Kazantzis and C. Kravaris, "Nonlinear observer design using Lyapunov's auxiliary theorem," *Systems & Control Letters,* vol. 34, pp. 241-247, 1998.

[5] I. Karafyllis and C. Kravaris, "From Continuous-Time Design to Sampled-Data Design of Observers," *IEEE Transactions on Automatic Control,* vol. 54, pp. 2169-2174, 2009.

[6] G. C. Walsh, H. Ye, and L. G. Bushnell, "Stability analysis of networked control systems," *IEEE Transactions on Control Systems Technology,* vol. 10, pp. 438-446, May 2002.

[7] A. Gani, P. Mhaskar, and P. D. Christofides, "Fault-tolerant control of a polyethylene reactor," *Journal of Process Control,* vol. 17, pp. 439-451, Jun 2007.

[8] K. B. McAuley and J. F. Macgregor, "Online Inference of Polymer Properties in an Industrial Polyethylene Reactor," *AIChE Journal,* vol. 37, pp. 825-835, Jun 1991.

[9] K. B. Mcauley, D. A. Macdonald, and P. J. Mclellan, "Effects of Operating-Conditions on Stability of Gas-Phase Polyethylene Reactors," *AIChE Journal,* vol. 41, pp. 868-879, Apr 1995.

469677 Comparison of Stochastic Fault Detection and Diagnosis Algorithms for Nonlinear Chemical Processes

Abstract:

Early detection of abnormal events and malfunctions defined as faults is of great interest, since faults may affect the product quality and lead to economic losses [1]. If a fault is detectable, the fault detection and diagnosis (FDD) system will provide symptomatic fingerprints, which in turn can be referred back to the FDD scheme to identify the root cause of the anomalous behaviour. Most of the available fault diagnosis algorithms can be broadly classified into three main classes [2, 3]: (i) Analytical methods that are solely based on first principlesâ€™ models of process; (ii) Empirical models that use the historical process data; and (iii) Semi-empirical algorithms that combine these aforementioned two classes. Each of these methods has its own advantages and disadvantages depending on the specific problem [4].

In terms of applications, many industrial processes are intrinsically nonlinear systems and they are operated at different operating conditions according to economic considerations [5]. Due to nonlinearity, the performance of linear FDD algorithms reported in literature [6] may be inaccurate and lead to missed detection of faults, since the process model will change from one operating conditions to another. It is critical to develop new methodologies for the detection of faults in the context of nonlinear chemical processes with multiple operating conditions [5].

Since most of the FDD schemes are invariably based on either first principle models or empirical models [3], a main restrictive factor of an efficient FDD system is the model uncertainty. Such uncertainty may originate from either intrinsic time varying phenomena of model parameters or may result from inaccurate measurements due to noise. Models with large uncertainties make the detection and isolation of small faults very difficult. However, the step of quantifying and propagating the uncertainties onto the measured quantities that can be used for fault detection is typically omitted in reported FDD studies, leading to a loss of useful information arising from these uncertainties [7]. Moreover, the quantitative analysis of the detectability of faults in the presence of uncertainty can provide more information to improve the development of FDD algorithm. For example, engineering effort can be saved, if it is impossible to detect a fault in reality due to uncertainties such as large measurement noise [8].

To evaluate the effect of uncertainty on FDD, one possibility is to propagate stochastic variations with Monte Carlo (MC) simulations [9], which involve drawing a large number of samples and running the models with each of these samples. However, approaches such as MC simulations are computationally prohibitive especially for complex processes as shown later in the manuscript. To improve the computational efficiency, this paper presents and compares two FDD algorithms in the presence of uncertainties. The uncertainty includes the parametric uncertainty of a process and measurement noise. In addition, the faults in this current work are stochastic perturbations superimposed on intermittent step changes in specific input variables for a nonlinear chemical plant. For the first FDD method, generalized polynomial chaos (gPC) [10, 11] in combination with first principlesâ€™ process models are used to quantify and propagate the uncertainty onto the measured quantities, which can be used for the detection of faults. For the second method, a surrogate metamodel is developed with Gaussian Process (GP) [12], which is calibrated with a minimal model adjustment algorithm and can be used estimate the value of fault.

The objective in this work is to address the capabilities of these methods and propose a possible strategy to overcome their limitation by combing their outcomes. For this purpose, the performance of each method is evaluated in terms of fault detection rate in the context of stochastic parametric input faults. These faults occur intermittently with stochastic perturbations, i.e., the mean value of faults switch between the non-faulty and faulty operating conditions in a random fashion. For simplicity, the stochastic perturbations are assumed to be time-invariant uncertainties. Thus, the key point is to identify and diagnose these step changes in the presence of the random perturbations in the parametric input faults, using available measurements corrupted with measurement noise. The performance of the above mentioned methods is evaluated in terms of fault detection rate by applying them to a chemical plant of two continuously stirred tank reactors (CSTRs) and a flash tank separator. The proposed methods are successful in detecting and diagnosing intermittent faults in the presence of uncertainty.

References

[1] J. Gerlter, Fault detection and diagnosis in engineering systems, NJ, USA: Taylor & Francis, 1998.
[2] R. Isermann, "Model based fault detection and diagnosis - status and applications," Annual reviews in control, vol. 29, pp. 71-85, 2005.
[3] V. Venkatasubramanian, R. Rengaswamy, K. Yin and S. Kavuri, "A review of process fault detecion and diagnosis Part I: Quantitative model-based methods," Computers and Chemical Engineering, vol. 27, pp. 293-311, 2003.
[4] R. Isermann, Fault diagnosis systems: An introduction from fault detection to fault tolerance, Berlin, Germany: Springer, 2006.
[5] A. Haghani, T. Jeinsch and S. X. Ding, "Quality related fault detection in industrial multimode dynamic processes," IEEE Transactions on Industrial Electronics, vol. 61, no. 11, pp. 6446 - 6453, 2014.
[6] X. Li and G. Yang, "Fault detection for linear stochastic systems with sensor stuck faults," Optimal control applications and methods, vol. 33, no. 1, pp. 61-80, 2012.
[7] R. J. Patton, P. M. Frank and R. N. Clark, Issues of fault diagnosis for dynamic systems, Springer, 2010.
[8] D. Eriksson, E. Frisk and M. Krysander, "A method for quantitative fault diagnosability analysis of stochastic linear descriptor models," Automatica, vol. 49, pp. 1591-1600, 2013.
[9] R. L. Harrison, "Introduction to Monte Carlo simulation," in AIP conference proceedings, Bratos;ava, 2010.
[10] R. Ghanem and P. Spanos, Stochastic finite elements: a spectral approach, 2nd ed., Berlin: Springer-Verlag, 1991.

[11] D. Xiu, Numerical Methods for Stochastic Computation: A Spectral Method Approach, Princeton University Press, 2010.
[12] C. E. Rasmussen and C. K. L. Williams, Gaussian processes for machine learning, Cambridge, MA: The MIT Press, 2006.

469768 An Ontology Supported Integration Framework for Models and Data in Biorefining

Abstract:

Process modelling and simulation are vital tools to plan, evaluate, assess and develop different alternatives in the biorefinery design for production of value added chemicals, materials, fuels, and energy from biomass. However, modelling for biorefining processes poses a number of technical challenges due to the complexity of the characteristics of biomass feedstock, and associated processing technologies evolving multiple pathways, including thermochemical-, catalytic-, biochemical- and hybrid-conversion routes. Due to the complexity and variety of these processes, numerous software tools have been developed to obtain the best-of-breed software for each application area, to fulfil the requirements of particular engineering tasks. During the process of developing a new model, the best suited modelling tools for different parts of the process are employed. However, to understand process design as a continuous work process from an integrated perspective requires the use of tools from a diverse range of sources and disciplines simultaneously. With the current increasing number of Computer Aided Chemical Engineering (CAPE) tools, the emergence of interoperability between heterogeneous modelling tools and process simulators was demonstrated by Braunschweig et al. (2000), which has ultimately led to the CAPE-OPEN initiative.

The CAPE-OPEN has conceptualised and developed a standard to define a set of interface specification as a method pertaining to the interoperability of the mainstream flowsheet applications (Pons 2010). The interoperability is facilitated by a standard middleware service, which enables the use of a single set of common interface as a communication tool (Braunschweig et al. 2004; Morales-Rodríguez et al. 2008). Having the common interface increases the potential reusability of existing models without having to invest in the software development, which further reduce the developing time, efforts, and the need for expertise (van Baten & Pons 2014). However, to fully exploit the potential reusability of existing models, the task to become aware of and have access to all of those models from a variety of sources is a challenge, yet required in order to identify the most suitable model to achieve the best solution for a particular engineering problem. As a consequence, Yang et al. (2008) indicated the potential misuse of the models that can further lead to inefficient or even wrong solution to the problem.

This paper is intended to address the challenges that previous research, CAPE-OPEN framework has faced and hence to minimise user intervention in selecting a model for the purpose of model integration. To enable automated discovery process in order to reduce the reliance of user intervention, the semantic representation is used to describe the models and data in terms of their input and output parameters, preconditions, and functionality (Koo & Cecelja 2015). Additionally, the Semantic Web Services (SWS) are employed to create a semantic profile of models and data representing biorefining processes. To this end, the domain ontology assures unified description of models by guiding the model registration and hence instantiation of ontology through ontology parsing, whereas the SWS assures the process of model discovery, its invocation and consequently integration. To demonstrate proposed approach, the framework is built for the models and data residing in web-based semantic repository in the domain of biorefining. In the domain ontology, the models are characterised in terms of functionality, modelling methods, model granularities, required software environment, etc. and their input and output parameters are characterised in terms of type (i.e. material, utility, economic and emissions), compositions and associated parameters, etc. The classification and characterisation of parameters used in matching process are not limited to numerical values (i.e. value for the flow rate), but also includes descriptive (i.e. type of material, energy) or even composite (i.e. range of flow rate with minimum and maximum values) types. The options of the discovered models to the requesting model, the input-output (I/O) matching technique is employed to identify the models that matches to a certain degree of request, which permits for a high degree of flexibility by allowing a partial matching rather than a full (exact) matching between I/O parameters. The discovered models from matching process are then ranked according to a set of preferences and the best option that satisfies the matching criteria is proposed to the user. To verify the performance of model discovery and selection, models representing biochemical- and thermochemical conversion processes are used to demonstrate as a case study with an aim to evaluate the best suited model whilst avoiding wrong design solution.

References:

Braunschweig, B.L. et al., 2000. Process modeling : The promise of open software architectures. Chemical engineering progress, 96(9), pp.65–76.

Braunschweig, B. et al., 2004. CAPE web services: The COGents way. Computer Aided Chemical Engineering, 18(C), pp.1021–1026.

L. Koo, F. Cecelja, "Model Integration Using Ontology Input-Output Matching," Computer Aided Chemical Engineering, 2015, volume 37, pp. 2567-2572.

Morales-Rodríguez, R. et al., 2008. Use of CAPE-OPEN standards in the interoperability between modelling tools (MoT) and process simulators (Simulis® Thermodynamics and ProSimPlus). Chemical Engineering Research and Design, 86(7), pp.823–833.

Pons, M., 2010. How to make use of CAPE-OPEN? In 2010 AIChE Annual Meeting, 10AIChE.

van Baten, J. & Pons, M., 2014. CAPE-OPEN: Interoperability in Industrial Flowsheet Simulation Software. Chemie Ingenieur Technik, 86(7), pp.1052–1064. Available at: http://doi.wiley.com/10.1002/cite.201400009.

Yang, a. et al., 2008. A multi-agent system to facilitate component-based process modeling and design. Computers and Chemical Engineering, 32(10), pp.2290–2305.

466202 A Biorefinery Boutique Made in Switzerland

Abstract:

Many challenges pave the road from an oil-based to a renewable economy. In the chemical industry, a substantial amount of work needs to be undertaken for renewable-based processes to compete with their petrochemical counterparts. Nowadays, the biggest hurdle is to evaluate which biorefinery concepts are environmental friendly and economic competitive at an early stage. Therefore, an innovative exploration was undertaken to create a biorefinery boutique for Switzerland and Europe for different future industrial scenarios.

Traditionally, there are two approaches to find biorefinery concepts. The first is to use overly simplified models, which can be analyzed according to their production portfolio and evaluated based on their profitability. In the second, detailed process simulations are used, which give detailed technological insight into each technology, but cannot be connected with each other due to their high complexity. This work employs a novel technique which connects detailed model biorefinery technologies in a network and finds tailor made biorefinery concepts for multiple future business scenarios.

Besides state of the art bioprocessing technologies, the biorefinery boutique incorporates life cycle assessment to indicate environmental issues and an economic assessment to project future profitability.

460997 New Short-Cut Tools for Early-Stage Investment Evaluation of Biorefineries

Abstract:

New short-cut tools for early-stage investment evaluation of biorefineries

Biorefineries offer a promising alternative to fossil-based processing industries and have met rapid development during the last years. Biorefinery processes employ state-of-the-art technologies and thus, represent high risk business decisions. Researchers undertaking the development of a biorefinery, often have to estimate capital and manufacturing costs based on minimum information, in order to decide wisely on project continuity and justify further funding for their research. Most R&D engineers rely on literature information to estimate the costs as historical cost data are proprietary information and are seldom announced. They also draw on various costing techniques and heuristics, which were developed for the needs of the petrochemical and the Oil & Gas Industry and require high level of process detail, which is not available at the early-stages of process conception.

We have evaluated some of these methods with means of both deterministic and statistical comparison from reported literature estimates and commercial biorefineries' cost data. As the majority were published during the 1970's-80's and were derived from petrochemical processes, they do not account for the technological progress often met within the state-of-the-art biorefinery processes and report discrepancies in their results. Fig. 1 illustrates the aforementioned discrepancies for dry corn mill ethanol biorefineries. We have been collecting capital cost data for commercial corn-to-ethanol plants (marked as circles). Only three out of the six cost methods fall within the 90% Confidence Intervals of the best regression line (marked as triangles), while the capital estimation reported in literature reference greatly underestimates the plant's final cost (marked as x). Our work reports similar figures for chemical and thermochemical biorefineries, as well CITATION Tsa16 \l 2057 [1].

Therefore, we decided to attempt new capital and manufacturing cost estimation methods to meet the needs of non-experienced cost estimators which undertake the design of first-of-a-kind biorefineries: the methods require information available at the start of the process conception. The first capital cost estimation method proposes new investment factors that draw on historical investment costs along with probabilistic estimation of the total investment. The second capital cost estimation method proposes cost estimation relationships for chemical, biochemical and thermochemical biorefineries, as well as simple techniques to define the uncertainty of the estimate. Finally, we have developed a new rapid manufacturing cost estimation methodology with uncertainty: it relies on factors requiring minimal user input, i.e. raw materials cost calculation, for determining the net production cost of the process, employing probabilistic means of determination.

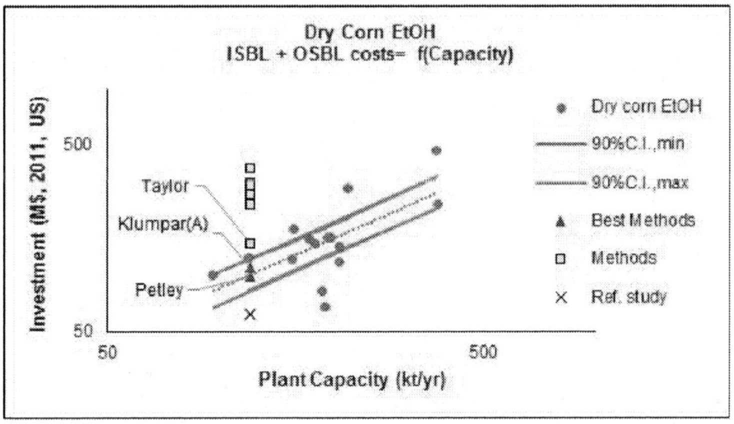

Fig. 1. Dry corn mill bioethanol production plants Investment Costs (M$, 2011,US) vs Plant Capacity (kt/yr)

Acknowledgements

The European Commission is gratefully acknowledged for funding the Renewable Systems Engineering project (RENESENG - FP7 Marie Curie project).

BIBLIOGRAPHY

[1] M. Tsagkari, J.-L. Couturier, A. Kokossis and J.-L. Dubois, "Early-stage capital cost estimation of biorefinery processes: a comparative study of heuristic techniques," *ChemSusChem*, 2016 *(under review)*.

466913 Global Sensitivity Analysis of Economic Assessment of Early Stage Process Design: The Case of the Glycerol Biorefinery

Abstract:

Nowadays it has become progressively important to achieve a sustainable process performance in order to keep a viable and competitive advantage in the international markets. Thus, the development of all-inclusive/holistic/comprehensive and systematic methods to accomplish this goal is the topic of this study. To this end, a multi-level framework for early-stage design and screening of process options is proposed for techno-economic and environmental sustainability analysis through risk assessment. The alternatives composing the design space are assessed by following the framework's steps, targeting to (i) quantify the economic risk; (ii) quantify the potential environmental risk under uncertainties by using monetary valuation of environmental impacts techniques; (iii) measure the concepts' eco-efficiency; and, finally (iv) facilitate the decision-making procedure by interpreting the trade-offs through a qualitative and quantitative sustainability risk assessment matrix. The framework application is highlighted through the critical assessment and screening of early stage glycerol-based biorefinery design concepts. It was found that the best potential options for glycerol valorisation is are obtained through the production of either: (i) lactic acid; (ii) succinic acid; or, (iii) 1,2-propanediol. To reduce the potential economic risk, a multi-product biorefinery is suggested, whose strategy is based on its capability of switching between the production of lactic acid and succinic acid.

The risk assessment tool described above provide valuable information on how uncertainties (technical versus external such as feed/product prices) impact the decision of optimal process concept and how to better mitigate these impacts by improving technology development. In this contribution, we take one step further and ask the question, a given a certain economic risk associated with a conceptual process alternative, what is the contribution of input uncertainties((e.g. feed composition, yield, feedstock prices,

product prices, etc)) to the risk? If risk is defined as variance (deviation from an expected value), what is the individual variance attributed to each input uncertainty? And how we can use this information to further minimize economic risk and improve technology advancement? To answer these questions, we use sensitivity analysis methods including local methods such as one-factor-at-a-time (OAT) as well as global methods such as variance decomposition to compute first order as well as total sensitivity indices.
We implemented algorithms in Matlab (R2015a) based on Monte Carlo sampling using Latin Hypercube Sampling with Iman Conover correlation control (Sin et al 2009) and sampling based on conditional probability of dependent variables proposed by Kucherenko et al 2012 to take into account the correlation structure of input uncertainties (especially price correlation matrix). The economic model used in this work is the discounted cash-flow rate of return (DCFR). The OAT sensitivity analysis has shown that deviations in the product's and feedstock prices, total production cost, fixed capital investment as well as discount rate, among others, have a high impact on the project's profitability (quantified with NPV and MSPA global sensitivity analysis of NPV indicated two major factors contributing to its variance (risk): uncertainty in fixed capital cost estimation (82%), and, uncertainty in product prices (14%). The NPV meta-model showed that improving the accuracy of fixed capital cost estimation (5% *std*) leads to a 95% reduction in the NPV's uncertainty. This global sensitivity analysis confirms the process engineering expectation that more detailed capital cost estimation supported through pilot scale studies or more rigorous model analysis is expected to decrease the economic risk and better judge the transferability of technology to the market. Global sensitivity analysis is an important and complementary tool to study and decompose impact of uncertainties to economic assessment of conceptual process design studies.

Keywords: techno-economic assessment, uncertainty analysis, glycerol-based biorefinery concepts, risk-based decision making, global sensitivity analysis.

References:

Kucherenko, S., Tarantola, S., & Annoni, P. (2012). Estimation of global sensitivity indices for models with dependent variables. *Computer Physics Communications*, *183*(4), 937-946.

Sin, G., Gernaey, K. V., & Lantz, A. E. (2009). Good modeling practice for PAT applications: Propagation of input uncertainty and sensitivity analysis. *Biotechnology progress*, *25*(4), 1043-1053.

Cheali, P., Gargalo, C., Gernaey, K. V., & Sin, G. (2015). A Framework for Sustainable Design of Algal Biorefineries: Economic Aspects and Life Cycle Analysis. In *Algal Biorefineries* (pp. 511-535). Springer International Publishing.

460678 General Bio-Separation Superstructure Optimization Framework

Abstract:

General Bio-Separation Superstructure Optimization Framework

Wenzhao Wu, Kirti Yenkie, Christos T. Maravelias*

Dept. of Chemical and Biological Engineering, University of Wisconsin-Madison, Madison, WI 53706

Recent advances in synthetic biology and metabolic engineering have enabled the production of a range of chemicals using engineered microorganisms[1,2]. However, despite the intriguing potential, the biological production of high-value chemicals is challenging because it is likely to have low titer and substantial amounts of excreted byproducts, and the purity specifications for high-value chemicals, as opposed to fuels, are rather strict, which means that separation costs are likely to represent a large fraction of the total production cost (more than 70% of total cost[3]). Thus, the efficient synthesis of bio-separation processes becomes a critical task. Although this synthesis problem has been studied for various chemicals in the past, these studies were mostly performed on a case-by-case basis. There has been limited research towards the development of systematic methods for bio-separations, applicable to all chemical targets. Accordingly, the goal of the present work is to develop a general bio-separation superstructure optimization framework, aiming to provide guidance on the preliminary synthesis of separation networks.

Based on general bio-separation principles and insights obtained from industrial separation processes for specific products, we first identify four separation stages: (1) cell treatment, (2) product phase isolation, (3) concentration and purification, and (4) final refinement. Specifically, cell treatment starts with the harvesting of cells (if the product is intracellular), which can be performed using solid-liquid separation technologies. Cell harvesting is followed by cell disruption to release the intracellular products. The first stage is skipped if the product is extracellular, and the separation starts from the second stage - product phase isolation, where the product-rich phase is isolated. In the concentration and purification stage, we remove large amount of water along with other impurities, by utilizing the differences between the product and the other components on volatility, molecular size, diffusivity, solubility in solvents, etc. In the final refinement stage, we further remove trace impurities and perform refined operations to satisfy special product specifications, such as colorlessness, complete dryness, and crystal form.

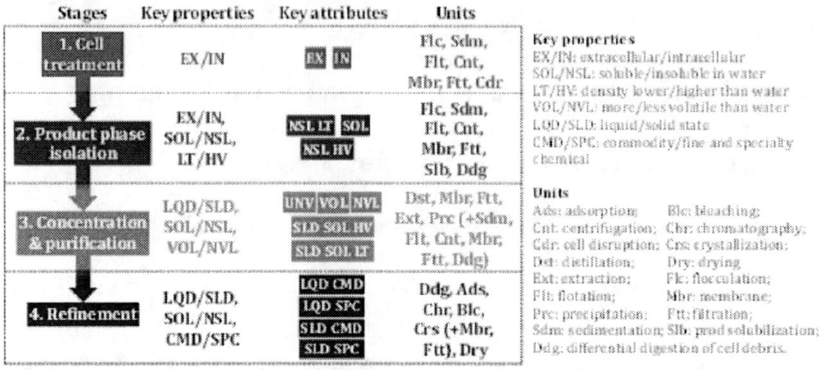

Figure 1. Stage-wise analysis of bio-separation processes, including key product properties, attributes, and units for each stage.

Next, we identify key product properties and key attributes that affect the separation processes in each stage, as shown in Figure 1. For example, in Stage 2, the different attributes imply different locations of the product (e.g., "NSL LT" indicates that the product is located at the top, while "SOL" indicates that the product is dissolved in water and evenly distributed in the system), and hence different separation processes. Major units in each stage are also identified.

Then for each stage, we systematically implement a set of connectivity rules[4] to develop attribute-specific superstructures. They are subsequently combined to generate a stage-specific superstructure. Finally, all the four stage-specific superstructures are integrated into a general superstructure (shown in Figure 2) that accounts for all types of bio-chemical products.

We further develop a superstructure reduction method to solve product-specific instances, based on product type, unit availability and suitability, case-specific considerations, and final product specifications. The reduced superstructure for an example case is shown in Figure 2, where an EX NSL LT NVL LQD CMD product (see abbreviations in Figure 1) is required to be completely colorless in its final product form, and all units in the general superstructure are available except for filtration.

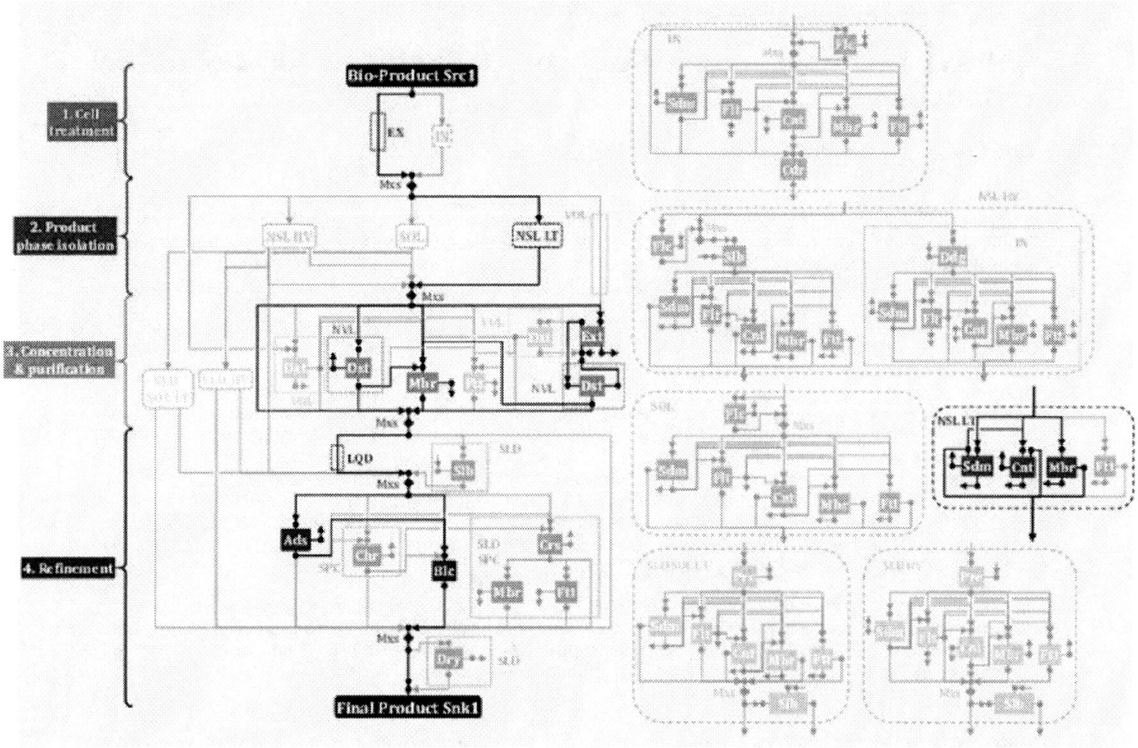

Figure 2. The general bio-separation superstructure (including the "dimmed" parts), and the reduced superstructure (excluding the dimmed parts) for an example case, where an EX NSL LT NVL LQD CMD product is required to be completely colorless in its final product form, and all units in the general superstructure are available except for filtration.

Finally, a general MINLP optimization model, including short-cut unit models (e.g., Fenske-Underwood equations for distillation unit models) for all types of separation units considered in the framework, is formulated. In the modeling of a specific reduced superstructure, only models of relevant units (which exist in the reduced superstructure) are included.

The proposed framework is applied to three case studies. Each case considers different set of product types, unit availability and suitability, case-specific considerations, and final product specifications. The common objective is to minimize total cost of purifying a dilute product stream (5 wt% concentration) to a product stream with at least 90 wt% purity.

References

[1] Gavrilescu, M. & Chisti, Y., 2005. Biotechnology- a sustainable alternative for chemical industry. *Biotechnology advances,* Volume 23, pp. 471-499.

[2] Bornscheuer, U. T. & Nielsen, A. T., 2015. Editorial overview: chemical biotechnology: interdisciplinary concepts for modern biotechnological production of biochemicals and biofuels. *Current Opinion in Biotechnology,* Volume 35, pp. 133-134.

[3] Kiss, A. A., Grievink, J. & Rito-Palomares, M., 2015. A systems engineering perspective on process integration in industrial biotechnology. *J Chem Technol Biotechnol,* Volume 90, pp. 349-355.

[4] Wu, W., Henao, C. & Maravelias, C., 2016. A Superstructure Representation, Generation and Modeling Framework for Chemical Process Synthesis. *AIChE J.*

459851 Technoeconomic Study of AB Synthesis for Biobutanol Production

Abstract:

Over the last 15 years there has been an important effort towards climate change mitigation. Among the different actions, one of the more challenging is reducing emissions in the transport sector. Ethanol and biodiesel, from different biomass sources, represent the typical alternatives to substitute crude based gasoline and diesel nowadays. However, there are other options such as FT fuels (Martín & Grossmann, 2011), glycerol ethers (Martín & Grossmann, 2014) or biobutanol. Biobutanol is considered an advanced fuel. It has similar properties to gasoline, it can be directly used in gasoline engines without modification, and it has higher energy density and boiling point than ethanol reducing evaporation losses. Moreover, it is less hydroscopic than ethanol (Durre, 2007).

The ethical issues involving first generation biofuels have led to focus on obtaining biofuels from agricultural residues. Corn being the major crop in the world, results in the fact that corn stover is widely available. However, in the case of biobutanol, the yield has been low due to the fact that together with biobutanol, ethanol and acetone are typically obtained. Thus, this work is divided in two parts: the first part is devoted to the experimental evaluation of the lignocellulosic biomass degradation. In the second one, a technoeconomic study of a new production process is described. The novelty of this work relies on the scale up of AB fermentations, that does not produce ethanol, (Qureshi et al., 1999). In addition, in-situ recovery techniques such as evaporation are also included at the process to make it economically competitive with conventional fuels and bioethanol.

In the first stage of the work, comprehensive evaluation of the experimental conditions to break down switchgrass and corn stover from Castille and Leon region (Spain) has been performed. The study considers both pretreatment and hydrolysis. With regards to pretreatment, several physio-chemical techniques such as acid (H2SO4), alkali (using KOH or NH4OH), ultrasounds processing, surfactants or organic solvents are tested. With regard to hydrolysis, various enzyme cocktails were used. It turns out that corn stover reaches the highest yield, 677.00 mg of sugars by gram of biomass, when it is pretreated using a diluted sulfuric acid solution 0.5% v/v, autoclaving the mixture during 90 min at 121 ºC with no sonication stage. The hydrolysis method is set up at 50 ºC and 150 rpm with a solid loading of 2.5 % w/v and a ratio of enzymes/biomass of 400 and 440 µl/g of NS50013 (cellulose complex) and NS50010 (β-glucosidase cocktail) respectively. The sugar recovery obtained is competitive with current best available technologies i.e. (Sundaram and Muthykumarappan, 2016).

In a second stage, we synthesize a process for the production of biobutanol from corn stover including the use of novel technologies such as pervaporation to improve the yield from biomass and reducing the by-products by means of AB fermentation. The processes consists of four sections, pretreatment, hydrolysis, fermentation-pervaporation system and product purification. We use the experimental results to determine the yield of the pretreatment and hydrolysis. A hybrid simulation using EXCEL to develop black box kind of models for the experimentally validated units and ChemCAD has been performed to evaluate the technoeconomic feasibility of this process.

The results show an investment of 186 MM€ and a butanol production cost of 1.09 €/kg for a 27 kt/y plant. The yield to butanol is 0.21 kg/kg of biomass resulting, altogether, in a fuel closer to become competitive versus second generation bioethanol and fossil based fuels in terms of yield and cost.

References

Durre, Biobutanol: an attractive biofuel Biotech. J., 2 (2007), pp. 1525–15234

M.S. Madihah, A.B. Ariff, K.M. Sahaid, A.A. Suraini, M.I.A. Karim (2001) Direct fermentation of gelatinized sago starch to acetone–butanol–ethanol by Clostridium acetobutylicum. World J. Microbiol. Biotechnol. 17 (2001), pp. 567–576

Martín, M., Grossmann, I.E. (2011) Process optimization of FT- Diesel production from biomass. Accepted Ind. Eng. Chem Res, 50 (23),13485–13499

Martín, M.; Grossmann, I.E. (2014) Simultaneous dynamic optimization and heat integration for the co-production of diesel substitutes: Biodiesel (FAME & FAEE) and glycerol ethers from algae oil. Ind. Eng. Chem. Res. 53, 11371-11383

Qureshi, N., Hans P. Blaschek. Production of Acetone Butanol Ethanol (ABE) by a Hyper-Producing mutant strain of Clostridium beijerinckii BA101 and recovery by Pervaporation. Biotechnol.Prog. 1999, 15, 594-602.

Sundaram, V., Muthukumarappan. K., Influence of AFEXTM pretreated corn stover and switchgrass blending on the compaction characteristics and sugar yields of the pellets. Industrial Crops and Products 83 (2016) 537–544.

465757 Effect of Market and Technical Parameter Uncertainties on the Optimal Design of Integrated Biorefineries

Abstract:

Effect of Market and Technical Parameter Uncertainties on the Optimal Design of Integrated Biorefineries

As research is exploring possibilities for developing renewable bio-based fuels and chemicals, there arises a need for development of strategies which can design sustainable value chains that can be scaled up efficiently and provide tangible net environmental benefits. Recent ventures in biofuel production have been fraught with corporate failures. A driving reason for these unsuccessful ventures, in part is governed by the lack of proper planning in designing plants and supply networks. Modeling strategic and operational level decision processes of an enterprise which produces renewable-based fuels and chemicals, using engineering and financial tools, can provide valuable insight into the inter-play of technology and process interactions for different product streams.

In this study we present the development of a comprehensive decision support tool, within which the business value of a multiproduct biorefinery is optimized by considering all types of uncertainties including uncertainties at strategic and operational level. A stochastic linear model is first developed to optimize production capacity of the plant for the desired planning horizon and then process simulation coupled with stochastic optimization algorithm is employed to optimize the operating condition of the plant under uncertainty. The proposed methodology is based on an iterative framework that utilizes systems-based optimization in conjunction with detailed mechanistic modelling and simulation of the process. Market uncertainties are taken into account at the strategic planning level, and uncertainties related to parameters characterizing the processing technologies are addressed in operational level optimization. Monte-Carlo based simulation and global sensitivity analysis are utilized to identify the most critical parameters and optimize the operating conditions of the plant accordingly. Additionaly, risk management strategies are introduced to the framework for explicit treatment of strategic and operational risks. The other key feature of our approach is the development of dynamic data exchange strategy embedded as part of the process simulation to incorporate the complex kinetics of bio-reactions in the simulation model. The final results of the proposed framework include strategic capacity plan, optimal expected NPV, and optimal operating conditions of the process.

464430 Centroidal Voronoi Tesselation Based Model Order Reduction for a Moving Boundary Problem: Application to a Hydraulic Fracturing System

Abstract:

More often than not a chemical process can be accurately described using a mathematical model which is generally comprised of nonlinear partial differential equations (PDE). One arrives at these equations by applying conservation laws and/or transport equations. Many chemical engineering processes are, owing to their nature, characterized by a moving boundary problem. Two such examples would be the propagation of fractures during a hydraulic fracturing operation [1] and the evolution of interface boundaries during a reaction-diffusion process [2, 3].

In order to accurately capture the dynamics of systems described by PDEs, a large number of state variables are required and this makes it computationally difficult to design online control systems. Although a model reduction technique, Proper orthogonal decomposition (POD), has proved to perform effectively, it often fails to capture the process dynamics in nonlinear systems, since it assumes that data belong to a linear space and therefore relies on the Euclidean distance as the metric to minimize. Hence, we can deal with the above mentioned problem by applying POD locally (with respect to time or spatial coordinates) to clusters instead of applying it globally because each cluster contains snapshots that show relatively close-in-distance behavior within itself, and considerably far with respect to other clusters. Also, the dominant behavior of each cluster can be captured by using less empirical Eigenfunctions as compared to that of global POD. Sahyoun and Djouadi [4] introduced different clustering schemes such as time snapshots clustering (TSC) and space vectors clustering (SVC), and applied them to a nonlinear convective PDE system governed by the Burgers' equation for fluid flows over 1D and 2D domains. Following these contributions, the idea here is to develop an optimal clustering technique using Centroidal Voronoi Tesselation (CVT) scheme and apply POD locally to these clusters which extracts a relevant set of basis

vectors for each cluster.

In this work, we first transformed the PDE with a time-dependent spatial domain into the one with an appropriate time-invariant spatial coordinate, and a representative ensemble of solutions is constructed by solving a high-order discretization of the PDE. Then we divide the ensemble into clusters using CVT and apply POD locally to these clusters to derive a small set of empirical eigenfunctions (dominant spatial patterns) that capture almost all the energy of the ensemble of solutions. The empirical eigenfunctions are subsequently used as basis functions within the Galerkin's model reduction framework to derive low-order ODE systems that accurately describe the dominant dynamics of the PDE system. These ODE systems are used as a basis for the synthesis of a low-dimensional nonlinear controller based on model predictive control theory. We applied the framework to the hydraulic fracturing problem and a reaction-diffusion process.

References:

[1] Qiuying Gu and Karlene A. Hoo, Evaluating the Performance of a Fracturing Treatment Design, *Ind. Eng. Chem. Res.*, 2014, *53* (25) : 10491–10503

[2] A. Armaou and P. D. Christofides, Nonlinear Feedback Control of Parabolic Partial Differential Equation Systems with Time-dependent Spatial Domains, J. Math. Anal. Appl., 1999, 239:124-157.

[3] M. Izadi and S. Dubljevic, Backstepping output-feedback control of moving boundary parabolic PDEs, European J. of Control, 2015, 21:27-35.

[4] S. Sahyoun and S. Djouadi, Time, Space, and Space-Time Hybrid Clustering POD with Application to the Burgers' Equation, *53rd IEEE Conference on Decision and Control December 15-17*, 2014. Los Angeles, California, USA

469168 Output Feedback Control of Transport-Reaction Processes with Unknown Parameters Using Adaptive Model Reduction with Minimum Feedback Information

Abstract:

We propose an adaptive output feedback framework to address the output feedback control problem for transport-reaction processes with unknown parameters. The mathematical description of these processes by nonlinear parabolic partial differential equation (PPDE) systems from the basis for our effort. Such systems frequently appear in chemical and advanced materials production industries, e.g., packed and fluidized bed reactors in refinery and petrochemical processes, polymerization and crystallization processes, and chemical vapor deposition in microelectronic fabrication processes. The infinite-dimensional representation of PPDE systems in appropriate functional spaces can be decomposed to a finite-dimensional slow (and possibly unstable) and infinite-dimensional fast (and stable) subsystems [1]. Due to the existence of a time-scale separation between slow and fast subsystems we can approximate the PPDEs by a finite number of ordinary differential equations (ODEs). This low-dimensional approximation (that can be employed as the basis for model-based adaptive output feedback control design) is constructed by discretizing the PPDEs using Galerkinâ€™s method. The required basis functions for such discretization are obtained by applying adaptive proper orthogonal decomposition (APOD) to the spatial profiles of the system states during process evolution. Obtaining such highly informative profiles might not be feasible owing to limited availability of sensors which frequently sample the entire process domain. To overcome such limitation, we use a PDE-based Luenberger-type dynamics observer to reconstruct the spatial profiles by employing measurement outputs from a set of moving sensors [3]. The effectiveness of the proposed adaptive control approach is illustrated on the thermal regulation problem of a tubular reactor where an irreversible reaction takes place in the presence of unknown diffusivity coefficient and reaction rate constant.

[1] Christofides, P.D., Nonlinear and robust control of PDE systems, Birkhauser, New York, 2000.

[2] Pourkargar, D.B., Armaou, A., Design of APOD-based switching dynamic observers and output feedback control for a class of nonlinear distributed parameter systems, Chem. Eng. Sci., 136 (2015), 62-75

[3] Demetriou, M.A., Hussein, I.I., Estimation of spatially distributed processes using mobile spatially distributed sensor network. SIAM J. Control Optim., 48(2009), 266-291

458097 Output Regulation for Boundary Controlled Linear Coupled Hyperbolic Pides: Application to a Parallel-Flow Heat Exchanger System

Abstract:

Transport equations described as first order hyperbolic linear partial differential equations (PDEs) are utilized to demonstrate different complex physical processes. To mention representative engineering applications, we consider heat exchanger [1], production systems [2], oil well [3] and road traffic [4]. Due to the wide area of applications, the control problems of such systems has been active research topics. We refer [5], [6], [7] for significant contributions related hyperbolic PDE systems.

The backstepping method known from nonlinear control theory [8], has been extended to PDEs solving stabilization problems. Many efforts and results have been established that can help to simplify the formulation of control problems, e.g. output regulation.

In this work, the output regulation is addressed for boundary controlled hyperbolic partial integro-differential equations (PIDE) systems. Thereby, the outputs to be measured do not belong to the set of outputs to be controlled. The change of variables is applied such that the completely separated first order hyperbolic PIDE systems can be obtained. Consequently, the backstepping approach is applied in such a way that the state feedback regulator problem is solved and regulator equations with a simple structure are obtained. In output feedback regulator design, in order to find observer gains, triangularization is performed and the observer error systems are transformed into cascade systems with a simple structure. Moreover, it is shown that for the resulting output feedback regulator, the separation principle holds implying output regulation for the exponentially stable closed-loop systems.

Finally, the output regulation results are demonstrated by means of a parallel-flow heat exchanged system with in-domain pointwise controlled outputs.

[1] Xu, X., and Dubljevic, S. (2016). The state feedback servo-regulator for countercurrent heat-exchanger system modelled by system of hyperbolic PDEs. European Journal of Control.

[2] Marca, M. L., Armbruster, D., Herty, M., and Ringhofer, C. (2010). Control of continuum models of production systems. IEEE Transactions on Automatic Control, , 55(11), 2511-2526.

[3] Landet, I. S., Pavlov, A., and Aamo, O. M. (2013). Modeling and control of heave-induced pressure fluctuations in managed pressure drilling, IEEE Transactions on Control Systems and Technoloy, 21(4), 1340–1351.

[4] Amin, S., Hante, F. M., and Bayen (2008). A. M., On stability of switched linear hyperbolic conservation laws with reflecting boundaries, in Hybrid Systems: Computation and Control. Springer-Verlag, 602–605.

[5] Coron, J.-M., d'Andrea Novel, B., and Bastin, G. (2007). A strict lyapunov function for boundary control of hyperbolic systems of conservation laws, IEEE Transactions on Automatic Control, 52(1), 2–11.

[6] Di Meglio, F., Vazquez, R., and Krstic, M. (2013). Stabilization of a system of n + 1 coupled first-order hyperbolic linear PDEs with a single boundary input, IEEE Transactions on Automatic Control, 58(12), 3097–3111.

[7] Aamo, O. M. (2016). Leak detection, size estimation and localization in pipe flows, IEEE Transactions on Automatic Control, 61(1), 246–251.

[8] P. V. Kokotovic (1992). The joy of feedback: nonlinear and adaptive, Control Systems, IEEE, 12(3), 7–17.

465247 A Novel Approach for Mechanistic Modeling and Simulation of Convection-Diffusion-Reaction Systems: Application to Nanoparticle Transport in Tumor Tissues

Abstract:

We present a novel Brownian Dynamics (BD) simulation approach to model convection-diffusion-reaction systems. The simulation approach combines the Method of Regularized Stokeslets (MRS) with a time-adaptive BD algorithm. The combined method has unique ability to mechanistically model convection and diffusion-driven transport of reactive particles in heterogeneous porous medium. The BD algorithm takes intelligent decisions to advance particle position based on distance from their reactive partners. In the bulk fluid, particles away from their reactive partners are treated as point objects, and advanced with larger step sizes. In contrast, in the vicinity of any reactive partner, particles are treated as circular objects of defined size, and advanced with sub-nanometer step sizes. Such adaptive approach enables efficient and robust computation, while capturing particle interaction with their partners with high resolution details. We have applied this approach to model drug-delivery nanoparticle transport in the heterogeneous interstitial space of tumor tissues. The model takes into account both diffusion and convection transport of nanoparticles, and their interaction with the cell boundaries. Using the model, we have analyzed tissue penetration efficacy of nanoparticles as function of particle size. Our analysis provides new insights contradicting the findings of several recent publications. We show that experiments based on artificial tissue systems, which are usually carried out in the absence of convective transport and particle-cell interaction, may yield misleading conclusion on size-dependent tissue penetration efficacy of nanoparticles. We have validated our model analysis and predictions using *in vitro* breast tumor tissue grown on a microfluidic system.

469307 Equation-Free Control of Distributed Parameter Systems Using Discrete Empirical Interpolation Method and Proper Orthogonal Decomposition

Abstract:

Diffusion, convection and reaction exist in many chemical and material industry processes. These phenomena lead to the spatial variation of the state variables. Their mathematical description can be derived using conservation equations and is in the form of distributed parameter system (DPS). Controlling these processes is of significant importance to achieve product quality, safety and economic profit. One of the challenges in the controller design for DPS is due to the infinite dimensional state space dynamics when the system is considered in a functional space. A standard way to deal with this issue is to use Galerkin's method to construct a reduced order model. The prerequisite basis functions in Galerkin's method can be generated by proper orthogonal decomposition (POD) [1]. After the reduced order model is constructed, established control methods, including feedback linearization, Lyapunov based ones and back-stepping can be used to design the controller. To implement these methods though, the information of the system state is required. The state of distributed parameter systems can be estimated by Luenberger-based observers, which relaxes the requirement on sensors. However, both the Luenberger-based observers and aforementioned controller design methods assume that a mathematical model of the system is available; furthermore unmodeled dynamics and model uncertainty always exits [2].

This issue motivates us to propose an approach to control systems when the knowledge of the chemical and physical law that describes the systems is unavailable or incomplete but the effect of actuators is known. First, POD is applied to generate two sets of basis functions that are used to estimate the system dynamics and state. Then discrete empirical interpolation method (DEIM) [3] is employed to determine the location of sensors. DEIM was proposed to reduce the computational cost associated with the nonlinear term in the reduced order model generated by Galerkin-POD methodology. Because DEIM has the property that the selection of the interpolation indices can limit the growth of the error of the approximation, it is modified to determine sensor location in our method. Compared with other sensor network design methods, this approach does not require the *a priori* knowledge of the governing equation. Having continuous measurements from these sensors, the state in the ROM is estimated by a static observer. Using a similar approach as the static observer, a mapping from the measurement of the velocity sensors onto the projection of the dynamics is generated. The estimation of the system dynamics and state make the model redundant in controller design.

Compared with other equation free methods, including divide-and-conquer techniques (such as memory based local modeling), subspace identification, neural network, and autoregressive moving average with exogenous inputs (ARMAX), the proposed method is computationally cheaper. The method performance is illustrated on a diffusion reaction process.

REFERENCES

[1] Lawrence Sirovich. Turbulence and the dynamics of coherent structures .1. coherent structures. Quarterly of Applied Mathematics, 45(3):561–571, 1987.

[2] Zhong Sheng Hou and Zhuo Wang. From model-based control to data-driven control: Survey, classification and perspective. Information Sciences, 235:3–35, 2013.

[3] Saifon Chaturantabut and DC Sorensen. Nonlinear model reduction via discrete empirical interpolation. SIAM Journal on Scientific Computing, 32(5):2737–2764, 2010.

465624 Sampled-Data Event-Triggered Control of Distributed Parameter Systems with Networked Sensors and Actuators

Abstract:

With the emergence of the Smart Plant paradigm in recent years [1] and the shift in industrial practice towards sensor and control systems that are accessed over shared communication networks instead of dedicated links, a significant and growing body of research work aimed at addressing the fundamental challenges associated with the design and analysis of networked control systems has emerged. A close examination of the literature on this topic, however, shows that the bulk of existing studies have focused on spatially homogenous systems whose underlying dynamics are modeled by systems of ordinary differential or difference equations. In contrast, spatially distributed systems whose underlying dynamics are captured by systems of Partial Differential Equations (PDEs) have received less attention. This is an important limitation given the fact that many industrial processes are characterized by spatial variations (e.g., transport-reactions and fluid flows), and that the control of these systems typically involves the use of spatially distributed control actuators and measurement sensors.

A number of recent studies have been carried out to bridge this gap (e.g., see [2]–[3] for some efforts in this direction). Bringing together tools from infinite-dimensional systems, model reduction and model-based control, the focus of these studies has been primarily on the development of systematic networked control methodologies that address various sensor-controller communication constraints, such as network resource limitations, communication delays, and real-time scheduling constraints. The problem is generally addressed by embedding a reduced-order model of the infinite-dimensional system in the control system. The model is used to generate the necessary control action when sensor-controller communication is suspended and is updated when communication is permitted using a time-triggered sensor-controller communication policy with a constant update rate.

An alternative approach for dealing with communication constraints in networked control systems is the use of event-triggered control strategies, where communication over the network is suspended or restored in response to certain events which are tied to the breach of certain desired closed-loop stability and/or performance thresholds. The aim is to keep network utilization to a minimum while guaranteeing desired levels of control performance. This is an appealing goal in the context of sensor/actuator networks where reducing network utilization can also reduce the energy expenditures of battery powered wireless devices. This approach has been widely studied in the context of lumped parameter systems (e.g., see [4]–[6]), and has also received some attention in the distributed parameter system setting [7]-[8]. In these studies, however, practical implementation issues such as the discrete availability of sampled state measurements, were not considered, and no assessment of the stability or performance properties of the networked closed-loop system under discretely-sampled measurements was made. In practice, state measurements are often sampled at discrete times. The limited access to state measurements during sampling intervals poses an important obstacle for the implementation of event-triggered control systems as it restricts the system's ability to properly monitor and evaluate the communication-triggering events, which typically involve state-dependent thresholds. A common approach for dealing with sampled-data control systems is the use of zero-order-hold schemes, which hold the last available measurement static until the next sampled measurement arrives. However, this approach does not take the system dynamics into account and can therefore lead to closed-loop instability when used to implement the event-triggered communication strategy.

Motivated by these considerations, we present in this work an event-triggered control strategy for spatially distributed systems with low-order uncertain dynamics, sensor-controller communication constraints and discretely-sampled state measurements. Based on a suitable finite-dimensional model that approximates the dominant dynamics of the infinite-dimensional system, an event-triggered networked control system that enforces closed-loop stability with minimal sensor-controller communication is initially designed for the case when state measurements are sampled continuously. The model is used by the controller to generate the necessary control action when communication is suspended, and its state is updated using the real-time measurements when communication is restored. Communication is triggered when a state-dependent threshold on the model estimation error is breached. The communication threshold is designed using Lyapunov techniques and is explicitly characterized in terms of the model and controller design parameters, as well as the actuator and sensor locations. To address the implementation of the control strategy under discretely-sampled measurements, a forecasting scheme that provides a worst-case bound on the evolution of the model estimation error over each sampling interval is devised and incorporated into the control system. This bound is used to obtain a modified (tighter) communication threshold which ensures that the actual stability threshold is not breached during periods of measurement unavailability. It is shown that if the underlying dynamics of the system are linear, the forecasting scheme can be carried out off-line leading to an explicit characterization of the modified stability threshold. On the other hand, for systems with nonlinear dynamics, an on-line forecasting strategy is required. Finally, the developed methodology is illustrated using a diffusion-reaction process example.

References:

[1] P. D. Christofides, J. F. Davis, N. H. El-Farra, D. Clark, K. R. Harris, and J. N. Gipson, "Smart plant operations: Vision, progress and challenges," *AIChE Journal*, vol. 53, no. 11, pp. 2734–2741, 2007.

[2] Z. Yao and N. H. El-Farra, "Resource-aware scheduled control of distributed process systems over wireless sensor networks," *Proceedings of American Control Conference*, pp. 4121–4126, 2010.

[3] Z. Yao and N. H. El-Farra, "Model-based networked control of spatially distributed systems with measurement delays," *Proceedings of American Control Conference*, pp. 2990–2995, 2012.

[4] M. Mazo and P. Tabuada, "Decentralized event-triggered control over wireless sensor/actuator networks," *IEEE Transactions on Automatic Control*, vol. 56, no. 10, pp. 2456–2461, 2011.

[5] X. Wang and M. D. Lemmon, "Event-triggering in distributed networked control systems," *IEEE Transactions on Automatic Control*, vol. 56, no. 3, pp. 586–601, 2011.

[6] E. Garcia and P. J. Antsaklis, "Model-based event-triggered control for systems with quantization and

time-varying network delays," *IEEE Transactions on Automatic Control*, vol. 58, no. 2, pp. 422–434, 2013.

[7] Z. Yao and N. H. El-Farra, "Resource-aware model predictive control of spatially distributed processes using event-triggered communication," *Proceedings of 52th IEEE Conference on Decision and Control*, pp. 3726–3731, 2013.

[8] D. Xue and N. H. El-Farra, "Networked event-triggered control of spatially distributed processes using a dual-mode communication strategy," *Proceedings of 54th IEEE Conference on Decision and Control*, pp. 1899–1904, 2015.

467944 Data-Driven Model Predictive Control of a Seeded Batch Crystallization Process

Abstract:

Particulates production has undergone phenomenal growth in the past few decades. They are used for manufacturing of numerous high-value industrial products such crystals, polymers etc. The products are often manufactured using batch processes to provide the flexibility of achieving different products by changing recipes.Â The physio-chemical and mechanical properties of the end product are known to be linked to the particle size distribution (PSD). For instance, in case of emulsion polymerization, PSD strongly impacts the polymerâ€™s rheology (viscosity), adhesion, drying characteristics etc. [1]

Traditional approach for control of batch processes (particulates or otherwise) have been to use open loop, recipe based policy. In this approach, a predefined input trajectory is applied to each batch [2]. The approach assumes that the desired product quality can be obtained by repeating historically successful input profiles. Although these approaches are easy to implement and do not require a model for the process, they are incapable of rejecting disturbances that affect the process. This has motivated the use of feedback control strategies.

A classical closed loop control approach in batch process is trajectory tracking. In this approach, a predefined set-point trajectory for a measured process variable, often obtained from past successful batches, is tracked. PI controllers are often used for implementing trajectory tracking control. When implementing trajectory tracking approaches (with PI or otherwise) even with perfect tracking, the desired quality may not be obtained, as the relationship between the measured/tracked variable and final quality may change significantly with changes in the process conditions.

These challenges are addressed by control strategies that are cognizant of the causal relationship not only between the manipulated and the (online) measured variables, but also between the manipulated and the final quality variable. A popular model based control approach, model predictive control (MPC), has increasingly been studied for the control of batch processes [3-6]. The reasons for popularity of these control schemes are twofold: first, the feedback controller can counter the model uncertainties that are associated with model simplifications and measurement errors, the other being their ability to handle the ever present input/output constraints. In implementing these formulations, where possible, good first principles models are preferred due to the ability to predict process behavior beyond the data set used to estimate the model parameters. However, first principles models are difficult to develop as it is often impractical to estimate the parameters through experiments, particularly because of the associated cost. Further, these first principles models, particularly in case of distributed parameter systems such as batch particulate processes, pose computational challenges for direct use in predictive control formulation.

These issues have motivated the use of simpler, often linear, models derived from past batches database. A variety of approaches for development of data-driven models have been proposed. One of the most celebrated and widely used approach is partial least squares (PLS), which models the process in a projected latent space [7]. These models are essentially time-varying linear models, linearized around mean past trajectories, and therefore require the batches to be of same length, or to recognize an appropriate alignment variable. To account for these limitations, a multi-model approach was proposed in [6]. These models were based on the 'current measurements' of the process instead of the 'time'. These developments were followed by contributions in the area of integration of these data-driven models with the advanced control formulations [5-6]. More recently a subspace identification based batch control approach was proposed in [8] where a LTI state-space model of the batch process is estimated.

The existing formulations have, however, dealt generally with lumped parameter systems and the applications of data-driven approaches for PBE systems have been limited. Some variation of PLS approach have been proposed for distributed parameter processes. In one approach [9] a PLS model is used in conjunction with the predictions obtained from PBEs for batch to batch optimization. A similar strategy is presented in [10] where only a PLS model is used for predicting the final PSD. As with other latent variable approaches, a limitation with the use of PLS models for batch processes is the need to align the batch lengths.

Motivated by these considerations, this work presents a subspace identification based state-space modeling and control approach for a seeded batch crystallization process. The proposed control design is compared with an open loop policy as well as a traditional trajectory tracking policy using classical control. The proposed MPC is shown to achieve 27.22% and 29.90% improvements in the desired product quality, and the ability to respect tighter product quality constraints compared to existing approaches.

[1] C. D. Immanuel, Y. Wang, N. Bianco, Feedback controllability assessment and control of particle size distribution in emulsion polymerisation, Chemical Engineering Science 63 (5) (2008) 1205-1216.

[2] D. Bonvin, B. Srinivasan, D. Hunkeler, Control and optimization of batch processes, IEEE Control Systems Magazine 26 (6) (2006) 34-45.

[3] D. Shi, N. H. El-Farra, M. Li, P. Mhaskar, P. D. Christofides, Predictive control of particle size distribution in particulate processes, Chemical Engineering Science 61 (1) (2006) 268-281.

[4] D. Bonvin, Optimal operation of batch reactors- a personal view, Journal of Process Control 8 (5-6) (1998) 355-368.

[5] S. Aumi, P. Mhaskar, Integrating Data-Based Modeling and Nonlinear Control Tools For Batch Process Control, Aiche Journal 58 (2012) 2105-2119.

[6] S. Aumi, B. Corbett, P. Mhaskar, and T. Clarke-Pringle. Data-based modeling and control of nylon-6, 6 batch polymerization. Control Systems Technology, IEEE Transactions on, 21(1):94-106, Jan 2013.

[7] J. Flores-Cerrillo, J. F. MacGregor, Control of particle size distributions in emulsion semibatch polymerization using mid-course correction policies, Industrial and Engineering Chemistry Research 41 (7) (2002) 1805-1814.

[8] B. Corbett, P. Mhaskar, Subspace identification for data-driven modelling and quality control of batch processes,

AIChE Journal (2016) n/a-n/a.

[9] F. J. Doyle, C. A. Harrison, T. J. Crowley, Hybrid model-based approach to batch-to-batch control of particle size distribution in emulsion polymerization, Computers & Chemical Engineering 27 (8-9) (2003) 1153-1163.

[10] M. Park, M. Dokucu, Regulation of the emulsion particle size distribution to an optimal trajectory using partial least squares model-based predictive control, Industrial & engineering (2004) 7227-7237.

471296 Model Predictive Control for Linear Distributed Parameter Systems

Abstract:

Distributed parameter systems (DPS) are ubiquitously present as models of fundamental conservation laws and in process control, manufacturing, transport systems and/or human society. The major drawback of DPS models is that they take form of partial differential equations containing higher order derivatives in space and time. The complexity of a partial differential equation **(PDE)** in the case of linear PDE models lies in necessity of modellers to account for model spatial characteristics by an approximating underlying model through some spatial approximation arriving to a finite dimensional model representation amenable for subsequent control, observer and/or monitoring device design [1], [2].

This work provides foundation for systematic development of modelling framework for a linear DPS system which uses a finite and low dimensional setting for the controller/observer/estimator design without application of any spatial approximation or order reduction. In particular, we are interested in formulating control design methodology for a general class of linear DPS systems which in this work account for an optimal constrained optimization based setting. Therefore, we propose to develop a linear model predictive controller design for a class of linear distributed parameter system. In this works, we present our results applied to the DPS emerging from the chemical transport-reaction systems varying from the convection dominated models of a plug flow reactor to diffusion dominated models of an axial dispersion reactor. In addition to classical chemical process systems, we also address wave and beam equation system which accounts for a large class of distributed parameter systems. In this work, the discrete model of a distributed parameter system is obtained by using energy preserving Cayley-Tustin discretization [1]. Discrete DPS models are low dimensional, energy preserving and do not dissipate numerically. In particular, discrete setting is amenable to an explicit, economic and/or a classical model predictive control setting realization, with emphasize on the different slight variations in realization of constrained finite dimensional controllers. Having this in mind, the model predictive control [4] is designed by utilizing standard optimal control law with input or/and state/output constraints. The issues of stabilization, optimality and constrained stabilization are addressed for an infinite-dimensional system in this work. Finally, the controller performance is assessed by numerical simulation with application on different distributed parameter systems.

[1] Ray, W. Harmon, Advanced Process Control, McGraw-Hill Inc.,USA, 1980

[2] Curtain, Ruth F and Zwart, Hans, An introduction to infinite-dimensional linear systems theory, Springer, 1995.

[3] V. Havu, J. Malinen, The Cayley transform as a time discretization scheme, Numerical Functional Analysis and Optimization 28 (7-8) (2007) 825-851.

[4] K. R. Muske, J. B. Rawlings, Model predictive control with linear models, AIChE Journal 39 (2) (1993) 262-287.

465350 Autocovariance-Based Plant-Model Mismatch Estimation for Linear MPC with Measurable Disturbances

Abstract:

Model predictive control (MPC) has become the standard approach for multivariable, constrained control in the chemical process industries [4]. MPC is capable of handling large-scale, non-square and highly interacting plants, while accounting for constraints on inputs, outputs and state variables. However, the performance of MPC controllers is highly dependent on the quality and accuracy of the process model. Phenomena such as corrosion, catalyst deactivation and fouling will cause the operation of a chemical process to drift away from the point where the model was identified. Compensating for such plant-model mismatch via (frequent) model update experiments is costly and inefficient, and it is highly desirable to develop an approach for quantifying plant-model mismatch using readily available closed-loop operating data.

A large portion of available MPC performance assessment techniques focus on characterizing controller performance and its degradation. These approaches include multivariate statistical process control (MSPC) [1] and controller performance benchmarking concepts [3]. However, neither of these approaches is able to locate or quantify plant-model mismatch. Recently, some contributions proposed techniques to locate the input/output pair(s) where mismatch exists and quantify the magnitude of mismatch using external excitations [2]. In our previous work, we proposed a novel autocovariance-based plant-model mismatch estimation approach for control loops under unconstrained MPC [6]. We showed that the mismatch in model parameters (where the model was represented as a transfer function matrix) can be estimated using steady state output data. We also proposed a partition technique that extended this approach to control loops with constrained MPC [5].

In this work, we rely on our previous results to consider the generic case of a plant operating under setpoint changes and account for measurable disturbance in the feedback control loop. Our framework is

based on a transformation that converts the raw, noisy closed-loop process output into a mean-centered variable. We then establish an explicit relation between the autocovariance matrices of the new mean-centered output and the magnitude of the mismatch. Finally, we formulate a least squares optimization problem to estimate the plant-model mismatch. The proposed approach is illustrated with a case study considering high-dimensional MIMO system featuring dynamic complexities such as higher order dynamics, time delays and inverse response.

References

[1] A. AlGhazzawi and B. Lennox. Model predictive control monitoring using multivariate statis- tics. *J. Proc. Contr.*, 19:314–327, 2009.

[2] Abhijit S. Badwe, Ravindra D. Gudi, Rohit S. Patwardhan, Sirish L. Shah, and Sachin C. Patwardhan. Detection of model-plant mismatch in MPC applications. *Journal of Process Control*, 19(8):1305–1313, 2009.

[3] T. J. Harris. Assessment of closed loop performance. *Can. J. Chem. Eng.*, 67:856–861, 1989.

[4] S. J. Qin and T. A. Badgwell. A survey of industrial model predictive control technology. *Control Engineering Practice*, 11:733–764, 2003.

[5] S. Wang, J. Simkoff, M. Baldea, L. Chiang, I. Castillo, R. Bindlish, and D. Stanley. Data- driven plant-model mismatch quantification in input-constrained linear mpc. In *Proceedings of The 11th IFAC Symposium on Dynamics and Control of Process Systems*, Trondheim, Norway, 2016.

[6] S. Wang, J.M. Simkoff, M. Baldea, L. Chiang, I. Castillo, R. Bindlish, and D. Stanley. Autocovariance-based plant-model mismatch estimation for unconstrained linear MPC. *Sys. & Contr. Lett.*, page submitted, 2016.

466990 Adaptive Multiple-Model Stochastic Predictive Control

Abstract:

Model predictive control (MPC) of uncertain systems is a challenging problem that has received considerable attention. A majority of the MPC methods for uncertain systems are focused on parametric and/or exogenous uncertainty, and are broadly classified as either *stochastic* (describing uncertainty with probability masses or densities) [1] or *robust* (describing uncertainty in terms of bounded, deterministic sets) [2]. Some of these approaches are adaptive, meaning the controller adjusts the control law as new information is revealed (e.g., [3,4]). A problem that has received less attention is that of an uncertain model structure; this situation may arise in a variety of problems, including when there is uncertainty regarding which physical phenomena are taking place, and when it is unclear what model order is required to adequately describe the process dynamics.

In this talk we present an adaptive multiple-model predictive control strategy with active learning of model structure for stochastic nonlinear systems. The controller considers a set of model hypotheses, each with different structural form. The uncertain time-invariant parameters in the model hypotheses are described with probability density functions (PDFs), and the temporal evolution of the state probability distributions is predicted using the generalized polynomial chaos framework [5]. The uncertain parameters in each hypothesis are estimated online [6] and their distributions are updated as new measurements become available. The measurements are also used to estimate the validity probability of each model hypothesis using Bayesian statistics, providing the controller with a metric for the degree to which each hypothesis agrees with the data. In order to effectively determine which hypothesis best describes the system, we include a measure of model-hypothesis similarity in the control formulation and specify that the controller attempts to decrease the predicted output distribution overlap, thereby facilitating the task of reconciling the next data point with one of the models. In the talk, we review some dissimilarity metrics and argue that the Kolmogorov distance [7] is best suited for our purposes. This results in actively adaptive control with respect to structural uncertainty, or suboptimal dual control in the sense of Feldbaum [8]. That is, the MPC algorithm simultaneously controls the process and performs active model discrimination.

We demonstrate the performance of the proposed adaptive multiple-model stochastic predictive control strategy through implementation on an atmospheric pressure plasma jet (based on [9]). The problem includes two model hypotheses, one for laminar flow and another for turbulent flow in the jet, and three uncertain parameters in each model hypothesis. The controller inputs will not only regulate the system dynamics, but also excite the system to effectively identify the flow regime (i.e., model hypothesis) in the presence of model parameter uncertainty.

[1] Mesbah, A. (2016). Stochastic model predictive control: An overview and perspectives for future research. *IEEE Control Systems Magazine*. Accepted.

[2] Bemporad, A., & Morari, M. (1999). Robust model predictive control: A survey. In *Robustness in identification and control*, pp. 207–226. Springer.

[3] Tanaskovic, M., Fagiano, L., Smith, R., & Morari, M. (2014). Adaptive receding horizon control for constrained MIMO systems. *Automatica*, 50(12), 3019–3029.

[4] Heirung, T.A.N., Ydstie, B.E., & Foss, B. (2015). Dual MPC for FIR systems: Information anticipation. In *Advanced Control of Chemical Processes*, pp. 1034–1039. Whistler, Canada.

[5] Xiu, D., & Karniadakis, G.E. (2002). The Wiener-Askey polynomial chaos for stochastic differential equations. *SIAM Journal on Scientific Computing*, 24(2), 619–644.

[6] Bavdekar, V.A., & Mesbah, A. (2016). A polynomial chaos-based nonlinear Bayesian approach for estimating state and parameter probability density functions. In *American Control Conference*. Boston, MA. To appear.

[7] Gibbs, A.L., & Su, F.E. (2002). On choosing and bounding probability metrics. *International Statistical Review*, 70(3), 419–435.

[8] Feldbaum, A.A. (1961). Dual-control theory. I. *Automation and Remote Control*, 21(9), 874–880.

[9] Gidon, D., Graves, D.B., & Mesbah, A. (2016). Model predictive control of thermal effects of an atmospheric pressure plasma jet for biomedical applications. In *American Control Conference*. Boston, MA. To appear.

472998 A Bayesian Filter Switching Strategy for Simultaneous State and Parameter Estimation

Abstract:

Recent advances in high-speed computing technology have lead to the frequent use of stochastic nonlinear models to represent complex system dynamics. The design and implementation of advanced control or monitoring strategies using such complex models require real-time estimation of the key system states and parameters that are either unmeasured or unknown. In situations, where the parameters are precisely known, the states can be estimated un- der the Bayesian framework by computing the state posterior density. The state posterior density is often computed by solving a state filtering problem [1]. A closed-form solution to the fi ering problem exists for linear state space models (SSMs) under the Gaussian noise settings or when the state space is finite [1]. Unfortunately, in many engineering systems, the model is often nonlinear and the parameters are not known or time-varying, and therefore need to be estimated before the states can be estimated. In practical settings, simultaneous state and parameter estimation is often the only realistic solution for it avoids processing of large data set and also allows for adaptation to the time-varying system behavior [2, 3].

The choice of an efficient Bayesian filter for simultaneous state and parameter estimation in nonlinear stochastic systems is still an open problem. This is because there is no single tractable Bayesian filter that is guaranteed to provide a consistent performance for a given system under all operating conditions [4]. A practitioner is thus left with no clear substitute for the optimal Bayesian filter.

This paper develops a filter switching strategy for simultaneous state and parameter estimation in systems represented by nonlinear, stochastic, discrete-time state space models (SSMs). The proposed strategy considers a bank of plausible Bayesian filters for simultaneous state and parameter estimation, and then switches between them based on their performance. The performance of a Bayesian filter is assessed using a performance measure derived from the posterior Cramer-Rao lower bound (PCRLB). The efficacy of

the filter switching strategy is illustrated on a practical simulation example.

References

[1] M. Arulampalam, S. Maskell, N. Gordon, and T. Clapp, "A tutorial on particle filters for online nonlinear/non-gaussian bayesian tracking," *IEEE Transactions on Signal Processing*, vol. 50, no. 2, pp. 174–188, 2002.

[2] H. He, R. Xiong, X. Zhang, F. Sun, and J. Fan, "State-of-charge estimation of the lithium-ion battery using an adaptive extended kalman filter based on an improved thevenin model," *IEEE Transactions on Vehicular Technology*, vol. 60, no. 4, pp. 1461–1469, 2011.

[3] C. Kravaris, J. Hahn, and Y. Chu, "Advances and selected recent developments in state and parameter estimation," *Computers & chemical engineering*, vol. 51, pp. 111–123, 2013.

[4] P. Minvielle, A. Doucet, A. Marrs, and S. Maskell, "A Bayesian approach to joint tracking and identificafition of geometric shapes in video sequences," *Image and Vision Computing*, vol. 28, no. 1, pp. 111–123, 2010.

455266 Fast Estimation of Plant Steady State, with Application to Static RTO

Abstract:

Fast Estimation of Plant Steady State, with Application to Static RTO

D. Rodrigues, M. Amrhein, J. Billeter and D. Bonvin

Laboratoire d'Automatique, Ecole Polytechnique Fédérale de Lausanne

EPFL – Station 9, 1015 Lausanne, Switzerland

diogo.mateusrodrigues@epfl.ch

In the operation of continuous processes, many tasks require the knowledge of plant steady state at various operating points. This is for example the case in the context of kinetic modeling, response surface modeling and real-time optimization. If the computational techniques are in principle straightforward, the time needed to reach steady state represents the main limiting factor. This work proposes a novel way of speeding up the *estimation* of plant steady state through the use of feedback control and rate estimation. It must be emphasized here that rate estimation requires only some structural information of the plant and no rate model. Generally speaking, the context of the present investigation corresponds to industrial practice, where there is significant plant-model mismatch, which typically calls for the use of measurements to feed data-driven techniques.

We will illustrate the fast estimation of plant steady state in the context of static optimization of continuous reactors. Real-time optimization (RTO) is typically implemented via some iterative scheme that uses steady-state plant measurements. The cost and constraints of the optimization problem are functions of the input and output steady-state values. At the kth iteration, the constant inputs $\mathbf{u}_k$ are usually applied to the plant in open loop and, once steady state is reached, the outputs $_k$ are measured and the cost and constraint values are evaluated. However, depending on the dominant time constant of the plant, the time necessary to reach steady state may be rather long. Hence, it would be useful to be able to speed up convergence to steady state, or at least speed up the *estimation* of plant steady state. This will be done in this work through combination of feedback control (to speed up a specific part of the plant) and rate estimation (to estimate the steady-state values of the remaining part of the plant so that the cost and constraint functions can be evaluated).

For this, let us consider a time-invariant dynamic system with the inputs **u** and the state variables **x** and **z** that can be described by the differential equations:

$$(t) = \mathbf{f}(\mathbf{x}(t),\mathbf{u}(t)) \quad \mathbf{x}(0) = \mathbf{x}_0$$

$$(t) = \mathbf{h}(\mathbf{x}(t),\mathbf{u}(t)) - \boldsymbol{\Omega}(\mathbf{x}(t),\mathbf{u}(t))\ \mathbf{z}(t) \quad \mathbf{z}(0) = \mathbf{z}_0$$

The particular structure of this dynamic system can be exploited when the *slow* and *fast* states **z** and **x** are associated with slow and fast dynamics, respectively, and the slow states do not affect the fast states. The idea is to use feedback control to speed up the convergence of the fast states **x** to and the inputs **u** to , provided that the states **x** are accessible, and then compute the steady-state values of the slow states as:

$$= \boldsymbol{\Omega}^{-1}(,)\ \mathbf{h}(,)$$

This computation relies on nonparametric estimation of the rates **h**(,) and **Ω**(,) using measurements and structural information of the plant. Note that can be estimated long before the state variables **z** converge to their steady-state values.

RTO is implemented via a two-layer approach. In the inner layer, feedback control is used to rapidly drive **x** to the setpoints $\mathbf{x}_{sp}$ by manipulating the inputs **u** and rate estimation is used to compute from and . In the outer (optimization) layer, the RTO algorithm computes optimal values for the setpoints $\mathbf{x}_{sp}$. The inner layer is described in more detail next for chemical reactors:

- The objective of the control scheme is to drive the fast states **x**, typically some reactant concentrations, to their constant setpoints in the shortest possible time after a step change in the setpoints $\mathbf{x}_{sp}$. This time is shorter than the time needed by the open-loop plant to reach steady state after a step change in the inputs **u**, typically inlet flowrates. Multivariable control is implemented via input-output feedback linearization [1], pole placement or optimal control and typically involves the measurement or estimation of **x**. Note that the gains to control the fast states (reactant concentrations) **x** are typically lower than those necessary to control both the fast (reactant) concentrations **x** and the slow (product) concentrations **z**, thereby making the control scheme with **x** less sensitive to measurement noise. Also note that the separation into slow and fast states assumes that the rates **f**(**x**,**u**) are independent of the slow (product) concentrations **z**.

- The rate estimation relies on the knowledge of the stoichiometry and the inlet concentrations, and on measurements of flowrates, reactor temperature, volume, and some of the concentrations *at steady state*. The rates are computed through numerical differentiation of the reaction variants using a Savitzky-Golay filter [1]. These reaction variants are easily computed from measured concentrations [2]. It will also be shown that the proposed Savitzky-Golay filter is the optimal convolution filter for rate estimation.

A simulated homogeneous CSTR [3] is used to illustrate the implementation of "fast static RTO" and address the following questions: (i) how to best combine feedback control and rate estimation, (ii) which controlled and manipulated variables to choose, (iii) how to eliminate the slow states in the objective and constraint functions, and (iv) how to deal with measurement noise in this measurement-based RTO algorithm.

[1] D. Rodrigues, J. Billeter, and D. Bonvin. Control of reaction systems via rate estimation and feedback linearization. In *PSE-12/ESCAPE-25*, Copenhagen, 2015.

[2] D. Rodrigues, S. Srinivasan, J. Billeter, and D. Bonvin. Variant and invariant states for chemical reaction systems, *Comput. Chem. Engng*, 73:23-33, 2015.

[3] B. Srinivasan, L. T. Biegler, and D. Bonvin. Tracking the necessary conditions of optimality with changing set of active constraints using a barrier-penalty function. *Comput. Chem. Engng*, 32(3):572–579, 2008.

462275 The Extended and Unscented Kalman Filtering Methods for Real-Time Plasma Insulin Concentration Estimation in an Artificial Pancreas Control System for Patients with Type 1 Diabetes

Abstract:

Background and objective: Artificial pancreas (AP) systems use continuous glucose measurements (CGM) to calculate optimum amount of insulin to be infused with an insulin pump [1, 2]. Real-time estimations for plasma insulin concentration (PIC) rather than generalized curve based insulin-on-board (IOB) estimates can be beneficial for improving the performance of AP control algorithms to calculate more realistic insulin infusion rates and prevent hypoglycemia that would be caused by overdosing of insulin [3]. Our objective is to fulfill a real-time estimation of PIC from CGM data by using two different mathematical models.

Methods: Two different glucose-insulin compartmental models, Hovorka's model and extended Bergman's minimal model, which were developed to describe glucose-insulin dynamic in different parts of the human body, have been incorporated into a continuous-discrete extended Kalman filter (CD-EKF) and an unscented Kalman filter (UKF), respectively, to provide an estimate of the plasma insulin concentration [4,5]. Furthermore, because of variability in system dynamics, uncertain parameters have been considered as new states in Hovorka's model and Bergman's minimal model to be estimated by CD-EKF and UKF, correspondingly. Thirteen datasets from nine different subjects with type 1 diabetes are used. Two euglycemic clamps were performed on separate mornings with (seven datasets) and without (six datasets) an insulin infusion site warming device. Subjects had one CGM sensor and blood was collected to measure plasma insulin levels for up to 5 hr. On both days, a bolus of 0.2 U/kg aspart insulin was infused at the beginning of each experiment and the basal infusion was suspended, and glucose levels were maintained between 90 and 100 mg/dl by infusion of variable rate of dextrose [6]. In addition, partial least squares models based on each patient's demographic information such as body weight, height, BMI and total daily insulin dose are developed for the initial guess of the time-varying unknown model parameters used in the nonlinear CD-EKF and UKF estimators.

Results: According to the CD-EKF estimator, the average root mean square errors are found to be 12.36 and 11.56 mU/L for the clamps with and without IP, respectively. The average mean absolute relative error with respect to the measured samples are 0.14 for the two clamps. The average Pearson's R coefficient are 0.95 and 0.91 for the clamps with and without IP, correspondingly. For UKF estimator, estimated plasma insulin concentrations are found to be linearly correlated with the measured insulin samples (R= 0.94 and 0.89 for the clamps with and without IP, respectively). The average root mean square errors are found to be 12.69 and 12.09 mU/L for the clamps with and without IP, respectively. The average mean absolute relative error with respect to the measured samples are 0.12 and 0.10 for the clamps with and without IP, correspondingly.

Conclusions: The proposed methods are able to estimate the plasma insulin concentration in real time by using only CGM measurements. These methods will may be beneficial for an AP system in terms of real time estimation of insulin concentration for preventing excessive insulin infusions if plasma insulin levels are already elevated.

References

[1] K. Turksoy, L. Quinn, E. Littlejohn, A. Cinar, Multivariable adaptive identification and control for artificial pancreas systems, Biomedical Engineering, IEEE Transactions on, 61 (2014) 883-891.

[2] K. Turksoy, E.S. Bayrak, L. Quinn, E. Littlejohn, A. Cinar, Multivariable adaptive closed-loop control of an artificial pancreas without meal and activity announcement, Diabetes technology & therapeutics, 15 (2013) 386-400.

[3] J. Walsh, R. Roberts, Pumping insulin, Everything you need for success on a smart insulin pump, 4th edition. Torrey Pines, San Diego, USA, (2006).

[4] A. Roy, R.S. Parker, Dynamic modeling of free fatty acid, glucose, and insulin: An extended" minimal model", Diabetes technology & therapeutics, 8 (2006) 617-626.

[5] R. Hovorka, V. Canonico, L.J. Chassin, U. Haueter, M. Massi-Benedetti, M.O. Federici, T.R. Pieber, H.C. Schaller, L. Schaupp, T. Vering, Nonlinear model predictive control of glucose concentration in subjects with type 1 diabetes, Physiological measurement, 25 (2004) 905.

[6] E. Cengiz, S.A. Weinzimer, J.L. Sherr, E.M. Tichy, L. Carria, D. Cappiello, A. Steffen, W.V. Tamborlane, Faster in and faster out: accelerating insulin absorption and action by insulin infusion site warming, Diabetes technology & therapeutics, 16 (2014) 20-25.

471213 Noise Covariance Estimation for an Air Separation Plant

Abstract:

Noise is ubiquitous in systems engineering. Therefore, process models often consist of a deterministic part describing the underlying physics and a stochastic part describing the noises affecting a system. Identifying the noise model is important because it is necessary in order to build an optimal state estimator for a system. Two assumptions are common in modeling noises:

1) There are two distinct types of noise: process noise and measurement noise.
2) Noises can be modeled as zero-mean Gaussian random variables.

Under these assumptions, the task of identifying the noise model reduces to identifying the process and measurement noise covariance matrices.

Given the deterministic part of a system model, there are several methods to estimate noise covariances from process data. Maximum likelihood estimation (MLE), often using the expectation maximization (EM) algorithm, is perhaps the most common. The major drawback of this approach is that the objective function of the MLE problem scales with the amount of data being processed and it can quickly become intractable. A more recent approach is the autocovariance least-squares (ALS) algorithm. The ALS problem is generally easier to solve than the MLE problem, but yields solutions that lack the clear statistical interpretation of the MLE approach.

In this work we present a case study using a model from an industrial air separation plant. We investigate the relevance and utility of noise identification in chemical engineering systems and the issues that arise when using these methods with industrial models.

470248 Multi-Rate Moving Horizon Estimation for an Electric Arc Furnace Steelmaking Process

Abstract:

Electric arc furnaces (EAFs) are commonly used in the steel industry for production of steel by melting down the scrap metal and altering its chemistry.Â The highly energy intensive steelmaking operation is a complex batch process, and involves limited automation. Both electrical and chemical energy sources are used to melt the scrap steel. The electrical energy is added in the form of electric arc through electrodes. The burners inject natural gas and oxygen in the furnace, which provide chemical energy through combustion reactions. As the melting progresses a flat bath of molten steel is formed. The oxygen reacts with metals to form oxides which become constituents of the slag layer floating on top of the molten steel. Slag chemistry is adjusted by direct addition of carbon, lime and dolomite through the furnace roof and by varying oxygen and carbon lancing. The decisions related to the timing and amount of additions rely heavily on the operators running the EAF. Although operatorâ€™s experience is invaluable for operation of the EAF, the subtle multivariable interactions and relationships can easily be overlooked. Additionally, considering the high energy consumption of the process, development and implementation of estimation and control strategies for EAFs to take economically optimal decisions is envisaged to generate significant savings through reduced consumption of electric power, natural gas, oxygen, carbon and fluxes such as limestone and dolomite.

Due to the harsh operating conditions, EAFs lack measurements and most of the states are not directly measured. Another issue arises due to the measurements being taken at different sampling rates. State knowledge is essential in real-time control applications. Very limited applications of state estimation for EAF operation have been reported. The extended Kalman filter (EKF) was applied in [1] to the refining stage of EAF operation but the model had 4 states, which are insufficient to capture the detailed process dynamics. An EKF was used in [2] to identify the arc current parameter for obtaining the electrical properties of the EAF load. However, the EAF model only involved the power system. A constrained multi-rate EKF was implemented in [3] to estimate the states of EAF system using plant measurements. The EKF showed an acceptable performance in tracking the true states of the process, even in the presence of parametric plant-model mismatch. Although different versions of the Kalman filter such as the EKF, unscented Kalman filter etc. are employed by some researchers, moving horizon estimation (MHE) is gaining popularity due to the ability to handle constraints and to use computationally efficient numerical optimization algorithms [4]. In a previous work, we presented a parameter estimation based multi-rate MHE framework for EAF operation [5]. The MHE problem consists of solving a nonlinear dynamic optimization problem subject to the nonlinear model under consideration and some other constraints specified by the user. It uses a finite set of past available measurements to reconstruct the full state of the process, thus keeping the optimization problem numerically tractable. The use of a finite size window of measurements by MHE provides a straightforward way to include measurements with different sampling rates [6].

In this work, we present a rigorous multi-rate MHE solution strategy for the EAF process. A novel initialization method for MHE problems based on a background solve to improve solution time is also introduced and implemented. The optimization-based strategy is particularly suitable for applications incorporating large-scale differential-algebraic equation (DAE) systems. The study is conducted using a first-principles dynamic EAF model developed originally in [7], in which the EAF is partitioned into four zones (gas, solid scrap, slag and molten metal). Chemical equilibrium is assumed within the slag and gas zones, with mass and energy transfer across the zone interfaces. The MHE problem built around the highly nonlinear model is solved using direct dynamic optimization approaches to estimate states for an EAF heat (batch). We provide an overview and a description of the MHE application that includes the multi-rate measurements handling, initialization of the online MHE problem, and investigation of different DAE optimization paradigms in the solution of the MHE optimization problem. The performance of the MHE under different conditions and using different optimization strategies is illustrated through application to several case studies. Additionally, avenues for future research will be identified and some perspective provided on the real-time application of the proposed strategy as a decision support tool for the operators of EAFs.

References

1. S. Billings, F. Boland and H. Nicholson, "Electric arc furnace modelling and control," *Automatica,* vol. 15, no. 2, pp. 137-148, 1979.

2. F. Wang, Z. Jin, Z. Zhu and X. Wang, "Application of extended Kalman filter to the modeling of electric," *Neural Networks and Brain, 2005. ICNN\&B'05. IEEE International Conference,* vol. 2, pp. 991-996, 2005.

3. Y. Ghobara, "Modeling, Optimization and Estimation in Electric Arc Furnace," *Master's thesis,* no. McMaster University, 2013.

4. F. Allgöwer, T. A. Badgwell, J. S. Qin, J. B. Rawlings and S. J. Wright, "Nonlinear predictive control and moving horizon estimation. An introductory overview," *Advances in Control,* no. Springer, pp. 391--449, 1999.

5. S. Shyamal and C. L. E. Swartz, "A Multi-rate Moving Horizon Estimation Framework for Electric Arc Furnace Operation," in *Proc. 11th DYCOPS-CAB, (accepted)*, Trondheim, Norway, 2016.

6. R. López-Negrete and L. T. Biegler, "A moving horizon estimator for processes with multi-rate measurements: A nonlinear programming sensitivity approach," *Journal of Process Control,* vol. 22, no. 22, pp. 677-688, 2012.

7. R. D. MacRosty and C. L. E. Swartz, "Dynamic modeling of an industrial electric arc furnace," *Ind.Eng.Chem.Res.,* vol. 44, pp. 8067-8083, 2005.

464244 A Simplified Dynamic Model of a Continuous Subsea Oil-Water Gravity Separator for Estimation of Unmeasurable Oil and Water Purities

Abstract:

Gravity separation is a widely used separation process in industry for separation of liquids based on difference in densities. In oil and gas industry, gravity separators are often used for subsea oil-water separation. However, a strict control over the quality of the outlet streams does not exist because subsea sensors for sufficiently accurate measurement of oil in water and water in oil concentrations are not available. Hence, there is a need for dynamic models for gravity separator that can predict the concentration of outgoing streams sufficiently accurately.

A common gravity separator design is a horizontally oriented cylindrical vessel, in which mixed fluids enter from one end and the separated fluids are removed from the other. Inside the separator, the fluids separate into layers. The layers from bottom to top are water layer, emulsion layer and oil layer. The emulsion layer consists of two sections: sedimentation layer and dense-packed layer. At the end of the separator, a weir prevents the emulsion layer and the water layer from flowing over the weir. The liquid levels inside the separator are controlled in such a way that only oil flows over the weir to the oil outlet section, while the water is removed continuously from the bottom of the separator.

In previous years, the modeling of gravity separation has focused on studying mechanisms that govern batch separation of two liquids, such as oil and water. The models for gravity separation presented in the literature have been summarized by Frising [1]. The models can be classified as sedimentation-based models and coalescence-based models. Sedimentation based models [2-6] consider sedimentation of droplets and interfacial coalescence between droplets and interface. These models are typically tuned to match the results from the batch experiments that demonstrate four layers: clear oil, sedimentation zone, dense-packed zone and clean water. On the other hand, coalescence based models [7,8] focus on the dense-packed zone considering binary as well as interfacial coalescence of the droplets of dispersed phase. They predict mean droplet size based on lifetime of a single film and assume that the presence of surfactant only scales the coalescence time. A model by Henschke [8], which is a coalescence-based model, is, perhaps, the most relevant for modeling continuous gravity separation because it predicts drop size in the dense-packed zone using a single coalescence parameter, which is independent of experimental equipment, mixing intensity and volume fraction of dispersed phase.

Some of these mechanisms of oil-water separation have been used for modeling continuous gravity separation. Sayda *et al.*[9] developed a simplified dynamic model for a continuously operating gravity separator based on material balances, and phase equilibrium equations. Their model takes into consideration simplified hydrodynamics, thermodynamics and spatial evolution of separation profiles. They used a standard form of Stoke's law for vertical settling velocity of droplets based on a constant average droplet diameter of oil droplets in water. Hence, their model does not consider coalescence effects that are typically very important in oil-water separation. Another attempt to model a continuous oil-water separation was conducted in [10], in which they model a horizontal pipe separator. They combine some of the principles from sedimentation based models [11] and Henschke's model [8]. Their approach is quite useful for modeling a gravity separator because horizontal pipe separator is quite similar to the conventional gravity separator. However, they model a pipe separator that is completely filled with fluids and does not have a weir. On the other hand, gravity separators often include a weir and are operated with the separator only partially filled with liquids.

In this work, we present a dynamic model for a continuous cylindrical gravity separator for oil-water separation to predict the purity of outlet streams based on operating conditions by assuming that there exists three layers of liquids: oil, emulsion and water inside the separator. It is based on lumped total mass balances and partial mass balances for each layer, in which the inter-layer mass transfer equations are derived using principles from the previous work in sedimentation based models and Henschke's model.

The model is solved in MATLAB using solver ode15s. It captures the effect of total inlet flow rate, inlet oil cut and inlet droplet size on purities of outlet streams and thickness of each layer of liquid. We use the model in a Kalman-based non-linear estimator for estimating unmeasurable variables, such as oil and water purities using level and density measurements and other measurements of inlet conditions, such as inlet flow rate, oil cut, droplet size and water outflow rate. Further, we study the effects of the variations to water level set point on the performance of the separator.

[1] Frising, T., Noïk, C. and Dalmazzone, C., 2006. The liquid/liquid sedimentation process: from droplet coalescence to technologically enhanced water/oil emulsion gravity separators: a review.*Journal of Dispersion Science and Technology*,*27*(7), pp.1035-1057.

[2] Jeelani, S.A.K. and Hartland, S., 1986. Prediction of dispersion height in liquid-liquid gravity settlers from batch settling data. *Chemical engineering research & design*, *64*(6), pp.450-460.

[3] Jeelani, S.A.K., Pandit, A. and Hartland, S., 1990. Factors affecting the decay of batch liquid liquid dispersions. *The Canadian Journal of Chemical Engineering*, *68*(6), pp.924-931.

[4] Jeelani, S.A.K. and Hartland, S., 1993. The continuous separation of liquid/liquid dispersions.*Chemical engineering science*, *48*(2), pp.239-254.

[5] Jeelani, S.A.K., Panoussopoulos, K. and Hartland, S., 1999. Effect of turbulence on the separation of liquid-liquid dispersions in batch settlers of different geometries. *Industrial & engineering chemistry research*, *38*(2), pp.493-501.

[6] Jeelani, S.A.K., Hosig, R. and Windhab, E.J., 2005. Kinetics of low Reynolds number creaming and coalescence in droplet dispersions. *AIChE journal*, *51*(1), pp.149-161.

[7] Lobo, L., Ivanov, I. and Wasan, D., 1993. Dispersion coalescence: Kinetic stability of creamed dispersions. *AIChE journal*, *39*(2), pp.322-334.

[8] Henschke, M., Schlieper, L.H. and Pfennig, A., 2002. Determination of a coalescence parameter from batch-settling experiments. *Chemical Engineering Journal*, *85*(2), pp.369-378.

[9] Sayda, A.F. and Taylor, J.H., 2007, July. Modeling and Control of Three-Phase Gravilty Separators in Oil Production Facilities. In *American Control Conference, 2007. ACC'07* (pp. 4847-4853). IEEE.

[10] Pereyra, E., Mohan, R.S. and Shoham, O., 2013. A Simplified Mechanistic Model for an Oil/Water Horizontal Pipe Separator. *Oil and Gas Facilities*,*2*(03), pp.40-46.

[11] Hartland, S. and Jeelani, S.A.K., 1988. Prediction of sedimentation and coalescence profiles in a decaying batch dispersion. *Chemical engineering science*, *43*(9), pp.2421-2429.

470841 Schuripopt: A Parallel Optimization Package for Structured Nonlinear-Programming Problems

Abstract:

Optimization plays an important role in a variety of areas, including process design, operation, and control. Large-scale nonlinear optimization problems that commonly arise in these areas need to be solved efficiently. These problems are characterized by having a large number of equations and variables and can become computationally prohibited. However, due to the nature of the problems, these problems often have a structure that can be exploited using parallel computing capabilities offered by modern computers. In this work, we developed SchurIpopt, a parallel extension to Ipopt that solves structured NLP problems efficiently with shared memory and distributed memory parallel architectures.

Interior-point methods have proven to be effective for solving large-scale nonlinear programming problems. The dominant computational steps in an interior-point algorithm are the solution of the KKT system and the computation of NLP functions and derivatives in every interior-point iteration. Our implementation uses a Schur-complement decomposition strategy to exploit the structure of the NLP problems that come from multi-scenario and dynamic optimization applications. In both cases, the inherited structure can be exploited by decomposing the problems to overcome computing memory and time limitations that commonly arise when dealing with large-scale problems. To achieve high parallel efficiencies, the implementation not only focuses on parallelizing the solution of the KKT system, but it also parallelizes function evaluations and scale-dependent operations like vector-vector and matrix-vector operations. This algorithm has been interfaced with PySP, an extension of Pyomo for modeling stochastic programming problems. To Illustrate the performance of the implementation we present two case studies in stochastic programming and dynamic optimization.

461713 A Scalable Design of Experiments Framework for Infinite-Dimensional Systems

Abstract:

The sensor placement problem seeks to determine the optimal number, location, and types of sensors that would maximize information about a dynamical system. Information is often expressed in terms of the posterior covariance matrix of the states of parameters of the system. The problem can thus be cast as an optimal design of the experiments problem. Such a problem is computationally challenging, particularly in the infinite-dimensional case, because one must solve a mixed-integer and bilevel optimization problem constrained by differential algebraic equations (DAEs) or by partial differential equations (PDEs).

The sensor placement problem has been addressed by using mixed-integer programming techniques in the context of contaminant detection in water networks. In these studies, an optimal set of sensor locations is selected from a set of candidate locations to minimize a certain engineering metric such as contaminant detection time, population exposea, or likelihood of detection. Likelihoods are assigned based on contamination scenarios and not on information content of the sensor data recorded, as in a traditional experimental design setting. Consequently, these approaches fail to provide statistically meaningful sensor network designs. Moreover, because the formulations capture flow dynamics by using surrogate representations such as transportation delays, they fail to capture nonlinear effects.

Sensor placement problems also have been addressed in a more general control setting where one weeks a measure of observability such as the covariance matrix, Kalman estimator gain, or so-called observability Grammian matrix. This problem again is a bilevel optimization problem. The covariance matrix approach in [1] bypasses this by assuming that the dynamic model is linear, thus allowing the inner minimization problem to be formulated as a linear matrix inequality. The approach in [2] models the dynamics of the covariance matrix directly as a Ricatti differential equation, which implicitly assumes linearity and thus enables the use of semidefinite programming algorithms. However, the work in [2] is focused on control

policy design to extract maximum information, and not on sensor placement design. Consequently, the authors do not consider mixed-integer formulations. A rigorous treatment of nonlinear dynamics is presented in [3] by casting the problem as a mixed-integer nonlinear programming. The authors, however, use a genetic algorithm to deal with the inner minimization problem that computes the observability metric. A similar approach is used in [4] to address the inner minimization problem. Mixed-integer techniques have also been used in the context of information maximization for Gaussian processes and for designing Latin hypercube samples [5]. These approaches do not capture physical models.

Recently, the sensor placement problem for systems described by PDEs has been cast as an A-experimental design problem in which the number of sensors (i.e., the design cost) is controlled by using an l0 regularization norm that is in turn approximated using a smoothing function [6]. This approach was shown to be scalable and applicable to infinite-dimensional systems, but it requires tuning and can be numerically unstable. One can also formulate and solve the problem as a mixed-integer programming problem directly, but this approach can become computationally intractable because the PDEs are in general nonconvex and because the problem is bilevel in nature.

In this work we present a scalable design of an experiments framework for sensor placement that seeks to compute optimal sensor locations where observational data are collected, by minimizing the uncertainty in parameters estimated from Bayesian inverse problems, which are governed by PDEs. The resulting problem is a mixed- integer infinite-dimensional optimal control problem. We approach this problem using two efficient heuristics that have a potential to be scalable for such problems: a sparsity-inducing approach [6] and a sum-up rounding approach [7]. We investigate two objectives: the total flow variance and the A-optimal design criterion. We conclude that the sum-up rounding approach produces shrinking gaps with increased meshes. We also observe that convergence for the white noise measurement error is slower than for the colored noise case. For the A-optimal design, the solution is close to the uniform distribution, but for the total flow variance the pattern is noticeably different.

[1] D. J. Chmielewski, T. Palmer, and V. Manousiouthakis. On the theory of optimal sensor placement. AIChE Journal, 48(5):1001–1012, 2002.

[2] D. Telen, F. Logist, R. Quirynen, B. Houska, M. Diehl, and J. Impe. Optimal experiment design for nonlinear dynamic (bio) chemical systems using sequential semidefinite programming. AIChE Journal, 60(5):1728–1739, 2014.

[3] A. K. Singh and J. Hahn. Sensor location for stable nonlinear dynamic systems: Multiple sensor case. Industrial & Engineering Chemistry Research, 45(10):3615–3623, 2006.

[4] E. Musulin, C. Benqlilou, M. J. Bagajewicz, and L. Puigjaner. Instrumentation design based on optimal kalman filtering. Journal of Process Control, 15(6):629–638, 2005.

[5] C. M. Harris, K. L. Hoffman, and L.-A. Yarrow. Using integer programming techniques for the solution of an experimental design problem. Annals of Operations Research, 58(3):243–260, 1995.

[6] A. Alexanderian, N. Petra, G. Stadler, and O. Ghattas. A-optimal design of experiments for infinite-dimensional bayesian linear inverse problems with regularized l0-sparsification. SIAM Journal on Scientific Computing, 36(5):A2122–A2148, 2014.

[7] S. Sager. Sampling decisions in optimum experimental design in the light of Pontryagin's maximum principle. SIAM Journal on Control and Optimization, 51(4):3181–3207, 2013.

475173 A Logical Benders Decomposition Algorithm for Binary-Constrained Quadratic Programs with Complementarity Constraints

Abstract:

An algorithm is presented for solving nonlinear optimization problems with chance constraints, i.e., those in which a constraint involving an uncertain parameter must be satisfied with at least a minimum probability. In particular, the algorithm is designed to solve cardinality-constrained nonlinear optimization problems that arise in sample average approximations of chance-constrained problems. No convexity assumptions are made. The algorithm employs a novel exact penalty function, which is minimized sequentially by solving quadratic optimization subproblems with linear cardinality constraints. Properties of minimizers of the penalty function in relation to minimizers of the corresponding nonlinear optimization problem are presented. The proposed algorithm converges to a stationary point of the penalty function. The effectiveness of the algorithm is demonstrated through numerical experiments. This is work in collaboration with Frank Curtis and Victor Zavala.

467664 Accelerating the Simplex Algorithm Via Novel Crash Procedures

Wednesday, November 16, 2016: 4:14 PM
Monterey I (Hotel Nikko San Francisco)

Nikolaos Ploskas, Carnegie Mellon University, Pittsburgh, PA, Nikolaos Samaras, University of Macedonia, Thessaloniki, Greece and **Nick Sahinidis**, Department of Chemical Engineering, Carnegie Mellon University, Pittsburgh, PA

Dantzig's simplex algorithm was initially proposed in the late 1940s [1]. Since then, variants of this algorithm have been studied extensively by numerous researchers [2] and have become core routines in many optimization algorithms for linear, integer and nonlinear programming [3, 4]. Leading implementations such as CPLEX [5] are currently capable of routinely solving linear programs with millions of constraints and variables using the simplex algorithm [3, 4]. The algorithm is in use by most Fortune 500 companies and was recently included in the list of the top-10 algorithms of the twentieth century [6]. Given this level of activity in the field and presumed maturity of the simplex algorithm, one would think that it would be very difficult, if at all possible, to speed up current implementations of the simplex algorithm. In this paper, we show that a novel initialization strategy provides a means of considerable acceleration of the simplex algorithm.

We propose novel crash procedures for finding an initial basis for the simplex algorithm. The main idea is to obtain a starting basis that is sparse and will reduce the fill-ins in LU factorization that forms the core operation of simplex iterations. We present extensive computational comparisons on standard benchmarks by supplying our initial basis to the state-of-the-art linear programming code CPLEX. The computations demonstrate that, in comparison to the CPLEX default crash procedure, our initialization strategy on average reduces the running time of CPLEX by 60% in the case of primal simplex and by 27% in the case of dual simplex. For large and difficult instances, our algorithm reduces the running time of CPLEX by an order of magnitude.

References cited:
[1] Dantzig, G. B. (1963). Linear programming and extensions. Princeton, NJ: Princeton University Press.
[2] D. Bertsimas and J. Tsitsiklis (1997). Introduction to Linear Optimization. Athena Scientific, Boston, MA.
[3] Bixby, R. E. (2002). Solving real-world linear programs: A decade and more of progress. Operations research, 50, 3–15.
[4] Bixby, R. E. (2012). A brief history of linear and mixed-integer programming computation. Documenta Mathematica, 107–121.
[5] IBM (2016). IBM ILOG CPLEX V12.6. Available online: http://www.ibm.com/support/knowledgecenter/SSSA5P_12.6.0 /ilog.odms.cplex.help/CPLEX/homepages/CPLEX.html
[6] Dongarra, J., Sullivan, F. (2000). Guest editors' introduction: The top 10 algorithms. Computing in Science & Engineering, 2(1), 22-23.

471112 Discreetly Discrete: Quietly Adding Integer-Valued Actuators to Model Predictive Control

Abstract:

In large-scale industrial processes, there are often both continuous ("how much") and discrete ("which one") control decisions that must be made. For the continuous decisions, a successful paradigm has been model predictive control (MPC), which chooses process inputs at regular intervals by solving an optimization problem considering the evolution of the system over some finite horizon. Owing to the historical difficulty of nonlinear and mixed continuous/discrete optimization, MPC theory and implementation almost always require that all control inputs take values in some convex set, as (in the case of linear models) the resulting convex optimization can be readily solved to (global) optimality. Because this framework precludes discrete optimization variables, such discrete decisions must be removed from the process control layer and are instead made using heuristics or other computational methods on a slower timescale, which can yield suboptimal control performance. The end result is an odd composite system that requires the controller to react to the discrete decisions instead of proactively making them. More desirable would be a single optimizing controller that has at its disposal all of the discrete and continuous actuators of the system.

With advances in computing hardware and optimization methods, it is now possible to quickly solve mixed continuous/discrete optimization problems provided that the formulation is sufficiently tight. For example, commercial solvers such as Gurobi or CPLEX can be used for "convex" mixed-integer quadratic programming (MIQP) problems, which consist of linear constraints and a convex quadratic objective function. This class of problems includes standard linear systems, as well as linear switched systems and piecewise-affine systems, both of which can be modeled using MIQPs. For nonlinear systems, spatial branch-and-bound algorithms can be applied; while such instances are often challenging, they are not significantly more difficult than global optimization for nonconvex continuous-variable problems, which is a requirement of many nonlinear MPC schemes proposed in the literature. As powerful optimization software becomes increasingly available, optimization-based control methods such as MPC become increasingly applicable.

While there have been significant improvements to computational capabilities, control theory for this class of problems has not kept pace, and there is a noticeable gap in the literature for MPC applied to general nonlinear systems with discrete actuators. However, it turns out that when assumptions are carefully chosen, MPC theory for systems with only continuous actuators can be extended without major modification to cover discrete-valued inputs. In this presentation, we discuss the extension of nominal stability theory to systems with both continuous and discrete actuators. We show that the key requirement is that the admissible input set not be assumed to contain the origin on its interior. Then, we discuss some of the computational difficulties of discrete actuators and show how they can be partially alleviated by suboptimal MPC, which relaxes the necessity of finding the global solution to each optimization problem. We conclude with a brief discussion of inherent robustness properties. While practical issues still remain, we feel that these results serve to break down some of the barriers toward industrial implementation of MPC with discrete actuators and allow practitioners to leverage advances in optimization methods to improve control capabilities.

461004 Global Optimization of Natural Gas Liquefaction Processes through Differentiable McCormick Relaxations

Abstract:

We present a new methodology and framework for the global optimization of liquefied natural gas production processes. Liquefaction processes are inherently challenging to optimize (even locally), largely due to the difficulty in modeling the complex heat transfer behavior and phase transitions in multistream heat exchanger unit operations. Much of the state-of-the-art work found in the literature involves the use of either mixed-integer or complementarity-constrained programming formulations and associated solution methods.[1],[2] However, building on a recent nonsmooth modeling approach for multiphase multistream heat exchangers,[3],[4] we demonstrate that these complex flowsheet optimization problems can instead be formulated as much simpler nonsmooth nonlinear programs. These problems can then be solved effectively in a branch-and-bound global optimization framework with the use of newly developed continuously-differentiable McCormick relaxations.[5] This relaxation technique calculates differentiable convex underestimators, differentiable concave overestimators and their respective gradients for complex compositions of simple intrinsic functions, and can be performed automatically in a procedure analogous to the forward mode of automatic differentiation. Moreover, these relaxations are based on the multivariate McCormick relaxation framework,[6] and so they are particularly well suited for this application, owing to the frequent occurrence of bivariate product, min and max terms in the nonsmooth multistream heat exchanger constraints. Additionally, the local solvers used to calculate lower bounds on the global solution can readily provide dual multiplier information because the relaxations are differentiable, meaning effective range-reduction strategies can be implemented. In combination with the comparatively small problem size relative to other approaches, these range-reduction methods significantly enhance the convergence rate of the branch-and-bound algorithm. Comparisons with state-of-the-art global optimization software are provided on problems involving both simplified and industrially-relevant liquefaction processes to demonstrate the favorable performance of the new modeling and optimization framework.

---

[1] M. M. F. Hasan, I. A. Karimi, H. E. Alfadala and H. Grootjans, "Operational modeling of multistream heat exchangers with phase changes," ***AIChE Journal***, Vol. 55, No. 1, pp. 150–171, 2009.

[2] R. S. Kamath, L. T. Biegler and I. E. Grossmann, "Modeling multistream heat exchangers with and without phase changes for simultaneous optimization and heat integration," ***AIChE Journal***, Vol. 58, No. 1, pp. 190–204, 2012.

[3]H. A. J. Watson, K. A. Khan and P. I. Barton, "Multistream heat exchanger modeling and design," ***AIChE Journal***, Vol. 61, No. 10, pp. 3390-3403, 2015.

[4] H. A. J. Watson and P. I. Barton, "Modeling phase changes in multistream heat exchangers." Submitted.

[5] K. A. Khan, H. A. J. Watson and P.I. Barton, "Differentiable McCormick relaxations." Submitted.

[6] A. Tsoukalas and A. Mitsos, "Multivariate McCormick relaxations," ***Journal of Global Optimization***, Vol. 59, pp. 633–662, 2014.

464294 Model Predictive Control and Materials Property Estimation of an Industrial Heat Treating Furnace

Abstract:

The energy consumption of metal processing industries, primarily related to the fuel required to operate furnaces and boilers, has reached around 2 quadrillion BTU (quads) per year in the United States [1]. The high energy demands are intensified due to the inherent losses in furnace operation (with 20% to 60%) and ineffective control strategies [2]. Motivated by the need to improve the efficiency in such operations, in this paper we study the use of model predictive control (MPC) to minimize the energy use in the plant without compromising product quality. MPC uses a dynamical model of the process to predict its future evolution and optimizes the control input signal, subject to the process, inputs and outputs constraints, based on both model predictions and real-time process measurements [3].

The system under consideration is a heat treating furnace, currently in operation at an industrial partner. During heat treatment (austenitization), the metal parts are heated to a very high temperature in an inert atmosphere (i.e., without oxidizing the surface), then rapidly quenched in an oil bath where desired mechanical properties such as ductility, hardness and shear strength are induced. The furnace is operated in a continuous manner. The metal parts travel on a conveyor belt through the furnace, which is heated by ceiling and floor radiant tube burners. For control purposes, the furnace is divided into four temperature zones. Part temperatures, especially the inside part regions, cannot be sensed and controlled directly. Hence, they are indirectly controlled by controlling the zone temperatures.

We develop a two-dimensional physics-based model of the furnace in order to balance computational efficiency with the ability to capture long range radiation interactions. The model provides the exit temperature profiles of the treated steel parts and the energy usage of the furnace by solving conservation equations. As a base case, we adopt a simple linear control strategy, wherein a PI controller controls the temperature of each zones by manipulating the fuel flow rate to the burners of the respective zone. We investigate the energy consumption and part temperature distributions for

a batch of forty parts, under the heuristic zone temperature set points suggested by the operators of the plant.

Then, the objective of model predictive control scheme is to control the minimum part temperature at furnace exit by adjusting the zone temperature set points of the PI controller. For this purpose, we develop simplified step response models that capture the dynamics of the system. These models are then used for anticipating future events and optimal control action is taken at each time step based on the simplified model predictions and the actual measured values from the detailed 2*D* model. Preliminary results show theoretical energy savings of 6.3%, when comparing the MPC operation mode with the aforementioned heuristic case.

Furthermore, we study the metallurgical property changes as a function of temperature and rate of heating. This is essential since changes in operating conditions of upstream heat treating can have ramifications downstream in machining. We model the extent of material transformation from Ferrite to Austenite using the Johnson-Mehl-Avrami-Kolmogorov (JMAK) equation [4] and ensure that there are no untransformed portions of the parts when operating under model predictive control.

References

[1] V Viswanathan, R Davies, and J Holbery. Opportunity analysis for recov- ering energy from industrial waste heat and emissions. Pacific Northwest National Laboratory, 2005.

[2] A Thekdi. Energy efficiency improvement opportunities in process heating for the forging industry. E3M, 2010.

[3] D Seborg, TF Edgar, and D Mellichamp. Process dynamics & control. John Wiley & Sons, 2004.

[4] Melvin Avrami. Kinetics of phase change. ii transformation-time relations for random distribution of nuclei. The Journal of Chemical Physics, 8(2):212â€" 224, 1940.

456469 A Multidimensional Population Balance Model for Predicting Crystal Size and Aspect Ratio in the Production of Phosphogypsum

Abstract:

A Multidimensional Population Balance Model for Predicting Crystal Size and Aspect Ratio in the Production of Phosphogypsum

Zhilong Zhu

You Peng

Kamal Samrane

Allan S. Myerson

Richard D. Braatz

Crystals of inappropriate crystal size distribution (CSD) and shape can cause problems in downstream processes [1]. Many recent efforts have been directed towards modeling CSD and shape using a multidimensional population balance model (mD-PBM) that can be used to optimize crystal shape [2]–[7]. The model equation is a hyperbolic integro-partial differential equation, which must be solved numerically except for some simple cases. Due to the advection term in the mD-PBM, most discretization methods experience numerical diffusion. Though more advanced schemes such as the high-resolution finite-volume method have demonstrated the ability to solve mD-PBM with much higher numerical accuracy [8], the computational cost of these methods is still not favorable for applications in process optimization and model predictive control. The method of characteristics (MOC) has demonstrated to be computationally efficient and accurate in handling the advection term in the 1D-PBM [9]–[11]. This presentation extends the MOC for solving mD-PBM with both growth and nucleation for the reactive crystallization of phosphogypsum.

During the production of phosphoric acid by the Wet Process, gypsum crystal is produced as a byproduct in the crystallizer. The phosphoric acid is separated from the gypsum by passing the slurry to the filter, which is the bottleneck process due to the undesired gypsum CSD and shape. A two-dimensional PBM with simultaneous growth and secondary nucleation is developed to model the crystal size and its varying aspect ratio in a continuous crystallizer. The solution supersaturation is estimated using the state-of-art Mixed Solvent Electrolyte (MSE) thermodynamic model [12], which models both the solution nonideality and speciation. In the presence of metal impurities, the growth rate in each direction is inhibited differently and results in a change in aspect ratio. The relative change in growth rate due to the added impurity is modeled by growth inhibition model [13]. The presented model is the first to combine the 2D-PBE, an impurity inhibition model, and the MSE model in the study of reactive crystallization under high ionic strength as found in the industrial process.

The proposed MOC method solves the 2D-PBM combined with the inhibition and the MSE model. Compared to existing PBM solvers, the MOC algorithm offers a computationally efficient and highly resolved method, especially in higher dimensions. The simulation results with experimentally measured crystal kinetic parameters show a good agreement with the measured CSD and mean aspect ratio. These simulation results also suggest that the presence of metal ion impurities cannot be ignored as they can modify crystal shape. This information would not be captured in a 1D-PBM for gypsum crystallization.

References:

[1] J. W. Mullin, *Crystallization*. Woburn, MA: Butterworth-Heinemann, 2001.

[2] R. Gunawan, D. L. Ma, M. Fujiwara, and R. D. Braatz, "Identification of kinetic parameters in multidimensional crystallization processes," *Int. J. Mod. Phys. B*, vol. 16, pp. 367–374, Jan. 2002.

[3] Z. K. Nagy and R. D. Braatz, "Advances and new directions in crystallization control," *Annu. Rev. Chem. Biomol. Eng.*, vol. 3, pp. 55–75, Jan. 2012.

[4] F. Puel, G. Févotte, and J. P. Klein, "Simulation and analysis of industrial crystallization processes through multidimensional population balance equations. Part 1: A resolution algorithm based on the method of classes," *Chem. Eng. Sci.*, vol. 58, no. 16, pp. 3715–3727, Aug. 2003.

[5] M. Jiang, X. Zhu, M. C. Molaro, M. L. Rasche, H. Zhang, K. Chadwick, D. M. Raimondo, K.-K. K. Kim, L. Zhou, Z. Zhu, M. H. Wong, D. O'Grady, D. Hebrault, J. Tedesco, and R. D. Braatz, "Modification of crystal shape through deep temperature cycling," *Ind. Eng. Chem. Res.*, vol. 53, no. 13, pp. 5325–5336, Apr. 2014.

[6] C. Borchert, N. Nere, D. Ramkrishna, A. Voigt, and K. Sundmacher, "On the prediction of crystal shape distributions in a steady-state continuous crystallizer," *Chem. Eng. Sci.*, vol. 64, no. 4, pp. 686–696, Feb. 2009.

[7] K. Sato, H. Nagai, K. Hasegawa, K. Tomori, H. J. M. Kramer, and P. J. Jansens, "Two-dimensional population balance model with breakage of high aspect ratio crystals for batch crystallization," *Chem. Eng. Sci.*, vol. 63, no. 12, pp. 3271–3278, Jun. 2008.

[8] R. Gunawan, I. Fusman, and R. D. Braatz, "High resolution algorithms for multidimensional population balance equations," *AIChE J.*, vol. 50, no. 11, pp. 2738–2749, Nov. 2004.

[9] F. Févotte and G. Févotte, "A method of characteristics for solving population balance equations (PBE) describing the adsorption of impurities during crystallization processes," *Chem. Eng. Sci.*, vol. 65, no. 10, pp. 3191–3198, 2010.

[10] E. Aamir, Z. K. Nagy, C. D. Rielly, T. Kleinert, and B. Judat, "Combined quadrature method of moments

and method of characteristics approach for efficient solution of population balance models for dynamic modeling and crystal size distribution control of crystallization processes," *Ind. Eng. Chem. Res.*, vol. 48, no. 18, pp. 8575–8584, Sep. 2009.

[11] S. Qamar and G. Warnecke, "Numerical solution of population balance equations for nucleation, growth and aggregation processes," *Comput. Chem. Eng.*, vol. 31, no. 12, pp. 1576–1589, Dec. 2007.

[12] P. Wang, A. Anderko, and R. D. Young, "A speciation-based model for mixed-solvent electrolyte systems," *Fluid Phase Equilib.*, vol. 203, no. 1–2, pp. 141–176, Dec. 2002.

[13] N. Kubota and J. W. Mullin, "A kinetic model for crystal growth from aqueous solution in the presence of impurity," *J. Cryst. Growth*, vol. 152, no. 3, pp. 203–208, 1995.

459234 Optimal Operation of Batch Enantiomer Crystallization: From Ternary Diagrams to Predictive Control

Abstract:

The compounds making up many useful molecules such as the active ingredients in pharmaceuticals appear in nature as enantiomers [1]. To produce a pure enantiomer, a variety of techniques can be used, including asymmetric synthesis and methods of chiral resolution such as chromatographic methods and crystallization [2]. When a set of enantiomers forms a racemic compound, production of pure crystals of one enantiomer in a batch operation requires that the original liquid mixture purity be above that of the eutectic composition due to thermodynamic restrictions [3]. These thermodynamic restrictions can be visualized through a ternary diagram for the mixture of the two enantiomers and a solvent [3]-[4].

In this work, we investigate the operation and control of a batch crystallization process for a racemic compound forming system with a fines trap and dissolution tank, and seeding of the initial liquid within the system (which is enriched in the desired enantiomer) with crystals of the desired enantiomer. We use the ternary diagram in the development of operating conditions for the system and for the development of constraints for an optimal control design for the system. We develop model equations for the batch crystallization process based on a population balance model for the crystals coupled with mass balances. We then apply a model predictive control strategy to the system to maximize the volume of crystals developed from the seeds while reducing the volume of crystals developed from nucleation, and show how the closed-loop trajectories are related to the ternary diagram thermodynamic data.

[1] Nguyen LA, He H, Pham-Huy C. Chiral drugs: An overview. International Journal of Biomedical Science. 2006;2:85-100.

[2] Lorenz H, Seidel-Morgenstern A. Processes to separate enantiomers. Angewandte Chemie International Edition.

2014;53:1218-1250.

[3] Kaspereit M. Separation of Enantiomers by a Process Combination of Chromatography and Crystallisation. PhD Thesis, Max Planck Institute for Dyanamics of Complex Technical Systems. Aachen, Germany: Shaker, 2006.

[4] Jacques J, Collet A, Wilen SH. Enantiomers, Racemates, and Resolutions. New York: John Wiley & Sons, 1981.

461227 Using Semidefinite Programming to Calculate Bounds on Particle Size Distributions

Abstract:

Many chemical engineering processes involve a population of particles with a distribution of sizes that changes over time. For example, crystallization, colloidal suspension formation, catalyst attrition, polymerization, and aerosol formation all fit this general framework [1]. In each of these processes, the particle size distribution (PSD) can have a large effect on macroscopic properties of engineering interest. For example, for pharmaceutical crystals, the PSD affects the ease with which the crystals can be filtered and compacted into tablets, thereby affecting the cost and processing time of the pharmaceutical product [2, 3]. Moreover, once the drug has been introduced to a patient's system, the PSD affects it's bioavailability. Thus, the PSD is tied to both the pharmaceutical's efficacy and safety [4,5].

Because of the importance of the PSD in these diverse chemical engineering applications, many methods have been developed to model how a PSD changes over time. Usually, this model is a PDE known as a population balance model [1]. In some cases, this PDE can be solved analytically. However, it often must be solved numerically. Solving the PDE has the advantage that the result is a full description of the final PSD in terms of a number density function; the disadvantage is that obtaining this solution numerically can be computationally expensive. For this reason, it is very common to instead model only finitely many *moments* of the PSD, which amounts to solving a system of ODEs [6,7]. Modeling only the moments certainly reduces the computational burden, but this comes at a cost: moments are only a summary description of the PSD, i.e, they do not contain enough information to reconstruct all of its details. This is because there are, in general, many PSDs corresponding to a given finite set of moments [8]. Thus, given only finitely many moments of an unknown distribution, there is no clear answer to industrially relevant questions such as:

- How many particles have size in the range *a* to *b*?
- What is the D10 of the distribution?
- What is the qualitative shape of the distribution?

Faced with these questions, one might be tempted to apply one of the various methods available for constructing a number density function with a specified finite set of moments [9,10,11,12]. With the resulting number density function, answering the above questions would be trivial. However, the problem with this strategy should be clear from the foregoing discussion: the calculated number density function describes just one of the many PSDs with the specified moments. Accordingly, it would provide just one of the many possible (valid) answers to each of the above questions, giving us a false sense of certainty in our knowledge of the distribution.

We take a more rigorous approach. Acknowledging that reconstructing a PSD from finitely many moments is an ill-posed inverse problem, we make no attempt to answer the above questions exactly. Instead, we calculate provable *bounds* on the answers. These bounds require no a priori knowledge of the shape of the distribution, no experimental data, and no regularity assumptions on the number density function describing the PSD.

The bounding algorithms we will present are a natural application of results from the mathematical literature regarding moments of positive finite Borel measures (i.e. generalized distributions) [13]. In particular, we will calculate the proposed bounds using Semidefinite Programs (SDPs) [14]. While SDPs have been applied in chemical engineering in the context of optimal control [15,16], to the best of the authors' knowledge, their natural application to particle size distributions has, until now, gone unnoticed.

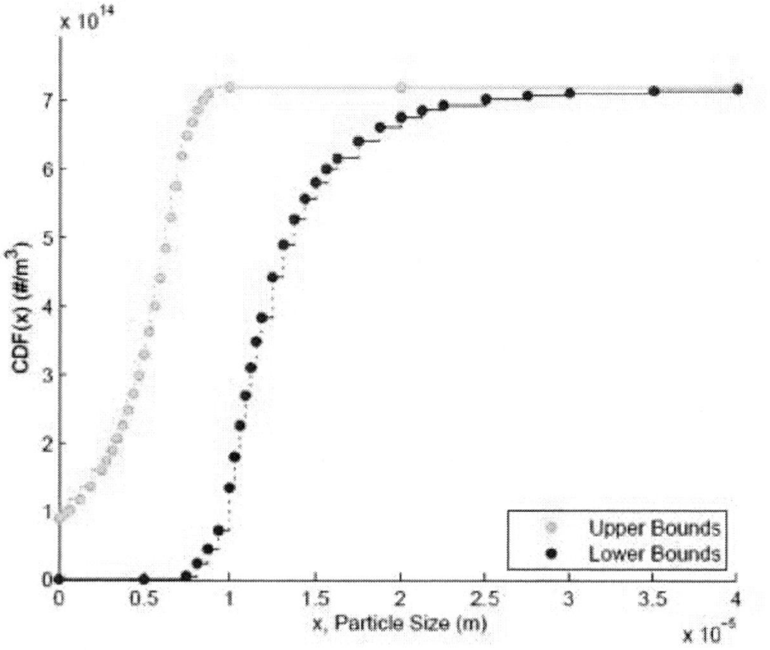

Figure 1: Example of bounds calculated on the PSD cumulative distribution function (CDF)

References

[1] Ramkrishna D. Population Balances. San Diego: Academic Press. 2000.

[2] Myerson A. Handbook of Industrial Crystallization. Woburn: Butterworth-Heinemann. 2002.

[3] Wibowo C, Chang W, Ng K. Design of integrated crystallization systems. AIChE Journal. 2001;47:2474-2492.

[4] Braatz RD. Advanced Control of Crystallization Processes. Annual Reviews in Control. 2002; 26(1):87-99.

[5] Nowee S, Abbas A, Romagnoli J. Model-Based Optimal Strategies for Controlling Particle Size in Antisolvent Crystallization Operations. Crystal Growth Des. 2007;8:2698-2706.

[6] Hulburt H, Katz S. Some problems in particle technology: A statistical mechanical formulation. Chemical Engineering Science. 1963;19:555-574.

[7] Randolph AD, A LM. Theory of Particulate Processes. New York: Academic Press Inc. 1988.

[8] McGraw R, Nemesure S, Schwartz SE. Properties and evolution of aerosols with size distributions having identical moments. Journal of Aerosol Science. 1998;29(7):761-772.

[9] de Souza LGM, Janiga G, John V, Thevenin D. Reconstruction of a distribution from a finite number of moments with an adaptive spline-based algorithm. Chemical Engineering Science. 2010;65(9):2741-2750.

[10] Hutton K, Mitchell N, Frawley PJ. Particle size distribution reconstruction: The moment surface method. Powder Technology. 2012;222:8-14.

[11] Cogoni G, Frawley PJ. Particle Size Distribution Reconstruction Using a Finite Number of Its Moments through Artificial Neural Networks: A Practical Application. Crystal Growth & Design. 2015; Published online.

[12] Woodbury AD. A FORTRAN program to produce minimum relative entropy distributions. Computers and Geosciences. 2004;30(1):131-138.

[13] Lasserre JB. Moments, positive polynomials and their applications, vol. 1. World Scientific. 2009.

[14] Vandenberghe L, Boyd S. Semidefinite programming. SIAM Review. 1996;38(1):49-95.

[15] VanAntwerp JG, Braatz RD, Sahinidis NV. Globally optimal robust process control. Journal of Process Control. 1999;9(5):375-383.

[16] VanAntwerp JG, Braatz RD, Sahinidis NV. Globally optimal robust control of large scale sheet and film processes. In: American Control Conference, 1997. Proceedings of the 1997, vol. 3. IEEE. 1997; pp. 1473-1477.

462030 Characterization of an Integrated Continuous Cooling Crystallization Process with an in-Situ Wet Mill System for Particle Size Reduction

Abstract:

Crystallization is one of the critical unit operations during the production of an active pharmaceutical ingredient or a pharmaceutical intermediate. Control of crystal attributes such as size and size distribution are critical not only to the efficiency of the downstream operations but also to the final drug product. Size reduction can be achieved by dry or wet milling operations. The integration of an insitu wet mill and continuous mixed suspension mixed product removal crystallizer (MSMPR) is studied in this work for cooling crystallization of Paracetamol from an aqueous isopropanol mixture. During this study, the impact of various operational conditions of both continuous *insitu* mill and MSMPR unit on the product properties is evaluated. The paper demonstrates that the MSMPR operational conditions (i.e. residence time, operational temperature) significantly impact the achievable steady state product characteristics. In addition, the influence of insitu mill operating conditions (i.e. rotor speed) on critical product and process qualities attributes such as particle size distribution, mean particle size, and yield. Towards this end, breakage kernels were evaluated and validated through a parameter estimation routine using the experimental data. The specific breakage rate considering magma density and rotor speed showed higher accuracy between the various kernels.

462046 Tailoring the Crystal Size Distribution By Controlling the Crystallization Trajectory in Mass-Count Space

Abstract:

Batch cooling crystallization is widely applied to isolate high-purity chemicals and pharmaceuticals. Control over the size of crystals produced by such operations can be critically important for efficient downstream processes. In this presentation we describe an approach for tailoring the crystal size distribution.

Previously, we proposed an alternative perspective on crystallization dynamics. Instead of understanding crystallization dynamics through the lens of the population balance framework, we viewed crystallization dynamics as movement in a 2D space mapped out by measured properties: the crystal mass and the chord count [1]. This, we found, revealed a simple, empirical model of crystallization dynamics and also facilitated the development of feedback schemes for controlling the mean crystal size [1, 2].

Here, a further use of the mass-count perspective will be described. By carefully controlling the mass-count *trajectory* we can rationally tailor the crystal size *distribution*. This will be demonstrated experimentally for paracetamol crystallization from ethanol.

[1] Griffin, D. J., Grover, M. A., Kawajiri, Y. and Rousseau, R. W. 2015. Mass-Count Plots for Crystal Size Control. *Chemical Engineering Science* 137, 338-351.

[2] Griffin, D. J., Grover, M. A., Kawajiri, Y. and Rousseau, R. W. 2016. Data-Driven Modeling and Dynamic Programming Applied to Batch Cooling Crystallization. *Industrial & Engineering Chemistry Research* 55, 1361-1372.

463021 Attainable Regions of Particle Sizes for Continuous Milling-Crystallization Processes

Abstract:

The properties of particulate products are not only defined by their composition and crystal structure (polymorph, salt form, solvates, etc.), but also by their size and shape. The influence of particle size and shape is particularly felt in downstream processing operations, such as filtration, drying, granulation, and blending of one particulate product with another. For instance, one can surmise that thin needle-like crystals are more prone to break in an agitated dryer than compact crystals of the same material or that the time required to separate mother liquor from crystals by cake filtration must – among other factors – depend on how tightly crystals in the filter cake are packed, which in turn depends on the crystal size and shape. Therefore, crystallization processes should be designed to deliver the high purity and yield, the right crystal form, as well as to deliver a reproducible particle size distribution that avoids undesired effects in downstream processing.

Within the quality by design (QbD) initiative[1], there is an increased drive towards a deeper understanding of manufacturing processes, aiming at identifying a design space for a given combination of product and process in which the product can be reliably produced to specification. Recent works within the crystallization community established a design methodology, which allows to deduce what particle sizes are attainable from a given crystallization process based on knowledge of the thermodynamics and kinetics[2]. This was formulated in regions of mean particle sizes against processing time (residence time for continuous crystallizers or batch time for batch crystallizers), while fulfilling constraints such as yield considerations and maximum/minimum allowed temperatures. The methodology was subsequently extended to take into account additional product characteristics, such as the width of the particle size distribution[3], as well as additional operational constraints[4]. The framework's robustness against uncertainty in the kinetic parameters was assessed as well[5]. However, these efforts were exclusively focused on crystallization processes that are dominated solely by crystal growth and nucleation, i.e., secondary kinetic processes, such as crystal breakage and agglomeration were neglected.

In this work, it will be shown that attainable regions can also be obtained for processes that include extensive breakage processes. The study is motivated by the realization that processes that intricately combine crystallization and milling devices often allow to tune particle size distributions in a more extensive way than crystallization processes alone[6]. For instance, they allow to obtain smaller particles, or allow reaching a desired yield in shorter processing times (through an increase in surface area of the crystals). This work specifically investigates the combination of a continuous mixed suspension mixed product removal crystallizer (MSMPRC) with a high shear wet suspension mill that is connected to the crystallizer in a recycle loop. After determining breakage kinetics for this mill were determined, attainable regions for this process were calculated using a population balance equation model including breakage kinetics. The methodology is applied to the cooling crystallization of paracetamol crystallized from a mixture of isopropanol and water (1:4 w/w).

References:

[1] Food and Drug Administration, "Pharmaceutical Quality for the 21st Century A Risk-Based Approach Progress Report", http://www.fda.gov/aboutfda/centersoffices/officeofmedicalproductsandtobacco/cder/ucm128080.htm, accessed: May 6th, 2016.

[2] Vetter, T., Burcham, C.L., Doherty, M.F., „Regions of attainable particle sizes in continuous and batch crystallization processes", Chem. Eng. Sci., 2014, 106, 167-180.

[3] Yang, Y., Nagy, Z.K., "Advanced control approaches for combined cooling/antisolvent crystallization in continuous mixed suspension mixed product removal cascade crystallizers", Chem. Eng. Sci., 2015, 127, 362-373.

[4] Power, G., Hou, G., Kamaraju, V.K., Morris, G., Zhao, Y., Glennon, B., „Design and optimization of a multistage continuous cooling mixed suspension, mixed product removal crystallizer", Chem. Eng. Sci., 2015, 133, 125-139.

[5] Vetter, T., Burcham, C.L., Doherty, M.F., „Designing robust crystallization processes in the presence of parameter uncertainty using attainable regions", Ind. Eng. Chem. Res., 2015, 54, 10350-10363.

[6] Yang, Y., Song, L., Zhang, Y., Nagy, Z.K., „Application of Wet Milling-Based Automated Direct Nucleation Control in Continuous Cooling Crystallization Processes", Ind. Eng. Chem. Res., 2016, 55, 4987-4996.

471499 Attainable Regions for Critical Quality Attributes in Drug Manufacture and End Performance

Abstract:

The attainable regions for Critical Quality Attributes (CQA) in the manufacture, such as Particle Size Distribution (PSD), and performance, such as fraction absorbed, of Solid Oral Dosage Forms (SODF) can be highly dependent upon the types and operation of the end-to-end production steps. This is particularly the case for low-dose drug products, where other factors, such as Content Uniformity (CU) of SODF can also be of a major concern. The CU of SODF is known to be highly dependent upon the particle size distribution of the Active Pharmaceutical Ingredient (API), with control of API particle size being a common approach to ensuring the content uniformity of drug products. In this work, we examine a continuous cascade cooling crystallization process and explore the impact of varying process configurations, both drug substance and drug product, on the attainable processing regions, subject to various processing and performance constraints. The continuous cooling crystallization of paracetamol and ethanol solutions was employed as an example case. Additional drug substance and drug product manufacturing steps were also considered in a modular fashion, in order to investigate their impact on the attainable regions, including filtration, wet and dry milling. The final performance of the drug product was determined via a segmented model of the Gastro-Intestinal (GI) tract. This model describes the dissolution and permeation of the API in the GI tract, yielding the fraction of the API absorbed in the patient as function of time and hence the bioavailability of the drug *in-vivo*.

A number of optimization objectives were posed, which included increasing throughput of the API and the number of crystallizer stages, whilst maintaining a target CU of the final product, sufficient bioavailability and other processing constraints, in order to evaluate these attainable regions. Once the attainable region of a given CQA versus total production time was determined, the robustness within the attainable region was evaluated and explored via Global Sensitivity Analysis (GSA). This holistic approach to the modelling of API manufacture and end performance, allows the interdependent processing steps of drug substance and drug product manufacture and their impact upon *in-vivo* drug performance to be considered in a more cohesive way.

461728 Surrogate-Based Optimization Methodology for Pharmaceutical Tablet Manufacturing Processes

Abstract:

The pharmaceutical manufacturing industry is going through a transformation from traditional batch processes to advanced continuous processes with the aim of increasing economic efficiency and process robustness [1]. Such a revolutionary change can further be facilitated by the enhanced process understanding with the help of Process Systems Engineering (PSE) tools [2]. Recently an increasing amount of research has been focused on the modeling of unit operations of continuous pharmaceutical processes, including feeders, blenders, roller compactors, the tablet press, etc. [3] [7], which can then be integrated into a flowsheet model to predict dynamic behaviors of the process under specific operating conditions and material properties. The flowsheet models can then be used in the process optimization problems.

There are several challenges faced by the pharmaceutical manufacturing processes in terms of the process optimization. First, the randomness in nature of the numerous input material properties can make the optimization problem rather complicated. Some inevitable variations in the process (e.g. refill in the feeder) and the propagation of such variations in the subsequent unit operations must be considered in order to insure desired product qualities [4]. Second, due to the strict standards of final pharmaceutical products, it is usually desired to design a rather flexible process that can withstand the potential uncertainty in the inlet materials [5]. Finally, depending on the different modeling techniques that are adopted, the flowsheet model can be computationally expensive and may give noisy output predictions, which makes it difficult to use traditional derivative-based solvers to find the optimal solution [6].

To address the difficulties outlined above, we develop a surrogate-based optimization methodology in order to solve the optimization problems in pharmaceutical manufacturing processes. To prevent considering unnecessary input factors, which have very little impact on the main output variables, a global sensitivity analysis is initially conducted to identify the subset of most significant input factors. Surrogate models (e.g. Kriging, RBF) are built to smooth out the noise in the output predictions, and used as an efficient approximation of the original flowsheet model. The surrogate models are adaptively updated both considering the feasibility of the constraints and the optimality of the objective function, in order to reduce the total sampling cost and efficiently find the optimal solution. The methodology is illustrated using a direct compression continuous manufacturing line.

References

[1] Plumb K. Continuous processing in the pharmaceutical industry: changing the mind set[J]. Chemical Engineering Research and Design, 2005, 83(6): 730-738.

[2] Gernaey K V, Cervera-Padrell A E, Woodley J M. A perspective on PSE in pharmaceutical process development and innovation[J]. Computers & Chemical Engineering, 2012, 42: 15-29.

[3] Rogers A J, Hashemi A, Ierapetritou M G. Modeling of particulate processes for the continuous manufacture of solid-based pharmaceutical dosage forms[J]. Processes, 2013, 1(2): 67-127.

[4] Boukouvala F, Ierapetritou M G. Surrogate-based optimization of expensive flowsheet modeling for continuous pharmaceutical manufacturing[J]. Journal of Pharmaceutical Innovation, 2013, 8(2): 131-145.

[5] Rogers A, Ierapetritou M. Feasibility and flexibility analysis of black-box processes part 2: Surrogate-based flexibility analysis[J]. Chemical Engineering Science, 2015, 137: 1005-1013.

[6] Sen M, Rogers A, Singh R, et al. Flowsheet optimization of an integrated continuous purification-processing pharmaceutical manufacturing operation[J]. Chemical Engineering Science, 2013, 102: 56-66.

[7] Hsu S H, Reklaitis G V, Venkatasubramanian V. Modeling and control of roller compaction for pharmaceutical manufacturing. Part I: Process dynamics and control framework[J]. Journal of Pharmaceutical Innovation, 2010, 5(1-2): 14-23.

462497 Probabilistic Robust Optimization and *a Posteriori* Bounds

Abstract:

Robust optimization is an attractive method for optimization under uncertainty in multi-scale systems engineering applications due to its deterministic handling of uncertain parameter realizations [1-8]. Key to the concept of robust optimization is the a priori simultaneous imposition of multiple possible parameter realization vectors onto uncertain constraints or an uncertain objective function [9]. Given bounded uncertainty, restricting robust optimization as simply a method for obtaining worst-case solutions [2] by imposing all possible parameter realizations onto the model will produce very conservative results. By imposing a subset of possible parameter realizations, contained in what is known as an uncertainty set [5-9], onto uncertain constraints, significantly less conservative solutions can be obtained given bounded uncertainty, and the method can be extended to unbounded uncertainty. This will yield a nonzero probability of constraint violation; the quality of a robust solution will rely heavily on the quality of applicable probabilistic expressions. In traditional one-pass robust optimization, probabilistic bounds are applied a priori in order to guide the size of uncertainty sets [10]. Fundamental theoretical advances have produced new a priori probabilistic bounds which are superior to previously existing bounds and significantly improve solutions in traditional one-pass robust optimization for models with uncertain parameters subject to unknown or known, and symmetric or asymmetric, probability distributions [11,12].

We present new a posteriori probabilistic bounds and expressions for robust optimization [11,13]. These expressions characterize robust solutions a posteriori and apply to models with uncertain parameters subject to (i) bounded but unknown probability distributions with limited information on their expected values, (ii) bounded and symmetric but otherwise unknown probability distributions, and (iii) certain known distributions. Algorithmic methods are discussed which are critical to the viability of two of the new methods. These new advances yield significantly less conservative characterizations of robust solutions when compared with existing a posteriori bounds [10,14,15]. The new a posteriori methods yield further significant improvements to traditional robust optimization techniques when applied in tandem with recent a priori bounds and within the context of iterative algorithms [16,17], as will be demonstrated with example problems.

References:

[1] Floudas, C. A., Niziolek, A. M., Onel, O., & Matthews, L. R. (2016). Multiâ€Scale Systems Engineering for Energy and the Environment: Challenges and Opportunities. AIChE Journal.

[2] Soyster, A. L. (1973). Convex programming with set-inclusive constraints and applications to inexact linear programming. Operations Research, 21(5), 1154-1157.

[3] El Ghaoui, L., & Lebret, H. (1997). Robust solutions to least-squares problems with uncertain data. SIAM Journal on Matrix Analysis and Applications, 18(4), 1035-1064.

[4] El Ghaoui, L., Oustry, F., & Lebret, H. (1998). Robust solutions to uncertain semidefinite programs. SIAM Journal on Optimization, 9(1), 33-52.

[5] Ben-Tal, A., & Nemirovski, A. (1998). Robust convex optimization. Mathematics of Operations Research, 23(4), 769-805.

[6] Ben-Tal, A., & Nemirovski, A. (1999). Robust solutions of uncertain linear programs. Operations Research Letters, 25(1), 1-13.

[7] Ben-Tal, A., & Nemirovski, A. (2000). Robust solutions of linear programming problems contaminated with uncertain data. Mathematical Programming, 88(3), 411-424.

[8] Bertsimas, D., & Sim, M. (2004). The price of robustness. Operations Research, 52(1), 35-53.

[9] Li, Z., Ding, R., & Floudas, C. A. (2011). A comparative theoretical and computational study on robust counterpart optimization: I. Robust linear optimization and robust mixed integer linear optimization. Industrial & Engineering Chemistry Research, 50(18), 10567-10603.

[10] Li, Z., Tang, Q., & Floudas, C. A. (2012). A comparative theoretical and computational study on robust counterpart optimization: II. Probabilistic guarantees on constraint satisfaction. Industrial & Engineering Chemistry Research, 51(19), 6769-6788.

[11] Guzman, Y. A., Matthews, L. R., & Floudas, C. A. (2016). New a priori and a posteriori probabilistic bounds for robust counterpart optimization: I. Unknown probability distributions. Computers & Chemical Engineering, 84, 568-598.

[12] Guzman, Y. A., Matthews, L. R., & Floudas, C. A. (2016). New a priori and a posteriori probabilistic bounds for robust counterpart optimization: II. A priori bounds for known symmetric and asymmetric probability distributions. In preparation.

[13] Guzman, Y. A., Matthews, L. R., & Floudas, C. A. (2016). New a priori and a posteriori probabilistic bounds for robust counterpart optimization: III. A posteriori bounds for known probability distributions. In preparation.

[14] Paschalidis, I., Kang, S., & Li, K. (2008). Distribution-dependent robust linear optimization with asymmetric uncertainty and application to optimal control. In Proceedings of the 17th World Congress IFAC, 10069-10074.

[15] Kang, S. C., Brisimi, T. S., & Paschalidis, I. C. (2015). Distribution-dependent robust linear optimization with applications to inventory control. Annals of operations research, 231(1), 229-263.

[16] Li, Z., & Floudas, C. A. (2014). A comparative theoretical and computational study on robust counterpart optimization: III. Improving the quality of robust solutions. Industrial & engineering chemistry research,53(33), 13112-13124.

[17] Matthews, L. R., Guzman, Y. A., Onel, O., Niziolek, A. M., & Floudas, C. A. (2016). Natural gas to liquid transportation fuels: Process synthesis under feedstock and product price uncertainty using robust optimization. In preparation.

462499 A Comparative Analysis of Robust Optimization and Scenario-Based Approaches for Optimization Under Uncertainty

Abstract:

It is common in mathematical optimization problems to assume that parameters will retain constant values, when in reality, these parameters may have potential to vary greatly. This uncertainty can have a tremendous impact on the objective function of the optimal solution, or on overall model feasibility. As noted in a recent perspective article on multi-scale engineering for energy and the environment, the inclusion of uncertainty in process systems engineering problems will be crucial in addressing major challenges ahead [1]. To address these issues, different approaches for optimization under uncertainty have been proposed for a range of applications [2,3]. Two widely used approaches include robust optimization [4,5] and stochastic programming using scenarios. The former has widely been considered overly conservative, as the optimal solution must remain feasible for an uncertainty set of parameter values. Scenario-based approaches, on the other hand, generate scenarios that include a subset of possible parameter values, and find the expected objective value over that set of scenarios, albeit with large computational times and limited information on the probability of constraint violation.

Through multiple applications in process systems engineering, the two approaches for handling uncertain parameters are compared for optimization models with uncertainty in the objective function. For problems with uncertainty in the objective function, the probability of constraint violation can be defined for the scenario-based approach in order to compare the solutions from each approach at various levels of conservatism. In order to cover the entire uncertain space with probabilistic guarantees, scenarios will be generated based on discretizations between the bounds of the uncertain parameters. Multiple discretization levels will be assumed in order to measure the impact of discretization on the performance of a scenario-based approach. Robust solutions will also be generated using *a priori* and *a posteriori* probabilistic bounds and iterative methods for improving the quality of robust solutions [6].

Due to recent advances in probabilistic bounds for robust optimization [7-9], it will be shown that robust optimization is a competitive, or even superior, approach for optimization under uncertainty. Optimal solutions and computational time will be compared for the two approaches using example problems and will be discussed in the context of large-scale optimization problems using robust results. These results will demonstrate that robust optimization provides robust solutions in a fraction of the computational time required for the scenario-based approach while giving as good or better objective function values at the same probabilities of constraint violation, based on the assumptions made regarding the uncertain parameter distributions.

[1] Floudas, C. A.; Niziolek, A. M.; Onel, O.; Matthews, L. R. Multi-scale systems engineering for energy and the environment: Challenges and opportunities. AIChE Journal 2016, 62 (3), 602-623.

[2] Verderame, P. M.; Elia, J. A.; Li, J.; Floudas, C. A. Planning and scheduling under uncertainty: A review across multiple sectors. Industrial & Engineering Chemistry Research 2010, 49, 3993-4017.

[3] Grossmann, I. E.; Apap, R. M.; Calfa, B. A.; Garcia-Herreros, P.; Zhang, Q. Recent advances in mathematical programming techniques for the optimization of process systems under uncertainty. Computers and Chemical Engineering (2016). In Press.

[4] Li, Z.; Ding, R.; Floudas, C. A. A Comparative Theoretical and Computational Study on Robust Counterpart Optimization: I. Robust Linear Optimization and Robust Mixed Integer Linear Optimization. Industrial & Engineering Chemistry Research 2011, 50, 10567-10603.

[5] Li, Z.; Tang, Q.; Floudas, C. A. A Comparative Theoretical and Computational Study on Robust Counterpart Optimization: II. Probabilistic Guarantees on Constraint Satisfaction. Industrial & Engineering Chemistry Research 2012, 51 (19), 6769-6788.

[6] Li, Z.; Floudas, C. A. A Comparative Theoretical and Computational Study on Robust Counterpart Optimization: III. Improving the Quality of Robust Solutions. Industrial & Engineering Chemistry Research 2014, 53 (33), 13112-13124.

[7] Guzman, Y. A; Matthews, L. R; Floudas, C. A New a priori and a posteriori probabilistic bounds for robust counterpart optimization: I. Unknown probability distributions. Computers & Chemical Engineering 2016, 84, 568-598.

[8] Guzman, Y. A; Matthews, L. R; Floudas, C. A New a priori and a posteriori probabilistic bounds for robust counterpart optimization: II. *A priori* bounds for known symmetric and asymmetric probability distributions. 2016, In Preparation.

[9] Guzman, Y. A; Matthews, L. R; Floudas, C. A New a priori and a posteriori probabilistic bounds for robust counterpart optimization: III. *A posteriori* bounds for known probability distributions. 2016, In Preparation.

463032 A Hybrid Data-Driven/Model-Based Framework for the Dynamic Characterization and Refinement of Model Uncertainty in Batch Systems: Application to Robust Online Optimization and Control

Abstract:

Stochastic optimization has emerged as a powerful means of solving a variety of optimization problems under uncertainty, which originate from different areas including supply-chain optimization, optimal design, robust dynamic optimization/optimal control, process scheduling and so on. However, the formulation of this type of optimization problems requires good quality information on the underlying characteristics of the uncertainty, which can affect the objective function as well as the problem constraints. Therefore, in recent times, the definition of strategies for extracting statistical information from historical economic/process data has progressively gained importance.

For the specific case of robust dynamic optimization/optimal control applied to batch processes, we can envisage the general strategy for "extracting statistical information" as a two-step (iterative) procedure:

1. First, historical process data (data from previous successful batches) and the process model are used to select a proper set of model parameters to be the uncertain parameters (UPs) and estimate their joint probability distribution (PDF);
2. Then, every "n" batch cycles ($n \geq 1$), the PDF is re-estimated as new data become available as well as the UPs set is updated by adding, removing or replacing parameters.

Notice that this dynamic PDF of the dynamic UPs set comprises an input to the optimization phase of many online optimization/control frameworks. Three specific features make the aforementioned dynamic estimation of PDF and UPs set non-trivial:

1. The available process data are usually measurements of process states and not measurements of the UPs themselves;
2. Some of the UPs might exhibit temporal dynamics (on a longer time scale than that of a single batch).
3. Proper metrics must be developed to manage the dynamic UPs set.

There are virtually no literature contributions concerning this specific type of problem, even though there exist frameworks for similar problems. In fact, there exist methods for online optimization/control coupled with run-to-run update of the process model ([1], [2] and [3]) but these frameworks allow only constant UPs sets and very often do not explicitly estimate the PDF of the UPs (except for [3], where the PDF is approximated using polynomial chaos expansion). Consequently, the aforementioned dynamic estimation problem appears to be an interesting research challenge.

In this paper, we tackle this problem using an iterative multi-step strategy that we can categorize as a Bayesian approach and involves a combination of Maximum Likelihood Estimation (MLE), Fitting (FG) and Sensitivity Analysis (SA). The methodology comprises two phases, referred to as phase I and phase II. Phase I is executed only before the very first batch cycle and allows estimating an initial UPs set as well as the corresponding PDF by means of a combination of FG, residuals analysis and SA. Necessary input information include historical process data along with the model of the target batch process. Phase II is performed every "n" batches ($n \geq 1$) and aims at updating the UPs set as well as the corresponding PDF, based on fresh process data and the process model. Its underlying logic is complex and combines FG, residuals analysis, SA and MLE in the following way:

1. New statistical information is acquired from the newly available process data via combination of FG and residuals analysis;
2. This information, combined with the results of a proper SA, is used to re-estimate a new set of UPs;
3. MLE is applied to merge the same newly available statistical information with the previous PDF of the UPs towards improved uncertainty characterization (a new PDF).

In contrast to conventional strategies ([1], [2] and [3]), this new method for "extracting statistical information" allows to dynamically and conveniently select the model parameters to consider as UPs, based on fresh process data. Moreover, it also allows estimating explicitly and dynamically the probability distribution of the UPs, which is necessary input to many robust dynamic optimization/optimal control strategies. Therefore, it is more general and flexible than other conventional strategies.

The new methodology proposed here is combined with a robust dynamic optimization/optimal control framework and applied to a fed-batch version of the Williams Otto process. The results achieved from the case study clearly demonstrate the benefit in applying the new approach, also compared to other conventional strategies ([1], [2] and [3]).

1. Nagy ZK. Model based robust batch-to-batch control of particle size and shape in pharmaceutical crystallization. *IFAC Proceedings Volumes.* 2009;7:195-200.
2. Chen T, Liu Y, Chen J. An integrated approach to active model adaptation and on-line dynamic optimisation of batch processes. *Journal of Process Control.* 2013;23:1350–1359.
3. Mandur J, Budman, H. A Robust algorithm for Run-to-run Optimization of Batch Processes. *10th IFAC International Symposium on Dynamics and Control of Process Systems.* 2013:541-546.

465581 Jointly Robust Optimization for Multiple Uncertain Constraints

Abstract:

Optimization under uncertainty is important for many decision making problems. Among the various approaches dealing with uncertainty in optimization problem, the advantages of robust optimization include the computational tractability as well as the absence of the distribution information requirement. It deals with the uncertainty in the parameters by introducing an uncertainty set which covers part of the whole uncertainty space. The target is to select the best solution that is feasible for any realizations of the uncertain parameters in the uncertainty set [1]. In the open literature, most of the robust optimization methods assumed that the uncertainties in the parameters are independent [2,3]. But in practice, correlation may exist among different uncertain parameters. As a result, the traditional robust optimization methods that ignore the correlation may lead to a solution that is too conservative. The major issue of robust optimization is the design of the uncertainty set. Different types of uncertainty set lead to different robust counterpart formulations [2]. Recently, robust optimization under correlated uncertainties within a single constraint has been studied by incorporating the covariance matrix in the uncertainty set design, and corresponding robust counterpart formulations were derived [4].

In this work, a novel jointly robust formulation for optimization under independent or correlated uncertainties across multiple constraints is proposed. First, for a set of inequality constraints with uncertainty $y_0^i + (y^i)^T$*Î¶â‰¤0* (where, y_0^i and y^i are variables and *Î¶* is uncertain parameter, *i*is the index for different constraints), it is converted to a single constraint by introducing a maximizing operator. Second, the selection of the largest left-hand-side is realized by introducing binary variables. Next, the mixed-integer constrained problem is equivalently transferred to a linear programming problem by relaxed the binary variables as continuous variables. Third, by incorporating the covariance matrix of the uncertainties in the uncertainty set as a general form, the robust counterpart can be formulated based on conic duality. With specified type of uncertainty set, the final robust counterpart formulations can be obtained. Lastly,

the proposed jointly robust optimization formulation can be improved by introducing adjustable coefficient to each individual constraints, which can lead to a robust solution of better quality. Numerical examples and applications in process operations problems are studied to illustrate the proposed jointly robust optimization framework for uncertainty in multiple constraints.

Reference:

[1] Ben-Tal, A., & Nemirovski, A. (1999). Robust solutions of uncertain linear programs. Operations research letters, 25(1), 1-13.

[2] Li, Z., Ding, R., & Floudas, C. A. (2011). A comparative theoretical and computational study on robust counterpart optimization: I. Robust linear optimization and robust mixed integer linear optimization. Industrial & engineering chemistry research, 50(18), 10567-10603.

[3] Li, Z., Tang, Q., & Floudas, C. A. (2012). A comparative theoretical and computational study on robust counterpart optimization: II. Probabilistic guarantees on constraint satisfaction. Industrial & engineering chemistry research, 51(19), 6769-6788.

[4] Yuan, Y., Li, Z., & Huang, B. (2016). Robust optimization under correlated uncertainty: Formulations and computational study. Computers & Chemical Engineering, 85, 58-71.

466626 Multi-Parametric Linear Programming with Global Uncertainty

Abstract:

Multi-parametric programming (mp-P) has proven to be an invaluable tool for optimisation problems liable to uncertainty [1]. Uncertainty is an inherent feature in all mathematical models either because of fluctuations in extrinsic data such as demand of products, prices, availability of raw material etc. or intrinsic parameters such as stoichiometric coefficients, transition times in a scheduling problem etc.[2] While the first type of uncertainty typically results in variations either in the objective function's coefficients (OFC) or on the right hand side (RHS) of the constraints of the problem the latter mostly appears as left hand side (LHS) uncertainty as varying parameters in the technology matrix. When uncertain parameters are present in the entries of the technology matrix the resulting multi-parametric programming problem is in general non-convex and hard to solve. To address this issue, hybrid and/or approximate solution strategies[3,4] and inversion of perturbed matrices based approaches[6] have been proposed. Recently, symbolic manipulation software has gained a considerable amount of attention and platforms like Mathematica© have created their own environments that allow users to efficiently perform symbolic calculations to provide exact solutions for certain classes of problems. The use of symbolic manipulation software for the solution of optimization problems was studied by Dua (2015) [5] where an algorithm for Mixed Integer Polynomial problems was proposed.

In this work, a new general framework is introduced for the exact solution of multi-parametric linear programming (mp-LP) problems with varying uncertain parameters in the OFC, the RHS and the LHS simultaneously. The proposed methodology is based on the analytical solution of the system of equations derived from the Karush-Kuhn-Tucker (KKT) conditions for general linear programming (LP) problems using symbolic manipulation software. Key contribution of this work is the ability of the proposed methodology to efficiently handle the LHS uncertainty by computing exactly the corresponding non-convex critical regions (CRs). The algorithm has been tested for a number of case studies which further underlines its advantages when compared with existing approaches.

References

1. Pistikopoulos, E. N., Georgiadis, M. C., & Dua, V. (2007). Multi-parametric programming. Volume I. Weinheim: WileyVCH.
2. Grossmann, I. E., Apap, R. M., Calfa, B. A., Garcia-Herreros, P., & Zhang, Q. (2015) "Recent Advances in Mathematical Programming Techniques for the Optimization of Process Systems under Uncertainty". 12th International Symposium on Process Systems Engineering and 25th European Symposium on Computer Aided Process Engineering (Vol. 37, p. 1). Elsevier.
3. Li, Z., & Ierapetritou, M. G. (2007). A new methodology for the general multiparametric mixed-integer linear programming (MILP) problems. Industrial & engineering chemistry research, 46(15), 5141-5151.
4. Wittmann-Hohlbein, M., & Pistikopoulos, E. N. (2013). On the global solution of multi-parametric mixed integer linear programming problems. Journal of Global Optimization, 57(1), 51-73.
5. Dua, V. (2015). Mixed integer polynomial programming. Computers & Chemical Engineering, 72, 387-394.
6. Khalilpour, R., & Karimi, I. A. (2014). Parametric optimization with uncertainty on the left hand side of linear programs. Computers & Chemical Engineering, 60, 31-40.

466879 An Efficient Branch-and-Bound Strategy for Multistage Stochastic Programs with Endogenous and Exogenous Uncertainties

Abstract:

An efficient branch-and-bound strategy for multistage stochastic programs with endogenous and exogenous uncertainties

Robert M. Apap & Ignacio E. Grossmann

Multistage stochastic programming with both endogenous and exogenous uncertain parameters presents many unique challenges in terms of the development of efficient models and solution methods. Apap and Grossmann (2015) recently addressed some of these challenges; specifically, by developing theoretical reduction properties that can eliminate all redundant non-anticipativity constraints (NACs), and by proposing a sequential scenario decomposition heuristic that can quickly determine high-quality feasible solutions. Efficient decomposition methods such as this heuristic are essential for problems with endogenous and exogenous uncertainties, as the corresponding models are often extremely large, even after redundant NACs have been removed.

Although the sequential scenario decomposition heuristic is capable of providing both upper and lower bounds, it is not an exact solution method that can be used in an iterative procedure to converge to the optimal solution. One rigorous decomposition method that satisfies such criteria, and has also been applied successfully to this class of problems (Apap and Grossmann (2015)), is Lagrangean decomposition. In general, however, there will be a duality gap between the best Lagrangean bound and the best feasible solution that may be very difficult (or impossible) to sufficiently close with standard techniques. This is not uncommon in large stochastic programs where optimal, or near-optimal, solutions are required. To close the duality gap, one can employ a branch-and-bound strategy that uses Lagrangean-relaxation bounds rather than LP bounds (Guignard, 2003). A well-known example of this in stochastic programming is Carøe and Schultz (1999). Goel and Grossmann (2006) and Goel et al. (2006) later proposed a similar Lagrangean-duality-based branch-and-bound algorithm that can also handle endogenous uncertainties.

In this work, we update the algorithm of Goel and Grossmann (2006) and Goel et al. (2006) to take advantage of recent advances in the modeling of endogenous and exogenous uncertainties. Two of these advances include the scenario grouping approach of Gupta and Grossmann (2014) to strengthen the Lagrangean-dual bound, and the heuristic of Apap and Grossmann (2015) to provide a warm-start for the branch and bound. As an illustrative example, we apply this procedure to a large instance of an oilfield development planning problem.

References:

Apap, R. M. & Grossmann, I. E. (2015). Models and computational strategies for multistage stochastic programming under endogenous and exogenous uncertainties. Submitted for publication.

Carøe, C. C. & Schultz, R. (1999). Dual decomposition in stochastic integer programming. Operations Research Letters, 24, 37-45.

Goel, V. & Grossmann, I. E. (2006). A class of stochastic programs with decision dependent uncertainty. Mathematical Programming, 108 (2-3, Ser. B), 355-394.

Goel, V., Grossmann, I. E., El-Bakry, A. S., & Mulkay, E. L. (2006). A novel branch and bound algorithm for optimal development of gas fields under uncertainty in reserves. Computers and Chemical Engineering, 30, 1076-1092.

Guignard, M. (2003). Lagrangean Relaxation. TOP, 11 (2), 151-228.

Gupta, V. & Grossmann, I. E. (2014). A new decomposition algorithm for multistage stochastic programs with endogenous uncertainties. Computers and Chemical Engineering, 62, 62-79.

467810 Financially Risk-Aware Plant Maintenance Turnaround Planning Incorporating Reliability in Integrated Chemical Sites

Abstract:

Maintenance turnarounds in processing plants ensure reliable operation and improve plant productivity. A turnaround may include large sets of preventive maintenance tasks, routine component replacements and additions, capacity expansions as well as discovery work and corrective actions. As a result, site-wide disruption from planned shutdowns are common. In practice, turnaround projects are managed by dedicated steering committees, possibly with additional help from third-party turnaround experts in planning and execution, but often with suboptimal coordination across the network. However, in a large chemical site network, planning and coordinating turnarounds for the various units from a high-level standpoint has potential benefits [1,2].

In our recent works [3,4], the focus has been quantifying the risk of loss in rescheduling a turnaround, which offers flexibility in strategic medium-term planning of turnarounds. Here, uncertainties in plant or unit reliability in the form of unplanned outages are considered. In [3], we presented a reactive or a myopic planning strategy and an anticipative planning strategy to quantify the risk of loss in an alternative turnaround schedule instead of a prior base schedule. The multistage stochastic linear programming (MSP) based anticipative model was shown to capture production planning better in the decision-making process via cumulative profit probability distribution profiles. In [4], we compared a portfolio of risk-functionals for the MSP objective on simple case studies along with a few sensitivity study for the timing of the turnaround reschedule.

In this work, we extend the stochastic programming model in [3,4] to a mixed-integer linear programming (MIP) model for medium-term turnaround planning in integrated sites that also takes into account plant reliability. The timing of the turnaround determines the unplanned outage scenarios, and thus, gives rise to endogenous uncertainty. We enforce equivalence of appropriate scenarios via logic constraints, which are analogous to nonanticipativity constraints, using the binary turnaround decision variables. The new formulation also considers multiple reliability-driven turnaround units. We consider different risk-averse objectives in the new anticipative model, and compare profit distributions for the optimal schedule with the myopic risk evaluation strategy. We also consider a simulation strategy to generate a profit profile that is synonymous with value of perfect information in two-stage stochastic programming within the current setting.

References

[1] S. Amaran, N. V. Sahinidis, B. Sharda, M. Morrison, S. J. Bury, S. Miller, and J. M. Wassick. Long-term turnaround planning for integrated chemical sites. Computers & Chemical Engineering 72, 145-158, 2015.

[2] S. Amaran, T. Zhang, N. V. Sahinidis, B. Sharda, S. J. Bury. Medium-term maintenance turnaround planning under uncertainty for integrated chemical sites. Computers & Chemical Engineering, 84, 422-433, 2016.

[3] S. Rajagopalan, N. V. Sahinidis, B. Sharda, S. Amaran, and S. J. Bury. Flexible turnaround planning in integrated chemical site networks. 2015 AIChE Annual Meeting, Salt Lake City.

[4] S. Rajagopalan, S. Amaran, A. Agarwal, N. V. Sahinidis, B. Sharda, S. J. Bury and J. M. Wassick. Financially Risk-Aware Strategies for Rescheduling Plant Maintenance Turnarounds in Integrated Sites. 2016 AIChE Spring Meeting and 12 th Global Congress on Process Safety, Houston.

466262 On the Systematic Integration of Different Generation Biorefineries

Abstract:

As the world strives to create a more sustainable environment, the biorefinery community fights to minimize the environmental impact of bioprocesses and build a green economy. The use of biomass as raw material for production of energy carriers and chemicals has emerged as a promising alternative to fossil resources for mitigating climate change and enhancing energy security. Second generation biorefineries (2G) came up as the natural evolution of the first generation (1G) bioprocesses to preserve food resources, while, recently, seaweed and microalgae processes have also started to hatch as the third generation (3G) biorefieneries. As different generation biorefineries treat different feedstock, their integration can reduce energy consumption and result in a more economical design with lower environmental impacts than the stand-alone.

To date, first and second generation processes have been integrated by the adoption of grassroots design methodologies [1-4], while, in reality, their coupling is far more complex. The upgrade of first generation plants constitutes a retrofit problem, constrained to make the best use of the invested equipment of the existing plant. Second and third generation upgrades account for new technology and investment very flexible to adjust so that the upgrades are best, but the portfolios of products need also to be investigated. Apart from integrating energy, there is also significant potential to exchange (by-) products. For example, bagasse can be used as feedstock in the second generation process, and, similarly, lignin can be used as fuel for the co-generation unit, while wastewater could be used for algae cultivation. Moreover, the processes may share common sections, like the downstream processes. Finally, the co-generation (CHP) unit, after modifications, can support both plants. Thus, the integration of first and second generation biorefineries is a combined grassroots–retrofit design problem, much larger and complex than most conventional applications, that hides inner trade-offs and proves the deficiency of present-day practices in proposing realistic. The size and the complexity of the problem impose the need for a systematic approach

in the form of mathematical models.

This work studies the different symbiotic options in the integration of different generation biorefineries, scoping for the development of a new systematic methodology, which can combine different process systems engineering tools for the integration of grassroots-retrofitted problems. Detailed designs for sugarcane (1G), bagasse (2G), and seaweed (3G) ethanol production are developed in Aspen Plus®, validated by the industrial partner (CIMV Process™) and laboratory experiments, and the potential of cultivating microalgae in wastewater is also investigated. The detailed models feed the mathematical models, developed in GAMS®. The processes are integrated following a mixed mathematical model. The model applies the methodology of retrofit integration on first generation [5] and grassroots integration on second and third generation technology [6, 7], following the transhipment analysis. Preliminary results indicate a significant potential in internal trade-offs savings, both in operating and capital cost.

Acknowledgment: The financial support from Marie Curie European Research Program, RENESENG (FP-607415) is gratefully acknowledged.

References

[1] R. Palacios-Bereche, A. Ensinas, M.Modesto, and S. Nebra, 2012, Ethanol production by enzymatic hydrolysis from sugarcane biomass–the integration with the conventional process. In Proceedings of the Ecos 2012–25th international conference on efficiency, cost, optimization, simulation and environmental impact of energy systems (pp. 26-29).

[2] M. Dias, O. Cavalett, R. Maciel Filho, and A. Bonomi, 2014, Integrated first and second generation ethanol production from sugarcane, Chemical Engineering Transactions, 37, 445-450.

[3] P.R. Lennartsson, P. Erlandsson, and M.J. Taherzadeh, 2014, Integration of the first and second generation bioethanol processes and the importance of by-products, Bioresource technology, 165, 3-8.

[4] J. Moncada, J. A.Tamayo, and C.A. Cardona, (2014). Integrating first, second, and third generation biorefineries: Incorporating microalgae into the sugarcane biorefinery. Chemical Engineering Science, 118, 126-140.

[5] R. Smith, M. Jobson, and L. Chen, 2010, Recent development in the retrofit of heat exchanger networks, Applied Thermal Engineering, 30(16), 2281-2289.

[6] A. Kokossis and A. Yang, 2010, On the use of systems technologies and a systematic approach for the synthesis and the design of future biorefineries, Computers & Chemical Engineering, 34, 1397-1405.

[7] A. Kokossis, M. Tsakalova, and K. Pyrgakis, 2015, Design of integrated biorefineries, Computers and Chemical Engineering, 81, 40–56.

470769 Use of Multi-Objective Optimization for Selecting Optimally Integrated Biorefinery Processes

Abstract:

Biorefineries are integrated processing facilities that convert biomass into value added products and energy after many physical, chemical and thermal conversion steps, aiming the maximization of the economic value of biomass while minimizing the waste streams. Recently, the attention has been focusing on developing feasible integrated biorefinery concepts as well as on the identification of the most promising bio-based chemicals. This is due to rising needs for renewable feedstocks as alternatives to petroleum and high environmental impacts of petroleum-derived products. The design of a biorefinery brings a significant number of challenges due to the large number of processes and products available. Hence, it requires the use of advanced process synthesis and optimization methods to ensure that these systems are economically viable, energy efficient, have minimum environmental impacts and realize the best use of the biomass resource.

In this study, a systematic process design methodology which incorporates process optimization and energy integration is applied to assess the thermo-economic performance, the environmental impact and the energy requirement for several processing pathway options. A superstructure of process models in a lignocellulosic biorefinery converting the feedstock in sugars (biochemical conversion) and syngas (thermochemical conversion) platforms for different bio-based fuels and valuable chemicals is developed. The heat recovery is represented by characterizing the process units energy requirement using pinch analysis.

Multi-objective optimization is implemented, using different objective functions such as economic and environmental criteria that are simultaneously considered to show the trade-offs between these conflicting objectives. Different performance targets are established to compare the alternative designs, to increase the utilization of biogenic carbon and to understand the best combination of products and the synergies between them. The overall design problem is

formulated as mixed integer linear programming (MILP) optimization problem and epsilon-constraint method is combined with integer cut constraints (ICC) algorithm to systematically generate the list of competing options in a Pareto front. The results of the proposed approach provide a set of non-dominated solutions (i.e., Pareto front), and each solution shows different biorefinery configuration. Finally, these trade-off solutions can be used as a decision support for engineers and decision-makers to identify the best biorefinery pathways.

Keywords: Integrated Biorefinery Systems, Process Design, Process Integration, Multi-Objective Optimization

465681 Process Models to Assess the Co-Processing of Gasoil and Bio-Based Feedstocks in Hydroprocessing Unit Operations

Abstract:

As the bioenergy industry focuses on the near-term development of conversion pathways from biomass feedstocks to infrastructure-compatible fuel blendstocks, a major area of interest is the integration of partially upgraded, biomass-derived intermediates into existing refinery upgrading unit operations. Development of analysis capabilities and process models are needed to assess these new ideas for refinery integration strategies on the basis of potential economic viability, impact on quality of refinery fuel products and benefits versus ill effects to reliability of petroleum refinery operations. One major area of focus for the modeling effort, enabled by funding from DOEâ€™s Bioenergy Technologies Office (BETO), is the development of hydroprocessing models in Aspen Plus and Aspen HYSYS to assess the co-processing of fossil-derived gasoils and bio-based feedstocks. One of the case studies explored through this initiative is the co-hydrotreating of gasoil and triglycerides.

This presentation will focus on (1) an overview of the analysis approaches and process models developed under this effort, (2) the benefits for utilizing the tools for assessing co-processing scenarios for biomass-derived intermediates, (3) insights gathered from model development and analysis for gasoil and triglyceride co-processing and (4) future work for this effort in assessing blending of oxygenated compounds.

469498 Design and Economic Analysis of a Macroalgae-to-Butanol Process Via a Thermochemical Route

Abstract:

Design and Economic Analysis of a Macroalgae-to-Butanol Process via a Thermochemical Route

Chinedu O. Okoli[1], Thomas A. Adams II[1][*], Boris Brigljević[2] and Jay J. Liu[2]

[1]Department of Chemical Engineering, McMaster University, 1280 Main Street West, Hamilton, Ontario, L8S 4L7

[2]Department of Chemical Engineering, Pukyong National University, 365 Sinseon-ro, Nam-gu, Busan 608-739,Republic of Korea

-

Biofuels produced from macroalgae are termed "third generation" biofuels and are becoming of increasing interest in the renewable fuels community. The advantages macroalgae have over first and second generation biofuel feedstocks include fast growth rates, which enable up to 4 - 6 harvest cycles per year, and the ability to be grown in the sea thus eliminating issues relating to land use and irrigation water [1]. Despite these advantages, question marks around their economic and environmental potential remain because of uncertainties around optimal macroalgae cultivation and harvest methods as well as the technologies to convert them to fuel. The objective of this presentation is to address the question of macroalgae conversion to fuel.

One such fuel receiving increased attention is biobutanol, as it is a better alternative to bioethanol as a gasoline replacement in current automobile engines and fuel pipeline networks [2]. Two major routes exist for biobutanol production; the biochemical route which primarily proceeds via fermentation of biomass feedstock to butanol, and the thermochemical route which proceeds via gasification of biomass feedstock to syngas and the conversion of syngas over catalysts to butanol. This work assesses a first-of-its-kind process for butanol production from macroalgae through a thermochemical pathway.

The design and simulation of different configurations was performed in Aspen Plus. In addition, the potential of the different configurations were assessed using economic and environmental metrics such as the minimum butanol selling price (MBSP), and cost of CO_2 equivalent emissions (CO_2e) avoided under different market scenarios. The resulting values of the metrics were then compared amongst the different configurations and also against standard literature references of similar processes. Finally, the impact that variations in key parameters have on the metrics were assessed using sensitivity analysis.

The results showed that the lowest MBSP was obtained with configurations which import natural gas and electricity as utility sources alongside the macroalgae feedstock. However these options perform poorly when the cost of CO_2e avoided is factored in. On the other hand, macroalgae-only configurations provide the best potential for cost of CO_2e avoided but perform poorly for the MBSP metric. Finally, evaluation of the sensitivity analyses results show that gasoline price changes have a high impact on the South Korean based plant configurations as shown in Fig. 1 [3].

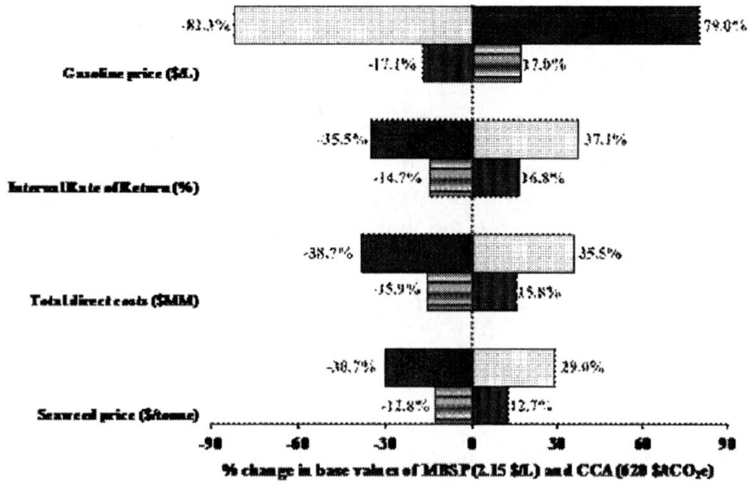

Figure 1: Sensitivity of key parameters from their base case values on MBSP (bottom bars with hatched fill) and CCA (top bars with solid fill) of the South Korea - self-sufficient plant scenario.

REFERENCES:

[1] G. Roesijadi, S. B. Jones, L. J. Snowden-Swan, and Y. Zhu, "Macroalgae as a biomass feedstock: a preliminary analysis," Pacific Northwest National Laboratory, Richland, 2010.

[2] M. Kumar and K. Gayen, "Developments in biobutanol production: New insights," *Appl. Energy*, vol. 88, no. 6, pp. 1999–2012, Jun. 2011.

[3] C. Okoli, T. A. Adams, 2016. "Design and economic analysis of a macroalgae-to-butanol process via a thermochemical route". Manuscript submitted for publication, 2016.

[*] Corresponding author: Email: tadams@mcmaster.ca; Ph: (905) 525-9140 x24782

462493 Robust Optimization of Biomass and Natural Gas to Liquid Transportation Fuel Refineries: Process Synthesis Under Uncertainty in Feedstock and Product Prices

Abstract:

Biomass and natural gas are two important feedstocks for the future production of liquid transportation fuels in the United States [1]. Biomass, as a renewable feedstock, can provide significant greenhouse gas (GHG) reductions for refineries, and the increased abundance of natural gas has significantly lowered its purchase price. Through process synthesis techniques and utilization of deterministic global optimization algorithms, optimum refinery topologies can be found that utilize both biomass and natural gas as feedstocks [2-7]. There are significant benefits to hybrid refineries in which both biomass and natural gas are converted to fuels, often through synthesis gas as an intermediate before upgrading through methanol or Fischer-Tropsch synthesis routes. These hybrid systems can offset the potentially high cost of biomass through low-cost natural gas, while providing significant GHG emission reductions based on the utilization of biomass.

In most process synthesis approaches, model parameters are assumed to be known and retain static parameter values. In reality, this is often not the case for parameters such as prices for feedstocks and products, among others; uncertain parameter realizations can have drastic impacts on the objective function values of the optimal solution, or even on overall model feasibility. Robust optimization is a framework for incorporating uncertainty which allows the optimization of large-scale process synthesis models which have uncertain parameters participating linearly in model constraints [8,9]. By imposing uncertainty sets for the parameters onto the model constraints, optimal solutions are found that ensure feasibility with known probabilities of constraint violation. Recent advances in robust optimization theory have greatly reduced the conservatism of robust solutions and allow the determination of competitive refinery topologies at known levels of risk through an iterative method utilizing *a priori* and *a posteriori* probabilistic bounds

[10-13].

The process synthesis superstructure for converting biomass, specifically hardwood or switchgrass, and natural gas to liquid transportation fuels is a non-convex, nonlinear mixed-integer optimization problem, and is solved to global optimality using a rigorous branch-and-bound algorithm [14]. Robust solutions will be presented when robust optimization model counterparts are incorporated into the process synthesis superstructure for a refinery with uncertain price parameters. These parameters appear in the objective function, allowing probabilistic guarantees on the level of profit to be provided. The impact of uncertainty on the objective function value, feedstock utilization, product distribution, and investment costs will be discussed. Case studies will be conducted using the box, interval + ellipsoidal, and interval + polyhedral uncertainty sets, and an iterative method will be utilized in order to provide high quality robust solutions at low probabilities of constraint violation.

[1] Floudas, C. A.; Niziolek, A. M.; Onel, O.; Matthews, L. R. Multi-scale systems engineering for energy and the environment: Challenges and opportunities. AIChE Journal 2016, 62 (3), 602-623.

[2] Baliban, R. C.; Elia, J. A.; Floudas, C. A. Biomass to liquid transportation fuels (BTL) systems: process synthesis and global optimization framework. Energy Environ. Sci. 2013, 6 (1), 267-287.

[3] Baliban, R. C.; Elia, J. A.; Floudas, C. A. Biomass and Natural Gas to Liquid Transportation Fuels: Process Synthesis, Global Optimization, and Topology Analysis. Industrial & Engineering Chemistry Research 2013, 52 (9), 3381-3406.

[4] Baliban, R. C.; Elia, J. A.; Floudas, C. A.; Gurau, B.; Weingarten, M. B.; Klotz, S. D. Hardwood Biomass to Gasoline, Diesel, and Jet Fuel: 1. Process Synthesis and Global Optimization of a Thermochemical Refinery. Energy & Fuels 2013, 27 (8), 4302-4324.

[5] Onel, O.; Niziolek, A. M.; Elia, J. A.; Baliban, R. C.; Floudas, C. A. Biomass and Natural Gas to Liquid Transportation Fuels and Olefins (BGTL+C2_C4): Process Synthesis and Global Optimization. Industrial & Engineering Chemistry Research 2015, 54 (1), 359-385.

[6] Niziolek, A. M; Onel, O.; Hasan, M. M. F.; Floudas, C. A Municipal solid waste to liquid transportation fuels - Part II: Process synthesis and global optimization strategies. Computers & Chemical Engineering 2015, 74 (0), 184-203.

[7] Matthews, L. R; Niziolek, A. M; Onel, O.; Pinnaduwage, N.; Floudas, C. A Biomass to Liquid Transportation Fuels via Biological and Thermochemical Conversion: Process Synthesis and Global Optimization Strategies. Industrial & Engineering Chemistry Research 2016, 55 (12), 3203-3225.

[8] Li, Z.; Ding, R.; Floudas, C. A. A Comparative Theoretical and Computational Study on Robust Counterpart Optimization: I. Robust Linear Optimization and Robust Mixed Integer Linear Optimization. Industrial & Engineering Chemistry Research 2011, 50, 10567-10603.

[9] Li, Z.; Tang, Q.; Floudas, C. A. A Comparative Theoretical and Computational Study on Robust Counterpart Optimization: II. Probabilistic Guarantees on Constraint Satisfaction. Industrial & Engineering Chemistry Research 2012, 51 (19), 6769-6788.

[10] Li, Z.; Floudas, C. A. A Comparative Theoretical and Computational Study on Robust Counterpart Optimization: III. Improving the Quality of Robust Solutions. Industrial & Engineering Chemistry Research 2014, 53 (33), 13112-13124.

[11] Guzman, Y. A; Matthews, L. R; Floudas, C. A New a priori and a posteriori probabilistic bounds for robust counterpart optimization: I. Unknown probability distributions. Computers & Chemical Engineering 2016, 84, 568-598.

[12] Guzman, Y. A; Matthews, L. R; Floudas, C. A New a priori and a posteriori probabilistic bounds for robust

counterpart optimization: II. A priori bounds for known symmetric and asymmetric probability distributions. 2016, In Preparation.

[13] Guzman, Y. A; Matthews, L. R; Floudas, C. A New a priori and a posteriori probabilistic bounds for robust counterpart optimization: III. A posteriori bounds for known probability distributions. 2016, In Preparation.

[14] Baliban, R. C; Elia, J. A; Misener, R.; Floudas, C. A. Global optimization of a MINLP process synthesis model for thermochemical based conversion of hybrid coal, biomass, and natural gas to liquid fuels. Computers & Chemical Engineering 2012, 42, 64-86.

465132 Dynamic Modeling of a Continuous Acetone-Butanol-Ethanol (ABE) Fermentation Process

Abstract:

Continuous acetone-butanol-ethanol (ABE) fermentation has drawn interest for large-scale production of biobutanol due to biobutanolâ€™s superior fuel properties over bio-ethanol and biodiesel [1]. Compared to batch fermentation, continuous ABE fermentation enables better control over process parameters and the possibility of achieving higher biobutanol productivity owing to the continuous product removal [1,2]. Dynamic modeling and analysis of integrated ABE fermentation and separation processes offer opportunities for optimal process design and operation. Dynamic models have been reported for fermentation processes integrated with product recovery units such as hollow fiber membrane, liquid-liquid extraction, and pervaporation [3]. In this work, we present a dynamic model for an integrated continuous ABE fermentation process that consists of a continuous fermentation unit [4], a tangential filtration unit, a membrane vapor extraction (MVE) unit [5], and a recycle stream. MVE is a novel product recovery technique that separates the vaporized solvent products from fermentation via solvent transport through a microporous, omniphobic membrane followed by isothermal absorption into a nonvolatile solvent [5]. The MVE unit enables continuous product removal with minimal energy consumption. An extensive dynamic sensitivity analysis has been performed to systematically investigate the effects of various kinetic and process parameters on the butanol productivity. The analysis provides guidelines for future process integration and optimization.

[1] P. Dürre. Biobutanol: An attractive biofuel. *Biotechnology Journal*. 2007; 2:1525.

[2] E. M. Green. Fermentative production of butanol- The industrial perspective. *Current Opinion in Biotechnology*.2011; 22:337.

[3] R. Mayank, A. Ranjan, and V. S. Moholkar. Mathematical models of ABE fermentation: Review and analysis.

Critical Reviews in Biotechnology. 2013; 33: 419.

[4] E. A. Buehler and A. Mesbah. Kinetic study of acetone-butanol-ethanol fermentation in continuous culture. *PLoS One*. 2016; Submitted.

[5] D. E. Liu, C. Cerretani, R. Tellez, A. P. Scheer, S. Sciamanna, P. F. Bryan, C. J. Radke, and J. M. Prausnitz. Analysis of countercurrent membrane vapor extraction of a dilute aqueous biosolute. *AIChE Journal*. 2015; 61: 2795.

458489 A Multi-Objective Optimization Approach to Optimal Sensor Location Problem in IGCC Power Plant in the Face of Uncertainties

Abstract:

Integrated Gasification Combined Cycle (IGCC) power plants provide a cleaner and more efficient way to obtain energy from coal. In order to operate an IGCC power plant in a safe and stable manner, many input and output process parameters need to be monitored. However, due to economic and operational constraints it is infeasible to place sensors at locations pertaining to all of these parameters. Hence, it becomes important to select the most effective sensor locations which can lead us to gain maximum information about the plant conditions. Drawing on the work done by Lee & Diwekar, 2012, this work attempts to broaden the realm of the optimal sensor location problem to address simultaneous optimization of multiple objective functions. In addition to "Fisher Information", two other objective functions, viz., "thermal efficiency of the power plant" and "cost", will be considered. Practical issues present in an IGCC power plant such as harsh physical conditions and variability in process parameters make the optimal sensor location problem an especially complicated one. Advanced simulation software and data from a set of virtual sensors are used to collect information about the output parameters which do not have any physical sensors to gauge them yet. In order to solve this real world large scale problem, we use a novel algorithm called Better Optimization of Nonlinear Uncertain Systems (BONUS). BONUS works in probability distribution space and avoids sampling for each optimization and derivative calculations iterations. In order to avoid infeasibilities, we derived a new 2-tier constraint method specific to this multi-objective optimization problem. The results of this nonlinear stochastic multi-objective problem is the non-dominated or Pareto set which provides trade-offs between various objectives like observability, cost and thermal efficiency.

459218 MILP Formulation for Optimal Planning of Electric Power Infrastructure

Abstract:

Energy systems planning models allow the evaluation of alternate scenarios of future growth, providing information to support the decision-making process and the selection of technology pathways in the power sector. Although different models [1,2] have been developed to plan the expansion of generation capacity and transmission infrastructure for long-term time horizons, there are major challenges that have not been fully addressed yet, such as the aging of the infrastructure, and the spatial and temporal multi-scale detail required to adequately model the increasing contribution of intermittent renewable generation to the power grid.

In this paper we propose an optimization modeling framework to evaluate the changes in generation and transmission infrastructure required to meet the projected demand for electricity over the next few decades while taking into account detailed operational constraints (e.g. unit commitment) and the variability and intermittency of renewable generation sources. The modeling framework, which is based on mixed-integer linear programming (MILP), takes the viewpoint of a central planning entity whose goal is to identify the source (nuclear, coal, natural gas, wind and solar), type, location and capacity of future power generation technologies and transmission infrastructure that can meet the projected electricity demand while minimizing the amortized capital investment of all new generating units and transmission lines, the operating and maintenance costs of both new and existing units, and suitable environmental costs (e.g. carbon tax).

The proposed formulation is applied to a case study in the region managed by the Electric Reliability Council of Texas (ERCOT) for a 30 year planning horizon, resulting in a large-scale MILP model with more than 100,000 discrete variables and constraints. Through a combination of judicious model formulation strategies such as time sampling [3] and generator clustering [4] as well as the use of specialized decomposition strategies [3], this work demonstrates that these large-scale MILP problems can be solved in a reasonable amount of time. The results for hourly and sub-hourly level of information are compared and, in both cases, it shows that the future growth will be met by a portfolio of different generation technologies.

[1] ReEDS model by NREL http://www.nrel.gov/analysis/reeds/

[2] EPAâ€™s Power Sector Modeling https://www.epa.gov/airmarkets/power-sector-modeling-platform-v515

[3] Mitra, S., Pinto, J. M., & Grossmann, I. E. (2014). Optimal multi-scale capacity planning for power-intensive continuous processes under time-sensitive electricity prices and demand uncertainty. Part I: Modeling.Â *Computers & Chemical Engineering,*Â *65*, 89-101.

[4] Palmintier, B. S., & Webster, M. D. (2014). Heterogeneous Unit Clustering for Efficient Operational Flexibility Modeling.Â *IEEE Trans. Power Syst. IEEE Transactions on Power Systems,*Â *29*, 1089-1098.

461089 A Stochastic Optimization Approach for Improving Power Systems Resilience through Operations and Planning

Abstract:

Electricity is one of the three components of the energy sector, one of sixteen critical infrastructure sectors defined by U.S. Presidential Policy Directive 21. Resilience of the electricity grid is vital for both the economy and public safety and health. Economic operation of power systems alone is difficult. Improving the resilience of the system to a wide range of scenarios (i.e., potential future events) is even more challenging due to the inherent uncertainty in the problem. Additionally, extreme events that disable many system components must be considered as opposed to the traditional N-1 contingency analysis where the system loses one transmission line per scenario [1]. Nominal operations of the system, where the primary manipulated variables are generator setpoints, is generally not enough to make systems resilient to such events. Other operational and planning decisions, such as transmission switching or physical protection of critical components, must be considered, adding even more complexity to the problem through the introduction of binary variables [2].

In this work, we present two-stage stochastic programming formulations for minimizing the consequence of extreme weather events to the electricity grid. While the second stage decisions are made in response to a particular event, the number and flexibility of options for recourse are limited. Therefore, the first stage decisions must be made carefully to make the system resilient to all scenarios. We utilize generator dispatch, transmission switching, and physical protection of transmission lines in the first stage, and we compare the effectiveness of each, along with combinations of the three. A linearized AC transmission model is used in the optimization, resulting in a large-scale stochastic mixed-integer linear program. The effectiveness of the approach is validated with the full nonlinear AC transmission model. The optimization problem is formulated with Pyomo, a flexible, python-based optimization modeling language. The analysis is performed on a real system with thousands of buses and transmission lines, and the scenarios used for the stochastic programming problem are generated with historical outage data.

[1] Jia Kang, John Siirola, Jean-Paul Watson, and Carl Laird,Â **"Parallel Solution of Nonlinear Contingency-constrained Network Problems"**, Proceedings of, Foundations of Computer-Aided Process Design 2014 (FOCAPD), July, 2014.

[2] Emily B. Fisher, Richard P. Oâ€™Neill, and Michael C. Ferris. **"Optimal transmission switching."**Â *Power Systems, IEEE Transactions on*Â 23.3 (2008): 1346-1355.

468930 An Opportunistic Hybrid Communications System for Distributed PV Coordination and Control

Abstract:

Widespread distributed solar power penetration requires the ability to monitor and communicate with distributed photovoltaic (PV) systems on residential and commercial buildings. However, the sheer number of possible devices presents challenges for effective communications and state estimation. An opportunistic hybrid communications system has been designed that combines both dedicated and non-dedicated communications infrastructure with robust algorithms for state estimation to help ensure reliable and cost effective power system operations with millions of distributed energy resources. This work investigates the joint design of monitoring and communication protocols, with the objective of enabling the fast, ubiquitous, and reliable collection of precise and accurate PV-state-related measurements for use at the distribution system operator (DSO) and transmission system operator (TSO) levels. It results in a comprehensive view of the present state of transmission and distribution systems with high distributed PV penetrations at any given moment. A new type of communications system architecture is needed, where information from local distributed PV systems is shared through WiFi and ZigBee type wireless access networks, information is fed up to hybrid distribution networks which leverage the dedicated power system communication channels, such as LTE networks. Furthermore, the information exchange for the nationwide power delivery is realized through a hybrid core network that utilizes fiber-optic and satellite communications. To accomplish this, a robust middleware system that can overcome availability and performance issues associated with any single PV system has been designed. To allow for this information to be best utilized without sampling from millions of generators, state estimation techniques are applied to both distributed PV output, and the power system implications. The communications system described will enable the cost effective and reliable integration of millions of distributed energy resources. The overall system design, and the implications of the middleware performance, will be presented. Future work includes rigorous validation of the design through both hardware-in-the-loop testing and integrated transmission-distribution-communications system simulation.

469449 Simulation and Analysis of Power Systems Using Non-Intrusive Appliance Load Monitoring

Abstract:

With the increase in world population and constant improvement in people's living standard, world electricity demand is increasing every year. In 2013, 67.4 % of electricity was generated through fossil fuels. Electricity generation through fossil fuel is generated in bulk and is transmitted to long distances by stepping up the voltage. The conventional grid's infrastructure is made considering bulk generation from fossil fuels. But fossil fuel are finite in quantity and lead to environmental pollution. This has motivated us to shift to a grid which will have rapid monitoring and transmit great share of electricity generated from renewable energy sources (RES). Also the renewable energy sources (RES) have non uniform spatial distribution and the generation is either DC or at variable frequency. So the future grid's source will have intermittent and distributed characteristics of the load.

The future grid with high penetration of RES, electric vehicle and storage will need of an optimal approach to match generation and demand. A significant amount of total generated electricity is consumed by residential sector. The residential sector has the potential to reduce the electricity waste and shift the demand on time scale. But to know when and what to shift we need proper monitoring and modelling of residential grid. An approach to monitor the residential load using information available from smart meter connected at utility entry point in a house is called Non-intrusive Load Monitoring (NILM) [2,3] The monitoring done through smart meter requires measurement of the aggregate electricity consumed in a house at a frequency of 1 Hz or higher. Non-intrusive load monitoring (NILM) technique identifies the appliance level electricity consumption pattern from the aggregate monitored data of a house. At industrial level technique was used for real time power factor correction [1]. NILM can also identify power quality offending loads.

This paper will discuss a Non-Intrusive Load Monitoring (NILM) technique which will identify the appliances operating at any time from a house hold aggregate energy meter data. Depending on the basic electrical components, the front end circuit and the task performed by an appliance, each appliance will have distinct power consumption level, power consumption trend and transient behaviour. The proposed approach uses multiple algorithms, each one specific to a class of appliances, with active & reactive power, states of operation in an appliance, power consumption levels and other features to detect the on/off state of various appliances [3,4,5]. The proposed approach is to be tested on various household aggregate energy datasets for identification of operating state of various appliances. Results from application of this approach to various Indian households will help identify the device utilization profile along with its power consumption, a key ingredient for modelling and simulation of residential electric systems. The information extracted form NILM will help in development of bottom up models to study the demand profile with vary with time and the region specific capability of AC/DC grid.

References

[1] G. W. Hart, "Nonintrusive Appliance Load Monitoring," 1992.

[2] J. Liang, S. K. K. Ng, G. Kendall, and J. W. M. Cheng, "Load Signature Study — Part I : Basic Concept , Structure , and Methodology," *IEEE Trans. POWER Deliv.*, vol. 25, no. 2, pp. 551–560, 2010.

[3] P. Ducange, F. Marcelloni, and M. Antonelli, "A novel approach based on finite-state machines with fuzzy transitions for nonintrusive home appliance monitoring," IEEE Trans. Ind. Informatics, vol. 10, no. 2, pp. 1185–1197, 2014

[4] D. Srinivasan, S. Member, W. S. Ng, A. C. Liew, and S. Member, "Neural-Network-Based Signature Recognition for Harmonic Source Identification," *IEEE Trans. Power Deliv.*, vol. 21, no. 1, pp. 398–405, 2006.

[5] D. He, S. Member, L. Du, S. Member, Y. Yang, R. Harley, and T. Habetler, "Front-End Electronic Circuit Topology Analysis for Model-Driven Classi fi cation and Monitoring of Appliance Loads in Smart Buildings," vol. 3, no. 4, pp. 2286–2293, 2012.

471279 Effective Variance Reduction and Gradient Estimation Techniques for Stochastic Simulation and Optimization of Microgrids

Abstract:

It is well known that renewable energy resources such as wind and solar have many economic and environmental benefits over conventional energy sources. However, efficiently and reliably generating power from renewable resources is extremely challenging due to their distributed and intermittent availability. Microgrids are small, autonomous power systems that address the geographical distribution of renewable power sources by intelligently pairing local loads with locally available resources. However, managing intermittency remains a key challenge, often necessitating complex systems that must coordinate multiple generation and storage technologies in real-time. In turn, this leads to challenging design and control problems that require efficient methods for both simulating and optimizing detailed microgrid models subject to the highly stochastic behavior of the renewable generators.

This presentation concerns the problem of jointly optimizing the generator/storage capacities and dispatching rules for microgrids under uncertainties in both renewable resources and loads. In existing literature, this problem has been addressed using: (1) mixed-integer programming formulations that must assume linear component models and deterministic load/resource profiles; (2) two-stage stochastic programming formulations that do not enforce non-anticipativity; (3) multistage stochastic programming formulations that are severely limited in the number of scenarios that can be addressed, even for linear models; and (4) trial-and-error or heuristic optimization approaches applied to black-box microgrid simulation codes. The later category is very attractive because it enables the use of much higher-fidelity models than the other approaches, particularly with respect to nonlinear component models and the essential discrete behavior introduced by dispatching rules. However, the heuristic methods currently used to optimize such models suffer from slow convergence and a lack of theoretical performance guarantees.

Our recent work has been addressing these shortcomings by developing gradient-based approaches for optimizing detailed microgrid simulation models that provide theoretical convergence guarantees and exhibit much faster convergence than black-box approaches in practice. As a first step, we have recently proven conditions under which a cost function computed through a stochastic microgrid simulation is differentiable, and hence amenable to efficient gradient-based optimization. This is nontrivial because of the discrete behavior introduced by component dispatching decisions in microgrid simulations, and interestingly our results show that only the true expected value of the cost is differentiable, whereas a finite sample-average is likely to be highly discontinuous. This result lays the theoretical groundwork for a gradient-based approach, but leaves a number of significant implementation challenges to be addressed.

This presentation will address the outstanding challenge of efficiently and accurately estimating the gradient of the expected cost from a finite number of stochastic simulations. As a starting point, we consider a straightforward finite difference (FD) approach based on direct Monte Carlo sampling, and demonstrate that this leads to very high variance estimates due to artificial 'rare events' created by the differencing scheme. Compensating for this high variance requires very large numbers of samples, which significantly deteriorates the performance of our gradient-based optimization algorithm. To address this, we next present three novel sample path estimators with substantially reduced variance compared to FD. Respectively, these estimators are based on (1) a custom conditional expectation expansion for the FD estimator; (2) an automated change-of-variables technique combined with automatic differentiation (AD); and (3) a method for evaluating a derivative formula in terms of surface integrals using so-called `phantom simulations'. We discuss conditions under which each estimator is preferable to the others, and present performances comparisons for some representative microgrid examples. Our results show much lower variance estimates relative to naïve FD, leading to very substantial speed-ups in the overall optimization scheme.

473060 Operation of Grid Scale Energy Storage Systems: Comparison of Multi-Stage Stochastic Programming and Empc

Abstract:

It is widely recognized that a major concern with renewable energy is the fact that wind and solar sources are non-dispatchable. That is, the power produced from renewable sources is dependent on environmental conditions and is likely uncorrelated with the power demand from load centers. While fossil based sources are dispatchable and currently have the ability to respond to the full range of consumer loads, the additional range imposed by renewable sources is expected to exceed the dispatch capability of these fossil plants at the point of 20% renewable power. Thus, many have advocated the use of massive energy storage systems to provide the additional level of dispatch capability required to maintain grid solvency.

Due to the uncertainty of consumer demand as well as that of renewable generation, the problem of optimal placement of these storage units within the grid along with the selection of equipment sizes must be formulated as a stochastic program. However, rather than being a fairly simple two-stage stochastic program, the dynamics imposed by the storage devices requires the formulation to be of the far more challenging multistage class. Due to the computationally intense nature of multistage stochastic programming, many have advocated the use of Economic MPC. The objective of this work is to provide a head-to-head comparison of feedback policies generated from multi-stage stochastic programming and EMPC. This comparison of policies provides a validation of EMPC based equipment design procedures.

453478 A Mid-Term, Market-Based Power Systems Planning Model

Abstract:

The integration of all the European national power markets into a single one is of top priority on the European Commission's political agenda. The security of the supply was one of the key drivers in the early stages of the power market integration in Europe. Indeed, the first electricity interconnections among neighbouring countries were constructed to cover emergency situations. The advantages of the electricity markets' integration include the diversification of the power generation mix, the reduction in the electricity prices by creating more competitive and transparent markets, and the fact that it can provide access to additional power generation capacity in case of a shortage in any one country.

Moreover, the expansion of grid interconnections is of paramount importance in order to facilitate the transition towards a power generation mix with very high levels of renewables penetration and lower CO_2 emission content (Hasan et al., 2015). As a direct consequence, the increased transmission capacity enables the more efficient utilization of the flexibility of the power plant fleet by increasing the flexibility to balance the intermittent wind and solar output (Hadera et al., 2015). Not surprisingly, the deployment of renewable energy at regions with weak interconnections infrastructure and far from the main load consuming centres often results in overloading of transmission lines (Xiao et al., 2011).

In that context, this work presents a generic Mixed Integer Linear Programming (MILP) model integrating a Mid-term Energy Planning model, which implements generation and transmission system planning at a yearly level (Koltsaklis et al., 2014), with a Unit Commitment model (Koltsaklis and Georgiadis, 2015), which simulates the Day-Ahead Electricity Market. The proposed modelling framework identifies the implementation (or not) of the interconnection of specific autonomous power networks to the mainland electric system, as well as the optimum interconnection capacity. It also systematically quantifies the effects in the Day-Ahead electricity market and on the energy mix. The proposed model can provide useful insights into the strategic and challenging decisions to be determined by investors and/or policy makers at a national and/or regional level, by providing the optimal energy roadmap and management, as well as clear price signal on critical energy projects under real operating and design constraints.

Keywords: Mixed integer linear programming; Unit commitment problem; Mid-term energy planning; Electric interconnections; Power markets

References

Hadera, H., Harjunkoski, I., Sand, G., Grossmann, I. E. & Engell, S. 2015. Optimization of steel production scheduling with complex time-sensitive electricity cost. *Computers & Chemical Engineering*, 76, 117-136.

Hasan, M. M. F., First, E. L., Boukouvala, F. & Floudas, C. A. 2015. A multi-scale framework for CO2 capture, utilization, and sequestration: CCUS and CCU. *Computers & Chemical Engineering*, 81, 2-21.

Koltsaklis, N. E., Dagoumas, A. S., Kopanos, G. M., Pistikopoulos, E. N. & Georgiadis, M. C. 2014. A spatial multi-period long-term energy planning model: A case study of the Greek power system. *Applied Energy*, 115, 456-482.

Koltsaklis, N. E. & Georgiadis, M. C. 2015. A multi-period, multi-regional generation expansion planning model incorporating unit commitment constraints. Applied Energy, 158, 310-331.

Xiao, J., Hodge, B.-M. S., Pekny, J. F. & Reklaitis, G. V. 2011. Operating reserve policies with high wind power penetration. *Computers & Chemical Engineering*, 35, 1876-1885. References

466075 Identification Methodology for Stirred and Plug-Flow Reactors

Abstract:

Nowadays, optimal operation of chemical reactors is a critical element in the economics of several processes of the chemical industry as, for instance, oil refining and petrochemistry. This optimal operation can be achieved if an accurate model of the chemical kinetics is available allowing us to simulate the reactor operation. For this purpose a major issue is the identification of the reaction mechanisms and their kinetics among different suitable functional expressions, by using experimental measurements and inverse problems methodology.

The present work deals with the identification problem of kinetic models and their corresponding parameters which better fit a set of experimental data, when the reaction system is supposed to be known. As said above, the optimal kinetic models are selected from a set of allowed functional expressions. For solving this inverse problem, several techniques can be considered, such as the so-called differential, integral and incremental methods (see [1]). When dealing with a stirred tank reactor (STR) our choice is the combination of two methods in cascade: the incremental method and the integral one (see [2] or [3]). Whereas in the case of plug flow reactors (PFR) we propose a multi-start strategy and then the integral method combined with the variable neighborhood search (VNS) to move towards a better solution (see [4] or [5]). In all cases, the adjoint method is used to compute the gradient of the objective function.

Often, information on temperature is not available and then the energy equation has to be considered in the model. Moreover, when experimental data of temperature are provided, differences between calculated values and experimental ones can be included into the objective function to be minimized. In our work, we consider both situations: either the temperature is given as a data and does not change inside the reactor or it is computed from the energy equation.

Finally, the applicability and performance of the proposed methodology will be shown through different numerical simulations for STRs and for PFRs.

Acknowledgements

Part of this research was developed as an activity in the Joint Research Unit Repsol-ITMATI (code file: IN853A 2014/03) which is funded by FEDER, the Galician Agency for Innovation (GAIN) and the Ministry of Economy and Competitiveness in the framework of the Spanish Strategy for Innovation in Galicia. It was also supported by the Spanish MICINN project MTM2013-43745-R and by the Xunta de Galicia under research project GRC13-014.

References

[1] A. Bermúez, E. Carrizosa and N. Esteban. A two steps identification process in stirred tank reactors. The incremental and integral methods. In preparation.

[2] A. Bermúdez, N. Esteban, J.L. Ferrín, J.F. Rodríguez-Calo, and M.R. Sillero-Denamiel, Identification problem in plug-flow chemical reactors using the adjoint method. In preparation.

[3] N. Bhatt, N. Kerimoglu, M. Amrhein, W. Marquardt, and D. Bonvin. Incremental identification of reaction systems-A comparison between rate-based and extent-based approaches. *Chemical Engineering Science*, 84: 24-38, 2012.

[4] M. Brendel, D. Bonvin and W. Marquardt. Incremental identification of kinetic models for homogeneous reaction systems. *Chemical Engineering Science*, 61:5404-5420, 2006.

[5] D. Rodrigues, J. Billeter and D. Bonvin. Incremental Model Identification of Distributed Two-phase Reaction Systems. *IFAC-PapersOnLine*, 48(8):266-271, 2015.

468307 A Mixed-Integer Formulation for Online Design of Model Discrimination Experiments

Abstract:

The optimal design of experiments (DoE) for model discrimination and parameter estimation is a crucial component of process systems engineering; DoE is essential for characterizing reaction networks and new products, and thus constitutes the cornerstone for reactor and new process design. Of particular interest is the design of dynamic experiments, whereby the system states (and potentially experimental conditions) can change in time.

The design of dynamic experiments aims to maximize the information gained about the structure (for model discrimination) and parameters (for parameter estimation) of the system under consideration. This goal is typically formulated in terms of optimizing a metric of the Fisher information matrix (equivalent to minimizing the parameter covariance matrix) by changing the experimental set-up and conditions [1] [2] [3]. Several metrics have been explored for parameter estimation purposes [4] [5]: the determinant (D-optimal design) minimizing the volume of the joint confidence region, the smallest eigenvalue (E-optimal design) minimizing the size of the largest axis, the trace (A-optimal design), minimizing the average axis, etc.

The design of dynamic experiments for model discrimination has fewer options, with most designs requiring a set of model candidates as a starting point. The experiment is then designed to maximize the difference between model predictions at the sampled points [6] [7]. Model discrimination and parameter estimation criteria have also been combined by summing the two weighted objective functions for an experimental design that includes both objectives [8]. Previous results have generally been reported for discrimination between alternate models, where the true solution is one of the candidate models.

In this work, we propose a new dynamic DoE approach, whereby we represent the (unknown) model as a combination of submodels [9] using binary variables to determine whether the submodels contribute to the final model structure. We examine the formulation of the optimal DoE problem for this new class of models using Fisher information matrix criteria, focusing on both model discrimination and parameter estimation. Further, we extend this approach with an online adaptation strategy, with the goal of accelerating the convergence of parameter estimates [10] [11].

We illustrate our results with a set of case studies concerning, i) model discrimination for a bioreactor (where we compare our results with previous literature), and ii) the identification of the mechanism and parameters of a complex reaction system under constraints related to avoiding undesirable side reactions throughout the experimental plan.

References

[1] SP Asprey and S Macchietto. Statistical tools for optimal dynamic model building. *Computers & Chemical Engineering*, 24(2):1261–1267, 2000.

[2] GEP Box and HL Lucas. Design of experiments in non-linear situations. *Biometrika*, 46(1/2):77–90, 1959.

[3] LC Zullo. *Computer aided design of experiments: an engineering approach*. PhD thesis, Imperial College London (University of London), 1991.

[4] SP Asprey and S Macchietto. Designing robust optimal dynamic experiments. *Journal of Process Control*, 12(4):545–556, 2002.

[5] Y Zhang and TF Edgar. PCA combined model-based design of experiments (doe) criteria for differential and algebraic system parameter estimation. *Industrial & Engineering Chemistry Research*, 47(20):7772–7783, 2008.

[6] BH Chen and SP Asprey. On the design of optimally informative dynamic experiments for model discrimination in multiresponse nonlinear situations. *Industrial & Engineering Chemistry Research*, 42(7):1379–1390, 2003.

[7] GB Ferraris, P Forzatti, G Emig, and H Hofmann. Sequential experimental design for model discrimination in the case of multiple responses. *Chemical Engineering Science*, 39(1):81–85, 1984.

[8] AC Atkinson. DT-optimum designs for model discrimination and parameter estimation. *Journal of Statistical Planning and Inference*, 138(1):56–64, 2008.

[9] K Edwards, TF Edgar, and VI Manousiouthakis. Reaction mechanism simplification using mixed-integer nonlinear programming. *Computers & Chemical Engineering*, 24(1):67–79, 2000.

[10] T Barz, DC Lopez Cardenas, H Arellano-Garcia, and G Wozny. Experimental evaluation of an approach to online redesign of experiments for parameter determination. *AIChE Journal*, 59(6):1981-1995, 2013.

[11] F Galvanin, M Barolo, F Bezzo. Online model-based redesign of experiments for parameter estimation in dynamic systems. *Industrial & Engineering Chemistry Research*, 48(9):4415-4427, 2009.

Gray-box modeling of 300 mm Czochralski single-crystal Si production process

Tatsuru Seto*, Sanghong Kim*, Manabu Kano†, Toshiyuki Fujiwara‡, Masahiko Mizuta‡, and Shinji Hasebe*

Abstract

The Czochralski (CZ) process is a dominant method for manufacturing large cylindrical single-crystal ingots for the electronics industry. To understand the phenomena in CZ process and efficiently operate it, many models and control methods have been proposed, but most of them have not been applied to the industrial processes. To the authors' best knowledge, only two papers reported modeling of industrial CZ processes. Lee et al. developed linear step response models which predict the crystal radius from the crystal pulling rate and the heater temperature. Zheng et al. constructed a nonlinear first-principle model which calculates the crystal radius and the crystal growth rate from the heater input, the crystal pulling rate, and the crucible rise rate. In this study, to improve the accuracy of the model used in Zheng's research, a gray-box model is developed. Moving window partial least squares (MW-PLS) is used to develop the statistical model since it can cope with the time-varying characteristics of CZ process, which originate from changes in the crystal length and the crucible position. The developed gray-box model was applied to a 300 mm CZ single-crystal silicon production process. It was confirmed that the accuracy of the developed model was much better than that of the model proposed by Zheng et al.

1 Introduction

The single-crystal silicon ingot is a key raw material of electronic devices. The progress of electronic industries increases the demand of single-crystal silicon. In addition, to improve performance of electronic devices, the request to the quality of single-crystal silicon becomes severe. To satisfy the needs for the single-crystal silicon, it is crucial to efficiently operate the Czochralski (CZ) process, which produces more than 95 % of single-crystal silicon ingots in 2009 [1]. The CZ process produces single-crystal silicon ingots as shown in Figure 1. The polycrystalline silicon is first charged into a silica crucible. Then the charged silicon is melted by the graphite heater, and a seed crystal is brought onto the melt surface (Figure 1 left). The silicon is crystalized on the solid-liquid interface, and by pulling up the puller the single-crystal silicon ingot is grown (Figure 1 middle). During the crystallization operation, the crucible and the silicon crystal are rotated so as to unify the temperature condition around the crystal. The position of the crucible is changed to keep the relative position of melted surface constant to the heater. The single-crystal silicon ingot is finally obtained as shown in Figure 1 right.

In CZ processes, it is required to produce silicon ingot with constant radius and small number of crystal defects. The crystal radius and the number of the crystal defects are affected by many

*Kyoto University, Dept. of Chemical Engineering
†Kyoto University, Dept. of Systems Science
‡SUMCO Corporation

THE GRAPHIC ON THIS PAGE WAS NOT AVAILABLE ON THE E-MEDIA VERSION OF THIS PAPER.

Figure 1: Schematic diagram of a Czochralski process.

factors such as the crystal pulling rate, the heater input, the crystal growth rate, the meniscus shape, and the temperature. In many cases, the crystal radius and the crystal growth rate are chosen as the controlled variables since the number of crystal defects is difficult to measure during the operation, and keeping the crystal growth rate constant ensures the small number of the defects. The crystal pulling rate and the heater input are the main manipulated variables.

To properly design and operate CZ processes, many modeling and control methods have been proposed. Gevelber and Stephanopoulos [2] developed a lumped-parameter first-principle model of the heat transfer in the CZ process, and analyzed the properties of CZ process by using the developed model. Gevelber [3, 4] later designed a PID control system based on the developed model. Irizarry-Rivera and Seider [5, 6] built a model predictive control system by using the heat transfer and the meniscus shape models. Winkler and coworkers [7, 8, 9] demonstrated the control results of a laboratory-scale real CZ process by using a hydrodynamic and geometrical model, in which a differential equation for the crystal slope angle is formulated. In addition to differential-algebraic equations used in these papers, the partial differential equations are used to develop a model of the CZ process. For example, Ng and coworkers [10, 11, 12] and Abdollahi et al. [13, 14] developed distributed-parameter first-principle models that take account of the crystal growth and the heat transfer. They calculated the temperature distribution inside the crystals and derived the optimal heater input based on the developed model. However, the applicability to the industrial process were not verified in these papers. Lee et al. [15] developed linear step response models for an industrial process that produces bulk silicon crystal with the diameter of 200 and 300 mm. In the developed models, the crystal radius is calculated by the crystal pulling rate and the heater temperature. The developed model was used for batch-to-batch updating of the feedforward trajectory of the heater temperature. However, the linear models cannot describe the process nonlinearity caused by various factors such as the radiative heat transfer. In addition, the specifications of bulk silicon crystal and single-crystal silicon are significantly different. Zheng et al. [16] developed a nonlinear first-principle model which estimates the crystal radius and the crystal growth rate, and applied it to the industrial data obtained from a 300 mm CZ single-crystal silicon production process. However, they showed the results in limited range of the operation condition and limited number of the batch.

In this research, to further improve the model accuracy and validate the model in many

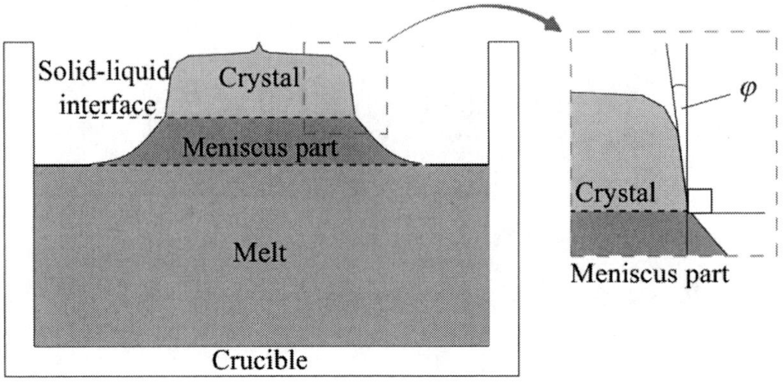

Figure 2: Definition of the crystal slope angle φ.

conditions, we developed a gray box model of the crystal radius. The statistical model is embedded into the first-principle model used in [16], and used for estimation of a key parameter of the first-principle model.

2 first-principle model of the crystal radius

Zheng et al. [16] used the first-principle model of the crystal radius proposed by Winkler et al. [7]. The first-principle model is called hydrodynamic and geometrical model, which is represented by the following equations.

$$\frac{\mathrm{d}r_{\mathrm{cry}}}{\mathrm{d}t} = v_{\mathrm{g}} \tan(\varphi) \tag{1}$$

$$\frac{\mathrm{d}\varphi}{\mathrm{d}t} = \frac{v_{\mathrm{c}} + \psi_1(r_{\mathrm{cry}}, \varphi)v_{\mathrm{g}} - v_{\mathrm{p}}}{-\psi_2(r_{\mathrm{cry}}, \varphi)} \tag{2}$$

Here, r_{cry} [m] is the crystal radius, v_{g} $[\mathrm{m} \cdot \mathrm{s}^{-1}]$ is the crystal growth rate, v_{c} $[\mathrm{m} \cdot \mathrm{s}^{-1}]$ is the crucible rise rate, v_{p} $[\mathrm{m} \cdot \mathrm{s}^{-1}]$ is the crystal pulling rate, and t [s] is the time. The crystal slope angle φ [rad] is shown in Figure 2, which determines the sign of the change in the crystal radius in Equation (1). ψ_1 [$-$] and ψ_2 [m] are the function of the crystal radius r_{cry} and the crystal slope angle φ, which are defined by the following equations.

$$\psi_1 = 1 - \frac{\rho_{\text{cry}} r_{\text{cry}}^2}{\rho_{\text{mel}} r_{\text{c(in)}}^2} + \left[\left(1 - \frac{r_{\text{cry}}^2}{r_{\text{c(in)}}^2} \right) \frac{\partial h_{\text{men}}}{\partial r_{\text{cry}}} - \frac{2 r_{\text{cry}} h_{\text{men}}}{r_{\text{c(in)}}^2} - \frac{a^2}{r_{\text{c(in)}}^2} \cos(\varphi_0 + \varphi) \right] \tan(\phi) \quad (3)$$

$$\psi_2 = \left(1 - \frac{r_{\text{cry}}^2}{r_{\text{c(in)}}^2} \right) \frac{\partial h_{\text{men}}}{\partial \varphi} + \frac{a^2 r_{\text{cry}}}{r_{\text{c(in)}}^2} \sin(\varphi_0 + \varphi) \quad (4)$$

$$h_{\text{men}} = a \sqrt{\frac{1 - \sin(\varphi_0 + \varphi)}{1 + \frac{a}{\sqrt{2} r_{\text{cry}}}}} \quad (5)$$

$$a = \sqrt{\frac{2\sigma_{\text{men}}}{\rho_{\text{mel}} g}} \quad (6)$$

$$\frac{\partial h_{\text{men}}}{\partial r_{\text{cry}}} = \frac{a^2 (1 - \sin(\varphi_0 + \varphi))^{1/2}}{2^{3/4} r_{\text{cry}}^{1/2} (\sqrt{2} r_{\text{cry}} + a)^{3/2}} \quad (7)$$

$$\frac{\partial h_{\text{men}}}{\partial \varphi} = -\frac{a r_{\text{cry}}^{1/2} \cos(\varphi_0 + \varphi)}{2^{3/4} (\sqrt{2} r_{\text{cry}} + a)^{1/2} (1 - \sin(\varphi_0 + \varphi))^{1/2}} \quad (8)$$

Here, ρ_{cry} $[\text{kg} \cdot \text{m}^{-3}]$ is the density of silicon, ρ_{mel} $[\text{kg} \cdot \text{m}^{-3}]$ is the density of the melt, $r_{\text{c(in)}}$ [m] is the inner diameter of the crucible, h_{men} [m] is the meniscus height, a [m] is the Laplace constant, σ_{men} $[\text{N} \cdot \text{m}^{-1}]$ is the surface tension of the meniscus, φ_0 [rad] is the wetting angle, and g $[\text{m} \cdot \text{s}^{-2}]$ is the acceleration of gravity. ψ_1 is always positive and ψ_2 is always negative.

3 Gray-box model of the crystal radius

In the hydrodynamic and geometrical model, the height of meniscus part h_{men} is calculated by Equation (5), which is the approximated solution of the Euler-Laplace equation. Although the Euler-Laplace equation assumes the steady-state, this assumption may no be satisfied since the crucible position and the crystal growth rate change during a batch. To cope with this problem, the desirable value of ψ_1 ($\psi_{1,\text{opt}}$) is derived by a data-based method, then, $\psi_{1,\text{opt}}$ and the operation condition is correlated with a statistical model. The statistical model is embedded into the first-principle model

The schematic diagram of the developed gray-box model is shown in Figure 3, and it can be represented by the following equations.

$$\frac{\mathrm{d} r_{\text{cry}}}{\mathrm{d} t} = v_{\text{g}} \tan(\varphi) \quad (9)$$

$$\frac{\mathrm{d} \varphi}{\mathrm{d} t} = \frac{v_{\text{c}} + \hat{\psi}_{1,\text{opt}} v_{\text{g}} - v_{\text{p}}}{-\psi_2(r_{\text{cry}}, \varphi)} \quad (10)$$

$$\hat{\psi}_{1,\text{opt}} = f(v_{\text{p}}, v_{\text{c}}, P, h_{\text{mel}}, \Delta r_{\text{cry}}) \quad (11)$$

The statistical model estimates $\psi_{1,\text{opt}}$ from the crystal pulling rate v_{p}, the crucible rise rate v_{c}, the heater input P, the height of meniscus part h_{mel} and the deviation Δr_{cry} of the crystal radius from its set-point. $\psi_{1,\text{opt}}$ is expected to be time-varying, thus it is calculated for every sub data shown in Figure 4 by using the following equations.

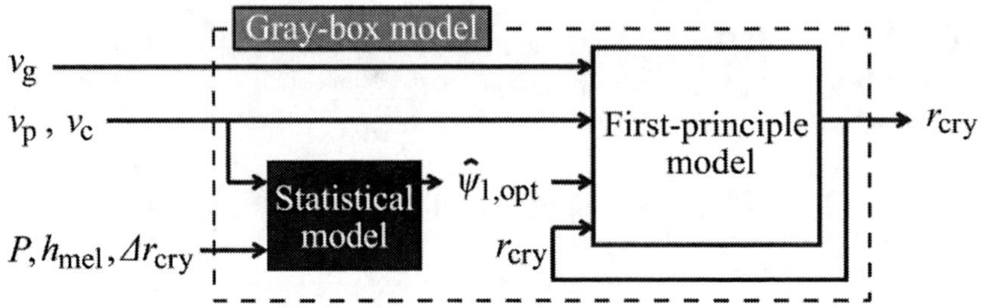

Figure 3: Schematic diagram of the developed gray-box model.

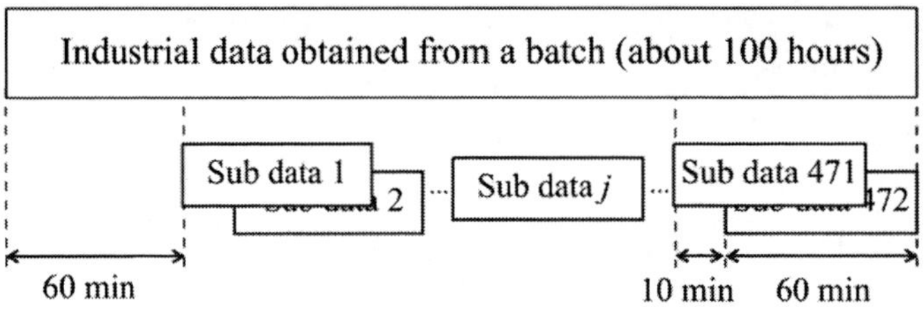

Figure 4: The definition of the sub data.

$$\psi_{1,\mathrm{opt}} = \arg \min_{\tilde{\psi}_1} J \tag{12}$$

$$J = \sum_{n=0}^{60} (r_{\mathrm{cry,calc}}(n\Delta t) - r_{\mathrm{cry,meas}}(n\Delta t))^2 \tag{13}$$

$$\frac{\mathrm{d}r_{\mathrm{cry}}}{\mathrm{d}t} = v_{\mathrm{g}} \tan(\varphi) \tag{14}$$

$$\frac{\mathrm{d}\varphi}{\mathrm{d}t} = \frac{v_{\mathrm{c}} + \tilde{\psi}_1 v_{\mathrm{g}} - v_{\mathrm{p}}}{-\psi_2(r_{\mathrm{cry}}, \varphi)} \tag{15}$$

Here, the subscripts "calc" and "meas" denote the calculated and measured values. J means the estimation error of the gray-box model, and Δt is the sampling interval, which is 1 minute in this study. To construct a statistical model, moving window PLS (MW-PLS) is used to cope with the time-varying characteristics of the CZ process. In this study, 50 most similar sets of the inputs ($v_{\mathrm{p}}, v_{\mathrm{c}}, P, h_{\mathrm{mel}} and \Delta r_{\mathrm{cry}}$) and the output $\psi_{1,\mathrm{opt}}$ are selected. The similarity is calculated on the basis of the crystal length corresponding to each sub data and the operation condition for which the estimation of $\psi_{1,\mathrm{opt}}$ is required. The number of the latent variables is determined as 2 by trial and error.

4 Industrial Application

To validate the developed gray-box model, 6 batches of industrial data is prepared. 2 batches data is used for model construction, and the other 4 batches data is used for model validation. The model validation data was divided into 1 hour sub data, and the estimation was conducted for each sub data, separately. Figure 5 shows the scaled estimation error in 4 validation data. The dimensionless crystal length L^* means the progress of the batch. When L^* is 0, the crystal growth is beginning, and it finishes when L^* is 1. In Figure 5, the estimation error of the gray-box model is smaller than that of the hydrodynamic and geometrical model. The gray-box model estimates the crystal radius through the batch while the hydrodynamic and geometrical model failed to estimate the crystal radius when $L^* \geq 0.8$. This result shows that the developed model can cope with the time-varying characteristics of the CZ process. In addition, it may cope with batch-to-batch variability since the estimation performance in every batch is small.

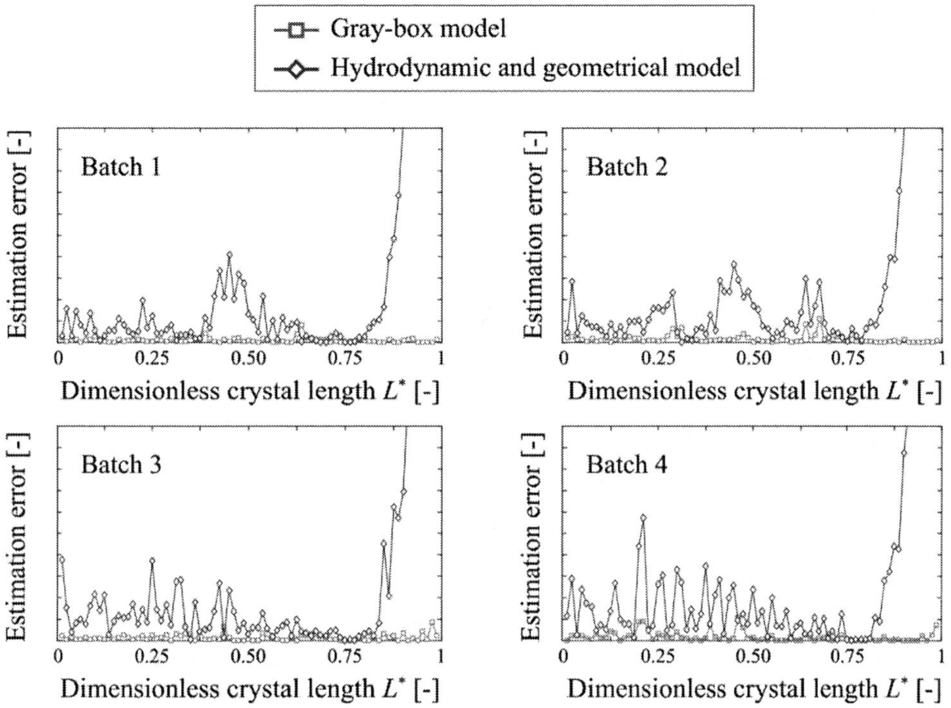

Figure 5: Estimation error of the crystal radius. Batch 1 (top-left), batch 2 (top-right), batch 3 (bottom-left), and batch 4 (bottom-right).

5 Conclusions

In this research, a gray-box model of a 300 mm Czochralski single-crystal silicon production process is developed to estimate the crystal radius, which is a important quality variable of single-crystal silicon. A statistical model is embedded into an existing first-principle model to improve the accuracy of the first-principle model. The accuracy of the gray-box model and the first-principle model was compared by using industrial data. The result showed that the gray-box model is superior to the first-principle model, and the accuracy of the gray-box model is acceptable in different batches and processing times.

References

[1] L. Wu. *Modeling and simulation of Czochralski bulk crystal growth process: investigation of transport effects in melt and gas phases.* VDM Publishing, 2009.

[2] M. A. Gevelber and G. Stephanopoulos. Dynamics and control of the Czochralski process I. Modelling and dynamic characterization. *J. Cryst. Growth*, 84:647–668, 1987.

[3] M. A. Gevelber. Dynamics and control of the Czochralski process III. Interface dynamics and control requirements. *J. Cryst. Growth*, 139:271–285, 1994.

[4] M. A. Gevelber. Dynamics and control of the Czochralski process IV. Control structure design for interface shape control and performance evaluation. *J. Cryst. Growth*, 139:286–301, 1994.

[5] R. Irizarry-Rivera and W. D. Seider. Model-predictive control of the Czochralski crystallization process part I. Conduction-dominated melt. *J. Cryst. Growth*, 178:593–611, 1997.

[6] R. Irizarry-Rivera and W. D. Seider. Model-predictive control of the Czochralski crystallization process part II. Reduced-order convection model. *J. Cryst. Growth*, 178:612–633, 1997.

[7] J. Winkler, M. Neubert, and J. Rudolph. Nonlinear model-based control of the Czochralski process I: Motivation, modeling and feedback controller design. *J. Cryst. Growth*, 312:1005–1018, 2010.

[8] J. Winkler, M. Neubert, and J. Rudolph. Nonlinear model-based control of the Czochralski process II: Reconstruction of crystal radius and growth rate from the weighing signal. *J. Cryst. Growth*, 312:1019–1028, 2010.

[9] M. Neubert and J. Winkler. Nonlinear model-based control of the Czochralski process III: Proper choice of manipulated variables and controller parameter scheduling. *J. Cryst. Growth*, 360:3–11, 2012.

[10] J. Ng and S. Dubljevic. Optimal control of convection-diffusion process with time-varying spatial domain: Czochralski crystal growth. *J. Proc. Cont.*, 21:1361–1369, 2011.

[11] J. Ng, S. Dubljevic, and I. Aksikas. Model predictive control of Czochralski crystal growth process. *19th Mediterranean Conference on Control and Automation, Corfu, Greece*, pages 825–831, 2011.

[12] J. Ng and S. Dubljevic. Optimal boundary control of a diffusion-convection-reaction PDE model with time-dependent spatial domain: Czochralski crystal growth process. *Chem. Eng. Sci.*, 67:111–119, 2012.

[13] J. Abdollahi, M. Izadi, and S. Dubljevic. Model predictive temperature tracking in crystal growth processes. *Comput. Chem. Eng.*, 71:323–330, 2014.

[14] J. Abdollahi, M. Izadi, and S. Dubljevic. Temperature distribution reconstruction in Czochralski crystal growth process. *AIChE J.*, 60:2839–2852, 2014.

[15] K. Lee, D. Lee, J. Park, and M. Lee. MPC based feedforward trajectory for pulling speed tracking control in the commercial Czochralski crystallization process. *Int. J. Cont. Autom.*, 3:252–257, 2005.

[16] Z. Zheng, T. Seto, S. Kim, M. Kano, T. Fujiwara, M. Mizuta, and S. Hasebe. Development of a first principle model of czochralski process and its verification with real industrial data. *APCChE 2015, Melbourne, Australia*, 2015.

470100 On the Identification of Meta-Models for the Optimization of Grade Transition in Polymerization Processes

Abstract:

An accurate mathematical model is essential in optimizing the grade transition in polymerization processes. Due to the complexity of the processes, large-scale knowledge-driven models consisting of Differential Algebraic Equations (DAE) have been developed to describe the process behaviours [1, 2]. Industrial models are expected to be more complex than the ones in the open literature. The grade transition problem is usually solved using model-based optimization approaches [3, 4] explicitly utilizing the available knowledge-driven models. There are two major limitations in this approach. Firstly, it might require substantial computational resources, especially when the detailed knowledge-driven models are very complex and when one needs to simultaneously schedule multiple plants and multiple grade transitions. Secondly, there is always a model-plant mismatch that needs to be taken into account. Adapting the highly complex and nonlinear knowledge-driven model through parameter re-estimation to match the plant 100% does not appear to be an easy task.

We approach the problem by first developing a meta-model of the detailed knowledge-driven model and quantify its performance in the grade transition optimization problem, from the prediction accuracy and the needed computational time points of view. A set of systematically selected *in silico* experiments, using the Design of Experiments approach [5, 6], is used to generate data for estimating the parameters of the Response Surface Model, the meta-model. The parameter estimation task aims to match the detailed model's predictions to those of the meta-model in the desired regions of operation. Analysis of variance concepts are used to assess the accuracy of the meta-model.

The optimal operating conditions, depending on the Hydrogen feeding profile, and leading to minimum amount of the off-spec product are determined using the detailed knowledge-driven model and its meta-model approximation. We compare the meta-model predictions against those of the knowledge-driven model concerning the amount of off-spec product and the needed time for transition. This is done in six grade transition examples covering the entire range of desired polymer grades. In all six transitions, the differences between the meta-model predictions and the knowledge-driven model's calculations are within 3%. More importantly, the computational time for the optimization using the meta-model is, on the average, 50 times smaller than the computational time required for solving the same problem using the knowledge-driven model. We also discuss how such a meta-model can be complemented to fit the plant's behaviour more accurately utilizing historical data on the expected and achieved plant behaviours.

References

1. Zacca, J.J. and W.H. Ray, *Modeling of the liquid-phase polymerization of olefins in loop reactors.* Chemical Engineering Science, 1993. **48**(22): p. 3743-3765.

2. Dunnebier, G., et al., *Optimization and control of polymerization processes.* Chemical Engineering & Technology, 2005. **28**(5): p. 575-580.

3. Shi, J., L.T. Biegler, and I. Hamdan, *Optimization of grade transitions in polyethylene solution polymerization processes.* Aiche Journal, 2016. **62**(4): p. 1126-1142.

4. Cervantes, A.M., et al., *Large-scale dynamic optimization for grade transitions in a low density polyethylene plant.* Computers & Chemical Engineering, 2002. **26**(2): p. 227-237.

5. Box, G.E.P. and N.R. Draper, *Response Surfaces, Mixtures, and Ridge Analysis.* 2007, Hoboken, NJ: Wiley.

6. Montgomery, D.C., *Design and Analysis of Experiments.* 8 ed. 2013, New York: Wiley. 729.

459085 Error-Triggered on-Line Model Identification for Model-Based Feedback Control

Abstract:

Empirical models obtained from model identification techniques are used in the chemical process industry when a first-principles model is unknown or too complex to use. Once a process model is obtained, it can be used in model-based control strategies like model predictive control (MPC) [1]-[2]. Many methods for the identification of linear empirical models based on input/output data exist such as the canonical variate algorithm (CVA) [3] and the multivariable output error state-space algorithm (MOESP) [4]. Linear empirical models can be used to model nonlinear process dynamics within a limited region. If plant variations occur or the process state moves out of the region for which the linear empirical model was valid for the nonlinear process, the linear model may need to be updated.

In this work, we propose a prediction error metric that can be used to trigger on-line model identification to update a linear empirical model for use in model-based control. A moving horizon error detector is developed based on the total relative error between the measured states and the states predicted using the current linear empirical model throughout the time period corresponding to the length of the moving horizon. Once a threshold value of the relative error metric is exceeded, model re-identification is triggered using the most recent input/output data, and the model-based controller is updated to include the new model. This prevents constant updating of the model and the controller, but allows for updates when the model no longer captures the nonlinear dynamics of the process. The method is demonstrated using Lyapunov-based economic model predictive control [5] and shown to improve the profit and to reduce plant-model mismatch compared to using one linear empirical model for the entire period of operation both for the case of plant variations (modeled through catalyst deactivation) and an expansion of the region of process operation.

[1] Alanqar A, Ellis M, Christofides PD. Economic model predictive control of nonlinear process systems using empirical models. AIChE Journal. 2015;61:816-830.

[2] Qin SJ, Badgwell TA. A survey of industrial model predictive control technology. Control Engineering Practice. 2003;11:733-764.

[3] Larimore WE. Canonical variate analysis in identification, filtering, and adaptive control. In: Proceedings of the 29th IEEE Conference on Decision and Control. Honolulu, HI, 1990;596-604.

[4] Verhaegen M, Dewilde P. Subspace model identification Part 1. The output-error state-space model identification class of algorithms. International Journal of Control. 1992;56:1187-1210.

[5] Heidarinejad M, Liu J, Christofides PD. Economic model predictive control of nonlinear process systems using Lyapunov techniques. AIChE Journal. 2012;58:855-870.

459702 A MINLP Approach to Model-Based Data Mining for the Quick Development of Nonlinear Dynamic Models

Abstract:

Simple and reliable phenomenological models always represent an attractive and powerful instrument in several chemical and biological industrial processes. A trustworthy model can potentially predict the response of a system outside the investigated range of experimental conditions and can be fruitfully exploited for the purposes of process design and non-empirical process optimization. Following the seminal work by Box and Lucas [1] a number of works appeared in the scientific literature regarding the discrimination among candidate models and the precise identification of the model parameters through model-based design of experiments (MBDoE) techniques for model discrimination and parameter precision (PP) in nonlinear dynamic systems [2,3]. However, the application of these techniques always starts from the availability of an existing set of candidate models while direct guidelines for subsequent model improvement are not obvious. A well-established systematic technique for quick model development and enhancement has not been proposed yet.

A model may be affected by two types of weaknesses: *i*) a structural weakness, intrinsically associated to the mathematical structure of the equations and to model identifiability [4]; *ii*) a descriptive weakness (i.e. the model can be weak on representing the system under certain experimental conditions because of an incorrect or incomplete set of equations). In this work a new approach is presented to guide the improvement of the model through the introduction of a method for the automated detection of the second-type weaknesses. The proposed method is based on the solution of a MINLP problem whose aim is the maximization of the likelihood function [5] acting on the following variables: *i*) the set of parameters of the candidate model (parameters are treated as continuous variables); *ii*) a set of user-defined binary variables. The user-defined binary variables act like switchers, including or removing experimental data from the parameter estimation problem, on single or groups of measurements (e.g. one could be interested in evaluating the model's capability in the prediction of certain experimental conditions grouping the measurements gathered in the same experiment under the same switcher). The method works as a model-based data mining (MBDM) filter for parameter estimation (PE) simultaneously removing the experimental results that the model is not able to fit taking into account the measurement uncertainty. It allows for a quick, automated mapping in the space of experimental conditions in terms of good and bad model predictive capabilities, preventing the thoughtless use of fake optimal process points located in regions where the model is not reliable. The presented MBDM-PE technique is proposed as part of a wider framework for a systematic approach to model identification which is synthetically shown in Figure 1a. The procedure starts with a candidate model and preliminary statistical information on measurement errors. The distribution of the residuals associated to the measurements removed by the MBDM-PE filter identifies the main weaknesses of the proposed model, highlighting both critical experimental conditions and specific measured variables whose associated prediction is very poor. This provides a feedback to the gradual improvement of the model itself. Once the enhanced model is known to give a satisfactory description of the phenomenon in the investigated experimental conditions, statistically unsatisfactory parameter estimates are amended performing additional experiments designed through known MBDoE methodologies for improving the parameter precision [1,2,3].

The MBDM-PE filtering technique has been successfully applied in a case study on the identification of a simplified reaction mechanism for the partial oxidation of methanol to formaldehyde over a silver catalyst in micro-reactor devices [6]. The investigation has been carried out adopting the optimization tool implemented in gPROMS Model Builder. The filter has automatically identified the poor predictive capabilities offered by a candidate kinetic model in the low temperature region of the experimental design space as shown in Figure 1b where the filtered (removed) experimental data have been highlighted with black circles. The underestimation of the conversion of methanol and the overestimation of selectivity for the formaldehyde indicate a parallel reaction of complete oxidation for methanol, non-negligible at low temperatures.

Future works will focus on the development of meaningful methods for assessing the model reliability in non-investigated intermediate experimental conditions and systematic approaches for model building, given the model structure, for the quick, automated identification of phenomenological models.

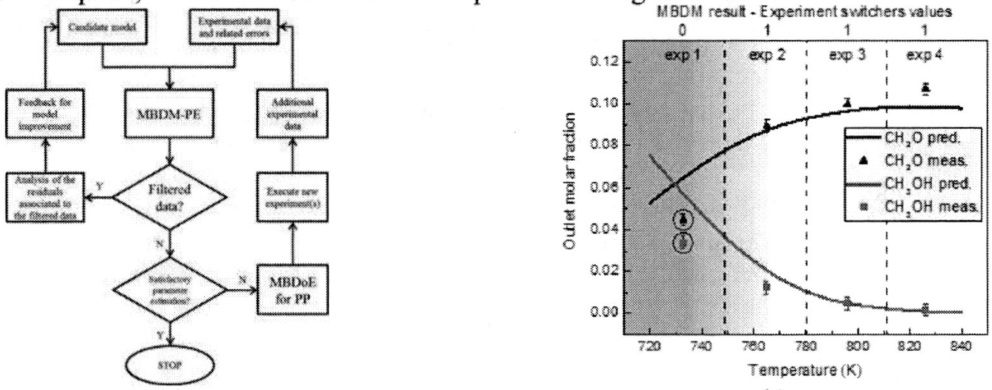

(a) (b)

Figure 1. (a) Systematic approach for model identification implementing the MBDM-PE filtering technique; (b) Results given by the MBDM filter applied to a simplified kinetic model for partial methanol oxidation. Solid lines represent model predictions; scattered triangles and squares represent measurements collected in four experiments at different temperature. The experiment switchers in the upper part of the graph represent the final values evaluated by the solver for the user-defined binary variables indicating which experiments have been included in the PE problem (1) and which have been removed (0). Removed experimental data are also highlighted with black circles. The grey-coloured side in the plot indicates the region of low model reliability.

References

[1] G.E.P. Box, H.L. Lucas (1959). Design of experiments in non-linear situations, Biometrika, 46, 77-90.

[2] D. Espie, S. Macchietto (1989). The optimal design of dynamic experiments. AIChE Journal , 35(2), 223-229.

[3] F. Galvanin, S. Macchietto and F. Bezzo (2007). Model-based design of parallel experiments. Ind. Eng. Chem. Res., 46, 871-882.

[4] F. Galvanin, C.C. Ballan, M. Barolo and F. Bezzo (2013). A general model-based design of experiments approach to achieve practical identifiability of pharmacokinetic and pharmacodynamic models. J. Pharmacokinet. Pharmacodyn., 40, 451-467.

[5] Yonathan Bard (1974), *Nonlinear parameter estimation*, ACADEMIC PRESS, New York and London.

[6] F. Galvanin, E. Cao, N. Al-Rifai, V. Dua and A. Gavriilidis (2015). Optimal design of experiments for the identification of kinetic models of methanol oxidation over silver catalyst. Chemistry Today, 33(3), 51-56.

463592 Parameter Estimation and Model Discrimination of Batch Solid-Liquid Reactors

Abstract:

Heterogeneous solid-liquid reactions are of great significance in metallurgical, agrochemical and pharmaceutical processes. This kind of industrial process is usually conducted in batch reactors, generating fine chemicals with high profits. Hence, understanding the mechanism and constructing qualified models are desired to facilitate process optimization and controller design. Motivated by this, we explore a model building exercise for a real industrial process that involves multiple solid-liquid reactions in a series of two batch reactors. Key activities are to elucidate the kinetic and transport models, estimate model parameters with limited plant data, and validate the models.

For our target application, the batch reactor operates sequentially through steps such as preparing, feeding, reacting, impurity removing, sampling and discharging. In the reacting step, solid particles react with liquid reactants and generate solid and/or liquid products. The kinetic mechanism is complicated through coupled mass and heat transfer and different models have been proposed by previous researchers, including two particular mechanisms -- solid surface reaction models [1] and dissolution models [2] with insoluble solid component reactions taking place on the interface and products sticking on or cracking away from the reaction surface. On the other hand, if solid components dissolve into the solvent, even in trace amounts, reactions may occur in liquid phase and the dissolution rate influences the production rate.

It is well known that model discrimination is difficult and requires carefully designed experiments. However, in this case only limited plant data are available to identify the model and uncertainties in the reaction process, such as solid particle sizes and shapes, must be resolved. Moreover, simple models with assumptions of uniform size and ideally nonporous spherical particles cannot fit data well. To deal with these issues, a generalized model is applied for both surface reaction and dissolution-controlled mechanisms. The particle morphology is considered similar to that described in [1] by introducing a shape factor. A particle size distribution is also implemented. This model requires estimation of several parameters, including shape factors, activation energies and diffusion coefficients. To enhance the model estimability, a number of parameters are lumped in the model.

Based on the above process characteristics, a dynamic model is derived with multiple stages. The parameter estimation is solved in an 'Errors-in-Variables-Measured' (EVM) optimization formulation [3]. Orthogonal collocation on finite elements is applied to discretize the problem, resulting in a large scale nonlinear programming problem. The problem is formulated in AMPL and solved by a nonlinear optimization solver IPOPT. The estimation performance is evaluated by calculating confidence intervals at the optimal solution. In our case, only reactor temperature profiles and end-point compositions are measured, with no concentration data available during reactions. Results show good fitting on process data and small confidence intervals, which indicates good model parameter selection and estimation. These results lead us to elucidate the likely mechanistic model for subsequent optimization studies.

References

[1] TapioSalmi, Henrik Grénman, Johan Wärnå, and Dmitry Yu Murzin. New modelling approach to liquid–solid reaction kinetics: From ideal particles to real particles. *ChemicalEngineering Research and Design*, 91(10):1876–1889, 2013.

[2] Claire L.Forryan, Oleksiy V.Klymenko, ColinM. Brennan, and Richard G. Compton. Heterogeneous kinetics of the dissolution of an inorganic salt, potassium carbonate, in an organic solvent, dimethylformamide. *The Journal of Physical Chemistry B*, 109(16):8263–8269, 2005.

[3] Victor M.Zavala, and Lorenz T. Biegler. "Large-scale parameter estimation in low-density polyethylene tubular reactors." *Industrial & engineering chemistry research* 45.23 (2006): 7867-7881.

462285 Development of a Model for a Continuous Ultra-Filtration System

Abstract:

Due to the wide applicability and simplicity of the ultra-filtration process, it is currently being used in a variety of commercial processes for the purpose of separation and concentration of valuable products and/or recovery of raw materials from dilute systems [1]. A predictive model for a system derived from first principles, is instrumental in evading the costs of conducting time-consuming experiments while also allowing one to not be dependent on a trial and error analysis approach. The validated final model can serve to understand the operational issues of the process and from it identify the important phenomena occuring in the process and/or the corresponding model parameters. Also, it can help to design experiments to collect useful measured data.

Therefore, a model was systematically developed in four phases: define the system to be modelled; derive the model equations representing the system; analyze and numerically solve the model equations; calibrate, validate and apply the model for different studies [2]. The objectives for the model are that it should be able to predict the feed and retentate flow-rates along with the permeate flux from a continuous ultra-filtration (UF) system consisting of different geometry (and 'N' membrane stacks) used for the concentration of enzyme solutions from a known inlet concentration to a desired (target) outlet concentration during the recovery and/or downstream processing of enzymes. The envisaged purpose of this model is to improve the efficiency of the UF separation process. The final model, saved as a model object, should be possible to use for the specified objectives as well as model parameter sensitivity, importance of the involved phenomena (mass transfer, permeability, etc.). Also, the model object could be used in an external simulation environment to guide process operators to run the system under different operaional scenarios, such as operate at flow-rates specifically needed to take specified feed solutions to the desired outlet concentrations; or, to optimise the feed concentration to complete the process in a fixed time and many more. This could reduce the energy requirements when compared to running the process at the maximum pump flow-rate.

The developed steady-state model equations are organized in terms of conservation equations, constitutive equations and connection equations. Changing the process geometry changes the conservation and connection equations. Chaning the model assumptions and/or the chemical system or involved phenomena, changes the constitutive equations. Therefore, from a master generic model, different versions of problem specific models are generated and used for different model scenarios. For example, different candidate models for filtration through UF membranes, which could describe the possible phenomenon occurring during the UF process (for example concentration polarization, pore blocking, cake formation, adsorption on the membrane) [3] have been generated and their performance compared and evaluated by using different sets of constitutive equations in the model. The connection equations were formulated to relate the retentate and permeate concentrations by the retentivity of the membrane.

Among the different models studied, the perofrmance of a modified gel polarisation model was identified and was found to yield good results. Experimental data from the literature was used to estimate the parameters of the model, i.e., the mass transfer coefficients and an operational parameter, the gelling concentration, could be estimated for each enzyme using a least-squares fit of the objective function. The mass transfer coefficient from theoretical relations is known to be a function of diffusion coefficient, velocity of flow and the membrane spacer dimensions [4]. The diffusion coefficient, the most sensitive parameter of this model, was found to be a function of the feed concentration or the gelling concentration of the enzyme, depending on the operational scenario. This result also confirmed observations reported in published papers on this subject [5]. Considering that, the velocity of flow and the membrane spacer dimensions are almost constant throughout the system, the mass transfer coefficient was found to vary only with the feed concentration. Lastly, the model was validated against experimental data and can predict the flow-rates within a difference of +-20% of the experimental data.

The presentation will provide details of the developed model; the modelling tool used to develop, analyse and solve the model equations; estimation and sensitivity analysis of important model parameters; and the comparison and validation of the model bevahior with published data.

References

[1] Trettin, D. and Doshi, M. (1980). Limiting flux in ultrafiltration of macromolecular solutions, *Chemical Engineering Communications*, 4(4-5), pp.507-522.

[2] Heitzig, M., Sin, G., Sales-Cruz, M., Glarborg, P. and Gani, R. (2011), Computer-Aided Modeling Framework for Efficient Model Development, Analysis, and Identification: Combustion and Reactor Modeling. *Industrial & Engineering Chemistry Research*, 50(9), pp.5253-5265.

[3] Van den Berg, G. and Smolders, C. (1990), Flux decline in ultrafiltration processes, *Desalination*, 77, pp.101-133.

[4] Baker, R. (2004). *Membrane technology and applications*. Chichester: J. Wiley.

[5] Van den Berg, G., Rácz, I. and Smolders, C. (1989). Mass transfer coefficients in cross-flow ultrafiltration. *Journal of Membrane Science*, 47(1-2), pp.25-51.

469700 Combined Use of the Sample Average Approximation and the Analytic Hierarchy Process to Support the Design of Chemical Processes Under Uncertainty

Abstract:

Chemical process design is typically performed under the assumption of fixed operating parameters (temperature, pressure, composition, flow rate, etc.) at nominal conditions. During plant operation, however, it is likely that the process will undergo changes in the operating parameters, thereby affecting the operability, controllability and feasibility of the design. Therefore, it is necessary to introduce some degree of flexibility in the design that will ensure that the process will be able to accommodate deviations from the nominal state during its operation. Flexibility considerations are typically incorporated in the conceptual design based on the engineersâ€™ experience and judgment and usually relying on empirical overdesign factors. These approaches may lead to significant extra costs and unintended consequences.

Published research in the process systems engineering has addressed both flexible process design and design under uncertainty. The former methods rely on flexibility indices that are difficult to implement in practice when dealing with complex systems and do not consider probabilistic information, while the latter may lead to large-scale stochastic programming models containing many scenarios and leading to large CPU times. In addition, the overwhelming majority of methods within the second group tend to maximize the expected value of the cost/profit distribution, thereby providing no control on the potential economic outcomes in every plausible scenario. To overcome these limitations, this paper introduces a novel methodology for the synthesis of chemical processes under uncertainty that combines the sample average approximation [1] and the analytic hierarchy process [2]. The approach presented comprises four main steps: i) uncertainty characterisation and generation of scenarios using sampling methods [3]; ii) process synthesis based on the sample average approximation algorithm to generate feasible potential designs, iii) evaluation of each design in the uncertain parameters space using both feasibility and probabilistic metrics to assess the performance of the designs; and iv) application of the analytic hierarchy process to identify the process design(s) that best reflects the decision-makersâ€™ preferences.

The capabilities of the proposed methodology are illustrated through its application to two problems: the design of heat exchanger networks and the optimisation of a crude oil distillation system of a petroleum refinery, considering uncertainties in stream inlet conditions. The former is originally formulated as a monolithic equation-oriented mixed-integer nonlinear programming problem, while the latter is implemented in a simulation-optimisation environment that combines rigorous model in Aspen HYSYS with an external optimiser coded in MatLab. The results obtained demonstrate that the proposed methodology is capable of identifying solutions that are flexible and operable under a wide range of operating conditions, and which shows better overall expected performance than the nominal design. One major advantage of the proposed method is that, compared to approaches based on flexibility indices, it can be easily implemented in process simulators without the need to write the mass, energy balances and thermodynamic equations in an explicit manner (a step required to derive the KKT conditions of the problem and determine the flexibility index).

References

[1] Kleywegt, A. J., Shapiro, A. & Homem-De-Mello, T. (2001). The sample average approximation method for stochastic discrete optimization. *Siam Journal on Optimization,* 12(2), 479-502.

[2] Saaty, T. L. (1990). How to Make a Decision - the Analytic Hierarchy Process. *European Journal of Operational Research,* 48(1), 9-26.

[3] Diwekar, U. M. & Kalagnanam, J. R. (1997). Efficient sampling technique for optimization under uncertainty. *Aiche Journal,* 43(2), 440-447.

470034 Using Resilience Principles for Prediction of Loss of Containment Events in Batch Operations

Abstract:

Increasing process safety and risk management challenges in the process industry and change in the perception of public about hazards and risk globally have necessitated exploring tools for efficient risk management [3]. Two types of factors: 1) technical (e.g., equipment malfunction, process parameter variation), and 2) social (e.g., regulations/policy, human and organizational factors) are important in assessing and managing risk of a process system. This need calls for the development of a holistic and integrated systems framework for risk management. The application of the resilience engineering perspective is gradually being explored as an approach for considering the dynamics of socio-technical aspects based on systems theory [2,4]. The resilience methodology emphasizes non-linear dynamics, consideration of new types of threats, uncertainty, and recovery from upset or catastrophic situations [1].

This paper presents a novel framework for incorporating both technical and social factors in an integrated approach - Process Resilience Analysis Framework and its four aspects: Early Detection, Error Tolerant Design, Plasticity and Recoverability. A combined framework for predictability, survivability and recoverability dynamic analysis is introduced with resilience metrics [4,5]. This work establishes and presents typical scenarios of Loss of Containment (LoC) events and resilience metrics for batch plant operations. The paper concludes with a discussion of the proposed integrated methodology for prediction of LoC events based on a mathematical model that explicitly accounts for process variations and safety constraints [6]. A detailed case study is utilized to illustrate the key ideas and a design optimization approach to obtain safer operational region at maximum average profit.

References

[1] Hollnagel, E. (2006). Resilience: the challenge of the unstable.

[2] Hollnagel, E., Nemeth, C. P., & Dekker, S. (2008). Resilience Engineering Perspectives, Volume 1: Remaining sensitive to the possibility of failure.

[3] Steen, R., & Aven, T. (2011). A risk perspective suitable for resilience engineering. Safety science, 49(2), 292-297.

[4] Jackson, S. (2009). Architecting resilient systems: Accident avoidance and survival and recovery from disruptions (Vol. 66). John Wiley & Sons.

[5] Dinh, L. T., Pasman, H., Gao, X., & Mannan, M. S. (2012). Resilience engineering of industrial processes: principles and contributing factors. Journal of Loss Prevention in the Process Industries, 25(2), 233-241.

[6] Thomaidis, T. V., & Pistikopoulos, E. N. (1995). Optimal design of flexible and reliable process systems. Reliability, IEEE Transactions on, 44(2), 243-250.

470221 A Demand Response Strategy for Continuous Processes Using Stochastic Optimal Scheduling

Abstract:

Renewable energy is dependent upon chaotic weather patterns beyond any engineerâ€™s control, increasing both the uncertainty and the variability of electricity produced from renewable energy sources. The equipment required to harness renewable energy also tends to have a high capital cost when built on a large scale, but individual consumers can often afford small scale wind turbines or solar panels themselves, leading to distributed energy generation. These distributed energy resources not only reduce the total amount of electricity sold from the grid, but also make the demand for grid-sourced electricity spike exactly in sync when any renewable electricity production owned by the grid plummets. An undesired but unavoidable result which comes as renewable energy sources are adopted is the increased difficulty of maintaining the 60 HZ frequency of the grid. It has already become impossible for suppliers to do entirely alone; consumers must learn to adapt the timing of their energy demands as well. To incentivize this, power suppliers are adopting pricing strategies wherein electricity prices vary depending on the instantaneous demand for electricity. [1] As electricity cost is a significant portion of daily expenditures for many industrial chemical processes, a reasonable way to reduce operating costs would be to adjust production in response to fluctuations in electricity price, an operating practice called Demand Response. [2]

Demand Response has been studied extensively in the literature for residential applications [3], however chemical processes nearly always obey nonlinear dynamics making their control notoriously difficult to optimize [4]. Adjusting the production set-point brings about a transition period during which product quality is uncontrolled. Unlike linear systems commonly modeled for classical control theory, the length of this transition period depends not only on the difference between the current state and the target but also on the current state itself. To work around the difficulty involved in determining arbitrary nonlinear transient profiles, a discrete set of stable operating points for the reactor are identified a priori by algebraically solving for values of control variables at as many equilibrium points as the user desires and tabulating the specific transition profiles between all possible pairs of equilibrium points. Once a sufficient number of the stable operating modes and their transitions have been tabulated, a clique wandering Dynamic Programming (DP) algorithm can be used to calculate the optimal schedule. [5] To do this, a clique (a fully connected graph) is generated for each time step using the stable operating points as the nodes, and weighting each edge according to the cost of the transition from the source to the set-point plus the cost of operating at the set-point for the remainder of the time after the transition. For each time step and each mode at each time step, every possible path from the previous feasible states is evaluated. Any of these new paths which both cost more and have less storage available than some other path or violate any constraint are then removed from the feasible set of this time step, preventing the number of computations from becoming exponential in time.

By using this clique wandering algorithm, the optimal scheduling policy was found for four case studies of CSTR operation. In the first case a CSTR was considered to be powered solely by the grid, and the envelope of pricing conditions which optimally lead to steady-state operation was found. The analysis was then repeated for the second case with on-site renewable energy generating units providing power to the plant. The results of using the DP algorithm on this case study matched or improved performance as compared to previous results of CSTR studied in the literature [5] [6]. Then the natural ability of Dynamic Programming to handle stochasticity was utilized for the third case, where uncertainty was introduced in the weather forecast and thus renewable power availability. In the fourth and final case the degree of stochasticity was increased further by adding random noise to the hourly electricity price. The results of the stochastic simulation for our case study show that the effects of taking the stochasticity are significant. Operating cost reductions above 10% compared to the steady state operation could be found in the best case, while savings of 4% are found for the average case.

References

[1] N. Li , L. Chen, S. H. Lowe, â€œOptimal demand response based on utility maximization in power networks," *Proc IEEE PES Gen Meeting,* pp. 1-8, 2011.

[2] I. Harjunkoski , E. Scholtz. X. Feng, "Industry Meets the Smart Grid," *CEP,* pp. 45-50, 2014.

[3] R. Deng , Z. Yang. M. Chow,Â J. Chen, "A Survey on Demand Response in Smart Grids: Mathematical Models and Approaches," *IEEE TRANSACTIONS ON INDUSTRIAL INFORMATICS,* vol. 11, no. 3, pp. 570-582, 2015.

[4] M.A. Javinsky & R. H. Kadlec, "Optimal Control of a Continuous Flow Stirred Tank Chemical Reactor," *AIChE Journal,* vol. 16, no. 6, pp. 916-924, 1970.

[5] C. Tong, N. El-Farra, A. Palazoglu, X. Yan, "Energy Demand Response of Process Systems through Production Scheduling and Control," *AIChE J.,* vol. 61, no. 11, pp. 3756-3769, 2015.

[6] T. Feital et al., "Modeling and Performance Monitoring of Multivariate Multimodal Processes," *AIChE,* vol. 59, no. 5, pp. 1557-1569, 2012.

470238 Rapid and Accurate Uncertainty Propogation for Nonlinear ODEs Using Nonlinear Solution Invariants

Abstract:

In this talk, we will present a new technique for computing tight interval enclosures of the solutions of nonlinear ordinary differential equations (ODEs) subject to a range of parameters and initial conditions. ODEs can be used to model a wide variety of dynamic chemical processes, and these models nearly always involve uncertain parameters or inputs. Quantifying the effects of such uncertainties on the model solutions provides essential information for many applications, including state and parameter estimation, controller synthesis, robust MPC, safety verification, and fault detection. Moreover, bounding the solutions of ODEs plays an essential role in all known methods for deterministic global optimization of dynamic models.

Despite substantial recent progress, for many systems of practical complexity, state-of-the-art ODE bounding methods still provide an unsatisfactory compromise between computational efficiency and the accuracy of the computed enclosures. In particular, classical methods based on interval computations are very efficient but often produce bounds that are too conservative to be useful. In contrast, modern methods based on high-order set enclosures (e.g., polytopes, zonotopes, and Taylor models) can provide highly accurate enclosures, but are much more computationally demanding and do not scale favorably in the number of uncertain model parameters. These issues severely limit the use of existing bounding methods for online control applications, and are also the predominant factors limiting the effectiveness of global dynamic optimization algorithms. Therefore, there is a critical need for an alternative approach that can provide bounds that are simultaneously sharp and highly efficient.

In 2013, Scott and Barton reported a fast, interval-based algorithm capable of producing very accurate bounds for a limited class of systems that obey affine solution invariants (i.e., affine equations satisfied by the system states for all time and all realizations of the uncertain parameters). These invariants are redundant with the original model, and can be exploited through a bounds-tightening procedure to greatly reduce the conservatism of fast interval bounding methods. More recently, we have made the crucial observation that arbitrary nonlinear systems (i.e., systems that do not obey any invariants) can be artificially endowed with affine solution invariants by introducing additional state variables defined as linear combinations of the original states. Our preliminary results, presented at last years AIChE meeting, show that this simple tactic can provide much more accurate bounds than standard interval methods with only a minor increase in CPU time. However, no general-purpose method has so far been proposed for identifying artificial invariants that will be effective in reducing conservatism. Moreover, this approach has so far been limited to invariants that are affine in the states, which proves to be a serious limitation for many nonlinear systems of interest.

In this talk, we will present an extension of this approach to permit the construction and exploitation nonlinear solution invariants, rather than only affine invariants. This is of interest first because many important systems naturally obey nonlinear invariants (e.g., Hamiltonian systems), and second because it provides much greater freedom in the design of effective artificial invariants for arbitrary systems that do not have pre-existing invariants. We will first outline our new theory and algorithms that enable the use of nonlinear invariants, and then present several case studies showing how nonlinear invariants can be constructed and exploited in order to produce highly accurate bounds using only fast interval computations.

470526 A Graph Theory Approach to Non-Anticipativity Constraint Generation in Multistage Stochastic Programs with Incomplete Scenario Sets

Abstract:

Multistage stochastic programs (MSSPs) with endogenous uncertainty arise from problem where the revelation of uncertainty is decision-dependent, such as pharmaceutical clinical trial planning problem (Colvin and Maravelias, 2008), the oil field infrastructure planning problem (Gupta & Grossmann, 2011), and the process synthesis problem (Goel & Grossmann, 2006). The MSSPs grow exponentially with the number of uncertain parameters and the number of realizations for each uncertain parameter. More specifically, the number of uncertain parameters and the number of realizations increase the number of scenarios in the model. The scenarios for the full-space model are generated using the Cartesian product of the realizations of the uncertain parameters. The size of the decision variables and the number of non-anticipativity constraints (NACs) increase with the increasing in the number of scenarios. As a result, most MSSPs considered in literature are limited to a few uncertain parameters with a few realizations for each parameter.

Often real-world size problems have many uncertain parameters with multiple realizations. Therefore, the MSSPs of these problems quickly become computationally intractable. One approach to address the complexity is to solve them using sample average approximation approaches where a subsets of scenarios generated from the full scenario set is considered. The solutions can be used to generate bounds for the full space problem, or can be implemented as heuristics. The challenge with this approach is that the traditional NAC reduction procedures are not applicable when only subsets of scenarios are used (Apap and Grossmann, 2015).

In this talk, we present a graph theory based algorithm, Sample Non-Anticipativity Constraint algorithm (SNAC), to determine the minimum number of NACs for the scenario subsets, and to efficiently generate at least one of the corresponding feasible NAC sets. The SNAC assumes that the scenarios are a subset of the full space scenario set, and that that the full space scenario set can be projected onto *n*-dimensional integer lattice, where *n* represents the number of uncertain parameters. This projection simply requires that the realizations of each uncertain parameter can be mapped to an integer range with a length equal to the number of realizations. The mapping allows the generation of an *n*-tuple representing the position of a scenario on the integer lattice.

The first step of the SNAC algorithm is to map the subset of scenarios to the full space integer lattice. The algorithm then adds edges to the lattice until the lattice forms a minimum spanning tree (MST) for the subset. The weight of each edge is set to its Euclidian distance for generating the MST. Using a MST connection, the procedure prioritizes connections of adjacent nodes and eliminates the need for redundant connections. The edges added in the generation of the MST represent the NACs needed to relate the selected scenarios. If the problem has only one stage of realizations (i.e., the problem is a two-stage stochastic problem), the added constraints also represent all the full set of NACs needed for the problem.

In MSSPs, when a realization occurs the scenario set is divided to form subsets. In order to achieve a feasible solution, all scenarios in each subset must be connected with proper NACs. The SNAC algorithm uses the MST generation procedure to generate the connections for each subset. To ensure that connections exist for any possible subset, the SNAC algorithm iterates over the possible subsets. To facilitate the enumeration of the subsets, the algorithm exploits the integer lattice mapping, which ensures that the division of the scenarios into subsets always occurs perpendicular to the plane of the realized uncertain parameter. For each subset, the algorithm reduces the number of NACs by using connections that were generated in previous subsets.

Figure 1 shows the results of applying the algorithm four times. In each application, samples of four scenarios selected from an MSSP with two uncertain parameters, and , and three realizations each. In Figure 1, the shaded dots represent the

ϕ_1

ϕ_2

sampled scenarios. The row and column in which each scenario appears corresponds to the realized values of the uncertain parameters. The lines connecting scenarios represent the NACs generated by the algorithm. In each of the cases the algorithm prioritizes connecting selected scenarios that are adjacent. In Figure 1(A), the initial MST connects scenarios in the first row because they are adjacent. The algorithm then connects the scenarios in the third column. When the algorithm generates all the possible subsets of scenarios, no additional NACs are needed. Similarly, in Figure 1(B), the SNAC algorithm yields NACs corresponding to connections of all adjacent scenarios. Then, a diagonal connection ensures non-anticipativity between the scenarios in the third column and the second column. Again, when subsets are generated no new NAC are needed. Figure 1(C) represents a subset where all scenarios are adjacent. The algorithm is able to generate all the NACs simply by connecting all adjacent scenarios. Figure

1(D) in contrast has only one pair of adjacent scenarios. The other two scenarios are connected to the adjacent pair with a horizontal line and a diagonal line to form the initial MST. When the algorithm iterates over the subsets, the two non-adjacent scenarios must be connected.

The talk will introduce the algorithm in detail, and demonstrate its application using the general two uncertain parameter problem, and, three uncertain parameter problem. We will present the generalization of the algorithm to an *n* uncertain parameter case, and compare the number of NACs generated using SNAC to the standard approaches. A preliminary analysis of the algorithm shows that the generation of the NACs occurs in polynomial time. In future, we plan to implement the SNAC algorithm for developing a sample average approximation approach to solve MSSPs.

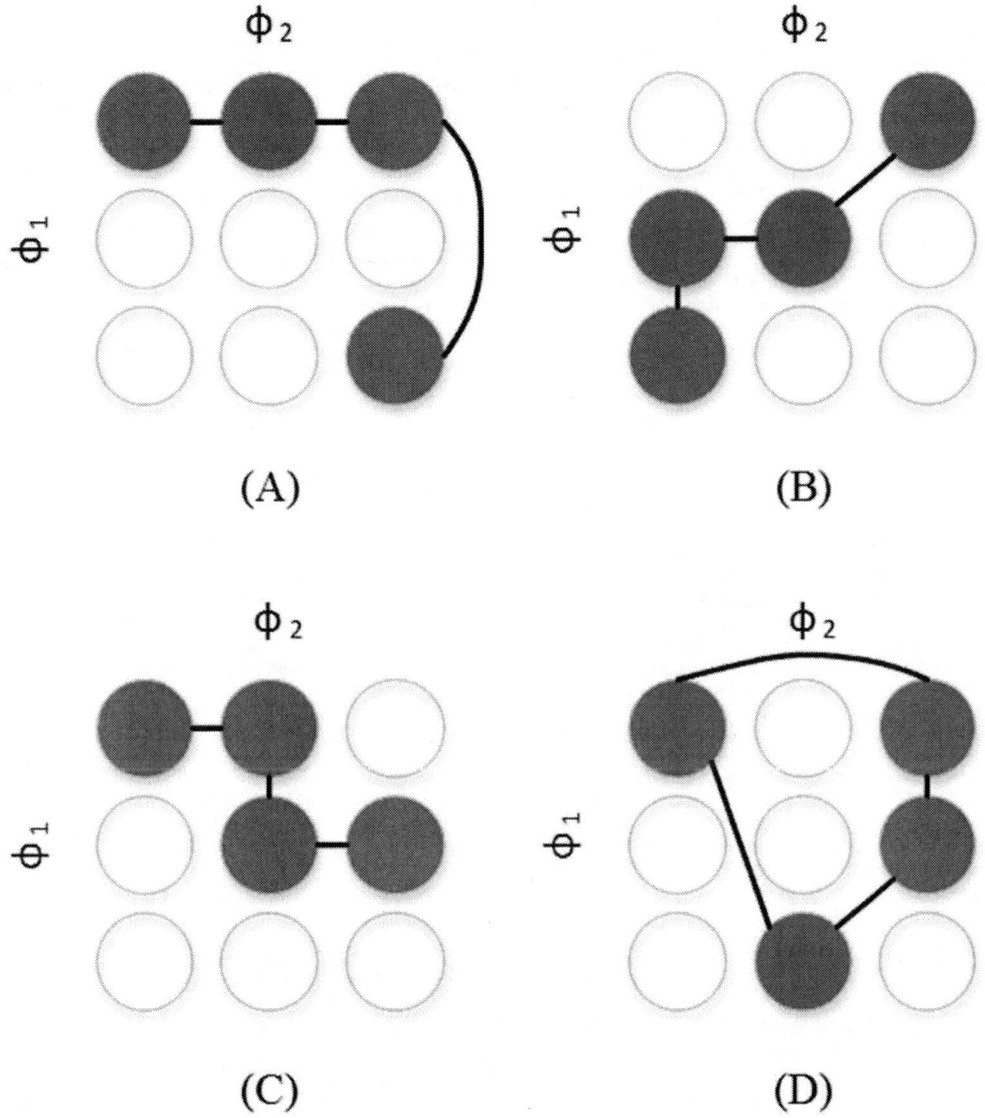

Figure 1 Non-anticipativity constraints generated using the SNAC algorithm

<>References

Colvin, M., & Maravelias, C. T. (2008). A stochastic programming approach for clinical trial planning in new drug development. *Computers & Chemical Engineering*, *32*, 2626–2642.

Goel, V., & Grossmann, I. E. (2006). A Class of Stochastic Programs with Decision Dependent Uncertainty. *Mathematical Programming*, *108*(2), 355–394.

Gupta, V., & Grossmann, I. E. (2011). Solution strategies for multistage stochastic programming with endogenous uncertainties. *Computers and Chemical Engineering*, *35*(11), 2235–2247.

Apap, R. M. & Grossmann, I. E. (2015). Models and computational strategies for multistage stochastic programming under endogenous and exogenous uncertainties. Manuscript in preparation.

470959 Towards Optimal Production of Industrial Gases with Uncertain Energy Prices

Abstract:

Nowadays, the inherent uncertainty in energy costs and the direct impact on industrial competitiveness have become a critical factor for efficient production decisions. Particularly, in high intensive energy processes, effective decision-making tools are urgently needed to simultaneously deal with alternative production modes and rates to lower power consumption. As a result, the development of optimal production scheduling strategies has emerged as a promising alternative to reduce the consumption of electricity.

In this work we consider air separation processes where the cost of energy changes hourly. The plant is assumed to participate in multiple energy markets for production: Day-ahead markets and Spot/Imbalance markets. In the day-ahead market, blocks of energy are nominated at an hourly level for the next day and are bought on a daily auction. The prices of energy are known after the auction closes, sometime around noon on the day before delivery. For a weekly time horizon, e.g. Monday to Sunday, the price and amount of power at every hour is known for Monday, whereas from Tuesday to Sunday neither the price nor the amount of power is known. However, forecasts are available for the day-ahead market for the next days. On the other hand, power is non-contracted in the imbalance market. It is the result of the imbalances in the physical power grid, and the attempt of the operators to match supply and demand. Prices of energy are known 15 minutes after the power is consumed. The challenge is to predict how long the price will remain profitable so the plant has time to react or even to forecast the spikes and valleys.

Therefore, we propose an efficient predictive and reactive solution strategy for real-world industrial scale problems to optimize participation in electricity markets under uncertainty in the operation of power-intensive air separation processes. A solution approach is developed based on a discrete-time scheduling formulation that allows modeling and optimizing operating decisions either in a fixed or a rolling horizon scheme. The objective is to compare both approaches in terms of computational efficiency and potential economical benefits. Accordingly, a deterministic MILP model is proposed to optimal production planning of continuous power-intensive air-separation processes to efficiently adjust production operation according to time-dependent electricity pricing. The main goal of this contribution is to find an optimal hourly schedule for next week, that minimizes total energy consumption cost while satisfying all operational constraints.

The model takes into account minimum and maximum production rates based on the plant state, storage capacity of the plant and minimum final tank level constraints, considering that minimum final tank levels must be fulfilled depending on the day of the week of last time period of the scheduling horizon. At the same time, detailed power consumption is taken into account for the different operating modes, which follows linear correlation. Expected daily demand and hourly electricity prices are used to generate and assess different scenarios.

An important aspect in this scheduling problem is to explicitly consider that there is an operational constraint on the minimum amount of time the plant should be running in the same operation mode. The plant has transition states to set-up and shut-down of equipment (ramp-up and ramp-down times) with minimum duration of 1 hour, and others states with minimum duration of 3 hours: uptime, standby time and downtime. Therefore, we propose an explicit modeling approach of feasible plant operational transitions and a systematic way of representing transition states.

A novel process state transition network is developed to model specific problem features. States with minimum duration of 3 hours are decomposed in 3 sub-states of 1 hour each and are called initial sequential transition states, Intermediate transition states, and critical transition states, respectively. The process state transition network model assigns states to time periods using proper binary variables denoting that process is operated at a given state at every time and ensuring that all operating constraints are satisfied. The model satisfies the start-up and shut-down restrictions, and also the constraints that concern the power consumption according to time-dependent electricity pricing schemes.

The proposed model was tested with real-world electricity price and demand input data. Scenarios were defined to assess how the model faces different situations. Scenarios tested with new formulation combines the input data (hourly and shift demand, and fixed, shift and hourly energy cost). In some cases the results are fixed to evaluate them with other data and to perform a fair comparison. The results show optimal solutions for the proposed methodology with a modest computational effort considering a one-hour time grid and one-week time horizon. Based on the preliminary results achieved, it can be concluded that the predictive MILP-based scheduling approach looks very efficient and robust. The model is able to consider all problem features and easy to adapt to reactive scheduling (a rolling horizon approach). Therefore, the developed model is a promising solution scheduling for the application to real-world air separation industrial plants. The model is also easy to adapt to other plant configurations, including the identified additional features. It allows the evaluation of daily and hourly reactive decisions based on energy price changes (day ahead market and imbalance market).

471504 MINLP Formulation and Global Solution Approach for Sensor Placement with Non-Uniform Failure Probabilities: Application to Gas Detection Systems

Abstract:

Sensor systems play a very important role in modern process safety, and in particular for gas and fire detection systems. To rapidly detect hazardous events and mitigate the corresponding damages, an efficient detection and mitigation system strongly depends on appropriate types, number, and placement of sensors. However, the optimal design of these systems is challenging because significant uncertainty must be taken into account. For example, to design a gas detection system, we must consider uncertainty including, but not limited to, gas compositions, leak locations, process conditions, and weather.

In the last decade, a large body of literature has focused on developing stochastic programming (SP) based approaches to solve many real-world sensor placement problems [1, 2, 4-7]. However, most of the previous problem formulations have assumed perfect sensors. In reality, detectors are imperfect and subject to unpredictable failures, which may be caused by poor maintenance, erroneous calibration, or power outage. In many cases, these imperfections can significantly impact the performance of the entire detector system. To design a sensor system with high reliability, therefore, it is necessary to explicitly consider sensor unavailability, i.e., the probability of a false-negative detection. Berry et al. [4] first proposed a SP-based imperfect-sensor model for water contamination warning systems. Inspired by this seminal work, Benavides-Serrano et. al. [1, 3] proposed SP formulations for flammable gas detection and mitigation systems considering sensor unavailabilities. However, the resulting sensor placement problems, formulated as large-scale mixed-integer nonlinear programming (MINLP) problems, are very challenging to solve due to strong nonlinearities and discrete variables. To solve these optimization problems efficiently, thus, there is a need for new problem formulations and solution strategies.

To reduce the inherent nonlinearities in the original MINLP formulation [4], in this work, we present a mathematically equivalent formulation based on log-transformation. Though new variables and constraints are introduced, this alternative formulation is mathematically preferable since all its nonlinearity comes from convex univariate terms. To solve this reformulation to global optimality, we propose a multitree method which depends on iteratively solving a sequence of upper bounding master problems and lower bounding subproblems. The master problem, which is formulated as a mixed-integer linear programming (MILP) problem, is a strict convex relaxation of our alternative formulation. To obtain a relatively tight and computationally efficient master problem, we introduce linear outer approximations and tight, problem specific, upper bounding constraints. The upper bounding problem is obtained by fixing all binary variables, which, in this case, results in a subproblem that can be directly computed with a single forward simulation. Our tailored global solution strategy is tested on a number of real data problems. The encouraging numerical results indicate that our solution framework is promising in solving sensor placement problems.

Bibliography

[1] Benavides-Serrano, A. J., Legg, S. W., Vázquez-Román, R., Mannan, M. S., & Laird, C. D. (2013). A Stochastic Programming Approach for the Optimal Placement of Gas Detectors: Unavailability and Voting Strategies. Industrial & Engineering Chemistry Research, 53(13), 5355-5365.

[2] Benavides-Serrano, A. J., Mannan, M. S., & Laird, C. D. (2015). A quantitative assessment on the placement practices of gas detectors in the process industries. Journal of Loss Prevention in the Process Industries, 35, 339-351.

[3] Benavides Serrano, A. J., Mannan, M. S., & Laird, C. D. (2016). Optimal placement of gas detectors: AP median formulation considering dynamic nonuniform unavailabilities. AIChE Journal.

[4] Berry, J., Carr, R. D., Hart, W. E., Leung, V. J., Phillips, C. A., & Watson, J. P. (2009). Designing contamination warning systems for municipal water networks using imperfect sensors. Journal of Water Resources Planning and Management, 135(4), 253-263.

[5] Berry, J., Carr, R. D., Hart, W. E., Leung, V. J., Phillips, C. A., & Watson, J. P. (2009). Designing contamination warning systems for municipal water networks using imperfect sensors. Journal of Water Resources Planning and Management, 135(4), 253-263.

[6] Legg, S. W., Wang, C., Benavides-Serrano, A. J., & Laird, C. D. (2013). Optimal gas detector placement under uncertainty considering Conditional-Value-at-Risk. Journal of Loss Prevention in the Process Industries, 26(3), 410-417.

[7] Legg, S. W., Benavides-Serrano, A. J., Siirola, J. D., Watson, J. P., Davis, S. G., Bratteteig, A., & Laird, C. D. (2012). A stochastic programming approach for gas detector placement using CFD-based dispersion simulations. Computers & Chemical Engineering, 47, 194-201.

473269 Data-Driven Adaptive Nested Robust Optimization: Modeling Framework and Solution Algorithm for Process Design and Operations Under Uncertainty

Abstract:

Optimization of process systems under uncertainty has attracted wide attention from both academia and industry [1, 2]. A plethora of optimization models and algorithms have been proposed to support decision making in the presence of uncertainty. Recently, adaptive robust optimization (ARO) offers a new paradigm for optimization under uncertainty by incorporating recourse decisions through a sequential decision-making framework [3-6]. However, the traditional ARO approach fails to take full advantage of the information embedded in uncertainty data. It does not take into account the correlation between uncertainty parameters or skewness of uncertainty data, which turn out to have substantial impacts on the ARO solutions. Besides, the traditional ARO is sensitive to outliers, yielding an optimized solution of low quality.

The aforementioned issues are relevant to process industries, in which a deluge of data has been collected and archived [7, 8]. In this work, we propose a data-driven adaptive nested robust optimization (DDANRO) approach for better decision-making from big data. A Bayesian nonparametric model – the Dirichlet process mixture model (DPMM) [9] – is adopted to learn an uncertainty set via a variational inference algorithm [10]. As a nonparametric model, DPMM is able to adjust its complexity to that of data. To be more specific, it utilizes potentially infinite mixtures of Gaussian distributions to characterize data. In this way, DPMM efficiently extracts valuable information from uncertainty data, including correlation and skewness. Following this DPMM approach, we propose a novel data-driven uncertainty set for ARO based on l_1 and l_∞ norms. We then integrate the statistical model and adaptive optimization model seamlessly in a four-level optimization framework (a min-max-max-min problem). Additionally, this DDANRO framework, as its name suggests, has two layers of robustness. The outer layer of the DDANRO framework is robust to outliers in uncertainty data. By using the weights of Gaussian components to hedge against outliers, components with smaller weights are "filtered out" as outliers by the outer layer. The inner layer of the DDANRO framework is robust to variations in the remaining – or "clean" – uncertainty data. The resulting DDANRO problem cannot be solved directly by any off-the-shelf optimization solvers due to its multi-level optimization structure. To address this computational challenge, we also propose a tailored column-and-constraint generation (C&CG [11]) algorithm. In this algorithm, each subproblem corresponds to a component of Gaussian mixtures. The algorithm iteratively solves a sequence of master problems and subproblems until the optimality gap reduces to a predefined tolerance.

The proposed modeling framework and solution algorithm are demonstrated using two applications on batch process scheduling [12] and on petrochemical complex planning [13], respectively. In the short-term batch process scheduling problem, processing time uncertainty and demand uncertainty are considered. The real batch processing time data are corrupted with outliers. The traditional ARO treats these outliers in processing time data as worst cases. Consequently, the scheduling strategy obtained from the traditional ARO performs less tasks over the same time horizon. In contrast, our method yields less conservative solutions, which means increased profits from batch processes. In the planning problem, the process network involves 38 processes and 28 chemicals. Both supply and demand are subject to uncertainty. The optimal process designs and process operations produced by our proposed method and the traditional ARO are quite different. In a case, the processes 1, 3, 8, 28 and 32 are planned to be expanded in the traditional ARO solution, while the processes 1, 3, 8, 14, 17, 28 and 32 are chosen to be expanded in the solution of our proposed model. Our proposed approach demonstrates superior performance in terms of net present value (NPV) over the traditional ARO. For example, in a case where product demands are correlated, the NPV of our proposed method is 7.5% higher than that of the traditional ARO.

References

[1] N. V. Sahinidis, "Optimization under uncertainty: state-of-the-art and opportunities," *Computers & Chemical Engineering,* vol. 28, pp. 971-983, 2004.

[2] I. E. Grossmann, R. M. Apap, B. A. Calfa, P. García-Herreros, and Q. Zhang, "Recent advances in mathematical programming techniques for the optimization of process systems under uncertainty,"*Computers &Chemical Engineering*,2016.DOI:10.1016/j.compchemeng.2016.03.002

[3] A. Ben-Tal, A. Goryashko, E. Guslitzer, and A. Nemirovski, "Adjustable robust solutions of uncertain linear programs," *Mathematical Programming,* vol. 99, pp. 351-376, 2004.

[4] J. Gong, D. J. Garcia, and F. You, "Unraveling Optimal Biomass Processing Routes from Bioconversion Product and Process Networks under Uncertainty: An Adaptive Robust Optimization Approach," *ACS Sustainable Chemistry & Engineering,* 2016. DOI: 10.1021/acssuschemeng.6b00188

[5] D. Yue and F. You, "Optimal supply chain design and operations under multi scale uncertainties: Nested stochastic robust optimization modeling framework and solution algorithm," *AIChE Journal*, 2016. DOI: 10.1002/aic.15255

[6] H. Shi and F. You, "A computational framework and solution algorithms for two-stage adaptive robust scheduling of batch manufacturing processes under uncertainty," *AIChE Journal,* vol. 62, pp. 687-703, 2016.

[7] C. Ning, M. Chen, and D. Zhou, "Hidden Markov Model-Based Statistics Pattern Analysis for Multimode Process Monitoring: An Index-Switching Scheme," *Industrial & Engineering Chemistry Research,* vol. 53, pp. 11084-11095, 2014.

[8] S. J. Qin, "Process data analytics in the era of big data," *AIChE Journal,* vol. 60, pp. 3092-3100, 2014.

[9] T. Campbell and J. P. How, "Bayesian nonparametric set construction for robust optimization," in *American Control Conference (ACC), 2015*, pp. 4216-4221.

[10] D. M. Blei and M. I. Jordan, "Variational inference for Dirichlet process mixtures," *Bayesian analysis,*

vol. 1, pp. 121-143, 2006.

[11] B. Zeng and L. Zhao, "Solving two-stage robust optimization problems using a column-and-constraint generation method," *Operations Research Letters,* vol. 41, pp. 457-461, 2013.

[12] Y. Chu, J. M. Wassick, and F. You, "Efficient scheduling method of complex batch processes with general network structure via agent-based modeling," *AIChE Journal,* vol. 59, pp. 2884-2906, 2013.

[13] F. You and I. E. Grossmann, "Stochastic inventory management for tactical process planning under uncertainties: MINLP models and algorithms," *AIChE Journal,* vol. 57, pp. 1250-1277, 2011.

473082 Value Based Sensor Network Design

Abstract:

In many control system hardware selection problems the objective in to minimize the capital cost of hardware, while satisfying pre-specified bounds with respect to operating performance [1, 2, 3, 4]. However, in the value based hardware selection problem the performance bounds are removed and replaced with an operating cost term in the objective function [5, 6, 7]. While this combined capital and operating cost formulation is more reflective of the industrial objectives of a chemical process, this class of design problems presents significant computational challenges. For example, in the data reconciliation based formulation it has been shown, [6], that the value based objective function contains a number of local minima and thus disrupts the global optimality of the tree search algorithm of [1]. In the case of closed-loop dynamic process, such a formulation will introduce a set of reverse-convex constraints along with the nonlinearities associated with the steady-state process model, [7]. In this work, we illustrate that a simple reformulation of the value based hardware selectin problem and subsequent application the generalized Benders decomposition will result in massive reductions in computational effort and in some cases lead to previously unachievable global solutions.

[1] Bagajewicz, MJ Design and retrofit of sensor networks in process plants, AIChE J, 9, pp 2300-2306 (1997).

[2] Chmielewski D. J., T. Palmer and V. Manousiouthakis, "On the Theory of Optimal Sensor Placement," AIChE J., 48(5), pp 1001-1012, (2002).

[3] Peng, J.K., and D.J. Chmielewski, "Covariance Based Hardware Selection-Part II: Equivalence Results for the Sensor, Actuator and Simultaneous Selection Problems," IEEE Trans. Cont. Sys. Tech., 14(2), pp 362-368, (2006).

[4] Ahmed S.K., J.-K. Peng, and D.J. Chmielewski, "Covariance-Based Hardware Selection-Part III:

Distributed Parameters Systems" AIChE J., 58(9), pp 2705-2713, (2012).

[5] Bagajewicz M, Markowski M, Budek A. Economic value of precision in the monitoring of linear system. AICHE J. 2005;51:1304– 615 1309. 616 27.

[6] Nguyen DQ, Bagajewicz MJ. New sensor network design and retrofit 617 method based on value of information. AICHE J. 2011;57:2136– 618 2148. 619 28.

[7] Peng JK, Chmielewski DJ. Optimal sensor network design using the 620 minimally backed-off operating point notion of profit. American 621 Control Conference, Portland, OR, USA, 2005.

450662 On the Robustness of Economic Nonlinear Model Predictive Control

Abstract:

Model predictive control (MPC) has seen a variety of applications, and its advantages include a natural way of handling inequality constraints and multiple-input-multiple-output systems due to the optimization formulation of the problem. A survey of industrial uses of MPC is given in [10], and a thorough treatment of MPC is given in [11]. Nonlinear model predictive control (NMPC) has the added advantage of being able to use a detailed first-principles dynamic model in order to provide accuracy across a wide range of states, and the fundamentals of NMPC are given in [4]. Furthermore, if a sensible initialization strategy is used, an exact solution of the associated nonlinear programming (NLP) problem is not required, as shown in [9,15,13,16].

Input-to-state stability (ISS) is used to extend stability analysis to systems with uncertainty. The property was originally described for continuous time (CT) systems in [12] and was extended to discrete time (DT) systems in [7]. Furthermore, ISS has been proposed as a framework for NMPC [8], and it provides a very convenient and natural way of thinking about robust stability. Previous work [3] provides a method for calculating ISS bounds. Input-to-state practical stability allows for a slight modification to ISS in order to be more widely applicable.

Economic NMPC allows for the use of more general objective functions in the controller. The difficulty is that standard stability results of tracking MPC cannot be used. Originally, asymptotic stability was shown by using strict dissipativity of the rotated cost function [2], but this property is not easy to show in general. However, it is possible to show a stronger property, strong convexity of the rotated cost function, by including regularization terms in the objective function. It has been shown that a sufficiently large regularization term (e.g., a tracking term) may be added to the objective in order to force the rotated cost function to be strongly convex, which leads to asymptotic stability [6]. The drawbacks of this method are that regularization weights may be difficult to calculate and very conservative, limiting economic gains. In [14] it is proposed to replace the regularization term in the objective with a stabilizing inequality constraint derived from the inherent robustness margin of an auxiliary, asymptotically stable MPC controller. It is demonstrated that this approach (that we call eMPC-sc) provides flexibility to optimize economic performance. Moreover, it is shown that this approach is related to regularization-based Lyapunov-based MPC approaches, but the cumbersome calculations required to find regularization terms are not required. The eMPC approach differs from the Lyapunov-based approach [5] in that feasibility of the stabilizing constraint can be guaranteed directly while in the Lyapunov approach the feasible set for the states needs to be adjusted. Also, a formulation for the tracking case with uncertainty with a similar stabilizing constraint is shown in [1]. Here, the stabilizing constraint is required to be feasible for a list of possible plant realizations, while we only require feasibility of the terminal constraint for the nominal plant.

In this work, we extend eMPC with a stabilizing constraint to the nonlinear case with uncertainty and state constraints. We show that a system to which this controller is applied is ISpS, which requires a slight modification to existing theory on ISpS. We show simulation results on a CSTR, as well as on a system of two distillation columns in series.

References

[1] T. Badgwell. A robust model predictive control algorithm for stable nonlinear plants. In *Preprints of the 1997 International Symposium on Advanced Control of Chemical Processes*, 1997.

[2] M. Diehl, R. Amrit, and J. Rawlings. A Lyapunov function for economic optimizing model predictive control. *IEEE Transactions on Automatic Control*, 56(3):703-707, 2011.

[3] D. Grith and L. Biegler. Trajectory bounds of input-to-state stability for nonlinear model predictive control. In *Preprints of the 9th International Symposium on Advanced Control of Chemical Processes*, pages 1028-1033, 2015.

[4] L. Grune and J. Pannek. *Nonlinear Model Predictive Control: Theory and Aglorithms*. Springer, London, 2011.

[5] M. Heidarinejad, J. Liu, and P. Christodes. Economic model predictive control of nonlinear process systems using Lyapunov techniques. *AICHE Journal*, 58(3):855-870, 2012.

[6] J. Jaschke, X. Yang, and L. Biegler. Fast economic model predictive control based on NLP-sensitivities. *Journal of Process Control*, 24:1260-1272, 2014.

[7] Z. Jiang and Y. Wang. Input-to-state stability for discrete-time nonlinear systems. *Automatica*, 37(6):857-869, 2001.

[8] D. Limon, T. Alamo, D. Raimondo, D. Peena, J. Bravo, A. Ferramosca, and E. Camacho. Input-to-state stability: A unifying framework for robust model predictive control. In L. Magni, D. Raimondo, and F. Allgower, Editors, *Nonlinear Model Predictive Control: Towards New Challenging Applications*. Springer, Berlin, 2009.

[9] G. Pannocchia, J. Rawlings, and S. Wright. Conditions under which suboptimal nonlinear MPC is inherently robust. *Systems & Control Letters*, 60:747-755, 2011.

[10] S. Qin and T. Bagdwell. A survey of industrial model predictive control technology. *Control Engineering Practice*, 1:733-764, 2003.

[11] J. Rawlings and D. Mayne. *Model Predictive Control: Theory and Design*. Nob Hill Publishing, Madison, WI, 2009.

[12] E. Sontag. Smooth stabilization implies coprime factorization. *IEEE Transactions on Automatic Control*, 34(4):435-443, Apr 1989.

[13] X. Yang and L. Biegler. Advanced-multi-step nonlinear model predictive control. *Journal of Process Control*, 23(8):1116-1128, 2013.

[14] V. M. Zavala. A multiobjective optimization perspective on the stability of economic MPC. *9th International Symposium on Advanced Control of Chemical Processes*, pages 975-981, 2015.

[15] V. M. Zavala and M. Anitescu. Real-time nonlinear optimization as a generalized equation. *SIAM Journal on Control and Optimization*, 48(8):5444-5467, 2010.

[16] V. M. Zavala and L. Biegler. The advanced-step NMPC controller: Optimality, stability, and robustness. *Automatica*, 45:86-93, 2009.

469259 A Carleman Approximation-Based Approach to Address the Performance Criteria Issue of Economic Model Predictive Control Via Lyapunov Method

Abstract:

The last decade witnessed a widespread growth of Economic-oriented Model Predictive Control (EMPC) in the chemical, pharmaceutical and petroleum industries. Built upon MPC, which predicts future system behavior over a finite horizon and takes actions accordingly, EMPC aims at maximizing the economic profit of the system while satisfying performance and safety constraints [1]. One of the most significant challenges faced with EMPC is the difficulty in maintaining the system in a desired performance criteria.

Lyapunov-based approaches have been reported in many EMPC designs for their provable stability and performance properties [123]. Defined as Lyapunov-based EMPC (LEMPC), it has been applied with success to constrained nonlinear systems, switched systems and fault-tolerant control schemes [45]. Carleman approximation (also known as Carleman linearization)-based formulations have been applied to nonlinear MPC and Moving Horizon Estimation (MHE) [678910]. Nonlinear systems are approached with polynomial expressions and then represented with extended bilinear forms. This formulation enables analytical prediction of future states and analytically providing the sensitivity of the cost function to the manipulated inputs as the searching gradient. Consequently, (i) the optimization is relieved from equality constraints; (ii) the computation of optimal control policy is accelerated.

In this presentation, we propose a method to construct Lyapunov functions via Carleman approximation in the design of EMPC controllers. After expressing the dynamic constraints with polynomial forms, we represent the dynamic system in an extended bilinear expression. With the extended bilinear expression, we incorporate the dynamic constraints directly in the construction of Lyapunov function and in the objective function. Hence, we relieve the optimization from dynamic constraints. Meanwhile, it becomes easier for us to prove the boundedness of the economic performance criteria with the extended bilinear expression. A case study example uses a polymerization system to illustrate the proposed method. We design a controller that bounds the economic performance of the system in a desired set and prove the boundedness of the economic performance criteria.

Key words: Carleman Approximation, economic-oriented model predictive control, Lyapunov-based control

Reference:

1. Ellis M, Durand H, Christofides PD. A tutorial review of economic model predictive control methods. *J Process Control*. 2014;24(8):1156-1178. doi:10.1016/j.jprocont.2014.03.010.

2. Mhaskar P, El-Farra NH, Christofides PD. Stabilization of nonlinear systems with state and control constraints using Lyapunov-based predictive control. *Syst Control Lett*. 2006;55(8):650-659. doi:10.1016/j.sysconle.2005.09.014.

3. Heidarinejad M, Liu J, Christofides PD. Economic Model Predictive Control of Nonlinear Process Systems Using Lyapunov Techniques Mohsen. *AIChE J*. 2012;58(3):855-870. doi:10.1002/aic.

4. El-Farra NH, Mhaskar P, Christofides PD. Output feedback control of switched nonlinear systems using multiple Lyapunov functions. *Syst Control Lett*. 2005;54:1163-1182. doi:10.1016/j.sysconle.2005.04.005.

5. Munoz de la Pena D, Christofides PD. Lyapunov-Based Model Predictive Control of Nonlinear Systems Subject to Data Losses. *IEEE Trans Automat Contr*. 2008;53(9):2076-2089. doi:10.1109/TAC.2008.929401.

6. Armaou A, Ataei A. Piece-wise constant predictive feedback control of nonlinear systems. *J Process Control*. 2014;24(4):326-335. doi:10.1016/j.jprocont.2014.02.002.

7. Fang Y, Armaou A. Nonlinear Model Predictive Control using a bilinear Carleman linearization-based formulation for chemical processes. In: *Proceedings of the American Control Conference*.Vol 1.; 2015:5629-5634. doi:10.1109/ACC.2015.7172221.

8. Fang Y, Armaou A. Carleman approximation based Quasi-analytic Model Predictive Control for Nonlinear Systems. *AIChE J*. 2016. doi:10.1002/aic.15298.

9. Hashemian N, Armaou A. Fast Moving Horizon Estimation of Nonlinear Processes via Carleman Linearization. In: *Proceedings of the American Control Conference*.Vol 1.; 2015:3379-3385. doi: 10.1109/ACC.2015.7171854.

10. Hashemian N, Armaou A. Simulation, Model-Reduction, and State Estimation of a Two-Component Coagulation Process. *AIChE J*. 2016;62(5):1557-1567. doi: 10.1002/aic.15146.

467835 Economic Nonlinear Model Predictive Control with a Path-Following Approach

Abstract:

Nonlinear Model Predictive Control (NMPC) with economic objective has been an active research field in recent years [1]. Especially its performance and closed-loop stability analysis have received much attention. The idea of economic NMPC is to integrate real time optimization (RTO) and supervisory layers in the process control hierarchy into a single economic NMPC layer in order to simultaneously optimize and control the process. Papers [2] and [3] provide tutorial and survey on recent developments.

In this work, we study a sensitivity-based advanced-step NMPC (asNMPC) as described in [4,5] with economic cost function. This type of controller consists of an online and an offline step. In the offline step, a full nonlinear programming problem (NLP) is solved for a predicted starting point in the future, while waiting for the measurement data of this point to become available. After the measurement data is obtained, a fast online correction is calculated based on the optimal sensitivity of the NLP problem (instead of solving the full NLP again). Thus, the computational delay between obtaining the measurements and implementing the computed inputs is significantly reduced.

When active set changes occur in the sensitivity update, special measures are required to ensure that the sensitivity update approximates the true NLP solution well enough. In [6] some methods are discussed for handling active set changes. Recently [7] proposed a path-following algorithm for the fast sensitivity update. The algorithm is capable of handling active set changes and non-unique-bounded Lagrange multipliers, which arise when the NLP satisfies MFCQ. This path-following algorithm is used in the online step of the asNMPC procedure. In addition to computing the optimal sensitivity along the path by solving a sequence of QPs, the algorithm solves a sequence of linear programming (LP) problems for computing the optimal changes of the Lagrange multipliers along the path.

In this work, we apply the path-following algorithm in the context of asNMPC for economic NMPC of a large-scale optimization problem consisting of a 41-tray distillation column case study. We employ orthogonal collocation to discretize the dynamic optimization problem and use automatic differentiation to obtain first and second order derivatives. We compare the performance of our method to ideal NMPC in which we solve the full NLP assuming negligible computational delay. We show that our method is able to preserve computational efficiency of the asNMPC in the presence of active set changes.

References: Â

1.Rawlings, J.B., Angeli, D., and Bates, C.N., 2012. Fundamental of economic model predictive control. In proceeding of IEEE 51st Annual Conference on Decision and Control (CDC), pp. 3851-3861.

2.Ellis, M., Durand, H., and Christofides, P.D., 2014. A tutorial review of economic model predictive control methods. Journal of Process Control, Volume 24, pp. 1156-1178.

3.Tran, T., Ling, K-V., and Maciejowski, J.M., 2014. Economic model predictive control â€" a review. In proceeding of The 31st International Symposium on Automation and Robotics in Construction and Mining (ISARC 2014).

4.Biegler, L.T., Yang, X., and Fischer, G.A.G., 2015. Advances in sensitivity-based nonlinear model predictive control and dynamic real-time optimization. Journal of Process control, Vol. 30. Pp. 104-116.

5.Zavala, V.M., and Biegler, L.T., 2009. The advanced-step NMPC controller: optimality, stability, and robustness. Automatica, Vol. 45. Pp. 86-93

6.Yang, X., and Biegler, L.T., 2013. Advanced-multi-step nonlinear model predictive control. Journal of Process Control, Vol. 23. Pp. 1116-1128.

7.Jäschke, J., Yang, X., and Biegler, L.T., 2014. Fast economic model predictive control based on NLP-sensitivities. Journal of Process Control, Vol. 24. Pp. 1260-1272.

472902 Economic MPC with Local Optimality

Abstract:

Model predictive control (MPC), or receding horizon control, refers to a control methodology that approximates the solution of a constrained infinite-horizon optimal control problem by solving finite-horizon optimal control problems in a receding horizon fashion. Generally, close-loop stability and performance of MPC cannot be guaranteed because of the finite horizon. A common approach to overcome the shortsightedness of MPC is to employ a terminal cost (often with terminal constraint). An ideal terminal cost to ensure stability and performance of MPC is the one characterizing the infinite-time cost-to-go of the system. However, for generic nonlinear systems, such a terminal cost can be very difficult to design, if it exists at all. An alternative to still ensure the stability of nonlinear MPC is to employ a terminal cost which is essentially an upper bound on the infinite-time cost-to-go (see e.g., [1-3]). Specifically, the terminal cost is constructed as a control Lyapunov function (CLF) which is compatible with the stage cost. Similar approach has been extended to the design of economic model predictive control (EMPC) which optimizes general economic cost functions to achieve better economic performance [4].

While the existing framework is superior for its stability property, with enlarged stability region compared to MPC/EMPC without terminal condition, it may not be favorable in terms of performance. In the effort to construct the CLF type terminal cost in which nonlinearity of the system is explicitly handled, local optimality of MPC is often lost. On the other hand, researches on the intrinsic properties of EMPC reveal that EMPC with a finite horizon is inherently stabilizing under certain conditions [5-6]. Thus it is conceivable that if a sufficiently large horizon is used or if the system state is close to the steady state, stability of MPC/EMPC is less of a concern. In these cases, the existing framework could be overly conservative.

In the present work, we provide an alternative framework for EMPC which is in favor of the performance. We design a terminal cost which preserves the local optimality for EMPC. First we show under certain conditions that for systems optimally operated at steady state, the optimal control is locally approximate to an LQR controller. The proposed terminal cost is constructed as the value function of the LQR controller

plus an extra linear term. For conventional NMPC with quadratic cost functions, the linear term can be dropped. We show that EMPC with the proposed terminal cost and a sufficiently large horizon is stabilizing and locally behaves like the LQR controller. Estimation of the region of attraction is also provided.

References

[1] H. Chen and F. Allgöwer. A quasi-infinite horizon nonlinear model predictive control scheme with guaranteed stability. Automatica, 34:1205–1217, 1998.

[2] A. Jadbabaie, J. Yu, and J. Hauser. Stabilizing receding horizon control of nonlinear systems: a control Lyapunov function approach. In Proceedings of the American Control Conference, San Diego, pages 1535–1539, 1999.

[3] D. Q. Mayne, J. B. Rawlings, C. V. Rao, and P. O. M. Scokaert. Constrained model predictive control: Stability and optimality. Automatica, 36:789–814, 2000.

[4] R. Amrit, J. B. Rawlings, and D. Angeli. Economic optimization using model predictive control with a terminal cost. Annual Reviews in Control, 35:178–186, 2011.

[5] L. Grüne. Economic receding horizon control without terminal constraints. Automatica, 49:725–734, 2013

[6] D. Limón, T. Alamo, F. Salas, and E. F. Camacho. On the stability of constrained MPC without terminal constraint. IEEE Transactions on Automatic Control, 51:832–836, 2006.

462068 Multi-Rate Subspace-Based System Identification and Economic Model Predictive Control of the Electric Arc Furnace

Abstract:

The electric arc furnace (EAF) process is widely employed in steelmaking, using electricity to melt post-consumer scrap metal to produce new steel. The process provides significant reductions in the labor, energy and environmental costs of steelmaking over conventional blast furnaces that forge steel out of iron ore, coal and gas. Despite these efficiency gains, the EAF process requires generating tremendous amounts of heat to recycle the scrap steel, while the industrial sector struggles with electricity costs that are exorbitantly high [1]. The energy efficiency and economics of the EAF process can hence be improved by leveraging the proliferation of new technologies in optimization and advanced control systems. The application of mathematical programming problems, such as model predictive control and optimization in general, however, rely heavily on good predictions from process models.

One approach for modeling and control of the EAF process involves the development of first-principles/mechanistic models that can then be used for the purposes of optimization [1]. While numerous variations of these model-based approaches have improved the computational tractability of such problems, the development and implementation of first-principles model-based controllers remains challenging. Compared to first-principles models, recent work has exploited the increased availability of historical process data to develop data-driven models that are simple enough to make model-based control design practical. The data-driven causal dynamic models identified using latent variable [2, 3] or subspace [4] methods can be readily integrated into an on-line optimization framework, where at each sampling instant a finite horizon optimal control problem is solved, yielding an optimal control sequence that achieves desired closed-loop performance. However, most of these data-driven models are ill-suited for the electric arc furnace application where output measurements are scheduled at different sampling rates because of hardware limitations on sensors. Specifically, the harsh environment and high corrosiveness of molten steel mean that on-line measurements of the molten steel temperature and chemical composition are often disrupted with unmeasured process variables and missing data, thus making existing data-driven modeling approaches inapplicable.

Even with the availability of good models, the optimization of the arc furnace, analogous to other batch processes, deals with the effective allocation of a set of limited resources over a finite time duration. When considering competing criteria and from an industrial perspective where the prevailing incentive is of an economic nature, the multi-objective optimization problem for the EAF process is to economically reach the desired product end-point target at the termination of the batch. While recent contributions have demonstrated the advantages of economic model predictive control under the assumption that all batches are of equal durations [5], using a data-driven model that is capable of handling inconsistent batch lengths opens new and beneficial possibilities of optimizing the operation of batches with variable durations.

Motivated by the above considerations, in this work we develop a system identification method and economic model predictive control framework for the EAF process. The system identification method is based on a subspace formulation and uses the reliable singular value decomposition to identify a dynamic model of the process from a finite number of noisy data samples that are disrupted with unmeasured process variables and asynchronous data. Furthermore, the proposed subspace-based system identification approach does not require the use of iterative computational procedures or nonlinear (and possibly non-convex) optimization. The resulting dynamic model is integrated into a tiered economic model predictive control (EMPC) formulation, where solutions to computationally tractable quadratic and linear programming problems achieve the desired final product end-point specification by batch termination while minimizing the operating costs. The multi-rate subspace-based system identification technique and EMPC framework is implemented on the EAF process subject to the limited availability of process measurements, missing data, measurement noise, and constraints.

[1] R. D. M. MacRosty and C. L. E. Swartz. Dynamic modeling of an industrial electric arc furnace. Ind. Eng. Chem. Res., 44:8067–8083, 2005.
[2] M. Golshan, J. F. MacGregor, M.-J. Bruwer, and P. Mhaskar. Latent variable model predictive control (LV-MPC) for trajectory tracking in batch processes. J. Proc. Cont., 20:538–550, 2010.
[3] S. Aumi, B. Corbett, T. Clarke-Pringle, and P. Mhaskar. Data-driven model predictive quality control of batch processes. AIChE J., 59:2852–2861, 2013.
[4] B. Corbett and P. Mhaskar. Subspace identification for data-driven modeling and quality control of batch processes. AIChE J., 62:1581–1601, 2016.
[5] M. M. Rashid, P. Mhaskar, and C. L. E. Swartz. Multi-rate modeling and economic model predictive control of the electric arc furnace. J. Proc. Cont., 40:50–61, 2016.

465360 Tracking Control of Boundary Controlled Continuum Models of Production Systems

Abstract:

Modelling of the average behaviour of the large scale production/manufacturing plant is described by nonlocal spatially distributed continuous transport models, see [1] and [2].
This description is suitable, e.g., for a semiconductor factory which produces a large number of items in a large number of steps. The considered continuum model satises conservation law, and the appropriate mathematical variable representing the production flow is a products density. Moreover, the production flow velocity of the conservative production system is given by hyperbolic PDE depending only on the products density and it is constant cross the entire system at a given time. Actually, this implies that in a real factory, all parts move through the factory with same speed at a given time.
Controlling the production rate of a manufacture production system is an important goal in manufacturing: producing too little of an item results in the lost sales and backlog costs while

producing too much leads to inventory and holding costs. In order to maximize protability, a production system has to be able match its projected demand. Although demand is stochastic over a given time period, a business typically generates a demand forecast for the next day, week, month, etc and runs its production system to match this demand accordingly and this demand can be modelled by a known signal process which is given by a so called exosystem.
In this work, we will develop a control law that controls the outflux of the continuum model solely by regulating the influx to achieve demanded tracking of targeted production. In [3], the theory about the output feedback stabilization of the considered continuum model has been investigated and important results in [3] assist to construct the boundary control law in such a way that the targeted demand tracking can be realized in the framework of internal model control. Furthermore, the solvability of resulting regulator equations is discussed and the solvability conditions is provided.
Finally, computer simulation will be presented to show the performance of the proposed controller, in other words, the demanded tracking production can be achieved.

[1] Armbruster, D., Marthaler, D. E., Ringhofer, C., Kempf, K., and Jo, T. C. (2006). A continuum model for a re-entrant factory. Operations research, 54(5), 933--950.

[2] Marca, M. L., Armbruster, D., Herty, M., and Ringhofer, C. (2010). Control of continuum models of production systems. IEEE Transactions on Automatic Control, 55(11), 2511--2526.

[3] Coron, J. M., and Wang, Z. (2013). Output feedback stabilization for a scalar conservation law with a nonlocal velocity. SIAM Journal on Mathematical Analysis, 45(5), 2646--2665.

458678 Model Predictive Control for Optimal Zone Tracking

Abstract:

Model predictive control (MPC) is an advanced control method that has been highly successful with thousands of applications in the chemical and petroleum industries alone (Qin & Badgwell, 2003). MPC relies on a model of the system to predict the process variables based on the actions taken by the controller (Rawlings & Mayne, 2009). In the traditional MPC paradigm, an online optimization problem is solved at each step to drive process variables to desired setpoints. However, in many industrial applications, the control objective is to keep process variables within a zone or target set. One such example is level control in a buffer tank between upstream and downstream process units. Another example is indoor air temperature control in commercial buildings. Since traditional MPC involves tracking to a setpoint, the theory needs to be extended to handle zone tracking. In this talk, we develop the MPC theory for tracking a target set, we propose an implementable framework for satisfying secondary control objectives, and we demonstrate the effectiveness of our formulation on two example problems.

In zone tracking, the stage cost is zero if the state is in the target set and positive otherwise. One such function that satisfies this property is the distance function from a point to a set. Under the stated assumptions, a Lyapunov function for the closed-loop system is derived to prove that the target set is asymptotically stable. A key practical consideration is that the solution to the optimal control problem when tracking zones is often not unique. There may exist a set of input sequences that can maintain the state within the target set. In this case, it is desirable to achieve secondary control objectives. To this end, other terms can be added to the objective function to minimize economic costs, input usage, or input movement to select the most desirable state trajectory within the target set for the given process conditions. A general framework is proposed to achieve these goals. After solving the zone tracking problem to determine the set of stabilizing input trajectories, a second optimization problem can be solved to determine the unique input trajectory from the stabilizing set that optimizes secondary control objectives. While this method requires solving two optimization problems at each MPC step, the procedure is simplified to solving a single optimization problem using exact penalties. Exact penalties on the deviation of the state from the target set can be used to ensure that the solution does not leave the target set if feasible. The resulting solution leads to stabilizing behavior if the penalty coefficient is chosen to be large enough.

The proposed formulation is applied to two example problems. First, a buffer tank between upstream and downstream process units is considered. The goal of the buffer tank is to prevent the upstream flow disturbance from propagating downstream by allowing the height of material in the tank to vary. While the height is allowed to vary, it should be kept within bounds to prevent emptying or overflowing. Through a simulation study, we show that the proposed MPC formulation accomplishes these goals. The controller changes the outflow of the tank to get the height within the bounds. When the height is within bounds, the controller keeps the outflow of the tank constant as long as the height stays in the target zone.

The second example problem considered is indoor air temperature control. The air temperature in commercial buildings is not required to be at a fixed setpoint. The air temperature should be maintained between upper and lower bounds to keep the occupants in the building comfortable. If the temperature is outside of these bounds, the controller should take action to drive the temperature to the comfort region. When the temperature is inside the bounds, the controller should minimize the amount of energy used by the heating, ventilation, and air conditioning (HVAC) system. We demonstrate that the proposed MPC formulation tracks the comfort region by rejecting the ambient disturbance. When the temperature is inside the comfort region, the controller avoids using the HVAC system to cool if possible by taking advantage of the disturbance while keeping occupants satisfied.

References

Qin, S. J., & Badgwell, T. A. (2003). A survey of industrial model predictive control technology. Control Engineering Practice, 11 (7), 733-764.

Rawlings, J. B., & Mayne, D. Q. (2009). Model Predictive Control: Theory and Design. Madison, WI: Nob Hill Publishing.

464459 A New Multitasking Continuous Time Formulation for Short-Term Scheduling of Operations in Multipurpose Plants

Abstract:

The problem of short-term scheduling for multipurpose plants has been the subject of significant research over the past two decades. Discrete and continuous formulations that take the form of a mixed-integer linear programming problem (MILP) have been proposed to address this emerging problem in the industry. Despite the burgeoning of continuous time formulations in the past decade, the existing approaches in the literature have not considered the case of machines capable of performing multiple tasks at the same time in multipurpose plants (i.e., multitasking). Plants in sectors such as semiconductor manufacturing and analytical services sectors consider machines capable of performing different tasks simultaneously; hence, the need to develop formulations that enable optimal scheduling of their operations. This paper aims at presenting a novel continuous time formulation for scheduling operations at multipurpose facilities with machines capable of multitasking. The proposed formulation uses the idea of Global event-based formulation to represent time as a continuous variable. Apart from accounting for operational constraints present at multipurpose facilities, e.g. sequential constraints, flow conservation constraint and capacity constraints, the formulation is capable of adopting different operational objective functions such as maximizing throughput or minimizing the turn-around time of the facility. These features make the formulation readily applicable to those multipurpose facilities that have machines with the multitasking feature. To assess the impact of incorporating the multitasking feature of the machines, the performance of the proposed formulation is compared against a continuous time formulation that is not capable of incorporating this feature (essentially, a single tasking formulation). The results, conducted through the solution of multiple instances, ranging from small to relatively large instances, shows that the quality of the solution obtained through the multitasking formulation is considerably better than its single-tasking counterpart. Furthermore, the CPU time obtained for both formulations is very similar for the instances considered in the comparative study. Thus, the proposed multitasking formulation is capable of generating considerably better quality solutions compared to its single-tasking counterpart while requiring similar solution times.

464173 Scheduling Straight Multiproduct Pipelines with Generalized Disjunctive Programming

Abstract:

Pipelines are the most cost-effective way of transporting large quantities of refined petroleum products (e.g. gasoline, kerosene, jet fuel, diesel, heating oil, liquefied petroleum gases) over large distances. Pipelines have grown considerably and have become structurally more complex. The main challenge of pipeline scheduling is to ensure that oil products form refineries arrive to the distribution centers serving local markets at the right time and preferably at the lowest cost. It involves batch sizing, sequencing and respecting minimum and maximum flowrate limits on the different segments that make the pipeline.

Pipeline scheduling models are typically of the mixed-integer linear type (MILP) unless an accurate representation of the pumping cost is required [1] that transforms the problem into a mixed-integer nonlinear program (MINLP). Similarly to general scheduling formulations, one may rely on a discrete or continuous representation, which in this case applies to both time and volume inside the pipeline. In this paper, we present a full-space continuous-time and volume formulation for straight pipelines that uses a single time grid to keep track of events taking place. The straight pipeline includes multiple intermediate nodes in addition to the first refinery and the last depot. The nodes can be single purpose (either input or output) or dual purpose nodes that ship and receive materials at different times.

We follow up on our recent work in [2] dealing with simultaneous injections and deliveries. The novelty is that we now allow for interacting pumping runs, where an intermediate refinery node can feed the downstream segment simultaneously with the upstream node, provided that the same product is involved. It allows to reduce the makespan as well as the number of time slots required to transfer the petroleum products and meet demand. Since the latter affects most model variables and constraints, the computational performance is also improved. Real-world case studies involving an Iranian pipeline are used to illustrate the advantages of the proposed model.

To ensure an efficient model by design, we rely on Generalized Disjunctive Programming (GDP) [3-5] and develop disjunctions that mostly allow for a compact convex hull reformulation [4]. Disjunctions relate binary and continuous variables, whereas the majority of constraints relating continuous variables are essentially mass balances. Finally, the logic propositions relate the different sets of binary variables. As such, a simpler and more systematic model is derived compared to our recent GDP efforts for pipeline scheduling [2, 6] (where a few constraints were developed ad-hoc), potentially making it easier to generalize to tree-like structures.

Acknowledgments: Financial support from Iranian Ministry of Science, Research and Technology and Fundação para a Ciência e Tecnologia (FCT) through the Investigador FCT 2013 program and project UID/MAT/04561/2013.

References:

[1] V.G. Cafaro, D.C. Cafaro, C.A. Mendéz, J. Cerdá, MINLP model for the detailed scheduling of refined products pipelines with flow rate dependent pumping costs, Comput Chem Eng 72 (2014) 210–221.

[2] H.Mostafaei, P.M. Castro, A. Ghaffari-Hadigheh, Short-term scheduling of multiple source pipelines with simultaneous injections and deliveries, Comput Oper Res 73 (2016) 27-42.

[3] R. Raman, I.E. Grossmann, Modeling and computational techniques for logic based integer programming, Comput Chem Eng 18 (1994) 563–578.

[4] P.M. Castro, I.E. Grossmann, Generalized disjunctive programming as a systematic modeling framework to derive scheduling formulations, Ind Eng Chem Res 51 (2012) 5781–5792.

[5] I.E. Grossmann, F. Trespalacios, Systematic modeling of discrete-continuous optimization models through generalized disjunctive programming, AIChE J 59 (2013) 3276–3295.

[6] H. Mostafaei, P.M. Castro, A. Ghaffari-Hadigheh, A novel monolithic MILP framework for lot-sizing and scheduling of multiproduct treelike pipeline networks, Ind Eng Chem Res 54 (2015) 9202–9221.

466880 Medium-Term Scheduling of Integrated Gasoline Blending and Delivery Operations Using Enhanced Rolling Horizon Decomposition Approach

Abstract:

Gasoline is one of the most profitable products of an oil refinery and can account for as much as 60-70% of total revenue [1-3]. Several gasoline cuts or fractions from various processes are typically blended together to produce different gasoline grades in order to meet customer orders of varying specifications and delivery dates. The entire operations involve handling a large numbers of orders, delivery dates, blenders, blend components, product tanks, and quality specifications, making scheduling highly complex. Optimal scheduling of such blending and delivery operations has several advantages such as avoiding ship demurrage, improving customer satisfaction, minimizing quality give-aways, reducing transitions, and slop generation, exploiting low-quality cuts, and reducing inventory costs. However, mathematical modeling of blending operations often introduces non-convex nonlinear terms, making the problem a large-scale complex non-convex mixed-integer nonlinear programming (MINLP). Scheduling based on industrial experiences or heuristics often leads to inferior schedules with costly quality give-aways, many transitions, and order demurrage. Thus, it is imperative to seek advanced techniques such as mixed-integer programming approach for simultaneous treatment of recipe, blending, scheduling and delivery operations.

The early work [4-5] mainly focused on finding optimal blending recipes of various intermediate fractions and some additives to meet product quality specifications. Scheduling decisions such as resources allocations and temporal decisions were not considered [2]. Although several efforts[6-8] have developed models and solution approaches to address the problem of scheduling of gasoline blending operation, an integrated treatment of recipe, blending, scheduling, and delivery is missing [2]. Li et al. [2], Li and Karimi [9], Castillo-Castillo and Mahalec [10-12], Li et al. [13] and Cerda et al. [14] addressed the problem of integrated gasoline blending and delivery operations. They incorporated many problem features such as non-identical parallel blenders, minimum run length and amount in a blend run, constant blending rates, changeover, one blender charging at most one tank at a time, multi-purpose product tanks, piecewise constant profiles for blend component qualities and feed rates, etc. It should be mentioned that recently Li et al. [13] did improve MILP relaxation of Li and Karimi [9] and proposed a hybrid global optimization approach to solve the MINLP model to e- (e.g., 1%) global optimality. All of these models and solution algorithms are restricted to short-term scheduling. When the scheduling horizon increases to medium term, they may need much more computational efforts or fail to find a feasible solution.

To address the problem of medium-term scheduling of gasoline blending and delivery operations, Castillo-Castillo and Mahalec [15] developed a multi-period inventory pinch-based algorithm to solve continuous-time scheduling models (MPIP-C algorithm). The entire problem was decomposed into three-level subproblems. The top-level model computed optimal recipes for aggregated blends over periods initially delineated by inventory pinch points. The second level model used fixed blend recipes to compute an approximate schedule. The last level model was the short-term continuous-time model that generated detailed schedules for the entire problem.

In this presentation, we use the model of Li et al. [13] and enhance rolling horizon decomposition approach to address the problem of medium-term scheduling of gasoline blending and delivery operations. The entire problem is decomposed into two-level sub-problems (i.e., Level-1 and Level-2 models). While the Level-1 model is utilized to determine the current time horizon and corresponding orders that should be included into the current horizon, the level-2 model is the same short-term scheduling model of Li et al. [13] to determine the detailed scheduling operations in the current horizon and minimize the total operating cost. We first solve 14 industrial-scale examples from Li et al. [13] and compare with those of Li et al. [13]. Then, we solve the examples from Castillo-Castillo and Mahalec [15] and compare with their solution algorithm MPIP-C [15]. The computational results demonstrate that we can generate better solutions using our enhanced rolling horizon decomposition than those of MPIP-C algorithm.

References

[1] Jia, Z. Y.; Ierapetritou, M. Mixed-Integer Linear Programming Model for Gasoline Blending and Distribution Scheduling. *Industrial and Engineering Chemistry Research* **2003**, *42*, 825-835.

[2] Li, J.; Karimi, I. A.; Srinivasan, R. Recipe Determination and Scheduling of Gasoline Blending Operations. *AIChE Journal* **2010**, *56*, 441-465.

[3] Shah, N. K.; Li, Z.; Ierapetritou, M. G. Petroleum Refining Operations: Key Issues, Advances, and Opportunities. *Industrial and Engineering Chemistry Research* **2011**, *50*, 1161-1170.

[4] Dewitt, C. W; Lasdon, L. S.; Waren, A. D.; Brenner, D. A.; Melhem, S. A. OMEGA: An Improved Gasoline Blending System for Texaco. *Interfaces* **1989**, *19*, 85-101.

[5] Rigby, B.; Lasdon, L. S.; Waren, A. D. The Evolution of Texacoâ€™s Blending Systems: From OMEGA to Starblend. *Interfaces* **1995**, 25, 64-83.

[6] Mendez, C. A.; Grossmann, I. E.; Harjunkoski, I.; Kabore, P. A. Simultaneous Optimization Approach for Off-Line Blending and Scheduling of Oil-Refinery Operations. *Computers and Chemical Engineering* **2006**, *30*, 614-634.

[7] Glismann, K.; Gruhn, G.; Short-Term Scheduling and Recipe Optimization of Blending Processes. *Computers and Chemical Engineering* **2001**, *25*, 627-634.

[8] Joly, M; Pinto, J. M. Mixed-Integer Programming Techniques for the Scheduling of Fuel Oil and Asphalt Production. *Trans IChemE. Part A.* **2003**, *81*, 427-447.

[9] Li, J.; Karimi, I. A. Scheduling Gasoline Blending Operations from Recipe Determination to Shipping using Unit Slots. *Ind. Eng. Chem. Res.* **2011**, *50*, 9156-9174.

[10] Castillo, P. A. C.; Mahalec, V. Inventory Pinch Based, Multiscale Models for Integrated Planning and Scheduling-

Part I: Gasoline Blend Planning. *AIChE Journal* **2014**, *60*, 2158-2178.

[11] Castillo, P. A. C.; Mahalec, V. Inventory Pinch Based, Multiscale Models for Integrated Planning and Scheduling-Part II: Gasoline Blend Scheduling. *AIChE Journal* **2014,** *60*, 2475-2497.

[12] Castillo-Castillo, P. A.; Mahalec, V. Improved Continuous-Time Model for Gasoline Blend Scheduling. *Computers and Chemical Engineering* **2016**, *84*, 627-646.

[13] Li, J.; Xiao, X.; Floudas, C. A. Integrated Gasoline Blending and Order Delivery Operations: Part I. Short-Term Scheduling and Global Optimization for Single and Multi-Period Operations. *AIChE Journal* **2016**, *62*, 2043-2070.

[14] Cerda, J.; Pautasso, P. C.; Cafaro, D. C. A Cost-Effective Model for the Gasoline Blend Optimization Problem. *AIChE Journal* **2016,** In press. DOI: 10.1002/aic.15208.

[15] Castillo-Castillo, P. A.; Mahalec, V. Inventory Pinch Gasoline Blend Scheduling Algorithm Combining Discrete- and Continuous-Time Models. *Computers and Chemical Engineering* **2016**, *84*, 611-626.

466962 Reducing Enterprise Wide Planning and Scheduling to Practice

Abstract:

Optimal planning and scheduling of significant maintenance activities such as plant turn-arounds grows in complexity and potential economic impact as the scale of the envelope grows from the local to global. Over the last few years we have invested significant research effort in the development of optimization approaches to handle these large scale problems. In this current paper we will present some accumulated wisdom from the successful implementation of planning and scheduling of turn-arounds from the short-term tactical scheduling to long-term to strategic planning of turn-arounds for integrated manufacturing sites for a global chemical manufacturer. We discuss the challenges of scaling the decision support tool as it moved from planning for a single integrated site to multiple sites across geographies and more network linkages are created with the inclusion of inventories, logistics and market conditions.

469251 Infeasibility Analysis for Scheduling Applications

Abstract:

Planning and scheduling software are crucial for industries and must provide feasible, robust, and operationally acceptable schedules. Our focus in this presentation is on tactical scheduling with a horizon of hours or days where real-time disturbances need to be accommodated into any proposed schedule. Equipment failures, resource storage policy change, resource availability are examples of such real-world uncertainties and in many cases, the optimal schedule obtained a priori is not robust to such changes. On-the-fly rescheduling algorithms try to repair such infeasibilities by employing quick heuristics.

In this work, we present use of infeasibility analysis in the context of rescheduling for a multi-purpose chemical batch process. We identify irreducible inconsistent sets (IISs), which are subsets of an infeasible model, such that the IIS itself is infeasible but any proper subset of the IIS is feasible. We isolate multiple IISs systematically and identify bottle-necks in process scheduling network by discovering a set of mutual inconsistencies that must be eliminated to achieve feasibility. We demonstrate this approach through an interactive and intuitive scheduling interface for variety of scheduling models using the well-known Westernberger-Kallrath case study.

469973 Mathematical Programming Models and Solution Methods for Online Scheduling of Central Heating/Cooling Plants

Abstract:

In large commercial buildings or campuses, the heating, ventilation, and air conditioning (HVAC) system is often served by a set of equipment operating in a "central plant." Use of this equipment must be scheduled so as to supply the time-varying heating and cooling demands of the overall system. While finding a feasible schedule is typically not challenging, it is desirable to operate the central plant at the lowest possible cost, which depends on a number of factors. First, utilities (primarily electricity) are often subject to time-varying prices or other rate structures that cause nominal operation (i.e., meeting system demand exactly on time) to be expensive. Second, equipment efficiency can vary highly with load, and so units should be operated at levels where they are most efficient. Third, to provide operational flexibility, thermal energy storage (TES) is often installed to temporally decouple chilled or hot water production and consumption. Fourth, equipment startup and shutdown may be accompanied by unmodeled transient dynamics, which requires that units not be switched rapidly. To address all of these considerations while minimizing operational costs, schedules can be determined by optimization using typical production scheduling methods. This approach avoids the use of heuristics and can lead to significant cost savings. However, a single schedule is not sufficient due to disturbances, and thus a practical implementation must address not offline scheduling but online scheduling.

Determining an optimal schedule requires forecasts for both utility prices (used in the objective function) and system thermal load (used in operational constraints). As the system evolves, these forecasts are updated to reflect measured system behavior, which can invalidate prior schedules. In addition, units may break down or be removed from service for maintenance. Thus, the schedule must be constantly updated to reflect the best available information, which imposes constraints on how much time can be spent determining an individual schedule. To meet these requirements, the scheduling formulation must be balanced so as to account for primary cost contributors (e.g., variable electricity price and equipment efficiency) while not considering too much detail so as to be computationally intractable. Although general production scheduling formulations can describe HVAC central plants, a specifically tailored formulation can provide more flexibility while yielding faster solution times.

In this talk, we present a mixed integer linear programming (MILP) formulation for online scheduling of HVAC central plants. Similar to a resource-task network (Pantelides, 1994) formulation, we use an abstract representation of the central plant in terms of "resources," which are mass or energy flows within the system, and "generators," which are pieces of equipment in the central plant. After adjusting equipment utilization and demand satisfaction constraints to more directly model the central plant, we embed piecewise-linear approximations of generator production/consumption models to accurately account for variable generator efficiency. Then, we employ symmetry-removal reformulations that greatly improve solution times for systems with multiple identical generators. The end result is a compact formulation that can accommodate a wide variety of central plant configurations and that can be optimized online for practically sized central plants.

Using this scheduling formulation, we also present a case study of closed-loop operation for a medium-size central plant consisting of chillers, cooling towers, pumps, and a storage tank. In this study, we discuss the effects of different prediction horizons and also inaccurate demand forecasts. We show that, even for moderate prediction horizons, the scheduling model can be solved to within 0.5% of optimality in 30 s. We also show that operating costs are not significantly impacted by uncertain forecasts due to closed-loop feedback. Finally, we discuss a cascaded approach that separates the long-term storage utilization decisions from the short-term unit commitments by using a simplified surrogate model with a long horizon at the upper level and a fully-detailed model with a short horizon at the lower level. Thus, rapid rescheduling can correct for inaccurate long-term forecasts, and existing central plants can be operated at significantly lower costs.

References

Pantelides, C. C. (1994). Unified frameworks for optimal process planning and scheduling. In Proceedings on the second conference on foundations of computer aided operations (pp. 253--274).

473931 A Unified Planning and Scheduling Platform for Refining Operations

Abstract:

A Unified Planning and Scheduling Platform for Petroleum Refining

Traditionally the functions of refinery planning and scheduling are separated both in terms of business organization structure as well as processes and technologies employed. Planning is typically done using a commercial optimization tool (such as Aspen PIMS) and executed by the economics planning function. Scheduling on the other hand, is done using primarily simulation-based tools at the refinery level (such as Aspen Petroleum Scheduler) augmented by optimization tools at the product blending level (such as Aspen MBO). This practice creates a disconnected set of workflows that primarily create an overly optimistic plan that is not constraint by detailed scheduling considerations. As a result, there is a mismatch between the plan and the schedule constraints that leaves a significant set of benefits unrealized. This is known in the refining industry as a margin leak.

In this paper, we review the main challenges in unifying refinery planning and scheduling and propose a framework for addressing those challenges. We also show the necessary key attributes for the new modeling and analysis environment and demonstrate how they can be brought together to enable the combined optimization of planning and scheduling.

469044 Continuous-Time Models for Scheduling of Parallel Batch Digesters for Pulp Production

Abstract:

The problem of short-term scheduling of batch digesters [1-2] involves development of a mathematical model for finding the optimal sequence of various operations, and their corresponding batch start and end times. The aim is to produce pulp of different qualities in a multi-digester facility comprising multiple production lines. In each production line there are parallel batch digesters along with shared resources (chip filling station, impregnation liquor pump, hot liquor pump, and blow down tank) for meeting the overall pulp demand in a specified time horizon. Unlike conventional batch digesters, displacement batch digesters offer improved energy efficiency through storage and reuse of the hot spent liquor in subsequent batches. Recently, Shaik and Bhat [3] proposed a mixed-integer linear programming (MILP) model for scheduling of displacement batch digesters using discrete-time formulation and State-Task-Network (STN) representation. In their model material balances were modified to include the flow to/from flow transfer tasks, and resource balances were proposed to monitor the utilization of digesters, and extensions to handle shared resources.

In this work, we consider scheduling of both conventional and displacement batch digesters using a resource-task-network (RTN) representation of production recipes, based on unit-specific event based continuous time representation resulting in a MILP problem. The basic framework is adapted from the unit-specific event based models [4-5] dealing with unification of STN/RTN representations and generalization of resource balances. The objective of the scheduling problem is either maximization of production or maximization of total duration of heating and cooking tasks which in turn reduces the overall energy consumption. Illustrative case studies will be presented on both conventional and displacement batch digesters for a digester house comprising of two parallel production lines with four digesters for producing pulp of two different qualities, to demonstrate the proposed modifications and comparison of results with discrete time model [3].

References

[1] Hvala, N.; Strmcnik, S.; Cernetic, J. (1993). Scheduling of batch digesters according to different control targets and servicing limitations, *Comp. Chem. Eng.*, 17, 739-750.

[2] Castro, P.; Matos, H.; Barbosa-Povoa, A.P.F.D. (2002). Dynamic modeling and scheduling of an industrial batch system, *Comp. Chem. Eng.*, 26, 671-686.

[3] Shaik, M.A.; Bhat, S. (2014). Scheduling of displacement batch digesters using discrete time formulation, *Chem. Eng. Res. & Des.*, 92, 318-339.

[4] Shaik, M.A.; Floudas, C.A. (2009). Novel unified modeling approach for short-term scheduling, *Ind. Eng. Chem. Res.*, 48, 2947-2964.

[5] Shaik, M.A.; Vooradi, R. (2013). Unification of STN and RTN based models for short-term scheduling of batch plants with shared resources, *Chem. Eng. Sci.*, 98, 104-124.

Model Predictive Control of a Rapid Pressure Swing Adsorption Medical Oxygen Concentrator

Presenting Author:
Matthew Urich

Contributing Authors:
Rama Rao Vemula
Mayuresh V. Kothare

Department of Chemical and Biomolecular Engineering, Lehigh University
Bethlehem, PA

Medical Oxygen Therapy and COPD

- Medical oxygen therapy used by patients with Chronic Obstructive Pulmonary Disease (COPD)
    - Film in lungs reduces oxygen uptake ability, decreased blood oxygen levels
    - High purity oxygen in lungs boosts mass transfer to blood
    - ~90% oxygen purity required
- Medical Oxygen Concentrators produce high purity oxygen from ambient air
    - Replace traditional oxygen cylinders
    - MOC size and weight a huge design factor
    - Typical design: 2 columns using pressure swing adsorption technology
- Various MOC sizes deliver different oxygen flow rates
    - Stationary Units: 1-5 L/min (at home only)
    - Ambulatory Units: 1-3 L/min (backpack-sized, portable)
    - Portable Units: <1 L/min (over the shoulder, portable)

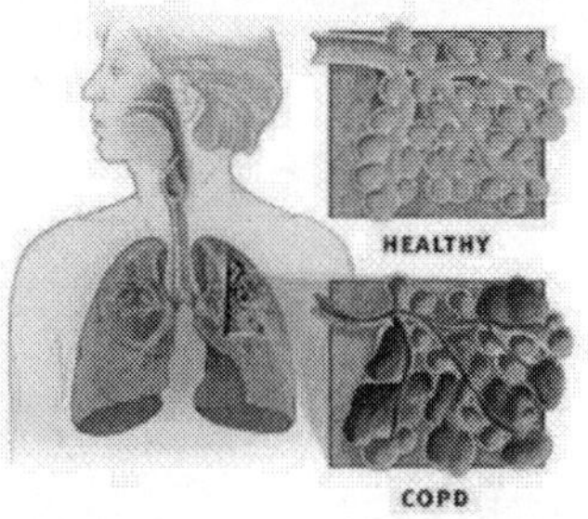

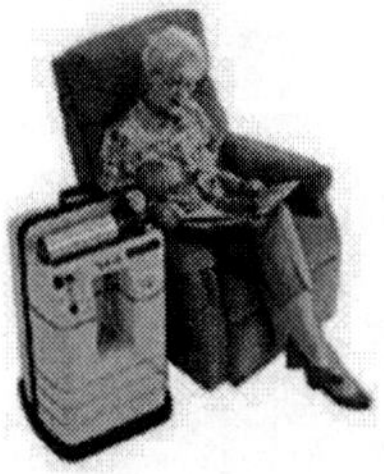

2

Novel Medical Oxygen Concentrator Design

Schematic of the RPSA MOC Apparatus [1]

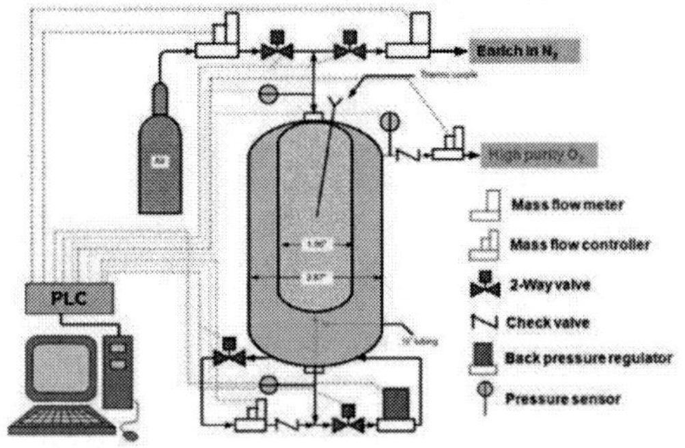

Optimal Cycle Time for the RPSA MOC [2]

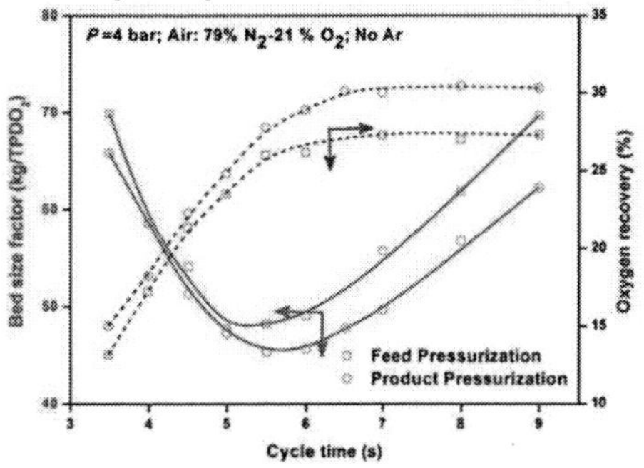

- The Rapid Pressure Swing Adsorption (RPSA) Medical Oxygen Concentrator (MOC) has a single adsorption column, inside a product storage tank
- Uses a zeolite adsorber material to continuously deliver ~90% oxygen to COPD patients
- RPSA is a cyclic process, and there is an optimal cycle that minimizes the bed size factor, and maximizes recovery
- The cycle timing plays a key role in RPSA operation
- RPSA currently operates without feedback control, and cannot reject process disturbances

[1] Rao VR, Kothare MV, Sircar S. Novel design and performance of a medical oxygen concentrator using a rapid pressure swing adsorption concept. AIChE J. 2014;60(9):3330-3335.
[2] Siew-Wah C, Sircar S, Kothare MV. *Miniature oxygen concentrators and methods*. 2012.

3

RPSA Cyclic Operation and Control Opportunities

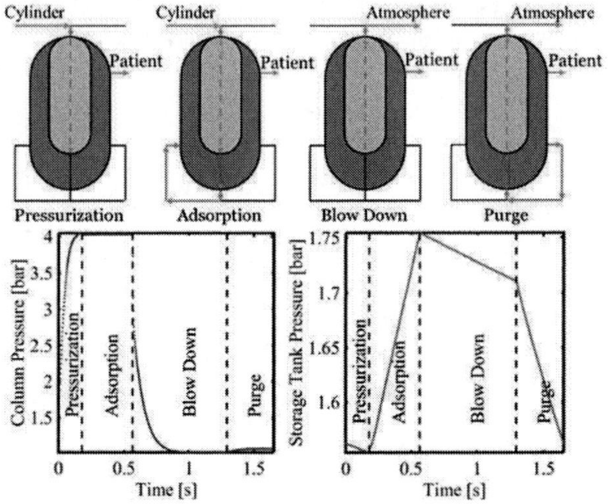

- Skarstrom-Like 4 Step RPSA Cycle:
  1. Pressurization: Adsorber Column Pressurizes with Feed Air
  2. Adsorption: Adsorber removes nitrogen, and delivers product oxygen to the storage tank
  3. Blow Down: Adsorbed Nitrogen Exhausts to Atmosphere from the column feed end
  4. Purge: Some oxygen product used to purge residual nitrogen from column
- "Traditional" Steady State not achieved, but Cyclic Steady State (CSS) does occur
- Storage Tank Pressure is Essential for Proper RPSA Operation
  - Delivery to patient and adequate purge
- MPC directly controls storage tank oxygen purity and pressure by manipulating the cycle step durations
- Feedback control necessary in all MOC designs

4

Unique Control Challenges in the RPSA System

- Challenge 1: Nonlinear System that Experiences Output Competition
  - Unique design requires controlling **both** oxygen purity and pressure in the storage tank
  - Multiple Outputs cannot be controlled by a single step duration
  - All four step durations can be independently adjusted to control both outputs

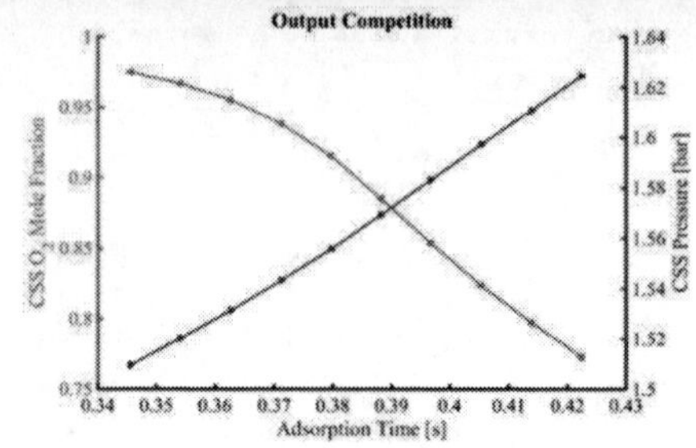

- Challenge 2: Rapid, Short Cyclic Operation
  - Total cycle time ~2-4sec, not much time for making control decisions
  - Controlling entire cycle profile not feasible
  - Use Cyclic Steady State, and control a reference point in cycle
  - Reference point choice: end of the purge step
  - Move from a continuous time to a discrete "cycle" time
  - Controller computation time is biggest challenge for implementation on the real RPSA system

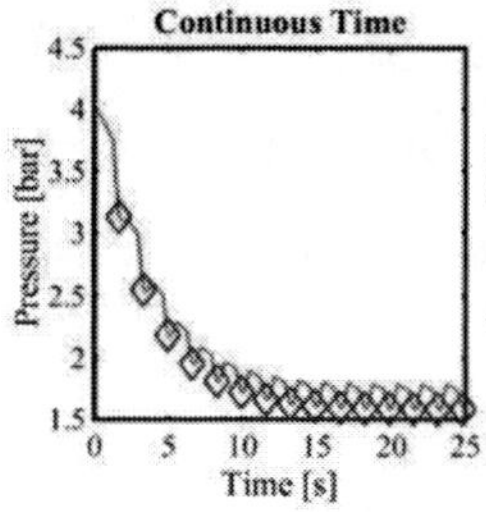

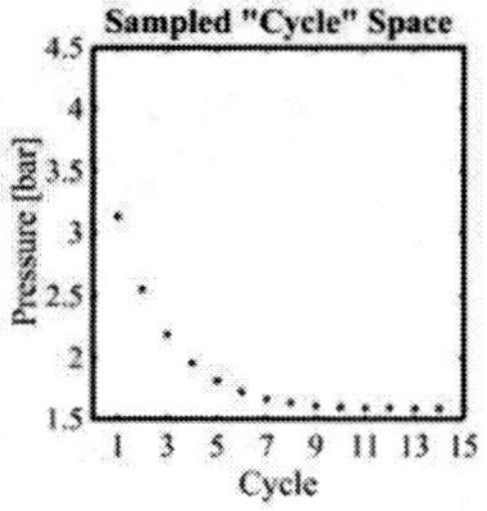

5

RPSA MOC Closed Loop System

- RPSA Plant Model
  - Mass Transfer Using LDF Model and Langmuir Isotherms
  - Heat Transfer between Solid and Gas Phases
  - Axially Dispersed Plug Flow in Adsorber Column
  - Storage Tank Mass Balances
  - Simulated with COMSOL Multiphysics and MATLAB
- Linear Model Identified from Plant Model Simulations
  - Pseudo-Random Binary Input Signals
  - Improve Computational Efficiency in MPC
  - Discrete Kalman Filter
- MPC Optimization Problem
  - Constrained Quadratic Optimization
  - Integral Action to Eliminate Offset
  - Optimal Inputs for Set Point Tracking and Disturbance Rejection

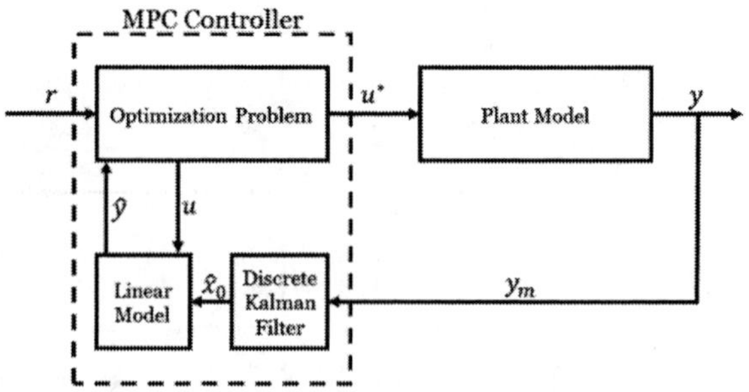

6

PRBS System Identification

- Computation Cost Biggest Disadvantage to MPC
  - Plant model too expensive, controller requires low-cost, simplified model
  - Linear models are least expensive, easiest to implement
- PRBS Input Signal Design Criteria:
  a) Each input signal has a lower and upper bound to ensure proper RPSA operation
  b) Input perturbations must cause large enough changes, but not too large
- Pseudo-Random Binary Signal System Identification
  - Relate the cycle step durations to the storage tank pressure and oxygen purity
  - Linear State space model replaces full plant model

$$\hat{x}(i+1) = A\hat{x}(i) + Bu(i)$$
$$\hat{y}(i) = C\hat{x}(i) + Du(i)$$

  - Identification Details: 8th Order Model
    92% Fit to Oxygen Purity
    85% Fit to Pressure

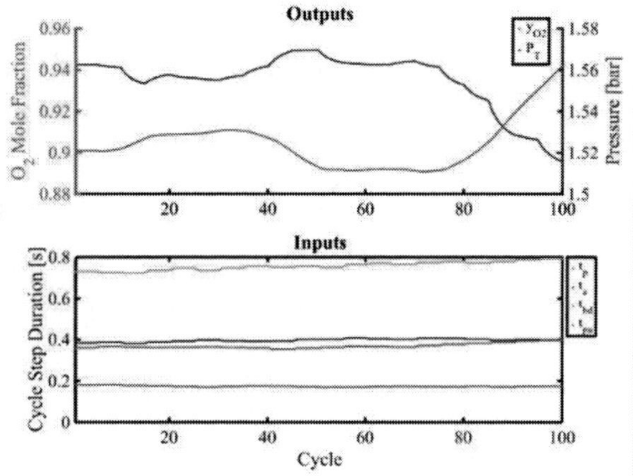

7

Model Predictive Controller Formulation

$$\min \sum_{i=1}^{N}[\hat{y}(k+i|k) - r(k+i|k)]^2_{\omega_1} + \sum_{i=2}^{M}\Delta u(k+i|k)^2_{\omega_2} + \sum_{i=1}^{M}[u(k+i|k) - u_b(k+i|k)]^2_{\omega_3}$$

$$s.t.\quad \hat{x}(k+i+1|k) = A\hat{x}(k+i|k) + Bu(k+i|k)$$
$$\hat{y}(k+i|k) = C\hat{x}(k+i|k) + Du(k+i|k)$$
$$\hat{y}_{min} \le \hat{y}(k+i|k) \le \hat{y}_{max}$$
$$u_{min} \le u(k+i|k) \le u_{max}$$

- Objective: set point tracking, minimize input changes
- Constraints: Identified model, physical output bounds, input operating limits
- MPC independently changes all step durations
- Standard MPC causes output oscillation even after tuning
- Integral action eliminates oscillations and offset at CSS

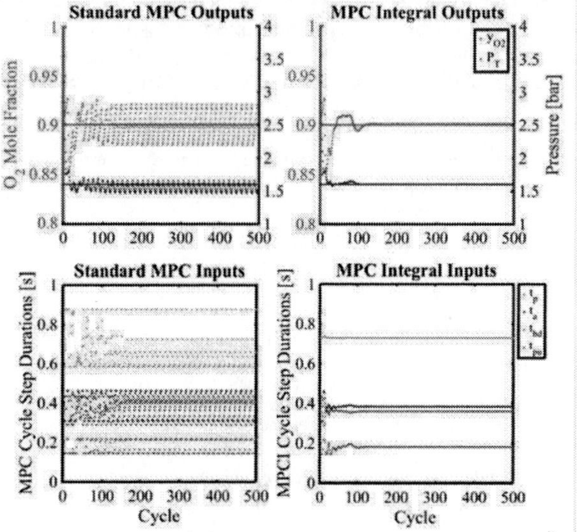

8

Disturbance Case Study: Changes in Feed Temperature

- Scenario: A patient takes a portable MOC outside during a cold winter month, feed air temperature drops
- Open Loop Case
  - Constant Cycle Step Durations (no feedback controller)
  - Delayed decrease in oxygen purity, negligible change in pressure
  - New CSS at 85% oxygen, not acceptable for patient therapy
- Closed Loop Case
  - Cycle Step Durations adjusted by MPC
  - Rejects disturbance, and follows set point at CSS
- Feed Temperature Delay Effects
  - Storage tank buffers adsorber column dynamics, causes some delay
  - Temperature affects adsorption kinetics slowly
- RPSA system *very* sensitive to changes in step durations

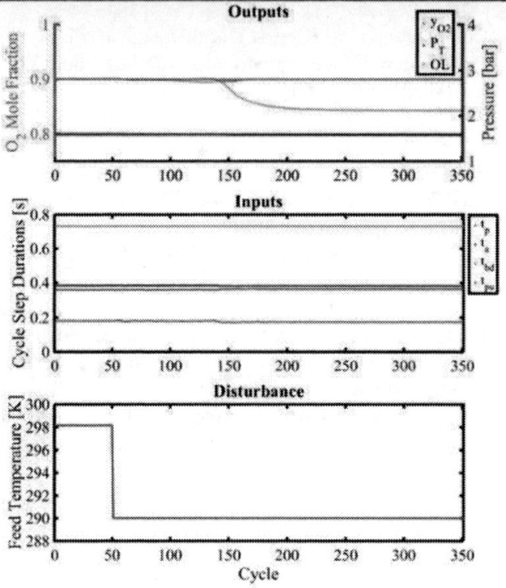

9

Comparison to PID Control

- PID Controller can only handle a SISO system
  - Control oxygen purity by manipulating the total cycle time
  - All cycle steps are a constant percentage of the total cycle time
  - Cannot directly control storage tank pressure
  - Step Durations must be constrained to ensure proper RPSA operation
- Perform the Same Feed Temperature Disturbance
- Large changes in total cycle time required to make any change in oxygen purity
  - Small changes by MPCI have much greater effects
- Closed loop response time is much longer in PID control
  - Long dynamics before returning to CSS
- Total cycle time less effective than independently manipulating the cycle step durations

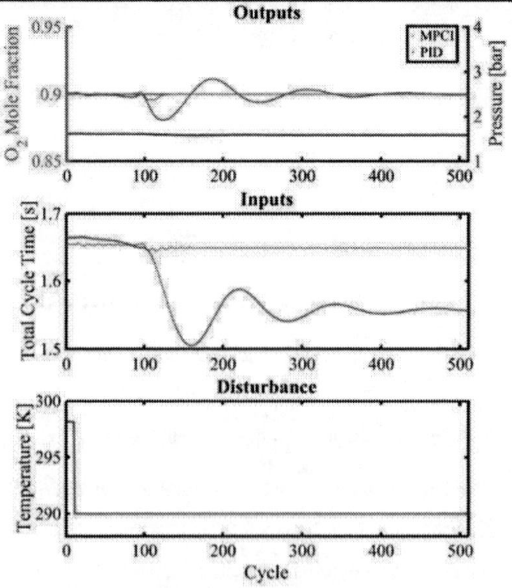

10

Disturbance Case Study: Changes in Product Flow Rate

- Typical MOCs deliver a range of oxygen flow rates to patients
  - Patient requirements lower at night
  - Higher for more severe illness
- When product rate changes, feed rate is adjusted to reduce power consumption and prolong device lifetime
- Scenario: The RPSA system experiences a step change in product flow rate from the storage tank
- Open Loop Response: With constant step durations, the purity drops and the RPSA device cannot respond
  - Offline experimentation needed to adjust RPSA back to producing 90% oxygen
- Closed Loop Response: Step times adjusted to maintain 90% oxygen
  - MPC removes offline experimentation and maintains required performance
- Demonstrates MPC is flexible enough for a MOC unit

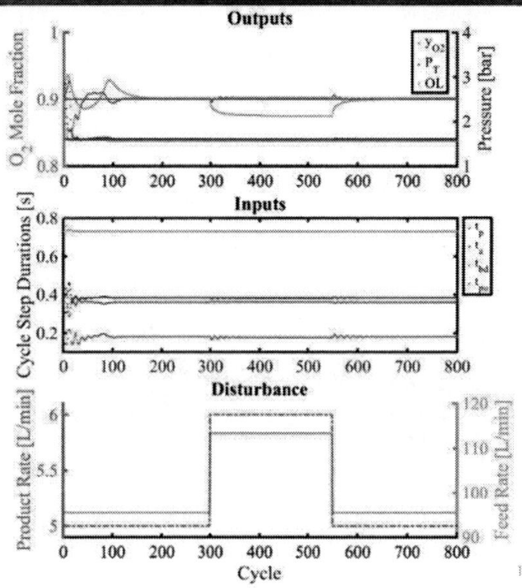

11

Disturbance Case Study: Changes in Oxygen Set Point

- MOCs designed for a single oxygen purity
  - RPSA technology can produce any oxygen purity
- MPC handles small changes (~5%) in oxygen purity
- Further from 90% means linear model approximation less accurate
- A single, linear MPC not robust enough
  - Multi-MPC using a collection of linear MPCs
  - Full Nonlinear MPC with nonlinear models
- Topic of Ongoing Research

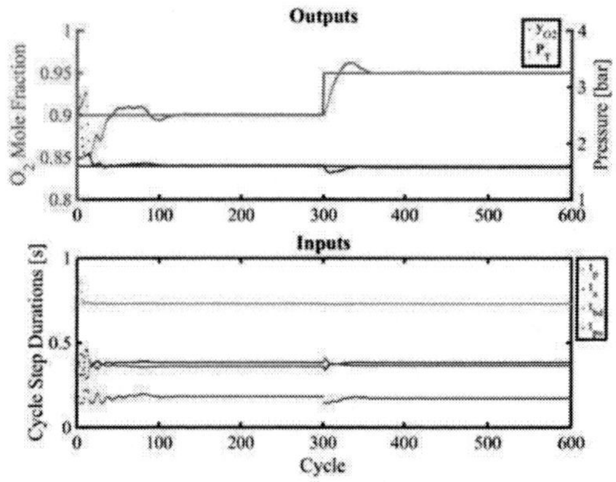

12

Conclusions and Future Work

- Novel design for a Rapid Pressure Swing Adsorption Medical Oxygen Concentrator with unique challenges
  - Output competition
  - Rapid, Cyclic Operation
- Model Predictive Controller manipulates cycle step durations, controls oxygen purity and pressure
- Rejects typical MOC disturbances: feed temperature and product flow rate
- Follows small changes in oxygen set point
  - More robust control needed for wider purity ranges
- Future work:
  - Implement MPC algorithm on lab-scale RPSA system
  - Develop more robust, nonlinear MPC algorithms

13

460428 Model Predictive Control of Atmospheric Pressure Plasma Jets for Biomedical Applications

Abstract:

Model Predictive Control of Atmospheric Pressure Plasma Jets for Biomedical Applications

D. Gidon, D. B. Graves, A. Mesbah

Department of Chemical and Biomolecular Engineering, University of California Berkeley

E-mail: dgidon@berkeley.edu

Atmospheric Pressure Plasma Jets (APPJs) are a class of cold, ambient pressure plasma devices that are increasingly used for biomedical purposes. Energetic electrons in these plasmas drive the production of therapeutically relevant components such as reactive oxygen and nitrogen species (RONs), UV-range photons, heat, and electric field. However, under atmospheric pressure, the characteristics of the plasma and the associated therapeutic effects can vary drastically with changing environmental conditions and disturbances [1]. Multiple modes observed in both radio frequency (RF) range [2] and kilohertz range [3] excited APPJs exemplifies this phenomenon. The change in plasma behavior is often associated with a heterogeneous, filamentary discharge structure and can lead to 'arcing' where all the current is concentrated in a thin conductive plasma channel. This sensitivity of the plasma behavior to external conditions makes reliable and reproducible operation of APPJs difficult, in particular for safety-critical applications. The large amount of heat and current delivered to the treated surface with arcing is also generally undesirable for biomedical applications due to safety considerations. The latter operational considerations and the requirements for high performance and strict constraint handling necessitate application of advanced control strategies for regulating the intricate dynamics of APPJs.

Dynamics of APPJs are commonly described by a set of coupled, high dimensional fluid equations. Small spatial discretization and small time steps are necessary to capture the non-equilibrium behavior of the plasma and the spatial distribution of variables of interest [1]. Consequently, the computational cost associated the solution of such fluid models becomes prohibitive for use in on-line optimal control strategies, such as model predictive control (MPC). Recent developments in analytical and numerical 'global' (zero-dimensional) modeling of APPJs [4] have revealed that simpler models can provide adequate descriptions of dynamical characteristics of plasmas. Moreover, for control, such global models must be extended to account for the full complexity of operation; including the electrical circuit associated with the device and the effects of the plasma on the treated surface. These requirements motivate our approach to control-oriented, lumped-parameter modeling of APPJs based on first principles.

In this work, a nonlinear MPC (NMPC) is designed based on a control-oriented lumped parameter model of the argon RF-APPJ reported in [5]. The control-oriented thermal model of the APPJ coupled with a biological surface, presented in [6], is extended by a lumped-parameter equivalent circuit model and a truncated reaction network to adequately represent system physics. The performance of the proposed NMPC approach is evaluated in terms of regulating the target surface temperature and the current delivered to the surface. The closed-loop simulation results of NMPC are compared to those of an internal model control system for various scenarios inspired from biomedical applications of the APPJ under study.

References

[1] H. W. Lee G. Y. Park, Y. S. Seo, Y. H. Im, S. B. Shim, and H. J. Lee, *Journal of Physics D: Applied Physics*, **44**, 053001 (2011).

[2] J. J. Shi, X. T. Deng, R. Hall, J. D. Punnett, and M. G. Kong, *Journal of Applied Physics,* **94:**10, 6303 (2003).

[3] J. Walsh, F. Iza, N. B. Jason, V. J. Law, and M. G. Kong, *Journal of Physics D: Applied Physics*, **43**, 075201 (2010).

[4] M. Lieberman, *Plasma Sources Science and Technology*, **24**, 025009 (2015)

[5] S. Hofmann, K. van Gils, S. van der Linden, S. Iseni. P. Bruggeman, *European Physical Journal D,* **68**:3 (2014)

[6] D. Gidon, D. B. Graves, A. Mesbah, *Proceedings of the American Control Conference*, Accepted, Boston (2016).

470383 Design and Implementation of Model Predictive Control Strategies for IGCC Power Plant Cycling with Carbon Capture

Abstract:

In recent years, as attention for energy and greenhouse gas emissions has raised, technologies for power plants with carbon capture have been studied worldwide. Integrated gasification combined cycle (IGCC) power plants with carbon capture, which have higher efficiencies than traditional pulverized coal-fired power plants, are a promising technology for power generation [1]. Additionally, with the increasing interest and focus on renewable energy sources, fossil-fueled power plants that were designed to operate at base-load conditions might need to cycle their load in the future. In this work, we analyze the implementation of model predictive control (MPC) strategies for base-load and cycling scenarios associated with an IGCC power plant equipped with a water gas shift membrane reactor (WGS-MR) unit for carbon capture. A model for this IGCC power plant represented by a system of differential algebraic equations was previously developed [2, 3]. Using this model, centralized control strategies based on linear and nonlinear MPC algorithms are presented to address different power generation scenarios. Given the computational time challenge for online implementation of nonlinear MPC to address high-dimensional systems, approaches for the MPC computational time reduction are also investigated.

Specifically, the linear MPC control strategy is based on a dynamic matrix control (DMC) method. For this method, a multiple-input multiple-output step response matrix is obtained by performing step tests in the IGCC plant. The proposed nonlinear MPC strategy is an extension of NLMPC [2, 4, 5], which is based on the conversion of the differential algebraic equation system into a large-scale nonlinear programming (NLP) problem. The resulting NLP is solved using IPOPT, an efficient interior point-based large-scale nonlinear optimization algorithm [6]. Also, the ADOL-C [7], an open-source automatic differentiation package, is employed for improved accuracy and speed of the calculations of system derivatives. For additional NLMPC computational improvements, sequential quadratic programming (SQP) algorithms are analyzed for solving the posed MPC NLP problem.

The proposed MPC algorithms are implemented for the IGCC-MR power plant as centralized plantwide control strategies. For such implementations, the selected control structure consists of the following controlled output variables (CVs): carbon capture, power generation, temperatures of cooled syngas and cooled permeate streams, steam to CO ratio at the WGS-MR inlet and hydrogen purity in the permeate stream. Also, the selected manipulated input variables (MVs) are: water flow rates for syngas and permeate cooling, steam injection flow to syngas for the facilitation of the WGS reaction, total coal/water slurry, oxygen enriched air flow for the gasifier, sweep flow for the WGS-MR and air flow for the gas turbine. A number of control scenarios for the IGCC power plant are addressed including: (i) trajectory tracking associated with a power generation demand cycling, in which the IGCC power plant demand follows a certain profile from its original power generation setpoint; (ii) disturbance rejection related to the variability in the quality of the coal feed. In this case, the power generation should still track the cycling profile while the carbon content in the coal slurry is reduced. In this presentation, different NLP solving methods will also be discussed when the nonlinear MPC controllers are implemented for this high-dimensional system. Results on the closed-loop responses for the different scenarios will be analyzed and compared considering the advanced linear and nonlinear MPC control strategies.

References

1. F.V. Lima, P. Daoutidis, M. Tsapatsis and J. J. Marano. Modeling and optimization of membrane reactors for carbon capture in Integrated Gasification Combined Cycle units. Ind. Eng. Chem. Res., 51(15):5480â€"5489, 2012.
2. F.V. Lima, R. Amrit, M. Tsapatsis and P. Daoutidis. Nonlinear model predictive control of IGCC plants with membrane reactors for carbon capture. In Proceedings of the American Control Conference. Washington, DC, June 2013.
3. F.V. Lima, P. Daoutidis and M. Tsapatsis. Modeling, optimization, and cost analysis of an IGCC plant with a membrane reactor for carbon capture. AIChE Journal, 62(5):1568-1580, 2016.
4. F.V. Lima, Xin He, R. Amrit and P. Daoutidis. Advanced control strategies for IGCC plants with membrane reactors for CO_2 capture. In Materials and process systems for CO_2 capture: modelling, design, control and integration, A.I. Papadopoulos and P. Seferlis (eds.), Wiley, 2016.
5. R. Amrit, J. B. Rawlings and L. T. Biegler. Optimizing process economics online using model predictive control. Computers and Chemical Engineering, 58:334-343, 2013.
6. A. Wächter and L.T. Biegler. On the implementation of an interior-point filter line-search algorithm for large-scale nonlinear programming. Mathematical programming, 106(1):25-57, 2006.
7. A. Walther and A. Griewank. Getting started with ADOL-C. Combinatorial Scientific Computing, 181-202, 2012.

470386 Optimal Controller Design Based on Efficient Ant Colony Optimization Algorithm. Case Study: Chemical Process Control

Abstract:

In this work, an advanced optimal control solution method inspired by biological systems is proposed. Biological systems have shown success of solving difficult problems encountered in nature and hence, serve as motivation for the formulation of optimization and advanced chemical process control approaches. For example, the behavior of the natural groups such as ants, bees and swarms demonstrates that self-organization and cooperation by following simple rules of interaction can result in a wide range of optimal patterns [1]. Inspired by ant's foraging behavior, this work proposes a biomimetic optimal control strategy for chemical process control specifically for non-linear fermentation processes. The proposed strategy employs the computationally efficient optimal control solver entitled efficient ant colony optimal control (EACOC) algorithm. The EACOC algorithm is a metaheuristic optimization technique inspired by the ants' foraging behavior which utilizes probabilistic and stochastic concepts for solving large-scale optimization problems. A brief description of this algorithm is presented below.

In basic ant colony optimization (ACO) algorithm, artificial ants are stochastic candidate solution construction procedures that exploit a pheromone model and possibly available heuristic information of the mathematical model. The artificial pheromone trails are used as means of communication among the artificial ants. Pheromone decay allows the artificial ants to forget the past history and focus on new promising search directions. The pheromone values are updated according to the information learned in the previous iterations and hence, the algorithmic procedure leads to accurate and potentially global optimal solution. EACO algorithm improves the performance of the conventional ACO algorithm for combinatorial, continuous and mixed-variable optimization problems by introducing Hamersley sequence sampling (HSS) for probabilistic elements of ACO. The initial solution archive diversity for continuous and mixed-variable optimization problems plays an important role in the performance of ACO algorithm. The uniformity property of the HSS technique is exploited to avoid clustering of the initial solution archive in a small region of the potential solution space. Moreover, ACO algorithm is a probabilistic method where several randomized probability functions are involved in the algorithm procedure. The distribution of these random numbers affects the performance of the ACO algorithm. At this step, the multidimensional uniformity property of HSS is also introduced to improve the computational efficiency of the ACO algorithm. The capabilities of the proposed methodology are illustrated using benchmark problems and real world case studies of computer aided molecular design problems. Specifically, these problems include the optimal design of solvent for extraction of acetic acid from waste process stream using liquid-liquid extraction [3] and adsorbent selection for naturally occurring radioactive materials of natural gas fracking produced water [4].

The EACOC algorithm is developed based on the efficient ant colony algorithm and orthogonal collocation methods. The proposed algorithm is implemented for optimal control of dynamic fermentation processes where the objective function is to track the product concentration set-point. The challenges in this fermentation process are the steady-state multiplicity and the oscillations in the concentration profile due to nonlinearities present in the process model [2, 5, 6, 7]. This approach successfully overcomes these challenges. The implementation results using EACO show potential improvement in the solution quality and reduction in computational time when compared to a gradient-based solver. Thus, the proposed algorithm can be used as an alternative to the gradient-based optimal control algorithms for optimization of large-scale dynamic problems.

References:

1. Hristu-Varsakelis D. and Shao C. "A bio-inspired pursuit strategy for optimal control with partially constraints final state". *Automatica* 2007; 43:1265-1273.
2. Sridhar L. "Elimination of oscillations in Fermentation Process". *AIChE Journal* 2011;57(9):2397-2405.
3. Gebreslassie BH, and Diwekar UM. Efficient ant colony optimization (EACO) for computer aided molecular design: Case study solvent selection problem. *Computers and Chemical Engineering* 2015: 78: 1-9.
4. Benavides PT, Gebreslassie BH, and Diwekar UM. Optimal design of adsorbents for NORM removal from produced water in natural gas fracking. Part 2: CAMD for adsorption of radium and barium. *Chemical Engineering Science* 2015: 37: 977-985.
5. Lima F. V., Li S., Mirlekar G. V., Sridhar L. N. and Ruiz-Mercado G. J., "Modeling and advanced control for sustainable process systems". Sustainability in the Analysis, Synthesis and Design of Chemical Engineering Processes, G. Ruiz-Mercado and H. Cabezas (eds.), Elsevier, 2016.
6. Mirlekar G. V., Gebreslassie B. H., Diwekar U. M. and Lima F. V., "Design and implementation of a biomimetic control strategy for chemical processes based on efficient ant colony optimization." Presented at AIChE Annual Meeting, Salt Lake City, Utah, November 2015.
7. Li S., Mirlekar G. V., Ruiz-Mercado G. J., and Lima F. V., "Development of Chemical Process Design and Control for Sustainability". Submitted for publication, 2016.

463113 Stochastic Predictive Control with Closed-Loop Model Adaptation: Application to a Cold Atmospheric Plasma Jet

Abstract:

Predictive controllers offer many advantages, including handling of multivariable dynamics, system constraints and competing control objectives [1]. However, the model quality often degrades over time due to system uncertainties or a limited validity range of the model as the system evolves (e.g., see [2] and the references therein). The key objective of this work is to address the problem of model uncertainty in the context of predictive control of stochastic systems. This work presents a stochastic MPC approach with integrated experiment design (iX-SMPC) for linear systems subject to parametric uncertainties and additive disturbances. The proposed approach allows for simultaneously controlling the system and probing the system dynamics to generate informative data for estimation of the uncertain model parameters. Hence, the designed control inputs have some form of dual control feature [3].

A control-oriented input design cost function is incorporated into a stochastic optimal control problem to allow for tuning the quality of model adaptation toward meeting a prespecified control performance level [4]. The closed-loop data generated by iX-SMPC can then be used for identifying the (posterior) probability distribution of model parameters and, consequently, facilitate online model maintenance during predictive control. The generalized polynomial chaos framework [5] is used to obtain a deterministic surrogate for the stochastic optimal control problem with integrated input design. For the case of expectation-type state constraints, a quadratic programming (QP) program is derived for the iX-SMPC approach. A gPC-based histogram filter [6], which relies on Bayesian estimation, is designed to re-estimate the probability density functions of the uncertain parameters using the closed-loop data obtained from the system.

The proposed iX-SMPC algorithm is implemented on a cold atmospheric plasma (CAP) jet. CAP jets have found extensive applications in materials processing and biomedical applications [7], [8]. This work considers the argon-

CAP jet presented in [9]. The control objective is to maintain the temperature of a target surface (e.g., tissue), in contact with the plasma jet, at a desired setpoint while fulfilling various system constraints in the presence of parametric uncertainties and system disturbances. The system constraints pertain to safe and reproducible operation of the device. The control inputs have a dual feature of controlling the system dynamics and generating informative data for parameter estimation. Monte Carlo simulations were carried out to evaluate the closed-loop performance of the control approach under different uncertainty realizations. The results indicate that the designed optimal inputs can generate informative closed-loop data such that the estimated probability distribution of the unknown model parameters converges to its true value. The simulation results suggest that the control approach can effectively realize the control objectives in the presence of the intrinsic system stochasticity.

References

[1] M. Morari and J. H. Lee, "Model predictive control: past, present and future," Computers and Chemical Engineering, vol. 23, pp. 667-682, 1999.
[2] A. Mesbah, X. Bombois, M. Forgione, H. Hjalmarsson and P. M. J. Van den Hof, "Least costly closed-loop performance diagnosis and plant re-identification," International Journal of Control, vol. 88, no. 11, pp. 2264-2276, 2015.
[3] A. A. Feldbaum, "Dual-control thoery I," Automatic Remote Control, vol. 21, pp. 874-880, 1961.
[4] C. A. Larsson, M. Annergren, H. Hjalmarsson, C. R. Rojas, X. Bombois, A. Mesbah and P. E. Moden, "Model predictive control with integrated experiment design for output error systems," in Proceedings of the European Control Conference, pp. 3790-3795, Zürich, 2013.
[5] D. Xiu and G. E. Karniadakis, "The Wiener-Asky polynomial chaos for stochastic differential equations," SIAM Journal on Computing, vol. 24, pp. 619-644, 2002.
[6] V. Bavdekar and A. Mesbah, "A polynomial chaos-based nonlinear Bayesian approach for estimating state and parameter probability distribution functions," In Proceedings of the American Control Conference, Accepted, Boston, 2016.
[7] A. Schutze, J. Y. Jeong, S. E. Babyan, J. Park, G. S. Selwyn and R. F. Hicks, "The Atmospheric-Preseure Plasma Jet: A Review and Comparison to Other Plasma Sources," IEEE Transactions on Plasma Science, vol. 26, no. 6, pp. 1685-1694, 1998.
[8] H. Lee, G. Park, Y. Seo, Y. Im, S. Shim and H. Lee, "Modelling of atmospheric pressure plasmas for biomedical applications," Jounal of Physics D: Applied Physics, vol. 44, no. 5, 2011.
[9] D. Gidon, D. B. Graves and A. Mesbah, "Model predictive control of thermal effects of an atmospheric plasma jet for biomedical applications," In Proceedings of the American Control Conference, Accepted, Boston, 2016.

469329 Robust Optimal Dosing Strategy for Bacterial Disinfection

Abstract:

Bacterial persistence is a bottleneck in designing an effective antibiotic dosing strategy for complete eradication of bacterial populations. Persister cells are not antibiotic-resistant mutants; rather they exhibit a different phenotype from normal cells, namely they lower their growth rate significantly when attacked by an antibiotic, thereby experiencing much lower death rates than normal cells. Persisters usually constitute only a small fraction of a total bacterial population. Because of dormancy of persister cells, an antibiotic has reduced efficacy when used to quickly eradicate a bacterial infection. The bactericidal action of the antibiotic is compromised due to phenotypic switching of normal cells to persister cells in the presence of the antibiotic and back-switching from persister cells to normal cells when the antibiotic is eventually depleted (Lewis 2007). Balaban, Merrin et al. (2004) proposed a simple model to study the bacterial phenotypic switching in the presence of an antibiotic. Subsequently, several investigators (Cogan 2007, Fasani and Savageau 2013) have proposed more complex models incorporating toxin hypotheses, to study the emergence of persistence and phenotypic switching. Based on these models, various dosing strategies, *such as piecewise constant dosing with alternate dose/withdrawal,* have been studied (Cogan 2006, De Leenheer and Cogan 2009, Cogan, Brown et al. 2012) in attempts to increase the bactericidal effect of an antibiotic. Some of these strategies are difficult to implement in practice, and many rely on the fidelity of the model describing the effect of an antibiotic on bacteria eradication. In this paper, we propose a robust optimal dosing strategy, which considers the effect of model parameter uncertainty on optimal dosing. To illustrate our approach, we first use experimental data to build a dynamic mathematical model, which we use next to study the population dynamics of normal and persister cells in the presence of an antibiotic. We finally use the model to find a robustly optimal dosing strategy for bacterial disinfection.

References

Balaban, N. Q., J. Merrin, R. Chait, L. Kowalik and S. Leibler (2004). "Bacterial persistence as a phenotypic switch."

Science **305**(5690): 1622-1625.

Cogan, N. G. (2006). "Effects of persister formation on bacterial response to dosing." Journal of Theoretical Biology **238**(3): 694-703.

Cogan, N. G. (2007). "Incorporating toxin hypothesis into a mathematical model of persister formation and dynamics." Journal of Theoretical Biology **248**(2): 340-349.

Cogan, N. G., J. Brown, K. Darres and K. Petty (2012). "Optimal Control Strategies for Disinfection of Bacterial Populations with Persister and Susceptible Dynamics." Antimicrobial Agents and Chemotherapy **56**(9): 4816-4826.

De Leenheer, P. and N. G. Cogan (2009). "Failure of antibiotic treatment in microbial populations." Journal of Mathematical Biology **59**(4): 563-579.

Fasani, R. A. and M. A. Savageau (2013). "Molecular mechanisms of multiple toxin-antitoxin systems are coordinated to govern the persister phenotype." Proceedings of the National Academy of Sciences of the United States of America **110**(27): E2528-E2537.

Lewis, K. (2007). "Persister cells, dormancy and infectious disease." Nature Reviews Microbiology **5**(1): 48-56.

469943 Utilizing Memory in Process Control

Abstract:

In the existing literature on process control, past experiences in the controller performance are rarely utilized. Therefore, if a controller had a poor performance in the past for a disturbance rejection or servo control problem, a similar performance would be expected even in the future if a similar control problem is faced. The main challenge in utilizing the past experience is to identify the similarity of the current problem with one or more previous control problem(s). In this work, we propose a novel sparse coding based approach that not only classifies the current control problem, but can also be useful as an adaptive approach for system identification.

In this work, a dynamic sparse coding (DSC) algorithm is developed for nonlinear time-varying systems for the purpose of using the individual sparse coefficients as classifiers. The algorithm developed here allows for evaluation of a linear combination of a subset of basis functions drawn from a candidate set by using an active-set likelihood maximization algorithm. The dynamic data are preprocessed and then used for obtaining the dynamically produced representations. The identification procedure is solved by an efficient sparse coding formulation using the Bayesian (Schwarz) Information Criterion (BIC) as a method for determining the minimal number of basis functions to fully describe the system being considered. This sparse representation is then used to infer the similarity of the control problem. Finally, if there is acceptable similarity to a past control problem, then the experience from the past controller performance is exploited to obtain improved performance by implementing a weighted scheme determined based on the similarity ratio. The sparse coding representation is also utilized for dynamically adapting the process model that is used in a model-based control technique. The sparse coding problem is formulated to ensure that if the process remains time-invariant, the sparse representation remains relatively invariant to changes in the inputs.

This approach is validated on an acid gas removal unit as part of an integrated gasification combined cycle plant. This unit consists of large number of equipment items, is highly nonlinear, and involves significant mass and heat integration along with considerable time delay. It is observed that even if the controller performance is poor to start with, repeated use of past experiences leads to almost perfect control.

465830 A Model-Based Framework for Advanced Optimal Operation of Polymerization Processes

Abstract:

A model-based framework for advanced optimal operation of polymerization processes

N. Ghadipasha, A. Geraili, J.A. Romagnoli

Department of Chemical Engineering

Louisiana State University

Baton Rouge, LA, 70809

Abstract:

As demand for new and more sophisticated polymers increases, the issue of controlling polymerization reactors becomes more important. The principal problems in achieving good control of polymerization processes are inadequate on-line measurement, lack of understanding of the dynamics of the process, nonlinearities arising during the reaction due to phenomena such as gelation and high exothermicity and lack of well-developed techniques for control of nonlinear processes. Hence, although there has been extensive work on controlling polymerization reactors, they have been limited to special case studies. Furthermore, due to lack of on-line measurement, most of the control works have been based on open loop methods.

This contribution discusses the formulation and implementation of a generic and flexible model-based framework for integrated simulation, estimation, optimization and feedback control of polymerization systems. The emphasis is on developing a comprehensive scheme which can be applied for the optimal operation of various polymeric systems. This goal was achieved by combining the powerful capabilities of the automatic continuous online monitoring of polymerization system, ACOMP, with a modern simulation, estimation and optimization software environment. ACOMP is a widely applicable platform for monitoring polymerization reactions. It combines simultaneous data from multiple detectors so continuous monitoring of salient reaction characteristics can be performed, such as kinetics, conversion of comonomers, evolution of molecular mass, intrinsic viscosity and detection of unusual phenomena, such as microgelation and runaway reactions. The proposed structure in this work will forge initial links between ACOMP and advanced modelling and control principles and demonstrate unprecedented feedback control of polymerization reactions.

The conceptual representation of the aforementioned framework is illustrated in Figure 1. The modelling work is carried out using gPROMS modelling language, providing a complete environment for modelling/analysis of complex systems. The parameter estimation entity makes use of the data gathered from the experimental runs. It has the ability to estimate various number of parameters, using data from multiple dynamic experiments and ability to specify different variance models among the variables as well as among different experiments. The optimization entity allows for the typical dynamic optimization problems arising from batch and/or semibatch operation to be formulated and implemented. Different off-line optimal strategies are developed and fully tested using experimental facilities. As an example, Figure 2 shows the validation results in terms of the model predictions and the experimental data when the obtained inputs trajectories are applied into the experimental systems. Finally, the issue of feedback control of polymerization reactors is considered. Development of a nonlinear input-output linearizing geometric approach to control the reactor concentration is explained and the performance is validated experimentally. Significant improvement compared with conventional controllers is observed which is due to the exploitation of nonlinear structure of the model in solving the control design problem.

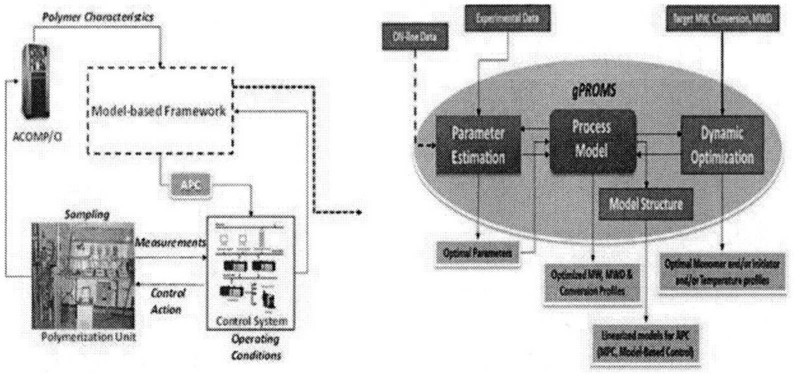

Figure 1: Schematic representation of the integrated simulation, estimation, optimization and feedback control of polymerization systems

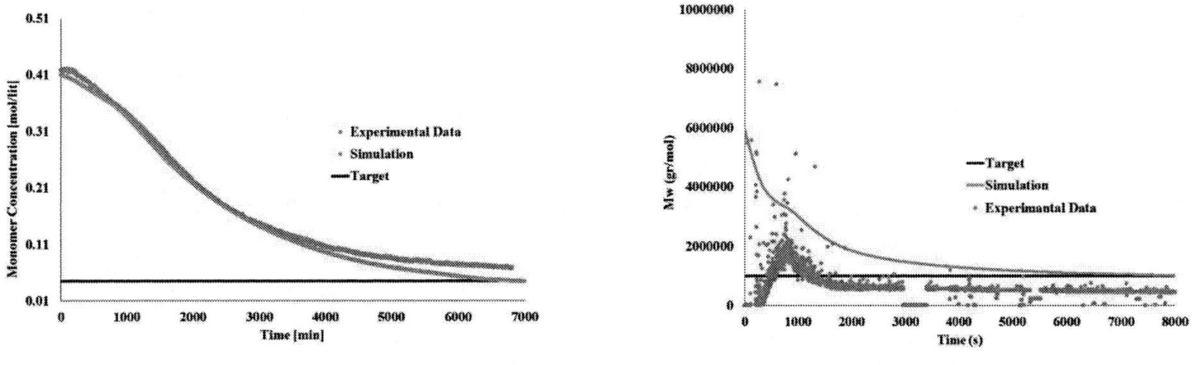

Figure 2: Validation of optimal runs for the monomer concentration and weight average molecular weight

467846 Towards Experimental Validation of a Multi-Resolution Approach for Directed Self-Assembly of Non-Periodic Structures:Â Spatial Control of Particle Densities

Abstract:

Self-assembly is the spontaneous association of components into patterns or structures.[1] As a manufacturing technology of advanced materials, directly self-assembly is exciting as it enables molecular precision. External fields can be introduced to better control the self-assembly process, which is often referred to as directed self-assembly. The key challenge for practical implementation of directed self-assembly is to engineer the interactions between particles and between particles and external fields such that directed self-assembly proceeds in a desired direction.

Various external fields have been investigated to direct self-assembly. Electric fields are of particular interest due to the tunable strength and direction.[2,3] For example, electric fields have been used as an actuator in automated feedback control methods to make advanced materials with low levels of defects.[4]

Self-assembly processes are prone to kinetic trapping and exhibit stochastic behavior, which complicates control of directed self-assembly to fabricate non-periodic structures.[5] Furthermore, although important progress is being made in this direction,[6] real-time observation of self-assembly processes is often not well possible in a non-invasive way, which complicates feedback control methods. Therefore, open-loop control methods are of particular interest for control of directed self-assembly. Several open-loop control methods have been proposed for control of directed self-assembly.[7] Solis et al.[7,8] proposed a novel multi-resolution approach to self-assemble non-periodic structures in a systematic way to avoid kinetic traps,[8,9] This multi-resolution approach is inspired by protein folding and decomposes the dynamic self-assembly into a number of sequential resolutions in which the required number of particles are directed to a desired part of a domain. By dividing the domain into more parts when proceeding from resolution to resolution, the differences in particle density on the domain start to resemble the final structure more closely until the final structure can be assembled in the final resolution. This approach avoids kinetic traps since the local particle densities match the requirements of the final non-periodic structure. From a different perspective, the approach systematically breaks the ergodicity of the system in a number of steps such that undesired structures are not accessible anymore when the final structure is assembled. Dynamic simulations have demonstrated the potential of this method to self-assemble non-periodic structures reliably.[10] However, the experimental validation of such multi-resolution approach has not been investigated to the best of our knowledge. Several challenges exist when implementing a multi-resolution approach experimentally. First, spatial control of the densities of the particles at different resolutions is needed. Second, the decomposition of the ergodicity of the system needs to be maintained when proceeding to lower resolutions. The former challenge may either require feed-back control or open-loop control when some form of predictive model is available.

The objective of this work is to validate an automated control strategy to control particle densities within different regions of a domain for a model system of directed self-assembly. Both open-loop and closed-loop control methods are investigated. Those methods can be used at a particular resolution to split a uniform particle population into two parts with desired densities. Ultimately, the ability to create different particle densities spatially and to maintain those divisions when proceeding to finer resolutions allows for the multi-resolution approach to be implemented experimentally.

Microscopic particles have been selected as the model system. The particles are suspended in a buffer solution and placed in a small cell under a microscope to observe the self-assembly process. Transparent electrodes on a glass substrate are patterned with non-conductive photoresist at the bottom of the cell to create an external field. A low-voltage AC electric field is used to direct the self-assembly of these microscopic particles. A dynamic analysis of this system has been conducted to quantify the diffusion and interaction of the particles with basic electrode shapes such as squares as function of electric field strength and frequency. Subsequently, an empirical input-output model is used to design an open-loop or closed-loop controller to separate a uniform particle population into various populations with different densities at specific regions of the domain. Figure 1 illustrates how the particles can be attracted to various electrodes and as such create density differences. Using automated image analysis (Matlab 2015b) for parallel experiments, the dynamics of the system can be quantified in terms of the density changes when applying step input changes in electric field properties, which are key for developing density control strategies to enable the experimental validation of a multi-resolution approach for control of directed self-assembly of non-periodic structures.

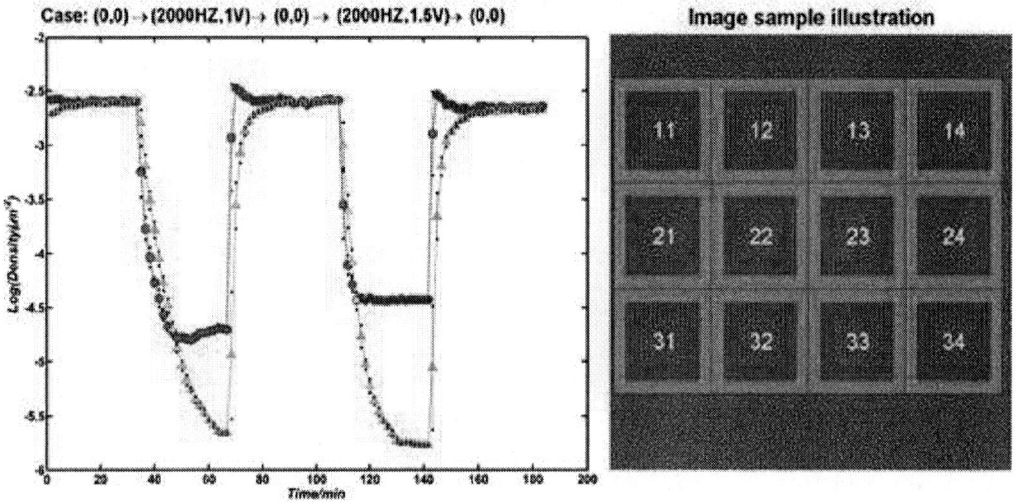

Figure 1: Dynamic analysis of the spatial particle densities in specific regions around square electrodes (indicated with row and column number) under AC electric fields with time-varying properties. The red line with circles represents the average particle density of the (inner) red area around the square electrodes computed from all 12 electrodes using image analysis. Similarly, the green line with triangles in the plotting represents the average particle density of the (outer) green area around the square electrodes.

Reference

1. Whitesides GM, Grzybowski B. Self-assembly at all scales. *Science (80-)*. 2002;295(5564):2418-2421.

2. Bhatt KH, Grego S, Velev OD. An AC electrokinetic technique for collection and concentration of particles and cells on patterned electrodes. *Langmuir*. 2005;21(14):6603-6612.

3. Xie R, Liu X-Y. Electrically directed on-chip reversible patterning of two-dimensional tunable colloidal structures. *Adv Funct Mater*. 2008;18(5):802.

4. Ju¨¢rez JJ, Bevan MA. Feedback Controlled Colloidal Self©\Assembly. *Adv Funct Mater*. 2012;22(18):3833-3839.

5. Paulson JA, Mesbah A, Zhu X, Molaro MC, Braatz RD. Control of self-assembly in micro-and nano-scale systems. *J Process Control*. 2015;27:38-49.

6. Weidman MC, Smilgies D-M, Tisdale WA. Kinetics of the self-assembly of nanocrystal superlattices measured by real-time in situ X-ray scattering. *Nat Mater*. 2016.

7. Komaee A, Barton PI. Directed Self-Assembly of Linear Nanostructures by Optimal Control of External Electrical Fields. *arXiv Prepr arXiv160300113*. 2016.

8. Solis EOP, Barton PI, Stephanopoulos G. Controlled formation of nanostructures with desired geometries. 1. Robust static structures. *Ind Eng Chem Res*. 2010;49(17):7728-7745.

9. Solis EOP, Barton PI, Stephanopoulos G. Controlled formation of nanostructures with desired geometries. 2. Robust dynamic paths. *Ind Eng Chem Res*. 2010;49(17):7746-7757.

10. Lakerveld R, Stephanopoulos G, Barton PI. A master-equation approach to simulate kinetic traps during directed self-assembly. *J Chem Phys*. 2012;136(18):184109.

471484 A Hybrid Control Technique to Achieve Spatial and Temporal Control of Droplets in a Microfluidic Device

Abstract:

Droplets moving in micro-channels act as micro reactors providing a unique environment to carry out reactions precisely. Spatial and temporal control of these droplets give advanced flexibility to perform high throughput screening and combinatorial chemistry. Such systems have huge potential in developing continous manufacturing technology for pharmaceutical industries. Control of these micro reactors can be achieved through active and passive control techniques. Although passive methods provide good control without any moving components, they are sensitive to experimental disturbances. In contrast, active control techniques can handle experimental uncertainties with ease; further, designing passive elements require several experimental trails. Therefore based on the broad applicability of the droplet systems, implementing advanced control techniques in the field of droplet microfluidics is essential.

Droplets affect the decisions, at bifurcation points, and velocities of the other droplets in the channel due to hydrodynamic interactions. The dynamics of these phenomena even in a simple bifurcating channel is astonishingly complex. The present dropletsâ€™ decisions affect past droplet decisions and droplet velocities. Droplet motion in microfluidic systems is highly nonlinear and interactive leading to chaos and presents a unique challenge for conventional control techniques. In this paper, we demonstrate the application of aÂ hybrid model predictive controller on a microfluidic loop device to produce droplets of desired sequence given an input sequence. A custom MINLP technique is formulated for droplet systems that efficiently predicts the actuations to control both decisions-discrete states- and velocities-continous states- of droplets. This technique is demonstrated in microfluidic loop device by generating a droplet sequence `AABB' from `ABABâ€™.

468259 Virtual Plant Environment Incorporating Data for Predictive Modeling and Control in Bio-Pharmaceutical Manufacturing

Abstract:

A virtual plant environment which incorporates data-driven models was developed for bio-pharmaceutical manufacturing. The objective of the virtual plant is replicate the control strategies implemented at bio-pharmaceutical plants in a virtual operating environment. This enables application and testing of data-driven model-based process simulation to gain insight into process and control system design while minimizing the number of wet experiments required. A DCS system was used as the platform to build the virtual plant, and Matlab was used to develop data-driven models. Data used in Matlab was taken from first principles simulations as well as empirical data. Bi-directional data communication was established between the DCS system and Matlab via OPC connection. A nested model scheme was developed to work within virtual plant. A case study was performed on a chromatography unit operation for drug substance manufacturing. The plug-and-play modular design of the virtual plant gives it the flexibility to simulate existing automation setups at many bio-pharmaceutical manufacturing plants.

469050 Model Predictive Control of Semicontinuous Distillation Process

Abstract:

Semicontinuous distillation is a process intensification technique introduced in 2000 [1]. It can purify ternary mixtures to desired specifications. Where conventional continuous distillation requires two distillation columns for ternary separation, semicontinuous system utilizes only one distillation column and a simple storage tank to perform the same task. Therefore, with lower total direct costs, semicontinuous systems have lower total annualized costs for low to intermediate production rates compared to continuous configuration. Semicontinuous distillation operates in stabilized cycles. Initially, the storage tank is charged with fresh feed. The tank feeds the column and the side stream of the column is recycled back to the tank. The light and the heavy boiling point components are removed in the distillate and bottom streams of the column, respectively, during the cycle. As these components are removed from the system, the concentration of the intermediate components reaches the desired purity in the tank. Finally, the tank is discharged and the third product is collected. Subsequently, the tank is recharged and a new cycle begins.

The system is dynamic and control driven and the cycle time is a function of initial tank hold up, column diameter, number of trays and controller parameters [2]. In the last 16 years, the main focus on semicontinuous research was on the design of the process. These studies have shown that the semicontinuous system can economically outperform the conventional continuous configuration for low production rates [3]. In all these studies, proportional integral (PI) controllers were used. PI controllers are easy to implement and show satisfactory performance for the system but tuning its parameters is tedious since the conventional tuning rules cannot be implemented for this dynamic system. On the other hand, due to high variable interactions and nonlinearities of the system, advanced control strategies such as model predictive control (MPC) can improve the performance and the economics of the system.

In this study, for the first time, the implementation of MPC on a semicontinuous system is studied. The separation of benzene, toluene and *o*-xylene was chosen as a case study. The first principal model of the semicontinuous system was simulated in gPROMS using its public model library (PML) [4]. The simulation results were sent to MATLAB (where the MPC algorithm was implemented) through the gO:MATLAB feature of gPROMS. The subspace identification method was adopted to identify a linear time invariant state-space model to be used in the MPC [5]. The model was identified using the input-output data collected from the process. Subsequently, a shrinking horizon MPC was implemented to obtain the desired purity of intermediate component in the tank and the average purities of the distillate and bottom streams by the end of the cycle while reducing the operating cost. In this implementation, the PI controllers drive the process and stabilize the cycle while the MPC determines the set-points of those PI controllers. Compared to the optimal PI control configuration (the best previously known), the MPC strategy resulted in an 11% improvement in the energy consumption of the system.

[1] Phimister JR, Seider WD. Semicontinuous, middle-vessel distillation of ternary mixtures. AIChE J. 2000;46(8):1508–1520.

[2] Meidanshahi V, Adams II TA. A new process for ternary separations: Semicontinuous distillation without a middle vessel. Chemical Engineering Research and Design 2015;93:100–112.

[3] Adams II TA, Pascall A. Semicontinuous thermal separation systems. Chem. Eng. Technol. 2012;35(7):1153–1170.

[4] Meidanshahi V, Adams II TA. Integrated design and control of semicontinuous distillation systems utilizing mixed integer dynamic optimization. Computers and Chemical Engineering, 2016;89:172–183.

[5] Corbett B, Mhaskar P. Subspace identification for data-driven modeling and quality control of batch processes, AIChE Journal, 2016:62(5):1581–1601.

461535 Regulation of Anemia in Chronic Renal Disease Using Zone Model Predictive Control

Abstract:

Regulation of Anemia in Chronic Renal Disease using Zone Model Predictive Control

Anemia is a state, characterized by a reduced mass of red blood cells (RBCs) or hemoglobin (Hgb), resulting in reduced oxygen carrying capacity and delivery to the body's tissues and organs. Anemia of chronic renal disease patients is primarily due to insufficient endogenous production of erythropoietin, as well as iron deficiency. Those patients with anemia are treated by the administration of recombinant human erythropoietin (rHuEPO) and iron. It has been recognized by the clinical community that while low Hgb levels lead to anemia, too large Hgb levels can increase the risk of mortality for the patient. Hence, effective methods are needed to determine the appropriate dose of rHuEPO and iron to maintain the target Hgb level. Many conventional methods for guiding rHuEPO and iron dosing used by clinicians rely on a set of rules based on past experiences or retrospective studies. Those methods for rHuEPO administration are generally imprecise, and as a result, the patient's Hgb level may move through the target range with large oscillations and overshoots. New effective anemia management methods are needed to avoid the adverse effects associated with increased Hgb levels while minimizing the effects of anemia. In this paper, one proposed solution to managing anemia in renal disease patients is the utilization of Zone Model Predictive Control (ZMPC).

To implement ZMPC, it is necessary to identify a suitable mathematical model to describe the system. Erythropoietin doses and hemoglobin data from 168 different patients over a span of up to 3 years were retrieved from medical records. Erythropoietin doses were typically given once per week, while hemoglobin values were taken from lab work which was done approximately every two weeks. To overcome the variance in the sampling time of Hgb and drug doses, the data was resampled into weekly intervals. A linear interpolation was used between Hgb values and the rHuEPO dose total for the entire trailing week was used. Due to the closed loop nature of the data acquisition, Direct Closed-loop System Identification was used to identify Autoregressive with Exogenous Input (ARX) models for individual patients. The model order and delay used in the model identification were selected based on priori medical knowledge of the life span of red blood cells and the response delay of Erythropoietin doses.

ZMPC is an advanced control system that uses a dynamic mathematical model of a system to optimize control inputs over a prediction horizon to guide the system to a target zone. For anemia management the target Hgb level selected was 9.5 to 11 g/dL which was based on physician recommendations. The ZMPC algorithm from [1] was adapted to incorporate the use of the identified ARX system models. Using a large target zone can introduce problems in which the output often has a tendency to settle near the boundaries of the target zone. This results in process noise and model mismatch often pushing the system outside of the desired zone. A viable approach to reducing the effect of this problem is to penalize the output in a funnel shape that reduces to a point along the prediction horizon. This is accomplished by introducing exponentially decaying constraints on the slack variable used within the cost function. In this manner, the controller guides the system towards the center of the zone with control actions that are far less aggressive than traditional model predictive control. ZMPC has the advantage of calculating less aggressive control inputs which can improve the robustness of the controller. The disadvantage is the requirement to tune the shape of the funnel constraints.

Computer simulations were used to explore the performance of MPC and ZMPC against the physician's protocol. Assessment of the controller output performance was based on the frequency and magnitude of zone violation. Although not minimized in the cost function, total erythropoietin doses were also compared. To test the performance of the controllers, both process noise and model mismatch were used. The process noise used in the simulations was determined from the residuals of the identified ARX model and the actual data. To represent model mismatch, one patient ARX model was used in the predictive controller while a different patient model was used as a patient simulator. It was found that both model predictive controllers consistently outperformed the physician protocol in simulations including process noise only. In simulations including process noise and model mismatch, both controllers outperformed the physician's protocol in the majority of cases. In some cases including model mismatch, the traditional MPC controller produced control actions that were too aggressive causing oscillations and overshoots in the Hgb levels. This phenomena is shown in Figure 1. The ZMPC controller with funnel constraints produced results consistently better than the physician's protocol in all simulations.

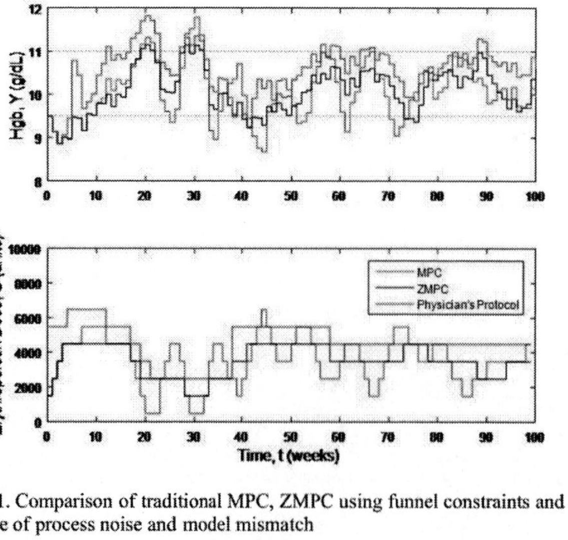

Figure 1. Comparison of traditional MPC, ZMPC using funnel constraints and Physician's Protocol performance in the presence of process noise and model mismatch

References:

[1] Roubal, Jirka and Vladimir Havlena. "Range Control MPC Approach for Two-dimensional System." (2005).

470856 A Scale-up Methodology for Continuous Bioreactor Systems Utilizing an Optimal Design Under Optimal Control Constraint Framework

Abstract:

Continuous operation is used in bioprocesses to allow for increased cell viability through the mitigation of substrate and product inhibition, thereby increasing the productivity of the desired product. However, the dilution rate, feed substrate composition, and reactor geometry play a key role as parameters affecting the productivity in a continuous bioreactor system. In addition, one of the many challenges found in the pharmaceutical industry is quantifying the effects of scale-up. Often, understanding of a biochemical process is gained through experimentation at the bench scale and it is common that these results cannot be replicated at production scale, where mass transfer becomes a more limiting factor[1]. Several heuristic techniques have been developed to try and mimic the bench scale system at the plant scale, including maintaining the fluid dynamics of the reactor via a constant impeller tip speed or Reynolds number or maintaining the mass transfer properties of the system by keeping the mass transfer coefficient k_La or dissolved oxygen level constant[2,3,4]. These scale-up methods are highly empirical, however, and more direct and accurate methodologies are needed.

This work proposes a method for the scale-up of continuous pharmaceutical systems utilizing a framework that pairs the optimal design and optimal control of the reactor to guarantee feasibility. The development of reliable bio-kinetic models is extended to the design of a continuous bioreactor that uses a multi-feed configuration allowing for the ability to use flow rate and substrate concentration of the feed stream as independent manipulated variables. The design problem for the bioreactor is formulated as a nonlinear program (NLP) for which the optimal control problem governing the dilution rate and substrate concentration and the transport and consumption of oxygen and nutrients are used as constraints. From this algorithm the optimal bioreactor geometry is determined for any bioreactor volume as well as the optimal control policy necessary to maximize the reactor productivity. A beta-carotene production process with kinetic models describing the glucose consumption and metabolic product formation and depletion in the *Saccharomyces cerevisiae*strain mutant SM14 is utilized as a case study for the described algorithm.

1. Garcia-Ochoa, F. and Gomez, E. Bioreactor scale-up and oxygen transfer rate in microbial processes: An overview. *Biotechnology Advances* 27, 153–176 (2009).
2. Puthli, M. S., Rathod, V. K. and Pandit, A. B. Gas–liquid mass transfer studies with triple impeller system on a laboratory scale bioreactor. *Biochemical Engineering Journal* 23, 25–30 (2005).
3. Wutz, J., Lapin, A., Siebler, F., Schäfer, J., Wucherpfennig, T., Berger, M. and Takors, R. Predictability of kLa in stirred tank reactors under multiple operating conditions using an Euler-Lagrange approach. *Eng. Life Sci.* (2016). doi:10.1002/elsc.201500135
4. Shuler, M.L and Kargi, F. Bioprocess Engineering: basic concepts. NJ: Prentice Hall PTR. (2002)

459124 Safety-Based Model Predictive Control with Data-Based Determination of Safety Level Sets

Abstract:

Maintaining safety of chemical processes is the most important operating objective in the chemical process industries. As a result, industry has developed a number of qualitative and quantitative means for assessing the safety of a chemical process, such as hazards and operability (HAZOP) studies, fault trees, and what-if scenarios. Process safety is promoted by designing a process to be inherently safe based on the chemistry and physics of the process; after the process has been designed, process control can be added to improve process safety [1]. The integration of safety with control [2] was pursued in [3], in which several Lyapunov-based economic model predictive controller (LEMPC) [4] formulations were developed with safety constraints. The safety constraints were designed to drive the process state into a smaller region of operation termed the safety region when a safety logic unit predicted that a component or software failure was possible if process operation was not moved to this smaller region of operation.

In this work, we improve the practical application of the safety-based LEMPC method proposed in [4] by developing a safety-based controller formulation in the context of Lyapunov-based model predictive control (LMPC) with a quadratic objective function for consistency with the tracking model predictive control (MPC) formulation commonly used in the chemical process industries [5]. In addition, we develop a data-based method for both the LMPC and LEMPC cases for determining the probability that the closed-loop state will leave a given safety level set, and the distance from the level set that the process states may move before returning to the safety level set, due to process disturbances and sensor noise. The knowledge gained from this data-based method can then be used by the safety logic unit as it evaluates safe regions of operation to quantify the probability that, for a given safety level set, the closed-loop state may move far enough from the safety level set to enter a region that the safety logic unit has determined to be unsafe. Furthermore, this knowledge may cause the safety logic unit to select smaller safety level sets for process operation if the disturbances and noise are known to move the process state significantly. In addition, closed-loop

simulations of a chemical process example showed that LMPC with safety-based constraints was able to drive the process state into a required safety region more quickly than tracking LMPC alone, showing that the proposed safety-based model predictive control formulation can enhance the safety of current chemical process systems.

[1] Crowl DA, Louvar JF. Chemical Process Safety: Fundamentals with Applications, 3rd ed. Upper Saddle River, NJ: Pearson Education, 2011.

[2] Leveson NG, Stephanopoulos G. A system-theoretic, control-inspired view and approach to process safety. AIChE Journal. 2014;60:2-14.

[3] Albalawi F, Alanqar A, Durand H, Christofides PD. A feedback control framework for safe and economically-optimal operation of nonlinear processes. AIChE Journal. 2016; in press.

[4] Heidarinejad M, Liu J, Christofides PD. Economic model predictive control of nonlinear process systems using Lyapunov techniques. AIChE Journal. 2012;58:855-870.

[5] Qin SJ, Badgwell TA. A survey of industrial model predictive control technology. Control Engineering Practice. 2003;11:733-764.

472996 Advanced Control Automation System of the Syngas Chemical Looping Process: Dynamic Model Simulation and Controller Development

Abstract:

Over the last two decades of development, chemical looping has become an attractive energy and/or chemical production technology with near-zero CO_2 emissions. In a chemical looping process, metal oxide particles serve as an oxygen carrier to indirectly convert carbonaceous fuels such as syngas, natural gas, coal or biomass into a concentrated stream of CO_2 without the need for energy intensive gas-gas separation techniques. The metal oxide-based oxygen carrier supplies oxygen to the fuel from air via a cyclic reduction-oxidation reaction without direct contact of the fuel with the nitrogen-laden air stream. Ohio State University has been developing an advanced chemical looping technology which utilizes a unique moving bed design to maximize fuel conversion to concentrated CO_2 and oxygen carrier conversion to minimize the solid circulation rate. Further, OSU has developed non-mechanical integrated reactor system that allows the oxygen carrier particles to circulate while preventing the gases present in each reactor from mixing. These concepts have been successfully demonstrated and scaled from laboratory testing to a $250kW_{th}$ â €" 3 MW_{tbyh} pilot plant constructed at the National Carbon Capture Center. One of the main challenges for operating the chemical looping unit at larger scale is the complexity of controlling and integrating multiple auxillary components simultaneously while maintaining proper operation of the chemical looping reactor system. Due to the interconnected nature of the reactor configuration and the high nonlinearity of the governing equations of the process dynamics, it is difficult to adjust or control the value of one process variable without manipulating multiple inputs.Â Therefore, the level of knowledge and experience required for the operators is relatively high to correctly perform state transitions between different operating conditions. The presented work seeks to develop a sophisticated control algorithm which is capable of performing automated action sequence to startup, maintain steady state and shutdown the chemical looping process. Multiple aspects of the project will be discussed, including the development of dynamic models of the SCL

process, the design of a hybrid sliding mode controller structure, the instrument and equipment design for automation, and the test results of concept demonstration performed on the sub-pilot and pilot scale SCL units integrated with the developed control algorithms.

471918 Ecosystems As Unit Operations: Designing Integrated Networks of Technological and Ecological Systems

Abstract:

One of the requirements to meet the goal of sustainable design is to ensure that impacts of engineering systems do not exceed the regenerative capacity of ecological systems [1], in order to prevent unintended harm. Industries striving towards sustainable development also aim to become carbon and emissions neutral, with zero water withdrawal. Many companies aim to go even beyond such goals by having "net positive" manufacturing systems. While the conventional approach to meeting this goal has been to increase the use of renewable fuels, optimize the system efficiency to reduce emissions and rely on end-of-pipe pollution control methods, such solutions tend to neglect the ecological overshoot resulting in some perverse outcomes.

In this work, we present an approach to designing coupled techno-ecological systems in order to minimize ecological overshoot within an "industrial-serviceshed", by including ecological systems in a manner analogous to unit operations within the engineering design framework. As a result, simultaneous decisions are made regarding the design of engineering systems to minimize the ecosystem service demand and the design of supporting ecological systems to supply the necessary ecosystems services demanded by industrial sites. Sustainability indices for such systems are quantified using the TES (Techno-Ecological Synergy) framework [1], as a function of the design variables and the objective is to identify optimal coupling between Technological and Ecological systems, to minimize the overshoot.

Traditional engineering unit-operation models capture the demand for ecosystem service, while the supply is quantified by spatially explicit unit-operation models of ecosystems. Similar to process level constraints regarding the plant operating conditions, feedstock availability etc. ecological constraints like land availability, type of ecological systems and other geo-spatial constraints based on remote-sensing and aerial survey are identified and included within the design problem. The objective of TES system design is to

maximize the net ecosystem service supply, quantified by the ecosystem service supply less the demand as opposed to the conventional approach of minimizing only the ecosystem service demand.

The TES design framework is applied to the design of a biofuel production system. Ecological systems included within the design framework include a forest ecosystem captured using the Urban Forest Effects model [2] and a horizontal subsurface flow wetland systems for treating the wastewater, designed using steady-state first order equations of an ideal plug flow reactor [3]. The problem is formulated as a multi-objective nonlinear program (Mo-NLP) and tradeoffs are identified between maximizing the net present value of the system and the net ecosystem service demand for multiple ecosystem services. The dynamic behavior of ecological systems and changes in the design and operation of systems are captured using a multi-period approach. Seasonal variations due to weather fluctuations are also bound to cause a change in ecological processes and thus robust designs are identified at each time-period accounting for these external disturbances. The problem formulation also accounts for the uncertainty associated with the critical design and operating variables.

Preliminary results from the case study demonstrate that systems designed by accounting for ecosystem service supply results in novel designs that have a higher economic value, and higher sustainability index, resulting in "win-win" designs. Thus, coupled TES systems are expected to have economic and environmental advantages over conventional systems design, mainly due to the reduction in external inputs by moving towards a closed loop production system.

References

[1] Bakshi, Bhavik R., Guy Ziv, and Michael D. Lepech. "Techno-ecological synergy: A framework for sustainable engineering." Environmental science & technology 49.3 (2015): 1752-1760.

[2] Nowak, David J., and Daniel E. Crane. "The Urban Forest Effects (UFORE) Model: quantifying urban forest structure and functions." (2000).

[3] Rousseau, Diederik PL, Peter A. Vanrolleghem, and Niels De Pauw. "Model-based design of horizontal subsurface flow constructed treatment wetlands: a review." Water research 38.6 (2004): 1484-1493.

459809 Economic, Environmental and Social Cost Optimization of Biomass Supply Chain for Electricity Generation in WI, USA

Abstract:

There is an increasing interest in identifying the optimal biomass supply chain for the renewable energy system, due to relatively low energy density and high procurement cost of biomass feedstock. The goal of this study is to develop an integrated framework to determine the optimal supply chain network for bioelectricity generation in Wisconsin, USA. First, forest residues collection sites, road and rail network across WI, MN and upper MI were mapped in Arc-GIS. Next, the optimal location of co-generation facility was identified by screening all potential sites by government policy, local community interest, land availability, distance to transmission lines and road/rail network. Finally, biomass supply chain was optimized by minimizing the economic, environmental and social cost of shipping forest residues from forest to the facility. Upstream (cradle-to-gate) GHG and criteria pollutants emissions of feedstock were also quantified for all potential facility sites. The result shows that Superior WI is the best location to build a bioelectricity facility in terms of economic, environmental, and social sustainability. In addition, the quantities, modes and distances of forest residues transportation were determined in order to meet the 0.5 million ton (current capacity) and 1 million ton (future capacity) demand of the bioelectricity facility.

472967 Renewable/Fossil Fuel Supply Selection and Supply Chain Optimization for a Synergized Upstream, Downstream and Power Economy System

Abstract:

The current environment of low oil prices constitutes a substantial headwind for much of the energy value chain, especially for energy systems where the upstream, downstream, and power sectors are synergized with the aid of governmental subsidies. In such cases, the economic viability of the downstream and power sectors depends heavily on the technology costs and the government regulation on oil prices, which are often fixed. In the current decade, energy systems with an abundant supply of oil were able to achieve sizable margins in the downstream sector within the high oil price environment. However, the uncertainty of future oil prices makes it difficult to put forward strategies for the energy and petrochemical sectors. An energy model that takes into account the incorporation of renewable resources mixed with fossil fuels, a dynamic selection of hydrocarbon prices, and a flexible downstream sector that is capable of liquid or gas feedstock inlets can lessen the burden that oil price fluctuations have. In this work, we present a model where static and dynamic optimization techniques are applied to the dynamic identification of prices that optimize supply decisions and minimize fuel and technology costs within a synergized upstream, downstream and power economy system. Emphasis is placed on understanding the role of modeling in the energy sector and understanding energy balances to develop a modern integrated energy planning model.

471652 Synthesis of Renewable Energy Supply Networks Considering Different Frequencies of Fluctuations in Supply and Demand

Abstract:

The demand for and dependency on fossil resources, coupled with growing environmental consciousness and new economic opportunities and developing/utilising clean and renewable energy sources on a larger scale, are among the major challenges and priorities of our society. In 2015 at least 164 countries had renewable energy targets, and most also had renewable energy support policies (Sawin et al., 2015). Among the most promising technologies to help achieve these goals are those using solar irradiance, wind and biomass (Davis and Martín, 2014).

In this study, a biorefinery supply network integrating wind and solar energy (Čuček et al., 2015), involving several biomass and waste sources for the production of biofuels and food (Čuček et al., 2014), is further extended by: i) applying more detailed models for wind and solar energy with more realistic data, and ii) better accounting for intermittent renewable energy sources, especially of wind and solar energy, and electricity demands. The dynamic model corresponds to a multi-period model, and considers different frequencies in supply and demand. In those cases where there is a more pronounced dynamic, the time interval selected is shorter. For supply and demand with longer-term fluctuations, the time interval is longer. Different time periods are applied leading to supply network models with a large number of variables and constraints, and to long computational times. For biomass and biofuels the time periods are longer, while for solar, wind and electricity consumption, the time intervals are shorter.

The proposed model is demonstrated on a case study of a relatively small region in Europe. Several scenarios are considered regarding possible percentage share of renewables in the future energy mix, preferably reaching 100% of renewable energy systems. Furthermore, a sensitivity analysis is performed

regarding the availability of energy sources, demand, and prices of biomass, energy technologies and products.

Acknowledgement

The authors acknowledge the financial support of the Slovenian Research Agency (programs P2-0032 and P2-0377), from a bilateral project between Slovenia and the United States entitled "Development of Methodology for the Synthesis of (Bio)chemical Supply Chains," from the Ministry of Science, Education and Sports of the Republic of Croatia and the EC under project CARBEN (NEWFELPRO Grant Agreement No. 39), and from a bilateral project between Slovenia and Croatia entitled SLOCRORES, "Planning of 100% RES communities by use and combination of Total Site Integration and Renewislands methodologies."

References

Čuček, L., Martín, M., Grossmann, I.E., Kravanja, Z., 2014. Multi-period Synthesis of Optimally-Integrated Biomass and Bioenergy Supply Network. Computers and Chemical Engineering 66, 57-70.

Čuček, L., Martín, M., Kravanja, Z., 2015. Integration of Wind and Solar Energy within Continental Biorefinery Supply Network. Proceedings of 2015 AIChE Annual Meeting, November 8 – 13, 2015, Salt Lake City, UT, USA.

Davis, W., Martín, M., 2014. Optimal year-round operation for methane production from $CO_2$and water using wind and/or solar energy. Journal of Cleaner Production 80, 252-261.

Sawin, J.L., Sverrisson, F., Rickerson, W., Lins, C., Musolino, E., Petrichenko, K., Rickerson, W., Sawin, J., L., Seyboth, K., Skeen, J., Sovacool, B., Williamson, L.E., 2015. Renewables 2015 - Global status report, Key findings 2015, Paris, France.

469724 A Holistic Look to Carbon Capture and Use in Sustainable and Economically Enhanced Bio-Based Supply Chains

Abstract:

Carbon capture and storage is considered worldwide as an alternative to take into account for cost-effective mitigation of CO_2 emissions. Notwithstanding, recent years have shown significant drawbacks associated with carbon capture and storage (CCS) to secure the CO_2 geological storage from industrial sources or power plants. Geological storage faces the possibility of leakage, liability issues, long-term problems with public acceptance of storage sites on land, and limited cost-effective storage capacity in some key regions (Cuéllar and Azapagic, 2015). Otherwise, sustainability concerns, evidence that a more holistic view is necessary. This includes the whole energy supply chain and a deeper insight on the protagonist role of carbon capture and utilization (CCU), specifically for the chemical industry, and as a part of the share of the environmentally friendly technological options of the future.

CCU provides a CO_2 management solution for a limited amount of CO_2. For instance, the CO_2 generated by a single coal power plant could potentially supply much of the CO_2 demand for CO_2-based methanol plants that supply a growing demand in Europe. This would avoid the use of fossil fuel, like natural gas, which is used nowadays in the conventional synthesis of methanol (Pérez-Fortes et al., 2016).

This work combines a bio-based energy supply chain with CO_2 capture and utilization, for economic optimisation based on a CO_2 emissions constraint. The problem is formulated as a mixed-integer linear program in GAMS. This study presents a complete and flexible supply chain superstructure, as well as the supporting framework developed to answer pending issues on the actual economic and environmental benefits achievable in this complex scenario for both, the biomass-coal plant and the CO_2 utilization plant; i.e. location of intermediate steps, fraction of CO_2 emissions captured per plant, or size of the CO_2 utilization plants, among others. Moreover, eventual incentives and environmental revenues are discussed, to make an economically viable business case. A large-size, real case study is used in this analysis. Two key scenarios are taken into account: (i) No use of biomass, neither capture nor utilization of CO_2 is contemplated; (ii) Biomass, capture and CO_2utilization are all contemplated. Finally, conclusions and guidelines are given.

References

Cuéllar-Franca R, Azapagic A. Carbon capture, storage and utilisation technologies: A critical analysis and comparison of their life cycle environmental impacts. *Journal of CO2 Utilization*. Elsevier. Volume 9, March 2015, Pages 82â€“102.

Available from: doi:10.1016/j.jcou.2014.12.001

Pérez-Fortes M, Laínez-aguirre JM, Puigjaner L. Economic and environmental assessment of CO_2 capture and utilisation technologies in bio-based supply chains. *Computers and Chemical Engineering*. 2016; in press.

.

459900 Optimal Integration of Renewable Based Processes for Fuels and Power Production: Case Study in Spain

Abstract:

Although renewable resources are plentiful, their sources are scattered. Therefore, to meet a certain demand of power and fuels using renewable sources, and mitigate the variability in their availability, process integration at local or greater scale has become an interesting engineering and technical challenge. Weekman (2010) and Yaun and Chen (2012) presented overviews regarding the integration possibilities as a perspective for the future combination of different sources of energy. In order to help with those decisions, process system engineering has the tools to compare sources and allocations in search for the optimal options at local and wider scales. Most of the studies either focus on fuels supply (Cucek et al 2014, Elia et al 2012) or power supply based on the unit commitment problem (Saravanan, et al 2013). However, power is most of the times produced by using a fuel, which is a chemical that can also be synthesized. Therefore, both supply networks are linked and must be addressed simultaneously. Over the last years plant size integration of renewable resources has been considered (Vidal and Martín, 2015, Martín and Davis, 2016, Martín and Grossmann, 2016). However, the integration at a larger scale, beyond the plant level, has not been attempted.

In this work we have developed a framework for the optimal integration of renewable sources of energy to produce fuels and power. A network is developed using surrogate models for various technologies that use solar energy, PV solar, CSP or algae to produce oil, wind technology, biomass based syngas to ethanol, methanol, FT-liquids and thermal energy, hydroelectric power and waste based power plant via biogas production. Hydrogen can be produced if there is a surplus of energy. It can be stored, by producing methanol or methane. Methane can be further used to produce power at need, while methanol is employed for oil transesterification to produced biodiesel. In both cases, the carbon source is CO_2, that has been captured during the gasification based processes or it imported from outside the network. The corresponding processes are modeled using an input- output approach based on the results of detailed optimization studies previously developed by the authors. However, we do not model an entire process from biomass to fuels as a black box, but we break it into sections that produce intermediates that can be used for a different purpose, i.e. syngas can be used to produced ethanol FT-liquids, etc, or because various technologies are available for that; i.e. dry or wet cooling for power plants. The model is formulated as an MILP that allows determining the optimal selection of technologies to meet certain demand. Both cost and environmental objective functions are considered.

We apply the network to evaluate process integration at different scales, from region level, where we also evaluate the effect of uncertainty in renewable resources, up to the level of an entire country (Spain). Solutions under uncertainty are more robust and expensive, about 25% higher requiring the use of further technologies to meet the demand. For Spain, and using 1% of the area, it is possible to substitute 20% of fossil fuels for transportation and 100% of the power demand. The solution suggests the use of Hydropower, and oil production in all regions, while bioethanol biodiesel plants are allocated close to demand points such as large population areas. Up to 44% of fuels and total power can be substituted with the current technology development status. However, we can only reach 90% substitution by using 20% area with the availability and efficiency of current processes. The investment required for this option is more than twice that to reach 20% fossil fuels substitution.

References

Cucek, L., MartAn, M., Grossmann, I.E., Kravanja, Z (2014) Large-Scale Biorefinery Supply Network â€“ Case Study of the European Union. Comp. Aid. Chem. Eng. Vol33 , 319-324.

Elia, J.A., Baliban R.C., Floudas, C.A. (2012) Nationwide energy supply chain analysis for hybrid feedstoch processes with significant CO2 emissions reduction. AIChE J. 58 (7) 2142-2154.

Martín, M., Grossmann, I.E. (2016) Optimal integration of a self sustained algae based facility with solar and/or wind energy. submitted *Energy*.

Martín, M, Davis, W. (2015) Integration of wind, solar and biomass over a year for the constant production of CH_4 from CO_2 and water Comp Chem Eng 84, 314-325,

Saravanan, B., Da, S., Sikri, S., Kpothari, D.P. (2013) A solution to the unit commitment problem-a review. Frointiers in Energy. 7(2) 223-236

Vidal, M., Martín, M.(2015) Optimal coupling of biomass and solar energy for the production of electricity and chemicals.Comp. Chem. Eng 72, 273-283

Weekman, V.W., 2010. Gazing into an energy crystal ball. CEP June 2010; 23-27

Yuan, Z., Chen, B., 2012. Process Synthesis for Addressing the Sustainable Energy Systems and Environmental Issues. AIChE J. 58 (11), 3370-3389

475314 Designing Functional Materials within the Materials Project

Abstract:

The Materials Project (www.materialsproject.org) â€" part of the broader Materials Genome Initiative - is an effort to compute the properties of all known inorganic materials and beyond, and offer that data to the community together with online analysis and design algorithms. The current release contains data derived from density functional theory (DFT) calculations for over 66,000 materials, each with searchable associated properties such as relaxed structure, electronic state, energy storage capability, aqueous and solid stability, and more. Furthermore, software algorithms are offered by the Materials Project and can be used by researchers for materials informatics, including both interactive web-based tools like the Phase Diagram App and the Pourbaix App, as well as open-source codebases and data access tools such as the pymatgen materials analysis library, FireWorks workflow software, and Materials API.Â Today â€" 4 years after launch, the Materials Project is driving materials innovation in broad chemical and structure spaces, for applications as varied as energy storage, energy production, thermoelectricity, transparent conductors, materials synthesis conditions etc. This talk will highlight the development of the Project, its growth attracting more than 20,000 users and a few of the many projects that have been successfully concluded or are being pursued using the capabilities and materials understanding that has emerged from our approach of data informed materials design. Such projects are i) stability of inorganic materials in aqueous electrolytes for battery, fuel cell and catalysis applications, and design principles for ii) novel multivalent intercalation cathode discovery and iii) electrolytes.

473787 Data Analytics and Machine Learning in Nanomaterials Discovery

Abstract:

The recent explosion of "big data" has positioned data analytics and machine learning techniques at the centre of social, marketing, medicine and manufacturing research. In nanomaterials discovery, data analytics tools are also playing a fundamental role to successfully tackle the exponential increase in size and complexity of functional nanomaterials. In this work, the intrinsic relationships between nanoscale features and functional properties of nanomaterials will be discussed for nanodiamonds[1], graphene nanoflakes[1-3] and nanoporous metal-organic frameworks (MOFs)[4-6]. Multivariate statistics are applied to identify the most relevant individual features in conjunction with the most representative nanostructures in nanoparticle ensembles and structural libraries. Meanwhile, machine learning techniques, from simple decision tree predictors to more complex suppert vector machines and neural networks, are utilized to predict the electronic properties of graphene nanoflakes and to identify "high-performing" nanoporous MOFs for carbon capture and storage applications. It will be demonstrated that simple predictors provide useful insights into the structure-property relations paradigm at the nanoscale, whilst more complex machine learning models are efficient tools to rapidly discriminate among potential candidate materials at a fraction of the traditional computational cost.

References

1. Fernandez, M. & Barnard, A. S. ACS Nano 9, 11980–11992 (2015).

2. Fernandez, M., Shi, H. & Barnard, A. S. J. Chem. Inf. Model. 55, 2500–2506 (2015).

3. Fernandez, M., Shi, H. & Barnard, A. S. Carbon 103, 142–150 (2016).

4. Fernandez, M., Boyd, P. G., Daff, T. D., Aghaji, M. Z. & Woo, T. K. J. Phys. Chem. Lett. 5, 3056–3060 (2014).

5. Fernandez, M. & Barnard, A. S. ACS Comb. Sci. (2016). doi:10.1021/acscombsci.5b00188

6. Aghaji, M. Z., Fernandez, M., Boyd, P. G., Daff, T. D. & Tom, K. Eur. J. Inorg. Chem. (2016) in press.

472740 Describing the Diverse Geometries of Gold from Nanoclusters to Bulk a First-Principles Based Hybrid Bond Order Potential

Abstract:

Gold nano structures are of intense interest for catalysis, sensing, plasmonic, and other applications, and gold nanoclusters (<200 atoms), in particular, exhibit unusual chemistry in the nano-scale. In gold nanoclusters, a fundamental understanding of the atomic structure, size-dependent dimensionality, surface chemistry, and atomic-scale dynamics over timescales spanning tens of nanoseconds holds the key to unraveling structure-property relationships. Gold nanoclusters undergo a fascinating size dependent transition in cluster dimensionality and display a wide range of structural motifs (e.g., planes, hollow cages, tubes, space-filled compact forms). The origin of such transitions has been largely unknown; previous density functional theory (DFT) and spectroscopy studies have speculated that while bond directionality effects predominantly govern structure/chemistry in small Au clusters, long-range dispersive forces probably become important in large clusters, surfaces and bulk condensed phases. A major roadblock has been our inability to perform long time molecular dynamics simulations, and global optimization of structures due to lack of accurate empirical force fields.

In this talk, we introduce a unified *ab initio*-based hybrid bond order potential (HyBOP), which captures the competition between short-range interactions (bond directionality) and long-range dispersions to accurately predict the size dependent diverse geometries of gold clusters, surfaces and bulk. Our work involves an important methodological development: we employed genetic algorithms to train our HyBOP potential against a large DFT dataset comprising (a) energies of a thousand 13-atom Au clusters, (b) equation of state for various bulk polymorphs of gold, and (c) low-index surfaces of bulk face-centered gold. Our newly developed HyBOP accurately predicts (a) evolution of various structural motifs in Au clusters with increasing size, (b) critical size for transition from planar to globular structures (*i.e.*, dimensionality changes), (c) global energy minimum configurations at various cluster sizes, and (d) thermodynamics, structure, elastic properties, and energetic ordering of bulk condensed phases as well as surfaces, in excellent agreement with DFT calculations and spectroscopic experiments. As a representative test, using long time HyBOP based MD simulations, we identified the atomic scale mechanisms governing the coalescence of icosahedral and planar clusters into pyramids.

469402 Topological Data Analysis of Nanoporous Materials Genome Using Pore-Geometry Recognition Technique

Abstract:

In most applications of nanoporous materials the pore structure is as important as the chemical composition as a determinant of performance. For example, one can alter performance in applications like carbon capture or methane storage by orders of magnitude by only modifying the pore structure. For these applications it is therefore important to identify the optimal pore geometry and use this information to find similar materials. However, the mathematical language and tools to identify materials with similar pore structures, but different composition, has been lacking. Here we develop a pore recognition approach to quantify similarity of pore structures and classify them using topological data analysis. Our approach allows us to identify materials with similar pore geometries, and to screen for materials that are similar to given top-performing structures. Using methane storage and carbon capture as a case study, we also show that materials can be divided into topologically distinct classes — and that each class requires different optimisation strategies. Inthis work we have focused on pore space, but our topological approach can be generalised to quantify similarity of any geometric object, which, given the many different Materials Genomics initiatives, opens many interesting avenues for big-data science.

468015 Unraveling the Role of Pore Topology and Chemical Functionality on the Carbon Capture Performance of Metal-Organic Frameworks

Abstract:

Metal-organic frameworks (MOFs) are novel porous materials obtained via assembly of inorganic and organic molecular building blocks into crystalline networks. Thousands of MOFs have been synthesized to date, based on different combination of building blocks. The evident structural and chemical versatility of MOFs has made them promising materials for applications such as CO_2 capture in fossil-fuel-based power stations. The challenge, however, is to efficiently identify high-performing materials within the vast MOF "structure space." Using molecular modeling, here we selectively explore this space aiming to elucidate the role of pore topology and chemical functionality in the CO_2 capture performance of MOFs to establish clear design rules that can guide the discovery and synthesis of optimal MOFs for this application.

The chemical functionality of MOFs has long been deemed key to enhance CO_2 adsorption in MOFs. However, since the incorporation of functional group also alters MOF textural properties, conflicting information on the positive or negative impact of a chemical moiety on CO_2 adsorption often arises. In addition, elucidating optimal functionalities experimentally has been difficult, because comprehensive sets of MOFs with identical structures but different functionality are not readily synthesizable or otherwise equally stable upon activation and testing. Similar issues are encountered in attempting to elucidate optimal pore topologies for CO_2 adsorption. The above roadblocks can be circumvented using molecular simulation to evaluate a suitable, comprehensive, diverse set of "perfect MOFs" but the challenge is to efficiently create and evaluate such predictably large set in the computer.

In this work, we created *in silico* 426 MOFs carrying different simple chemical functionalities, based on 28 "parent" MOFs of 16 different topologies, using our recently developed automated MOF construction algorithm. Since CO_2 adsorption is likely to be highly sensitive to Coulombic interactions, we explored different strategies to assign partial charges to MOF atoms, including a new building block-based strategy that we introduce in this work to be compatible with our MOF construction algorithm. Based on our simulations, we provide insights into how and under what conditions different chemical functionalities and topologies are likely to enhance CO_2 adsorption. For instance, from our data analysis we find that at pressures relevant to post-combustion CO_2 capture, alcohol and thiol functionalities tend to outperform other moieties often suggested to enhance CO_2 adsorption such as amino functionalities. Additionally, we present potentially attractive MOF synthesis targets based on CO_2 adsorption properties predicted during our computational screening.

460977 An Automated Approach for Developing Graph-Theoretical Cluster Expansions of the Total Energy of Adsorbed Layers

Abstract:

An Automated Approach for Developing Graph-Theoretical Cluster Expansions of the Total Energy of Adsorbate Layers

E. Vignola[1,2], S. N. Steinmann[2], M. Stamatakis,[3] P. Sautet[2,4]

[1] *TOTAL Petrochemicals, Route de la Chimie, 76700 Gonfreville-l'Orcher, France;* [2] *Univ Lyon, Ens de Lyon, CNRS, Université Lyon 1, Laboratoire de Chimie UMR 5182, F-69342, Lyon, France;* [3] *Department of Chemical Engineering, University College of London, Torrington Place, London WC1E7JE, United Kingdom;* [4] *Department of Chemical and Biomolecular Engineering, University Of California Los Angeles, CA, USA*

The accurate description of the total energy of adsorbate layers is crucial for the understanding of chemistry at interfaces. For catalysis applications in particular, adsorbate-adsorbate lateral interactions have been shown to significantly affect activation energies of reactions, thereby shaping experimentally observed trends. [1,2] Modelling the interactions of atomic adsorbates has traditionally been achieved using effective Ising-type Hamiltonians [3], whereby a set of spin-like values is attributed to the layer's lattice points describing the occupancy of the corresponding catalytic sites (vacant occupied by a species). Pairwise additive adsorbate-adsorbate lateral interactions in this model are captured by appropriate coupling constants.

Such a Hamiltonian is however limited, as it cannot account for adsorbates that bind to more than one sites (bi-dentate or even multi-dentate species), and it cannot capture many-body contributions to the total energy (3-body interactions – triplets). To overcome these limitations one has to adopt a cluster expansion Hamiltonian formalism [4], which has recently been implemented in a graph-theoretical scheme [5,6] to enable the representation of multi-dentate species. Automating the development of such cluster expansion Hamiltonians for catalytic systems is challenging. Existing approaches for such automation [7] can only tackle mono-dentate adsorbates and cannot account for the various binding modes that molecules exhibit on solid surfaces.

The current work develops a scheme for automating the development of cluster expansions applicable to molecular species on catalytic surfaces. The scheme has been implemented in a FORTRAN 95 program compatible with the graph-theoretical kinetic Monte Carlo code, *Zacros* [8].

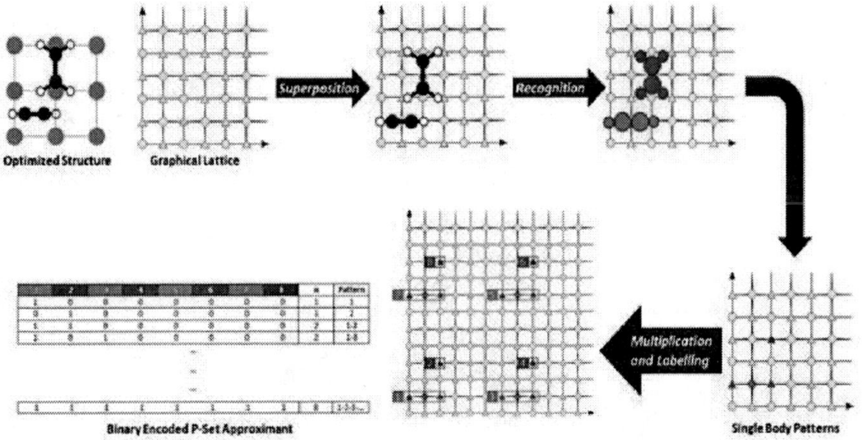

Figure 1. Pattern Recognition and Encoding Scheme

References

[1] Stamatakis, M.; Piccinin, S. *ACS Catalysis* **2016**, *6*, 2105

[2] Frey, K.; Schmidt, D. J.; Wolverton, C.; Schneider, W. F. *Catal. Sci. Technol*, **2014**, *4*, 4356

[3] Mussardo, G. *Statistical Field Theory. An Introduction to Exactly Solved Models in Statistical Physics*, Oxford University Press, New York, 2010.

[4] Sanchez, J. M.; Ducastelle, F.; Gratias, D. *Phys. A Stat. Mech. its Appl.* **1984**, *128 (1-2)*, 334.

[5] Stamatakis, M.; Vlachos, D. G. *J. Chem. Phys.* **2011**, 134, 214115

[6] Nielsen, J.; d'Avezac, M.; Hetherington, J.; Stamatakis, M. J. Chem. Phys. **2013**, *139 (22)*, 224706

[7] van de Walle, A.; Ceder, G. *J. Phase Equib.* **2002**, 23, 348.

[8] Stamatakis M. *Zacros: Advanced Lattice-KMC simulation Made Easy*, http://www.zacros.org/, 2013.

453762 Property Prediction of Crystalline Solids from Composition and Crystal Structure

Abstract:

Property Prediction of Crystalline Solids from Composition and Crystal Structure

Bruno A. Calfa[1] and John R. Kitchin[2]

Predicting properties of compounds has received considerable attention across different disciplines and has seen applications in diverse areas. The approach proposed in this paper is influenced by the advances in Group Contribution Methods (GCMs).

A key characteristic of GCMs in general is that the property descriptors (i.e., predictors) are explicitly represented by the molecular structure (and chemical composition) of a compound. One can pose an optimal *inverse* (design) problem, which attempts to obtain the original molecule (structure and composition) given a target property value. This is the main goal of the research area called Computer-Aided Molecular/Mixture Design (CAMD) as exemplified by several works in the literature (Eljack et al., 2007; Samudra and Sahinidis, 2013).

In the crystalline solids literature, several research groups have proposed statistical (machine learning) approaches for property prediction. Saad et al. (2012) investigated both unsupervised and supervised machine learning techniques to predict structure and properties of crystals with chemical formula AB. Ma et al. (2015) developed a machine-learning-augmented model based on artificial neural networks (ANNs) that captures nonlinear adsorbate-substrate interactions.

We propose using kernel regression (Li and Racine, 2007) as a data-driven and rigorous nonparametric statistical technique to predict properties of atomic crystals. A key feature of the proposed approach is the possibility of treating predictors not only as continuous, but also as categorical data. The latter specifically allows the predictive model to capture the discrete nature of crystals with regards to composition (number of atoms in the chemical formula) and spatial configuration (finite number of crystallographic space groups). Another important aspect of using kernel regression is the direct access to its explicit mathematical form, which can be directly embedded in optimal inverse problems to design new crystalline materials with given target properties. The property prediction approach is illustrated by training models to predict electronic properties of 746 binary metal oxides and elastic properties of 1,173 crystals. As a first approach to solving the inverse problem, we describe an exhaustive enumeration algorithm (Calfa and Kitchin, 2016).

References

Calfa, B. A.; and Kitchin, J. R. 2016. Property Prediction of Crystalline Solids from Composition and Crystal Structure. AIChE Journal. In press. DOI: 10.1002/aic.15251.

Eljack, F. T.; Eden, M. R.; Kazantzi, V.; Qin, X.; and El-Halwagi, M. M. 2007. Simultaneous Process and Molecular DesignÑA Property Based Approach. AIChE Journal. 53(5):1232Ð1239.

Li, Q., and Racine, J. S. 2007. Nonparametric Econometrics: Theory and Practice. Themes in Modern Econometrics. Princeton University Press. New Jersey, NJ. USA.

Ma, X.; Li, Z.; Achenie, L. E. K.; and Xin, H. 2015. Machine-Learning-Augmented Chemisorption Model for CO2 Electroreduction Catalyst Screening. Journal of Physi- cal Chemistry Letters. 6(18):3528Ð3533.

Saad, Y.; Gao, D.; Ngo, T.; Bobbitt, S.; Chelikowsky, J. R.; and Andreoni, W. 2012. Data Mining for Materials: Computational Experiments with AB Compounds. Physical Review B. 85(10):104104.

Samudra, A. P., and Sahinidis, N. V. 2013. Optimization-Based Framework for Computer- Aided Molecular Design. AIChE Journal. 59(10):3686Ð3701.

[1] Department of Chemical and Biological Engineering. University of Wisconsin-Madison. Madison, WI 53706. USA.

[2] Department of Chemical Engineering. Carnegie Mellon University. Pittsburgh, PA 15213. USA.

453137 Local Pattern Discovery for Uncovering Structure-Property Relationships of Materials

Abstract:

Machine learning applications to data from atomistic material simulations typically focus on the inference of a global prediction model for some physical or chemical property of interest, such as activation barriers or binding energies. Due to their global prediction-based objective, these models are not well suited for the discovery of physically interpretable characterizations describing sub-groups of material configurations that share common properties. In contrast to predictive global modeling, local pattern discovery techniques directly aim to uncover specific local properties of the data. Moreover, they describe these properties through models that are built from discrete symbols corresponding to meaningful notions of the discovery domain, e.g., find sub-groups of materials with large band gaps and low formation energies. Therefore, they constitute a complimentary part of the data mining toolbox for the automatic analysis of materials science data repositories, such as for the Novel Materials Discovery Repository. As an exemplary application of local pattern discovery for materials science, we consider the automatic discovery of structure-property relationships of gold clusters (size 7-14). Local patterns between geometrical (e.g., local bond coordination, cluster size, and radius of gyration) and physicochemical (e.g., HOMO-LUMO gap and intramolecular van der Waals energy) properties are found.

464732 Alloy Catalyst Discovery Using Computational Alchemy

Abstract:

There is great interest in finding identifying high-performance catalysts that are economical. Computational quantum chemistry schemes employing Kohn-Sham density functional theory (KS-DFT) can be used to screen catalyst materials on the basis of thermodynamic descriptors (see for example ref. [1]). Though usually considered reliable for descriptor-based analyses, KS-DFT calculations are computationally expensive and intractable for use when screening across the full chemical space of all possible alloy materials. Toward this goal, we employ a model Hamiltonian method, Ôcomputational alchemyÕ [2-4] to approximate KS-DFT energies at a fraction of the computational cost. We will introduce the theory of computational alchemy and how it can be used to facilitate descriptor-based screening on many thousands of alloys structures.

Reference:

1. Greeley, J., Jaramillo, T. F., Bonde, J., Chorkendorff, I. & N¿rskov, J. K. Computational high-throughput screening of electrocatalytic materials for hydrogen evolution. Nat Mater 5, 909Ð913 (2006).

2. Lilienfeld, O. A. von, Lins, R. D. & Rothlisberger, U. Variational Particle Number Approach for Rational Compound Design. Phys. Rev. Lett. 95, 153002 (2005).

3. Lilienfeld, O. A. von & Tuckerman, M. E. Molecular grand-canonical ensemble density functional theory and exploration of chemical space. The Journal of Chemical Physics 125, 154104 (2006).

4. Sheppard, D., Henkelman, G. & Lilienfeld, O. A. von. Alchemical derivatives of reaction energetics. The Journal of Chemical Physics 133, 084104 (2010).

470073 To Address Surface Reaction Network Complexity Using Machine Learning, Scaling Relations, and DFT Calculations

Abstract:

Surface reaction networks involving hydrocarbons exhibit enormous complexity with hundreds of species and thousands of reactions for all but the very simplest of chemistries. Among fields that regularly confront this challenge, such as catalysis, metabolic engineering, and combustion, catalysis is unique in that the most likely reaction pathway varies for each surface and active site and must be re-determined in every instance proving a fundamental bottleneck in new catalyst discovery. We present a framework to efficiently explore a reaction network from scratch using a hierarchy of well-established predictive methods such as group additivity and linear scaling relations as guides to traditional density functional theory (DFT) approaches. Unexplored regions of the reaction network are predicted using machine learning methods, and the mechanism is progressively refined using uncertainty-aware DFT calculations on the parts of the network that are likely rate-limiting. Applying these methods to the reaction of syngas on rhodium (111), we identify the most likely reaction mechanism along with bounds on the likelihood unexplored reaction pathways do not contribute, using DFT on just a fraction of the possible intermediates. Propagating uncertainty throughout this process allows for estimates that the final mechanism is complete given DFT-level uncertainty and measurements on only a subset of the entire network.

467637 Machine-Learning Acceleration of the Exploration of Potential Energy Surface within Amp

Abstract:

Novel nanostructures with enhanced properties are required to be computationally designed in various fields of science and engineering, e.g. nanomaterials with improved load-bearing properties in materials science [1] or catalysts to speed up reaction kinetics in chemistry [2]. In principle, accurate calculations can be carried out based on electronic structure methods. However, with the current status of computational facilities, such accurate calculations are feasible only on relatively small systems within limited time scale. In other words the accuracy might have to be sacrificed in favor of less computational expense as the length and time scales of simulations increase. As the result, emergence of new computational methods seems essential.

Pattern recognition of energetics from an *ab initio* database can help reducing the number of required quantum mechanical calculations, as the result making larger-length or longer-time scale simulations practical. This can be done inside our open-source modular Atomistic Machine-learning Package (Amp) [3], which has been recently developed by the authors of this paper. Amp has two main components, a descriptor for modeling local atomic environments and a regression model to establish the relationship between energies/forces with local atomic environment. Amp is developed compatible with the widely-used open-source Atomic Simulation Environment (ASE), facilitating interactive cooperation between the Amp machine-learning calculator and any *ab initio* calculator within ASE. This will be illustrated here in the context of two well-known problems in molecular sciences, a) determination of the structure of stable nanoclusters, and b) finding the minimum energy pathway of reactions. It will be shown that machine-learning can remarkably improve the exploration of the potential energy surface.

References

[1] Q. Qin et al.; Recoverable plasticity in penta-twinned metallic nanowires governed by dislocation nucleation and retraction, Nature Communication, 6, 2015.

[2] S. Freakley et al.; Palladium-tin catalysts for the direct synthesis of H2O2 with high selectivity, Science, 351, 2016.

[3] A. Khorshidi, A. A. Peterson; Amp: A modular approach to machine learning in atomistic simulations, Computer Physics Communications, 2016.

461315 Constructing a Framework for Measuring National Energy Security

Abstract:

Energy policy in many countries, driven by concerns about energy security and environmental damage, is promoting a shift from fossil fuels to a variety of renewable sources. We examine risks to energy security arising from resource discovery and fuel processing, conversion, and use. We analyse the whole of the supply chain for renewable and non-renewable fuels, both current and potential sources, in a national context. The result is a set of indicators summarising the issues which are important for energy security.

Sustainability and energy security are concepts which share common features, some of which are poorly defined or lacking good data. We use the Process Analysis Method for systematically selecting indicators to measure a nationâ€™s energy security which includes environmental, human/social, and economic elements within a single self-consistent framework. This is a triple-bottom-line method of assessment, particularly suitable for complex systems where quantitative numerical models cannot be constructed. To use the indicators in a particular case we require a description of the energy supply and demand system, and we show how this can be constructed to meet the needs of the analysis. Appropriate energy and economic data are needed, together with assessments of issues such as political stability, international relations, and demand management. Although ill-defined, these must be included because of their potential importance. The method underlines the need for assessments and data relating to many issues which are commonly not considered as part of energy security.

Sustainability analysis of the whole value chain of bio-syngas production and its potential as chemical platform for syngas fermentation

Baudine Gevers Deynoot, Delft University of Technology, Delft, the Netherlands
Eduardo Almeida Benalcázar, Delft University of Technology, Delft, the Netherlands
Henk Noorman, DSM Biotechnology Center, Delft, the Netherlands
Patricia Osseweijer, Delft University of Technology, Delft, the Netherlands
John A. Posada Duque[1], Delft University of Technology, Delft, the Netherlands

Introduction

As the world's energy demand is growing and fossil fuels are depleting, concerns are raising on how this demand can be continuously supplied. Solar and wind energy are renewable resources that may contribute to replace fossil energy, However, they cannot provide the carbon needed to produce fuels and chemicals. This fact, coupled to uncertain oil prices and need of lowering the atmospheric CO_2, have led the attention towards biomass and other short-cycled carbon raw materials, such as CO, CO_2 and CH_4 [1]. These feedstocks can yield fuels and chemicals with improved environmental performance (compared to fossil-based products), while preventing the impacts over food security that are argued against first generation feedstocks [2].

Commonly, bio-refineries are conceptualized to produce fuels and chemicals through two types of processing platforms that utilize biomass as feedstock [3,4]: i) the sugar platform, in which biochemical pre-treatment of cellulose yields a fermentable mixture of 5 and 6-carbon carbohydrates; and ii) the syngas platform, which through thermochemical treatment, at 750 – 1000 °C, decomposes biomass into a mixture of gases, commonly referred to as syngas, formed by carbon monoxide, carbon dioxide, hydrogen and small fractions of methane and higher hydrocarbons, hydrogen sulfide and ammonia [5]. The syngas platform is currently under use for heat and electricity generation, and as source of raw materials for the production of fuels via Fischer-Tropsch synthesis (FTS) [6]. Syngas, in recent years, has also been seen as possible feedstock for fermentation [7].

The thermochemical conversion of biomass offers potential advantages over the biochemical route, for instance: i) higher yields, ii) simpler process configuration, iii) more flexibility on feedstock admission and product composition, and iv) higher scale-up possibilities [8]. Moreover, fermentation also shows possible advantages over the catalytic conversion of syngas since microorganisms can: i) work under moderate temperatures and pressures; ii) be more tolerant than catalysts to poisoning by sulphur, tars and chlorine; iii) accept a wider range of gas compositions; and iv) achieve higher conversions and yields [6,7]. Therefore, a hybrid process, combining thermochemical conversion of biomass and syngas fermentation, could drive to reduction on capital and operation costs as well as environmental impacts to the produce fuels and chemicals.

Currently, only few studies have been found to simultaneously address the technical, economic and environmental performance of large-scale production of syngas, as potential feedstock for fermentation [7]. Therefore, in is study we have performed a sustainability

[1] Corresponding author at: van der Maasweg 9 (Room: C0.360), 2629 HZ, Delft, the Netherlands. E-mail address: J.A.PosadaDuque@tudelft.nl (J.A. Posada)

analysis (*i.e.* techno-economic and life cycle assessments) [9] of the syngas production process, from biomass production, its delivery and conversion into a theoretically fermentable gas, in order to evaluate and compare the effects of different processing configurations and operating conditions at multiple stages of the process.

Methodology

The syngas production scale was defined to supply the need of an average second generation bioethanol production plant of 20 million gallons/year, *i.e.* an average of 388 kton of syngas per year, since each type of biomass has different yields to syngas.

Biomass supply

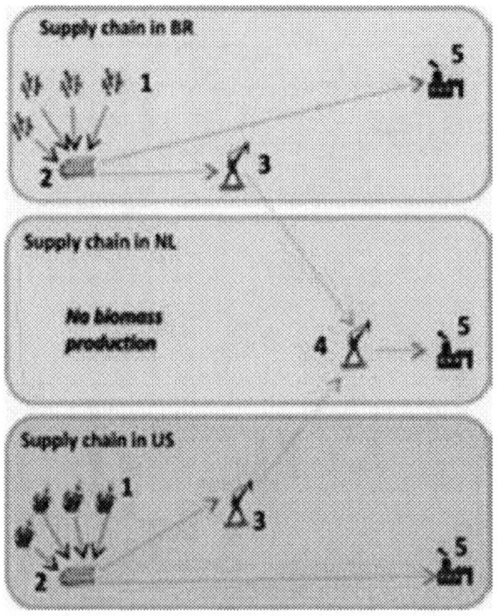

Figure 1. Configuration of supply chains according to the geographical location. Arrows represent transport by truck (black), train (red), and ship (blue). Transfer points: 1. biomass production site, 2. central gathering point, 3. export terminal, 4. import terminal, 5. conversion plant (biomass to syngas and syngas to chemicals)

Different biomass supply configurations were designed considering four different feedstocks: forestry residues (FR) and corn stover (CS), produced in the United States of America (US); and sugarcane bagasse (SB) and eucalyptus wood (EW), produced in Brazil (BR). The Netherlands (NL) has been included only as location for processing, which means biomass is imported from the other two locations and then processed first into syngas and later into one of the bio-based products. The four feedstocks were selected based on availability on the production site, price and composition.

Here, the approach proposed by Hamelinck et al. (2005) [10] was adopted to obtain supply costs, environmental impacts and energy requirements for the delivery of the lignocellulosic biomass. Three main process operations were considered to structure the (inter)national biomass supply, they are: i) biomass production (type of feedstock produced, land establishment, land maintenance and collection/harvesting); ii) conditioning (chipping, baling, pelletization and drying); and iii) transportation (truck, train and ship transport) [10]. The sequence of operations was defined based on generic theoretical chains which contain five transfer points [1] as shown in

Figure **1**: 1) the production site, 2) a central gathering point, 3) a transport terminal for export, 4) a transport terminal for import, and 5) the conversion plant. In the specific case of international transport, pelletizing of biomass is included to increase its density and makes long distance transportation more efficient. The location of the central gathering points was defined to keep minimum costs for transport between the sites with the highest biomass productivities and the next transfer points (considering road infrastructure and cost-efficiency of the transport type). Determination of the ports was mainly based on infrastructure and scale; ports were selected that can take up Panamax ships, assuring large scale biomass transportation [10]. The moisture content of the biomass entering the gasification unit, is considered to be 10 wt.%, which is pointed as the optimum to minimize size and operation costs of the gas-to-liquid plant. When biomass required partial drying before international transport, the costs were accounted for.

Thermochemical conversion

The gasification process here considered was based on the technology described by NREL [11], which is able to produce syngas suitable for fermentation. The process takes place at atmospheric pressure and uses three fluidized bed reactors: an indirectly heated (allothermal) gasification reactor, a combustion reactor (as heat source) and a gas conditioning reactor. The gasification and combustion reactors function with circulating fluidized beds (CFB) while the gas conditioning reactor uses a bubbling fluid bed (BFB). Olivine, which is used as bed material, circulates through the three reactors and is (re)heated in the combustion reactor where char is burned with air.

Table 1. Biomass data: physicochemical characterization, production requirements and production costs

Parameter	***Forestry residues (PW)***	***Agricultural residues***		***Energy crops (EW)***
		SB	***CS***	
Region of production	US	BR	US	BR
Costs in \$/ton$_{wet}$ (@year)	15 (2011)	17 (2012)	24 (2011)	31 (2011)
Moisture content$_{Green\ Biomass}$ (wt.%)*	45	51	24	45
C (wt.%)	49.7	45.2	47.3	46.0
H (wt.%)	7.8	5.5	5.1	5.8
O (wt.%)	41.8	40.7	40.6	44.5
N (wt.%)	0.1	0.1	0.8	0.3
S (wt.%)	0.05	0.05	0.22	0.0

* Green biomass is defined as wet biomass at its harvesting phase

Table 2. Gasification and tar reforming reactions

Gasification	
$C + H_2O \rightarrow CO + H_2$	(1)
$C + O_2 \rightarrow CO_2$	(2)
$2H_2 + O_2 \rightarrow 2H_2O$	(3)
$C + CO_2 \rightarrow CO$	(4)
$C + 2H_2 \rightarrow CH_4$	(5)
$CO_2 + H_2 \rightarrow CO + H_2O$	(6)
$N + 1.5H_2 \rightarrow NH_3$	(7)
$S + H_2 \rightarrow H_2S$	(8)
Reforming	
$CH_4 + H_2O \rightarrow CO + 3H_2$	(9)
$CH_4 + CO_2 \rightarrow 2CO + 2H_2$	(10)
$2CO + O_2 \rightarrow 2CO_2$	(11)
$2NH_3 \rightarrow 3H_2 + N_2$	(12)

The gasification process was simulated in Aspen Plus V8.8 using the physicochemical characterization obtained from literature data (see Table 1) for all considered feedstocks. The developed model is based on a general approach to represent a dual fluidized bed gasifier, known as fast internally circulating fluidized bed [12]. The model considers the following assumptions: ideal gas law applies for the gas phase; operation takes place isothermally and at steady state; distribution of all gases within the emulsion phase is homogenous; drying and pyrolysis are instantaneous; there are not pressure drops or heat losses; chemical equilibrium is assumed; since CH_4 represents all hydrocarbons; all nitrogen and sulfur in biomass are converted into NH_3 and H_2S; and char only consists of carbon and ash.

For the gasification process, steam (at 450 °C) enters at the gasification reactor. This steam has a double function, it acts as a fluidizing medium and as oxidizing agent for the gasification and tar reforming reactions, as shown in Table 2. The mass flow rate of steam is defined by the steam to biomass ratio which was set at 0.75 [12]. The gasification reactions (see Table 2) take place when high temperatures are reached (850 °C). The temperature of the air stream fed to the gasification reactor was set at 450 °C.

In the tar reforming process, the gas coming from the gasifier is sent to the catalytic tar reformer, where methane and higher hydrocarbons are oxidized into CO and H_2, CO into CO_2, and NH_3 into both: N_2 and H_2 (Table 2). Steam addition to the tar reformer was excluded from the model since the water content in the input gas was already sufficient to drive the reactions and avoid additional costs and gas dilution.

Economic evaluation

The economic assessment of the designed processes was made from the perspective of the syngas production costs, considering capital and operational expenditures (CAPEX and OPEX, respectively). CAPEX was constructed from several factors linked to the equipment purchase costs (EPC) [12]. The EPC's of the gasification plant were adapted from and NREL's report on syngas production [11]. The 6/10 scaling factor rule was used for capacity corrections [12]. Prices were updated to 2015 by using the Chemical Engineering Plant Cost Index (CEPCI). A project lifetime was fixed at 15 years. OPEX was assumed to be formed by [12]: facility-dependent costs (depreciation, maintenance, insurance, local taxes and overhead), process-related costs (raw materials and utilities) and labor.

Furthermore, the economic performance for each scenario was adjusted to the specific geographical locations (NL, US and BR) by using specific prices for fuels (used for defining prices of utilities) and for lignocellulosic biomass.

Life Cycle Assessment

The LCA was performed in agreement with the procedures stated in the ISO 14040 and 14044 [14,15] norms. System boundaries were defined from a cradle-to-gate approach. Therefore, the system includes from feedstock's production and supply until syngas production. The assessment was performed from the perspective of two environmental impact categories [16]: the global warming potential (GWP) and the primary energy use (calculated as non-renewable energy use, NREU). The functional units for the inventory analysis and the impacts assessment is the production of 1 kg of syngas and 1 MJ of its low heating value (see Table 3).

Individual impact characterization factors for materials and energy carriers were taken from the Ecoinvent V2.2 database, literature and internet references. The environmental impact data for sugarcane bagasse and corn stover were adopted from literature by applying economic allocation since they are considered as waste materials. Factors for electricity production were defined as site-specific for each country where location was considered.

Results and discussion

Process design

The supply chains were designed to produce a syngas that complies with all technical requirements for its further fermentation into chemicals. Table 3 shows the resulting yields of syngas on biomass for each feedstock as well as the composition of the resulting syngas for each case. The highest syngas yield is obtained from pine wood as a result of its high hydrogen content and low presence of ash and nitrogen compared to the other feedstocks. However, it is worth to notice that the syngas yield does not have significant variability among the four feedstocks; only a variation of 5.3 % is observed.

Table 3. Yields of biomass and final products, and composition of resulting syngas at a gasification temperature of 850 °C

Parameter	PW	SB	CS	EW
Syngas composition (vol. % $_{dry}$)				
H_2	60.9	57.0	55.4	56.5
CO	28.1	28.6	30.0	29.0
CO_2	9.9	13.1	13.0	13.2
N_2	1.0	1.2	1.5	1.2
CH_4	593.2 ppmv	379 ppmv	412 ppmv	386 ppmv
NH_3	12.4 ppmv	12 ppmv	13 ppmv	12 ppmv
H_2S	160.6 ppmv	189 ppmv	822 ppmv	0
H_2/CO	2.2	2.0	1.8	1.9
LHV (kJ/kg)	5101	5451	5378	5446
Yield (kg syngas/kg biomass$_{dry}$)	1.20	1.14	1.18	1.19

Economic evaluation

Figure 2 shows the costs distribution for syngas production, for the four feedstocks, with their respective region of origin and according to the location for biomass gasification. When syngas production takes place in the region of the feedstock origin, the lowest production costs are obtained from CS in the US. However, when syngas production takes place in NL after biomass import, PW led to the lowest syngas production costs of all feedstocks. The costs related to biomass delivery to the gasification plant represent the largest share of the total syngas production costs. For instance, when syngas is produced in NL, the cost of biomass represents around the 50 % of production costs; this makes production in NL is the most expensive option. Utilities are the second largest contributor to the syngas production costs; for example, when syngas is produced in NL, utilities represent between 21 and 24%; while for syngas production in US and BR these contributions range between 19 and 25%.

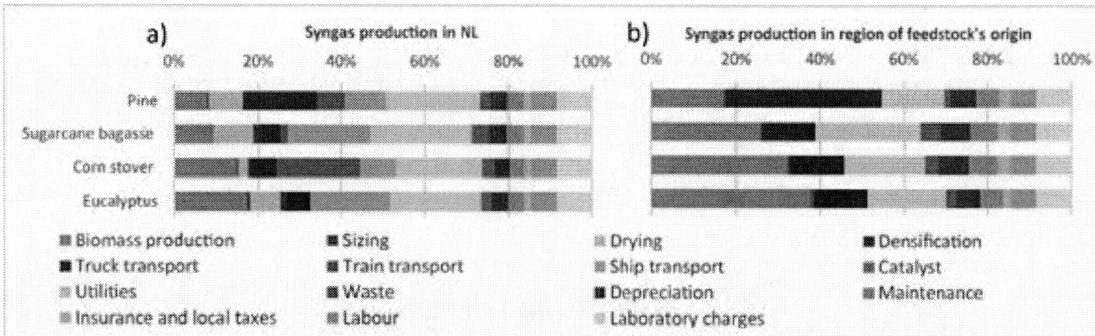

Figure 2. Cost breakdown of syngas production for: a) syngas production in NL, and b) syngas production at the region of the feedstock production

Generally, syngas production costs are found to be higher than the ones obtained by the present study. However, literature data consider syngas as feedstock for FTS, which has larger requirements on gas composition than fermentation. Nevertheless, production costs of bio-syngas from corn stover with a H_2/CO ratio of 2 for FTS, is reported to have a value of \$221.5/ton [17]. This is 1.3 – 85.7 % higher than the values obtained in this study.

Table 4. Total production costs of syngas: contribution analysis by feedstocks and location

Parameter	Costs contribution ($/ton syngas)							
	PW (US)		SB (BR)		SC (US)		EW (BR)	
	(NL)*	(local)**	(NL)	(local)	(NL)	(local)	(NL)	(local)
Production costs ($/ton)	190.7	144.5	198.2	130.0	218.7	119.3	198.0	150.0

* NL case: syngas production takes place in NL

** Local case: syngas production takes place in the region where the feedstock is produced

Life cycle assessment

The obtained environmental impacts (GWP and NREU) for the production of syngas in the three locations (NL, BR and US) from PW, SB, CS and EW are shown in Table 5. Of all feedstocks, eucalyptus returns the lowest impacts during syngas production, while corn stover yields, by far, the highest; additionally BR is the best location while NL produces the highest impacts, mainly due to international transport. Figure 3 shows a contribution analysis for GWP and NREU. Two key factors contributing to the environmental impacts of syngas production are steam for gasification and biomass production; since the latter includes the conditioning of biomass and its delivery, its contribution is larger for the cases in which gasification is carried out in NL.

Again for the environmental impacts, comparing the results of this study with literature data is rather difficult due to the differences in syngas requirements and process configurations. When comparing against averaged results obtained for syngas produced as feedstock for FTS [17], this study returned averaged 44% lower GWP. On the other hand, when impacts are compared against syngas derived from coal in India, this study returned 55 and 73 % lower GWP and NREU.

Table 5. Environmental impacts (GWP and NREU) of syngas production

Impact category	Functional unit	NL				BR		US	
		PW	SB	CS	EW	SB	EW	PW	CS
GWP	kg syngas	0.26	0.26	0.32	0.21	0.23	0.20	0.21	0.29
	MJ syngas	0.051	0.047	0.059	0.038	0.042	0.036	0.041	0.053
NREU	kg syngas	5.87	3.71	5.20	5.99	3.13	3.74	3.46	4.09
	MJ syngas	1.15	0.68	0.97	1.10	0.57	0.69	0.68	0.76

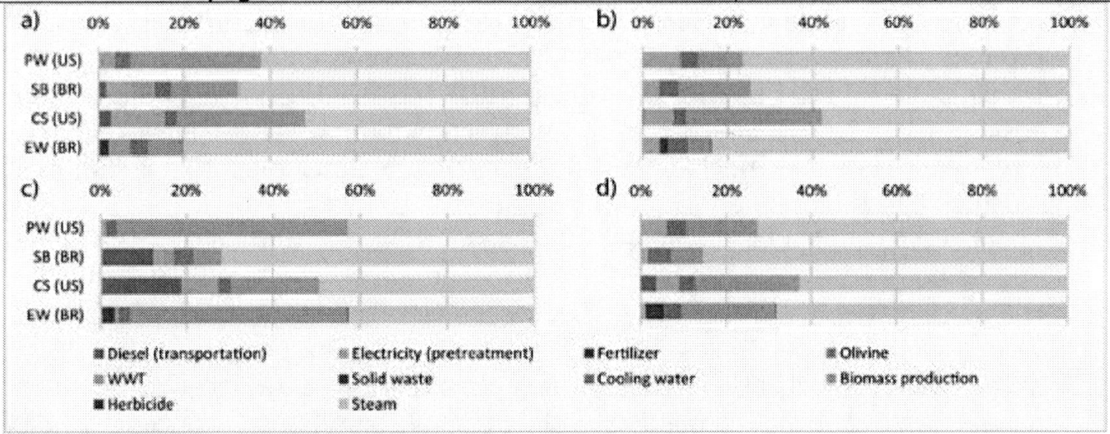

Figure 3. Contribution analysis for environmental impacts of syngas production: a) GWP in NL; b) GWP in the region of feedstocks origin; c) NREU in NL; d) NREU in the region of feedstocks origin

Conclusions

All four feedstocks (pine, sugarcane bagasse, corn stover and eucalyptus) proved to be able to produce syngas suitable for fermentation, with different production cost, environmental impacts and technical performance. Additionally, no specific feedstock-plant location combination performs best on all aspects: economic, environmental and technical performance. However, corn stover when gasification takes place in the US, proved to have the lowest production cost: $119.27/ton. This configuration also delivered syngas with the highest CO and H_2 concentration, plus the lowest nitrogen and ash content; this resulted in the highest syngas yield while keeping low CO_2 concentration. Environmentally, syngas production in Brazil returned the lowest GWP of 0.20 kgCO_2eq/kg when eucalyptus is used as feedstock, and the lowest NREU of 3.13 MJ/kg when sugarcane bagasse is the feedstock. In general, location of gasification plant in the same region where feedstock production takes place is, economically and environmentally, the most optimum configuration, mainly due to the large impacts of biomass transport.

Optimum gasification temperatures were obtained in terms of syngas composition (high CO and H_2, and low CH_4 and CO_2). These temperatures range between 850 and 1000 °C, depending on the feedstock. At these temperatures theoretically, the tar reformer could be excluded from the process configuration. This resulted in reduction of syngas production costs, GWP and NREU by 20.1, 14.0 and 16.0 %, respectively; however, the NH_3 concentration at these conditions is too high for syngas fermentation, according to literature. Finally, heat recovery through steam generation is highly recommended to increase the overall efficiency of the system: maximum 29 and 23 % reductions on syngas production costs and GWP, were respectively observed for pine gasification in the US.

The results of this analysis, among others [18], are used as input data to study the economic and GHG performance of the the syngas fermentation platform for the production of ethanol, 2,3-butanedion and hexanoic acid [19].

References

1. Daystar, Jesse; Gonzalez, Ronalds; Reeb, Carter; Venditti, Richard; Treasure, Trevor; Abt, Robert and Kelley, Steve (2014), "Economics, Environmental Impacts, and Supply Chain Analysis of Cellulosic Biomass for Biofuels in the Southern US: Pine, Eucalyptus, Unmanaged Hardwoods, Forest Residues, Switchgrass, and Sweet Sorghum", BioResources, pp. 393–444.
2. Posada, John A. and Osseweijer, Patricia (2016) "Socio-economic and environmental considerations for sustainable supply and fractionation of lignocellulosic biomass in a biorefinery context", In: Biomass fractionation technologies for a lignocellulosic feedstock based biorefinery, 1st ed. Mussatto SL, p. 674.
3. Cheali, Peam; Posada, John A.; Gernaey, Krist V. and Sin, Gürkan (2015), "Upgrading of lignocellulosic biorefinery to value added chemicals: Sustainability and economics of bioethanol-derivatives", Biomass and Bioenergy, April, pp.282–300.
4. Cheali, Peam; Posada, John A.; Gernaey, Krist V. and Sin, Gürkan (2016), "Economic risk analysis and critical comparison of optimal biorefinery concepts: Economic risk analysis and critical comparison of optimal biorefinery concepts", Biofuels, Bioproducts and Biorefining, May, pp. 435-445.

5. Munasinghe, Pradeep C. and Khanal, Samir K. (2010), "Biomass-derived syngas fermentation into biofuels: Opportunities and challenges", Bioresource Technology, July, pp. 5013–5022.
6. Liew, Fung M.; Köpke, Michael and Dennis, Séan (2013), "Gas Fermentation for Commercial Biofuels Production", In: Liquid, Gaseous and Solid Biofuels - Conversion Techniques, InTech.
7. Daniell, James; Köpke, Michael and Simpson, Séan D. (2012), "Commercial Biomass Syngas Fermentation", Energies, December, pp. 5372–5417.
8. Siedlecki, marcin; De Jong, Wiebren and Verkooijen, Adrian H. (2011), "Fluidized Bed Gasification as a Mature And Reliable Technology for the Production of Bio-Syngas and Applied in the Production of Liquid Transportation Fuels—A Review", Energies, March, pp. 389–434.
9. Palmeros-Parada, Mar; Osseweijer, Patricia; Posada, John A. (2016) Sustainable biorefineries, an analysis of practices for incorporating sustainability in biorefinery design. Industrial Crops and Products http://dx.doi.org/10.1016/j.indcrop.2016.08.052
10. Hamelinck, Carlo; Suurs, Roald S. and Faaij, André P. (2005), "International bioenergy transport costs and energy balance", Biomass Bioenergy, August, pp. 114-134.
11. Worley, M. and Yale, J. (2012), "Biomass Gasification Technology Assessment", NREL.
12. Doherty, Wayne; Reynolds, Anthony and Kennedy, David (2013), "Aspen Plus Simulation of Biomass Gasification in a Steam Blown Dual Fluidised Bed", In: Materials and processes for energy: communicating current research and technological developments, Formatex Research Centre.
13. Peters Max S. and Timmerhaus, Klaus D. (1991), "Plant design and economics for chemical engineers", 4. ed. McGraw-Hill.
14. ISO, (1997), "ISO 14040 Environmental management - Life cycle assessment - Principles and framework".
15. ISO, (2006), "ISO 14044 Environmental management - Life cycle assessment - Requirements and guidelines".
16. Posada, John A; Brentner, Laura B; Ramirez, Andrea; Patel, Martin K. (2016), "Conceptual design of sustainable integrated microalgae biorefineries: parametric analysis of energy use, greenhouse gas emissions and techno-economics", Algal Research, July, pp. 113-131.
17. Moncada, Jonathan; Posada, John A. and Ramírez, Andrea (2015), "Early sustainability assessment for potential configurations of integrated biorefineries. Screening of bio-based derivatives from platform chemicals", Biofuels, Bioproducts and Biorefining, November, pp. 722–748.
18. Posada, John A; Cardona, Carlos A. (2012). Propionic acid production from raw glycerol using commercial and engineered strains. Industrial & Engineering Chemistry Research 51:2354–2361.
19. Almeida-Benalcázar, Eduardo F.; Gevers-Deynoot, Baudine; Noorman, Henk; Osseweijer, Patricia; Posada, John A. (2016). Syngas fermentation for the production of bulk chemicals: early-stage techno-economic and life cycle assessment. 2016 AIChE Annual Meeting. November 13-18, 2016. San Francisco, CA

462372 Design and Operation of Biofuel Supply Chains with Variable Regional Depot and Biorefinery Locations

Abstract:

The transportation, handling and storage of bulky raw biomass is one of the key challenges in biofuel supply chains. To address this challenge, the concept of regional biomass processing depot [1] (referred to in this work as 'depot') has been introduced. At the depot, bulky biomass is pretreated and/or densified into a stable and dense intermediate, which can then be transported economically over long distances. Despite all the extensive research in the area [2,3], the design of biofuel supply chains with depots has received limited attention; depots have not been studied at all or assumed to be installed in predetermined locations. However, the efficiency of the biofuel supply chain (SC) can be improved substantially by optimizing the location of depots and biorefineries.

Accordingly, in this work, we propose a multi-period mixed-integer linear programming (MILP) model after briefly reviewing a mixed-integer non-linear programming (MINLP) model developed by Ng and Maravelias [4]. The new MILP model accounts for biomass selection and allocation, technology selection and capacity planning at depots and biorefineries, inventory and shipment planning, and, importantly, variable depot and biorefinery locations. Unlike other biofuel SC approaches, which precalculate the distances based on the predetermined depot and biorefinery locations, we develop a series of approximations and reformulations that allow us to overcome the bilinearities (shipments*transportation_distance) in the calculation of transportation cost, based on the following assumptions:

1) The biomass from a harvesting site is sent to only one downstream node, either a depot or a biorefinery; and, similarly, the intermediates from a depot are shipped to a single biorefinery. This is a realistic assumption because it is difficult to have multiple pickups from a harvesting site or a depot and also to "dynamically" change where and how much to ship.

2) Biomass and intermediates are stored at the origin (harvesting site and depot, respectively). This assumption has minimal effect because inventory unit costs and material deterioration coefficients are practically independent of location.

3) We approximate the exact timing of the shipment as all biomass from selected harvesting site is eventually shipped to the same depot or biorefinery. Thus, we can replace the variable shipments with the total availability of biomass. A correction factor though is introduced to account for the fact that the actual shipments will be slightly lower than the total availability of biomass due to biomass deterioration during storage.

Based on the above assumptions, we first precalculate the range of correction factors based on the possible inventory profiles. We can choose the average value or a factor that results from a reasonable inventory profile. Second, we disaggregate all variable distance components into "actual" and "dummy" distance variables, which are bounded by variable upper bound constraints using binary arc selection variables. The actual distance variable is equal to zero if the corresponding arc is not selected. Finally, we reformulate the transportation cost, which can be calculated based on the actual distance variables and approximated shipments. The latter are treated as parameters, thereby resulting in a linear transportation cost calculation.

We illustrate the applicability of the proposed model by considering a biofuel SC in South Central Wisconsin that includes Columbia, Dane, Dodge, Grant, Green, Iowa, Jefferson, Lafayette, Richland, Rock, and Sauk counties, with a total of 64 harvesting sites. We consider two types of biomass: corn stover and switchgrass, both of which are abundant biomass resources in Wisconsin. At most three depots can be installed in the studied region. We employ certain criteria for screening the potential biorefinery locations: Candidate biorefinery locations were restricted to cities/towns/villages with populations between 3,000 and 10,000 in 2014 for a maximum of one biorefinery per county [5] and total biomass availability more than 120,000 dry ton per year. Among the locations that met the screening criteria, we preselect two towns, which are located near the center of the studied region, because they are likely to lead to smaller transportation cost. The coordinates of the biorefinery are constrained by the convex hull of the selected town, therefore, the biorefinery can be installed at various points within the selected region, instead of at the centroid of the region.

The objective is to minimize the total annual cost (TAC), which includes biomass acquisition, inventory, operating and capital costs. The total capital investment for the ethanol production level of 340 million liters per year, is $ 592 million, leading to a TAC of $ 251.44 million. In this case study, two depots are selected, while the biorefinery is installed at West Oregon. The SC configuration with two depots has lower TAC as compared to the one without depots. The transportation cost in the configuration with three depots or more is lower than in the configuration with two depots, but the higher capital costs leads to higher TAC.

We also calculate the relative error of our approximation using the exact biomass/intermediate shipments obtained. The calculated relative error is found to be insignificant. Finally, for comparison purposes, we extend the MINLP model [4] to account for variable biorefinery locations and resolve the case study. The proposed model is shown to yield better solutions in a faction of the time required by the MINLP model.

References:

[1] P.L. Eranki, B.D. Bals, B.E. Dale, Advanced regional biomass processing depots: a key to the logistical challenges of the cellulosic biofuel industry, Biofuels, Bioprod. Biorefining. 5 (2011) 621–630.

[2] B. Sharma, R.G. Ingalls, C.L. Jones, A. Khanchi, Biomass supply chain design and analysis: Basis, overview, modeling, challenges, and future, Renew. Sustain. Energy Rev. 24 (2013) 608–627.

[3] D. Yue, F. You, S.W. Snyder, Biomass-to-bioenergy and biofuel supply chain optimization: Overview, key issues and challenges, Comput. Chem. Eng. 66 (2014) 36–56.

[4] R.T.L. Ng, C.T. Maravelias, Design of Cellulosic Ethanol Supply Chains with Regional Depots, Ind. Eng. Chem. Res. 55 (2016) 3420–3432.

[5] W. Alex Marvin, L.D. Schmidt, S. Benjaafar, D.G. Tiffany, P. Daoutidis, Economic Optimization of a Lignocellulosic Biomass-to-Ethanol Supply Chain, Chem. Eng. Sci. 67 (2012) 68–79.

462459 Process Synthesis of Natural Gas to Liquid Transportation Fuels Under Uncertainty: A Robust Optimization Approach

Abstract:

As discussed in a recent perspective [1], natural gas has been explored as an exciting feedstock for the production of liquid transportation fuels [2], olefins [3], and aromatics [4] through advanced process synthesis and global optimization techniques in recent years. The growing abundance of natural gas as a feedstock, coupled with the importance of these products in society, provides substantial motivation for the discovery of optimum refinery topologies and product distributions. Focusing on natural gas to liquid transportation fuels (GTL) specifically, there are many different natural gas conversion routes using synthesis gas as an intermediate and processes such as methanol synthesis or Fischer-Tropsch conversion for final fuel production in the form of gasoline, diesel, and kerosene. Identifying the most cost-effective method for fuel production is imperative.

Process synthesis as a whole, however, has many areas in which uncertainty can appear in model parameters. This uncertainty can be detrimental to the optimal solutions, which may become infeasible or give objective function values worse than expected due to uncertain parameter realizations. To include parameter uncertainty in the model, uncertain constraints are reformulated using robust optimization [5]. The robust counterparts ensure that constraints will feasible for an uncertainty set of parameter realizations; the size of the uncertainty set can be determined using probabilistic bounds, in which considerable advances have recently been made [6-9]. These probabilistic bounds are utilized *a priori* and *a posteriori*to give robust solutions with minimal conservatism and known levels of risk.

Uncertainty has been included in the GTL process synthesis superstructure in order to incorporate price uncertainty from the key feedstocks and products through reformulation of the objective function. Uncertain parameters such as investment costs are also considered and discussed. The non-convex mixed-integer nonlinear optimization problems are solved to global optimality to give optimal solutions at known probabilities of constraint violation [10]. As the uncertainty appears in the objective function alone, probabilistically guaranteed levels of profit can be found with conservative assumptions about the probability distributions for uncertain parameters. An iterative method will be utilized to provide high quality robust solutions at low probabilities of constraint violation using the box, interval + ellipsoidal, and interval + polyhedral uncertainty sets [11]. The guaranteed and expected profit levels, optimal product distributions, and overall investment costs of the robust solutions will be discussed at varying probabilities of constraint violation. These insights will provide key information on the best GTL refinery topologies moving forward under uncertainty.

[1] Floudas, C. A.; Niziolek, A. M.; Onel, O.; Matthews, L. R. Multi-scale systems engineering for energy and the environment: Challenges and opportunities. AIChE Journal 2016, 62 (3), 602-623.

[2] Baliban, R. C.; Elia, J. A.; Floudas, C. A. Novel Natural Gas to Liquids Processes: Process Synthesis and Global Optimization Strategies. AIChE Journal 2013, 59 (2), 505-531.

[3] Onel, O.; Niziolek, A. M; Floudas, C. A Optimal Production of Light Olefins from Natural Gas via the Methanol Intermediate. Industrial & Engineering Chemistry Research 2016, 55 (11), 3043-3063.

[4] Niziolek, A. M.; Onel, O.; Floudas, C. A. Production of benzene, toluene, and xylenes from natural gas via methanol: Process synthesis and global optimization. AIChE Journal 2016, 62(5), 1531-1556.

[5] Li, Z.; Ding, R.; Floudas, C. A. A Comparative Theoretical and Computational Study on Robust Counterpart Optimization: I. Robust Linear Optimization and Robust Mixed Integer Linear Optimization. Industrial & Engineering Chemistry Research 2011, 50, 10567-10603.

[6] Li, Z.; Tang, Q.; Floudas, C. A. A Comparative Theoretical and Computational Study on Robust Counterpart Optimization: II. Probabilistic Guarantees on Constraint Satisfaction. Industrial & Engineering Chemistry Research 2012, 51 (19), 6769-6788.

[7] Guzman, Y. A; Matthews, L. R; Floudas, C. A New a priori and a posteriori probabilistic bounds for robust counterpart optimization: I. Unknown probability distributions. Computers & Chemical Engineering 2016, 84, 568-598.

[8] Guzman, Y. A; Matthews, L. R; Floudas, C. A New a priori and a posteriori probabilistic bounds for robust counterpart optimization: II. A priori bounds for known symmetric and asymmetric probability distributions. 2016, In Preparation.

[9] Guzman, Y. A; Matthews, L. R; Floudas, C. A New a priori and a posteriori probabilistic bounds for robust counterpart optimization: III. A posteriori bounds for known probability distributions. 2016, In Preparation.

[10] Baliban, R. C; Elia, J. A; Misener, R.; Floudas, C. A. Global optimization of a MINLP process synthesis model for thermochemical based conversion of hybrid coal, biomass, and natural gas to liquid fuels. Computers & Chemical Engineering 2012, 42, 64-86.

[11] Li, Z.; Floudas, C. A. A Comparative Theoretical and Computational Study on Robust Counterpart Optimization: III. Improving the Quality of Robust Solutions. Industrial & Engineering Chemistry Research 2014, 53 (33), 13112-13124.

465584 Multi-Scale Operational Planning and Optimal Sizing of Hybrid Renewable Energy System

Abstract:

Decentralized and hybrid renewable energy system has recently received attention for providing energy in a small and remote area with its own grid (referred to as micro-grid). In this study, *operational planning and optimal sizing* of a micro-grid having large wind farm and battery storage device is studied. Operational planning is generally made in two time-scales; unit commitment decisions on which source should be on-line at what time are pre-determined one day ahead; and dispatch decisions on the output of on-line sources are then decided on hourly time basis. Two stage stochastic programming (2SSP) is conventionally formulated for this one-day operation with a number of intra-day uncertainty scenarios. However, the finite horizon setting is a main factor to limit sustainable operation due to end-effects. At the same time, the yearly investment problem for sizing energy generation and storage resources is typically performed to minimize future operating costs as well as capital cost. These argue yearly operation model beyond one-day is required. Therefore in this study, 2SSP model for one-day operation is combined with daily-evolving Markov decision process (MDP). That is, the value of being in a state of current commitment and battery at the end of a day (called value function) is included in the objective function to account for longer term implications of the decisions. In the MDP formulation, inter-day variation of wind is captured, and the value function is approximated with a linear model. The coefficient vector of the linear model is recursively updated with a sampled observation estimated from the one-day operation model. With this value function capturing all future operating costs, optimal sizing of the wind farm and battery devices is determined based on a surrogate function optimization. Meanwhile, for the integration of the decision hierarchy, a wind model consistent across multiple timescales from seasonal to hourly is developed to provide both intra-day scenarios capturing hourly ramping and inter-day transition information. The results of the proposed method are compared to those of the original 2SSP model through a case study and real wind data.

461974 Modelling the Gradual Expansion of Integrated Wind-Hydrogen-Electricity Networks That Decarbonise the Domestic Transport in Great Britain By 2050

Abstract:

The evolution of integrated wind-hydrogen-electricity networks that satisfy the demands of the domestic transport sector is optimised using a detailed, spatially-distributed mixed integer linear programming model that accounts for hourly operational issues, such as intermittency and dynamics of energy storage, simultaneously with long-term strategic decisions.

In our previous recent work [1], the design and operation of the network were determined using representative wind speed and hydrogen demand data for a whole year, to meet the full demand for hydrogen without considering the gradual transition from the current state with no hydrogen demand. However, in reality the full network will be established over many years as the penetration of fuel cell vehicles increases. Therefore, the model presented in Samsatli et al. (2016) [1] was extended to consider staged capacity investments and retirements, such that the demands for hydrogen, based on an assumed trajectory for the penetration of fuel cell vehicles from now to 2050, are met at all times and that the net present cost of doing so is minimised, subject to the available land area for wind turbines.

The potential technologies and infrastructures that may become part of the network include wind turbines; electrolysers, fuel cells, compressors and expanders; pressurised vessels and underground storage for hydrogen storage; hydrogen pipelines and electricity overhead/underground transmission lines; and fuelling stations and distribution pipelines.

The model determines the optimal spatial structure and the gradual expansion of the integrated networks across Great Britain, i.e. the optimal combinations of conversion and storage technologies invested in at every 5-year period, where to locate them, what their capacities should be and how to connect them by means of the transmission infrastructure networks. Simultaneously with this, the model also determines the hourly operation of each technology that forms the integrated network and the interactions between hydrogen and electricity networks.

In this conference, we will discuss the model and a decomposition method that enables this very large optimisation model to be solved. We will also present some case studies to illustrate the applicability of the model to scenarios in Great Britain.

References:

[1] S. Samsatli, I. Staffell, and N.J. Samsatli. Optimal design and operation of integrated wind-hydrogen-electricity networks for decarbonising the domestic transport sector in great britain. International Journal of Hydrogen Energy, 41:447 â€"475, 2016.

465831 A Computer-Aided Scenario Planning of Future Regional and National Energy Systems Based on Feasible Technology Options in Japan

Abstract:

The design of energy systems has become an issue all over the world. A single optimal system cannot be suggested because the availability of infrastructure and resources, and the acceptability of the system should be discussed locally with the involvement of all related stakeholders in the energy systems. In particular, researchers and engineers of technologies related to energy systems should be able to perform the forecasting and roadmapping of future energy systems and indicate quantitative results of scenario analyses. A simulation-based approach is strongly needed for such a purpose.

We developed a nation-wide energy flow model developed for analysing scenarios of future energy systems implementing a variety of feasible technology options [Kikuchi et al. 2014]. Decentralized power source technologies have been also modeled and analyzed with scenarios installing them in various countries [Shimizu et al. 2015; Kikuchi et al. 2016a]. A case study was conducted with a model for industrial symbiosis in a remote island in Japan, where excess energy and resources from cane sugar mills are effectively utilized in local energy systems [Kikuchi et al., 2016b]. These models were modularized and represented as functionals of appropriate technology options, which enables the aggregation and disaggregation of energy systems by defining functionals for individual technologies, packages integrating multi-technologies, and mini-systems. For scenario planning for actual energy systems, such a computer-aided approach should be able to address the issues related with system design and decision by stakeholders. It includes not only technological aspects, but also socio-economic ones such as the acceptability by energy consumers.

In this study, we are tackling with the scenario planning of energy systems in regional and national boundaries of Japan by developing and using computer-aided tools. Based on the latest energy technology roadmaps by the Society of Chemical Engineers, Japan (SCEJ) [Kato et al. 2016], feasible technology options are modeled for simulating micro- and macro energy grids in Japan. As case examples of microgrid simulation, the energy systems in remote islands were examined and scenarios were planned considering locally available resources. The introduction of energy carriers in Japan was also examined on the basis of energy demand/supply statistics. Regarding socio-economic aspects, questionnaire surveys were conducted to the energy consumers in remote islands and metropoles in Japan to scrutinize the consumers' preferences on energy. Based on the case studies, the requirements of a computer-aided approach for scenario planning of energy systems were defined.

The categorization of feasible technologies options listed in the latest energy technology roadmap [Kato et al. 2016] could be considered in the model framework quantifying energy demand/supply. In the regional grid case study, the energy integration in the rural area was examined on the systems centering on two types of power plant, the existing combined heating and power (CHP) in the sugar mill and new CHP for district heating and cooling in the central town area. The utilization of resources available in regional area can be an effective option for mitigating greenhouse gas (GHG) emission, which is also supported by the socio-economic analyses. Regarding the macro level introduction of energy carriers in Japan, their advantages have a strong connection with the site-specific condition on energy demand/supply, for example, the places where the unused or renewable resources are available have a priority in mitigation of GHG emission by introducing energy carrier-related technologies. Through case studies based on the developed models, it was demonstrated that the contribution of technologies to, e.g., the reduction in GHG emissions, should be carefully examined by quantitative analyses of interdependencies of the technology options for actual decision-making on energy systems. A simulation-based approach can contribute to the achievement of " be smart" by considering the benefit and risk of future energy-system options.

Acknowledgements:

Part of this study is financially supported by grants from the project JSPS KAKENHI Grant Numbers 16H06126 (Grant-in-Aid for Young Scientists A), 15H01750 (Scientific Research A) and the Environment Research and Technology Development Fund (1RF–1503). Activities of the Presidential Endowed Chair for "Platinum Society" in the University of Tokyo are supported by the KAITEKI Institute, Inc., Nippon Telegraph and Telephone Corporation, Fujifilm Holdings Corporation, Mitsui Fudosan Co., Ltd, LIXIL Corporation, and Shin-Etsu Chemical Co., Ltd..

References:

Kato, Y., Koyama, M., Fukushima, Y., Nakagaki, T. (Edited), Future Energy Systems Based on Feasible Technologies Beyond 2030, Springer: Tokyo, ISBN13: 9784431559498 (2016)

Kikuchi, Y., Kanematsu,Y., Sato, R., and Nakagaki, T., Distributed cogeneration of power and heat within an energy management strategy for mitigating fossil fuel consumption, Journal of Industrial Ecology, 20(2) 289-303 (2016)

Kikuchi, Y., Kanematsu,Y., Ugo, M., Hamada, Y., and Okubo, T., Industrial symbiosis centered on a regional cogeneration power plant utilizing available local resources: A case study of Tanegashima, Journal of Industrial Ecology, 20(2) 276-288 (2016)

Kikuchi, Y., Kimura, S., Okamoto, Y., and Koyama, M., A scenario analysis of future energy systems based on an energy flow model represented as functionals of technology options, Applied Energy, 132, 586-601 (2014)

Shimizu, T., Kikuchi, Y., Sugiyama, H., Hirao, M., Design method for a local energy cooperative network using distributed energy technologies, Applied Energy, 154 781-793 (2015)

AUTHOR INDEX

AUTHOR INDEX

AUTHOR INDEX

AUTHOR INDEX

AUTHOR INDEX

AUTHOR INDEX

AUTHOR INDEX

AUTHOR INDEX

AUTHOR INDEX

AUTHOR INDEX

AUTHOR INDEX